Lecture Notes in Computer Science 3378

Commenced Publication in 1973
Founding and Former Series Editors:
Gerhard Goos, Juris Hartmanis, and Jan van Leeuwen

Editorial Board

David Hutchison
   Lancaster University, UK
Takeo Kanade
   Carnegie Mellon University, Pittsburgh, PA, USA
Josef Kittler
   University of Surrey, Guildford, UK
Jon M. Kleinberg
   Cornell University, Ithaca, NY, USA
Friedemann Mattern
   ETH Zurich, Switzerland
John C. Mitchell
   Stanford University, CA, USA
Moni Naor
   Weizmann Institute of Science, Rehovot, Israel
Oscar Nierstrasz
   University of Bern, Switzerland
C. Pandu Rangan
   Indian Institute of Technology, Madras, India
Bernhard Steffen
   University of Dortmund, Germany
Madhu Sudan
   Massachusetts Institute of Technology, MA, USA
Demetri Terzopoulos
   New York University, NY, USA
Doug Tygar
   University of California, Berkeley, CA, USA
Moshe Y. Vardi
   Rice University, Houston, TX, USA
Gerhard Weikum
   Max-Planck Institute of Computer Science, Saarbruecken, Germany

Joe Kilian (Ed.)

Theory of Cryptography

Second Theory of Cryptography Conference, TCC 2005
Cambridge, MA, USA, February 10-12, 2005
Proceedings

 Springer

Volume Editor

Joe Kilian
Yianilos Labs
707 State Rd., Rt. 206, Suite 212, Princeton, NJ 08540, USA
E-mail: joe@pnylab.com

Library of Congress Control Number: 2005920136

CR Subject Classification (1998): E.3, F.2.1-2, C.2.0, G, D.4.6, K.4.1, K.4.3, K.6.5

ISSN 0302-9743
ISBN 3-540-24573-1 Springer Berlin Heidelberg New York

This work is subject to copyright. All rights are reserved, whether the whole or part of the material is concerned, specifically the rights of translation, reprinting, re-use of illustrations, recitation, broadcasting, reproduction on microfilms or in any other way, and storage in data banks. Duplication of this publication or parts thereof is permitted only under the provisions of the German Copyright Law of September 9, 1965, in its current version, and permission for use must always be obtained from Springer. Violations are liable to prosecution under the German Copyright Law.

Springer is a part of Springer Science+Business Media

springeronline.com

© Springer-Verlag Berlin Heidelberg 2005
Printed in Germany

Typesetting: Camera-ready by author, data conversion by Scientific Publishing Services, Chennai, India
Printed on acid-free paper   SPIN: 11390305   06/3142   5 4 3 2 1 0

Preface

TCC 2005, the 2nd Annual Theory of Cryptography Conference, was held in Cambridge, Massachusetts, on February 10–12, 2005. The conference received 84 submissions, of which the program committee selected 32 for presentation. These proceedings contain the revised versions of the submissions that were presented at the conference. These revisions have not been checked for correctness, and the authors bear full responsibility for the contents of their papers.

The conference program also included a panel discussion on the future of theoretical cryptography and its relationship to the real world (whatever that is). It also included the traditional "rump session," featuring short, informal talks on late-breaking research news.

Much as hatters of old faced mercury-induced neurological damage as an occupational hazard, computer scientists will on rare occasion be afflicted with egocentrism, probably due to prolonged CRT exposure. Thus, you must view with pity and not contempt my unalloyed elation at having my name on the front cover of this LNCS volume, and my deep-seated conviction that I fully deserve the fame and riches that will surely come of it. However, having in recent years switched over to an LCD monitor, I would like to acknowledge some of the many who contributed to this conference.

First thanks are due to the many researchers from all over the world who submitted their work to this conference. Lacking shrimp and chocolate-covered strawberries, TCC has to work hard to be a good conference. As a community, I think we have.

Shafi Goldwasser, the general chair, and Joanne Talbot Hanley, her administrative assistant, went far beyond the call of duty in their support for this conference. It is a matter of debate whether temporary insanity is a prerequisite for volunteering to be general chair, or a consequence. But, certainly, volunteering twice consecutively qualifies one for academic sainthood, if not martyr status. I wish them both several months of well-deserved peace and quiet.

Evaluating submissions requires deep knowledge of the literature, razor-sharp analytical skills, impeccable taste, wisdom and common sense. For my part, I have some pretty good Python scripts. The rest was filled in by my committee. I picked twelve people, and every last one of them did a great job. That just doesn't happen any more, not even in the movies. They supported me far more than I led them.

Like everyone else these days, we outsourced. Our deliberations benefited greatly from the expertise of the many outside reviewers who assisted us in our deliberations. My thanks to all those listed in the following pages, and my thanks and apologies to any I have missed.

I have had the pleasure of working with our publisher, Springer, and in particular with Alfred Hofmann, Ursula Barth, and Erika Siebert-Cole. Although this was my second time working with Springer, I am sure I have not lost my amateur status. It is wrong to prejudge based on nationality, so forgive me, but I did sleep easier knowing that in Germany people spell "Kilian" correctly.

I am grateful to Mihir Bellare, the steering committee chair, and the steering committee in general for making this conference possible.

The time I spent on this project was graciously donated by my places of employment and by my family. I thank NEC and Peter Yianilos for their support and understanding. I thank Dina, Gersh and Pearl for their support, understanding and love.

Finally, I wish to acknowledge the lives and careers of Shimon Even and Larry Stockmeyer, who left us much too soon. Looking at my own work, I can point to specific papers and research directions where their influence is direct. On a deeper level, both shaped their fields by their work and by their interactions with others. Many are their heirs without knowing it. Thank you.

December 2004   Joe Kilian
                                                        Program Chair
                                                        TCC 2005

TCC 2005

February 10–12, 2005, Cambridge, Massachusetts, USA

General Chair

Shafi Goldwasser, Massachusetts Institute of Technology, USA
Weizmann Institute, Israel
Administrative Assistant: Joanne Talbot Hanley

Program Committee

Boaz Barak	IAS and Princeton University, USA
Amos Beimel	Ben-Gurion University, Israel
Rosario Gennaro	IBM, USA
Joe Kilian (Chair)	Yianilos Labs, USA
Anna Lysyanskaya	Brown University, USA
Tal Malkin	Columbia University, USA
Rafail Ostrovsky	UCLA, USA
Erez Petrank	Technion Institute, Israel
Tal Rabin	IBM, USA
Leonid Reyzin	Boston University, USA
Alon Rosen	MIT, USA
Amit Sahai	UCLA, USA
Louis Salvail	Aarhus University, Denmark

Steering Committee

Mihir Bellare (Chair) (UCSD, USA), Ivan Damgård (Aarhus University, Denmark), Oded Goldreich (Weizmann Institute, Israel), Shafi Goldwasser (MIT, USA and Weizmann Institute, Israel), Johan Håstad (Royal Institute of Technology, Sweden), Russell Impagliazzo (UCSD, USA), Ueli Maurer (ETHZ, Switzerland), Silvio Micali (MIT, USA), Moni Naor (Weizmann Institute, Israel), Tatsuaki Okamoto (NTT, Japan)

Table of Contents

Hardness Amplification and Error Correction

Optimal Error Correction Against Computationally Bounded Noise
  Silvio Micali, Chris Peikert, Madhu Sudan, David A. Wilson 1

Hardness Amplification of Weakly Verifiable Puzzles
  Ran Canetti, Shai Halevi, Michael Steiner 17

On Hardness Amplification of One-Way Functions
  Henry Lin, Luca Trevisan, Hoeteck Wee 34

Graphs and Groups

Cryptography in Subgroups of $\mathbb{Z}_n$
  Jens Groth 50

Efficiently Constructible Huge Graphs That Preserve First Order
Properties of Random Graphs
  Moni Naor, Asaf Nussboim, Eran Tromer 66

Simulation and Secure Computation

Comparing Two Notions of Simulatability
  Dennis Hofheinz, Dominique Unruh 86

Relaxing Environmental Security: Monitored Functionalities and
Client-Server Computation
  Manoj Prabhakaran, Amit Sahai 104

Handling Expected Polynomial-Time Strategies in Simulation-Based
Security Proofs
  Jonathan Katz, Yehuda Lindell 128

Security of Encryption

Adaptively Secure Non-interactive Public-Key Encryption
  Ran Canetti, Shai Halevi, Jonathan Katz 150

Adaptive Security of Symbolic Encryption
Daniele Micciancio, Saurabh Panjwani 169

Chosen-Ciphertext Security of Multiple Encryption
Yevgeniy Dodis, Jonathan Katz 188

Steganography and Zero Knowledge

Public-Key Steganography with Active Attacks
Michael Backes, Christian Cachin 210

Upper and Lower Bounds on Black-Box Steganography
Nenad Dedić, Gene Itkis, Leonid Reyzin, Scott Russell 227

Fair-Zero Knowledge
Matt Lepinski, Silvio Micali, Abhi Shelat 245

Secure Computation I

How to Securely Outsource Cryptographic Computations
Susan Hohenberger, Anna Lysyanskaya 264

Secure Computation of the Mean and Related Statistics
Eike Kiltz, Gregor Leander, John Malone-Lee 283

Keyword Search and Oblivious Pseudorandom Functions
*Michael J. Freedman, Yuval Ishai, Benny Pinkas,
Omer Reingold* 303

Secure Computation II

Evaluating 2-DNF Formulas on Ciphertexts
Dan Boneh, Eu-Jin Goh, Kobbi Nissim 325

Share Conversion, Pseudorandom Secret-Sharing and Applications to
Secure Computing
Ronald Cramer, Ivan Damgård, Yuval Ishai 342

Toward Privacy in Public Databases
*Shuchi Chawla, Cynthia Dwork, Frank McSherry, Adam Smith,
Hoeteck Wee* 363

Quantum Cryptography and Universal Composability

The Universal Composable Security of Quantum Key Distribution
*Michael Ben-Or, Michał Horodecki, Debbie W. Leung,
Dominic Mayers, Jonathan Oppenheim* 386

Universally Composable Privacy Amplification Against Quantum Adversaries
Renato Renner, Robert König 407

A Universally Composable Secure Channel Based on the KEM-DEM Framework
Waka Nagao, Yoshifumi Manabe, Tatsuaki Okamoto 426

Cryptographic Primitives and Security

Sufficient Conditions for Collision-Resistant Hashing
Yuval Ishai, Eyal Kushilevitz, Rafail Ostrovsky 445

The Relationship Between Password-Authenticated Key Exchange and Other Cryptographic Primitives
Minh-Huyen Nguyen 457

On the Relationships Between Notions of Simulation-Based Security
Anupam Datta, Ralf Küsters, John C. Mitchell, Ajith Ramanathan .. 476

Encryption and Signatures

A New Cramer-Shoup Like Methodology for Group Based Provably Secure Encryption Schemes
*María Isabel González Vasco, Consuelo Martínez,
Rainer Steinwandt, Jorge L. Villar* 495

Further Simplifications in Proactive RSA Signatures
Stanisław Jarecki, Nitesh Saxena 510

Proof of Plaintext Knowledge for the Ajtai-Dwork Cryptosystem
Shafi Goldwasser, Dmitriy Kharchenko 529

Information Theoretic Cryptography

Entropic Security and the Encryption of High-Entropy Messages
Yevgeniy Dodis, Adam Smith 556

Error Correction in the Bounded Storage Model
  Yan Zong Ding 578

Characterizing Ideal Weighted Threshold Secret Sharing
  Amos Beimel, Tamir Tassa, Enav Weinreb 600

Author Index 621

Optimal Error Correction Against Computationally Bounded Noise

Silvio Micali, Chris Peikert, Madhu Sudan, and David A. Wilson

MIT CSAIL, 77 Massachusetts Ave,
Building 32, Cambridge, MA, 02139
{silvio, cpeikert, madhu, dwilson}@mit.edu

Abstract. For computationally bounded adversarial models of error, we construct appealingly simple, efficient, cryptographic encoding and unique decoding schemes whose error-correction capability is much greater than classically possible. In particular:

1. *For binary alphabets, we construct positive-rate coding schemes which are uniquely decodable from a $1/2 - \gamma$ error rate for any constant $\gamma > 0$.*
2. *For large alphabets, we construct coding schemes which are uniquely decodable from a $1 - \sqrt{R}$ error rate for any information rate $R > 0$.*

Our results are qualitatively stronger than related work: the construction works in the public-key model (requiring no shared secret key or joint local state) and allows the channel to know everything that the receiver knows. In addition, our techniques can potentially be used to construct coding schemes that have information rates approaching the Shannon limit. Finally, our construction is qualitatively optimal: we show that unique decoding under high error rates is *impossible* in several natural relaxations of our model.

1 Introduction

The theory of error correction is concerned with sending information reliably over a "noisy channel" that introduces errors into the transmitted data. In this setting, a *sender* starts with some *message*, which is a fixed-length string of symbols over some alphabet. The sender *encodes* the message into a longer string over the same alphabet, then transmits the block of data over a *channel*. The channel introduces *errors* (or *noise*) by changing some of the symbols of the transmitted block, then delivers the corrupted block to the *recipient*. Finally, the recipient *decodes* the block (hopefully to the intended message). Whenever the sender wants to transmit a new message, the process is repeated.

Two quantities are of special interest in this setting: the *information rate* (i.e., the ratio of the message length to the encoded block length) and the *error rate* (i.e., the ratio of the number of errors to the block length). Coding schemes having high information rate and tolerating high error rate are, of course, the most desirable. Small alphabets are desirable too, and in particular most natural channels are indeed best at transmitting only bits.

But the question remains: *how should we model a noisy channel?*

Standard Channels. There are two historically popular ways to model a noisy channel. Shannon's *symmetric channel* independently changes each symbol to a random different one, with some fixed probability. Hamming's *adversarial channel* changes symbols in a worst-case fashion, subject only to an upper bound on the number of errors per block of data. In particular — though this is not often stated explicitly — the adversarial channel is computationally *unbounded*. Working with this "pessimistic" model certainly ensures the robustness of the resulting coding scheme, but it also severely restricts the information and error rates. For instance, when the alphabet is binary, the error rate must be less than $1/4$ for unique decoding to be possible (unless the blocks are exponentially longer than the messages).

One way to recover from a higher error rate is to relax the task of decoder, allowing it to output a short *list* of messages which contains the intended one. To tolerate adversarial channels with high error rates, list decoding seems to be the best one can do — but under a more *"realistic"* model of an adversarial channel, is it possible to *uniquely* decode under high error rates?

Computationally Bounded Channels. In 1994, Lipton [9] put forward the notion of a computationally bounded channel, which is essentially a Hamming channel restricted to *feasible* computation. That is, the channel still introduces errors adversarially (always subject to a given error rate), but must do so in time polynomial in the block length.

We posit that natural processes can be implemented by efficient computation, so *all* real-world channels are, in fact, computationally bounded. We therefore have confidence that results in this model will be as meaningful and applicable as classical codes. Indeed, the nature of the model is such that if some malicious (or natural!) process is capable of causing incorrect decoding, then that process can be efficiently harnessed to break standard hardness assumptions. In contrast to coding schemes which are only guaranteed to work against channels that are modelled by *very specific, limited* probabilistic processes, results in this model apply *universally* to any channel which can be modelled by efficient computation.

Remarkably, under standard cryptographic assumptions and assuming that sender and receiver share secret randomness, Gopalan, Lipton, and Ding [3] proved that for such a bounded channel, it is possible to decode correctly from higher error rates. Unfortunately, their result requires the communicating parties to share a secret key which is unknown to the channel.

More significantly, though the bounded-channel model was first envisioned over a decade ago, nobody has yet shown an *essential* use of this assumption to yield any unique benefits over an unbounded channel. That is, previous constructions still work when the channel is computationally unbounded, as long as the sender and receiver share some secret randomness. The bounded-channel assumption is used to reduce the *amount* of shared randomness that is needed, but not to eliminate it altogether. This computational assumption is thus an *additional* one, and does not supplant the assumption of secret randomness shared between the sender and receiver.

Our goal is to provide a general method for optimal error correction, exploiting the bounded-channel assumption in an essential way.

1.1 Our Contributions

Our Setting. We work in a very simple cryptographic setting: we assume that a one-way function exists (the "minimal" cryptographic assumption) and that the sender has a public key known to the receiver (and, perhaps, to the channel as well).

The sender (but *not* the receiver) keeps a small amount of state information, which he uses when encoding messages. Because the sender keeps state, our constructions are actually *dynamic* coding schemes, in which the same message results in a different encoding each time it is sent.

Our Results. Our setting yields great benefits in error correction for both binary and large alphabets. Namely,

1. *For binary alphabets, we construct positive-rate dynamic coding schemes which are uniquely decodable from a $1/2 - \gamma$ error rate for any constant $\gamma > 0$.*
   Classically, a $1/4 - \gamma$ error rate is the best possible for unique decoding (and positive information rate). We stress that *in any reasonable model*, decoding of *any kind* (even list decoding) is impossible under an error rate of $1/2$. Therefore this result is optimal in a very strong sense, and matches the best possible error rates in the weaker Shannon model.
2. *For large alphabets, we construct dynamic coding schemes which are uniquely decodable from a $1 - \sqrt{R}$ error rate for any information rate $R > 0$.*
   The $1 - \sqrt{R}$ error rate is actually a consequence of known list decoding algorithms, and not imposed by our technique. Note that when $R < 1/4$, we can uniquely decode from error rates much greater than $1/2$, which is impossible in the Hamming model.

To achieve these results, we actually prove a very general *reduction*, namely,

> If one-way functions exist, *(dynamic) unique decoding* from e errors in the bounded-channel model reduces to efficient *(static) list decoding* from e errors in the Hamming model (with no asymptotic loss in information rate).

We obtain results 1 and 2 above by applying this reduction to the classical Guruswami-Sudan [7] and Reed-Solomon codes.

Optimality of Our Model. There are three defining characteristics of our model: (1) the sender is stateful (the amount of state required is minimal; either a single counter value or a local clock would suffice) while the receiver is stateless, (2) the sender keeps a secret key which is unknown to the channel, and (3) the channel is assumed to be computationally bounded.

We show that our model is qualitatively optimal: relaxing any of these three requirements makes the task of unique decoding under high error rates *impossible*. Thus our construction can be seen as the "best possible" use of the bounded-channel assumption for error correction. See Section 4.3 for details.

Overview of the Construction. Starting with any static code, we specify a *cryptographic sieving* procedure, which only certain "authentic" codewords will pass. Authentic words are hard for the adversary to compute (even after seeing other authentic codewords), but easy for the sender to generate and for the recipient to sieve out.

Upon receiving a corrupted word, the recipient first *list decodes* it. Of course, list decoding only provides a set of candidate codewords. In order to uniquely decode, the recipient next uses the cryptographic sieve to filter out only the authentic word(s). Provided that the number of errors is suitably limited, the intended codeword is guaranteed to appear in the decoded list and pass the sieve. However, it may not be alone: though the bounded channel cannot produce any *new* authentic codewords, it may be able to cause *prior* ones to appear in the decoded list. This is where the sender's state comes into play: dynamic encoding allows the receiver to choose the "freshest" word that passes the sieve, resulting (with overwhelming probability) in correct, unique decoding.

2 Related Work

We wish to contrast our results with several other models and techniques for tolerating high error rates.

2.1 List Decoding

One of the best-known methods of decoding beyond classical limits under adversarial error is known as *list decoding*. In list decoding, a received word is not decoded to a unique message, but rather to a short *list* of possible messages. If the number of errors is within the *list-decoding radius*, the original message will appear in the list.

There exist codes with rate approaching the Shannon capacity of the channel and yielding constant-size lists (cf. [5]); however, no efficient list decoding algorithms are known for such codes. Still, many popular codes have efficient list-decoding algorithms that can decode significantly beyond the half-the-distance bound.

The obvious drawback of list decoding is that one typically desires to know the *unique* message that was sent, rather than a list of possible messages. The works presented below, as well as our cryptographic sieve, use list decoding as a tool to extend the decoding radius, then employ additional assumptions in order to identify the correct, unique message in the list.

2.2 List Decoding with Side Information

Guruswami [4] achieves unique decoding under high error rates for binary alphabets. However, he makes two strong assumptions: first, the communicating parties must share a *side channel* which is *noise-free*, and must use it every time a message is sent. (Note that the side channel does not trivialize the problem, because it is only used to send strings that are much shorter than the messages.) Second, the adversary must not know what is sent over the side channel when introducing errors in the main channel; this imposes either a privacy or timing constraint on the side-channel communication.

2.3 Code Scrambling Using Shared Randomness

The code-scrambling method of Gopalan, Lipton, and Ding [3] assumes that the communicating parties share some secret random (or pseudorandom) data. The randomness is used to "scramble" codewords, which reduces adversarial noise to (random) Shannon noise. Under such a random-error model, and for certain properly-chosen codes, maximum-likelihood decoding yields the correct word with high probability.

Code scrambling and our cryptographic sieve are both based on the minimal cryptographic assumption of the existence of one-way functions.[1] But our underlying model compares favorably to that of code scrambling in some important ways:

Cryptographic Setup. The code-scrambling method requires a random secret key to be shared between the communicating parties and kept secret from the channel. Such a setup requires either the parties to meet in advance (which may be unrealistic), or some *interactive* protocol to establish the private key (in addition, such a protocol would have to deal with the noisy channel that separates the two parties!).

In contrast, our cryptographic sieve works in the *public key* setting: we only require the sender to have a single public key that is known to the recipient. In fact, our results hold even when the channel possesses all the information available to the receiver, and is potentially even more computationally powerful than the receiver. Previous results certainly do not allow the channel to be this powerful.

Local State. In reality, two communicating parties usually send and receive many messages over time. Using a classical (static) code, this is no problem: each message is simply encoded and decoded on its own, with no implications for

---

[1] The two employ different cryptographic primitives, both of which are implied by one-way functions. Gopalan *et al* use a pseudorandom generator, while our solution uses an existentially unforgeable signature scheme [2, 10].

correctness. However, when shared randomness and state are introduced, one must be more careful.

The first observation is that in the code-scrambling method, the shared key (or portion thereof) must only be used *one time*. If any part is re-used, the adversary gains information about how the codewords are scrambled, and may be able to introduce "non-random" errors in the future. Therefore, the code-scrambling method requires both parties to keep *synchronized state*. That is, they must always agree on what fresh portion of their shared (pseudo)random key should be used for scrambling and unscrambling the next message. If the parties fall out of sync (say, due to an unexpected transmission or hardware failure), then future messages will decode incorrectly.[2]

In contrast, our cryptographic sieve only requires the sender to maintain a small amount of local state, independent of the recipient (who is stateless). If a message is dropped or ignored, there is no effect on the correctness of future messages.[3]

We also compare our quantitative results with those of Gopalan *et al*:

Binary Alphabets. For binary alphabets, the code-scrambling method can yield coding schemes that handle the optimal error rate of $\epsilon = 1/2 - \gamma$ for any $\gamma > 0$. In addition, the information rate is optimal, because it meets the Shannon limit of $1 - H(\epsilon)$.

Our method also provides unique decoding from a $1/2-\gamma$ error rate. We stress that while our technique yields positive asymptotic information rate, it does not yet match the Shannon limit, because the information rate is dictated by the underlying efficiently list-decodable code. While list-decodable codes matching the Shannon limit for any error rate are known to exist, it is not known how to efficiently list decode them. Fortunately, improvements in list decoding techniques automatically carry over to our construction.

Large Alphabets. Implemented with Reed-Solomon codes (which require large alphabets), code scrambling allows unique decoding from a $\min(1 - \sqrt{R}, 1 - 2R)$ error rate (where R is the information rate), while classical unique decoding of RS codes only allows a $(1 - R)/2$ error rate. Therefore, for $R \geq 1/3$ the code-scrambling method offers no advantage over classical decoding; for $R \in (1/4, 1/3)$ there are some benefits but they are not as pronounced as in the low-rate case.

---

[2] To relax the synchronization requirements, one might imagine sending some "synchronizing information" along with each message. However, the synchronizing information is also subject to errors, so it must be protected by some encoding, and also be recoverable separately from the message. (Using a pseudorandom function for synchronization suffers from a similar problem.) Eliminating the synchrony requirement, while retaining desirable information and error rates, seems to be quite difficult.

[3] Of course, we cannot provide such a guarantee if the channel is allowed to arbitrarily *delay* messages and *swap* their order — however, neither can any scheme that uses synchronized state.

In contrast, our method meets or exceeds the asymptotic error correction rate of the code-scrambling method, at all information rates. In particular, it allows unique decoding from a $1 - \sqrt{R}$ error rate, for all values of R.

Universality. Because it relies on the randomness of the errors, the analysis of the code-scrambling method depends essentially upon the properties of Reed-Solomon codes.[4] The authors also point out that a similar analysis can be applied to certain algebraic-geometry codes, and that experimental simulations using expander codes have shown positive results. However, each application of code scrambling to a new family of codes requires a new analysis (and yields, potentially, a new set of workable parameters).

Our construction, instead, is fully general: it uses an efficient list-decoding algorithm as a black-box, and requires no other special properties of the code. It reduces *unique decoding* against a *bounded* adversary to *list decoding* against an *unbounded* adversary, and retains all the asymptotic parameters of the code.

2.4 Private Codes

In a recent independent work, Langberg [8] describes "private codes" in which the sender and recipient use a shared secret random key (which is not known to the channel) to uniquely decode under high error rates.

Langberg assumes a computationally unbounded channel and focuses mainly on existential (rather than efficiently constructible) results, and on tight bounds for the length of the secret key. The construction uses certain combinatorial set systems to define a "secret subcode" C_r of a given static code C, based on the secret key r. Unique decoding is performed by maximum-likelihood decoding within C_r. The analysis and security proof of the scheme are somewhat complex and difficult to penetrate.

Compared to our cryptographic sieving, private codes share two main drawbacks with code scrambling. Namely,

1. *They require secret randomness to be shared between the sender and receiver and kept secret from the channel.* By contrast, in our model the channel is on "equal footing" with the receiver: it knows everything that is known to the latter.
2. *They require sender and receiver to keep synchronized state.* (Else they may not be able to understand each other whenever multiple messages are sent.) No such requirement exist in our model, and multiple messages can be safely sent.

Finally, in our view, we contribute a conceptually cleaner framework and simpler security proof. In retrospect, it is possible to cast private codes as a *specific*

---

[4] For Lemma 3.2 in Gopalan *et al*, the *maximum distance separability* (MDS) property of Reed-Solomon codes is the key ingredient. The MDS property is true of RS codes but not true in general.

construction in our general framework of *message authentication* and *cryptographic sieving*.[5] (We thank Adam Smith for pointing out this relationship.)

3 Preliminaries

3.1 Notation

Messages and words, which are just vectors over some alphabet, are written in bold: e.g., **m**, **x**, **r**. The concatenation of two vectors **x**, **y** is denoted $\mathbf{x} \circ \mathbf{y}$. The set $\{1, \ldots, n\}$ is denoted by $[n]$. A negligible function in n is one which vanishes faster than $1/p(n)$ for any fixed polynomial p, and is denoted by $\nu(n)$.

3.2 Relevant Coding Theory

Basic Concepts. The *message* is the information to be sent before it is encoded; it is a vector of k symbols from some finite *alphabet* Σ (by convention, we define $q = |\Sigma|$; a binary alphabet is one where $q = 2$). The message is encoded as a *codeword* of n symbols from Σ (n is called the *block length*). The *information rate* R of the code is defined as $R = k/n$; this is a measure of how much meaningful information is carried by each transmitted symbol.

After passing through the channel, a (potentially corrupted) word is received and decoded, ideally back to the intended k-symbol message. However, in order for this to be the case, the number of errors must be suitably limited. The *Hamming distance* $\Delta(\mathbf{x}, \mathbf{y})$ between two words **x** and **y** is the number of symbols that differ between the two; we wish to decode the received word to the message whose codeword is nearest in Hamming distance.

Definition 1 (Hamming distance). *For any* $\mathbf{x}, \mathbf{y} \in \Sigma^n$, *the* Hamming distance *between* **x** *and* **y**, *denoted* $\Delta(\mathbf{x}, \mathbf{y})$, *is the number of positions i in which x_i and y_i differ:* $\Delta(\mathbf{x}, \mathbf{y}) = |\{i \in [n] : x_i \neq y_i\}|$.

Definition 2 (Coding scheme, rate). *An* $(n, k)_q$-*coding scheme* $C = (E, D)$ *over alphabet* Σ *is an* encoding function $E : \Sigma^k \to \Sigma^n$ *and a* decoding function $D : \Sigma^n \to \Sigma^k$ *for some positive integers* $n \geq k$, $q = |\Sigma| \geq 2$. *The (relative) rate or information rate of the scheme, denoted R, is defined as $R = k/n$. The scheme tolerates error rate ρ if, for all* $\mathbf{m} \in \Sigma^k$ *and all* **r** *such that* $\Delta(E(\mathbf{m}), \mathbf{r}) \leq \rho n$, $D(\mathbf{r}) = \mathbf{m}$.

List Decoding. Even if the actual error rate exceeds the rate ρ tolerated by a coding scheme, in some contexts it may be sufficient to decode to a short *list*

---

[5] One can interpret private codes as using an (information-theoretically secure) secret-key *message authentication code* (MAC), rather than a (computationally secure) digital signature scheme. In this interpretation, the "secret subcode" C_r consists of encoded message-tag pairs, where the tag is a valid MAC of the message under secret key r. Maximum-likelihood decoding within C_r can be accomplished by first list decoding within C, then sieving out the decoded message-tag pairs that are authentic relative to r.

of messages containing the intended one. *List decoding* finds such a list of all messages which encode to within some distance ϵn of the received word, where ϵ may be significantly greater than ρ.

Definition 3 (List decodability). *An $(n,k)_q$ coding scheme $C = (E,D)$ over Σ is $(\epsilon n, L)$-list decodable if, for any $\mathbf{r} \in \Sigma^n$, there exist $\ell \leq L$ distinct messages $\mathbf{m}_1, \ldots, \mathbf{m}_\ell \in \Sigma^k$ such that $\Delta(\mathbf{r}, E(\mathbf{m}_j)) \leq \epsilon n$ for all $j \in [\ell]$.*

Asymptotics. In order to make meaningful asymptotic statements about the rate of a coding scheme or the efficiency of encoding and (list) decoding, we must consider infinite *families* having increasing block lengths.

Definition 4 (Family of coding schemes). *An infinite family $\mathbb{C}$ of coding schemes is a set $\mathbb{C} = \{C_i\}_{i=1}^\infty$ where $C_i = (E_i, D_i)$ is an $(n_i, k_i)_{q_i}$ coding scheme, and $\lim_{i \to \infty} n_i = \infty$.*

The (asymptotic) information rate *(often just abbreviated* rate*) of $\mathbb{C}$, denoted $R(\mathbb{C})$, is defined to be $R(\mathbb{C}) = \liminf_{i \to \infty} k_i/n_i$.*

If C_i is an $(\epsilon_i n_i, L_i)$-list decodable coding scheme, then we say that $\mathbb{C}$ is list decodable *under error rate $\epsilon(\mathbb{C}) = \liminf_{i \to \infty} \epsilon_i$.*

If $\{E_i\}$ and $\{D_i\}$ (respectively) can be computed by two uniform polynomial-time algorithms, we say that the coding scheme is efficient.

Definition 5 (List decoding algorithm). *If $\mathbb{C} = \{C_i\}$ is a family of $(n_i, k_i)_{q_i}$ coding schemes $C_i = (E_i, D_i)$ over alphabet Σ_i, and each C_i is $(\epsilon_i n_i, L_i)$-list decodable, then an efficient* list decoding algorithm *for these parameters is a polynomial-time algorithm LD such that for all i and any $\mathbf{r} \in \Sigma_i^{n_i}$, $LD(\mathbf{r})$ outputs all $\ell \leq L_i$ messages $\mathbf{m}_1, \ldots, \mathbf{m}_\ell \in \Sigma_i^{k_i}$ such that $\Delta(\mathbf{r}, E_i(\mathbf{m}_j)) \leq \epsilon_i n_i$ for all $j \in [\ell]$.*

Note that LD must run in polynomial time in the size of its input, so in particular the list size L_i must be polynomially related to n_i. Many families are indeed efficiently list decodable for high error rates.

3.3 Relevant Cryptography

We require signature schemes which are existentially unforgeable under chosen message attack [2]. Such schemes were first shown to exist under a hardness of factoring assumption, and later under the assumption that one-way functions exist [10].

4 Dynamic Coding Schemes

4.1 The Formal Model

The issues of a bounded adversary, stateful players, and chosen-message attacks are not captured by the classical coding theory definitions. Here we formally define a bounded noisy channel and the requirements for the sender and receiver.

For ease of notation, we provide definitions modeling the channel as a uniform algorithm; a non-uniform treatment is easily adapted.

A *dynamic coding scheme* for a family of parameters $\{(n_i, k_i, q_i)\}_{i=1}^{\infty}$ (with $\lim_{i \to \infty} n_i = \infty$) is a triple of probabilistic polynomial-time algorithms (G, S, R) such that:

- $G(1^{k_i}, 1^{n_i})$ outputs a pair (pk, sk);
- $S(\mathbf{m}, sk, aux)$, where sk was produced by $G(1^{k_i}, 1^{n_i})$, $\mathbf{m}$ is of length k_i over a q_i-ary alphabet Σ_i, and aux is some local state, outputs $(\mathbf{x}, aux')$ where $\mathbf{x} \in \Sigma_i^{n_i}$, and aux' is the updated local state that will be provided on the next invocation of S;
- $R(\mathbf{r}, pk)$, where pk was produced by $G(1^{k_i}, 1^{n_i})$ and $\mathbf{r} \in \Sigma_i^{n_i}$, outputs some $\mathbf{m}' \in \Sigma_i^{k_i}$.

The *information rate* of such a scheme is $\liminf_{i \to \infty} k_i/n_i$.

An (adversarial) *channel* $\mathcal{C}$ with *error rate* ϵ is a probabilistic poly-time algorithm which interacts with a sender S and receiver R in a chosen-message attack, which proceeds as follows:

1. $G(1^{k_i}, 1^{n_i})$ produces (pk, sk).
2. On input pk to $\mathcal{C}$, the following process is repeated until $\mathcal{C}$ terminates:
   - On the jth iteration, $\mathcal{C}$ chooses a message $\mathbf{m}_j \in \Sigma_i^{k_i}$ and hands it to the sender.
   - The sender encodes $\mathbf{m}_j$ using $S(\mathbf{m}_j, sk, aux_j)$, yielding aux_{j+1} and some $\mathbf{x}_j \in \Sigma_i^{n_i}$, which is given to $\mathcal{C}$.
   - $\mathcal{C}$ produces a word $\mathbf{r}_j$ such that $\Delta(\mathbf{x}_j, \mathbf{r}_j) \leq \epsilon n_i$ with probability $1 - \nu(n_i)$, and hands $\mathbf{r}_j$ to the recipient.
   - The recipient runs $R(\mathbf{r}_j, pk)$ and outputs a message $\mathbf{m}'_j$.

We say that $\mathcal{C}$ *succeeds* at causing an incorrect decoding if, for any j in the above experiment, $\mathbf{m}'_j \neq \mathbf{m}_j$. We say that a dynamic code *uniquely decodes from error rate ϵ* if, for any channel $\mathcal{C}$ of error rate ϵ, $\Pr[\mathcal{C} \text{ succeeds}] \leq \nu(n_i)$, where the probability is taken over the random choices of G, S, R, and $\mathcal{C}$.

Remark 1. In contrast to many cryptographic definitions of an adversary, our channel is *not* allowed to drop or re-order messages. That is, the channel must deliver a corrupted message before requesting a new message from the sender.

Against a more powerful channel which *can* drop and re-order messages, we are still able to construct coding schemes which provide similar guarantees about message integrity — provided that the receiver also keeps state. (However, the receiver's state is independent of the sender's.) We omit the details in this version of the paper.

4.2 The Construction

Intuition. The first attempt at a cryptographic sieve is to just sign each message and send the signature along with it. This obviously doesn't work because the signature is also subject to errors. The natural fix is to protect the signature in transit by also encoding it, using appropriate parameters. Unfortunately, a careful analysis (even using list decodable coding schemes for both the message and signature) shows that this approach yields no improvement in overall rate.

The key insight is that the message and signature should be encoded *together*, rather than separately. To communicate a message **m**, the sender first signs **m**, then encodes the message-signature pair. To decode a received word **r**, the recipient applies the list decoding algorithm to **r**, yielding a list of potential message-signature pairs. If the number of errors is suitably bounded, then the original pair will appear in the list, and the signature will verify. And by the unforgeability of the signature scheme, the other pairs will not verify and can be thrown out. Therefore the original message can be recovered uniquely.

There is one hidden difficulty in the above description: the list may actually include several pairs that verify, but which correspond to messages sent *in the past*.[6] This event cannot be used to break the signature scheme, because a valid forgery must involve a *new*, unsigned message. Therefore we must find a way to disambiguate among potentially several valid message-signature pairs in the list.

The solution is for the sender to maintain a short *counter t*. The current value of the counter is appended to each message (and signed along with it), then incremented. Among all valid message-signature pairs in the decoded list, the recipient chooses the one with the largest counter. In other words, the recipient always decodes to the "freshest" message in the list. We note that the recipient is *stateless*, while the sender need only maintain the value of the counter.

Indeed, there are only two essential requirements for the counter values: they must not be reused, and the receiver must be able to recognize the most recent counter value in a list. Any monotonically increasing sequence satisfies these requirements; in particular, a timestamp is sufficient. (Note that the sender's clock need not be synchronized with any other clock, nor is relative clock drift a concern.) Our construction may be viewed as confirmation of some conventional wisdom: one should always date and sign one's correspondence!

The Formal Construction. Let $\mathbb{C}$ denote some family $\mathbb{C} = \{C_i\}$ of $(n_i, k_i)_{q_i}$, (e_i, L_i) list decodable coding schemes that has an efficient encoding procedure E and an efficient list decoding procedure LD for parameters (e_i, L_i). Note that the efficiency constraints imply that n_i grows polynomially with k_i (in fact, we are most interested in the case where this growth is linear), and that the decoded list sizes are polynomial in n_i.

Also assume we are given some signature scheme (GEN, SIG, VER) which is existentially unforgeable under an adaptive chosen-message attack. Recall that the key generator GEN requires a security parameter k'; for simplicity we assume wlog that the corresponding signature length is k'. We require the signature size to be small relative to the message size, therefore when using code C_i we use a security parameters of, say, $k'_i = \sqrt{k_i}$.

---

[6] This can happen if two prior message-signature pairs map to two codewords separated by, say, the minimum distance of the code.

Additionally, the sender S maintains a state variable t, which is a counter initialized to 0. This counter will also appended to the message, so again we want it to be short. When using code C_i, we append the value of the counter using $k'_i = \sqrt{k_i}$ symbols in some canonical way; this provides for $q_i^{k'_i}$ unique counters, which is super-polynomial in n_i.

We now describe the dynamic coding scheme. The codes will have message lengths of $k_i - 2k'_i = k_i - 2\sqrt{k_i}$ and corresponding block lengths of n_i. The algorithms are as follows:

- $G(1^{k_i}, 1^{n_i})$: let $k'_i = \sqrt{k_i}$. Compute and output $(pk, sk) \leftarrow \text{GEN}(1^{k'_i})$.
- $S(\mathbf{m}, sk, aux = t)$: compute $\sigma \leftarrow \text{SIG}_{sk}(\mathbf{m} \circ t)$. Output $(E(\mathbf{m} \circ t \circ \sigma), aux' = t + 1)$.
- $R(\mathbf{r}, pk)$: list decode $\mathbf{r}$: $(\mathbf{m}_1 \circ t_1 \circ \sigma_1), \ldots, (\mathbf{m}_\ell \circ t_\ell \circ \sigma_\ell) \leftarrow LD(\mathbf{r})$. Consider all pairs $(\mathbf{m}_i, t_i)$ such that $\text{VER}_{pk}(\mathbf{m}_i \circ t_i, \sigma_i) = 1$ and $t_i = \max_j t_j$. If at least one such $\mathbf{m}_i$ exists and are all the same message, output that message. Otherwise, output $\bot$.

Theorem 1. *Assuming one-way functions exist, the above dynamic code (G, S, R) uniquely decodes from error rate $\epsilon(\mathbb{C})$ and has information rate $R(\mathbb{C})$.*

Proof. Clearly the rate of the scheme is $R(\mathbb{C})$, because $\liminf_{i \to \infty} (k_i - 2\sqrt{k_i})/n_i = \liminf_{i \to \infty} k_i/n_i = R(\mathbb{C})$.

Now suppose for contradiction there exists a channel $\mathcal{C}$ with error rate ϵ that causes R to incorrectly decode some message with probability $1/p(n_i)$ for some polynomial $p(\cdot)$ and for infinitely many n_i. We will use $\mathcal{C}$ to construct a forger F for the signature scheme. F will receive message requests from $\mathcal{C}$ and use its signing oracle to simulate the sender, then pass authentic codewords to the channel. The channel will produce corrupted words, and the forger will simulate the recipient on them. If decoding is incorrect at any point (which is detectable), the incorrect output can be used to construct a forgery. We now proceed more formally.

First note that for infinitely many n_i, $\mathcal{C}$ makes fewer than $q_i^{k'_i}$ queries, and the counter values are all distinct. From now on, we consider only those n_i. Now $F^{\mathcal{O}}(pk)$ works as follows: let $k_i = (k'_i)^2$, and n_i be the block length corresponding to k_i, and $t \leftarrow 0$. Run $\mathcal{C}(pk)$. When $\mathcal{C}$ requests a message $\mathbf{m}$ to be sent, query $\sigma \leftarrow \mathcal{O}(\mathbf{m} \circ t)$, and return $\mathbf{w} = E(\mathbf{m}, t, \sigma)$ to the channel. Receive $\mathbf{r}$ (a corrupted version of $\mathbf{w}$) from the channel, where $\Delta(\mathbf{r}, \mathbf{w}) \leq \epsilon n_i$ (except with probability $\nu(n_i)$), and list decode: $(\mathbf{m}_1 \circ t_1 \circ \sigma_1), \ldots, (\mathbf{m}_\ell \circ t_\ell \circ \sigma_\ell) \leftarrow LD(\mathbf{r})$, where $\ell \leq L_i$. By assumption on $\mathcal{C}$, $(\mathbf{m} \circ t \circ \sigma)$ appears in the list (except with probability $\nu(n_i)$). If there exists some $\mathbf{m}_i$ such that $\text{VER}_{pk}(\mathbf{m}_i \circ t_i, \sigma_i) = 1$ and $t_i > t$, or such that $m_i \neq m$ and $t_i = t$, then output $(\mathbf{m}_i \circ t_i, \sigma_i)$ as a forgery. Otherwise, increment t and repeat with the next message chosen by $\mathcal{C}$, until it aborts.

Note that R decodes incorrectly if and only if some σ_i is a valid signature of $m_i \circ t_i$, and either $t_i > t$, or $t_i = t$ and $m_i \neq m$ (because all counters in the experiment are unique). In the first case, $\mathcal{O}$ was never queried on $m_i \circ t_i$, because t_i is too large. In the second case, $\mathcal{O}$ was never queried on $m_i \circ t_i$ because only $m \neq m_i$ was queried with counter $t = t_i$. Therefore $(m_i \circ t_i, \sigma_i)$ constitutes a forgery. The success probability of $F^{\mathcal{O}}$ on security parameter k'_i is negligibly smaller than $p(n_i)$ (the success probability of $\mathcal{C}$ on block lengths of size n_i), and by assumption n_i grows polynomially with $k_i = (k'_i)^2$. This contradicts the unforgeability of the signature scheme, as desired.

Remark 2. A similar construction works in the *private-key setting*, in which the sender and recipient share a private key that is unknown to the channel. Instead of signing each message, the sender uses a *message authentication code* (MAC) that is existentially unforgeable under chosen-message attack.[7] We stress that the receiver remains stateless in this modified scheme. The only difference in the proof is that the forger cannot detect a successful forgery, but instead chooses a random element of the decoded list during a random query by the channel. This alters the concrete security analysis of the scheme, but still leads to a non-negligible chance of forgery.

Corollary 1. *Assuming one-way functions exist, there exist binary dynamic coding schemes with error rate $1/2 - \gamma$ for any $\gamma > 0$ and positive asymptotic information rate.*

Proof. Apply Theorem 1 to the concatenated codes of Guruswami and Sudan [7]. The Reed-Solomon concatenated codes are efficiently encodable, have asymptotic rate $\Omega(\gamma^8)$, and can be efficiently list decoded under error rate $1/2 - \gamma$, for any constant $\gamma > 0$. The algebraic-geometry concatenated codes also have efficient algorithms and asymptotic rate $\Omega(\gamma^6 \log 1/\gamma)$.

Alternatively, one may use the efficiently list-decodable codes of Guruswami et al [5], which have asymptotic rate $\Omega(\gamma^4)$ for error rate $1/2 - \gamma$.

Corollary 2. *Assuming one-way functions exist, there exist large-alphabet dynamic coding schemes with error rate $1 - \sqrt{R}$ for any information rate $R > 0$.*

Proof. Apply Theorem 1 to Reed-Solomon codes. These codes are efficiently encodable and list decodable under error rate $1 - \sqrt{R}$ using the Guruswami-Sudan algorithm [6].

4.3 Optimality of the Model

In our model, the sender keeps a secret key and maintains some local state between each encoding, and the channel is computationally bounded. Propositions 1, 2, and 3 establish that these features are *essential*: under several natrual relaxations of the model, decoding from high error rates is impossible.

The following well-known lemma will be useful in our proofs:

Lemma 1 (Plotkin bound). *For any $2n+1$ strings $\mathbf{x}_1, \ldots, \mathbf{x}_{2n+1} \subset \{0,1\}^n$, there exist i, j such that $i \neq j$ and $\Delta(\mathbf{x}_i, \mathbf{x}_j) < n/2$.*

---

[7] Such a MAC can be constructed using a pseudorandom function family, which exists if and only if one-way functions exist [1].

Proposition 1. *Consider any dynamic coding scheme for parameters $\{(n_i, k_i, q_i)\}_i$ where both the sender and receiver are stateless.*

If $q_i = 2$ and $2^{k_i} \geq 2n_i + 1$ for infinitely many i, an adversarial channel with any error rate $> 1/4$ can cause incorrect decoding with probability non-negligible in n_i.

For any values of q_i, an adversarial channel with any error rate $> 1/2$ can cause incorrect decoding with probability non-negligible in n_i.

These results hold even if the sender and receiver are randomized and have secret or public keys, and the channel is polynomially bounded.

Proof. We first prove the result for binary alphabets (i.e., $q_i = 2$). (We assume $2^{k_i} \geq 2n_i + 1$ simply to guarantee that at least $2n + 1$ distinct messages can be sent.)

For any n and $\mathbf{x}, \mathbf{x}' \in \{0,1\}^n$, if $\Delta(\mathbf{x}, \mathbf{x}') < n/2$, define $M(\mathbf{x}, \mathbf{x}')$ to be some canonical $\mathbf{r}$ such that $\Delta(\mathbf{x}, \mathbf{r}) < \lceil n/4 \rceil$ and $\Delta(\mathbf{r}, \mathbf{x}') \leq \lceil n/4 \rceil$. Note that M is easily computable. We now describe a channel $\mathcal{C}$ which will cause an incorrect decoding with non-negligible probability.

$\mathcal{C}$ chooses two distinct messages $\mathbf{m}, \mathbf{m}'$ at random and queries the sender on $\mathbf{m}$, yielding $\mathbf{x} \in \{0,1\}^n$, then passes it to the receiver without error. $\mathcal{C}$ then queries the sender on $\mathbf{m}'$, yielding $\mathbf{x}' \in \{0,1\}^n$. If $\Delta(\mathbf{x}, \mathbf{x}') < n/2$, $\mathcal{C}$ sends either $M(\mathbf{x}, \mathbf{x}')$ or $M(\mathbf{x}', \mathbf{x})$ to the recipient, each with probability $1/2$ (this requires introducing at most $\lceil n/4 \rceil$ errors to $\mathbf{x}'$). Otherwise, $\mathcal{C}$ sends $\mathbf{x}'$ uncorrupted.

Conditioned on $\Delta(\mathbf{x}, \mathbf{x}') < n/2$, the receiver's view *of the second message* is distributed identically to its view in a world where $\mathbf{m}'$ is queried *first* and $\mathbf{m}$ is queried second. (This relies on the statelessness of both sender and receiver.) Therefore, $\mathcal{C}$ will cause incorrect decoding with probability at least $1/2$.

It remains to bound $\Pr[\Delta(\mathbf{x}, \mathbf{x}') < n/2]$. Consider a thought experiment in which the channel additionally queries $2n - 1$ more distinct random messages before querying $\mathbf{m}$ and $\mathbf{m}'$. By the Plotkin bound, the encodings of some two messages will have Hamming distance less than $n/2$. Since all messages are random and each encoding is independent (due to the sender's statelessness), $\Pr[\Delta(\mathbf{x}, \mathbf{x}') < n/2] \geq 1/\binom{2n+1}{2} = \Omega(1/n^2)$. This completes the proof for binary alphabets.

For the large-alphabet case, we apply a similar (but simpler) argument. Since *all* pairs of codewords of length n are within Hamming distance n for any alphabet, an adversarial channel with any error rate $> 1/2$ can cause incorrect decoding with probability $1/2$.

Proposition 2. *Consider any dynamic coding scheme for parameters $\{(n_i, k_i, q_i)\}_i$ where all the sender's inputs are known to the channel.*

If $q_i = 2$ and $2^{k_i} \geq 2n_i + 1$ for infinitely many i, an adversarial channel with any error rate $> 1/4$ can cause incorrect decoding with probability at least $\Omega(1/n)$.

For any values of q_i, an adversarial channel with any error rate $> 1/2$ can cause incorrect decoding with probability at least $1/2$.

These results hold even if the sender and receiver are randomized and stateful, the receiver has secret inputs, and the channel is polynomially bounded.

Proof. Because the sender is a polynomial-time algorithm with only public inputs, it can be "simulated" by the channel. That is, the channel $\mathcal{C}$ can simply use the encoding function as a subroutine. Thus, instead of making queries to the sender, $\mathcal{C}$ can simply simulate the process, encoding messages itself. This difference is essential: since the sender is not actually encoding these messages, his internal state remains unchanged. The channel can thus simulate the execution of the sender on a variety of inputs with the same internal state.

Once again we start with the binary case. As in the proof of Proposition 1, for any n and $\mathbf{x}, \mathbf{x}' \in \{0,1\}^n$, if $\Delta(\mathbf{x}, \mathbf{x}') < n/2$, define $M(\mathbf{x}, \mathbf{x}')$ to be some canonical $\mathbf{r}$ such that $\Delta(\mathbf{x}, \mathbf{r}) < \lceil n/4 \rceil$ and $\Delta(\mathbf{r}, \mathbf{x}') \leq \lceil n/4 \rceil$. We now describe an adversarial channel that can cause incorrect decoding:

The channel first makes a *real* query to the sender on a random message $\mathbf{m}_1$ and receives $\mathbf{x}_1$, its encoding. For block length n, the channel then *simulates* the encoding $2n$ more random, distinct messages $\mathbf{m}_2, \ldots, \mathbf{m}_{2n+1}$. (If the sender is stateful, the channel encodes with the sender's state *at the time $\mathbf{m}_1$ was encoded*.) Denote the encoding of $\mathbf{m}_i$ as $\mathbf{x}_i$. By the Plotkin bound, there exist $\mathbf{x}_i, \mathbf{x}_j$ such that $i \neq j$ and $\Delta(\mathbf{x}_i, \mathbf{x}_j) < n/2$. By symmetry and without loss of generality, $\Pr[i=1] \geq 2/(2n+1)$. In the event that $i=1$, the channel corrupts $\mathbf{x}_1$ in the following way: it sends $M(\mathbf{x}_1, \mathbf{x}_j)$ with probability $1/2$, and $M(\mathbf{x}_j, \mathbf{x}_1)$ otherwise.

Conditioned on $i=1$, the recipient's view is distributed identically to the case where the roles of $\mathbf{x}_1$ and $\mathbf{x}_j$ are switched, and $\mathbf{x}_j$ was the real query. Therefore the recipient will decode incorrectly with probability $\Omega(1/n)$.

For large alphabets, a similar argument works with only one simulated message encoding.

Proposition 3. *Consider any dynamic coding scheme for parameters $\{(n_i, k_i, q_i)\}_i$ where and receiver is stateless and has only public inputs.*

If $q_i = 2$ and $2^{k_i} \geq 2n_i + 1$ for infinitely many i, a computationally unbounded adversarial channel with any error rate $> 1/4$ can cause incorrect decoding with probability at least $1/2$.

For any values of q_i, a computationally unbounded adversarial channel with any error rate $> 1/2$ can cause incorrect decoding with probability at least $1/2$.

These results hold even if the sender and receiver are randomized, and the sender has a public key.

Proof. We again start with the case of binary alphabets. Note that because the receiver is stateless, its output distribution on a given word $\mathbf{r}$ is always the same, regardless of what transmissions have preceded $\mathbf{r}$. We now describe an unbounded adversarial channel that can cause incorrect decoding:

For block length n, the channel will make up to $2n+1$ arbitrary distinct message queries $\mathbf{m}_1, \ldots, \mathbf{m}_{2n+1}$. Denote the sender's encoding of $\mathbf{m}_i$ as $\mathbf{x}_i$. When receiving $\mathbf{x}_j$, the channel exhaustively searches all $\mathbf{r}$ such that $\Delta(\mathbf{x}_j, \mathbf{r}) \leq \lceil n/4 \rceil$. Because the receiver is stateless and only has *public* inputs, the unbounded channel can compute the receiver's output distribution for word $\mathbf{r}$. There are two cases: (1) if for some $\mathbf{r}$ the receiver would fail to output $\mathbf{m}_j$ with probability $\geq 1/2$, the channel corrupts $\mathbf{x}_j$ as $\mathbf{r}$, sends it to the receiver, and halts; (2) otherwise, the channel sends $\mathbf{x}_j$ uncorrupted and makes the next query.

We now need only argue that the case (1) eventually occurs for some query. By the Plotkin bound, for some j there exists $i < j$ and an $\mathbf{r}$ such that $\Delta(\mathbf{x}_i, \mathbf{r}) \leq \lceil n/4 \rceil$ and $\Delta(\mathbf{r}, \mathbf{x}_j) < n/4$. Suppose that for queries $1, \ldots, j-1$, case (1) did not occur. Then by this assumption, on input $\mathbf{r}$ the receiver outputs $\mathbf{m}_i$ with probability $\geq 1/2$. Therefore $\mathbf{r}$ is close enough to $\mathbf{x}_j$ but fails to decode to $\mathbf{m}_j$ with probability at least $1/2$, and we are done.

For large alphabets, a similar argument works with only two distinct message queries.

Acknowledgements

We are grateful to Adam Smith for helpful discussions, including his explanation of the results of Langberg [8] in our framework. We also thank the anonymous reviewers for their helpful and thorough comments.

References

1. O. Goldreich, S. Goldwasser, and S. Micali. How to construct random functions. *J. ACM*, 33(4):792–807, 1986.
2. S. Goldwasser, S. Micali, and R. L. Rivest. A digital signature scheme secure against adaptive chosen-message attacks. *SIAM J. Comput.*, 17(2):281–308, 1988.
3. P. Gopalan, R. J. Lipton, and Y. Z. Ding. Error correction against computationally bounded adversaries. Manuscript, October 2004.
4. V. Guruswami. List decoding with side information. In *18th IEEE Annual Conference on Computational Complexity*, pages 300–312, 2003.
5. V. Guruswami, J. Håstad, M. Sudan, and D. Zuckerman. Combinatorial bounds for list decoding. In *Proceedings of the 38th Annual Allerton Conference on Communication, Control and Computing*, 2000.
6. V. Guruswami and M. Sudan. Improved decoding of reed-solomon and algebraic-geometric codes. In *IEEE Symposium on Foundations of Computer Science*, pages 28–39, 1998.
7. V. Guruswami and M. Sudan. List decoding algorithms for certain concatenated codes. In *Proceedings of the thirty-second annual ACM symposium on Theory of computing*, pages 181–190. ACM Press, 2000.
8. M. Langberg. Private codes or succinct random codes that are (almost) perfect. In *Proceedings of the forty-fifth annual IEEE Symposium on Foundations of Computer Science*, 2004.
9. R. J. Lipton. A new approach to information theory. In *Proceedings of the 11th Annual Symposium on Theoretical Aspects of Computer Science*, pages 699–708. Springer-Verlag, 1994.
10. J. Rompel. One-way functions are necessary and sufficient for secure signatures. In *Proceedings of the Twenty-Second Annual ACM Symposium on Theory of Computing*, pages 387–394. ACM Press, 1990.

Hardness Amplification of Weakly Verifiable Puzzles

Ran Canetti*, Shai Halevi, and Michael Steiner

IBM T.J. Watson Research Center, Hawthorne, NY, USA
{canetti, msteiner}@watson.ibm.com, shaih@alum.mit.edu

Abstract. Is it harder to solve many puzzles than it is to solve just one? This question has different answers, depending on how you define puzzles. For the case of inverting one-way functions it was shown by Yao that solving many independent instances simultaneously is indeed harder than solving a single instance (cf. the transformation from weak to strong one-way functions). The known proofs of that result, however, use in an essential way the fact that for one-way functions, verifying candidate solutions to a given puzzle is easy. We extend this result to the case where solutions are efficiently verifiable *only by the party that generated the puzzle*. We call such puzzles **weakly verifiable**. That is, for weakly verifiable puzzles we show that if no efficient algorithm can solve a single puzzle with probability more than ε, then no efficient algorithm can solve n independent puzzles simultaneously with probability more than ε^n. We also demonstrate that when the puzzles are not even weakly verifiable, solving many puzzles may be no harder than solving a single one.

Hardness amplification of weakly verifiable puzzles turns out to be closely related to the reduction of soundness error under parallel repetition in computationally sound arguments. Indeed, the proof of Bellare, Impagliazzo and Naor that parallel repetition reduces soundness error in three-round argument systems implies a result similar to our first result, albeit with considerably worse parameters. Also, our second result is an adaptation of their proof that parallel repetition of four-round systems may not reduce the soundness error.

1 Introduction

This work is concerned with the fundamental question of hardness amplification via parallel repetition. Suppose we knew that no efficient device succeeds in some computational task with probability much better than ε. Then what can we say about the success probability of efficient devices in performing n such tasks in parallel? The answer clearly depends on the type of task at hand. For the simple case where the task is inverting an efficiently computable function, the answer implicit in the groundbreaking work of Yao [13] is that the success probability cannot be much more than ε^n. A proof can be found in Goldreich's book [6– Chapter 2.3]. However, this proof relies heavily on the ability to efficiently verify

---

* Supported by NSF CyberTrust Grant #0430450.

the correctness of a candidate solution: That is, on input y in the range of the function f and candidate x from its domain, it is possible to efficiently verify that indeed $y = f(x)$. A natural question is whether parallel repetition amplifies hardness also for other types of puzzles, and in particular for the case where the entity posed with the puzzle cannot efficiently verify on its own the correctness of candidate solutions. (Following [12, 8], we use the term "puzzles" to denote somewhat-hard automatically-generated computational problems.)

We identify a more general class of puzzles for which parallel repetition indeed amplifies hardness. Specifically, we show that the same hardness amplification result holds even in the case where only the entity generating the puzzles can efficiently verify correctness of candidate solutions. More precisely, we consider the case where puzzles are generated (by some efficient algorithm) together with some "secret check information". Efficient verification of correctness of candidate solutions for a puzzle is guaranteed only if the corresponding secret check information is known. In particular, the entity posed with the puzzle may not be able to efficiently verify correctness of candidate solutions. We call such puzzles **weakly verifiable**. We show that, even in this setting, if no efficient algorithm can solve a single puzzle with probability much more than ε, then no efficient algorithm can simultaneously solve n puzzles with probability much more than ε^n, which is essentially optimal.[1] We also show that the weak verifiability property is essential for obtaining such a general hardness amplification result: We exhibit an example of puzzles that are not even weakly verifiable, and where the probability of solving multiple instances is the same as the probability of solving a single instance.

One example of weakly verifiable puzzles is the notion of computer-generated inverse Turing tests, or CAPTCHAs [11, 1]. These are distribution of puzzles that are easily solvable by humans, but are assumed to be solvable by computers only with small (albeit noticeable) probability. Automatically verifying a solution to a given CAPTCHA is typically just as hard as solving it, since the space of solutions is fairly small. Still, CAPTCHAs are weakly verifiable, as it is possible to efficiently generate a CAPTCHA together with its unique solution. In the work of von-Ahn et al. [1], they suggests *sequential* repetition as a method of hardness amplification for CAPTCHAs. Our work indicates that parallel repetition can be used as well. We note that puzzles that are only weakly verifiable can be constructed from one-way functions (e.g., the puzzle is $f(x)$ and the solution is a hard-core bit of x). On the other hand, our definition in Section 2 does not imply one-way functions. See Section 2.2 for more discussion on the definition of puzzles and relations to other notions of computational hardness.

1.1 Soundness Amplification of Argument Systems

Bellare, Impagliazzo, and Naor [2] investigated the problem of reducing the soundness error of interactive argument systems (i.e., proof systems with compu-

---

[1] As usual, the analysis incurs slackness of negligible quantities. This in particular means that the amplification is only meaningful when ε is not negligible.

tational soundness). They showed that for three-round systems, n-fold parallel repetition reduces the soundness error exponentially in n, whereas for four-round systems parallel repetition may not reduce the soundness error at all. This problem turns out to be closely related to ours: For three-round systems, once the first prover message is fixed, the remaining two messages can be regarded as a weakly verifiable puzzle sent by the verifier to the prover, followed by a solution candidate sent by the prover. When there are more than three rounds, this puzzle may not be even weakly verifiable without additional communication. Indeed, the result in [2] for three-round argument systems implies that parallel repetition of weakly verifiable puzzles reduces the success probability exponentially. Similarly, their example of a four-round system whose soundness error is not reduced by parallel repetition can be translated to a family of puzzles (that are not weakly verifiable) where parallel repetition does not amplify hardness.

In terms of concrete parameters, however, the result in [2] is far from optimal. Specifically, they show that if no algorithm can solve a single puzzle much better than ε, then no algorithm can solve n independent puzzles simultaneously with probability much better than δ^n, where $\delta \approx \exp\left(\frac{-(1-\varepsilon)^2}{128}\right)$. Note that δ is quite close to one: we always have $\delta > \exp(-1/128) > 0.99$, regardless of ε. In particular, if ε is small (say, $\varepsilon = 1/\text{poly}$), then it may take many repetitions before any amplification whatsoever is guaranteed. Hence, although the bound in [2] suffices for an asymptotic result, it is not very useful for the case of amplifying hardness of moderately hard weakly verifiable puzzles. We remark that our improved bounds apply also to the problem of parallel repetition of argument systems. Also there, we obtain optimal bounds.

1.2 Our Techniques

To show hardness amplification, we need to transform an algorithm A that solves n puzzles with probability ε^n into another algorithm A' that solves a single puzzle with probability ε. We consider the following matrix M that represents n-vectors of puzzles: The columns are labeled by all the possibilities for the first puzzle, and rows are labeled by all the possibilities for puzzles $2..n$, so each entry in the matrix corresponds to a particular n-vector of puzzles. Each entry consists of two bits, where the first bit is 1 if the answer-vector that A returns for these n puzzles includes a correct solution to the first puzzle, and the second bit is 1 if it includes correct solutions to all the puzzles $2..n$.

We make the following combinatorial observation. Assume that the fraction of (1,1) entries in M is at least some positive γ, and let α, β be positive numbers such that $\gamma = \alpha \cdot \beta$. Then, either M has some column with α-fraction of entries of the form $(\star, 1)$, or else the conditional probability of a (1,1) entry given that the entry is of the form $(\star, 1)$ is at least β.

We use this combinatorial observation with $\alpha = \varepsilon^{n-1}$, $\beta = \varepsilon$ and $\gamma = \alpha \cdot \beta = \varepsilon^n$, and indeed we know that the fraction of $(1, 1)$ entries in the matrix is at least ε^n. We thus conclude that either there exists a particular puzzle x, such that when x is the first puzzle, A correctly solves all the other puzzles with probability ε^{n-1}, or the conditional probability of a correct answer to the first puzzle when the answers to all the others are correct is at least ε. In the former case we trivially get an algorithm that solves $n-1$ puzzles with probability ε^{n-1} and we can continue by induction. In the latter case we directly get an algorithm that solves a single puzzle with probability ε: Simply choose many random $(n-1)$-vectors of puzzles, insert the input puzzle at the beginning to form an n-vector, and run A on the resulting vector. Repeat this process until we get correct answers to the last $(n-1)$ puzzles. Then we guess that we also have the right answer to the input puzzle and output that answer. The heart of the analysis is proving that this strategy indeed yields success probability (close to) ε.

Relations to xor-lemma proofs. We comment that there is some parallel between proofs of hardness amplification and proofs of xor lemmas, and indeed our proof is reminiscent of the xor-lemma proof of Myers [10]. In particular, he too has two asymmetric cases, where in one case we only get dimension reduction by one but it is essentially "for free", and in the other case we directly go to dimension one but pay some polynomial factor in complexity.

2 Notations and Definitions

Below we use the term *efficient algorithms* as a synonym to (probabilistic) polynomial-time Turing machines. A function (from positive integers to positive real numbers) is *negligible* if it approaches zero faster than any inverse polynomial. (Also, we informally say that something is *noticeable* when it is larger than some inverse polynomial.) We use $\mathsf{negl}(\cdot)$ to denote an unspecified negligible function. We also use the notation $\tilde{O}(x)$ as a shorthand for $O(x \log^c x)$ for some constant c.

2.1 Puzzles

A system for weakly verifiable puzzles consists of algorithms for generating random puzzles and for verifying solutions to these puzzles. Specifically, it consists of a pair of efficient algorithms $\mathcal{Z} = (G, V)$, such that

- The puzzle-generator algorithm G, on security parameter k, outputs a random puzzle p along with some "check information" c, $(p, c) \xleftarrow{\$} G(1^k)$.
- The "puzzle verifier" V is a deterministic efficient algorithm that on input a puzzle p, check-information c, and answer a, outputs either zero or one, $V(p, c, a) \in \{0, 1\}$.

A *solver* for this puzzle system is an efficient algorithm S that gets a puzzle p as input and outputs an answer a. The *success probability* of S is the probability that the answer is accepted by the puzzle verifier,

$$\mathsf{succ}_{\mathcal{Z}}[S] \stackrel{\text{def}}{=} \Pr_{G,S}\left[(p, c) \xleftarrow{\$} G(1^k),\ a \xleftarrow{\$} S(p)\ :\ V(p, c, a) = 1\right]$$

where the probability is taken over the randomness of G and S. (Note that $\mathsf{succ}_{\mathcal{Z}}[S]$ is a function of the security parameter k.) The *hardness* of the puzzle system $\mathcal{Z}$ is a bound on the success probability of any efficient solver.

Definition 1 (Hardness of puzzles). *Let $\varepsilon : \mathbb{N} \to [0,1]$ be an arbitrary function. A puzzle-system $\mathcal{Z}$ is said to be $(1-\varepsilon)$-hard if for any efficient solver S, there is a negligible function* negl *such that* $\mathsf{succ}_{\mathcal{Z}}[S] \leq \varepsilon + \mathsf{negl}$.

Repetition. Let $\mathcal{Z} = (G, V)$ be a puzzle system, and let $n : \mathbb{N} \to \mathbb{N}$ be an arbitrary function. We denote by G^n the algorithm that on security parameter k runs $G(1^k)$ for $n(k)$ times and outputs all the n puzzles with their check information, $(\langle p_1, \ldots, p_n \rangle, \langle c_1, \ldots, c_n \rangle) \stackrel{\$}{\leftarrow} G^n(1^k)$. Similarly, we denote by V^n the function that gets three n-vectors $\boldsymbol{p}, \boldsymbol{c}, \boldsymbol{a}$ and outputs one if and only if $V(p_i, c_i, a_i) = 1$ for all $i \in \{1, ..., n\}$. The n-fold repetition of $\mathcal{Z}$ is the puzzle system $\mathcal{Z}^n = (G^n, V^n)$.

2.2 Discussion

We discuss some aspects of the definition of puzzles, and relate it to other notions of computational hardness. First, note that our definition of weakly-verifiable puzzles allows the veracity of answer a to puzzle p to depend on the check-information c. Namely, we allow the possibility of two outputs (p, c), and (p, c') in the support of G (with the same p but different c's), such that for some answer a it holds that $V(p, c, a) = 1$ but $V(p, c', a) = 0$. This extra generality may seems somewhat non-intuitive at first. In particular, it allows the hardness of solving the puzzle to be information-theoretic. (For example, consider a system where the puzzle is always the all-zero string, the check information is a random k-bit string, and an answer is accepted if it equals the check information.) We chose this more general formulation since defining things this way is slightly simpler, and because it captures also the soundness in proof systems where soundness is argued unconditionally. Also, it makes our result a bit stronger (since it works even for this wider class of puzzles).

Still, the interesting cases for hardness amplification are usually the ones where the veracity of the solution does not depend on the check information. Notice that in such systems the hardness can only be computational. (Indeed, an infinitely powerful solver, on input puzzle p, can exhaustively search for a pair (c, a) where (p, c) is in the support of G and $V(p, c, a) = 1$.) Below we therefore call such systems weakly-verifiable *computational* puzzle systems.

It may be instructive to relate the notion of weakly-verifiable computational puzzles to the notion of average-case hardness due to Levin [9]. Recall that according to Levin, a *distributional problem* is a pair $(\mathcal{P}, \mathcal{D})$, where $\mathcal{P}$ is a (search or decision) problem and $\mathcal{D}$ is a distribution on the instances of $\mathcal{P}$. Hence, weakly-verifiable puzzles are a special case of distributional search problems. Most of the literature concerning Levin's theory is focused on the study of the case where $\mathcal{P}$ is an NP-problem (either search or decision). From the perspective of the current work, this means that candidate solutions are always efficiently verifiable. Hence the previous proofs of Yao's theorem can be used just as well to prove hardness amplification for all these prior notions.

In contrast, in this work we consider search problems that *are not even in NP*. When cast as distributional search problems, a weakly-verifiable computational puzzle system is a distributional search problem $(\mathcal{P}, \mathcal{D})$ where $\mathcal{D}$ is efficiently sampleable and the relation $R_{\mathcal{P}} = \{(p, a) : a \text{ is a correct solution for } p\}$, is *not necessarily efficiently computable*. (When viewed as a language, $R_{\mathcal{P}}$ is itself in NP, with the witness roughly being the check information c.)[2] Hence, the class of distributional search problems that result from weakly-verifiable computational puzzle systems is a superset of the class $\langle$NP, P-sampleable$\rangle$ of Ben-David et al. [3], in which $R_{\mathcal{P}}$ in an NP-relation (where pairs can be recognized in polynomial time).

The class $\langle$NP, P-sampleable$\rangle$ is itself a superset of the class DistNP from Levin's work [9], in which the distribution $\mathcal{D}$ is P-computable. However, it was shown by Impagliazzo and Levin [7] that if $\langle$NP, P-sampleable$\rangle$ contains hard problems, then so does DistNP. It is also easy to see that a hard weakly-verifiable computational puzzle can be transformed into a hard problem in $\langle$NP, P-sampleable$\rangle$ (by changing the goal of the search problem from finding a to finding (a, c)). Hence, if any of these classes contains hard problems, then they all do. We finally comment that the existence of hard problems in these classes is not known to imply the existence of one-way functions.

3 Hardness Amplification of Puzzles

Theorem 1. *Let $\varepsilon : \mathbb{N} \to [0, 1]$ be an efficiently computable function, let $n : \mathbb{N} \to \mathbb{N}$ be efficiently computable and polynomially bounded, and let $\mathcal{Z} = (G, V)$ be a weakly verifiable puzzle system. If $\mathcal{Z}$ is $(1 - \varepsilon)$-hard, then $\mathcal{Z}^n$, the n-fold repetition of $\mathcal{Z}$, is $(1 - \varepsilon^n)$-hard.*

The core of the proof is a transformation that turns an algorithm A that solves (G^n, V^n) with probability δ^n (for some δ) into an algorithm A' that solves (G, V) with probability $\delta(1 - \frac{1}{q})$, where q is some "slackness parameter".[3] The running time of A' is polynomial in $n, q, 1/\delta^n$, and the running times of A, G, and V.

Lemma 1. *Fix efficiently computable functions, $n, q : \mathbb{N} \to \mathbb{N}$, and $\delta : \mathbb{N} \to (0, 1)$. Also fix a puzzle system $\mathcal{Z} = (G, V)$, and denote the running times of G, V, by T_G, T_V, respectively. If there exists a solver A for $\mathcal{Z}^n$ with success probability δ^n and running time T, then there exists also a solver A' for $\mathcal{Z}$ with success probability $\delta(1 - \frac{1}{q})$ and running time $T' = \tilde{O}\left(\frac{nq^3}{\delta^{2n-1}}(T + nT_G + nT_V)\right)$.*

---

[2] Technically, the witness has to be randomness of the generator when generating (p, c). The reason is that even a computational puzzle system can have "invalid check information" c' such that $V(p, c', a) = 1$ for an incorrect answer a, as long as the generator G has probability zero of outputting (p, c').

[3] The parameter q is introduced in order to achieve an "optimal" hardness amplification result, from ε to ε^n. One can instead set, say, $q = 2$, in which case you can only prove amplification from ε to $(2\varepsilon)^n$.

3.1 The Solver A'

Having input puzzle p the algorithm A' consists of two phases. Roughly, in a pre-processing phase, A' tries to find a puzzle p_1^*, such that when p_1^* is placed as the first puzzle in a vector, the algorithm A correctly solves all the other puzzles with probability at least δ^{n-1}. This pre-processing is the most time-consuming operation in the execution of A'. If such p_1^* is found, then A' makes a "recursive call to itself", using as a solver for $\mathcal{Z}^{n-1}$ the algorithm A with p_1^* hard-wired as the first puzzle.

If A' fails to find such puzzle p_1^*, it moves to the on-line phase, where it actually tries to solve its input puzzle p. This is done by repeatedly sampling $(n-1)$-vectors $(\boldsymbol{p}, \boldsymbol{c}) \leftarrow G^{n-1}(1^k)$, and running A on the n-vector $(p, \boldsymbol{p})$, getting the n-vector of answers $(a, \boldsymbol{a}) \leftarrow A(p, \boldsymbol{p})$. If the answers 2..$n$ are correct (i.e., $V^{n-1}(\boldsymbol{p}, \boldsymbol{c}, \boldsymbol{a}) = 1$) then A' "hopes that the solution to p is also valid", and outputs the first answer a. (If too many trials have passed without getting a correct answers for puzzles 2..n, then A' aborts.) A detailed description of A' follows, and the code for A' can be found in Figure 1.

Pre-processing phase: In the pre-processing phase, A' tries to find a prefix of $v \leq n-1$ puzzles that has high probability of residual success. That is, conditioned on this prefix, A solves the suffix of $n-v$ puzzles with probability at least δ^{n-v}. The pre-processing phase consists of iterations, where in the iteration i, A' already has a prefix of $i-1$ puzzles and it tries to add to it the i'th puzzle. This is done in a straightforward manner: let prefix be the prefix of length $i-1$ from the previous iteration. A' repeatedly chooses candidates to extend the prefix, and for each candidate p^* it estimates the residual-success probability of prefix $\circ\, p^*$. (I.e., the probability that A, on input (prefix, p^*, $n-i$ random puzzles), solves correctly the last $n-i$ puzzles.) If A' finds a candidate p^* for which the estimated probability is at least δ^{n-i}, then it adds it to the prefix and continues to the next iteration. We stress that since A' generates these last "$n-i$ random puzzles" by itself, it also has the corresponding check information so it can verify the solutions to these puzzles.

In iteration i, A' tries at most $N_i = \lceil \frac{6q}{\delta^{n-i+1}} \ln\left(\frac{18qn}{\delta}\right) \rceil$ candidates. If none of them yields estimated probability of δ^{n-i} then A' terminates the pre-processing and moves to the on-line phase. For each candidate, A' estimates the probability up to additive accuracy of $\delta^{n-i}/6q$ with confidence of $\delta/18qnN_i$. Namely, $\Pr[|\text{estimated} - \text{actual}| > \delta^{n-i}/6q] < 2\delta/18qnN_i$. Using Chernoff bound, one can see that it is sufficient to sample $M_i = O(\frac{q^2}{\delta^{n-i}} \ln(\frac{qn}{\delta}))$ points to get these accuracy and confidence bounds.

On-line phase: Going into the on-line phase, A' has an input puzzle p, and a prefix of $v \leq n-1$ puzzles, and we know the residual-success probability of that prefix was estimated to be at least δ^{n-v}. Due to the accuracy and confidence bounds that were used in the estimation above, we can assume that the actual probability is at least $\delta^{n-v}(1 - 1/6q)$ (and this assumption holds expect with very small probability).

> SOLVER $A'(p)$:   // Parameters: k, n, q, δ
> PREPROCESSING PHASE:
> 0. initialize prefix $\leftarrow$ empty-vector
> 1. **for** $i = 1$ **to** $n-1$
> 2. $p^* \leftarrow$ EXTEND-PREFIX(prefix, i)
> 3.   **if** $p^* = \perp$ **then** $v \leftarrow i$, **goto** ONLINE PHASE
> 4.   **else** prefix $\leftarrow$ prefix $\circ\, p^*$
> 5. $v \leftarrow n$, **goto** ONLINE PHASE
>
> ONLINE PHASE:
> 10. **if** $v = n$ // Base case, prefix has $n-1$ puzzles
> 11. $a \leftarrow A(\text{prefix}, p)$
> 12.   **return** a_n
> 13. **else**   // prefix has $v - 1 \leq n - 2$ puzzles
> 14.   **repeat** $\left\lceil \frac{6q \ln(6q)}{\delta^{n-v+1}} \right\rceil$ times:
> 15. $(\langle p_{v+1}, \ldots, p_n \rangle, \langle c_{v+1}, \ldots, c_n \rangle) \leftarrow G^{n-v}(1^k)$
> 16. $a \leftarrow A(\text{prefix}, r, p_{v+1}, \ldots, p_n)$
> 17.   **if** $V(p_i, c_i, a_i) = 1$ for all $i \in \{v+1, \ldots, n\}$ **then return** a_v
> 18.   **if** none of the repetitions succeeded **then abort**
>
> EXTEND-PREFIX(prefix, i):   // prefix has $i - 1$ puzzles
> 21. $N_i \leftarrow \left\lceil \frac{6q}{\delta^{n-i+1}} \ln\left(\frac{18qn}{\delta}\right) \right\rceil$
> 21. **repeat** N_i times:
> 22. $(p^*, c^*) \leftarrow G(1^k)$
> 23. $\bar{\mu}_{p^*} \leftarrow$ ESTIMATE-RES-SUCC-PROB(prefix $\circ\, p^*$, i)
> 24.   **if** $\bar{\mu}_{p^*} \geq \delta^{n-i}$ **then return** p^*
> 25. **return** $\perp$ // No good extension found
>
> ESTIMATE-RES-SUCC-PROB(prefix, i):   // prefix has i puzzles
> 30. $M_i \leftarrow \left\lceil \frac{84q^2}{\delta^{n-i}} \ln\left(\frac{18qn \cdot N_i}{\delta}\right) \right\rceil$
> 31. **repeat** M_i times:
> 32. $(\langle p_{i+1}, \ldots, p_n \rangle, \langle c_{i+1}, \ldots, c_n \rangle) \leftarrow G^{n-i}(1^k)$
> 33. $a \leftarrow A(\text{prefix}, p_{i+1}, \ldots, p_n)$
> 34.   this sample is successful if $V(p_j, c_j, a_j) = 1$ for all $j \in \{i+1, \ldots, n\}$.
> 35. **return** number-of-successes$/M_i$

Fig. 1. The solver A' for $\mathcal{Z}$

If the prefix is of length $n - 1$, then we know that A solves any single puzzle with probability at least $\delta(1 - 1/6q)$, so A' directly uses it to solve the input puzzle p. Otherwise, we know that the last iteration of the pre-processing phase failed to find an extension to the prefix that has estimated residual-success probability of δ^{n-v}. As we will show later, this means that with the given prefix, the conditional probability that A solves the v'th puzzle *given that it solves puzzles* $v + 1, \ldots, n$ is very close to δ. Thus, A' samples many random vectors of $n - v$ puzzles (with their check information) and use $A(\text{prefix}, p, \text{random puzzles})$

to try and solve the random puzzles. Since the overall residual-success probability with the given prefix is close to δ^{n-v+1} we expect to succeed after not much more than $1/\delta^{n-v+1}$ trials (but for technical reasons A' tries as many as $6q\ln(6q)/\delta^{n-v+1}$ times). Once A' gets an answer vector $\boldsymbol{a}$ that contains correct solutions to the random puzzles $v+1,\ldots,n$, it outputs the answer a_v (i.e., the one that corresponds to the input puzzle). If all the trials fail, then A' aborts.

3.2 Analysis of A'

In the analysis we refer directly to the code of A' from Figure 1. We begin with the running time of A'. The pre-processing phase consists of at most $n-1$ calls to EXTEND-PREFIX, where the i'th call makes at most $N_i = \left\lceil \frac{6q}{\delta^{n-i+1}} \ln\left(\frac{18qn}{\delta}\right) \right\rceil$ calls to ESTIMATE-RES-SUCC-PROB. The routine ESTIMATE-RES-SUCC-PROB (when called during the i'th iteration) goes through its loop for $M_i = \left\lceil \frac{84q^2}{\delta^{n-i}} \ln\left(\frac{18qnN_i}{\delta}\right) \right\rceil$ times, and each loop makes one call to A and $n-i < n$ calls to G and V. Thus, the total time of the pre-processing phase is less than

$$n \cdot \frac{6q}{\delta^n} \cdot O\left(\ln(\frac{qn}{\delta})\right) \cdot \frac{84q^2}{\delta^{n-1}} \cdot O\left(\ln(\frac{qn}{\delta})\right) \cdot (T + nT_G + nT_V) = \tilde{O}\left(\frac{nq^3}{\delta^{2n-1}}(T + nT_G + nT_V)\right)$$

The on-line phase of A consists of at most $6q\ln(6q)/\delta^{n-v+1} \leq 6q\ln(6q)/\delta^n$ repetitions of a loop, and each repetition makes one call to A and $n-v < n$ calls to G and V, so the time of this phase is $O(\frac{q\ln q}{\delta^n}(T + nT_G + nT_V))$. The total running time is therefore $\tilde{O}\left(\frac{nq^3}{\delta^{2n-1}}(T + nT_G + nT_V)\right)$, as stated in Lemma 1.

Next we analyze the success probability, and we begin with a few notations. For a vector $\boldsymbol{p}$ with $i \leq n-1$ puzzles, we denote the residual-success probability of $\boldsymbol{p}$ by

$$\mathsf{rsp}_i[\boldsymbol{p}] \stackrel{\mathrm{def}}{=} \Pr\left[\begin{array}{l}(\langle p_{i+1},\ldots,p_n\rangle,\langle c_{i+1},\ldots,c_n\rangle) \stackrel{\$}{\leftarrow} G^{n-i}(1^k),\ \boldsymbol{a} \stackrel{\$}{\leftarrow} A(\boldsymbol{p},\ p_{i+1},\ldots,p_n) \\ : V(p_j, c_j, a_j) = 1 \text{ for all } i+1 \leq j \leq n\end{array}\right].$$

(In this notation, it is assumed that the security parameter k is implicit in the puzzles in $\boldsymbol{p}$.) For a vector $\boldsymbol{p}$ with $i-1 \leq n-2$ puzzles, we denote by $\mathsf{Ext}_i(\boldsymbol{p})$ the set of "good extensions" of $\boldsymbol{p}$, namely those puzzles p^* such that the residual-success probability of $\boldsymbol{p} \circ p^*$ is noticeably more than δ^{n-i},

$$\mathsf{Ext}_i(\boldsymbol{p}) \stackrel{\mathrm{def}}{=} \{\ p^* :\ \mathsf{rsp}_i(\boldsymbol{p} \circ p^*) \geq \delta^{n-i}(1 + \frac{1}{6q})\ \}$$

Next, we let prefix_i be the random variable describing the prefix of length i after the i'th iteration in the pre-processing phase, if there is one (otherwise we let $\mathsf{prefix}_i = \perp$). By convention, $\mathsf{prefix}_0 = \Lambda \neq \perp$.

We say that iteration i in the pre-processing phase *makes the wrong decision*, either if it returns an extension to prefix_{i-1} with residual-success probability that is too low, or if it fails to find an extension even though many good extensions exist. Formally, the event Wrong_i is defined when $\mathsf{prefix}_{i-1} \neq \perp$, and one of the following holds:

(a) either $\text{prefix}_i \neq \bot$ (i.e., the routine EXTEND-PREFIX(prefix_{i-1}, i) returned some $p^* \neq \bot$), but $\text{rsp}(\text{prefix}_i) < \delta^{n-i}(1 - \frac{1}{6q})$,

(b) or $\text{prefix}_i = \bot$ (i.e., the routine EXTEND-PREFIX(prefix_{i-1}, i) returned $\bot$), but it holds that $\Pr[(p, c) \xleftarrow{\$} G(1^k) : p \in \text{Ext}_i(\text{prefix}_{i-1})] \geq \frac{\delta^{n-i+1}}{6q}$.

Claim. For all $i \leq n-1$, $\Pr[\text{Wrong}_i] \leq \frac{\delta}{6qn}$.

Proof. This claim essentially follows from the Chernoff bound. Recall that the Chernoff bound asserts that for any 0-1 random variable with mean μ, is we choose M independent samples of that variable and let $\bar{\mu}$ be their average, then for any $\gamma > 0$ we have

$$\Pr[\mu - \bar{\mu} > \gamma] < \exp\left(\frac{-M\gamma^2}{2\mu(1-\mu)}\right) < \exp\left(\frac{-M\gamma^2}{2\mu}\right)$$

(and the same expression also bounds $\Pr[\bar{\mu} - \mu > \gamma]$). Fix some i and assume that $\text{prefix}_{i-1} \neq \bot$, which means that the pre-processing phase indeed calls EXTEND-PREFIX(prefix_{i-1}, i). For any possible value of prefix_{i-1}, we now bound the probability of Wrong_i conditioned on this value of prefix_{i-1}. For each puzzle p^* that can be chosen by EXTEND-PREFIX in line 22, let $\mu_{p^*} = \text{rsp}_i(\text{prefix}_{i-1} \circ p^*)$. Note that the estimation routine uses $M_i = \left\lceil \frac{84q^2}{\delta^{n-i}} \ln\left(\frac{18qnN_i}{\delta}\right) \right\rceil$ samples to provide an estimate $\bar{\mu}_{p^*}$ for μ_{p^*}. Using the Chernoff bound, we have for any puzzle p^* such that $\mu_{p^*} \leq \delta^{n-i}(1 - \frac{1}{6q})$

$$\Pr\left[\bar{\mu}_{p^*} \geq \delta^{n-i}\right] \leq \Pr\left[\bar{\mu}_{p^*} - \mu_{p^*} > \delta^{n-i}/6q\right] < \exp\left(-M_i \cdot \frac{\delta^{2(n-i)}}{72q^2 \mu_{p^*}}\right)$$
$$< \exp\left(-M_i \cdot \frac{\delta^{n-i}}{72q^2}\right) < \exp\left(-\ln\left(\frac{18qnN_i}{\delta}\right)\right) = \frac{\delta}{18qnN_i}$$

and since the routine EXTEND-PREFIX examines at most N_i candidates p^*, the probability that it gets an estimate $\bar{\mu}_{p^*} \geq \delta^{n-i}$ for any candidate p^* with $\mu_{p^*} \leq \delta^{n-i}(1 - 1/6q)$ is at most $\delta/18qn$. Hence the probability of sub-case (a) is at most $\delta/18qn$. Similarly, for a puzzle p^* such that μ_{p^*} exactly equals $\delta^{n-i}(1 + \frac{1}{6q})$ we have

$$\Pr\left[\bar{\mu}_{p^*} < \delta^{n-i}\right] \leq \Pr\left[\bar{\mu}_{p^*} - \mu_{p^*} < -\delta^{n-i}/6q\right] < \exp\left(-M_i \cdot \frac{\delta^{2(n-i)}}{72q^2 \mu_{p^*}}\right)$$
$$< \exp\left(-M_i \cdot \frac{\delta^{n-i}}{72q^2(1 + \frac{1}{6q})}\right) < \exp\left(\frac{-84}{72(1 + \frac{1}{6q})} \ln\left(\frac{18qnN_i}{\delta}\right)\right) \leq \frac{\delta}{18qnN_i}$$

where the last inequality holds since $\frac{84}{72(1+1/6q)} \geq \frac{84}{72(1+1/6)} = 1$. Clearly, if $\mu_{p^*} > \delta^{n-i}(1 + 1/6q)$ then the probability of $\bar{\mu}_{p^*} < \delta^{n-i}$ is even smaller. We see that when the routine EXTEND-PREFIX picks any $p^* \in \text{Ext}(\text{prefix}_{i-1})$ in line 22, it returns that p^* in line 24 with probability more than $1 - \frac{\delta}{18qnN_i}$. On the other hand, if the probability weight of $\text{Ext}_i(\text{prefix}_{i-1})$ is more than $\frac{\delta^{n-i+1}}{6q}$, then the probability that none of the N_i candidates that EXTEND-PREFIX picks belongs to $\mathsf{Ext}_i(\mathsf{prefix}_{i-1})$ is at most

$$\left(1 - \frac{\delta^{n-i+1}}{6q}\right)^{N_i} = \left(1 - \frac{\delta^{n-i+1}}{6q}\right)^{\frac{6q}{\delta^{n-i+1}} \ln(18qn/\delta)} < \exp(-\ln(\frac{18qn}{\delta})) = \frac{\delta}{18qn}$$

We conclude that the probability of sub-case (b) is at most $\frac{2\delta}{18qn}$, and therefore the overall probability of the event Wrong_i is at most $\frac{\delta}{18qn} + \frac{2\delta}{18qn} = \frac{\delta}{6qn}$.

In the analysis below of the online phase, we therefore assume that the pre-processing phase never makes the wrong decision, and this assumption effects the overall error probability of A' by at most $(n-1)\frac{\delta}{6qn} < \frac{\delta}{6q}$.

The Online Phase of A'. Consider now the on-line phase of A'. Recall that this phase gets a prefix with $v - 1$ puzzles, prefix_{v-1}, and an input puzzle p, and that these two are independent (since the pre-processing phase is independent of the input). Below we denote the input puzzle by p_v^*, and we denote by c_v^* the corresponding check information (that A' never actually sees, but may determines the veracity of A''s answer.) Assuming that the pre-processing phase did not make a wrong decision, we know that

$$\mathsf{rsp}(\mathsf{prefix}_{v-1}) \geq \delta^{n-v+1}(1 - \frac{1}{6q}) \tag{1}$$

(since iteration $v - 1$ of the pre-processing returned some $p_{v-1} \neq \bot$). If $v = n$, this means that $\mathsf{rsp}(\mathsf{prefix}_{v-1}) \geq \delta(1 - 1/6q)$, so running $A(\mathsf{prefix}_{v-1}, p_v^*)$ and taking the last answer yields a correct solution to p_v^* with probability at least $\delta(1 - 1/6q)$.

The more interesting case to analyze is when $v < n$, which means that iteration v in the pre-processing phase failed to extend the prefix. Assuming again that this was not a wrong decision, it means that

$$\Pr[(p, c) \xleftarrow{\$} G(1^k) : p \in \mathsf{Ext}_v(\mathsf{prefix}_{v-1})] < \frac{\delta^{n-v+1}}{6q} \tag{2}$$

From now on, we fix some value for prefix_{v-1} for which Equations 1 and 2 hold. For convenience in the discussion below, we let E be the set of pairs (p, c) such that $p \in \mathsf{Ext}_v(\mathsf{prefix}_{v-1})$. Namely,

$$E \stackrel{\text{def}}{=} \{ (p, c) : \mathsf{rsp}_v(\mathbf{prefix}_{v-1} \circ p) \geq \delta^{n-v}(1 + \frac{1}{6q}) \}$$

and from Equation 2 we know that

$$\Pr[(p, c) \xleftarrow{\$} G(1^k) : (p, c) \in E] < \frac{\delta^{n-v+1}}{6q}.$$

Consider the experiment where we choose at random a single vector of $n - v + 1$ puzzles, $(p_j, c_j) \leftarrow G(1^k)$ for $j = v, \ldots, n$, and then run A to get $\boldsymbol{a} \leftarrow$

$A(\mathsf{prefix}_{v-1}, p_v, \ldots, p_n)$. We are interested in the "success" event where $\boldsymbol{a}$ contains the right answers to all the puzzles $p_v, \ldots, p_n$, and in the "almost success" event where we only know that the answers to $p_{v+1}, \ldots, p_n$ are right. For any pair (p_v, c_v) we let $w(p_v, c_v)$, $s(p_v, c_v)$, $a(p_v, c_v)$, respectively, be the probability weight of that pair, and the probabilities of "success" and "almost success" conditioned on it.

$$w(p_v, c_v) \stackrel{\text{def}}{=} \Pr[G(1^k) = (p_v, c_v)]$$

$$s(p_v, c_v) \stackrel{\text{def}}{=} \Pr\left[\begin{array}{l} (\langle p_{v+1}, \ldots, p_n \rangle, \langle c_{v+1}, \ldots, c_n \rangle) \leftarrow G^{n-v}(1^k), \\ \boldsymbol{a} \leftarrow A(\mathsf{prefix}_{v-1}, p_v, p_{v+1}, \ldots, p_n) \\ : V(p_j, c_j, a_j) = 1 \text{ for all } j \in \{v, \ldots, n\} \end{array}\right]$$

$$a(p_v, c_v) \stackrel{\text{def}}{=} \Pr\left[\begin{array}{l} (\langle p_{v+1}, \ldots, p_n \rangle, \langle c_{v+1}, \ldots, c_n \rangle) \leftarrow G^{n-v}(1^k), \\ \boldsymbol{a} \leftarrow A(\mathsf{prefix}_{v-1}, p_v, p_{v+1}, \ldots, p_n) \\ : V(p_j, c_j, a_j) = 1 \text{ for all } j \in \{v+1, \ldots, n\} \end{array}\right]$$

Clearly, the overall probability of "success" is exactly the residual-success probability of prefix_{v-1}, namely,

$$\sum_{(p_v, c_v)} w(p_v, c_v) s(p_v, c_v) = \mathsf{rsp}(\mathsf{prefix}_{v-1}) \geq \delta^{n-v+1}(1 - \frac{1}{6q}). \tag{3}$$

Also, from what we know about E

$$a(p_v, c_v) \leq \delta^{n-v}(1 + \frac{1}{6q}) \text{ for any } (p_v, c_v) \notin E, \text{ and } \sum_{(p_v, c_v) \in E} w(p_v, c_v) \leq \frac{\delta^{n-v+1}}{6q} \tag{4}$$

Recall that A', on input p_v^*, chooses at random many continuations $(p_{v+1}, \ldots, p_n)(c_{v+1}, \ldots, c_n)$ until it finds an answer vector $\boldsymbol{a}$ that is an "almost success". Then A' outputs a_v, and this is correct only if $\boldsymbol{a}$ is also a "success". This means that conditioned on not aborting, the success probability of A' is exactly $s(p_v^*, c_v^*)/a(p_v^*, c_v^*)$. That is, for any fixed pair (p_v^*, c_v^*) we have

$$\Pr[V(p_v^*, c_v^*, A'(p_v^*)) = 1 \mid A'(p_v^*) \text{ does not abort}] = s(p_v^*, c_v^*)/a(p_v^*, c_v^*) \tag{5}$$

Next, let B (for $\underline{B}$ad) be the set of input puzzles (and their associated check information) on which A' is unlikely to succeed. More specifically,

$$B \stackrel{\text{def}}{=} \{(p_v, c_v) \ : \ s(p_v, c_v) < \delta^{n-v+1}/6q\}. \tag{6}$$

It is easy to see that when $(p_v^*, c_v^*) \notin B$, then $A(p_v^*)$ almost never aborts. Indeed A' aborts only if it does not find puzzles $p_{v+1}, \ldots, p_n$ that A solves correctly after $\frac{6q \ln(6q)}{\delta^{n-v+1}}$ trials. As each trial success with probability $a(p_v^*, c_v^*) \geq s(p_v^*, c_v^*) \geq \frac{\delta^{n-v+1}}{6q}$, the probability that they all fail is at most $1/6q$,

$$\Pr_{A'}[A'(p_v^*) \text{ aborts}] \leq \frac{1}{6q}, \text{ for all } (p_v^*, c_v^*) \notin B \tag{7}$$

The last crucial observation that we need is that the sets B and E together cannot contribute too much to the probability of success. Namely,

$$\sum_{(p_v,c_v)\in B\cup E} w(p_v,c_v)s(p_v,c_v) \leq \sum_{(p_v,c_v)\in B} w(p_v,c_v)s(p_v,c_v) + \sum_{(p_v,c_v)\in E} w(p_v,c_v)s(p_v,c_v)$$

$$\leq \sum_{(p_v,c_v)\in B} w(p_v,c_v)\delta^{n-v+1}/6q + \sum_{(p_v,c_v)\in E} w(p_v,c_v)$$

$$\leq \frac{\delta^{n-v+1}}{6q} + \frac{\delta^{n-v+1}}{6q} = \frac{\delta^{n-v+1}}{3q}$$

and combined with Equation 3 we get

$$\sum_{(p_v,c_v)\notin B\cup E} w(p_v,c_v)s(p_v,c_v) \geq \delta^{n-v+1}(1-\frac{1}{6q}) - \frac{\delta^{n-v+1}}{3q} = \delta^{n-v+1}(1-\frac{1}{2q}) \quad (8)$$

Putting everything together, we have

$$\Pr[A' \text{ answers correctly}] = \sum_{(p_v^*,c_v^*)} w(p_v^*,c_v^*) \cdot \Pr[V(p_v^*,c_v^*,A'(p^*))=1]$$

$$\geq \sum_{(p_v^*,c_v^*)\notin B\cup E} w(p_v^*,c_v^*) \cdot (1-\Pr[A'(p_v^*) \text{ aborts}])$$

$$\cdot \Pr[V(p_v^*,c_v^*,A'(p_v^*))=1 \mid A'(p_v^*) \text{ does not abort}]$$

$$\stackrel{(a)}{\geq} \sum_{(p_v^*,c_v^*)\notin B\cup E} w(p_v^*,c_v^*) \cdot (1-\frac{1}{6q}) \cdot \frac{s(p_v^*,c_v^*)}{a(p_v^*,c_v^*)}$$

$$\stackrel{(b)}{\geq} \sum_{(p_v^*,c_v^*)\notin B\cup E} w(p_v^*,c_v^*) \cdot (1-\frac{1}{6q}) \cdot \frac{s(p_v^*,c_v^*)}{\delta^{n-v}(1+1/6q)}$$

$$= \frac{1-1/6q}{\delta^{n-v}(1+1/6q)} \sum_{(p_v^*,c_v^*)\notin B\cup E} w(p_v^*,c_v^*)s(p_v^*,c_v^*)$$

$$\stackrel{(c)}{\geq} \frac{1-1/6q}{\delta^{n-v}(1+1/6q)} \cdot \delta^{n-v+1}(1-1/2q) = \delta \cdot \frac{(1-1/6q)(1-1/2q)}{1+1/6q} > \delta(1-\frac{5}{6q})$$

where inequality (a) is due to Equations 5 and 7, inequality (b) is due to (the first part of) Equation 4, and inequality (c) is due to Equation 8. We conclude that the probability of a wrong decision in the preprocessing phase is at most $\delta/6q$, and that if no wrong decisions were made then the online phase of A' solves the input puzzle with probability at least $\delta(1-5/6q)$, hence the overall success probability of A' is at least $\delta(1-5/6q) - \delta/6q = \delta(1-1/q)$. This completes the proof of Lemma 1. □

3.3 Proof of Theorem 1

All that is left now is to provide an asymptotic interpretation to the concrete bounds from Lemma 1. Let $\mathcal{Z}$ be a $(1-\varepsilon)$-hard puzzle system, and assume toward contradiction that there exists a T-time solver S^n that (for infinitely many k's) solves (G^n, V^n) with probability at least $\varepsilon(k)^n + 1/r(k)$, where both $T(\cdot)$, $r(\cdot)$ are polynomials.

Let us denote $q = 4nr$, and let δ be the solution to $\delta^n = \varepsilon^n + 1/r$. We note that since δ^n is noticeably larger than ε^n, then also δ is noticeably larger than ε, specifically $\delta > \varepsilon + 1/2rn$. To see this, denote $\gamma = \delta - \varepsilon$, and assume toward contradiction that $\gamma < 1/2rn < 1/n$. Then we have

$$(\varepsilon + \gamma)^n = \varepsilon^n + \sum_{t=1}^{n} \binom{n}{t} \varepsilon^{n-t} \gamma^t = \varepsilon^n + \gamma \sum_{t=1}^{n} \binom{n}{t} \varepsilon^{n-t} \gamma^{t-1}$$

$$< \varepsilon^n + \gamma \sum_{t=1}^{n} \binom{n}{t} \frac{\varepsilon^{n-t}}{n^{t-1}} < \varepsilon^n + \gamma \sum_{t=1}^{n} \frac{n^t}{t!} \cdot \frac{\varepsilon^{n-t}}{n^{t-1}}$$

$$= \varepsilon^n + \gamma n \sum_{t=1}^{n} \varepsilon^{n-t}/t! < \varepsilon^n + 2\gamma n$$

Thus, $\varepsilon^n + 2\gamma n > (\varepsilon + \gamma)^n = \delta^n = \varepsilon^n + 1/r$, and therefore $\gamma > 1/2rn$, contradiction.

Applying Lemma 1 with the given n, q, δ and T, we get an algorithm S for solving (G, V), with success probability

$$\delta(1 - 1/q) \geq (\varepsilon + 1/2rn)(1 - 1/4rn) > \varepsilon + 1/8rn,$$

which is noticeably larger than ε (since r, n are both polynomial in k). The running time of S is polynomial in n, q, T, T_G, T_V and $1/\delta^n$, which are all polynomial in k. (Note that $\delta^n = \varepsilon^n + 1/r$ is noticeable since r is polynomial in k.) This contradicts the $(1 - \varepsilon)$ hardness of $\mathcal{Z}$, concluding the proof of Theorem 1. □

4 The Weak Verification Property (Informal)

As discussed in the introduction, the proof of Theorem 1 relies on the fact that we have efficient generation and verification algorithms, so that A' can generate puzzles and recognize correct solutions to these puzzles. We now show that without this property, hardness amplification is not guaranteed. In particular, we describe a "puzzle system" where the verification algorithm is not efficient, and show that solving a few puzzles in parallel is not any harder than solving just one.

The example is essentially the one that was used by Bellare, Impagliazzo and Naor, except that we do not need the last two flows of their protocol. Assume that we have a non-interactive, perfectly-binding commitment scheme for one bit, $C(\cdot)$, and fix some parameter n. The generator for the puzzle system picks at random a bit b, and outputs as puzzle a random commitment $c = C(b)$. A potential solution to this puzzle is a vector of $n - 1$ commitments $c'_1, \ldots, c'_{n-1}$ such that (i) each c'_i is indeed a valid commitment to some bit b'_i, (ii) $c'_i \neq c$ for all i, and (iii) $b \oplus b'_1 \oplus \cdots \oplus b'_n = 0$. (Note that the last condition is well defined since the commitment scheme is perfectly binding.)

It is easy to see that if $C(\cdot)$ is non-malleable [5], then no efficient solver can solve this system with probability noticeably more than $1/2$. Informally, since the solver cannot return $c'_i = c$, then the bits b'_i must be almost independent of b, so their sum must also be almost independent of b, and therefore $\Pr[b \oplus b'_1 \oplus \cdots \oplus b'_n = 0] \approx 1/2$.

On the other hand, a solver that gets n random puzzles $c^1, \ldots, c^n$ can return as a solution to puzzle c^i all the other puzzles $c^j, j \neq i$. To analyze the success probability of the solver, let b_i denote the committed bit defined by commitment c_i. (b_i is well defined since the commitment is perfectly binding.) Then, with probability one half we have that $b_1 \oplus \cdots \oplus b_n = 0$. In this case, the solver has solved *all* the puzzles. (Indeed, the probability that any two of the c^i's are the same is negligible.)

The only problem with this example is that there are no known provable constructions of non-interactive, perfectly-binding, non-malleable commitment schemes in the "bare model". Such schemes are only known to exist in the common-random-string model [4] (or the common-reference-string model, or the random-oracle model, etc.) At the current state of affairs, this example is therefore only valid with respect to one of these models. However, since we only need non malleability with respect to one specific relation, it may be possible to devise such scheme in the bare model. In particular, assuming that such scheme exist seems like a rather reasonable assumption. (We comment that the negative result of Bellare, Impagliazzo, and Naor, for four-round proofs also requires non-interactive, non-malleable commitment schemes.)

References

1. L. von Ahn, M. Blum, N. Hopper, and J. Langford. CAPTCHA: Using hard AI problems for security. In *Advances in Cryptology - EUROCRYPT'03*, volume 2656 of *Lecture Notes in Computer Science*, pages 294–311. Springer-Verlag, 2003.
2. M. Bellare, R. Impagliazzo, and M. Naor. Does parallel repetition lower the error in computationally sound protocols? In *38th Annual Symposium on Foundations of Computer Science* (FOCS'97), pages 374–383. IEEE, 1997.
3. S. Ben-David, B. Chor, O. Goldreich, and M. Luby. On the theory of average case complexity. *Journal of Computer and System Sciences*, 44(2):193–219, 1992. Preliminary version in STOC'89.
4. G. Di Crescenzo, Y. Ishai, and R. Ostrovsky. Non-interactive and non-malleable commitment. In *Proceedings of the thirtieth annual ACM symposium on theory of computing (STOC'98)*, pages 141–150. ACM Press, 1998.
5. D. Dolev, C. Dwork, and M. Naor. Non-malleable cryptography. *SIAM J. on Computing*, 30(2):391–437, 2000. Preliminary version in STOC'91.
6. O. Goldreich. *Foundations of Cryptography, Basic tools*. Cambridge University Press, 2001.
7. R. Impagliazzo and L. A. Levin. No better ways to generate hard NP instances than picking uniformly at random. In *31st Annual Symposium on Foundations of Computer Science (FOCS'90)*, pages 812–821. IEEE, 1990.
8. A. Juels and J. Brainard. Client puzzles: A cryptographic defense against connection depletion attacks. In *Proceedings of the 1999 Network and Distributed System Security Symposium (NDSS'99)*, pages 151–165. Internet Society (ISOC), 1999.

9. L. A. Levin. Average case complete problems. *SIAM Journal of Computing*, 15(1):285–286, 1986. Preliminary version in STOC'84.
10. S. Myers. Efficient amplification of the security of weak pseudo-random function generators. *Journal of Cryptology*, 16(1):1–24, 2003. Extended Abstract appeared in EUROCRYPT 2001.
11. M. Naor. Verification of a human in the loop or identification via the Turing test. Manuscript, available on-line from http://www.wisdom.weizmann.ac.il/~naor/PAPERS/human_abs.html, 1996.
12. R. L. Rivest, A. Shamir, and D. A. Wagner. Time-lock puzzles and time-released crypto. Technical Report MIT/LCS/TR-684, MIT laboratory for Computer Science, 1996.
13. A. C. Yao. Theory and applications of trapdoor functions. In *23rd Annual Symposium on Foundations of Computer Science*, pages 80–91. IEEE, Nov. 1982.

A Results for Proof Systems

A computationally-sound proof system for a language $L \subseteq \{0,1\}^*$, is a pair $\mathcal{P} = (P, V)$ of polynomial time interactive Turing machines, who gets a common input string x, whose length is considered the security parameter. (The honest prover may also get some additional input that the verifier does not see.) Since we only care about soundness, then it is enough to consider only the verifier V. The success probability of a "cheating prover" B on input x is the probability that V accepts when interacting with B on common input x. The probability is taken over the randomness of B and V.

Definition 2 (Soundness error). *Fix a language L and a verifier V, and let $\varepsilon : \mathbb{N} \to (0,1)$ be some function. We say that V has soundness error at most ε if for any efficient prover B there is a negligible function* negl, *such that for any $x \notin L$, the success probability of $(B,V)(x)$ is at most $\varepsilon(|x|) + \mathsf{negl}(|x|)$.*

We comment that to be of interest, the proof system also has to satisfy some completeness property (say, for a prover that is given an NP witness for the membership of x in L). In this work, however, we are only interested in soundness.

The n-fold parallel repetition of a proof system $\mathcal{P} = (P, V)$ is defined in a straightforward manner: there are n independent copies of the verifier (and the honest prover), all running in "lock steps" in parallel on the same common input x. The n-fold parallel repetition of $\mathcal{P}$ is denoted $\mathcal{P}^n = (P^n, V^n)$.

A.1 Hardness Amplification for Three-Round Proof Systems

A rather straightforward adaptation of our result from Section 3 yields:

Theorem 2. *Let $\varepsilon : \mathbb{N} \to (0,1)$, $n : \mathbb{N} \to \mathbb{N}$ be efficiently computable functions, where n is polynomially bounded, and let $\mathcal{P} = (P,V)$ be an interactive proof system with three flows (or less) and soundness error at most ε. Then the proof system $\mathcal{P}^n$ has soundness error at most ε^n.*

Proof. (sketch) We again assume that there exists a cheating prover B such that for infinitely many common inputs $x \notin L$, the success probability of $(B, V^n)(x)$ is at least $\varepsilon(|x|)^n + 1/r(|x|)$ for some polynomial r. We show a cheating prover B' such that for those x'es, the success probability of $(B', V)(x)$ is noticeably more than ε.

Assume w.l.o.g. that the proof system has exactly three flows of communication, which means that the first flow is from the prover to the verifier. The prover B' first samples (polynomially) many first-flow messages of B, and for each one it estimates the success probability of $(B, V^n)(x)$ conditioned on that message. B' tries $N = poly(r, n)$ candidate first messages, and for each candidate it estimates the conditional success probability with accuracy $1/4r$ and confidence $1/32rnN$. Since the overall success probability of P is at least $\varepsilon^n + 1/r$, then P' can find a first-flow message y with estimated conditional success probability of at least $\varepsilon^n + 3/4r$ after trying only $N = poly(r, n)$ candidates. With the given accuracy and confidence bounds, it follows that except with probability $1/32rn$, P' indeed finds such first-flow message y, and the conditional success probability of y is at least $\varepsilon^n + 1/2r$.

Now P' consider the puzzle system with the generating algorithm defined by the verifier V^n on input x and y, and whose verifying function is the final verification procedure of V^n. (This does not quite satisfy our definition of the n-fold repetition of a puzzle system, since each copy now has a different generating and verifying procedures, because of the different first-flow messages. However, this difference does not change anything in the proof of Lemma 1.) P' applies the transformation from Lemma 1 with $\delta^n = \varepsilon^n + 1/2r$, and slackness $q = 8rn$, thus obtaining a strategy that convinces V with probability at least $\varepsilon + 1/16rn$. Since we have an error probability $1/32rn$ for choosing y, then the overall success probability of $(P', V)(x)$ is at least $\varepsilon + 1/32rn$, which is noticeably more than ε.

A comment about the common-reference-string model Computationally-sound proofs were define in the work of Bellare et al. [2] somewhat more generally than above: essentially they defined proofs in the common-reference-string model. We note, however, that their "positive result" (as well as ours) does not extend to this model. Indeed, they only show hardness amplification when the common reference string is fixed (so the soundness error is defined with respect to a fixed reference string, rather than with respect to a random choice of that string).

For example, one may think of a cheating prover B that on ε^n fraction of the reference strings is able to convince the verifier with probability one, and on other strings it always fails. It is not hard to see that no black-box reduction can transform this prover to one that succeeds in a single proof with probability more than ε^n.

On Hardness Amplification of One-Way Functions

Henry Lin, Luca Trevisan[*], and Hoeteck Wee[*]

Computer Science Division, UC Berkeley
{henrylin, luca, hoeteck}@cs.berkeley.edu

Abstract. We continue the study of the efficiency of black-box reductions in cryptography. We focus on the question of constructing strong one-way functions (respectively, permutations) from weak one-way functions (respectively, permutations). To make our impossibility results stronger, we focus on the weakest type of constructions: those that start from a weak one-way permutation and define a strong one-way function.

We show that for every "fully black-box" construction of a $\epsilon(n)$-secure function based on a $(1 - \delta(n))$-secure permutation, if $q(n)$ is the number of oracle queries used in the construction and $\ell(n)$ is the input length of the new function, then we have $q \geq \Omega(\frac{1}{\delta} \cdot \log \frac{1}{\epsilon})$ and $\ell \geq n + \Omega(\log 1/\epsilon) - O(\log q)$. This result is proved by showing that fully black-box reductions of strong to weak one-way functions imply the existence of "hitters" and then by applying known lower bounds for hitters. We also show a sort of reverse connection, and we revisit the construction of Goldreich et al. (FOCS 1990) in terms of this reverse connection.

Finally, we prove that any "weakly black-box" construction with parameters $q(n)$ and $\ell(n)$ better than the above lower bounds implies the unconditional existence of strong one-way functions (and, therefore, the existence of a weakly black-box construction with $q(n) = 0$). This result, like the one for fully black-box reductions, is proved by reasoning about the function defined by such a construction when using the identity permutation as an oracle.

1 Introduction

We continue the study of efficiency of reductions in cryptography, and we focus on the question of constructing strong one-way functions or permutations from weak one-way functions or permutations.

1.1 Efficiency of Cryptographic Reductions

Several fundamental results in the foundations of cryptography, most notably the proof that pseudorandom generators exist if one-way functions exist [HILL99], are proved via constructions and reductions that are too inefficient to be used

---

[*] Work supported by US-Israel BSF Grant 2002246.

in practice. It is natural to ask whether such inefficiency is a necessary consequence of the proof techniques that are commonly used, namely "black-box" constructions and reductions.

The first proof of a lower bound to the efficiency of a reduction for constructing a cryptographic primitive from another was by Kim, Simon and Tetali [KST99], in the context of constructing one-way hash functions from one-way permutations. Later work by Gennaro and Trevisan [GT00] and by Gennaro, Gertner and Katz [GGK03] has focused on constructions of pseudorandom generators from one-way permutations and of signature schemes and encryption schemes from trapdoor permutations.

The study of limitations of black-box reductions was initiated by Impagliazzo and Rudich [IR89], who showed that key agreement and public-key encryption cannot be based on one-way functions or one-way permutations using black-box reductions. Several other impossibility results for black-box reductions are known, including a result of Rudich [Rud88] and Khan, Saks, and Smyth [KSS00] ruling out constructions of one-way permutations based on one-way functions, a result of Rudich [Rud91] ruling out round-reduction procedures in public key encryption, and results of Gertner et al. [GKM+00] and Gertner, Malkin and Reingold [GMR01] giving a hierarchy of assumptions in public key encryption that cannot be proved equivalent using black-box reductions.

1.2 Black-Box Constructions

To illustrate the definition of a black-box construction, consider for example the notion of black box construction of a key agreement protocol based on one-way functions formalized in [IR89]. In this model, the one-way function $f()$ is given as an oracle, and the protocols A and B for Alice and Bob are oracle procedures with access to $f()$. This, for example, means that a protocol where the code (or circuit) of $f()$ is used in the interaction is not black-box as defined above. The security of the protocol is also defined in a black-box way as follows: we assume that there is a security reduction R (a probabilistic polynomial time oracle algorithm) such that if E is a procedure for Eve (of arbitrary complexity) that breaks (A^f, B^f), then $R^{E,f}$ inverts f on a noticeable fraction of inputs. Notice that a proof of security in which the code (or circuit) of the adversary E is used in the reduction would not fit the above model. This model, in which both the "one-way function" $f()$ and the adversary E are allowed to be of arbitrary complexity, is called the *fully black-box* model in [RTV04]. In all the above cited papers [IR89, Rud88, Rud91, KST99, KSS00, GT00, GKM+00, GMR01, GGK03], as well as in the results of this paper, fully black-box reductions are ruled out unconditionally.

A less restrictive model of black-box construction, introduced in [GT00] and formally defined in [RTV04], is the *weak black-box* model. As before, in a weakly black-box construction of key agreement from one-way functions, the algorithms for Alice and Bob are oracle algorithms that are given access to a function $f()$. A proof of security, however, only states that if $f()$ is hard to invert for efficient procedures that are given oracle access to $f()$, then the protocol is secure in the standard sense (that is, for adversaries that are ordinary probabilistic polynomial time algorithms with no oracles). In this model one is still not allowed to use the code of $f()$ in the construction or in the security analysis. The code of the adversary, however, may be used in the security analysis. Note that if we have a provably secure construction of, say, a key agreement protocol, then we also have a weakly black-box construction of key agreement based on one-way functions: just make the algorithms for Alice and Bob be oracle algorithms that never use the oracle. For this reason, one cannot unconditionally rule out weakly black-box constructions: at most, one can show that a weakly black-box construction implies the unconditional existence of some cryptographic primitive and possibly complexity theoretic separations that we do not know how to prove. Negative results for weakly black-box constructions are proved in [GT00] and [GGK03], where the authors show that weakly black-box constructions that make a small number of oracle queries imply the existence of one-way functions. The other lower bounds cited above [IR89, Rud88, Rud91, KST99, KSS00, GKM+00, GMR01], however, do not rule out weakly black-box reductions.

The reason for this lack of negative results about weakly black-box reductions is partly explained in [RTV04]. For example, Reingold et al. [RTV04] prove that, unless one-way functions exist and key agreement is impossible in the real world, then there is a weak black-box construction of key agreement based on one-way functions. In other words, from the existence of a weakly black-box construction of key agreement from one-way functions it is impossible to derive any other consequence besides the obvious one that the existence of one-way functions implies the existence of key agreement schemes. See [RTV04] for a precise statement of this result and for a discussion of its interpretation.

In this paper, we are able to prove unconditional lower bounds for fully black-box reductions, and to show that a weakly black-box reduction improving on our lower bounds implies the unconditional existence of one-way functions.

1.3 Amplification of Hardness

We say that a function $f()$ is $\alpha(n)$-secure[1] if for every family of polynomial-size oracle circuits $\{C_n\}$ and for all n, the probability that $C_n^f(f(x))$ outputs a preimage of $f(x)$ is at most $\alpha(n)$, where the probability is taken over the uniform choice of x from $\{0,1\}^n$. We say that a function $f()$ is a *strong one-way function* if it is computable in polynomial time and is also $\epsilon(n)$-secure, for $\epsilon(n) = n^{-\omega(1)}$. We say that $f()$ is a *weak one-way function* if it is computable in polynomial time and is also $(1-\delta(n))$-secure, for $\delta(n) = n^{-O(1)}$.

The problem of "amplification of hardness" is to deduce the existence of strong one-way functions (respectively, permutations) from the existence of weak one-way functions (respectively, permutations).

---

[1] We avoid definitions and statements in terms of concrete security as they do not directly apply to adversaries of arbitrary complexity in fully black-box reductions. The results in Section 5 may be restated with concrete security parameters in a straight-forward manner.

The "direct product" construction is a simple approach to prove amplification of hardness results. Given a weak one-way function $f()$ we define a new function $f'()$ as $f'(x_1, x_2, \ldots, x_{q(n)}) = (f(x_1), f(x_2), \ldots, f(x_{q(n)}))$, where $q(n)$ is a polynomial. The function f' is still computable in polynomial time, and a non-trivial analysis shows that if f is weak one-way then f' is strong one-way.[2] Furthermore, if $f()$ is a permutation then $f'()$ is a permutation. See for example [Gol01–Sec 2] for more details.

The direct product construction is not, however, "security-preserving," in that the input length of the new function is polynomially larger than the input length of the original function. (In a security-preserving construction, the input length of the new function would be linear in the input length of the original one.) See for example [Lub96] for a discussion of "security preserving" reductions and the importance, in a cryptographic reduction, of not increasing the input length of the new primitive by too much.

For one-way permutations, we do have a security-preserving construction due to Goldreich et al. [GIL$^+$90] based on random walks on expanders. We stress that our results do not rule out fully black-box security-preserving hardness amplification for one-way functions, and we hope that the connections presented in this paper will help resolve this open problem.

1.4 Our Results

We say that a polynomial time computable oracle function $F^{(\cdot)}$ is a fully black-box construction of $\epsilon(n)$-secure functions from $(1 - \delta(n))$-secure functions if there is a probabilistic polynomial time oracle algorithm $R^{(\cdot,\cdot)}$ such that for every function $f : \{0,1\}^n \to \{0,1\}^n$ and adversary $A()$ with the property that $A()$ inverts F^f on a $\geq \epsilon(n)$ fraction of inputs, then $R^{A,f}()$ inverts $f()$ on a $\geq 1 - \delta(n)$ fraction of inputs. From this definition it is immediate to see that if $f()$ is polynomial time computable and no polynomial time adversary can invert it on more than a $1 - \delta(n)$ fraction of inputs, then it follows that F^f is polynomial time computable and no polynomial time adversary can invert it on more than a $\epsilon(n)$ fraction of inputs. The definition, however, requires the reduction R to transform an adversary $A()$ of arbitrary complexity that inverts $F^f()$ into an adversary that inverts $f()$ in polynomial time given oracle access to $A()$ and $f()$.

A polynomial time computable oracle function $F^{(\cdot)}$ is a weak black-box construction of $\epsilon(n)$-secure functions from $(1 - \delta(n))$-secure functions if for every function $f : \{0,1\}^n \to \{0,1\}^n$ and polynomial time adversary $A()$ with the property that $A()$ inverts F^f on a $\geq \epsilon(n)$ fraction of inputs, then there is a polynomial time oracle adversary $R^{(\cdot)}$ such that R^f inverts $f()$ on a $\geq 1 - \delta(n)$ fraction of inputs.

Impossibility of Fully Black-Box Constructions. Our first main result is as follows.

---

[2] This approach is typically credited to [Yao82].

Theorem 1. *Let $F^{(\cdot)}$ be a fully black-box construction of $\epsilon(n)$-secure functions from $(1-\delta(n))$-secure permutations, let ℓ be the input length of $F^{()}$, n the length of inputs of the oracle function, and q be the number of oracle queries. Then $q \geq \Omega(\frac{1}{\delta}\log\frac{1}{\epsilon})$ and $\ell \geq n - O(\log q) + \Omega(\log\frac{1}{\epsilon})$.*

In comparison, the direct product construction has $q = O(\frac{1}{\delta}\log\frac{1}{\epsilon})$, which is tight, but $\ell = nq$. The construction of [GIL+90], that only works if the oracle is a permutation, has $q = O(\frac{1}{\delta}\log\frac{1}{\epsilon})$, which is tight, and $\ell = O(n + \log\frac{1}{\epsilon})$ for $\delta = n^{-O(1)}$, which is nearly tight.

It should be noted that our result applies even to constructions that require the oracle to be a permutation and that do not guarantee the new function to be a permutation. In particular, it applies as a special case to constructions that map permutations into permutation and functions into functions.

To prove Theorem 1, we first show that a fully black-box reduction of strong to weak one-way functions or permutations implies the existence of a disperser, or a "hitter" in the terminology of [Gol97] with efficiency parameters that depend on the efficiency of the reduction. A hitter is a randomized algorithm that outputs a small number of strings in $\{0,1\}^n$ such that for every sufficiently dense subset of $\{0,1\}^n$, the output of the hitter hits the set (that is, at least one of these strings is contained in the set) with high probability. In a hitter, we would like to use a small number of random bits, to generate a small number of strings, and we would like the density of the sets to be low and the probability of hitting them to be high. Various impossibility results are known for hitters and, in particular, if ℓ is the number of random bits, q is the number of strings, δ is the density of the sets and $1-\epsilon$ is the hitting probability, then it is known that $q \geq \Omega(\frac{1}{\delta}\cdot\log\frac{1}{\epsilon})$ and $\ell \geq n - \log q + \Omega(\log\frac{1}{\epsilon})$. Our negative results for fully black-box constructions will follow from the connection between such constructions and hitters and from the above negative results for hitters.

The intuition for the connection is, with some imprecision, as follows: a function $f: \{0,1\}^n \to \{0,1\}^n$ may be $(1-\delta)$-secure and still be extremely easy to invert on a $1-\delta$ fraction of inputs, while only a subset H of density δ of inputs is very hard to invert. If the computation of $F^f(z)$ involves only oracle queries to $f()$ on inputs outside H, then $F^f()$ is not "using" the hardness of $f()$, and $F^f(z)$ will be "easy" to invert. Considering that at most an ϵ fraction of inputs of $F^f()$ can be easy to invert, it follows that, for at least a $1-\epsilon$ fraction of the choices of z, the oracle queries of the computation $F^f(z)$ hit the set H. In conclusion, using ℓ random bits (to choose z) we have constructed q strings in $\{0,1\}^n$ (the oracle queries in the computation $F^f(z)$) such that a set of density δ (the set H) is hit with probability at least $1-\epsilon$. Of course none of this is technically correct, but the above outline captures the main intuition.

Moving on to a more precise description of our proof, we show that if $F^{(\cdot)}$ is a fully black-box construction of ϵ-secure functions from $(1-\delta)$-secure ones, where ℓ is input length of F, n the input length of the oracle, and q the number of oracle queries, then we can derive a hitter that uses randomness ℓ, produces q strings, and has hitting probability $1-\sqrt{\epsilon}$ for sets of density 2δ. The hitter, given an ℓ-bit random string z, simply outputs the oracle queries in the computation of $F^{\text{id}}(z)$, where id is the identity permutation.

Suppose that the construction is not a hitter as promised, then there is a set H of density 2δ such that $F^{\text{id}}(z)$ avoids querying elements of H for a $\sqrt{\epsilon}$ fraction of the z. Let $\pi : \{0,1\}^n \to \{0,1\}^n$ be a permutation that is the identity on elements not in H and that is a random permutation on H. Then F^{id} and F^π agree on at least a $\sqrt{\epsilon}$ fraction of inputs. We can also show that if A is a uniform (possibly, exponential time) algorithm that inverts F^{id} everywhere then A inverts F^π on at least a fraction ϵ of the inputs. (We would not lose this quadratic factor if we insisted that F^{id} and F^π be permutations.) In fact, such an A exists as long as F is polynomial time computable. Now we have that $R^{A,\pi}$ is a uniform algorithm that inverts π on at least a $1 - \delta$ fraction of inputs, using polynomial number of oracle queries into π. Restricting ourselves to H, we get that $R^{A,\pi}$ inverts π on at least $1/2$ of the elements of H, which is impossible because π is a random permutation over H, and it cannot be inverted on many inputs by a uniform procedure (regardless of running time) that makes a polynomial number of oracle queries [IR89, Imp96, GT00]. We have reached a contradiction, and so our construction was indeed a hitter as promised.

Impossibility of Weakly Black-Box Constructions. For weakly black-box constructions, we show that improving beyond our lower bounds is possible only by constructing strong one-way functions from scratch.

Theorem 2. *Let $F^{(\cdot)}$ be a weakly black-box construction of $\epsilon(n)$-secure permutations from $(1 - \delta(n))$-secure permutations, let ℓ be the input length of $F^{()}$, n the length of inputs of the oracle function, and q be the number of oracle queries.*

There are constants c_1, c_2, c_3 such that if $q \leq c_1 \frac{1}{\delta} \log \frac{1}{\epsilon}$ or $\ell \leq n - c_2 \log q + c_3 \log \frac{1}{\epsilon}$, then one-way permutations exist unconditionally and, in particular, F^{id} is a $(1 - \epsilon(n))$-secure permutation.

The proof is similar to the one of Theorem 1. We define a hitter based on the computation of F^{id} as before. If q and ℓ are too small, then the hitter must fail, and there must be some set H of density 2δ that is avoided with probability at least 2ϵ. Then we define a permutation π that is random on H and the identity elsewhere, and we note that F^{id} and F^π have agreement at least 2ϵ. If there were a polynomial time algorithm that inverts F^{id} on a $1 - \epsilon$ fraction of inputs, the same algorithm would invert F^π on a ϵ fraction of inputs. This would yield a polynomial time oracle algorithm that given oracle access to π, inverts π on a $1 - \delta$ fraction of inputs, which is again a contradiction. Therefore F^{id} is a (weak) one-way permutation.

A Reverse Connection. We also point out a reverse connection, namely that special types of hitters yield fully black-box amplification of hardness (F, R). Specifically, we require that the F satisfy two additional properties (apart from computing a hitter). Suppose F^f on input z queries f on $x_1, \ldots, x_q$. The first property tells us that inverting F^f on $F^f(z)$ is at least hard as inverting f on all of $f(x_1), \ldots, f(x_q)$. Next, given a challenge $f(x)$, the second property allows us to sample a challenge $F^f(z)$ (with the appropriate distribution) by substituting $f(x)$ for one of $f(x_1), \ldots, f(x_q)$. We may view both hardness amplification via direct product and via random walks on expanders [GIL+90] in this framework, which yields a more modular and arguably simpler presentation of both results.

1.5 Perspective

The new connection between fully black-box hardness amplification and hitters makes explicit the construction of hitters in previous results on hardness amplification (namely a hitter from independent sampling in amplification via direct product and from random walks on expanders in [GIL+90]) and shows that such a construction is in fact necessary. In addition, we see from [GIL+90] in order to address the major open problem in this area of research - whether we can achieve security-preserving hardness amplification for one-way functions, it would be sufficient to give a hardness amplification procedure based on (δ, ϵ)-hitters with randomness complexity $O(n + 1/\delta \log 1/\epsilon)$ (which is optimal up to constant factors for constant δ but not sub-constant δ). There are simple and direct constructions of hitters achieving such parameters, and reviewing these constructions may prove to be a fruitful starting point for resolving this open problem.

2 Preliminaries

2.1 Notation

We use U_n to denote the uniform distribution over $\{0,1\}^n$. Given a function $G : \{0,1\}^m \to (\{0,1\}^n)^k$, $G_i : \{0,1\}^m \to \{0,1\}^n$, for $i = 1, 2, \ldots, k$, is the function that on input z, outputs the i'th block of $G(z)$. In probability expressions that involve a probabilistic computation (of a probabilistic algorithm, say), the probability is also taken over the internal coin tosses of the underlying computation.

2.2 Notions of Reducibility

Here, we are only interested in hardness amplification wherein the construction of the strong one-way function f' uses black-box access to a weak one-way permutation f. However, we distinguish between fully black-box and weakly black-box constructions, following the work of [RTV04], depending on whether the proof of security is black-box.

Definition 1 (Fully Black-Box Amplification of Hardness). *A fully black-box construction of $\epsilon(n)$-secure functions from $(1 - \delta(n))$-secure permutations is a pair of polynomial time computable oracle procedures F and R (where F is deterministic whereas R may be randomized) such that, for every permutation $f : \{0,1\}^n \to \{0,1\}^n$, F^f is a function mapping $\ell(n)$ bits into $\ell(n)$ bits, and for every function $A : \{0,1\}^{\ell(n)} \to \{0,1\}^{\ell(n)}$, if*

$$\Pr_{z \sim U_{\ell(n)}}[A(F^f(z)) = z' : F^f(z') = F^f(z)] \geq \epsilon(n)$$

then
$$\Pr_{x \sim U_n}[R^{A,f}(f(x)) = x] \geq 1 - \delta(n) .$$

By requiring that F and R be polynomial time computable, we guarantee that if f and A are polynomial time computable, then F^f and $R^{A,f}$ are also polynomial time computable. However, (F, R) must also satisfy the stated property even when given oracle access access to some function f and A that may not be polynomial time computable.

Definition 2 (Weakly Black-Box Amplification of Hardness). *A weakly black-box construction of $\epsilon(n)$-secure functions from $(1 - \delta(n))$-secure permutations is a (deterministic) polynomial time computable oracle procedure F such that, for every permutation $f : \{0,1\}^n \to \{0,1\}^n$, F^f is a function mapping $\ell(n)$ bits into $\ell(n)$ bits, and if there is a probabilistic polynomial time algorithm A such that*
$$\Pr_{z \sim U_{\ell(n)}}[A(F^f(z)) = z' : F^f(z') = F^f(z)] \geq \epsilon(n)$$
then there is a probabilistic polynomial time oracle algorithm I such that
$$\Pr_{x \sim U_n}[I^f(f(x)) = x] \geq 1 - \delta(n) .$$

Remark 1. In both definitions, the new function defined by the construction is $\epsilon(n)$-secure on inputs of length $N = \ell(n)$, and so, according to our definition, it would be more precise to call it a $\epsilon(\ell^{(-1)}(N))$-secure function.

2.3 Hitters

Definition 3 (Hitter [Gol97]). *A function $G : \{0,1\}^m \to (\{0,1\}^n)^k$ is a (δ, ϵ)-hitter if all for sets $H \subseteq \{0,1\}^n$ of density at least δ,*
$$\Pr_{z \sim U_l}[\forall\, i = 1, 2, \ldots, k, G_i(z) \notin H] \leq \epsilon$$

We refer to m and k as the randomness complexity and sample complexity of G respectively.

An equivalent, and more common, notion is that of a *disperser*. Using a notation consistent with the one above, a function $D : \{0,1\}^m \times \{0,1\}^\kappa \to \{0,1\}^n$ is a (b, δ)-disperser if for every distribution X over $\{0,1\}^m$ of min-entropy at least b, and for every set $H \subseteq \{0,1\}^n$ of density at least δ, there is a non-zero probability that $D(X, U_\kappa)$ hits H. Such an object is easily seen to be equivalent to a $(2^{b-n}, \delta)$-hitter with $k = 2^\kappa$. We will use the hitter notation because it is more convenient for our purposes. The following lower bounds for hitters are proved in [Gol97, RTS97].

Theorem 3 (Lower Bounds for Hitters). *If $G : \{0,1\}^m \to (\{0,1\}^n)^k$ is a (δ, ϵ)-hitter, then:*

$$\text{(sample complexity)} \qquad k \geq \frac{1}{2\delta} \ln \frac{1}{2\epsilon} \qquad \text{provided } \epsilon \leq 1/8$$

$$\text{(randomness complexity)} \quad m > n - \log k + \log \frac{1}{\epsilon} + \log \log \frac{1}{\delta}$$

Efficient constructions of hitters are known that match these lower bounds up to constant factors.

Theorem 4. *[Gol97] There exists a polynomial time computable (δ, ϵ)-hitter with sample complexity $O(\frac{1}{\delta} \log \frac{1}{\epsilon})$ and randomness complexity $2n + O(\log \frac{1}{\epsilon})$.*

The construction of dispersers of Ta-Shma [TS98] give even tighter bounds.

2.4 Hardness of Inverting Random Permutations

We begin by establishing that a permutation that is a random permutation on a subset of $\{0,1\}^n$ of density 2δ and is the identity everywhere else is $(1-\delta)$-secure. We will be using this permutation as a weak one-way function for establishing lower bounds for black-box hardness amplification.

Lemma 1. *Fix $T(n) = n^{\log n}$. For all sufficiently large n, for all $\delta > \frac{1}{T(n)}$, for all sets $H \subseteq \{0,1\}^n$ of density 2δ, there exists a permutation π_H on $\{0,1\}^n$ such that π_H is the identity on $\{0,1\}^n - H$, and for all oracle Turing machines M with description at most $\log n$ bits that makes at most $T(n)$ oracle queries,*

$$\Pr_{x \sim U_n}[M^{\pi_H}(\pi_H(x)) = x] < 1 - \delta$$

Proof. Let Π_H denote the set of permutations that is the identity on $\{0,1\}^n - H$. Fix an oracle Turing machine M. Then,

$$E_{\pi \sim \Pi_H}[\#\{y \in H : M^\pi(y) = \pi^{(-1)}(y)\}] \leq 2\delta \cdot 2^n \left(\frac{n^{\log n} + 1}{2\delta \cdot 2^n - n^{\log n}} \right) < \frac{\delta}{4n} \cdot 2^n$$

Hence,

$$\Pr_{\pi \sim \Pi_H}[\#\{x \in H : M^\pi(\pi(x)) = x\} \geq \delta 2^n] \leq \frac{1}{4n}$$

This allows us to take a union bound over all oracle Turing machines M with description at most $\log n$ bits. □

Note that we could also derive a non-uniform analogue of this lemma using the counting argument of [GT00]:

Lemma 2. *Fix $T(n) = n^{\log n}$. For all sufficiently large n, for all $\delta > \frac{1}{T(n)}$, for all sets $H \subseteq \{0,1\}^n$ of density 2δ, there exists a permutation π_H on $\{0,1\}^n$ such that π_H is the identity on $\{0,1\}^n - H$, and for all probabilistic oracle Turing machines M with description at most $\log n$ bits that makes at most $T(n)$ oracle queries and uses at most $T(n)$ bits of non-uniformity and at most $T(n)$ random coin tosses,*

$$\Pr_{x \sim U_n}[M^{\pi_H}(\pi_H(x)) = x] < 1 - \delta$$

Remark 2. We stress that in both Lemma 1 and Lemma 2, we allow the machine M to have arbitrary (possibly exponential) running time, but we require that M has bounded non-uniformity, makes a bounded number of oracle queries and uses a bounded number of random coins; in this sense, M still has "low complexity".

By the weakly black-box property of $F^{()}$, there exists an efficient oracle algorithm B such that

$$\Pr_{x \sim U_n} [B^{\pi_H}(\pi_H(x)) = x] \geq 1 - \delta$$

a contradiction. □

5 Revisiting the Direct Product Construction and [GIL+90]

We present a simple, modular and unified view of the analysis for previous results for fully black-box hardness amplification. We stress that the analysis is not novel, and is based largely on the exposition of [GIL+90] in [Gol01–Sec 2.6].

Theorem 5. *Let F be a (deterministic) polynomial time computable oracle procedure such that for every function (resp permutation) $f : \{0,1\}^n \to \{0,1\}^n$, F^f is a function mapping $\ell(n)$ bits to $\ell(n)$ bits and makes at most $q(n)$ oracle queries. Let $G^{()} : \{0,1\}^\ell \to (\{0,1\}^n)^q$ be the function that computes the sequence of q oracle queries that F makes. Suppose F also satisfies the following properties for every function (resp permutation) $f : \{0,1\}^n \to \{0,1\}^n$:*

1. *(consistent) for any $z, z' \in \{0,1\}^\ell$ such that $F^f(z') = F^f(z)$, we have $f(G_i^f(z')) = f(G_i^f(z))$ for all i.*
2. *(restrictable) there exists a polynomial time oracle algorithm that given oracle access to f, and given input $i \in [k]$ and $y \in f(\{0,1\}^n)$, outputs a random sample from the distribution[3] $\{F^f(U_l) \mid f(G_i^f(U_l)) = y\}$.*
3. *(hitting) G^f is a $(\delta/2, \epsilon/2)$-hitter.*

Then, there exists a probabilistic polynomial time oracle procedure R such that (F, R) constitute a fully black-box construction of a ϵ-secure function from a $(1 - \delta)$-secure function (resp permutation). In addition, $R^{A,f}$ makes $O(\frac{q^2}{\epsilon} \log \frac{1}{\delta})$ oracle queries to A.

Consider what happens in a black-box reduction for a proof of security of hardness amplification. We are given oracle access to an algorithm A that inverts F^f on a ϵ fraction of input and a challenge $f(x)$. The "consistent" property tells us (informally) that inverting F^f on $F^f(z)$ is at least as hard as inverting f on all of the $f(G_i^f(z))$'s ($i = 1, 2, \ldots, q$), and the "restrictable" property allows us to construct from $f(x)$ a challenge $F^f(z)$ for A by substituting $f(x)$ for one of the $f(G_i^f(z))$'s. Note that the "consistent" property is trivially satisfied if F^f is injective. We also do not need to make any assumptions about the distributions $G_i^f(U_\ell)$, $i = 1, 2, \ldots, q$.

---

[3] The distribution may be described more precisely by the following two-step experiment: pick z uniformly at random from $\{z \in \{0,1\}^\ell : f(G_i^f(z)) = y\}$ and output $F^f(z)$. We stress that the sampling algorithm may not compute z explicitly.

Proof. Let A be a function that inverts F^f on an ϵ fraction of input. Now, consider an oracle procedure I that given oracle access to A, f and on input $y \in \{0,1\}^n$, does the following: for each $i = 1, 2, \ldots, q$,

1. samples $y^{(i)}$ from $\{F^f(U_l) \mid f(G_i^f(U_l)) = y\}$, and computes $z^{(i)} = A(y^{(i)})$;
2. checks whether $f(G_i^f(z^{(i)})) = y$, and if so, outputs $G_i^f(z^{(i)})$.

Define the set H (for "hard") by:

$$H = \{x \in \{0,1\}^n \mid \Pr[I^{A,f}(f(x)) \in f^{(-1)}(f(x))] < \epsilon/2q\}$$

It is easy to see that $x \in H$ iff $f(x) \in f(H)$.

Claim. $|H| < \delta/2 \cdot 2^n$

Proof. (of claim) Suppose otherwise. Then,

$$\Pr_{z \sim U_\ell}[A \text{ inverts } F^f(z)]$$

$$\leq \Pr_{z \sim U_\ell}[\forall i, G_i^f(z) \notin H] + \sum_{i=1}^{q} \Pr_{z \sim U_\ell}[A \text{ inverts } F^f(z) \text{ and } G_i^f(z) \in H]$$

$$\leq \epsilon/2 + \sum_{i=1}^{q} \Pr_{z \sim U_\ell}[A \text{ inverts } F^f(z) \text{ and } f(G_i^f(z)) \in f(H)] \quad \text{(by "hitting")}$$

$$\leq \epsilon/2 + \sum_{i=1}^{q} \max_{y \in f(H)} \Pr_{z \sim U_\ell}[A \text{ inverts } F^f(z) \mid f(G_i^f(z)) = y]$$

$$\leq \epsilon/2 + \sum_{i=1}^{q} \max_{y \in f(H)} \Pr_{z \sim U_\ell}[I^{A,f} \text{ inverts } y] \quad \text{(by "consistent")}$$

$$< \epsilon/2 + q \cdot \epsilon/2q \leq \epsilon$$

a contradiction. □

Consider the oracle procedure R that given oracle access to A and f, runs $I^{A,f}$ $O(\frac{q}{\epsilon} \log \frac{1}{\delta})$ times. This allows us to amplify the success probability of inverting values not in H to $1 - \delta/2$. Hence,

$$\Pr_{x \in U_n}[R^{A,f}(f(x)) \notin f^{(-1)}(f(x))]$$

$$\leq \Pr_{x \in U_n}[x \in H] + \Pr_{x \in U_n}[R^{A,f}(f(x)) \notin f^{(-1)}(f(x)) \mid x \notin H] < \delta$$

The result follows. □

Next, we review previous results on hardness amplification in our framework:

Direct Product. [Yao82] Here, we start with a $(1 - \delta)$-secure function f, and we define $F^f : (\{0,1\}^n)^q \to (\{0,1\}^n)^q$ is given by $F(x_1, \ldots, x_q) = (f(x_1), \ldots, f(x_q))$, where $q = O(1/\delta \log 1/\epsilon)$. $G^f : \{0,1\}^{nq} \to (\{0,1\}^n)^q$ is then the identity function for all f. It is easy to check that F satisfies all of the 3 properties, from which hardness amplification via direct product follows.

Random Walk on Expanders. [GIL+90] Here, we start with a $(1-\delta)$-secure permutation π and a family of d-regular explicitly constructible expanders $\{\Gamma_n\}$ with vertex set $\{0,1\}^n$, where d is a constant. We define $G^\pi : \{0,1\}^n \times [d]^t \to (\{0,1\}^n)^{t+1}$ as follows:

$$G_1^\pi(x, \sigma_1, \ldots, \sigma_t) = x$$
$$G_{i+1}^\pi(x, \sigma_1, \ldots, \sigma_t) = g_{\sigma_i}(\pi(G_i^\pi(x, \sigma_1, \ldots, \sigma_t))) \quad i = 1, 2, \ldots, t$$

where $g_\sigma(x)$ for $x \in \{0,1\}^n$ and $\sigma \in [d]$ denotes the σ'th neighbor of vertex x in Γ_n. Note that the output of G is the set of vertices visited in a random walk on G started at x along the path $\sigma_1, \ldots, \sigma_t$, interspersed with an application of π before each step. Since π is a permutation, applying π does not affect the mixing properties of the random walk, and therefore if we take $t = O(1/\delta \log 1/\epsilon)$, then G^π yields a $(\delta/2, \epsilon/2)$-hitter. The new function $F^\pi : \{0,1\}^n \times [d]^t \to \{0,1\}^n \times [d]^t$ is given by

$$F^\pi(x, \sigma_1, \ldots, \sigma_t) = (G_{t+1}^\pi(x, \sigma_1, \ldots, \sigma_t), \sigma_1, \ldots, \sigma_t)$$

It is easy to see that F^π is injective, and thus F is "consistent" and F^π is a permutation. The "restrictable" property is satisfied using the following algorithm: given $i \in [t+1]$ and $y \in \{0,1\}^n$, pick $\sigma_1, \ldots, \sigma_t$ independently at random from $[d]$, and output $(\pi(g_{\sigma_t}(\ldots g_{\sigma_i}(y)\ldots)), \sigma_1, \ldots, \sigma_t)$. This constitutes the basic building block: a fully black-box construction of ϵ-secure permutations from $(1-\delta)$-secure permutations with $\ell = n + O(1/\delta \log 1/\epsilon)$ and $q = O(1/\delta \log 1/\epsilon)$.

To obtain a security-preserving construction of an ϵ-secure permutation on $\{0,1\}^{O(cn+\log 1/\epsilon)}$ from a $(1-1/n^c)$-secure permutation on $\{0,1\}^n$, we compose the basic building block $c+1$ times as follows: we first construct a $(1-1/2n^{c-1})$-secure permutation, then a $(1-1/2n^{c-2})$-secure one, and right up to $1/2$-secure permutation. In the last composition, we construct a ϵ-secure permutation from a $1/2$-secure one.

6 Conclusion

Our negative result for weakly black-box constructions is less general than the one for fully black-box constructions: in the former case we restrict ourselves to constructions that define a permutation if the original primitive is a permutation. It should be noted that both the direct product construction and the construction of [GIL+90] satisfy this property. It would be possible to strengthen Lemma 5 to hold under the assumption that $F^{()}$ is a construction of ϵ-secure *functions*, and with the conclusion that either G^{id} is a $(2\delta, 2\epsilon)$-hitter or that one-way functions exist uncondtionally. The proof would have followed along the lines of the proof of Lemma 3, using a result of Impagliazzo and Luby [IL89] to construct a polynomial time algorithm that approximates algorithm A in the proof of Lemma 3 assuming that one-way functions do not exist. We will give more details in the full version of this paper.

The main open problem that is still unresolved is whether there is a fully black-box security-preserving hardness amplification for one-way functions. From the work of [GIL+90], we know that it would suffice to construct a "restrictable" and "consistent" hitter (see the statement of Theorem 5 for the terminology) with randomness complexity $O(n + 1/\delta \log 1/\epsilon)$.

References

[GGK03] Rosario Gennaro, Yael Gertner, and Jonathan Katz. Lower bounds on the efficiency of encryption and digital signature schemes. In *Proceedings of the 35th ACM Symposium on Theory of Computing*, pages 417–425, 2003.

[GIL+90] Oded Goldreich, Russell Impagliazzo, Leonid Levin, Ramarathnam Venkatesan, and David Zuckerman. Security preserving amplification of hardness. In *Proceedings of the 31st IEEE Symposium on Foundations of Computer Science*, pages 318–326, 1990.

[GKM+00] Yael Gertner, Sampath Kannan, Tal Malkin, Omer Reingold, and Mahesh Viswanathan. The relationship between public key encryption and oblivious transfer. In *Proceedings of the 41st IEEE Symposium on Foundations of Computer Science*, pages 325–335, 2000.

[GMR01] Yael Gertner, Tal Malkin, and Omer Reingold. On the impossibility of basing trapdoor functions on trapdoor predicates. In *Proceedings of the 42nd IEEE Symposium on Foundations of Computer Science*, pages 126–135, 2001.

[Gol97] Oded Goldreich. A sample of samplers - a computational perspective on sampling. Technical Report TR97-020, Electronic Colloquium on Computational Complexity, 1997.

[Gol01] Oded Goldreich. *The Foundations of Cryptography - Volume 1*. Cambridge University Press, 2001.

[GT00] Rosario Gennaro and Luca Trevisan. Lower bounds on the efficiency of generic cryptographic constructions. In *Proceedings of the 41st IEEE Symposium on Foundations of Computer Science*, pages 305–313, 2000.

[HILL99] J. Håstad, R. Impagliazzo, L. Levin, and M. Luby. A pseudorandom generator from any one-way function. *SIAM Journal on Computing*, 28(4):1364–1396, 1999.

[IL89] Russell Impagliazzo and Michael Luby. One-way functions are essential for complexity based cryptography. In *Proceedings of the 30th IEEE Symposium on Foundations of Computer Science*, pages 230–235, 1989.

[Imp96] Russell Impagliazzo. Very strong one-way functions and pseudo-random generators exist relative to a random oracle. Unpublished manuscript, 1996.

[IR89] Russell Impagliazzo and Steven Rudich. Limits on the provable consequences of one-way permutations. In *Proceedings of the 21st ACM Symposium on Theory of Computing*, pages 44–61, 1989.

[KSS00] J. Kahn, M. Saks, and C. Smyth. A dual version of Reimer's inequality and a proof of rudich's conjecture. In *Proceedings of the 15th IEEE Conference on Computational Complexity*, 2000.

[KST99] J.H. Kim, D.R. Simon, and P. Tetali. Limits on the efficiency of one-way permutations-based hash functions. In *Proceedings of the 40th IEEE Symposium on Foundations of Computer Science*, pages 535–542, 1999.

[Lub96] M. Luby. *Pseudorandomness and Cryptographic Applications*. Princeton University Press, 1996.

[RTS97] J. Radhakrishnan and A. Ta-Shma. Tight bounds for depth-two superconcentrators. In *Proceedings of the 38th IEEE Symposium on Foundations of Computer Science*, pages 585–594, 1997.

[RTV04] Omer Reingold, Luca Trevisan, and Salil Vadhan. Notions of reducibility between cryptographic primitives. In *Proceedings of the 1st Theory of Cryptography Conference*, pages 1–20. LNCS 2951, 2004.

[Rud88] S. Rudich. *Limits on the provable consequences of one-way functions*. PhD thesis, University of California at Berkeley, 1988.

[Rud91] S. Rudich. The use of interaction in public cryptosystems. In *Proceedings of CRYPTO'91*, pages 242–251, 1991.

[TS98] A. Ta-Shma. Almost optimal dispersers. In *Proceedings of the 30th ACM Symposium on Theory of Computing*, 1998.

[Yao82] A.C. Yao. Theory and applications of trapdoor functions. In *Proceedings of the 23th IEEE Symposium on Foundations of Computer Science*, pages 80–91, 1982.

Cryptography in Subgroups of $\mathbb{Z}_n^*$

Jens Groth*

jg@brics.dk

Abstract. We demonstrate the cryptographic usefulness of a small subgroup of $\mathbb{Z}_n^*$ of hidden order. Cryptographic schemes for integer commitment and digital signatures have been suggested over large subgroups of $\mathbb{Z}_n^*$, by reducing the order of the groups we obtain quite similar but more efficient schemes. The underlying cryptographic assumption resembles the strong RSA assumption.

We analyze a signature scheme known to be secure against known message attack and prove that it is secure against adaptive chosen message attack. This result does not necessarily rely on the use of a small subgroup, but the small subgroup can make the security reduction tighter.

We also investigate the case where $\mathbb{Z}_n^*$ has semi-smooth order. Using a new decisional assumption, related to high residuosity assumptions, we suggest a homomorphic public-key cryptosystem.

Keywords: RSA modulus, digital signature, homomorphic encryption, integer commitment.

1 Introduction

Consider an RSA-modulus $n = pq$, where p and q are large primes. Many cryptographic primitives take place in the multiplicative group $\mathbb{Z}_n^*$ and use the assumption that even if n is public, the order of the group $\varphi(n) = (p-1)(q-1)$ is still unknown. Concrete examples of such primitives are homomorphic integer commitments [FO97, DF02], public key encryption [RSA78, Rab79, Pai99, CF85, KKOT90, NS98] and digital signatures that do not use the random oracle model [BR93] in their security proofs [CS00, CL02, Fis03].

In order to speed up cryptographic computations it is of interest to find as small groups of hidden order as possible. We suggest using a small subgroup of $\mathbb{Z}_n^*$. More precisely, if we have primes $p'|p-1, q'|q-1$, then we look at the unique subgroup $\mathbb{G} \leq \mathbb{Z}_n^*$ of order $p'q'$. We make a strong root assumption for this group, which roughly states that it is hard to find a non-trivial root of a random element in $\mathbb{G}$. We call this the strong RSA subgroup assumption.

Following Cramer and Shoup [CS00] several very similar signature schemes have been suggested. One variation is the following: We publish n and elements $a, g, h \in \mathrm{QR}_n$. To sign a 160-bit message m, select at random a 161-bit randomizer r and a 162-bit prime e. Compute y so $y^e = ag^m h^r \bmod n$. The signature is

---

* Work done while at Cryptomathic, Denmark and BRICS, Dept. of Computer Science, University of Aarhus, Denmark.

(y, e, r). A natural question to ask is whether we really need the randomizer r. We analyze this question and show that indeed it is not needed provided we are willing to accept a weaker security reduction.

Restricting ourselves to an even more specialized group, we look at $n = pq = (2p'r_p + 1)(2q'r_q + 1)$, where r_p, r_q consists of distinct odd prime factors smaller than some low bound B. We can form a new cryptosystem using modular arithmetic in $\mathbb{Z}_n^*$. Let g have order $p'q'r_g$ and h have order $p'q'$. Assuming random elements of $\mathbb{G}$ are indistinguishable from random elements of QR_n we can encrypt m as $c = g^m h^r \bmod n$. To decrypt we compute $c^{p'q'} = g^{p'q'm} h^{p'q'r} = (g^{p'q'})^m \bmod n$. Since $g^{p'q'}$ has order $r_g | r_p r_q$, which only has small prime factors, it is now possible to extract $m \bmod r_g$. This cryptosystem is homomorphic, has a low expansion rate $\frac{|c|}{|m|}$ and fast encryption. The decryption process is slow, yet as we shall see, there are applications of this kind of cryptosystem. A nice property of the cryptosystem is that under the strong RSA assumption it serves at the same time as a homomorphic integer commitment scheme. This comes in handy in verifiable encryption where we want to prove that the plaintext satisfies some specified property.

2 Subgroup Assumptions

As mentioned in the introduction, it is of interest to find a small group, where some sort of strong root assumption holds. Obviously, a prerequisite for a strong root assumption is that the order of the group is hidden. Otherwise we have for any $g \in \mathbb{G}$ that $g = g^{1+ord(\mathbb{G})}$ giving us a non-trivial root. We suggest using a subgroup of $\mathbb{Z}_n^*$, where n is some suitable RSA modulus.

A small RSA subgroup of unknown order. Throughout the paper we shall work with RSA moduli on the form $n = pq$, where p, q are primes. We choose these moduli in a manner such that $p = 2p'r_p + 1, q = 2q'r_q + 1$, where p', q' are primes so there is a unique subgroup $\mathbb{G} \leq \mathbb{Z}_n^*$ of order $p'q'$. Let g be a random generator for this group. We call (n, g) an RSA subgroup pair.

Definition 1 (Strong RSA subgroup assumption). *Let K be a key generation algorithm that produces an RSA subgroup pair (n, g). The strong RSA subgroup assumption for this key generation algorithm is that it is infeasible to find $u \in \mathbb{Z}_n^*, w \in \mathbb{G}$ and $d, e > 1$ such that $g = uw^e \bmod n$ and $u^d = 1 \bmod n$.*

In comparison with the strong RSA assumption [BP97] we have weakened the assumption by only worrying about non-trivial roots of elements from $\mathbb{G}$. On the other hand, we have strengthened the assumption by generating the RSA modulus in a special way and publicizing a random generator g of a small subgroup of $\mathbb{Z}_n^*$.

We write $\ell_{p'}, \ell_{q'}$ for the bit-length of the primes p', q'. The order of $\mathbb{G}$ then has bit-length $\ell_G = \ell_{p'} + \ell_{q'}$. A possible choice of parameters is $\ell_{p'} = \ell_{q'} = 100$. We shall also use a statistical hiding parameter ℓ_s. Given some number a and a random $|a| + \ell_s$-bit integer r, the idea is that $a + r$ and r should be statistically indistinguishable. A reasonable choice is $\ell_s = 60$.

Lemma 1. *Consider a subgroup pair (n,g) generated in a way such that the strong RSA subgroup assumption holds. Let $g_1, \ldots, g_k$ be randomly chosen generators of $\mathbb{G}$. Give $(n, g_1, \ldots, g_k)$ as input to an adversary $\mathcal{A}$ and let it produce $(y, e, e_1, \ldots, e_k)$ such that $y^e = g_1^{e_1} \cdots g_k^{e_k} \bmod n$. If $e = 0$, then $e_1, \ldots, e_k = 0$. Else we have $e|e_1, \ldots, e|e_k$ and $y = u \prod_{i=1}^{k} g_i^{e_i/e} \bmod n$, where $u^e = 1 \bmod n$.*

Proof. Pick $\gamma_1, \ldots, \gamma_k \leftarrow \{0,1\}^{\ell_G + \ell_s}$ and set $g_i = g^{\gamma_i} \bmod n$. We give $(n, g_1, \ldots, g_k)$ to $\mathcal{A}$ that with noticeable probability produces $(y, e, e_1, \ldots, e_k)$. We have $y^e = g^{\sum_{i=1}^{k} \gamma_i e_i} \bmod n$. If $e = 0$ then $g = g^{1 + \sum_{i=1}^{k} \gamma_i e_i}$. Unless $e_1, \ldots, e_k = 0$ this is likely to be a breach of the strong RSA subgroup assumption.

Assume from now on $e \neq 0$. Let $d = \gcd(e, \sum_{i=1}^{k} \gamma_i e_i)$ and choose α, β such that $d = \alpha e + \beta \sum_{i=1}^{k} \gamma_i e_i$. We have $g^d = g^{\alpha e + \beta \sum_{i=1}^{k} \gamma_i e_i} = (g^\alpha y^\beta)^e \bmod n$. If $p'q'|d$ then $g = g^{1+e}$ and a breach of the strong RSA subgroup assumption has been found. If $1 < \gcd(d, p'q') < p'q'$, then we have $1 < \gcd(g^d - 1, n) < n$ giving us a non-trivial factorization of n, and indirectly a breach of the strong RSA subgroup assumption. Therefore, d is invertible modulo $p'q'$ and we have $g = u(g^\alpha y^\beta)^{e/d} \bmod n$, where $u^d = 1 \bmod n$. Unless $d = \pm e$, this breaks the strong RSA subgroup assumption.

So $e | \sum_{i=1}^{k} \gamma_i e_i$. Write $\gamma_i = \kappa_i p'q' + \lambda_i$. We have $e | p'q' \sum_{i=1}^{k} \kappa_i e_i + \sum_{i=1}^{k} \lambda_i e_i$. Since κ_i is completely hidden to the adversary and randomly chosen this implies $e|e_i$ for all i. We now have $y = u \prod_{i=1}^{k} g_i^{e_i/e} \bmod n$, where $u^e = 1 \bmod n$. □

Definition 2 (Decisional RSA subgroup assumption). *Let K be a key generation algorithm that produces an RSA subgroup pair (n, g). The decisional RSA subgroup assumption for this key generation algorithm K is that it is hard to distinguish elements drawn at random from $\mathbb{G}$ and elements drawn at random from QR_n.*

The assumption is related to high-residuosity assumptions made by other authors [GM84, CF85, KKOT90, NS98]. These assumptions are on the form: Given (n, r), where $r | r_p r_q$, it is hard to distinguish a random element and a random element on the form $z^r \bmod n$. In comparison, the decisional RSA subgroup assumption is weaker in the sense that we do not publish $r = r_p r_q$. On the other hand, it is stronger in the sense that we may have a much smaller group $\mathbb{G}$.

Under the decisional RSA subgroup assumption, the strong RSA subgroup assumption implies the standard strong RSA assumption. To see this, consider choosing $r \leftarrow \{0,1\}^{\ell_G + \ell_s}$ at random and feeding g^r to the strong RSA assumption adversary. Under the decisional RSA subgroup assumption this looks like a random element and the SRSA assumption adversary might return $w, e > 1$ so $g^r = w^e$. Write $r = \kappa p'q' + \lambda$, then κ is perfectly hidden from the adversary. There is at least 50% chance of $\gcd(e, \kappa p'q' + \lambda) \neq e$. This contradicts Lemma 1, which states $e|r$.

2.1 RSA with Semi-smooth Order

In Section 7, we restrict the way we generate the RSA subgroup pair. Consider choosing p, q so $p = 2p'p_1 \cdots p_{t_p} + 1, q = 2q'q_1 \cdots q_{t_q} + 1$, where

$p_1, \ldots, p_{t_p}, q_1, \ldots, q_{t_q}$ are distinct odd primes smaller than some small bound B. We call (n, g) a semi-smooth RSA subgroup pair.

Define $P_B = \prod_{1 < p < B, p \text{ is prime}} p$. Choosing h at random and setting $g = h^{P_B}$ we have overwhelming probability of g generating $\mathbb{G}$. In other words, given n it is easy for anybody to find a generator for $\mathbb{G}$. We can therefore save specifying g and just make n public.

Typical parameters would be $\ell_{p'} = \ell_{q'} = 160$ and $B = 2^{15}$. Setting $\ell_{p_i} = 15$, we choose $t = t_p + t_q$ distinct odd primes $p_1, \ldots, p_{t_p}, q_1, \ldots, q_{t_q}$ such that $p = 2p'p_1 \cdots p_{t_p} + 1, q = 2q'q_1 \cdots q_{t_q} + 1$ are primes.

Lemma 2. *Let n be a semi-smooth RSA subgroup modulus generated with parameters as described above. Pick g at random from QR_n and let d be an arbitrary non-negative integer smaller than t. With probability at least $1 - 1/p' - 1/q' - \frac{(t2^{1-\ell_{p_i}})^{d+1}}{(1-t2^{1-\ell_{p_i}})(d+1)!}$ the order of g is greater than $p'q'2^{(t-d)(\ell_{p_i}-1)}$.*

Proof. Consider a generator h of QR_n. Pick at random $x \in \mathbb{Z}_{p'q'r_p r_q}$. Then $g = h^x$ is uniformly distributed in QR_n. Consider the prime factors $p_1, \ldots, p_t$ of $r_p r_q$. We will consider the probability that $x = 0 \bmod p_i$ for more than d of these prime factors. Each event is independent of the others and has at most probability $2^{1-\ell_{p_i}}$ of occurring. Therefore, from Lemma 3 we get a probability lower than $\frac{(t2^{1-\ell_{p_i}})^{d+1}}{(1-t2^{1-\ell_{p_i}})(d+1)!}$. Combine this with the probabilities $1/p'$ and $1/q'$ for respectively $x = 0 \bmod p'$ and $x = 0 \bmod q'$ to conclude the proof. □

Lemma 3. *Consider n independent Bernoulli-trials with probability p, where $np < 1$. The probability of having at least k successes out of n trials is lower than $\frac{(np)^k}{(1-np)k!}$.*

Proof.

$$\sum_{i=k}^{n} \binom{n}{i} p^i (1-p)^{n-i} \le \sum_{i=k}^{n} \binom{n}{i} p^i \le \sum_{i=k}^{n} \frac{n^i}{i!} p^i \le \frac{1}{k!} \sum_{i=k}^{n} (np)^i$$

$$= \frac{1}{k!} \frac{(np)^k - (np)^{n+1}}{1 - np} \le \frac{(np)^k}{(1-np)k!}.$$

□

3 Factorization Attacks

If we can factor n we know $p-1, q-1$ and it is easy to break the strong RSA subgroup assumption. In the case of a semi-smooth RSA subgroup modulus, the factorization would also tell us the factors p', q' and we can break the decisional RSA subgroup assumption. We do not know of any non-factorization attacks that could be used to break either the strong RSA subgroup assumption or the decisional RSA subgroup assumption, therefore we will focus on the possibility of factoring n.

Pollard's rho method. Consider a semi-smooth RSA subgroup pair (n, g). We can use the following variation of Pollard's ρ-method [Pol75] to factor n. We define f by $f(0) = g$ and $f(i+1) = (f(i)+1)^{P_B} \mod n$. Intuitively this corresponds to taking a random walk on $\mathbb{G}$ starting in g. Actually, modulo p, it corresponds to taking a random walk on a group of size p', and modulo q, it corresponds to taking a random walk on a group of size q'. We now hope to find points i, j such that $f(i) = f(j) \mod p$ or $f(i) = f(j) \mod q$. This would give us $\gcd(n, f(i) - f(j)) > 1$ and most likely a non-trivial factor of n. Using Brent's [Bre80] cycle finding method we expect to find a factorization using $\mathcal{O}(\min(\sqrt{p'}, \sqrt{q'}) \log(P_B)) = \mathcal{O}(2^{\ell_{p'}/2} B)$ modular multiplications.

In case (n, g) is simply a normal RSA subgroup modulus it seems hard to find a function f that always ends up inside $\mathbb{G}$. It therefore seems like Pollard's ρ-method is of little use.

Other factorization methods. Other methods such as the baby-step giant-step algorithm of Shanks [Sha71], Pollard's λ-method [Pol78] or Pollard's $p-1$ method [Pol74] seem to use at least $2^{\ell_{p'}}$ modular multiplications.

While the above mentioned algorithms take advantage of a special structure of the divisors of n, other algorithms such as the elliptic curve method (ECM) [Len87] or the general number field sieve (GNFS) [CP01] do not. We therefore believe that the best one can do here is to run the general number field sieve with heuristic running time $\exp((1.92 + o(1)) \ln(n)^{1/3} \ln \ln(n)^{2/3})$.

Dangers. It is of course important not to give away too much information about the factorization of $p-1$ and $q-1$. An adversary knowing p' could compute $\gcd(n, g^{p'} - 1) = p$. For this reason, we do not release $p'q'$.[1]

Likewise, if we were to release $\sigma | (p-1)(q-1)$ with $|\sigma| > |n|/4$ then we may risk the factorization attack described in [NS98]. Therefore, we must make sure that there is enough entropy in the primes $p_i | (p-1)(q-1)$ that the adversary cannot guess a significant portion of them. Unlike other high-residuosity schemes, we cannot publicize the value $\prod_{p_i|(p-1)/2} p_i \prod_{q_i|(q-1)/2} q_i$.

4 Signature

Cramer and Shoup [CS00] suggest an efficient signature scheme based on the strong RSA assumption where security can be proved in the standard model without using random oracles. Subsequently, Fischlin [Fis02] has proposed efficient schemes for both the case of a statefull signer and a stateless signer. Koprowski [Kop03] points out a minor flaw in the statefull signature scheme and an easy correction of it. Camenisch and Lysyanskaya [CL02] have suggested a variant that is more suitable as a building block in larger protocols such as group signatures. Finally, Zhu [Zhu03] suggests a variation that combines the

---

[1] Actually, such a factorization attack is possible on scheme 3 of the Paillier cryptosystem [Pai99], since it uses an element $g = 1 \mod q$. In a subsequent variant [PP99] this has been corrected and they work in a subgroup of the same nature as we do.

efficiency of the stateless version of Fischlin's scheme with the suitability of the Camenisch and Lysyanskaya signature scheme. All these signature schemes use safe-prime product moduli. We will suggest similar looking signature schemes for both the statefull and the stateless case and prove security under the strong RSA subgroup assumption. We are not the first to use RSA moduli that are not safe-prime products, Damgård and Koprowski [DK02, Kop03] have generalized the Cramer-Shoup signature approach to basing signature schemes on general groups with a strong root assumption. RSA subgroups as we suggest using can be seen as an example of such a group.

Key generation: We generate an RSA subgroup $\mathbb{G}$ and pick $a, g, h \leftarrow \mathbb{G}$. The private key is $p'q'$, the order of $\mathbb{G}$. We select a positive integer t so $t(\ell_e - 1) + 1 > \ell_m$.
Public verification key $vk = (n, a, g, h, t)$. Private signature key $sk = p'q'$.
Signature: To sign a message $m \in \{0, 1\}^{\ell_m}$, choose an ℓ_e-bit prime e that has not been used before. Choose at random $r \in \mathbb{Z}_{e^t}$. Compute $y = (a g^m h^r)^{e^{-t} \bmod p'q'} \bmod n$.
The signature on m is (y, e, r).
Verification: Given a purported signature (y, e, r) on $m \in \{0, 1\}^{\ell_m}$, check that e is an ℓ_e-bit number and $r \in \mathbb{Z}_{e^t}$. It is not necessary to check specifically that e is a prime. Accept if $y^{e^t} = a g^m h^r \bmod n$.

For a stateless signature scheme it would be reasonable to choose $\ell_m = 160, \ell_e = 161$ and $t = 1$. We can use the method from [CS00] to pick the primes e, this way it is still unlikely that we run into a collision where we use the same prime in two different signatures. For a statefull signature scheme we can pick $\ell_m = 160, \ell_e = 28$ and $t = 6$ and keep track of the last prime we used. Whenever we wish to sign, we pick the subsequent prime and use that in the signature. For so small primes, the exponentiation is the dominant computational cost.

Theorem 1. *If the strong RSA subgroup assumption holds for the key generation algorithm, then the signature scheme described above is secure against existential forgery under adaptive chosen message attack.*

Proof. There are three cases to consider. The first case is where the adversary forges a signature using a prime e that it has not seen before. The second case is where the adversary reuses a prime, i.e., $e = e_i$, where e_i is the prime from query i but $r \neq r_i$. The third case is where the adversary reuses both e_i and r_i for some i.

Case 1: $e \neq e_i$. With non-negligible probability, we can guess the number k of signing queries the adversary is going to make. Choose according to the signature algorithm distinct ℓ_e-bit primes $e_1, \ldots, e_k$. Set $E = \prod_{j=1}^{k} e_j^t$. Given random elements $\alpha, \gamma, \eta \in \mathbb{G}$ we set $a = \alpha^E, g = \gamma^E, h = \eta^E$. We give (n, a, g, h) to the adversary. We can answer the ith query since we know e_i^t-roots of a, g, h. Consider now the adversary's signature (y, e, r). We have $y^{e^t} = a g^m h^r = \alpha^E \gamma^{Em} \eta^{Er}$ so by Lemma 1 we have $e^t | E$. This means, $e = e_i$ for some i, i.e., the first case only occurs with negligible probability.

Case 2: $e = e_i, r \neq r_i$. Consider next the case of an adversary that reuses e_i. We guess the query i, where the adversary is going to make the forgery. We pick r_i at random and set up $a = \alpha^E h^{-r_i}, g = \gamma^E, h = \eta^{E/e_i^t}$. We can easily answer queries $j \neq i$, and for query i we return the answer (y_i, e_i, r_i), where $y_i = \alpha^{E/e_i^t} \gamma^{m_i E/e_i^t}$. Consider now the adversary's signature (y, e_i, r) on message m. We have $(y/y_i)^{e_i^t} = g^{m-m_i} h^{r-r_i} = \gamma^{(m-m_i)E} \eta^{(r-r_i)E/e_i^t}$. By Lemma 1, we have $e_i^t | (r - r_i) E/e_i^t$. Since e_i does not divide E/e_i^t and $|r - r_i| < e_i^t$ this means $r = r_i$, so the second case occurs with negligible probability.

Case 3: $e = e_i, r = r_i$. Consider finally the case where the adversary reuses both e_i and r_i. We make the following setup. Pick at random $r_m \in \mathbb{Z}_{e_i^t + 2^{\ell_m}}$. Set $a = \alpha^E g^{-r_m}, g = \gamma^{E/e_i^t}, h = g\eta^E$. On query m_i we pick $r_i = r_m - m_i$, which enables us to compute y_i. r_i is uniformly distributed over $\mathbb{Z}_{e_i^t + 2^{\ell_m}} - m_i$ and has more than 50% chance of being inside $\mathbb{Z}_{e_i^t}$. Conditioned on $r_i \in \mathbb{Z}_{e_i^t}$, we have a correctly distributed signature. Suppose now the adversary forms a new signature (y, e_i, r_i) on message m. We get $(y/y_i)^{e_i^t} = g^{m-m_i} = \gamma^{(m-m_i)E/e_i^t}$. By Lemma 1 we have $e_i^t | (m - m_i) E/e_i^t$ so $m = m_i$. □

To form a signature we make an exponentiation with $e^{-t} \bmod p'q'$. In comparison, the other schemes use an exponent of size ℓ_n. Especially for the statefull signature scheme, we obtain a significant reduction in computation.

Strong signature. A signature scheme is strong if it impossible to form a new signature on a message m, even if we have already seen many signatures on this message under the chosen message attack. If we ensure that no ℓ_e-bit primes divide $\varphi(n)$, then it is impossible to find a non-trivial u such that $u^e = 1$, where e is an ℓ_e-bit prime. Generating the modulus like this makes the signature scheme strong, since this way the adversary can only use y belonging to $\mathbb{G}$ because $((ag^m h^r)^{e^{-t}} y^{-1})^{e^t} = 1$.

Applications. An advantage of the signature scheme is that it allows us to sign a committed message without knowing the content. The receiver creates a commitment $c = ug^m h^r \bmod n$ and proves knowledge of an opening $(m, (u, e, r))$ of c. We then choose a prime e and return (y, e) where $y = (ac)^{e^{-t} \bmod p'q'} \bmod n$. The receiver now has a signature $(y, e, r \bmod e^t)$ on m.

This kind of committed signature can be set up in a safe-prime product modulus as suggested in [CL02]. To hide m this requires a large r. We gain an advantage by working in a small group and thus needing a much shorter r. One application of this is to speed up group signatures such as [CG04].

5 Simplified Signature

It is well known that if a signature scheme secure against known message attack suffices, then we can drop the r in the scheme described in the previous section. I.e., a stateless signature can look like (y, e), where $y^e = ag^m \bmod n$. The public key is also shorter since we do not need h any more. We shall investigate whether this signature scheme is actually secure against adaptive chosen message attack.

Key generation: We generate an RSA subgroup $\mathbb{G}$ and pick $a, g \leftarrow \mathbb{G}$.
Public verification key $vk = (n, a, g)$. Private signature key $sk = p'q'$.
Signature: To sign a message $m \in \{0,1\}^{\ell_m}$ choose a random ℓ_e-bit prime e.
Compute $y = (ag^m)^{e^{-1} \bmod p'q'} \bmod n$.
The signature on m is (y, e).
Verification: Given a purported signature (y, e) on $m \in \{0,1\}^{\ell_m}$ check that e is an ℓ_e-bit number. It is not necessary to check specifically that e is a prime. Accept if $y^e = ag^m \bmod n$.

In practice, there may be more convenient ways to choose the prime e than completely at random. Consider for instance the method of Cramer and Shoup for generating 161-bit primes [CS00]. It is important for the proof of Theorem 2 that the primes have a distribution that is somewhat close to uniform though.

Choosing parameters for the signature scheme is not straightforward. We do certainly need $\ell_e > \ell_m$, as well as ℓ_n to be large enough to make factoring n hard. We also want the group $\mathbb{G}$ to be large enough to make it hard to break the strong RSA subgroup assumption. To simplify notation we will assume p', q' both are $\ell_{p'}$-bit primes, i.e., $\ell_G = 2\ell_{p'}$. On the other hand, for reasons that will become apparent in the proof of Theorem 2 we must be able to factor $\ell_e + \ell_{p'}$-bit numbers.

Consider a rigorous factorization algorithm such as the class-group-relations method. Lenstra and Pomerance [LP92] prove that it takes time $L(2^\ell) = \exp((1 + o(1))\sqrt{\ln(2^\ell)\ln\ln(2^\ell)})$ to factor an ℓ-bit number. We want $L(\ell) < \ell_n^d$ for some degree d, i.e., a running time that is polynomial in the security parameter. This is satisfied if ℓ is chosen such that $\ell \ln(2) \ln(\ell \ln(2)) \leq (d \ln(\ell_n)/(1 + o(1)))^2$. With this choice of ℓ we also have $\ell \leq \frac{\ln^2(\ell_n)}{\ln(2)} \frac{d^2}{(1+o(1))^2 \ln(\ell \ln(2))}$. Letting $\ell = \ell_e + \ell_{p'}$ we have an upper bound on the length of $\ell_{p'} = \ell - \ell_e$.

For the strong RSA subgroup assumption to hold, we need that it is hard to guess the order of the group $\mathbb{G}$. Known algorithms that compute this order use at least time $2^{\ell_{p'}}$. We therefore want $2^{\ell_{p'}}$ to be superpolynomial in the security parameter. Suppose we choose the parameters so $\ell/3 \leq \ell_{p'}$, then we want to choose ℓ as large as possible so $2^{\ell/3}$ is superpolynomial. To see whether there is room for that consider choosing ℓ so $\ell = \frac{\ln^2(\ell_n)}{\ln(2)} \frac{d^2}{(1+o(1))^2 \ln(\ell \ln(2))}$. We then have

$$2^{\ell/3} = \ell_n^{\frac{\ln(\ell_n)}{\ln(\ell \ln(2))} \frac{d^2}{3(1+o(1))^2}} \geq \ell_n^{\frac{\ln(\ell_n)}{\ln((d \ln(\ell_n)/(1+o(1)))^2)} \frac{d^2}{3(1+o(1))^2}}.$$

This is a superpolynomial function of ℓ_n. So we do have reasonable hope to have wriggle-room for choosing $\ell_e, \ell_{p'}$ so that the strong RSA subgroup assumption holds and at the same time, it takes polynomial time to factor $\ell_e + \ell_{p'}$-bit numbers.

Theorem 2. *If the strong RSA subgroup assumption holds for the key generation algorithm and factoring of $\ell_e + \ell_{p'}$-bit numbers can be done in polynomial time then the signature scheme described above is a strong signature scheme secure against existential forgery under adaptive chosen message attack.*

Proof. We consider two cases. In the first case the adversary forges a signature using a prime e that it has not seen before in an adaptive chosen message attack. In the second case the adversary reuses a prime e_i that it has received in an answer to query i.

Case 1: $e \neq e_i$. Consider first a variation where we choose α, γ at random from $\mathbb{G}$. We guess the number of signature queries the adversary will make and choose at random corresponding primes $e_1, \ldots, e_k$. Let $E = \prod_{i=1}^{k} e_i$. Then $a = \alpha^E, g = \gamma^E$ look like random elements from $\mathbb{G}$ and we can answer the k queries. After having asked the queries the adversary must produce a message m and a signature (y, e) so $y^e = ag^m$. I.e., $y^e = \alpha^E \gamma^{mE}$, which by Lemma 1 implies that $e|E$. Since e must be an ℓ_e-bit number this means $e = e_i$ for some i. Case 1 occurs with negligible probability, a successful forger must reuse a prime e_i from one of the oracle queries.

Case 2: $e = e_i$. Consider a different way to set up the signature scheme. We choose z at random from $\mathbb{G}$, and $\alpha \leftarrow \{0,1\}^{\ell_{p'}}, \gamma \leftarrow \{0,1\}^{\ell_{p'}}, \eta \leftarrow \{0,1\}^{\ell_e}$. We guess the number of signing queries k that the adversary will make and an index i for which it will make a forgery. Set $E = \prod_{i \neq j} e_j$. We set $a = z^{E(\alpha 2^{\ell_e} + \eta)}, g = z^{E\gamma}$ and give the public key (n, a, g) to the adversary.

The probability of $\alpha < p', \gamma < q'$ is at least 25%. Conditioned on $\alpha < p', \gamma < q'$ our key looks like a real public key. If we work modulo p, then we have $a = (z^{E\eta})(z^{E 2^{\ell_e}})^\alpha \mod p$, which is distributed as a random element. If we work modulo q, then we have $g = (z^E)^\gamma \mod q$, which is distributed as a random element too. Overall, it therefore looks like the discrete logarithm x so $a = g^x$ is perfectly random. Since z is chosen at random from $\mathbb{G}$ we also have g is randomly distributed. So a, g are perfectly indistinguishable from two randomly chosen elements from $\mathbb{G}$.

It is easy to answer signature queries $j \neq i$ by returning $y = z^{(\alpha 2^{\ell_e} + \eta + \gamma m_j)E/e_j}$ together with e_j. Remaining is the question of answering query i. Suppose signature query i ask for a signature on m_i. Consider $\alpha 2^{\ell_e} + \eta + \gamma m_i$. Since η is statistically hidden to the adversary, it must choose m_i independently of η. $\alpha 2^{\ell_e} + \eta + \gamma m_i \mod 2^{\ell_e}$ therefore looks like a random number. By assumption we can factor $\alpha 2^{\ell_e} + \eta + \gamma m_i$ in polynomial time. With some luck it contains an ℓ_e-bit prime factor e_i, if not we give up in the simulation. We can now return $y_i = z^{E(\alpha 2^{\ell_e} + \eta + \gamma m_i)/e_i}$.

With the method presented above a given ℓ_e-bit prime has either probability $2^{1-\ell_e}$ or probability $2^{2-\ell_e}$ of being chosen. In the real signature scheme, the distribution of primes may be different. Consider for instance the method of Cramer and Shoup [CS00] for picking primes, this distribution is very different from what we have. However, we can consider our distribution of primes as a weighted sum of two distributions: The correct distribution and a residual distribution. We include in the residual distribution all the cases where we simply do not find any prime-factor of $\alpha 2^{\ell_e} + \eta + \gamma m_i$. I.e., we have $\text{Dist}_{\text{our}} = w \text{Dist}_{\text{correct}} + (1-w) \text{Dist}_{\text{residual}}$. In [CS00] they suggest using 161-bit primes and get a distribution where none of the possible primes has more than probability 2^{-144} of being chosen. In our distribution each prime, and thus each of those primes, has at least probability 2^{-160} of being chosen. Thus, w can be chosen to be at least 2^{-16}.

With probability w, we end up in a case where we give the adversary a signature that is statistically indistinguishable from a real signature. Consider now a signature (y, e_i) on message m produced by this adversary. We have $y^{e_i} = ag^m$ so $(y/y_i)^{e_i} = g^{m-m_i} = z^{\gamma(m-m_i)E/e_i}$. By Lemma 1 it must be the case that $e_i|(\gamma(m - m_i)E/e_i)$. However, e_i is a prime and $e_i > \gamma, e_i > |m - m_i|$ and e_i does not divide E/e_i. Therefore, $m = m_i$. We can therefore not produce a signature on a new message if w is non-negligible.

Strong signature. We still need to argue that the signature scheme is strong. Consider the adversary's signature (y, e_i) on $m = m_i$, where the signature oracle returned (y_i, e_i). We then have $y^{e_i} = y_i^{e_i} = ag^m$. This means $y = uy_i$, where $u^{e_i} = 1$. However, with overwhelming probability $\gcd(e_i, p'q'r_p r_q) = 1$ so $u = 1$. □

In the proof we need to factor $\alpha 2^{\ell_e} + \eta + \gamma m_i$. We discussed the class-group-relations method earlier since this has a rigorously proved run-time. Other possible choices include the GNFS, which is not relevant for practical parameters but gives good asymptotics, and the QS [CP01], which works better than the class-group-relations method in practice. The best option would probably be to use the ECM, which has a heuristic run-time of $L(p)^{\sqrt{2}+o(1)}$, with p being the smallest prime factor. This prime factor should be no larger than $\ell_{p'}$ bit in our case.

If we use the ECM we can also consider tackling the original safe-prime setting of this type of signature schemes, where $p = 2p' + 1, q = 2q' + 1$. In this case α, η, γ are so large that we cannot reasonably hope to factor $\alpha 2^{\ell_e} + \eta + \gamma m_i$, however, all we need is an ℓ_e-bit prime factor. As long as ℓ_e is small enough, it is feasible to get out such a small prime factor using the ECM.

Applications. Consider a tag-based simulation sound trapdoor commitment scheme as defined by MacKenzie and Yang [MY04]. It takes as input a message and a tag and forms a commitment. With the trapdoor, it is possible to open the commitment with this tag to any message. The hiding property is defined as usual, however, the binding property is strengthened in the following way: Even if we have seen arbitrary trapdoor openings of commitments with various tags, it is still hard to open a commitment to two different messages using a tag for which no commitment has been equivocated.

[MY04] construct a simulation sound trapdoor commitment scheme based on the Cramer-Shoup signature scheme. Essentially, a commitment to message m using tag tag is a simulated honest verifier zero-knowledge argument of knowledge of a signature on tag using challenge m. We can simplify this trapdoor simulation sound commitment scheme by instead simulating an honest verifier zero-knowledge argument of a signature on tag using challenge m, where we use the simplified signature scheme. I.e., we pick a prime e, pick at random r

and set $c = r^e(ag^{tag})^{-m} \mod n$. The commitment is (c, e, tag), while the opening is (r, m). A double opening would give us $(r/r')^e = (ag^{tag})^{m'-m}$. Since $\gcd(e, m' - m) = 1$, this gives us an e-root of ag^{tag}, i.e., a signature on tag.

[MY04] use 5 exponentiations to form their simulation sound trapdoor commitment and remark that using the Fischlin signature scheme it can be reduced to 4 exponentiations. In comparison, we only use 3 exponentiations.

6 Commitment

Homomorphic integer commitments based on the strong RSA assumption were first suggested by Fujisaki and Okamoto [FO97]. Later Damgård and Fujisaki [DF02] corrected a flaw in the security proof of the former paper and generalized the commitment scheme to abelian groups satisfying some specific assumptions. In this section, we suggest a similar integer commitment scheme based on the strong RSA subgroup assumption.

Key generation: We generate an RSA subgroup $\mathbb{G}$ and choose at random two generators g, h.
The public key is $pk = (n, g, h)$.
Commitment: To commit to integer m using randomizer (u, e, r), where $u^e = 1 \mod n, e > 0$ and $r \in \mathbb{Z}$ we compute

$$c = commit_{(n,g,h)}(m; (u, e, r)) = ug^m h^r \mod n.$$

When making a commitment from scratch we choose $r \leftarrow \{0, 1\}^{\ell_G + \ell_s}$ and use the randomizer $(1, 1, r)$.
Opening: To open commitment c we reveal $(m, (u, e, r))$ such that $c = ug^m h^r \mod n$, where $u^e = 1 \mod n, e > 0$.

Theorem 3. *The commitment scheme is statistically hiding and if the strong RSA subgroup assumption holds for the key generation algorithm then it is computationally binding.*

Proof. It is easy to see that the commitment is statistically hiding since h^r is almost uniformly distributed on $\mathbb{G}$.

To see that the commitment scheme is binding consider a commitment c and two openings $(m, (u, e, r))$ and $(m', (u', e', r'))$ produced by the adversary. We have $c = ug^m h^r = u'g^{m'} h^{r'}, u^e = 1, (u')^{e'} = 1$. We must have $\gcd(e, p'q') = \gcd(e', p'q') = 1$, since otherwise we can as in the proof of Lemma 1 break the strong RSA subgroup assumption. This means $u, u' \in \mathbb{Z}_n^*/\mathbb{G}$ and therefore $u = u'$. We then have $1^0 = g^{m-m'} h^{r-r'}$. By Lemma 1 we get $m = m'$. □

The commitment scheme has several nice properties. It is homomorphic in the sense that for all $(m, (u, e, r)), (m, (u', e', r'))$ we have $commit_{(n,g,h)}(m + m'; (uu', ee', r + r')) = commit_{(n,g,h)}(m; (u, e, r)) commit_{(n,g,h)}(m'; (u', e', r'))$. It is a trapdoor commitment scheme, if we know both $p'q'$ and x such that

$g = h^x$ and an opening $(m, (u, e, r))$ of c, then we can open c to m' by revealing $(m', (u, e, r'))$, where r' is picked at random from $\{0,1\}^{\ell_G+\ell_s}$ such that $r' = (m-m')x + r \bmod p'q'$. Finally, it has the following root extraction property: Consider an adversary that produces $(c, m, (u, e, r), d)$ so $c^e = ug^m h^r, u^d = 1$, then we can find a valid opening of c. Notably, we have $c^{de} = g^{dm}h^{dr}$ so from Lemma 1 we get $e|m, e|r$ and $c = (u')g^{\frac{m}{e}}h^{\frac{r}{e}}$, where $(u')^{ed} = 1$. The homomorphic property combined with the root extraction property means that we can form efficient honest verifier zero-knowledge arguments (Σ-protocols [CDS94]) for many interesting properties of the message inside the commitment.

The commitment schemes of [FO97, DF02] pick the randomness from $\{0,1\}^{\ell_n+\ell_s}$ while we pick the randomness from $\{0,1\}^{\ell_G+\ell_s}$. This means that we have a much shorter exponentiation when computing the commitment.

7 Encryption

Recall that a semi-smooth RSA subgroup modulus $n = (2p'r_p + 1)(2q'r_q + 1)$ has B-smooth r_p, r_q. Suppose we have $h \in \mathbb{G}$ and g has order $p'q'r_g$. Given $c = g^m h^r$ we can compute $c^{p'q'} = g^{p'q'm} h^{p'q'r} = (g^{p'q'})^{m \bmod r_g}$. Since r_g is B-smooth, we can from this compute $m \bmod r_g$. This is the main idea in the following cryptosystem.

Key generation: Generate an RSA subgroup modulus $n = pq = (2p'r_p + 1)(2q'r_q + 1)$, where r_p, r_q are B-smooth and all prime factors are distinct. Select $g \leftarrow \mathrm{QR}_n$ and $h \leftarrow \mathbb{G}$.
The public key is (n, g, h). The secret key is the factorization of $\varphi(n)$.
Encryption: We wish to encrypt a message $m \in \{0,1\}^{\ell_m}$ using randomness $(u, r) \in \{-1, 1\} \times \mathbb{Z}$. The ciphertext is

$$c = E_{(n,g,h)}(m;(u,r)) = ug^m h^r \bmod n.$$

We usually choose $u = 1$ and $r \leftarrow \{0,1\}^{\ell_G+\ell_s}$.
Decryption: Given a ciphertext $c \in \mathbb{Z}_n^*$ we compute $C_p = c^{p'} = (g^{p'})^{m_p} \bmod p$. Since the order of $g^{p'}$ in $\mathbb{Z}_p^*$ is smooth, we can now find $m_p \bmod p_i$ for all $p_i | r_p, p_i | ord(g)$. Similarly, we can find $m_q \bmod q_i$ for $q_i | r_q, q_i | ord(g)$. Using the Chinese remainder theorem, we end up with $m \bmod \gcd(r_p r_q, ord(g))$. If $m \in \{0,1\}^{\ell_m}$ we output m, otherwise we output `invalid`.

Theorem 4. *If the decisional RSA subgroup assumption holds for the key generation algorithm then the cryptosystem is semantically secure against chosen plaintext attack.*

Proof. By the decisional RSA subgroup assumption, we can replace g in the public key with a randomly chosen element from $\mathbb{G}$ without the adversary noticing it. This leaves us with a statistically hiding commitment, which of course does not allow the adversary to distinguish plaintexts. □

It is worthwhile to observe that given a semi-smooth RSA subgroup modulus n an adversary can only produce trivial (u, e) so $u^e = 1, e > 1$. It is with overwhelming probability the case that $u = \pm 1$. To see this first note as in the proof of Lemma 1 that if $\gcd(e, p'q') > 1$, then we can break the strong RSA subgroup assumption. If there is a prime $p_i < B$ so $p_i |\gcd(e, ord(u))$ then we can find s so $U = u^{e/p_i^s} \neq 1 \bmod n$ and $U^{p_i} = 1 \bmod n$. This means $U = 1 \bmod p, U \neq 1 \bmod q$ or the other way around. I.e., $1 < \gcd(n, U-1) < n$ gives us a factorization of n.

The cryptosystem looks just like the integer commitment scheme, where we always choose $u = \pm 1$ and $e = 2$. As we argued above it is not possible for an adversary to find $u \neq \pm 1$ so this is not a problem. Since we cannot distinguish between a random g from QR_n and a random g from $\mathbb{G}$ we actually have all the nice properties of the commitment scheme we presented before. In particular, the cryptosystem is homomorphic as long as we are careful to avoid overflows where the messages are longer than ℓ_m bits. It also has the root extraction property that is useful in zero-knowledge arguments.

Let us consider the length of the messages ℓ_m. The ciphertext has length ℓ_n, however, ℓ_G bits are used for the randomization. Suppose d is chosen such that there is negligible probability that more than d of the primes p_i, q_i do not divide the order of g. We are then left with $\ell_m \leq (t-d)(\ell_{p_i} - 1)$.

In comparison with other cryptosystems such as [Pai99, NS98, OU98] the present scheme offers a better expansion rate. Generalized Paillier encryption [DJ01] has expansion rate $|c|/|m| = 1 + 1/s$, where s is some small positive integer. Their scheme, however, requires a modulus of size n^{s+1}. Okamoto-Uchiyama encryption uses a modulus n of about the same size as we do, however, the expansion rate is around 3. Our cryptosystem has an expansion rate as low as $\ell_n/\ell_m = \ell_n/((t-d)(\ell_{p_i}-1))$. With the parameters $\ell_n = 1280, \ell_{p'} = \ell_{q'} = 160, B = 2^{15}, t = 64, d = 7$ we get from Lemma 2 that the order of g has bit-length no smaller than $320 + (64 - 7)(15 - 1) = 1118$ with probability higher than $1 - 2^{-80}$, giving us an expansion rate of $1280/798 \approx 1.6$.

Applications. Strengthening the decisional RSA subgroup assumption a little, we could get away with picking g of full order $p'q'r_p r_q$. This way, we can increase the message space $\{0,1\}^{\ell_m}$ slightly. According to Lemma 2 a random g does have high order so the difference is not that big though.

The reason we prefer a random g is that part of the public key can be picked by coin-flipping. This property can be useful. Consider as an example the universally composable commitment scheme of Damgård and Nielsen [DN02, Nie03]. In their scheme, they first carry out a 2-move coin-flipping protocol to determine the key for what they call a mixed-commitment scheme. If a corrupt party is making a commitment, the coin-flipping protocol makes the key be a so-called X-key. The setup is such that a simulator knows the corresponding secret key, and thus can extract what the corrupt party committed to. On the other hand, if an honest party is making a commitment we can tweak the coin-flipping protocol to produce a so-called E-key. A commitment under an E-key is equivocable. The simulator can therefore make the commitment now, and later when learning the real value it can equivocate the commitment to this value.

Damgård and Nielsen suggest universally composable commitments based on the subgroup-p assumption [OU98] and based on the decisional composite residuosity assumption [Pai99]. Our cryptosystem provides an efficient alternative to these variations. We generate a (n, h) as in the key generation of the cryptosystem. The corresponding trapdoor is the factorization of $\varphi(n)$. Running a coin-flipping protocol we get a random element g. Using this g we can commit to $m \in \{0,1\}^{\ell_m}$ as $g^m h^r$. If g is random, then it is a ciphertext and we can extract m with our knowledge of the factorization. On the other hand, we could also select $x \leftarrow \{0,1\}^{\ell_G + \ell_s}, g = h^x$, which would make g an E-key. With this g we have set up the statistically hiding commitment scheme and with knowledge of the trapdoor x we can form commitments that can be opened to our liking.

Notice, we only use the decryption property in the simulation in the security proof. In a real run of the universally composable commitment protocol we never decrypt anything. Therefore, it does not hurt us that the decryption process is slow.

Consider further the universally composable threshold cryptosystem of Damgård and Nielsen [DN03]. Here the sender encrypts his message and at the same time makes a universally composable commitment to it. He also proves that the two messages are identical.

The cryptosystem itself needs to be a threshold cryptosystem. They suggest using a variation over the Paillier cryptosystem, which gives us a message space on the form $\mathbb{Z}_n$, with known n. However, the UC commitment scheme does not need to be a threshold scheme. Actually, it is only used in the security proof where the simulator can extract the message from the UC commitment rather than the ciphertext itself. Using our universally composable commitment scheme, we have the additional advantage that it serves as an integer commitment. This means, it is easy to make an efficient zero-knowledge argument of the ciphertext and the commitment containing the same message, even though the message spaces are different.

References

[BP97]   Niko Bari and Birgit Pfitzmann. Collision-free accumulators and fail-stop signature schemes without trees. In *proceedings of EUROCRYPT '97, LNCS series, volume 1233*, pages 480–494, 1997.

[BR93]   Mihir Bellare and Phillip Rogaway. Random oracles are practical: A paradigm for designing efficient protocols. In *ACM CCS '93*, pages 62–73, 1993.

[Bre80]   Richard P. Brent. An improved monte carlo factorization algorithm. *BIT*, 20:176–184, 1980.

[CDS94]   Ronald Cramer, Ivan Damgård, and Berry Schoenmakers. Proofs of partial knowledge and simplified design of witness hiding protocols. In *proceedings of CRYPTO '94, LNCS series, volume 893*, pages 174–187, 1994.

[CF85]   Josh D. Cohen and Michael J. Fischer. A robust and verifiable cryptographically secure election scheme. In *proceedings of FOCS '85*, pages 372–382, 1985.

[CG04]   Jan Camenisch and Jens Groth. Group signatures: Better efficiency and new theoretical aspects. In *proceedings of SCN '04, LNCS series*, 2004.
[CL02]   Jan Camenisch and Anna Lysyanskaya. A signature scheme with efficient protocols. In *SCN '02, LNCS series, volume 2576*, pages 268–289, 2002.
[CP01]   Richard Crandall and Carl Pomerance. *Prime Numbers - a Computational Perspective*. Springer Verlag, 2001.
[CS00]   Ronald Cramer and Victor Shoup. Signature schemes based on the strong rsa assumption. *ACM Transactions on Information and System Security (TISSEC)*, 3(3):161–185, 2000.
[DF02]   Ivan Damgård and Eiichiro Fujisaki. A statistically-hiding integer commitment scheme based on groups with hidden order. In *proceedings of ASIACRYPT '02, LNCS series, volume 2501*, pages 125–142, 2002.
[DJ01]   Ivan Damgård and Mads J. Jurik. A generalisation, a simplification and some applications of paillier's probabilistic public-key system. In *proceedings of PKC '01, LNCS series, volume 1992*, 2001.
[DK02]   Ivan Damgård and Maciej Koprowski. Generic lower bounds for root extraction and signature schemes in general groups. In *proceedings of EUROCRYPT '02, LNCS series, volume 2332*, pages 256–271, 2002.
[DN02]   Ivan Damgård and Jesper Buus Nielsen. Perfect hiding and perfect binding universally composable commitment schemes with constant expansion factor. In *proceedings of CRYPTO '02, LNCS series, volume 2442*, pages 581–596, 2002. Full paper available at http://www.brics.dk/RS/01/41/index.html.
[DN03]   Ivan Damgård and Jesper Buus Nielsen. Universally composable efficient multiparty computation from threshold homomorphic encryption. In *proceedings of CRYPTO '03, LNCS series, volume 2729*, pages 247–264, 2003.
[Fis02]   Marc Fischlin. On the impossibility of constructing non-interactive statistically-secret protocols from any trapdoor one-way function. In *proceedings of CT-RSA '02, LNCS series, volume 2271*, pages 79–95, 2002.
[Fis03]   Marc Fischlin. The cramer-shoup strong-rsasignature scheme revisited. In *proceedings of PKC '03, LNCS series, volume 2567*, pages 116–129, 2003.
[FO97]   Eiichiro Fujisaki and Tatsuaki Okamoto. Statistical zero knowledge protocols to prove modular polynomial relations. In *proceedings of CRYPTO '97, LNCS series, volume 1294*, pages 16–30, 1997.
[GM84]   Shafi Goldwasser and Silvio Micali. Probabilistic encryption. *J. Comput. Syst. Sci.*, 28(2):270–299, 1984.
[KKOT90] Kaoru Kurosawa, Yutaka Katayama, Wakaha Ogata, and Shigeo Tsujii. General public key residue cryptosystems and mental poker protocols. In *proceedings of EUROCRYPT '90, LNCS series, volume 473*, pages 374–388, 1990.
[Kop03]   Maciej Koprowski. Cryptographic protocols based on root extracting. Dissertation Series DS-03-11, BRICS, 2003. PhD thesis. xii+138 pp.
[Len87]   Hendrik W. Lenstra. Factoring integers with elliptic curves. *Ann. of Math.*, 126:649–673, 1987.
[LP92]   Hendrik W. Lenstra and Carl Pomerance. A rigourous time bound for factoring integers. *J. Amer. Math. Soc.*, 5:483–516, 1992.
[MY04]   Philip D. MacKenzie and Ke Yang. On simulation-sound trapdoor commitments. In *proceedings of EUROCRYPT '04, LNCS series, volume 3027*, pages 382–400, 2004. Full paper available at http://eprint.iacr.org/2003/252.

[Nie03]   Jesper Buus Nielsen. On protocol security in the cryptographic model. Dissertation Series DS-03-8, BRICS, 2003. PhD thesis. xiv+341 pp.

[NS98]   David Naccache and Jacques Stern. A new public key cryptosystem based on higher residues. In *ACM Conference on Computer and Communications Security*, pages 59–66, 1998.

[OU98]   Tatsuaki Okamoto and Shigenori Uchiyama. A new public-key cryptosystem as secure as factoring. In *proceedings of EUROCRYPT '98, LNCS series, volume 1403*, pages 308–318, 1998.

[Pai99]   Pascal Paillier. Public-key cryptosystems based on composite residuosity classes. In *proceedings of EUROCRYPT '99, LNCS series, volume 1592*, pages 223–239, 1999.

[Pol74]   John M. Pollard. Theorems of factorization and primality testing. *Proc. Cambridge Phil. Soc.*, 76:521–528, 1974.

[Pol75]   John M. Pollard. A monte carlo method for factorization. *BIT*, 15:331–334, 1975.

[Pol78]   John M. Pollard. Monte carlo methods for index computation (mod p). *Math. Comp.*, 32(143):918–924, 1978.

[PP99]   Pascal Paillier and David Pointcheval. Efficient public-key cryptosystems provably secure against active adversaries. In *proceedings of ASIACRYPT '99, LNCS series, volume 1716*, pages 165–179, 1999.

[Rab79]   Michael O. Rabin. Digitalized signatures and public-key functions as intractable as factorization. Technical Report MIT/LCS/TR-212, MIT Laboratory for Computer Science, 1979.

[RSA78]   Ronald L. Rivest, Adi Shamir, and Leonard M. Adleman. A method for obtaining digital signatures and public-key cryptosystems. *Commun. ACM*, 21(2):120–126, 1978.

[Sha71]   Daniel Shanks. Class number, a theory of factorization, and genera. In *1969 Number Theory Institute (Proc. Sympos. Pure Math., Vol. XX, State Univ. New York, Stony Brook, N.Y., 1969)*, pages 415–440. Amer. Math. Soc., Providence, R.I., 1971.

[Zhu03]   Huafei Zhu. A formal proof of zhu's signature scheme. Cryptology ePrint Archive, Report 2003/155, 2003. http://eprint.iacr.org/.

Efficiently Constructible Huge Graphs That Preserve First Order Properties of Random Graphs

Moni Naor[*], Asaf Nussboim[**], and Eran Tromer

Department of Computer Science and Applied Mathematics,
Weizmann Institute of Science, Rehovot 76100, Israel
{moni.naor, asaf.nussbaum, eran.tromer}@weizmann.ac.il

Abstract. We construct efficiently computable sequences of random-looking graphs that preserve properties of the canonical random graphs $G(2^n, p(n))$. We focus on first-order graph properties, namely properties that can be expressed by a formula ϕ in the language where variables stand for vertices and the only relations are equality and adjacency (e.g. having an isolated vertex is a first-order property $\exists x \forall y (\neg \text{EDGE}(x,y)))$. Random graphs are known to have remarkable structure w.r.t. first order properties, as indicated by the following 0/1 law: for a variety of choices of $p(n)$, any *fixed* first-order property ϕ holds for $G(2^n, p(n))$ with probability tending either to 0 or to 1 as n grows to infinity.

We first observe that similar 0/1 laws are satisfied by $G(2^n, p(n))$ even w.r.t. sequences of formulas $\{\phi_n\}_{n \in \mathbb{N}}$ with bounded quantifier depth, $depth(\phi_n) \leq \frac{n}{\lg(1/p(n))}$. We also demonstrate that 0/1 laws do not hold for random graphs w.r.t. properties of significantly larger quantifier depth. For most choices of $p(n)$, we present efficient constructions of huge graphs with edge density nearly $p(n)$ that emulate $G(2^n, p(n))$ by satisfying $\Theta(\frac{n}{\lg(1/p(n))})$-0/1 laws. We show both probabilistic constructions (which also have other properties such as K-wise independence and being computationally indistinguishable from $G(N, p(n))$), and deterministic constructions where for each graph size we provide a specific graph that captures the properties of $G(2^n, p(n))$ for slightly smaller quantifier depths.

1 Introduction

We deal with small families of graphs that resemble large ones. In general we think of our graphs as being huge so they are not represented explicitly, but rather by a procedure that evaluates edge-queries using a succinct representation (a seed) of the graph. Such small families are sampled by randomly picking the succinct representation.

---

[*] Partly supported by a grant from the Israel Science Foundation.
[**] Partly supported by the Minerva Foundation 2-8495.

We attempt to capture a large class of properties of truly random graphs $G(N,p)$ where $N = 2^n$ vertices are fixed and the edges are independently picked each with probability $p = p(n)$. A prominent class of properties is that of first order properties, namely those that can be expressed by a formula ϕ in the language where variables stand for vertices and the only relations are equality and adjacency (e.g containing a triangle is a first order property of quantifier depth 3 written as $\exists x \exists y \exists z \, (\text{EDGE}(x,y)) \wedge (\text{EDGE}(x,z)) \wedge (\text{EDGE}(y,z))$. Random graphs are known to exhibit remarkable structure w.r.t. first order properties, namely the famed 0/1 law: any fixed first-order property ϕ holds for $G(N,p)$ with probability tending either to 0 or to 1 as N grows to infinity[1]. Thus one can view this work as dealing with graphs that look random to distinguishers that are expressible as first order properties.

We show that for sufficiently large k, any *exact* k-wise independent graphs (defined below) preserve the 0/1 law of random graphs (this is not true for *almost* k-wise independent graphs). We also show a construction of computationally pseudo-random graphs that satisfy the 0/1 law of random graphs (note that in general, computational pseudo-randomness does not imply such combinatorial properties). Finally, we provide for each graph size a single graph that captures the first order properties of $G(N,p)$, and is efficiently computable. Those results can be extended to first-order properties of quantifier depth up to $\frac{n}{\log(1/p)}$.

On the other hand we show that no efficiently constructed family of graphs can achieve $D(n)$-equivalence to random graphs w.r.t. to an arbitrarily large polynomial $D(n)$. Before elaborating on our main results, we review other notions that capture aspects of the structure of random graphs.

1.1 Random-Looking Graphs

Several characterizations for the concept of a "random-looking" graph have been extensively studied and are known to have a wealth of applications in combinatorics and computer science:

$K(n)$-wise independent graphs. These are a relaxation of $G(N, p(n))$ in the sense that each edge appears w.p. $p(n)$, and the distribution of any fixed $K(n)$ potential edges is mutually independent. Efficient constructions of n^c-wise independent graphs are known for all fixed c and a wide variety of densities $p(n)$ (e.g., [2]).

Combinatorial pseudo-random graphs. This term refers to a collection of definitions that consider a single graph g_n for each size n and intend to capture the edge distribution of $G(N,p)$ by requiring that any induced subgraph of g_n has density $\approx p$. Two of the variants are Thomason's jumbled graphs where for each vertex set U, $\left||E(U)| - p\binom{|U|}{2}\right| \leq \alpha|U|$, where $\alpha = \sqrt{pN}$ is the desired accuracy achieved by $G(N,p)$ and $E(U)$ is the set of vertices in the subgraph induced by U (see [23]). A weaker (yet very useful) definition is *quasi-random*

---

[1] Note that despite the term "law", the 0/1 law is actually a characteristic that may or may not hold for specific families of graphs.

graphs, which requires only that $\forall U ||E(U)| - p\binom{|U|}{2}| \leq o(N^2)$. Quasi-random graphs were shown by Chung, Graham and Wilson ([6]) to be equivalent to the surprisingly innocent condition that the number of labeled cycles of length 4 is $(pN)^4(1\pm o(1))$ when $E(g_n) = (p\pm o(1))\binom{N}{2}$. Several deterministic constructions are known for such quasi-random and jumbled graphs (see a recent survey by Krivelevich and Sudakov [16]).

Computationally pseudo-random graphs. These are defined as graphs which are computationally indistinguishable from random graphs [13], in the sense of [12]. Namely, no polynomial-time distinguishing algorithm that performs edge-queries of its choice can tell apart a pseudo-random graph from a random graph $G(N, p(n))$. Explicit constructions of computationally pseudo-random graphs are easily derived from pseudo-random functions. The latter are known to exist iff one-way functions exist [12, 15].

Graphs that preserve specific combinatorial properties of random graphs. Random graphs are known to exhibit a remarkable combinatorial structure (see Bollobás's survey [4]). For instance, consider $G(2^n, 1/2)$ which is the same as the uniform distribution on all 2^n-vertices graphs, and let $N = 2^n$. Then for some value $s(N) \approx 2 \lg N$, it holds that with overwhelming probability $G(2^n, 1/2)$ is:

1. Connected, Hamiltonian, and has a perfect matching.
2. Has clique number and independence number precisely $s(N) \pm 1$.
3. Has chromatic number $\frac{N}{s(N)}(1 \pm \frac{1}{\sqrt{\lg(N)}})$.
4. Has maximal and minimal degree $\frac{1}{2}N(1 \pm 2\sqrt{\frac{\lg(N)}{N}})$.
5. Has connectivity number $\frac{1}{2}N(1 \pm 2\sqrt{\frac{\lg(N)}{N}})$.

Some, of these properties are met by $poly(n)$-wise independent graphs, and by combinatorial pseudo-random graphs. It was shown in [13][17] that there are efficient constructions of graphs which are simultaneously: computationally pseudo-random (w.r.t. $G(2^n, 1/2)$), almost n^c-wise independent, preserve properties 1–3 above, and approximate properties 4 and 5.

Our work. While the constructions of [13][17] are tailor-made to preserve a *fixed* number of prescribed properties (some of which are provably more complex then first-order properties), the current work constructs small families of graphs that preserve *arbitrary* first-order properties of random graphs and in addition may be computationally pseudo-random (w.r.t. $G(N, p)$) and n^c-wise independent. Alternatively we construct a single graph that satisfies *arbitrary* first-order properties of random graphs.

1.2 Preserving First-Order Properties of Random Graphs

First-order properties are graph properties that can be expressed in first order language, where the variables stand for vertices and the only relations are equality and adjacency. For instance, having an isolated vertex can be written as $\exists x \forall y \neg \text{EDGE}(x,y)$ (see section 2 for definitions).

From the first-order lens, random graphs exhibit a remarkable structure (see Spencer's [19] for an excellent survey). The following 0/1 law is known to hold for $G(N,p)$: every first order property ψ holds with probability tending either to 0 or to 1 as the size of the graph grows to infinity. The case where p is constant is due to Fagin [9] and independently Glebskii et al [11]. The other known case where $p(n) = 2^{-\alpha n}$ for an irrational α is due to Shelah and Spenser [22].

Can one efficiently construct random-looking graphs that resemble $G(N, p(n))$ and satisfy this 0/1 law? The answer is positive, but we shall actually consider graphs that meet a much stronger requirement.

Generalized 0/1 Laws. Rather than fixing a single first-order formula, we shall consider sequences of formulas $\Phi = \{\phi_n\}_{n \in \mathbb{N}}$. Such a sequence can express much richer properties than a single formula. For instance, containing a clique of size $\lg n$ can be expressed by the sequence where $\phi_n = \exists x_1 ... \exists x_{\lg n} \bigwedge_{i \neq j}((x_i \neq x_j) \wedge \text{EDGE}(x_i, x_j))$, and the quantifier depth is $depth(\phi_n) = \lg n$ (quantifier depths are formally defined in section 2).

A natural generalization of the basic 0/1 law is the $D(n)$-0/1 law which is satisfied by huge graphs $\mathcal{G}$ if for any sequence Φ having quantifier depth $depth(\phi_n) \leq D(n)$ it holds that

$$\lim_{n \to \infty} \Pr[\mathcal{G}_n \models \phi_n] \in \{0,1\}. \tag{1}$$

Choosing the quantifier depth as the complexity measure for Φ, rather than the entire length of the formulas, will be well-motivated by the discussed results. Some relaxation of this definition is required, however, since for any sequence Φ satisfying the limit condition in (1), if we negate all formulas for odd n then the limit no longer exists. This shows that with the above definition can never be satisfied, even when $D(n) = 1$. This is overcome by requiring (instead of condition (1)) that for each sequence Φ satisfying $depth(\phi_n) \leq D(n)$ there exists a similar sequence Φ' s.t. $\phi'_n \in \{\phi_n, \neg \phi_n\}$, and $\Pr[\mathcal{G}_n \models \phi'_n] \stackrel{n \to \infty}{\longrightarrow} 1$.

Alas, it can be easily seen that with the above relaxation, the $D(n)$-0/1 laws no longer imply the basic 0/1 law. Thus, to reinstate this implication we explicitly also require that for any fixed formula $\phi \in \Phi$ the limit $\lim_{n \to \infty} \Pr[\mathcal{G}_n \models \phi]$ should exist. Note that with this final definition, satisfying the basic 0/1 law is identical to satisfying the $D(n)$-0/1 law for all $D(n) = \Theta(1)$.

Next, recall that we wish to formalize the notion of some huge graphs $\mathcal{G}^1$ preserving the first-order properties of $\mathcal{G}^2 = G(N,p)$. Having a 0/1 law hold for both $\mathcal{G}^1$ and $\mathcal{G}^2$ may not suffice as it might be the case that $\Pr[\mathcal{G}^1_n \models \phi_n] \stackrel{n \to \infty}{\longrightarrow} 1$, whereas $\Pr[\mathcal{G}^2_n \models \phi_n] \stackrel{n \to \infty}{\longrightarrow} 0$. Therefore the following definition is introduced: $\mathcal{G}^1$ and $\mathcal{G}^2$ are said to be $D(n)$-equivalent, if for any sequence Φ having quantifier depth $depth(\phi_n) \leq D(n)$, it holds that $\lim(\Pr[\mathcal{G}^1_n \models \phi_n] - \Pr[\mathcal{G}^2_n \models \phi_n]) \stackrel{n \to \infty}{\longrightarrow} 0$.

1.3 Our Results

Maximal 0/1 laws for random graphs. We start by establishing the maximal depth, $D(n)$, for which $G(N,p(n))$ satisfies $D(n)$-0/1 laws. For any choice of $p(n)$,[2] we set $D^* = D^*(n,p(n)) = \frac{n(1-o(1))}{\lg(1/p(n))}$ and show that $G(N,p(n))$ satisfies the D^*-0/1 law. On the other hand, we show that for any $p(n)$ there exists $p'(n) = p(n)(1-o(1))$ s.t. $G(N,p'(n))$ defies the $2D^*$-0/1 law as long as $p(n) \geq 2^{o(\sqrt{n})}$.

A probabilistic construction. For D^* as above, we show that arbitrary n^3-wise independent graphs satisfy the D^*-0/1 law and are D^* equivalent to $G(N,p(n))$. Since for any non-trivial[3] density $p(n)$ there are explicit efficient constructions of n^3-wise independent graphs $\mathcal{G}$ with density $p'(n) = p(n)(1-o(1))$, our goal is accomplished. A modification of the construction for $\mathcal{G}$ can guarantee (in addition to the above), the computational indistinguishability of $\mathcal{G}$ from $G(N,p(n))$, if one-way functions exist.

Deterministic construction using Paley graphs. We show that for every n and p there exists a specific efficiently computable graph of size $N' = 2^{\Theta(n)}$ and edge density $p' = p \pm \epsilon$, which is $D(n)$-equivalent to $G(N',p')$. Here $D(n)$ depends on ϵ; for example, for $\epsilon(n) > \Theta(1/n)$ we get $D(n) > \frac{n}{2\lg(1/\epsilon)}(1-o(1))$. For the special case $p = 1/2$ we obtain edge density exactly p and $D(n) = \Theta(n)$ which is optimal up to a factor of $4 + o(1)$.

Negative results. While the above positive results are close to optimal, one may still consider the case where $D(n)$ equivalence to random graphs is desired for $D(n)$ so large that $D(n)$-0/1 laws no longer hold for $G(N,p)$. We obtain the following negative result: efficiently constructed graphs $\mathcal{G}$ with seed length $m(n)$ are never $D(n)$-equivalent to $G(2^n, \frac{1}{2})$, for $D(n) = \omega(n + \sqrt{m(n)})$. If one wishes to separate $\mathcal{G}$ from $G(2^n, \frac{1}{2})$ by sequences that have $poly(n)$ total length, then a similar negative result holds for $D(n) = 2m + \omega(n + \sqrt{m(n)})n$. Similar results can be obtained for various choices of p.

1.4 Relationships Among Concepts of Random-Looking Graphs

Figure 1 summarizes the relationships between the main notions of random-looking graphs for a given density $p(n)$. A black arrow stands for implication, while a dotted one implies that implication fails to hold (the bottom left square refers to the conjunction of the properties). Interestingly, while no notion implies all the others, a single construction achieves all four requirements simultaneously (assuming that one-way functions exist).

We sketch the references to the information given in the table. The two following facts are well known. Any computationally pseudo-random graphs with

---

[2] Throughout this subsection we assume that $p(n) \leq \frac{1}{2}$. Otherwise each term $p(n)$ concerning quantifier depths should be replaced by $\min\{p(n), 1-p(n)\}$).

[3] A trivial density is one for which the graph is empty w.p. $1 - o(1)$.

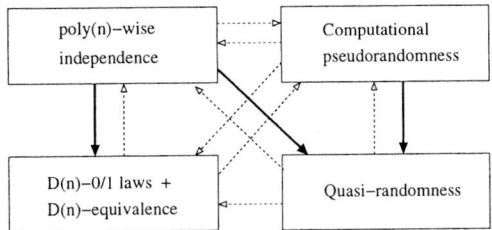

Fig. 1. Relation between notions of random-looking graphs

seed length n^c are statistically far from any n^{c+1}-wise independent graphs. On the other hand, n^c-wise independent graphs generated via polynomials of degree n^c, are easily distinguished from random graphs using only $n^c + 1$ edge queries.

Next, quasi-randomness, $D(n)$-0/1 laws and $D(n)$-equivalence to random graphs may hold even for a single graph per size, and consequently, these conditions do not imply neither K-wise independence nor computational pseudo-randomness.

Using the equivalent condition for quasi-randomness concerning the number of 4-cycles, it is easy to show that quasi-randomness is guaranteed by either computational pseudo-randomness or by $\Theta(1)$-wise independence.

Next, it can be seen that computational pseudo-randomness, and (consequently by the above) also quasi-randomness, both fail to imply even depth-2 0/1 laws and depth-2 equivalence to random graphs. Indeed, assuming the existence of one-way functions, by [13] there exist pseudorandom graphs that have an isolated vertex for odd n but are connected for even n.

Although we can provide graphs satisfying $D(n)$-0/1 laws without achieving quasi-randomness, it is not clear whether $D(n)$-0/1 laws combined with $D(n)$-equivalence to random graphs implies quasi-randomness or not.

Finally, our probabilistic construction shows that n^3-wise independence ensures optimal 0/1 laws and optimal equivalence to random graphs. When this construction is strengthen to maintain computational pseudo-randomness (assuming that one-way functions exist), we achieve a single construction which simultaneously meets all 4 criteria for a being random-looking graph.

2 Preliminaries

2.1 First Order Logic on Graphs

Formally, the alphabet of first order logic on graphs is made of:

1. Infinitely many variable symbols such as 'x','y','z'.
2. The binary relation symbols '$=$' and 'EDGE'.
3. The quantifier symbols '$\forall$' and '$\exists$', the connective symbols '$\neg$', '$\vee$', '$\wedge$', and the signs '(' and ')'.

A first order formula is a formula written in graphs' first order logic. A first order property is a graph property that can be expressed by a first order formula where the variables x, y, z stand for vertices, '=' stands for equality and 'EDGE' stands for adjacency.

The quantifier depth $depth(\phi)$ of a formula ϕ is defined inductively:

1. For atomic expressions, $depth(x = y) = depth(\text{EDGE}(x, y)) = 0$.
2. $depth(\neg \phi) = depth((\phi)) = depth(\phi)$.
3. $depth(\phi \bigvee \psi) = depth(\phi \bigwedge \psi) = max\{depth(\phi), depth(\psi)\}$
4. $depth(\exists x \phi) = depth(\forall x \phi) = depth(\phi) + 1$.

For instance, the property of being either an empty graph or containing a triangle is a first order property that can be expressed by the following formula of quantifier depth 3: $(\forall u \forall v \neg \text{EDGE}(u, v)) \bigvee (\exists x \exists y \exists z \ (\text{EDGE}(x, y)) \bigwedge (\text{EDGE}(x, z)) \bigwedge (\text{EDGE}(y, z))$.

2.2 Distributions on Huge Graphs

Definition 1 (Distributions on Huge Graphs). *Let $\ell : \mathbb{N} \longrightarrow \mathbb{N}$ be a $poly(n)$-bounded length function. Distributions on huge graphs with vertex sets $\{V_n\}_{n \in \mathbb{N}}$, $V_n \subseteq \{0,1\}^{\ell(n)}$ are a sequence of distributions $\mathcal{G} = \{\mathcal{G}_n\}_{n \in \mathbb{N}}$, where each $\mathcal{G}_n$ is taken over the set of simple, labeled undirected graphs over V_n.*

For our probabilistic constructions the vertex sets are simply $V_n = \{0,1\}^n$. For our deterministic constructions the distributions $\mathcal{G}_n$ are degenerate (i.e., have support of size 1). We often abbreviate the term "distributions over huge graphs" and refer to "huge graphs" instead.

Definition 2 (Efficiently constructible huge graphs). *Let $\ell_1, \ell_2 : \mathbb{N} \longrightarrow \mathbb{N}$ be $poly(n)$-bounded length functions. Distributions on huge graphs $\mathcal{G} = \{\mathcal{G}_n\}_{n \in \mathbb{N}}$ with vertex sets $\{V_n\}_{n \in \mathbb{N}}$, $V_n \subseteq \{0,1\}^{\ell_1(n)}$ are efficiently constructible if there exists a deterministic polynomial-time evaluation algorithm E such that for all $n \in \mathbb{N}$: for uniformly drawn $s \in \{0,1\}^{\ell_2(n)}$, the distribution of graph*

$$(V_n, \{(u, v) : E(s, u, v) = 1\})$$

is identical to $\mathcal{G}_n$.

Note that for our deterministic constructions, $\mathcal{G}_n$ is degenerate and $\ell_2(n) = 0$.

Definition 3 ($(p(n), K(n))$-wise independent graphs). *Let $p : \mathbb{N} \to [0,1]$, and $K : \mathbb{N} \to \mathbb{R}^+$. Huge graphs $\mathcal{G} = \{\mathcal{G}_n\}_{n \in \mathbb{N}}$ are $(p(n), K(n))$-wise independent if in $\mathcal{G}_n$ every potential edge appears w.p. $p(n)$, and the distribution of any $K(n)$ potential edges is mutually independent.*

Computational Indistinguishability between distributions over huge graphs is defined exactly like (standard) computational indistinguishability between distributions over functions, with function evaluation replaced by graph edge queries. (For more details the reader may consult [13] [17].)

2.3 New Definitions: Generalized 0/1 Laws

Definition 4 ($D(n)$-0/1 law). *Let $\mathcal{G}$ be huge graphs, and let $D : \mathbb{N} \to \mathbb{N}$. The $D(n)$-0/1 law holds for $\mathcal{G}$ if for any sequence of formulas Φ having quantifier depth $depth(\phi_n) \leq D(n)$ the following conditions are satisfied:*

- *There exist a sequence $\Phi' = \{\phi'_n\}_{n \in \mathbb{N}}$, such that $\phi'_n \in \{\phi_n, \neg \phi_n\}$, and $\Pr[\mathcal{G}_n \models \phi'_n] \stackrel{n \to \infty}{\longrightarrow} 1$.*
- *For any single formula $\phi \in \Phi$ the limit $\lim_{n \to \infty} \Pr[\mathcal{G}_n \models \phi]$ exists.*

For the motivation of this definition, see Section 1.2. Note that meeting the basic 0/1 law is precisely the same as satisfying the $D(n)$-0/1 law for all $D(n) = \Theta(1)$.

Definition 5 ($D(n)$-equivalence of huge graphs). *Let $D : \mathbb{N} \to \mathbb{N}$. Two huge graphs $\mathcal{G}^1, \mathcal{G}^2$ are $D(n)$-equivalent if for any sequence of formulas Φ having quantifier depth $depth(\phi_n) \leq D(n)$ it holds that $\lim(\Pr[\mathcal{G}^1_n \models \phi_n] - \Pr[\mathcal{G}^2_n \models \phi_n]) \stackrel{n \to \infty}{\longrightarrow} 0$.*

3 Extension Properties and 0/1 Laws

We now describe extension properties, which were used by Fagin as a sufficient condition for his basic 0/1 law [9]. These extension properties will be used for two purposes: first, to establish the maximal depth, $D(n)$, for which $D(n)$-0/1 laws are satisfied by $G(N, p(n))$, and later, for proving $D(n)$-0/1 laws for efficiently constructed graphs.

Definition 6 (Extension Properties).

- *A single graph g maintains the t-extension property P_t^{EXT} if for all distinct vertices $v_1, ..., v_t$ and any bits $b_1, ..., b_t$ there exists an extending vertex $u \notin \{v_1, ..., v_t\}$ s.t. the edge $\{u, v_i\}$ appears in g iff $b_i = 1$.*
- *Let $T : \mathbb{N} \to \mathbb{N}$. A sequence of huge graphs $\mathcal{G} = \{\mathcal{G}_n\}_{n \in \mathbb{N}}$ achieves the $T(n)$-extension property if $\Pr[\mathcal{G}_n \models P_{T(n)}^{EXT}] \stackrel{n \to \infty}{\longrightarrow} 1$.*

We first state the sufficiency of $D(n)$-extension to $D(n)$-0/1 laws. We remark that although Spencer considers only the case of a single formula (rather then a sequence of formulas), the following Theorem is actually proved in Spencer's [19–Section 2.5]:

Theorem 1. *Let $\mathcal{G}$ be huge graphs, and let $D : \mathbb{N} \to \mathbb{N}$ be an arbitrary increasing function. If $\mathcal{G}$ achieves $D(n)$-extension, then $\mathcal{G}$ satisfies the $D(n)$-0/1 law.*

We next prove that any pair of huge graphs that achieve $D(n)$-extension are $D(n)$-equivalent.

Theorem 2. *Let $\mathcal{G}^1$ and $\mathcal{G}^2$ be huge graphs, and let $D : \mathbb{N} \to \mathbb{N}$ be an arbitrary increasing function. If both $\mathcal{G}^1$ and $\mathcal{G}^2$ achieve $D(n)$-extension, then $\mathcal{G}^1$ and $\mathcal{G}^2$ are $D(n)$-equivalent.*

Proof. Assume towards contradiction that $\mathcal{G}^1$ and $\mathcal{G}^2$ (as above) are not $D(n)$-equivalent. By Theorem 1, both $\mathcal{G}^1$ and $\mathcal{G}^2$ satisfy the $D(n)$-0/1 law. Therefore our negation assumption implies that there exist an infinite subset $N \subset \mathbb{N}$ and a sequence $\Phi = \{\phi_n\}_{n \in \mathbb{N}}$ having quantifier depth $d_\Phi(n) \leq D(n)$, s.t. $\Pr[\mathcal{G}_n^1 \models \phi_n] \overset{n \in N, n \to \infty}{\longrightarrow} 1$, whereas $\Pr[\mathcal{G}_n^2 \models \phi_n] \overset{n \in N, n \to \infty}{\longrightarrow} 0$.

Consider a third distribution $\mathcal{G}^3 = \frac{1}{2}\mathcal{G}^1 + \frac{1}{2}\mathcal{G}^2$. Namely, we construct two graphs g_1, g_2 according to $\mathcal{G}^1, \mathcal{G}^2$ resp. and then toss a fair coin to choose the final graph $g_3 \in \{g_1, g_2\}$. We get $\Pr[\mathcal{G}_n^3 \models \phi_n] \overset{n \in N, n \to \infty}{\longrightarrow} 1/2$. On the other hand, $\mathcal{G}^3$ clearly achieves the $T(n)$-extension property, so Theorem 1 implies that $\mathcal{G}^3$ satisfies the $D(n)$-0/1 law. This contradiction completes the proof. ∎

We next claim that the maximal extension achieved by $G(N, p(n))$ is approximately $\frac{n}{\lg(1/p(n))}$ (the proof is omitted in this preliminary version):

Theorem 3. *For arbitrary $p : \mathbb{N} \to (0,1)$, set $p'(n) = \min\{p(n), 1 - p(n)\}$, and let $T(n) = \frac{n - 2\lg n}{\lg(1/p'(n))}$. Then $G(2^n, p(n))$ achieves the $T(n)$-extension property, and does not achieve the $(1+\Delta)T(n)$-extension property for any constant $\Delta > 0$.*

An interesting consequence of Theorem 3 is that from the lens of first order logic, very sparse graphs and very dense graphs look the same. Formally, this is expressed by the fact that by Theorem 3 $G(N, p)$ and $G(N, 1-p)$ have the same extension. This coincides with the intuition that for $p < 1/2$, finding an extending vertex for the hardest requirement that "all edges must appear" is just as hard as finding an extending vertex for the requirement that "all edges must not appear" when $p' = 1 - p$. For instance, we get that depth-$\frac{n}{10}$ properties can not distinguish between $G(2^n, 0.001)$ and $G(2^n, 0.999)$.

Is the $D(n)$-extension property not only a sufficient but also a *necessary* condition for $D(n)$-0/1 laws? While for general graphs the answer is no (we can show examples where $2^{\omega(n)}$-0/1 laws are satisfied without achieving even 2-extension), we now show that for $G(N, p(n))$ the maximal extension and the maximal depth of 0/1 laws are roughly the same in the following sense: for any choice of $p(n)$ there exists $p'(n) \approx p(n)$ s.t. $G(N, p'(n))$ cannot achieve $D(n)$-0/1 laws for $D(n)$ larger than twice its maximal extension:

Theorem 4. *Let $p : \mathbb{N} \to (0,1)$, s.t. $\frac{1}{p(n)}$ and $\frac{1}{1-p(n)} = 2^{o(\sqrt{n})}$. Then there exists $p' : \mathbb{N} \to (0,1)$ where $p'(n) = p(n)(1 \pm o(1))$, s.t. $G(2^n, p'(n))$ defies the $D(n)$-0/1 law for $D(n) = (2 \pm o(1))\frac{n}{\lg(\frac{1}{p'(n)})}$.*

Proof. The claim will follow by presenting $p'(n)$ as above and a sequence of first-order formulas $\Phi = \{\phi_n\}_{n \in \mathbb{N}}$ having $depth(\phi_n) = (2 \pm o(1))\frac{n}{\lg(1/p'(n))}$ s.t.

1. For sufficiently large n, $1/4 \leq \Pr[G(2^n, p'(n)) \models \phi_n] \leq 3/4$.
2. The limit $\lim_{n \to \infty} \Pr[G(2^n, p'(n)) \models \phi_n]$ does *not* exist.

We use formulas ϕ_n that state the existence of a clique of size $\approx 2\frac{n}{\lg(1/p(n))}$ in the graph. We assume w.l.o.g. that $p(n) \leq 1/2$ (otherwise, let $p(n) > 1/2$, ϕ_n states the existence of independent sets that size).

By the classical analysis of Bollobás and Erdös concerning cliques in random graphs [5], there exists an integer $S^* = S^*(n,p(n)) = (2-o(1))\frac{n}{\lg(1/p(n))}$ s.t. S^*-cliques appear in $G(2^n, p(n))$ almost surely. Namely, for $\phi_n = \exists v_1...v_{S^*} \bigwedge_{i \neq j}((v_i \neq v_j) \wedge \text{EDGE}(v_i, v_j))$, we have $\Pr[G(2^n, p(n)) \models \phi_n] = 1 - o(1)$.

Fix a sufficiently large n s.t. $\Pr[G(2^n, p(n)) \models \phi_n] \geq 3/4$, and define H as follows:

$$H(q) = \Pr[G(2^n, q) \models \phi] =$$

$$\Sigma_{g \models \phi} \Pr[G(2^n, q) = g] = \Sigma_{g \models \phi} q^{E(g)}(1-q)^{\binom{2^n}{2}-E(g)},$$

where $E(g)$ denotes the number of edges in g. Clearly, $H(\cdot)$ is continuous in q, and ϕ_n is a monotone property[4]. Thus, for any choice of $1/4 \leq \mu(n) \leq 3/4$ there exists (a unique) $p'(n) \leq p(n)$ s.t. $\Pr[G(2^n, p'(n)) = \phi_n] = \mu(n)$. In particular, we can take $\{\mu(n)\}_{n \in \mathbb{N}}$ s.t. the sequence has no limit. We thus get that $G(2^n, p'(n))$ defies the $2\frac{n}{\lg(1/p(n))}$-0/1 law.

We need to prove that the $\approx 2\frac{n}{\lg(1/p'(n))}$-0/1 law is also defied by $G(2^n, p'(n))$, so to complete the entire proof we will show that:

1. $\lg(1/p(n)) = \lg(1/p'(n))(1 \pm o(1))$, and
2. $p'(n) = p(n)(1 - o(1))$.

Indeed, fix n so $p = p(n), p' = p'(n), \mu = \mu(n)$ and $\phi = \phi_n$, and let $\delta = \delta(n)$ be defined by $p' = p(1-\delta)$. Let $\mathbb{E}_{S,n,q}$ denote the expected number of S-cliques in $G(2^n, q)$. Again, by [5] $\mathbb{E}_{S^*,n,p} \leq 2^{(2+o(1))n}$. Next, Markov's inequality gives:

$$1/4 \leq \mu \stackrel{\text{def}}{=} \Pr[G(2^n, p') \models \phi] \leq \mathbb{E}_{S^*,n,p'} = \binom{2^n}{S^*} \times p'^{\binom{S^*}{2}} = \binom{2^n}{S^*} p^{\binom{S^*}{2}}(1-\delta)^{\binom{S^*}{2}}$$

$$\leq 2^{(2+o(1))n}(1-\delta)^{\binom{S^*}{2}} = 2^{(2+o(1))n}e^{-\Theta(\delta(S^*)^2)} = 2^{(2+o(1))n}e^{-\Theta(\delta(\frac{n}{\lg(\frac{1}{p})})^2)}$$

Thus $\delta(n) = o(1)$ iff $(\lg\frac{1}{p})^2 = o(n)$ but the latter condition is met since the conditions of the theorem include $\frac{1}{p} = 2^{o(\sqrt{n})}$. This proves that $p'(n) \stackrel{\text{def}}{=} p(n)(1 - \delta(n)) = p(n)(1 - o(1))$.

Finally, as $0 < \delta \leq 1/2$ we get $\frac{1}{1-\delta} = 1 + \frac{\delta}{1-\delta} \leq 1 + 2\delta \leq e^{2\delta}$. Consequently,

$$\frac{\lg\frac{1}{p'}}{\lg(1/p)} = \frac{\lg\frac{1}{p(1-\delta)}}{\lg(1/p)} = \frac{\lg\frac{1}{p} + \lg\frac{1}{1-\delta}}{\lg(1/p)} \leq 1 + \frac{\lg e^{2\delta}}{\lg(1/p)} = 1 + \Theta(\frac{\delta}{\lg(1/p)}) = 1 + o(1),$$

since $\delta(n) = o(1)$. The claim follows. ∎

An immediate corollary of Theorems 1, 3, and 4 is that the maximal depth $D^*(n)$ for which $G(N, p)$ satisfies $D^*(n)$-0/1 laws is $\Theta(\frac{n}{\lg(\frac{1}{p})})$:

---

[4] Namely, if $g \models \phi$ and g' is obtained by adding edges to g, then $g' \models \phi$ as well.

Theorem 5. *Let $p : \mathbb{N} \to (0, 1)$. Then*

1. *$G(2^n, p(n))$ satisfies the $[\frac{n-2\lg n}{\lg(1/p(n))}]$-0/1 law.*
2. *If $\frac{1}{p(n)}, \frac{1}{1-p(n)} \leq 2^{o(\sqrt{n})}$, then there exists $p' : \mathbb{N} \to (0, 1)$ s.t. , $p'(n) = p(n)(1 \pm o(1))$, and $G(2^n, p'(n))$ defies the $[\frac{2n}{\lg(1/p'(n))}]$-0/1 law.*

In light of Theorem 5, our aim becomes to efficiently construct graphs that satisfy $\Theta(\frac{n}{\lg(\frac{1}{p})})$-0/1 laws and are $\Theta(\frac{n}{\lg(\frac{1}{p})})$-equivalent to $G(N, p)$.

4 Computational and k-wise Independent Graphs and Equivalence

Given the tight relationship between extensions and first-order graph properties, constructing computational and k-wise independent graphs satisfying the 0/1-laws is simple. The next theorem shows that n^3-wise independence in graphs guarantees the optimal $\frac{n}{\lg(\frac{1}{p})}$-0/1 laws and thus $\frac{n}{\lg(\frac{1}{p})}$-equivalence to random graphs.

Theorem 6. *Let $p : \mathbb{N} \to (0, 1)$, and set $p'(n) = \min\{p(n), 1 - p(n)\}$. Let $D(n) = \frac{\frac{n}{2} - 2\lg n}{\lg(1/p'(n))}$ and let $K(n) = 2nD^2(n)$. Let $\mathcal{G}$ be $(p(n), K(n))$-wise independent huge graphs (see definition 3). Then $\mathcal{G}$ satisfies the $D(n)$-0/1 law and is $D(n)$-equivalent to $G(2^n, p(n))$.*

The proof is via the extension property and we omit it in this preliminary version. Recall that for arbitrary $p(n)$, one can construct (based on [13, 17]), $poly(n)$-wise independent graphs that are also computationally pseudo-random w.r.t. $G(2^n, p(n))$. Combining this with Theorem 6 one can show the following:

Theorem 7. *Let $c > 0, p : \mathbb{N} \to [0, 1]$. Then there exist an explicit efficient construction of huge graphs $\mathcal{G}$ that for some $D(n) = \frac{n}{\lg(1/p(n))}(1 - o(1))$ are:*

1. *$(p'(n), n^c)$-wise independent for some $p'(n)$ s.t. $|p'(n) - p(n)| \leq 2^{-3n}$.*
2. *Satisfy the $D(n)$-0/1 law and are $D(n)$ equivalent to $G(2^n, p(n))$.*
3. *Computationally indistinguishable from $G(2^n, p(n))$ if one-way functions exist.*

5 A Single Graph Equivalent to Random Graphs

In this section we demonstrate a single huge graph (for each size) that is deterministically constructible and "behaves like $G(N, p(n))$": the sequence is $D(n)$-equivalent to $G(N, p(n))$ and have edge density $p(n) \pm \epsilon$. The construction is based on Paley graphs, which are known to preserve a variety of properties of random graphs [2]. We employ the following generalized definition:

Definition 7 (Paley graph). *Let $\mathcal{F}$ be a finite field of size N, let $M \in \mathbb{N}$ such that $2M \mid (N-1)$, and let $p \in \{\frac{1}{M}, \frac{2}{M}, \ldots, \frac{M-1}{M}\}$. Let $Z \subset \{a \in \mathcal{F} : a^M = 1\}$ with $|Z| = pM$. Then the Paley graph $G_{\mathcal{F},M,p,Z} = (\mathcal{F}, E_{\mathcal{F},M,p,Z})$ is given by*

$$E_{\mathcal{F},M,p,Z} = \left\{ \{u,v\} : u,v \in \mathcal{F}, \ (u-v)^{(N-1)/M} \in Z \right\} \qquad (2)$$

It is readily verified that every node has exactly $p(N-1)$ neighbors, and that the graph is undirected since the exponent in (2) is even.

The rest of this section is structured as follows. First, as a technical aid we define sets of linear equalities that contain certificates to "$x \not\equiv 0 \pmod{M}$", and observe that for certain M these sets can be small. Then, we show that the $D(n)$-0/1 properties of a Paley graph $G_{\mathcal{F},M,p,Z}$ are related to the size of the smallest such certifying set for M. Next, we show that for appropriate parameters we can efficiently compute edge queries in $G_{\mathcal{F},M,p,Z}$. Finally, we describe two concrete sequences of Paley graphs, and invoke the aforementioned lemmas to derive their efficient computability and $D(n)$-0/1 properties.

Definition 8 (nonzero-certifying set). *A set $C \subset \mathbb{N} \times \mathbb{Z}$ is nonzero-certifying modulo M if $\sum_{(y,z) \in C} y < M$ and for all $x \in \mathbb{Z}$:*

$$x \not\equiv 0 \pmod{M} \quad \text{iff} \quad \exists (y,z) \in C : y_j x \equiv z_j \pmod{M} \qquad (3)$$

For example, for any $M \in \mathbb{N}$ the set $\{(1,r)\}_{r \in \{1,\ldots,M-1\}}$ is nonzero-certifying modulo M. Smaller sets can be obtained by the following:

Lemma 1. *Let $M = q_1^{e_1} q_2^{e_2} \cdots q_s^{e_s}$ for distinct primes q_i and $e_i \in \mathbb{N}$. Then there exists a set C which is nonzero-certifying modulo M and $|C| = \sum_{t=1}^{s} e_t(q_t - 1)$.*

Proof (sketch). Denote $\pi_t = \prod_{t'=t+1}^{s} q_{t'}^{e_{t'}}$, and set $C = \left\{ \left(\pi_t q_t^i, \pi_t (M/q_t) r \right) \right\}_{t,i,r}$ where $t \in \{1,\ldots,s\}$, $i \in \{0,\ldots,e_t - 1\}$, $r \in \{1,\ldots,q_t - 1\}$. Then $|C| = \sum_{t=1}^{s} e_t(q_t - 1) < \sum_{t=1}^{s} (\lg q_t^{e_t})(B-1) = (B-1) \lg M$, and it is readily verified that $\sum_{(y,z) \in C} y = M - 1$. To show that (3) indeed holds, show that it holds modulo each $q_t^{e_t}$ by considering the q_t-ary representation of $z \bmod q_t^{e_t}$; then apply the Chinese Remainder Theorem.[5] ∎

The next lemma shows that Paley graphs satisfy $D(n)$-0/1 laws with $D(n)$ that is related to the size of nonzero-certifying sets. The analysis follows Graham and Spencer's proof of the connection between similar Paley graphs (restricted to $M = 2$) and tournament problems [14][3]. Recall that for a finite field $\mathcal{F}$, a character $\chi : \mathcal{F} \to \mathbb{C}$ of order M is a multiplicative homomorphism from $\mathcal{F}^*$ onto the M-th roots of unity, extended with $\chi(0) = 0$; such χ exist whenever $M \mid (N-1)$. We will invoke Weil's theorem:

---

[5] Essentially, we are forming a system of linear equations which expresses a special case of the additive analogue of the Pohlig-Hellman-Silver algorithm [18].

Theorem 8 (Weil). *Let $\mathcal{F}$ be a finite field, let $N = |\mathcal{F}|$, and let χ be a character of order M. Let $f(x) \in \mathcal{F}[x]$ be a monic polynomial that is not an M-th root of any polynomial in $\mathcal{F}[x]$. Then:*

$$\left| \sum_{u \in \mathcal{F}} \chi(f(u)) \right| < (\deg F - 1)\sqrt{N}$$

Lemma 2. *Let $\mathcal{G} = \{G_{\mathcal{F},M,p,Z}\}_n$ be a sequence of Paley graphs with $\mathcal{F} = \mathcal{F}(n)$, $M = M(n)$, $p = p(n)$, $Z = Z(n)$, $N = |\mathcal{F}(n)|$ such that $N > M^{\omega(1)}$. Let $\ell = \ell(n)$, and suppose that for every n there exist a set of size ℓ which is nonzero-certifying modulo M. Then $\mathcal{G}$ satisfies the $D(n)$-0/1 law for $D(n) = \frac{\lg N}{2\ell}(1-o(1))$.*

Proof. By Theorem 11, it suffices to show that $G_{\mathcal{F},M,p,Z}$ satisfies the $D(n)$-extension law. Denote $d = D(n)$, $\ell = \ell(n)$. Let $C = \{(y_j, z_j)\}_{j=1}^{\ell}$ be nonzero-certifying modulo M, and let $\chi : \mathcal{F} \to \mathbb{C}$ be a character of order M.

Let $v_1, \ldots, v_d \in \mathcal{F}$ be arbitrary vertices, and let $b_1, \ldots, b_d \in \{0,1\}$. We wish to show that there exists an extending vertex $u \in \mathcal{F}\setminus\{v_1,\ldots,v_d\}$ such that $\{u, v_i\} \in E_{\mathcal{F},M,p,Z}$ iff $b_i = 1$ for all $i = 1,\ldots,d$. Let $w_1, \ldots, w_d \in \mathcal{F}$ be chosen arbitrarily subject to $w_i^{(N-1)/M} \in Z$ iff $b_i = 1$, for $i = 1,\ldots d$. Then by definition of $E_{\mathcal{F},M,p,Z}$, it suffices to show that there exists a vertex $u \notin \{v_1,\ldots,v_d\}$ such that $(u-v_i)^{(N-1)/M} = w_i^{(N-1)/M}$ for all i. This further reduces to $\chi(u-v_i) = \chi(w_i)$, since in this case $\mu_i = (u-v_i)/w_i$ is in $\text{Ker}_\chi = \chi^{-1}(1)$ so the order of μ_i divides $|\text{Ker}_\chi| = (N-1)/M$, whence $(u-v_i)^{(N-1)/M}/w_i^{(N-1)/M} = \mu_i^{(N-1)/M} = 1$.

It thus suffices to show that there exists $u \in \mathcal{F}$ such that $\chi(u-v_i) = \chi(w_i)$ for all i. Let α be a generator of $\mathcal{F}^*$, and denote:

$$h(u) = \prod_{i=1}^d h_i(u) \quad \text{where} \quad h_i(u) = \prod_{j=1}^\ell \left(1 - \frac{\chi(u-v_i)^{y_j}}{\chi(w_i^{y_j}\alpha^{z_j})}\right) \quad (i = 1,\ldots,d)$$

Note that $h_i(u) = 0$ iff there exists $j \in \{1,\ldots,\ell\}$ such that $\chi(u-v_i)^{y_j}/\chi(w_i^{y_j}\alpha^{z_j}) = 1$. Since $\chi(\alpha)$ is a generator of the multiplicative group of M-th roots of unity in $\mathbb{C}$, which has order M, for $u \neq v_i$ we can take discrete logs to base $\chi(\alpha)$. Then:

$$h_i(u) = 0 \quad \text{iff} \quad \exists j \in \{1,\ldots,\ell\} : y_j \log_{\chi(\alpha)}((u-v_i)/w_i) \equiv z_j \pmod{M}$$

Since C_n is nonzero-certifying modulo M, by considering $x = \log_{\chi(a)}((u-v_i)/w_i)$ we get that $h_i(u) = 0$ iff $x \equiv 0 \pmod{M}$, i.e., iff $\chi(u-v_i) \neq \chi(w_i)$. Our task is thus reduced to showing the existence of an "extending vertex" $u \in \mathcal{F} \setminus \{v_1,\ldots,v_d\}$ such that $h(u) \neq 0$.

Denote $S = \sum_{u \in \mathcal{F}} h(u)$. By the triangle inequality:

$$|S| \leq \sum_{\substack{u \in \mathcal{F} \\ h(u) \neq 0}} \prod_{i=1}^d \prod_{j=1}^\ell \left(1 + \left|\frac{\chi(u-v_i)^{y_j}}{\chi(w_i^{y_j}\alpha^{z_j})}\right|\right) \leq \sum_{\substack{u \in \mathcal{F} \\ h(u) \neq 0}} 2^{d\ell} = d2^{d\ell} + \sum_{\substack{u \in \mathcal{F}\setminus\{v_1,\ldots,v_d\} \\ h(u) \neq 0}} 2^{d\ell}$$

(4)

Thus, if $|S| > d2^{d\ell}$ then there exists an extending vertex. To lower bound $|S|$, we first expand the product over i and j. Denote $\mathcal{I} = \{1, \ldots, d\} \times \{1, \ldots \ell\}$. Then:

$$S = \sum_{u \in \mathcal{F}} \prod_{i=1}^{d} \prod_{j=1}^{\ell} \left(1 + \frac{\chi(u - v_i)^{y_j}}{-\chi(w_i^{y_j} \alpha^{z_j})}\right) = \sum_{u \in \mathcal{F}} \sum_{I \subseteq \mathcal{I}} \prod_{(i,j) \in I} \frac{\chi(u - v_i)^{y_j}}{-\chi(w_i^{y_j} \alpha^{z_j})}$$

$$= \sum_{u \in \mathcal{F}} \sum_{I \subseteq \mathcal{I}} P_I \left(\prod_{(i,j) \in I} \chi(u - v_i)^{y_j}\right) \quad \text{where} \quad P_I = \prod_{(i,j) \in I} \frac{1}{-\chi(w_i^{y_j} \alpha^{z_j})}$$

By separating the case $I = \emptyset$ and, changing order of summation and using the multiplicativity of χ, we then obtain:

$$S = N + \sum_{\substack{I \subseteq \mathcal{I} \\ I \neq \emptyset}} P_I \sum_{u \in \mathcal{F}} \chi(f_I(u)) \quad \text{where} \quad f_I(u) = \prod_{(i,j) \in I} (u - v_i)^{y_j}$$

For all $I \subseteq \mathcal{I}$ with $I \neq \emptyset$, $f_I(u)$ has at least one root v_i and the multiplicity of any root v_i is at most $\sum_{j=1}^{\ell} y_i < M$ by Definition 8, so $f_I(u)$ is not an M-th power of any polynomial in $\mathcal{F}[u]$. Also, $\deg f_I \leq d(M-1)$. Invoking Weil's theorem, we obtain for all such I:

$$\left|\sum_{u \in \mathcal{F}} \chi(f_I(u))\right| \leq (d(M-1) - 1) \sqrt{N}$$

Then by the triangle inequality,

$$|S| \geq N - \sum_{\substack{I \subseteq \mathcal{I} \\ I \neq \emptyset}} P_I \left|\sum_{u \in \mathcal{F}} \chi(f_I(u))\right| > N - 2^{d\ell} d(M-1) \sqrt{N}$$

By (4), there remains to show that $2^{d\ell} d \geq N - 2^{d\ell} d(M-1) \sqrt{N}$. Indeed:

$$\left(N - 2^{d\ell} d(M-1) \sqrt{N}\right) - 2^{d\ell} d \geq \sqrt{N} \left(\sqrt{N} - 2^{d\ell} dM\right)$$

and the latter is greater than 0 when $\lg N > 2(d\ell + \lg d + \lg M)$, i.e., when $d > \frac{\lg N - 2 \lg M}{(2+o(1))\ell} > \frac{\lg N - 2 \lg N/\omega(1)}{(2+o(1))\ell} = \frac{\lg N}{2\ell}(1 - o(1))$. ∎

Remark 1. Since the choice $w_1, \ldots, w_d \in \mathcal{F}$ in the above proof was arbitrary, we have actually shown a stronger result: for the same parameters as in Lemma 2, there exists an edge labeling $L : \mathcal{F} \times \mathcal{F} \to \{1, \ldots, M\}$ of the full graph of size N, such that for any d vertices $v_1, \ldots, v_d$ and labels $a_1, \ldots a_d$ there exists a vertex $u \in \mathcal{F} \setminus \{v_1, \ldots, v_d\}$ such that $L(u, v_i) = a_i$ for all $i = 1, \ldots, d$. ∎

Recall that $M \in \mathbb{N}$ is called B-smooth if no prime divisor of M is larger than B.

Corollary 1. Let $\mathcal{G} = \{G_{\mathcal{F},M,p,Z}\}_n$ be a sequence of Paley graphs with $\mathcal{F} = \mathcal{F}(n)$, $M = M(n)$, $p = p(n)$, $Z = Z(n)$, $N = |\mathcal{F}(n)|$ such that $N > M^{\omega(1)}$ and M is B-smooth for $B = B(n)$. Then $\mathcal{G}$ satisfies the $D(n)$-0/1 law for

$$D(n) = \frac{\lg N}{2(B-1)\lg M}(1 - o(1))$$

Proof. Let $M = q_1^{e_1} q_2^{e_2} \cdots q_s^{e_s}$ for distinct primes $q_i \leq B$ and $e_i \in \mathbb{N}$. Then by Lemma 1, there exists a set C which is nonzero-certifying modulo M and $\ell(n) = |C| = \sum_{t=1}^{s} e_t(q_t - 1) < \sum_{t=1}^{s}(\lg q_t^{e_t})(B-1) = (B-1)\lg M$. The claim follows by Lemma 2. ∎

We now address the issue of efficient computability. The following lemma shows that there are sequences of Paley graphs in which edge queries can be computed efficiently, under constraints which will be addressed by the concrete sequences described later.

Lemma 3. *There exists a deterministic algorithm A which, for any $\mathcal{F}$, N, M and p as in Definition 7, evaluates edge queries in a Paley graph $G_{\mathcal{F},M,p,Z}$ in the following sense: given an oracle $\mathcal{O}_\mathcal{F}$ which computes the basic operations in $\mathcal{F}$, and given an element $g \in \mathcal{F}$ of order M in $\mathcal{F}^*$, there exists Z as in Definition 7 such that $A^{\mathcal{O}_\mathcal{F}}(N, M, p, g, u, v) = 1$ iff $(u, v) \in E_{\mathcal{F},M,p,Z}$. Moreover, if M is B-smooth then A runs in time $\mathrm{poly}(\log N, B)$.*

Proof. Note that $\langle g \rangle = \{a \in \mathcal{F} : a^M = 1\}$, and set $Z = \{a \in \langle g \rangle : \log_g a < pM\}$. For $u \neq v$, to test whether $a = (u-v)^{(N-1)/M}$ fulfills $a \in Z$, it suffices to compute discrete logarithms in the group $\langle g \rangle$, whose order is B-smooth. This can be done deterministically in time $\mathrm{poly}(\log N, B, |C_\mathcal{F}|)$ using the Pohlig-Hellman-Silver algorithm [18]. ∎

We can now proceed to describe two specific efficiently computable huge graphs based on sequences of Paley graphs. As we have seen, it suffices to find a deterministically computable sequence of pairs (N, M) such that N is a prime power, $2M|(N-1)$, M is highly smooth, and we can deterministically find an efficient representation of the finite field $\mathcal{F} = \mathrm{GF}(N)$ and an element $g \in \mathcal{F}^*$ of order M. Moreover, we wish the sequence to be dense: for every $n \in \mathbb{N}$ there should be (N, M) fulfilling $M = 2^{\Theta(n)}$.

Recall the following results about finite fields, from [20] and [21].

Theorem 9 (Shoup). *(a) Let q be prime and $m \in \mathbb{N}$. Then there exists a deterministic algorithm that computes an irreducible polynomial $I(X)$ of degree m in $\mathrm{GF}(q)[X]$ in time $\mathrm{poly}(q, m)$. (b) Let $I(X)$ be any an irreducible polynomial of degree m in $\mathrm{GF}(q)[X]$, and let $\mathcal{F} = \mathrm{GF}(q)[X]/(I(X))$. There exists a deterministic algorithm which, given $I(X)$, runs in time $\mathrm{poly}(q, m)$ and outputs a set of elements in $\mathcal{F}$ which contains at least one generator of $\mathcal{F}^*$.*

The following is an explicit construction which approximates any desired edge density $p(n)$ up to an additive term of $\epsilon(n) < \Theta(1/n)$, and achieves $D(n)$ which is optimal up to a constant. Here, we choose N and M using Euler's theorem.

Theorem 10. *Let $p = p(n) \in (0,1)$ and let $\epsilon = \epsilon(n) \geq c_0/n$ for a certain constant $c_0 > 0$. Then there exists a deterministically efficiently computable huge graph $\mathcal{G} = \{g_n\}_n$ which satisfies the $D(n)$-0/1 law for $D(n) = \frac{n}{2\log(1/\epsilon)}(1-o(1))$, and g_n has size $2^{\theta(n)}$ and edge density $p'(n)$ such that $|p'(n) - p(n)| < \epsilon(n)$.*

Proof. Set $c_0 = 2 \lg 3$. Let $N = 3^{n'}$ where $n' = 2^k$ and $k = \lceil \lg(n/\lg 3) \rceil$. Let $M = 2^{\lceil \lg(1/\epsilon) \rceil}$. Note that $2^n < N \leq 2^{2n}$, and that $M < 2^{\lg(1/\epsilon)+1} < 2^{\lg(n/2\lg 3)+1} = 2^{\lg(n/\lg 3)} \leq n'$, so $M \mid n'$. Since 3 is relatively prime to $2n'$, Euler's theorem yields $3^{\varphi(2n')} \equiv 1 \pmod{2n'}$, where $\varphi(2n') = n'$. Hence $2M \mid (N-1)$. We have $\epsilon/2 < \frac{1}{M} \leq \epsilon$, and can choose $p'(n) \in \{\frac{1}{M}, \frac{2}{M}, \ldots, \frac{M-1}{M}\}$ such that $|p'(n) - p(n)| \leq \frac{1}{M} \leq \epsilon$.

By Theorem 9(a), we can deterministically compute an irreducible polynomial of degree n' in $GF(3)[X]$ in time $poly(n') = poly(n)$, and can thus efficiently calculate in the field $\mathcal{F} = GF(3^{n'})$.[6] To deterministically find an element of order M in time $poly(n)$, run the algorithm of Theorem 9(b) and, for each output element β, directly test whether $\gamma = \beta^{(N-1)/S}$ has order M by computing the first M powers of γ. Note that when β generates $\mathcal{F}^*$, γ indeed has order M.

By the above and Lemma 3 there exists a set Z such that $G_{\mathcal{F},M,p,Z}$ is a Paley graph whose edge queries can be computed deterministically in time $poly(\log N) = poly(n)$. Then $\mathcal{G} = \{G_{\mathcal{F}(n),M(n),p'(n),Z(n)}\}_n$ is a deterministically efficiently computable huge graph with density $p' = p \pm \epsilon$. Since M is 2-smooth, by Corollary 1 $\mathcal{G}$ satisfies the $D(n)$-0/1 law for $D(n) = \frac{\lg N}{2 \lg M}(1-o(1)) \geq \frac{n}{2\lg(1/\epsilon)}(1-o(1))$. ∎

The above allows only $\epsilon(n) > \Theta(1/n)$, which means we cannot meaningfully approximate graphs with density $p \ll 1/n$. To enable better approximation ϵ, and also to obtain N closer to 2^n (albeit at some cost in the extension $D(n)$), we will replace Euler's totient function $\varphi(\cdot)$ with Carmichael's function $\lambda(\cdot)$, which likewise satisfies that $b^{\lambda(a)} \equiv 1 \pmod{a}$ for any relatively prime $a, b \in \mathbb{N}$. The benefit is that $\lambda(a)$ occasionally assumes much smaller values than $\varphi(a)$ (cf. [8]). For square-free $a \in \mathbb{N}$, $\lambda(a) = \text{lcm}\{q-1 : q \text{ prime}, q \mid a\}$. For $b \in \mathbb{N}$, let $\eta(b) = \prod_{q \text{ prime}, q-1 \mid b} q$. Note that $\lambda(\eta(\mid b)) = b$. Then by [1]:

Theorem 11 (Pomerance, Odlyzko). *There exists a constant $c_1 > 0$ such that for all sufficiently large A, there exists $b < (\ln A)^{c_1 \ln \ln \ln A}$ s.t. $\eta(b) > A$.*

Theorem 12. *Let $p = p(n) \in (0,1)$ and let $\epsilon > 2^{-n^{1/c_2 \ln \ln n}}$ for a constant $c_2 > 0$. Then there exists a deterministically efficiently computable huge graph $\mathcal{G} = \{g_n\}_n$ which satisfies the $D(n)$-0/1 law for $D(n) = n/\log(1/\epsilon)^{\Theta(\log \log \log(1/\epsilon))}$, and g_n has size $2^{n(1+o(1))}$ and edge density $p'(n)$ such that $|p'(n) - p(n)| < \epsilon(n)$.*

Proof. We first find appropriate N, M. Let $B = (\ln(6/\epsilon))^{c_1 \ln \ln \ln(6/\epsilon)}$. Then by Theorem 11, for sufficiently large n there exists $b < B$ such that $\eta(b) > 6/\epsilon$. We can deterministically find such b by exhaustive search in time $poly(B) < poly(n)$.

---

[6] Alternatively replace 5 with 3, and by [10], $X^{2^k} - 2$ is irreducible in $GF(3)[X]$.

Fix any c_2 larger than c_1. It is readily verified that $\log(6/\epsilon)^{c_1 \ln \ln n} < n/\sqrt{\ln n}$ for sufficiently large n, and since $n > \ln(6/\epsilon)$ we get $B < n/\sqrt{\lg n}$ and thus $b < n/\sqrt{\ln n} = o(n)$. Let n' be the smallest multiple of b that is larger than n, and let $N = 3^{n'}$. Then $2^n \leq N \leq 2^{n(1+o(1))}$.

Let $M = \prod_{\text{prime } q | \eta(b),\, q > \kappa} q$ where κ is the largest such that $M \geq 1/\epsilon$. Note that $M \mid \eta(b)$ and $2 \mid \eta(b)$ but $2 \nmid M$, so $2M \mid \eta(b)$, and and from the definition of λ we get $\lambda(2M) \mid \lambda(\eta(\mid b)) = b$. Thus $\lambda(2M) \mid n'$, and since $3 \nmid M$ we get $3^{2^{r}b} \equiv 1 \pmod{2M}$, i.e., $2M | (N-1)$. Also note that all prime factors of M are at most $b+1$, so M is $(B+1)$-smooth and $M < (B+1)/\epsilon = (1/\epsilon)^{1+o(1)}$. Since $\frac{1}{M} \leq \epsilon$, we can choose $p'(n) \in \{\frac{1}{M}, \frac{2}{M}, \ldots, \frac{M-1}{M}\}$ such that $|p'(n) - p(n)| \leq \frac{1}{M} \leq \epsilon$.

Conclude as in Theorem 10, with two differences. First, to test whether $\gamma = \beta^{(N-1)/M}$ is of order M, use the fact that M is $(B+1)$-smooth and square-free: by the Chinese Remainder, γ has order M iff $\gamma^{S/q} \neq 1$ (and thus $\gamma^{M/q}$ has order q) for every prime $q \mid M$, and this can be checked in time $\text{poly}(B \lg M) = \text{poly}(n)$. Second, M is $(B+1)$-smooth so we get $D(n) = \frac{n}{2B \lg M}(1-o(1)) = n/B^{1+o(1)}$. ∎

6 The Limits of Small Families

We now argue that no small and efficient family can be $D(n)$-equivalent to $G(2^n, 1/2)$ once $D(n)$ is an arbitrary polynomial in n. We can generalize the theorem to hold for various choices of p.

Theorem 13. *Let $\mathcal{G}$ be an efficiently constructed distribution on huge graphs with seed length $m(n)$, and let $D: \mathbb{N} \to \mathbb{N}$, s.t. $D(n) = 2m(n) + \omega(\sqrt{m(n)} + n)n$. Then $\mathcal{G}$ is not $D(n)$-equivalent to $G(2^n, 1/2)$.*

Proof. Intuitively, the theorem stems from the fact that any efficiently constructed graph has a low Kolmogorov complexity (KC), whereas random graphs exhibit a high KC. The claim will follow once we provide a sequence of separating formulas $\Phi = \{\phi_n\}_{n \in \mathbb{N}}$ which have depth $\text{depth}(\phi_n) = D(n)$ and length $|\phi_n| = n^{\Theta(1)}$ s.t. $\Pr[\mathcal{G}_n \models \phi_n] = 1$, but $\Pr[G(2^n, \frac{1}{2}) \models \phi_n] \xrightarrow{n \to \infty} 0$.

Fix n and let $m = m(n), d = D(n)$. Let E be the evaluating algorithm of $\mathcal{G}$. Namely, to each graph g in the support of $\mathcal{G}$ there corresponds a seed $s = s(g) \in \{0,1\}^m$ s.t. for any vertex pair $u, v \in \{0,1\}^n$, it holds that $E(s, u, v) = 1$ when the edge $\{u, v\}$ appears in g, and $E(s, u, v) = 0$ otherwise. The standard reduction from Turing machines to Boolean circuits implies the existence of a $\text{poly}(n)$-size Boolean formula $\psi_{E,n}$ s.t. $\psi_{E,n}(s, u, v) = E(s, u, v)$ for all inputs s, u, v of appropriate length.

We wish the separating formulas to hold for a graph g iff g is in the support of $\mathcal{G}$, namely, when there exists a seed $s = s(g)$ s.t. all the edges of g are correctly evaluated by $\psi_{E,n}(s, \cdot, \cdot)$. However, to reduce the quantifier depth of ϕ_n, we only attempt that ϕ_n expresses the following condition where $r = r(n)$ is specified later:

Condition 1. *Every subgraph on r vertices $v_1, \ldots, v_r$ is isomorphic to some subgraph correctly evaluated by $\psi_{E,n}$ using some seed $s \in \{0,1\}^m$.*

Condition 1 can be expressed as follows (here $u_{i_1}...u_{i_n}$ denote the bits of a vertex $u_i \in \{0,1\}^n$):

$$\psi_n = \forall v_1, v_2 \ldots v_r \exists u_{1_1} \ldots u_{1_n}, \ldots, u_{r_1}...u_{r_n} \exists s = s_1...s_m$$
$$\bigwedge_{i \neq j} \text{EDGE}(v_i, v_j) \Leftrightarrow \psi_{E,n}(s_1...s_m, u_{1_1}...u_{1_n}, u_{r_1}...u_{r_n}).$$

This expression is, however, not a first-order sentence on graphs. In first-order language the variables stand for vertices, whereas in the above expression $\psi_{E,n}(s_1...s_m, u_{1_1}...u_{1_n}, u_{r_1}...u_{r_n})$ actually refers to the bits s_i and u_{i_j}. This is resolved as each bit can be encoded using a single edge (or non-edge). Indeed, a string $x = x_1...x_\ell \in \{0,1\}^\ell$ is encoded using 2ℓ (not necessarily distinct) vertices $\bar{x}_1, ..., \bar{x}_\ell, \bar{x}'_1, ..., \bar{x}'_\ell$ s.t. $Enc(x) = \text{EDGE}(\bar{x}_1, \bar{x}'_1)...\text{EDGE}(\bar{x}_\ell, \bar{x}'_\ell)$. Note that for any string x, a valid encoding exists as long as the graph contains both edges and non-edges.[7]

We recall that all the encodings in ψ_n are valid as long as the graph is neither the complete nor the empty graph. Thus we define the separating formula ϕ_n by $\phi_n = \psi_n \lor \gamma \lor \gamma'$, where γ, γ' are two fixed formulas which state that the graph is either complete or empty ($\gamma = \forall u, v(u \neq v \Rightarrow \text{EDGE}(u,v))$, and $\gamma' = \forall u, v(\neg \text{EDGE}(u,v))$).

We finally prove that ϕ_n indeed separates $\mathcal{G}_n$ from $G(2^n, \frac{1}{2})$. We first note that $\Pr[\mathcal{G}_n \models \phi_n] = 1$. Indeed for any single graph g in the support of $\mathcal{G}_n$, if the graph is either complete or empty we are done. Otherwise, each vertex in g has a valid encoding. Since all the encodings in ψ_n are valid, clearly $g \models \psi_n$.

On the other hand $G(2^n, \frac{1}{2})$ is complete or empty with only vanishing probability. Hence it suffices to show that w.h.p. $G(2^n, \frac{1}{2}) \not\models \psi_n$. Indeed assume for a fixed graph g, that $g \models \psi_n$. This implies that for any subgraph on r vertices g_r of g the following holds: there exist strings $\bar{s} \in \{0,1\}^m$, and $\bar{v}_i \in \{0,1\}^n, i = 1,...,r$ s.t. when the evaluator $E = E_\mathcal{G}$ is given all $\binom{r}{2}$ inputs in lexicographic order, then $E(\bar{s}, \bar{v}_i, \bar{v}_j)$ is exactly the adjacency string of g_r. In particular this implies that g_r has Kolmogorov complexity $KC(g_r) \leq m + rn + \Theta(1)$. Since with overwhelming probability a r-subgraph of a random graph has $KC(g_r) \geq \Omega(r^2)$, we get that ψ_n rarely holds for random graphs when $m + rn \leq o(r^2)$, namely when we set $r = \omega(\sqrt{m} + n)$. As the depth of ϕ_n is clearly $r + 2nr + 2m = 2m + \omega(\sqrt{m} + n)n$ the claim follows and this concludes the proof. ∎

Remark 2. The above can be strengthened to show that $\mathcal{G}$ is not $D(n)$-equivalent to $G(2^n, 1/2)$ even for $D(n) = \omega(\sqrt{m(n)} + n)$, at the expense of using seperating formulas of size exponential in n.

---

[7] Note that one cannot write a first order expression that states the validity of the encoding, namely the requirement that indeed $\text{EDGE}(\bar{x}_i, \bar{x}'_i) = x_i$. Yet, this encoding will suffice for our needs.

Acknowledgments. The second author wishes to thank Nati Linial and Avi Wigderson for helpful discussions and Daniel Reichman for referring him to [16]. We thank Ronen Gradwohl, Eran Ofek, Guy Rothblum, Tal Sagiv and Udi Wieder for helpful comments on an earlier draft.

References

1. L. M. Adleman, C. Pomerance and R. S. Rumely, *On Distinguishing Prime Numbers from Composite Numbers*, Annals of Mathematics, vol. 117, no. 1, 173–206, 1983.
2. N. Alon and J. H. Spencer, *The Probabilistic Method*, John Wiley and Sons, 1992.
3. L. Babai, *Character Sums, Weil's Estimates, and Paradoxical Tournaments*, lecture notes, http://people.cs.uchicago.edu/~laci/reu02.dir/paley.pdf
4. B. Bollobás. *Random Graphs*, Academic Press, 1985.
5. B. Bollobás and P. Erdös, *Cliques in Random Graphs*, Cambridge Philosophical Society Mathematical Proc., vol. 80, 419–427, 1976.
6. F. R. K. Chung, R. L. Graham and R. M. Wilson, *Quasi-random graphs*, Combinatorica, vol. 9, 345–362, 1989.
7. A. Ehrenfeucht, *An Application of Games to the Completeness Problem for Formalized Theories*, Fundamenta Mathematicae, vol. 49, 129–141, 1961.
8. P. Erdös, C. Pomerance and E. Schmutz, *Carmichael's Lambda Function*, Acta Arithmetica, vol. 58, 363-385, 1991.
9. R. Fagin, *Probabilities in Finite Models*, Journal of Symbolic Logic, vol. 41, 50–58, 1969.
10. S. Gao and D. Panario, *Tests and Constructions of Irreducible Polynomials Over Finite Fields*, Foundations of Computational Mathematics (F. Cucker, M. Shub, Eds.), 346–361, Springer, 1997.
11. Y. V. Glebskii, D. I. Kogan, M. I. Liagonkii, V. A. Talanov, *Range and Degree of Realizability of Formulas in the Restricted Predicate Calculus*, Cybernetics, vol. 5, 142–154, 1976.
12. O. Goldreich, S. Goldwasser, S. Micali, *How to Construct Random Functions*, Journal of the ACM, vol. 33, no. 4, 276–288, 1985.
13. O. Goldreich, S. Goldwasser, A. Nussboim, *On the Implementation of Huge Random Objects*, proc. 44th IEEE Symposium on Foundations of Computer Science, 68–79, 2003.
14. R. L. Graham and J. H. Spencer, *A Constructive Solution to a Tournament Problem*, Canadian Math Bulletin, vol. 14, 45–48, 1971.
15. J. Håstad, R. Impagliazzo, L.A. Levin and M. Luby, *A Pseudo-Random Generator from any One-Way Function*, SIAM Journal on Computing, vol. 28, num. 4, 1364–1396, 1999.
16. M. Krivelevich and B. Sudakov, *Pseudo-random Graphs*, preprint, http://www.math.princeton.edu/~bsudakov/papers.html
17. A. Nussboim. *Huge Pseudo-Random Graphs that Preserve Global Properties of Random Graphs*, M.Sc. Thesis, Advisor: S. Goldwasser, Weizmann Institute of Science, 2003. http://www.wisdom.weizmann.ac.il/~asafn/psdgraphs.ps
18. S. C. Pohlig and M. E. Hellman, *An Improved Algorithm for Computing Logarithms Over* $GF(p)$ *and Its Cryptographic Significance*, IEEE Transactions on Information Theory, Vol. IT-24, 106–110, 1978.

19. J. H. Spencer. *The Strange Logic of Random Graphs*. Springer Verlag, 2001.
20. V. Shoup, *New Algorithms for Finding Irreducible Polynomials over Finite Fields*, Mathematics of Computation, vol. 54, 435–447, 1990.
21. V. Shoup, *Searching for primitive roots in finite fields*, Mathematics of Computation, vol. 58, 369–380, 1992.
22. J. H. Spencer and S. Shelah, *Zero-One Laws for Sparse Random Graphs*, Journal of the American Mathematical Society, vol. 1, 97–115, 1988.
23. A. Thomason, *Pseudo-random graphs*, Proceedings of Random Graphs, Annals of Discrete Mathematics 33, 307–331, 1987.

Comparing Two Notions of Simulatability

Dennis Hofheinz and Dominique Unruh

IAKS, Arbeitsgruppe Systemsicherheit, Prof. Dr. Th. Beth,
Fakultät für Informatik, Universität Karlsruhe, Am Fasanengarten 5,
76 131 Karlsruhe, Germany

Abstract. In this work, relations between the security notions *standard simulatability* and *universal simulatability* for cryptographic protocols are investigated.

A simulatability-based notion of security considers a protocol π as secure as an idealization τ of the protocol task, if and only if every attack on π can be simulated by an attack on τ.

Two formalizations, which both provide secure composition of protocols, are common: *standard simulatability* means that for every π-attack and protocol user H, there is a τ-attack, such that H cannot distinguish π from τ. *Universal simulatability* means that for every π-attack, there is a τ-attack, such that no protocol user H can distinguish π from τ.

Trivially, universal simulatability implies standard simulatability. We show: the converse is true with respect to perfect security, but not with respect to computational or statistical security.

Besides, we give a formal definition of a *time-lock puzzle*, which may be of independent interest. Although the described results do not depend on any computational assumption, we show that the existence of a time-lock puzzle gives an even stronger separation of standard and universal simulatability with respect to computational security.

Keywords: Reactive simulatability, universal simulatability, protocol composition.

1 Introduction

Recently, simulatability-based characterizations of security for cryptographic protocols received a lot of attention. In particular, several modelings of multi-party computation have been presented which allow for secure composition of protocols, cf. [PW00, Can00, PW01, Can01, BPW04b]. All these models share the idea of simulatability: a protocol is considered secure only relative to another protocol. That is, a protocol π is as secure as another protocol τ (usually an idealization of the respective protocol task), if every attack on π can be simulated by an attack on τ.

A little more formally, this means that for every adversary A_π attacking π, there is an adversary A_τ (sometimes referred to as the simulator) that attacks τ, such that from an outside view, both attacks and protocols "look the same." There are different interpretations of what "looking the same" means concretely.

Roughly, the interpretation of [Can01] is the following: π is as secure as τ, iff for every A_π, there is an A_τ such that no protocol user H (this entity is called the environment $\mathcal{Z}$ in [Can01]) is able to distinguish running with π and A_π from running with τ and A_τ.

Although [PW00, PW01, BPW04b] provide this criterion as "universal simulatability," the default notion of security in these works is that of "standard simulatability." Roughly, standard simulatability demands that for every A_π and every protocol user H, there is an A_τ such that H cannot distinguish π and A_π from τ and A_τ. So basically, the difference between these notions is that with standard simulatability, the simulator A_τ may depend on the user H, whereas universal simulatability requests the existence of "user-universal" simulators A_τ.

All presently known proofs (e.g., in [Can01, BPW04a]) that one can securely compose a *polynomial* number of concurrent protocols depend on the fact that the honest user/environment is chosen in dependence of the simulator. Consequently, we do not know how to prove such a composition theorem in the case of standard security.

1.1 Our Results

In this contribution, we study the relation between standard and universal simulatability. Therefore, we focus on the modeling of [BPW04b], which provides both flavors of simulatability. For a relation to the framework [Can01, CLOS02] of universal composability, see Section 1.2.

By definition, universal simulatability implies standard simulatability. We show that even the converse is true when requiring perfect security (i.e., equality of user-views in the definitions). Apart from giving structural insights, this result may be of practical interest: especially when dealing with idealized protocols, often perfect simulatability can be achieved. Our result enables to conduct a (potentially easier) proof of standard simulatability, and then to conclude universal simulatability using Theorem 1.

On the other hand, we can show that standard simulatability does *not* imply universal simulatability with respect to statistical or computational security. For this, we construct a protocol which is secure only with respect to standard simulatability (in the statistical or computational case). This result shows that proofs of universal simulatability can be stronger than proofs of standard simulatability.

Unfortunately, the constructed protocol is not strictly polynomial-time. So in the computational case, one may wish to have a stronger separation by means of a strictly polynomial-time protocol. We provide such a protocol, and prove that it separates standard and universal simulatability in the computational case. As a technical tool, we need the computational assumption of *time-lock puzzles*, cf. [RSW96]. So additionally, we provide a formal definition of a time-lock puzzle, which may be of independent interest.

1.2 Connections to Universal Composability

Although the framework [Can01, CLOS02] of Universal Composability (UC) does not directly provide an equivalent to the notion of standard simulatability, a formulation of standard simulatability there would seem straightforward. As our proofs below do not depend on specific model characteristics, we believe that our proofs can then be adapted to that framework; this would show that standard and universal simulatability can be separated there, too.

However, recently we have been told [Can04] by Ran Canetti, that in a slightly modified UC setting with a different formulation of polynomial-time, the two notions coincide. At a closer look, this is no contradiction to our results. Namely, Canetti proposes a different notion of standard simulatability than used in, e.g., [BPW04b]: in Canetti's formulation, the environment[1] has a runtime bounded polynomially in the length of its auxiliary input, which again is chosen in dependence of the simulator. So effectively, the (polynomial) runtime bound of the environment is chosen after the simulator, whence our proofs do not apply in that case.

However, since we show that our separating examples also hold for the case of honest users H with non-uniform auxiliary input (that does *not* affect H's runtime), they should be applicable to the notion of "Specialized-simulator UC"[2] defined in [Lin03].

1.3 Organisation

Section 2 establishes the equality of standard and universal simulatability for the case of perfect security; in Section 3, a separation of these notions for statistical and computational security is presented. The discussed stronger separation by means of a strictly polynomial-time protocol using time-lock puzzles is investigated in Section 4. This work ends with a conclusion in Section 5. In Appendix A, we briefly review the modeling of [BPW04b].

2 The Perfect Case

We start by relating standard and universal simulatability for the case of perfect security. Perfect security demands that the respective user-views in the compared protocol situations are completely equal. We show that with respect to perfect security, standard and universal simulatability are equivalent notions. For this, we only need to show that standard simulatability already implies universal simulatability—the other direction is trivial from the definitions.

The idea of our proof is to construct a "universal" protocol user H_u, that simply chooses all of its outputs at random, such that any finite sequence of outputs occurs with nonzero probability. In a sense, H_u incorporates all possible protocol users H.

Now standard simulatability implies that there is a simulator which is "good" with respect to this user H_u. But informally, anything H could do will be done by

---

[1] i.e., the UC counterpart of the honest user H
[2] This notion is identical to standard simulatability, except that a possible non-uniform auxiliary input for the environment is chosen after the simulator.

H_u with nonzero probability. Since H_u's views are completely identical in both protocols, this allows to conclude that this simulator is not only "good" with respect to H_u, but with respect to all possible users H.

Theorem 1. *With respect to perfect security, standard simulatability implies universal simulatability.*

Proof. As a prerequisite, let $\mathcal{D}$ be a probability distribution over Σ^* which satisfies $\Pr[s \leftarrow \mathcal{D}] > 0$ for all $s \in \Sigma^*$. (As in [PW01, BPW04b], Σ denotes the (finite) message alphabet over which messages sent by machines are formed.) Such a $\mathcal{D}$ necessarily exists, since Σ^* is countable.

Let $(\hat{M}_1, S)$ and $(\hat{M}_2, S)$ be structures with $(\hat{M}_1, S) \geq_{\text{sec}}^{\text{perf}} (\hat{M}_2, S)$. Then, let further $(\hat{M}_1, S, \mathsf{H}, \mathsf{A}_1) \in \mathsf{Conf}^{\hat{M}_2}(\hat{M}_1, S)$. That is, let H, A_1 be a valid pair of user and adversary for protocol $\hat{M}_1$. Without loss of generality, we assume H to have exactly one self-clocked self-connection (i.e., a connection from H to itself) with name loop, and to have its ports ordered lexicographically.[3] Then the sequence of ports of H only depends on A_1.

Let H_u be a machine with the same port sequence as H, but with a state set Σ^* and initial states $\{1\}^*$. H_u's transition function makes H_u switch as follows: independent of state and input, H_u's next state and all of its outputs, including outputs on clock ports, are drawn (independently) from $\mathcal{D}$.

Intuitively, H_u is universal in the following sense: for a fixed A_1, H_u is independent of H. H_u's construction guarantees that every finite prefix of H_u-outputs and -states has non-zero probability.

Clearly we have $(\hat{M}_1, S, \mathsf{H}_u, \mathsf{A}_1) \in \mathsf{Conf}^{\hat{M}_2}(\hat{M}_1, S)$, which means that $\mathsf{H}_u, \mathsf{A}_1$ is a valid pair of user and adversary for protocol $\hat{M}_1$. Then the standard security $(\hat{M}_1, S) \geq_{\text{sec}}^{\text{perf}} (\hat{M}_2, S)$ which we assumed ensures the existence of an A_2 with

$$view_{(\hat{M}_1, S, \mathsf{H}_u, \mathsf{A}_1)}(\mathsf{H}_u) = view_{(\hat{M}_2, S, \mathsf{H}_u, \mathsf{A}_2)}(\mathsf{H}_u). \tag{1}$$

We will show

$$view_{(\hat{M}_1, S, \mathsf{H}, \mathsf{A}_1)}(\mathsf{H}) = view_{(\hat{M}_2, S, \mathsf{H}, \mathsf{A}_2)}(\mathsf{H}), \tag{2}$$

which suffices to prove $(\hat{M}_1, S) \geq_{\text{sec}}^{\text{uni,perf}} (\hat{M}_2, S)$, since A_2 does not depend on H.

So let $k \in \mathbb{N}$ be an arbitrary security parameter.[4] The following notation for views of H in protocol runs with A_1 and $\hat{M}_1$, resp. A_2 and $\hat{M}_2$ will simplify the presentation: let $tr_{\mathsf{H}}^{(1)} := view_{(\hat{M}_1, S, \mathsf{H}, \mathsf{A}_1), k}(\mathsf{H})$, and $tr_{\mathsf{H}}^{(2)} := view_{(\hat{M}_2, S, \mathsf{H}, \mathsf{A}_2), k}(\mathsf{H})$. Analogously, we define $tr_u^{(1)}$ and $tr_u^{(2)}$ for views of H_u. For $n \in \mathbb{N}$, let $(tr)_n$ denote the n-th step in a view tr; $(tr)_{1..n}$ is the n-step prefix of tr. When it is clear that $I \in \Sigma^*$ is a vector of inputs, we write $I \in st$ to denote the event that in a step st, the machine input vector is I.

---

[3] Every H can be turned into an H' of this form, so that there is a probability-respecting identification of H-views and H'-views.

[4] Note that this already determines the initial state 1^k for both H and H_u.

We prove the following two statements simultaneously by induction over $n \in \mathbb{N}$:
$$A(n): \quad \left(tr_{\mathsf{H}}^{(1)}\right)_{1..n} = \left(tr_{\mathsf{H}}^{(2)}\right)_{1..n},$$
and
$$B(n): \quad \forall s: \Pr\left[\left(tr_{\mathsf{H}}^{(1)}\right)_{1..n} = s\right] > 0 \Rightarrow \Pr\left[\left(tr_{u}^{(1)}\right)_{1..n} = s\right] > 0.$$

$A(0)$ and $B(0)$ hold trivially. So assume that $A(n)$ and $B(n)$ hold. Let an arbitrary $(n+1)$-step prefix $(st)_{1..n+1}$ with
$$\alpha := \Pr\left[\left(tr_{\mathsf{H}}^{(1)}\right)_{1..n+1} = (st)_{1..n+1}\right] > 0 \tag{3}$$
be given. To show $A(n+1)$, it suffices to prove
$$\Pr\left[\left(tr_{\mathsf{H}}^{(2)}\right)_{1..n+1} = (st)_{1..n+1}\right] = \alpha. \tag{4}$$

To see (4), we first remark that for the machine input vector I_{n+1} in $(st)_{n+1}$, (1) implies
$$\Pr\left[I_{n+1} \in \left(tr_{u}^{(1)}\right)_{n+1} \mid \left(tr_{u}^{(1)}\right)_{1..n} = (st)_{1..n}\right]$$
$$= \Pr\left[I_{n+1} \in \left(tr_{u}^{(2)}\right)_{n+1} \mid \left(tr_{u}^{(2)}\right)_{1..n} = (st)_{1..n}\right]. \tag{5}$$

Here we also need $B(n)$ to be sure that the probabilities of the conditions are not only equal, but also positive. Furthermore, we have for $i \in \{1, 2\}$:
$$\Pr\left[I_{n+1} \in \left(tr_{\mathsf{H}}^{(i)}\right)_{n+1} \mid \left(tr_{\mathsf{H}}^{(i)}\right)_{1..n} = (st)_{1..n}\right]$$
$$= \Pr\left[I_{n+1} \in \left(tr_{u}^{(i)}\right)_{n+1} \mid \left(tr_{u}^{(i)}\right)_{1..n} = (st)_{1..n}\right] > 0, \tag{6}$$

because the distribution on the next user-inputs is completely determined by the history over all preceding user-outputs. The probabilities for the conditions are positive by (3), $A(n)$, and the construction of H_u. From here, $B(n+1)$ follows from the construction of H_u.

We continue proving $A(n+1)$. Combining (5) and (6) yields
$$\Pr\left[I_{n+1} \in \left(tr_{\mathsf{H}}^{(1)}\right)_{n+1} \mid \left(tr_{\mathsf{H}}^{(1)}\right)_{1..n} = (st)_{1..n}\right]$$
$$= \Pr\left[I_{n+1} \in \left(tr_{\mathsf{H}}^{(2)}\right)_{n+1} \mid \left(tr_{\mathsf{H}}^{(2)}\right)_{1..n} = (st)_{1..n}\right].$$

But since input and current state already determine the distribution on outputs and next states, we have

$$\Pr\left[\left(tr_{\mathsf{H}}^{(1)}\right)_{n+1} = (st)_{n+1} \mid \left(tr_{\mathsf{H}}^{(1)}\right)_{1..n} = (st)_{1..n}\right] \quad (7)$$
$$= \Pr\left[\left(tr_{\mathsf{H}}^{(2)}\right)_{n+1} = (st)_{n+1} \mid \left(tr_{\mathsf{H}}^{(2)}\right)_{1..n} = (st)_{1..n}\right].$$

Because the probabilities for the respective conditions in (7) are positive and equal by $A(n)$ and 3, this shows (4), and thus $A(n+1)$.

Summarising, $A(n)$ holds for all n, which in particular implies (2), and thus shows the theorem. □

This proof idea does not work in the computational or statistical case. Very informally, H_u behaves like a given user H too seldom; the resulting success to distinguish protocols would be much smaller than that of H.

So one may ask whether Theorem 1 also holds for computational or statistical security. The next section shows that this is not the case.

3 The Statistical and the Computational Case

Recall that simulatability with respect to *statistical security* demands that polynomial prefixes of the user-views in the real, resp. ideal model must be of "small" statistical distance. Here, "small" may denote a negligible or even exponentially small function in k. The following proof deals with negligible functions as those "small" functions. However, the proof carries over to other classes of "small" functions.

On the other hand, for simulatability with respect to *computational security*, users and adversaries are restricted to (strict) polynomial-time. In this case, the user-views in the real, resp. ideal model only need to be computationally indistinguishable.

Here we give a real and an ideal protocol such that the real protocol is as secure as the ideal one with respect to standard simulatability, but not with respect to universal simulatability. Roughly, the ideal protocol asks adversary and user for bitstrings and then outputs who of them gave the longest bitstring. The real protocol does the same, but always outputs "adversary".

A successful simulator must hence be able to give a longer input than the user with overwhelming probability. We show that such a simulator exists for any given user; we also show that there can be no such simulator which gives longer inputs than every user.

Theorem 2. *With respect to computational and statistical security, standard simulatability does not imply universal simulatability. This holds also if we allow non-uniform polynomial-time honest users for the case of computational security.*

Proof. Let $(\hat{M}_1, S)$ be a structure with machines $\hat{M}_1 = \{\mathsf{M}_1\}$ and service ports S, where $S^c = \{\mathsf{user}!, \mathsf{user}^{\triangleleft}!, \mathsf{out}?\}$. The machine M_1 is depicted in Figure 1. M_1

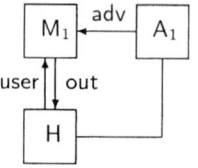

Fig. 1. Machines in the real case

waits for an input h on port user?, and an input a on port adv?; only the first respective input is considered. When having received both such inputs h and a, M_1 outputs and clocks the value $b = 0$ on out!.

Let $(\hat{M}_2, S)$ be a structure with $\hat{M}_2 = \{\mathsf{M}_2\}$. The machine M_2 is identical to M_1, except that the value b that is eventually output on out! is determined as $b = 1$ if $|h| > |a|$, and $b = 0$ otherwise. So intuitively, $b = 1$ (resp. $b = 0$) indicates that the user (resp. the simulator) delivered the longest bitstring.

We claim $(\hat{M}_1, S) \geq_{\text{sec}}^{NEGL} (\hat{M}_2, S)$. So let a real configuration $(\hat{M}_1, S, \mathsf{H}, \mathsf{A}_1) \in \mathsf{Conf}^{\hat{M}_2}(\hat{M}_1, S)$ be given. Denote by h_k the random variable that describes M_1's first user?-input in runs with security parameter k, or $\bot$, if there is no user?-input. Since H and A_1 are fixed, there is a function $f : \mathbb{N} \to \mathbb{N}$ for which

$$\Pr[h_k < f(k) \vee h_k = \bot] > 1 - 2^{-k}. \tag{8}$$

Thus, let A_2 be the combination of A_1 and a special machine S, cf. Figure 2. In this combination, the adv! and $\mathsf{adv}^\triangleleft$! ports of A_1 are renamed to $\overline{\mathsf{adv}}$! and $\overline{\mathsf{adv}^\triangleleft}$!, respectively. The special machine S converts every $\overline{\mathsf{adv}}$?-input into an adv!-output $1^{f(k)}$, which is clocked by S immediately. If we restrict to runs in

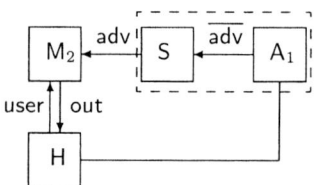

Fig. 2. The simulator for standard simulatability

which either $h_k = \bot$ or $h_k < f(k)$, then we get exactly the same distribution on H-views as in the real configuration. Using (8), we get

$$view_{(\hat{M}_1, S, \mathsf{H}, \mathsf{A}_1)}(\mathsf{H}) \approx_{NEGL} view_{(\hat{M}_2, S, \mathsf{H}, \mathsf{A}_2)}(\mathsf{H}).$$

This implies in particular $(\hat{M}_1, S) \geq_{\text{sec}}^{NEGL} (\hat{M}_2, S)$.

On the other hand, we claim $(\hat{M}_1, S) \not\geq_{\text{sec}}^{\text{uni}, NEGL} (\hat{M}_2, S)$. For this, consider the following real adversary A_1, which is master scheduler and has an additional token connection to the honest user. In its first activation, A_1 outputs and clocks the value $a^{(1)} = 1$ on adv!. In its second activation, it activates H by outputting and clocking 1 on port token!.

Furthermore, for a function $g : \mathrm{N} \to \mathrm{N}$, let H_g be the machine which writes and clocks $h = 1^{g(k)}$ onto out! in its first activation. The remaining ports of H_g are chosen to close the collection $\{\mathsf{M}_1, \mathsf{H}_g, \mathsf{A}_1\}$.

We show that for every simulator A_2, there is a function g for which H_g distinguishes M_1 and A_1 from M_2 and A_2. So let A_2 be a simulator for which $(\hat{M}_2, S, \mathsf{H}_1, \mathsf{A}_2) \in \mathsf{Conf}(\hat{M}_2, S)$. (Then $(\hat{M}_2, S, \mathsf{H}_g, \mathsf{A}_2) \in \mathsf{Conf}(\hat{M}_2, S)$ for all g.) Denote by $a_k^{(2)}$ the random variable that describes the first adv?-input that M_2 gets in runs with security parameter k, or $\bot$, if there is no adv? input. Since A_2 is fixed, there is a function g for which

$$\Pr\left[a_k^{(2)} < g(k) \vee a_k^{(2)} = \bot\right] > 1 - 2^{-k}. \tag{9}$$

In the configuration $(\hat{M}_1, S, \mathsf{H}_g, \mathsf{A}_1)$, H's view in its second activation contains out?-input 0. But by (9), and since the distribution of $a_k^{(2)}$ is independent of g, H_g's view in $(\hat{M}_2, S, \mathsf{H}_g, \mathsf{A}_2)$ contains out?-input 0 with only negligible probability. So $(\hat{M}_1, S) \not\gtrsim_{\mathsf{sec}}^{\mathsf{uni},NEGL} (\hat{M}_2, S)$, and the theorem follows for the statistical case.

The proof above carries over literally to the computational case (with respect to uniform as well as non-uniform honest users), because for polynomial-time H and A, there are polynomials f and g fulfilling (8) and (9). □

4 A Stronger Separation

The proof from the preceding section does not work, if we restrict to protocol machines (i.e., structures) that are strictly polynomial-time. Although the machines M_1 and M_2 used in the proof above are weakly polynomial-time (i.e., they are polynomial-time in the overall length of their inputs and the security parameter, cf. [BPW04b]), at least M_2 needs to accept arbitrarily long inputs.[5] For a separation of the computational simulatability notions by means of strictly polynomial-time structures, we have to work a little harder. As a technical tool, we use time-lock puzzles (see [RSW96]).

Definition 1. *A PPT-algorithm[6] $\mathcal{G}$ (called the problem generator) together with a PPT-algorithm $\mathcal{V}$ (the solution verifier) is called a* time-lock puzzle *iff the following holds:*

- *sufficiently hard puzzles: for every PPT-algorithm B and every $e \in \mathrm{N}$, there is some $c \in \mathrm{N}$ with*

$$\sup_{t \geq k^c, |h| \leq k^e} \Pr\left[(q,a) \leftarrow \mathcal{G}(1^k, t) : \mathcal{V}(1^k, a, B(1^k, q, h)) = 1\right] \tag{10}$$

negligible in k.

---

[5] However, intuitively, nothing "superpolynomial" happens: M_2 determines the length of its inputs.
[6] Probabilistic polynomial time algorithm

– *sufficiently good solvers: there is some $b \in \mathbb{N}$ such that for every $d \in \mathbb{N}$ there is a PPT-algorithm C such that*

$$\min_{t \leq k^d} \Pr\left[(q,a) \leftarrow \mathcal{G}(1^k, t); c \leftarrow C(1^k, q) : \mathcal{V}(1^k, a, c) = 1 \wedge |c| \leq k^b\right] \quad (11)$$

is overwhelming in k.

Intuitively, $\mathcal{G}(1^k, t)$ generates a puzzle q of hardness t, and a description a of valid solutions for q. $\mathcal{V}(1^k, a, b)$ verifies if b is a valid solution as specified by a.

First, we require that any given PPT-algorithm B can't solve sufficiently hard puzzles. Formally, we want B to be unable to solve puzzles which are of hardness t, $t \geq k^c$ for some c depending on B.

We add an auxiliary input h (of polynomial length) to prevent the following scenario: Bob (B) wants to show to Alice, that he is able to perform calculations of some hardness t, therefore he chooses t, and then Alice ($\mathcal{G}$) chooses the puzzle. It is now imaginable, that Bob may choose some auxiliary information h and the hardness t simultaneously, s.t. using h one can solve time-lock puzzles of hardness t.[7] This is prevented by our definition, since (10) is negligible even in presence of polynomially length-bounded auxiliary inputs h.[8]

Second, we demand that for every polynomial hardness value, there is an algorithm C solving puzzles of this hardness. It is sensible here to ask for short solutions (i.e., $|c| \leq k^b$): otherwise, the definition allows time-lock puzzles in which the solution of every t-hard puzzle is deterministically 1^t.[9]

[RSW96] promotes the following family of puzzles as candidates for time-lock puzzles. A puzzle of hardness t consists of the task to compute $2^{2^{t'}} \bmod n$ where $t' := \min\{t, 2^k\}$ and $n = pq$ is a Blum integer.[10] In our notation, this is denoted by $\mathcal{G}(1^k, t) = ((n, \min\{t, 2^k\}), (p, q, \min\{t, 2^k\}))$, where n is a random k-bit Blum integer with factorisation $n = pq$, and $\mathcal{V}(1^k, (p, q, t'), c) = 1$ if and only if $c \equiv 2^{2^{t'}} \bmod pq$. (Note that $2^{t'} \bmod \varphi(pq)$ can be efficiently computed with knowledge of p and q.)

We return to the problem of separating standard and universal simulatability by means of strictly polynomial-time structures. The idea is very similar to the one used in the proof of Theorem 2. In the ideal setting, we let a machine M_2 check and output whether H has more "computational power" than the adversary A_2. The corresponding real machine M_1 always outputs "no". Here we

---

[7] Imagine that, e.g., being able to find the pre-image of t under some function would already solve the puzzle.
[8] Note that for the proof of Theorem 3, this additional constraint is not necessary.
[9] In fact, using such a "degenerate" puzzle would yield a proof for Theorem 2, very similar to that given in Section 3.
[10] One might wonder why our formulation uses $t' = \min\{t, 2^k\}$ instead of t (in contrast to the original formulation in [RSW96]). It can be shown that for $t := (2^k)! + 1$ it is $2^{2^t} \equiv 4 \bmod n$ for all k-bit Blum integers. Therefore (10) would be violated (consider $B(1^k, q, h) = 4$). For practical purposes, this does not pose a threat, since the length of $t = (2^k)! + 1$ is not polynomial in k.

have standard simulatability, since A_2 can be chosen in dependence of H (and thus more powerful). On the other hand, the simulatability is not universal, since an A_2-dependent user H can be chosen so powerful that M_2 outputs "yes".

Now time-lock puzzles are exactly what M_2 needs to check the computational power of A_2 and H. Concretely, M_2 simply picks a puzzle for both A_2 and H and outputs "yes" if H is the *only* one to solve that puzzle. Our definition of time-lock puzzles guarantees that puzzles can be generated and solutions can be checked by a strictly polynomially bounded M_2.

An exact theorem statement and a proof follow:

Theorem 3. *Assume that time-lock puzzles exist. Then for computational security, standard simulatability and universal simulatability can be separated by two strictly polynomial-time structures. This holds also if we allow non-uniform polynomial-time honest users.*

Proof. First, let $\mathcal{D}$ denote a PPT-algorithm which, upon input 1^k, returns a uniformly chosen t from $\{2^1, 2^2, \ldots, 2^k\}$.

Let $(\mathcal{G}, \mathcal{V})$ be a time-lock puzzle. Let $(\hat{M}_1, S)$ be a structure with machines $\hat{M}_1 = \{M_1\}$ and service ports S, where $S^c = \{\mathsf{user!}, \mathsf{user}^{\triangleleft}!, \mathsf{puz\_user?}, \mathsf{out?}\}$.

The machine M_1 is depicted in Figure 3. All indicated connections (not counting the potential connections between A_1 and H) are clocked by the sending machine.

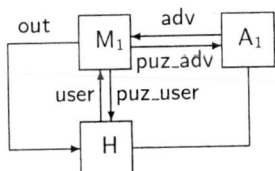

Fig. 3. Machines in the real case

Upon its first activation, M_1 chooses $t \leftarrow \mathcal{D}(1^k)$. Then it computes $(q, a) \leftarrow \mathcal{G}(1^k, t)$.

Upon the first input via $\mathsf{user?}$, M_1 sends and clocks q via $\mathsf{puz\_user}$ to H. Upon the first input via $\mathsf{adv?}$, M_1 sends and clocks q via $\mathsf{puz\_adv}$ to A_1.

The second input via $\mathsf{user?}$ (called c_H) is verified via $v_H \leftarrow \mathcal{V}(1^k, a, c_H)$, and analogously we set $v_A \leftarrow \mathcal{V}(1^k, a, c_A)$ upon the second $\mathsf{adv?}$-input c_A.

At the time both v_H and v_A are determined, M_1 outputs and clocks $b = 0$ on $\mathsf{out!}$.

Let $(\hat{M}_2, S)$ be a structure with machines $\hat{M}_2 = \{M_2\}$. The machine M_2 is identical to M_1, except that the value $b \in \{0, 1\}$ that is eventually output on $\mathsf{out!}$ is determined as an evaluation of the predicate $v_H \wedge \neg v_A$. Intuitively, $b = 1$ happens (i.e., H can distinguish) if and only if H is able to successfully solve harder puzzles than the adversary.

Note that by setting suitable length functions (i.e., $l(\mathsf{user?}) = l(\mathsf{adv?}) = k^b$ for the $b \in \mathbb{N}$ from (11), M_1 and M_2 can be made polynomial-time.

The rest of this proof follows an idea similar to that of the proof of Theorem 2. For showing that $(\hat{M}_1, S) \geq_{\text{sec}}^{\text{poly}} (\hat{M}_2, S)$, we construct a simulator A_2 that can solve any puzzle H can solve, and for showing $(\hat{M}_1, S) \not\geq_{\text{sec}}^{\text{uni,poly}} (\hat{M}_2, S)$ we construct a H which can solve some puzzles A_2 cannot solve.

For showing $(\hat{M}_1, S) \geq_{\text{sec}}^{\text{poly}} (\hat{M}_2, S)$, let a real configuration $(\hat{M}_1, S, \mathsf{H}, \mathsf{A}_1) \in \text{Conf}^{\hat{M}_2}(\hat{M}_1, S)$ be given.

We sketch how to view H as a PPT-algorithm B: $B(1^k, q', h)$ simulates a run of the collection $[\{\mathsf{H}, \mathsf{A}_1, \mathsf{M}_1\}]$ with security parameter k (auxiliary input h is ignored); only a possible message q from M_1 to H or A_1 on user_puz or adv_puz is substituted by q'. B outputs H's answer a_H (i.e., M_1's second user?-input), or $\bot$, if M_2 never receives such an a_H.

Let $f(k) = k^c$ for the $c \in \mathbb{N}$ that arises from (10) for this B. Let C be the PPT-algorithm from (11) when setting $d = c$. So intuitively, C is able to solve (except with negligible error) any puzzle that H can solve. Similar to the construction of the simulator A_2 in the proof of Theorem 2 (cf. also Figure 2), let S be a machine which places itself between A_1 and M_1. A_1's ports puz_adv?, adv!, and $\text{adv}^\triangleleft!$ are renamed to $\overline{\text{puz\_adv?}}$, $\overline{\text{adv!}}$, and $\overline{\text{adv}^\triangleleft!}$, respectively. Finally, A_2 is the combination of A_1 and this machine S.

The idea of S is simple: Whenever A_1 sends a (possibly wrong) solution of a puzzle of hardness $f(k)$ to M_2, S solves the puzzle and sends a correct solution to M_2. This will allow to show indistinguishability, since H will only notice that it runs with the ideal protocol if A_2 sends a wrong solution to M_2 for a puzzle H was able to solve.

More formally, S immediately forwards all messages from M_2 to A_2. However, the first message q on adv_puz is stored. When A_1 sends the second message to M_2 via adv, that message is replaced by a solution $c \leftarrow C(1^k, q)$ and then forwarded to M_2.

Using a suitable length function, S can be made polynomial-time. Similar to Figure 2, let A_2 be the combination of S and A_1 (with renamed ports). S only substitutes messages between M_1 and A_1, and does not change the scheduling.

We can now observe the following: First, if we can show that for the output $b = v_\mathsf{H} \wedge \neg v_\mathsf{A}$ by M_2, it is $b = 1$ with only negligible probability, then H's views when running with the real and the ideal protocol are indistinguishable, and thus $(\hat{M}_1, S) \geq_{\text{sec}}^{\text{poly}} (\hat{M}_2, S)$.

Second, consider the definition of B. We can define B' completely analogous, using A_2 and M_2 instead of A_1 and M_1. By noticing that H's answer to the puzzle is chosen before M_1 or M_2 outputs b, we see that B and B' have the same output distributions.

Let v_A and v_H denote the corresponding predicates calculated by M_2 in a run of the ideal protocol, where we set $v_\mathsf{A} = \bot$ and $v_\mathsf{H} = \bot$ if the respective variable predicate is never determined (this can happen only when no answer to the respective puzzle is made).

By (10), the following is negligible:

$$\sup_{t \geq k^c, |h| \leq k^e} \Pr\left[(q,a) \leftarrow \mathcal{G}(1^k, t) : \mathcal{V}(1^k, a, B(1^k, q, h)) = 1\right]$$

$$\geq \Pr\left[t \leftarrow \mathcal{D}(1^k), (q,a) \leftarrow \mathcal{G}(1^k, t) : \mathcal{V}(1^k, a, B(1^k, q, 0)) = 1 \wedge t \geq k^c\right]$$

$$\stackrel{(*)}{=} \Pr\left[v_\mathsf{H} = 1 \wedge t \geq k^c\right]$$

$$\geq \Pr\left[v_\mathsf{H} = 1 \wedge v_\mathsf{A} = 0 \wedge t \geq k^c\right]. \tag{12}$$

(Note that in the last two terms of (12), t denotes the t chosen by M_2.) These inequalities hold for all sufficiently large k. To see (∗), note that M_2 chooses t, q, a via $t \leftarrow \mathcal{D}(1^k)$, $(q, a) \leftarrow \mathcal{G}(1^k, t)$, and then calculates v_H via $\mathcal{V}(1^k, a, a_\mathsf{H})$. Further $B(1^k, q, 0)$ generates a_H by simulating the ideal configuration for given q, so (∗) follows.

Since A_2 solves the puzzle with overwhelming probability for $t < k^c$ (it uses C, which again does so by (11)), the following is negligible:

$$\Pr\left[v_\mathsf{A} = 0 \wedge t < k^c\right]$$

$$\geq \Pr\left[v_\mathsf{H} = 1 \wedge v_\mathsf{A} = 0 \wedge t < k^c\right]. \tag{13}$$

By (12,13), the probability $\Pr\left[v_\mathsf{H} = 1 \wedge v_\mathsf{A} = 0\right]$ is negligible, too, therefore we can conclude $(\hat{M}_1, S) \geq^{\mathsf{poly}}_{\mathsf{sec}} (\hat{M}_2, S)$.

If we allow non-uniform honest users, we have to modify the definition of B as follows: $B(1^k, q', h)$ runs a simulation as above, but now h is given to the non-uniform honest user as auxiliary input. Then, for some function $\tilde{h}$ (mapping the security parameter to the auxiliary input), $B(1^k, q', \tilde{h}(k))$ simulates the network containing the non-uniform honest user H. Replacing $B(1^k, q, 0)$ with $B(1^k, q, \tilde{h}(k))$ in (12) yields a valid proof for the non-uniform case.

Note that this construction even applies when the auxiliary input $\tilde{h}$ of the honest user H is chosen in dependence of the simulator (as with the "Specialized-simulator UC" formulation in [Lin03]).

The remaining statement $(\hat{M}_1, S) \not\geq^{\mathsf{uni,poly}}_{\mathsf{sec}} (\hat{M}_2, S)$ can be shown in a similar way: We define a family of honest users H_d, such that no simulator will be able to solve all puzzles these H_d can solve. Formally:

For any d, let C_d be the puzzle-solver C whose existence is guaranteed by (11) for that d. Then H_d is the user having ports $\{\mathsf{user!}, \mathsf{user}^{\triangleleft}!, \mathsf{out?}, \mathsf{puz\_user?}, \mathsf{init?}\}$ and running the following program: Upon the first activation via init?, send (and schedule) a non-empty message to M_1. When receiving q via puz_user? from M_1, let $c \leftarrow C_d(1^k, q)$ and send (and schedule) c to M_2.

The real adversary A_1 has ports $\{\mathsf{adv!}, \mathsf{adv}^{\triangleleft}!, \mathsf{init!}, \mathsf{init}^{\triangleleft}!, \mathsf{puz\_adv?}\}$ and runs the following program: Upon the first activation, activate H_d via init, and upon any further activation send and schedule 1 to M_2 via adv.

The resulting network is depicted in Figure 4.

We now assume for contradiction that $(\hat{M}_1, S) \geq^{\mathsf{uni,poly}}_{\mathsf{sec}} (\hat{M}_2, S)$. Because then $\mathrm{conf}_1^d := (\hat{M}_1, S, \mathsf{H}_d, \mathsf{A}_1) \in \mathrm{Conf}^{\hat{M}_2}(\hat{M}_1, S)$ for all d, there is a polynomial simulator A_2, s.t. for all d, $\mathrm{conf}_2^d := (\hat{M}_2, S, \mathsf{H}_d, \mathsf{A}_2) \in \mathrm{Conf}(\hat{M}_2, S)$ and

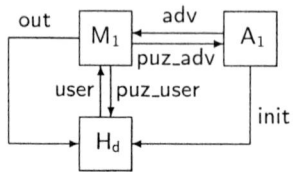

Fig. 4. The configuration conf_1^d

$$view_{\mathsf{conf}_1^d}(\mathsf{H}_d) \approx_{\mathsf{poly}} view_{\mathsf{conf}_2^d}(\mathsf{H}_d). \tag{14}$$

Similar to the construction of B in the first part of this proof, let $B_d(1^k, q', h)$ simulate a run of the collection $[\{\mathsf{H}_d, \mathsf{A}_2, \mathsf{M}_2\}]$ with security parameter k (auxiliary input h is ignored); only a possible message q from M_2 to H or A_1 on user_puz or adv_puz is substituted by q'. B_d outputs A_2's answer a_A (i.e., M_2's second adv?-input), or $\bot$, if M_2 never receives a_A.

It is easy to see that A_2's answers do not depend on H_d's answers, therefore B_d is independent of d.

Now, by (10), there is some c, s.t.

$$\sup_{t \geq k^c, |h| \leq k^e} \Pr\left[(q,a) \leftarrow \mathcal{G}(1^k, t) : \mathcal{V}(1^k, a, B_0(1^k, q, h)) = 1\right]$$

is negligible.

We will now examine the situation, where H_{c+1} runs with $\mathsf{A}_2, \mathsf{M}_2$; that is, we consider $run_{\mathsf{conf}_2^{c+1}}$. Let as above v_H and v_A denote the corresponding variables of M_2, with $v_\mathsf{H} = \bot$ or $v_\mathsf{A} = \bot$ if the corresponding variable is not set.

First note, that by definition of $\mathcal{D}$, it is

$$\Pr\left[t \leftarrow \mathcal{D}(1^k) : k^c \leq t \leq k^{c+1}\right] \geq \tfrac{1}{k} \tag{15}$$

for sufficiently large k.

By definition of c and (10), the following is overwhelming:

$$\min_{k^c \leq t \leq k^{c+1}} \Pr\left[(q,a) \leftarrow \mathcal{G}(1^k, t) : \mathcal{V}(1^k, a, B_0(1^k, q, 0)) \neq 1\right]$$
$$\leq \Pr\left[t \leftarrow \mathcal{D}(1^k), (q,a) \leftarrow \mathcal{G}(1^k, t) : \mathcal{V}(1^k, a, B_0(1^k, q, 0)) \neq 1 \mid k^c \leq t \leq k^{c+1}\right]$$
$$= \Pr\left[t \leftarrow \mathcal{D}(1^k), (q,a) \leftarrow \mathcal{G}(1^k, t) : \mathcal{V}(1^k, a, B_{c+1}(1^k, q, 0)) \neq 1 \mid k^c \leq t \leq k^{c+1}\right]$$
$$\stackrel{(**)}{=} \Pr\left[v_\mathsf{A} \neq 1 \mid k^c \leq t \leq k^{c+1}\right]. \tag{16}$$

Here $(**)$ is shown like $(*)$ in the first part of the proof.

Note further that by definition of H_d (in the particular case $d = c + 1$) and (11), and considering (15), we have that

$$\Pr\left[v_\mathsf{H} \neq 0 \mid k^c \leq t \leq k^{c+1}\right] \tag{17}$$

is overwhelming.

Combining (17) and (16), we conclude that

$$\Pr\left[v_\mathsf{H} \neq 0 \ \wedge \ v_\mathsf{A} \neq 1 \mid k^c \leq t \leq k^{c+1}\right] \leq \Pr\left[b=1 \ \vee \ b=\bot \mid k^c \leq t \leq k^{c+1}\right]$$

are both overwhelming (consider that M_2's output is $b=0$ only if v_H and v_A are defined and $\neg v_\mathsf{H} \vee v_\mathsf{A}$). Using (15), we finally have that

$$\Pr\left[b=1 \ \vee \ b=\bot\right]$$

is non-negligible. Since in the run of the real configuration conf_1^{c+1}, it is $b=0$ with overwhelming probability, and b shows up in H_{c+1}'s view, this is a contradiction to (14) and shows $(\hat{M}_1, S) \not\geq_{\mathsf{sec}}^{\mathsf{uni},\mathsf{poly}} (\hat{M}_2, S)$. □

5 Conclusion

We have separated standard and universal simulatability in the case of computational and statistical security. This shows that these security notions are indeed different. However, it would be nice to know whether there is a less "artificial" separating example than ours. In particular, it is not clear whether there is a more "cryptographic" example.

We have also shown that for perfect security, standard and universal simulatability coincide. This result may ease security proofs—showing standard simulatability automatically shows universal simulatability.

Acknowledgements

We thank Ran Canetti, Jörn Müller-Quade, and Rainer Steinwandt for interesting and valuable discussions. Furthermore, we thank the anonymous referees for helpful comments.

References

[Bac04] Michael Backes. E-mail communication with the authors, June 2004.

[BPW04a] Michael Backes, Birgit Pfitzmann, and Michael Waidner. A general composition theorem for secure reactive systems. In Moni Naor, editor, *Theory of Cryptography, Proceedings of TCC 2004*, number 2951 in Lecture Notes in Computer Science, pages 336–354. Springer-Verlag, 2004. Online available at http://www.zurich.ibm.com/security/publications/2004/BaPfWa2004MoreGeneralComposition.pdf.

[BPW04b] Michael Backes, Birgit Pfitzmann, and Michael Waidner. Secure asynchronous reactive systems. IACR ePrint Archive, March 2004. Online available at http://eprint.iacr.org/2004/082.ps.

[Can00] Ran Canetti. Security and composition of multi-party cryptographic protocols. *Journal of Cryptology*, 3(1):143–202, 2000. Full version online available at http://eprint.iacr.org/1998/018.ps.

[Can01] Ran Canetti. Universally composable security: A new paradigm for cryptographic protocols. In *42th Annual Symposium on Foundations of Computer Science, Proceedings of FOCS 2001*, pages 136–145. IEEE Computer Society, 2001. Full version online available at http://eprint.iacr.org/2000/067.ps.

[Can04] Ran Canetti. Personal communication with one of the authors at TCC, February 2004.

[CLOS02] Ran Canetti, Yehuda Lindell, Rafail Ostrovsky, and Amit Sahai. Universally composable two-party and multi-party secure computation. In *34th Annual ACM Symposium on Theory of Computing, Proceedings of STOC 2002*, pages 494–503. ACM Press, 2002. Extended abstract, full version online available at http://eprint.iacr.org/2002/140.ps.

[Lin03] Yehuda Lindell. General composition and universal composability in secure multi-party computation. In *44th Annual Symposium on Foundations of Computer Science, Proceedings of FOCS 2003*, pages 394–403. IEEE Computer Society, 2003. Online available at http://www.research.ibm.com/people/l/lindell/PAPERS/gc-uc.ps.gz.

[PW00] Birgit Pfitzmann and Michael Waidner. Composition and integrity preservation of secure reactive systems. In *7th ACM Conference on Computer and Communications Security, Proceedings of CCS 2000*, pages 245–254. ACM Press, 2000. Extended version online available at http://www.semper.org/sirene/publ/PfWa_00CompInt.ps.gz.

[PW01] Birgit Pfitzmann and Michael Waidner. A model for asynchronous reactive systems and its application to secure message transmission. In *IEEE Symposium on Security and Privacy, Proceedings of SSP 2001*, pages 184–200. IEEE Computer Society, 2001. Full version online available at http://eprint.iacr.org/2000/066.ps.

[RSW96] Ronald L. Rivest, Adi Shamir, and David A. Wagner. Time-lock puzzles and timed-release crypto. Technical Report MIT/LCS/TR-684, Massachusetts Institute of Technology, February 1996. Online available at http://theory.lcs.mit.edu/~rivest/RivestShamirWagner-timelock.ps.

A Review of Reactive Simulatability

In this section, we present the notion of reactive simulatability. This introduction only very roughly sketches the definitions, and the reader is encouraged to read [BPW04b] for more detailed information and formal definitions.

Reactive Simulatability is a definition of security which defines a protocol $\hat{M}_1$ (the *real protocol*) to be *as secure as* another protocol $\hat{M}_2$ (the *ideal protocol*, the *trusted host*), if for any adversary A_1 (also called the *real adversary*), and any *honest user* H, there is a *simulator* A_2 (also called the *ideal adversary*), s.t. the view of H is indistinguishable in the following two scenarios:

- The honest user H runs together with the real adversary A_1 and the real protocol $\hat{M}_1$
- The honest user H runs together with the simulator A_2 and the ideal protocol $\hat{M}_2$.

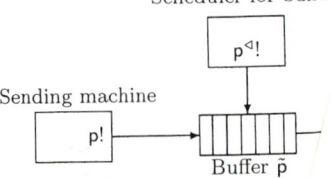

Fig. 5. A conne[ction]

(asymptotically smaller than the inverse of any polynomial). **buffer**: Stores message sent from a simple out- to a simple in-port. Needs an input from a clock port to deliver. **clock out-port p◁!**: A port used to schedule connection. **closed collection**: A collection is closed, if all ports have all their necessary counterparts. **collection**: A set of machines. **combination**: The combination of a set of machines is a new machine simulating the other machines. A set of machines can be replaced by its combination without changing the view of any machine. **composition**: Replacing sub-protocols by other sub-protocols. **computational security**: When in the security definition, honest user and adversary are restricted to machines running in polynomial time, and the views are computationally indistinguishable. **configuration**: A structure together with an honest user and an adversary. **free ports**: The free ports of a collection are those missing their counterpart. **honest user**: Represents the setting in which the protocol runs. Also called environment. **intended structure**: A structure from which a system is derived making a structure for every corruption situation. **master clock port clk◁?**: A special port by which the master scheduler is activated. **master scheduler**: The machine that gets activated when no machine would get activated. **perfect security**: When in the security definition, the real and ideal run have to be identical, not only indistinguishable. Further the machines are completely unrestricted.[11] **run**: The transcript of everything that happens while a collection is run. Formally a random variable over sequences. $run_{conf,k,l}$ is the random variable of the run when running the configuration conf upon security parameter k, restricted to its first l elements. If k is omitted, a family of random variables is meant. If l is omitted, we mean the full run. **service ports**: The ports of a structure to which the honest user may connect. They represent the interface of the protocol. As service ports are most often ports of a buffer, they are sometimes specified through the set S^c of their complementary ports; S^c consists of all ports which directly connect to a service port. **simple in-port p?**: A port of a machine, where it can receive messages from other machines. **simple out-port p!**: As simple in-port, but for sending. **statistical security**: When in the security definition the statistical distance of polynomial prefixes of the views have a statistical distance which lies in a set of small functions $SMALL$ (in the security parameter k). Usually $SMALL = NEGL$. Further the machines are completely unrestricted.11⁰ **structure**: A collection together with a set of service ports, represents a protocol. **view**: A subsequence of the run. The $view(M)$ of some collection or machine M consists of the run restricted to the ports and states of M. Possible indices are as with runs.

---

[11] In [BPW04b] a machine can in every activation for a given input and current state only reach one of a finite number of states (this convention has been chosen for simplicity [Bac04]). However, this cannot even model the simple Turing machine that tosses (within one activation) coins until a 1 appears, and then stores the number of coin tosses. Therefore we will here adopt the convention that each state can have a countable number of potential successor states, from which one is chosen following some distribution depending on the input and the current state.

Relaxing Environmental Security: Monitored Functionalities and Client-Server Computation

Manoj Prabhakaran* and Amit Sahai**

Princeton University and UCLA
{mp, sahai}@cs.princeton.edu

Abstract. Definition of security under the framework of Environmental Security (a.k.a Network-Aware Security or Universally Composable Security) typically requires "extractability" of the private inputs of parties running a protocol. Formalizing concepts that appeared in an earlier work [19], we introduce a framework of "Monitored Functionalities," which allows us to avoid such a requirement from the security definition, while still providing very strong composition properties. We also consider a specialization of the Environmental Security framework by designating one party as a "server" and all other parties as clients. Both these contributions in the work are aimed at being able to provide weaker Environmental Security guarantees to simpler protocols. We illustrate the usability of the Monitored Functionalities framework by providing much simpler protocols in the plain model than in [19] for some limited functionalities in the server-client model.

1 Introduction

At the onset of theoretical investigations into defining and achieving cryptographic security, idealized settings involving just one protocol were used. A highly successful theory was developed which gave satisfactory solutions to a multitude of cryptographic problems like encryption and multi-party computation, reducing the nebulous questions of security to concrete problems in computational complexity theory. Having successfully tackled these problems, this nascent field of computer science started focusing on more challenging problems, thrown its way by the new requirements of the fast changing world of information technology.

An important challenge was to enhance the theory so that it could handle less idealized settings relevant to the modern world – where there are many computers connected to an unreliable network. Rather annoyingly, this complicated the situation spectacularly. Definitions of security turned out to be inadequate as new attacks came into the picture. Protocols which were provably secure earlier either became insecure, or worse still, remained open challenges to be proven secure or insecure.

Two important works which explored these complications identified *Non-malleability* [8] and *Concurrent simulation* [9] as two most basic problems to be studied. Since then a significant amount of work went into tackling these basic aspects.

---

* Part of this work was done while the author was an intern at IBM TJ Watson Research Center.
** Research done while the author was at Princeton University. Research supported in part by NSF ITR, NSF Cybertrust, and the Alfred P. Sloan Foundation.

Combining (17) and (16), we conclude that

$$\Pr\left[v_H \neq 0 \wedge v_A \neq 1 \mid k^c \leq t \leq k^{c+1}\right] \leq \Pr\left[b = 1 \vee b = \perp \mid k^c \leq t \leq k^{c+1}\right]$$

are both overwhelming (consider that M_2's output is $b = 0$ only if v_H and v_A are defined and $\neg v_H \vee v_A$). Using (15), we finally have that

$$\Pr\left[b = 1 \vee b = \perp\right]$$

is non-negligible. Since in the run of the real configuration conf_1^{c+1}, it is $b = 0$ with overwhelming probability, and b shows up in H_{c+1}'s view, this is a contradiction to (14) and shows $(\hat{M}_1, S) \not\geq_{\mathsf{sec}}^{\mathsf{uni},\mathsf{poly}} (\hat{M}_2, S)$. □

5 Conclusion

We have separated standard and universal simulatability in the case of computational and statistical security. This shows that these security notions are indeed different. However, it would be nice to know whether there is a less "artificial" separating example than ours. In particular, it is not clear whether there is a more "cryptographic" example.

We have also shown that for perfect security, standard and universal simulatability coincide. This result may ease security proofs—showing standard simulatability automatically shows universal simulatability.

Acknowledgements

We thank Ran Canetti, Jörn Müller-Quade, and Rainer Steinwandt for interesting and valuable discussions. Furthermore, we thank the anonymous referees for helpful comments.

References

[Bac04]   Michael Backes. E-mail communication with the authors, June 2004.

[BPW04a] Michael Backes, Birgit Pfitzmann, and Michael Waidner. A general composition theorem for secure reactive systems. In Moni Naor, editor, *Theory of Cryptography, Proceedings of TCC 2004*, number 2951 in Lecture Notes in Computer Science, pages 336–354. Springer-Verlag, 2004. Online available at http://www.zurich.ibm.com/security/publications/2004/BaPfWa2004MoreGeneralComposition.pdf.

[BPW04b] Michael Backes, Birgit Pfitzmann, and Michael Waidner. Secure asynchronous reactive systems. IACR ePrint Archive, March 2004. Online available at http://eprint.iacr.org/2004/082.ps.

[Can00]   Ran Canetti. Security and composition of multi-party cryptographic protocols. *Journal of Cryptology*, 3(1):143–202, 2000. Full version online available at http://eprint.iacr.org/1998/018.ps.

[Can01]   Ran Canetti. Universally composable security: A new paradigm for cryptographic protocols. In *42th Annual Symposium on Foundations of Computer Science, Proceedings of FOCS 2001*, pages 136–145. IEEE Computer Society, 2001. Full version online available at http://eprint.iacr.org/2000/067.ps.

[Can04]   Ran Canetti. Personal communication with one of the authors at TCC, February 2004.

[CLOS02]   Ran Canetti, Yehuda Lindell, Rafail Ostrovsky, and Amit Sahai. Universally composable two-party and multi-party secure computation. In *34th Annual ACM Symposium on Theory of Computing, Proceedings of STOC 2002*, pages 494–503. ACM Press, 2002. Extended abstract, full version online available at http://eprint.iacr.org/2002/140.ps.

[Lin03]   Yehuda Lindell. General composition and universal composability in secure multi-party computation. In *44th Annual Symposium on Foundations of Computer Science, Proceedings of FOCS 2003*, pages 394–403. IEEE Computer Society, 2003. Online available at http://www.research.ibm.com/people/l/lindell/PAPERS/gc-uc.ps.gz.

[PW00]   Birgit Pfitzmann and Michael Waidner. Composition and integrity preservation of secure reactive systems. In *7th ACM Conference on Computer and Communications Security, Proceedings of CCS 2000*, pages 245–254. ACM Press, 2000. Extended version online available at http://www.semper.org/sirene/publ/PfWa_00CompInt.ps.gz.

[PW01]   Birgit Pfitzmann and Michael Waidner. A model for asynchronous reactive systems and its application to secure message transmission. In *IEEE Symposium on Security and Privacy, Proceedings of SSP 2001*, pages 184–200. IEEE Computer Society, 2001. Full version online available at http://eprint.iacr.org/2000/066.ps.

[RSW96]   Ronald L. Rivest, Adi Shamir, and David A. Wagner. Time-lock puzzles and timed-release crypto. Technical Report MIT/LCS/TR-684, Massachusetts Institute of Technology, February 1996. Online available at http://theory.lcs.mit.edu/~rivest/RivestShamirWagner-timelock.ps.

A Review of Reactive Simulatability

In this section, we present the notion of reactive simulatability. This introduction only very roughly sketches the definitions, and the reader is encouraged to read [BPW04b] for more detailed information and formal definitions.

Reactive Simulatability is a definition of security which defines a protocol $\hat{M}_1$ (the *real protocol*) to be *as secure as* another protocol $\hat{M}_2$ (the *ideal protocol*, the *trusted host*), if for any adversary A_1 (also called the *real adversary*), and any honest user H, there is a *simulator* A_2 (also called the *ideal adversary*), s.t. the view of H is indistinguishable in the following two scenarios:

– The honest user H runs together with the real adversary A_1 and the real protocol $\hat{M}_1$
– The honest user H runs together with the simulator A_2 and the ideal protocol $\hat{M}_2$.

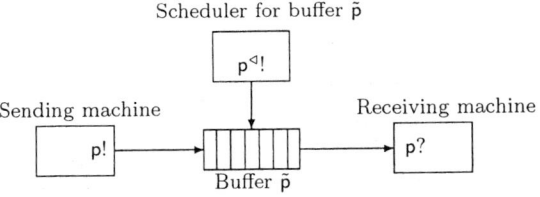

Fig. 5. A connection

Note that there is a security parameter k common to all machines, so that the notion of indistinguishability makes sense.

This definition allows to specify some trusted host—which is defined to be a secure implementation of some cryptographic task—as the ideal protocol, and then to consider the question, whether a real protocol is as secure as the trusted host (and thus also a secure implementation of that task). In order to understand the above definitions in more detail, we have to specify what is meant by machines "running together". Consider a set of machines (called a *collection*). Each machine has so-called *simple in-ports* (written p?), *simple out-ports* (written p!), and *clock out-ports* (written $p^{\triangleleft}!$). Ports with the same name (p in our example) are considered to belong together and are associated with a *buffer* $\tilde{p}$. These are then interconnected as in Figure 5 (note that some or all ports may originate from the same machine). Now, when a collection runs, the following happens: At every time, exactly one machine is activated. It may now read its simple in-ports (representing incoming network connections), do some work, and then write output to its simple out-ports. After such an activation the contents of the simple out-ports p! are appended to the queue of messages stored in the associated buffer $\tilde{p}$. However, since now all messages are stored in buffers and will not be delivered by themselves, machines additionally have after each activation the possibility to write a number $n \geq 1$ to at most one clock out-port $p^{\triangleleft}!$. Then the n-th undelivered message of buffer $\tilde{p}$ will be written to the simple in-port p? and deleted from the buffer's queue. The machine that has the simple in-port p? will be activated next. So the clock out-ports control the scheduling. Usually, a connection is clocked by (i.e., the corresponding clock out-port is part of) the sender, or by the adversary. Since the most important use of a clock out-port is to write a 1 onto it (deliver the oldest message in the buffer), we say a machine clocks a connection or a message when a machine writes a 1 onto the clock port of that connection.

At the start of a run, or when no machine is activated at some point, a designated machine called the *master scheduler* is activated For this, the master scheduler has a special port, called the *master clock port* $\mathsf{clk}^{\triangleleft}?$.

Note that not all collections can be executed, only so-called *closed* collections, where all connections have their simple in-, simple out-, and clock out-port. If a collection is not closed, we call the ports having no counterpart *free ports*.

In order to understand how this idea of networks relates to the above sketch of reactive simulatability, one has to get an idea of what is meant by a protocol. A protocol is represented by a so-called *structure* $(\hat{M}, S)$, consisting of a collection $\hat{M}$ of the protocol participants (parties, trusted hosts, etc.), and a subset of the free ports of $\hat{M}$, the so-called *service ports* S. The service ports represent the protocol's interface (the connections to the protocol's users). The honest user can then only connect to the service ports (and to the adversary), all other free ports of the protocol are intended for the communication with the adversary (they may e.g. represent side channels, possibilities of attack, etc.). Since usually a protocol does not explicitly communicate with an adversary, such free non-service ports are more commonly found with trusted hosts, explicitly modelling their imperfections.

With this information we can review the above "definition" of security. Namely, the honest user H, the adversary, and the simulator are nothing else but machines, and the protocols are structures. The view of H is then the restriction of the run (the transcripts of all states and in-/output of all machines during the protocols execution, also called trace) to the ports and state of H.

The definition, as presented so far, still has one drawback. We have not introduced the concept of a corruption. This can be accommodated by defining so-called systems. A *system* is a set of structures, where to each "corruption situation" (set of machines, which are corrupted) one structure corresponds. That is, when a machine is corrupted, it is not present anymore in the corresponding structure, and the adversary takes its place. For a trusted host, the corresponding system usually consists of structures for each corruption situation, too, where those connections of the trusted host, that are associated with a corrupted party, are under the control of the adversary.

We can now refine the definition of security as follows: A *real system* Sys_1 is as secure as an *ideal system* Sys_2, if every structure in Sys_1 is as secure as the corresponding structure in Sys_2.

A major advantage of a security definition by simulatability is the possibility of *composition*. The notion of composition can be sketched as follows: If we have on structure or system A (usually a protocol) implementing some other structure or system B (usually some primitive), and we have some protocol X^B (having B as a sub-protocol, i.e. using the primitive), then by replacing B by A in X^B, we get a protocol X^A which is as secure as X^B. This allows to modularly design protocols: first we design a protocol X^B, and then we find an implementation for B.

A.1 Glossary

In this section we explain the technical terms used in this paper. Longer and formal definitions can be found in [BPW04b].

$[\hat{C}]$: Completion of the collection $\hat{C}$. Results from adding all missing buffers to $\hat{C}$. $\mathsf{Conf}_x(\hat{M}_2, S)$: Set of ideal configurations that are possible for structure $(\hat{M}_2, S)$. $\mathsf{Conf}_x^{\hat{M}_2}(\hat{M}_1, S)$: Set of real configurations possible for structure $(\hat{M}_1, S)$. $\mathsf{ports}(M)$: The set of all ports, a machine or collection M has. **to clock**: To write 1 onto a clock out-port. EXPSMALL : The set of exponentially small functions. NEGL: The set of negligible functions (asymptotically smaller than the inverse of any polynomial). **buffer**: Stores message sent from a simple out- to a simple in-port. Needs an input from a clock port to deliver. **clock out-port $p^{\triangleleft}!$**: A port used to schedule connection. **closed collection**: A collection is closed, if all ports have all their necessary counterparts. **collection**: A set of machines. **combination**: The combination of a set of machines is a new machine simulating the other machines. A set of machines can be replaced by its combination without changing the view of any machine. **composition**: Replacing sub-protocols by other sub-protocols. **computational security**: When in the security definition, honest user and adversary are restricted to machines running in polynomial time, and the views are computationally indistinguishable. **configuration**: A structure together with an honest user and an adversary. **free ports**: The free ports of a collection are those missing their counterpart. **honest user**: Represents the setting in which the protocol runs. Also called environment. **intended structure**: A structure from which a system is derived making a structure for every corruption situation. **master clock port $clk^{\triangleleft}?$**: A special port by which the master scheduler is activated. **master scheduler**: The machine that gets activated when no machine would get activated. **perfect security**: When in the security definition, the real and ideal run have to be identical, not only indistinguishable. Further the machines are completely unrestricted.[11] **run**: The transcript of everything that happens while a collection is run. Formally a random variable over sequences. $run_{conf,k,l}$ is the random variable of the run when running the configuration conf upon security parameter k, restricted to its first l elements. If k is omitted, a family of random variables is meant. If l is omitted, we mean the full run. **service ports**: The ports of a structure to which the honest user may connect. They represent the interface of the protocol. As service ports are most often ports of a buffer, they are sometimes specified through the set S^c of their complementary ports; S^c consists of all ports which directly connect to a service port. **simple in-port $p?$**: A port of a machine, where it can receive messages from other machines. **simple out-port $p!$**: As simple in-port, but for sending. **statistical security**: When in the security definition the statistical distance of polynomial prefixes of the views have a statistical distance which lies in a set of small functions $SMALL$ (in the security parameter k). Usually $SMALL = NEGL$. Further the machines are completely unrestricted.11⁰ **structure**: A collection together with a set of service ports, represents a protocol. **view**: A subsequence of the run. The $view(M)$ of some collection or machine M consists of the run restricted to the ports and states of M. Possible indices are as with runs.

---

[11] In [BPW04b] a machine can in every activation for a given input and current state only reach one of a finite number of states (this convention has been chosen for simplicity [Bac04]). However, this cannot even model the simple Turing machine that tosses (within one activation) coins until a 1 appears, and then stores the number of coin tosses. Therefore we will here adopt the convention that each state can have a countable number of potential successor states, from which one is chosen following some distribution depending on the input and the current state.

Relaxing Environmental Security: Monitored Functionalities and Client-Server Computation

Manoj Prabhakaran* and Amit Sahai**

Princeton University and UCLA
{mp, sahai}@cs.princeton.edu

Abstract. Definition of security under the framework of Environmental Security (a.k.a Network-Aware Security or Universally Composable Security) typically requires "extractability" of the private inputs of parties running a protocol. Formalizing concepts that appeared in an earlier work [19], we introduce a framework of "Monitored Functionalities," which allows us to avoid such a requirement from the security definition, while still providing very strong composition properties. We also consider a specialization of the Environmental Security framework by designating one party as a "server" and all other parties as clients. Both these contributions in the work are aimed at being able to provide weaker Environmental Security guarantees to simpler protocols. We illustrate the usability of the Monitored Functionalities framework by providing much simpler protocols in the plain model than in [19] for some limited functionalities in the server-client model.

1 Introduction

At the onset of theoretical investigations into defining and achieving cryptographic security, idealized settings involving just one protocol were used. A highly successful theory was developed which gave satisfactory solutions to a multitude of cryptographic problems like encryption and multi-party computation, reducing the nebulous questions of security to concrete problems in computational complexity theory. Having successfully tackled these problems, this nascent field of computer science started focusing on more challenging problems, thrown its way by the new requirements of the fast changing world of information technology.

An important challenge was to enhance the theory so that it could handle less idealized settings relevant to the modern world – where there are many computers connected to an unreliable network. Rather annoyingly, this complicated the situation spectacularly. Definitions of security turned out to be inadequate as new attacks came into the picture. Protocols which were provably secure earlier either became insecure, or worse still, remained open challenges to be proven secure or insecure.

Two important works which explored these complications identified *Non-malleability* [8] and *Concurrent simulation* [9] as two most basic problems to be studied. Since then a significant amount of work went into tackling these basic aspects.

---

* Part of this work was done while the author was an intern at IBM TJ Watson Research Center.
** Research done while the author was at Princeton University. Research supported in part by NSF ITR, NSF Cybertrust, and the Alfred P. Sloan Foundation.

While there has been quite some success in resolving many challenges, the new protocols kept getting more complicated and less efficient. A fresh look at the problem was taken in [3], which offered a comprehensive definition of security, namely that of *Environmental Security* (ES, for short) or *Network-Aware Security*. (It was introduced in [3] under the name Universally Composable (UC) Security; hence we shall refer to this version of Environmental Security as ES/UC.) ES/UC security, at once subsumes non-malleability and concurrency. This allowed simpler and intuitive compositions of protocols. Using a setup called the "common reference string," (where all parties are provided a string produced by a trusted party), or alternatively using a trust assumption of "honest majority" (where majority of players among a pre-defined subset of parties are trusted to be honest), it was shown how to do "secure multi-party computation," arguably the ultimate cryptographic task [3, 7]. However it fell short of offering a viable solution to the protocol designers: it was shown that to achieve provable security under this new model, some "trust assumption" would be necessary in most cases of interest [3, 4, 6]. Recently [19] showed a way to get around the need for trust assumption by modifying the ES/UC framework to obtain what is called the *generalized Environmental Security* (gES), or *generalized Network-Aware Security*, framework.

This Work. Environmental Security (ES/UC as well as gES) addresses the concern that protocols are run in an arbitrary environment along with other processes and network activity. However the comprehensive definitions of security offered by these frameworks tend to require complex protocols. The thesis of this work is that we need to develop relaxed notions of Environmental Security which will help us prove some level of security for simpler protocols, at least for certain limited applications. We explore two separate directions simultaneously.

First, we develop a model intended to remove some of the concerns of universal composability from environmental security. The model restricts the protocol executions for which security analysis will be applied, by requiring "fixed roles" for the participants across all protocol executions. (This can be viewed as a generalization of the setup introduced in concurrent zero knowledge [9], to the ES setting.) This restriction frees us from concerns of certain "man-in-the-middle" attacks (or malleability of the protocols). Our interest in the client-server model is as a useful theoretical construct in itself – a platform for tackling some of the Environmental Security issues without having to deal with all the composition issues. For the protocols in this work, use of this model is *not crucial*, but it leads to somewhat simpler protocols, simpler assumptions and simpler analysis.

Second – and this is the main focus in this work – we introduce a significant relaxation of the security requirements in the ES framework, in a new framework of *Monitored Functionalities*. Indeed, [19] shows how to relax ES/UC, without much loss in applicability.[1] gES removes the restriction in the ES/UC framework that the "simulation" used to define security should be computationally bounded, and still manages to retain Universal Composability. However the gES protocols in [19] are still complex.

---

[1] Technically, security in gES framework is not a relaxation of security in ES/UC framework, but involves a relaxation followed by a strengthening, which in general makes it incomparable with the latter.

We go one step further, and redefine security in such a way that one more requirement from the ES/UC framework is removed – namely that of "extractability." We show how to define meaningful notions of security without the extractability requirement, and yet obtain Environmentally Secure and Universally Composable protocols. This is achieved by introducing a new entity alongside the "trusted party" in the ideal world: it is a computationally unbounded *"monitor"* which can watch over the ideal world execution and raise an alarm if some specific condition is violated.

Two of our protocols (for commitment and zero-knowledge proofs) are adapted from [19], with simplifications allowed by the client-server model. The results in [19] do show that those protocols can be used to obtain full-fledged gES security for these tasks (against static adversaries, without the client-server restriction, and without resorting to monitored functionalities). However it is far from a direct use: they are used as subroutines within a larger protocol for commitment. Our attempt is to use these protocols in a more direct fashion and obtain much simpler protocols.

Our Results. We introduce a new framework of Environmental Security, where the correctness and security requirements of a protocol are separately defined (unlike [3, 19]). Further, we consider a model, called "client-server model," which considers restricted patterns of executions of the protocols analysed. Both are aimed at getting relaxations of the existing ES frameworks, so that possibly weaker levels of Environmental Security can be defined and proven for simpler protocols.

Then we show how to realize tools like commitment, zero knowledge proof and commit-and-prove functionalities in this setting. We illustrate the use of these tools in implementing a special class of functionalities called the "server-client" functionalities. All our protocols are very simple and relatively efficient (compared to protocols in ES/UC and gES models). The protocols are all in the "plain-model" (no common reference string, or other trust assumptions), and are much more efficient than even the ones in [19] (which solve a more difficult problem).

We point out that the 4 message zero knowledge protocol we give is, in particular, a *concurrent* zero knowledge argument with only 4 messages, wherein the simulator (and corrupt verifiers) are allowed some super polynomial computational power. Previous results (which worked with polynomial time simulators) gave either protocols with a large number of rounds (dependent on the security parameter), or were dependent on the number of verifiers that the protocol was secure against. Further, our protocol is a simple variant of a well-known simple protocol which has been around for many years, but for which no such strong composability has been proven till now.

Limitations of Our Results. There are some serious limitations to our current results. It is not clear if the approach in this work can directly yield protocols for the most general kind of functionalities. Firstly, our 2-party protocol is for a very special kind of multi-party computations only, which we term the server-client computation. (In a server-client computation, the client receives as output some function of its input and the server's input. But the server receives as output, the client's input.)

But a more serious limitation lies with the nature of security guarantee provided. Along with correctness and secrecy guarantees, one would like to have a guarantee that the server's input to the function is independent of the client's input. The security definition provided in this work does not make this last guarantee. Nevertheless, we sketch how this can be remedied under the condition that the client never *uses* its input previously (the full technical details of which will be published elsewhere).

Despite the limitations, our new framework is a step in the direction of formalizing relaxed notions of security (relaxed, but still accounting for a general environment), which will help prove security guarantees for simpler and more efficient protocols.

Previous Results. As mentioned above [3, 17] introduced the ES/UC framework, as a model to consider general composability and complex environments. But in the plain-model very little was available in terms of positive results. Recently, [19] introduced a modified notion of security, by allowing the IDEAL world environment and adversary access to super-polynomial powers. This made it possible to achieve secure multi-party computation in the plain model. However the protocols in [19] are still much more involved than the ones presented here. An earlier attempt at reducing the requirements of the ES/UC framework was in [5], which also introduced a semi-functionality-like notion in the context of secure Key-Exchange.

Work on concurrent model stretches back to [9], who introduced it in the context of Zero Knowledge proofs, followed by a sequence of works in the same context [20, 11, 18], where an arbitrary polynomial number of concurrent executions can be handled, but with the number of rounds in the protocol growing with the security parameter. [1] gave a constant round protocol for *bounded* concurrent ZK, in which the communication complexity depended on the number of concurrent sessions that can be handled. A similar result, but with similar limitations, was shown for *general* 2-party computations (general, as opposed to our Client-Server Computation) recently [12, 16]. All these protocols are somewhat complicated and conceptually involved. Relaxing the requirement of polynomial time simulation in the definition of security was used in some earlier works [13, 14] too, in the context of zero knowledge proofs.

Connections with [19]. Our starting point is the two "semi-functionalities" for commitment and zero-knowledge proofs, introduced in [19]. There they are used for the specific purpose of implementing a (full-fledged gES) commitment functionality. However we seek to directly use them for "end uses." Our new framework for monitored functionalities lets us extend the approach there to a formal definition of security. We introduce two more semi-functionalities, namely commit-and-prove and server-client computation. We give protocols for these semi-functionalities and also prove that these semi-functionalities have the required correctness property in our framework. We then observe that for such functionalities to be more useful, it would help if the correctness guarantees on the semi-functionalities are strengthened. We show that such a strengthening can indeed be formalized and proven (see Section 6.2).

2 Basic Ideas

The next few subsections introduce the novel tools and concepts we employ. All these are new, except the ideas in Section 2.3 (which were recently introduced in [19]).

2.1 Client-Server Model

We present the *Client-Server model* as a simplified setting to investigate some of the Environmental Security issues, without having to deal with all the composition issues. In this model, the security guarantees will be given only to sessions of protocols in all of which the participants have the same "fixed roles." The inspiration for this model is the model used for concurrent zero knowledge proofs [9].

The specific fixed role restriction in our model is as follows. There is a special party called the *server* $\mathcal{S}$. All the other parties are considered *clients*. We shall use $\mathcal{K}$ to denote a generic client. We allow only static adversaries, i.e., parties can be corrupted only at the beginning of the system. (Recall that the concurrent ZK-model also has a static adversary.) In this model we shall typically investigate functionalities where $\mathcal{S}$ plays a special role: for instance, a commitment functionality where $\mathcal{S}$ is the committing party (and a client receives the commitment), or a zero knowledge proof where $\mathcal{S}$ is the prover. We also consider a class of multi-party computation problems where $\mathcal{S}$ has a special role. Universal composition in the client-server model is limited to concurrent sessions of the protocols where the same party plays the server in all sessions. Thus, in particular, we do not offer general non-malleability, just as in the case of concurrent zero knowledge. However unlike there, *we require environmental security*: i.e., the security definition incorporates the presence of an arbitrary environment.

Note that the client-server model does not have a different definition of security, but rather inherits it from the underlying ES model. (The new security definition we introduce in this work is given in Section 2.2). It only specifies restrictions on the protocol executions for which security will be defined.

The main purpose of introducing the client-server model is to allow simplification of protocols, by exploiting the asymmetry in the model. It allows us to use simpler assumptions, as will be described in Section 2.3.

2.2 A New Framework for Specifying Security

In the concurrent ZK-model security requirement (for concurrent ZK proofs) is specified by the two properties: zero-knowledge (secrecy) and soundness (correctness). In contrast, in the ES/UC model security requirements are specified by giving an *ideal functionality*, which totally captures both these requirements. We propose a new framework, where we still require Environmental Security for the more subtle secrecy requirement. But the correctness requirement is specified separately, as in the concurrent ZK model. Below we elaborate on this.

"Semi-Functionalities" and "Monitors." In the ES/UC-model, a 2-party IDEAL functionality usually interacts with both the parties in an ideal way. For instance the IDEAL commitment functionality would involve receiving a value from the *sender* secretly, and notifying the *receiver* (and adversary) of the receipt, and later on receiving a command to reveal from the sender, sending the original value to the receiver. This functionality makes sure that the sender is bound to a value on committing (this is the "correctness guarantee") and that the value remains secret ("secrecy guarantee"). The idea behind defining a *semi-functionality* is to free one of these requirements from the IDEAL functionality, and somehow enforce that requirement separately.

Commitment: $\langle \mathcal{F}_{\widetilde{\text{COM}}} \rangle$

Semi-Functionality $\mathcal{F}_{\widetilde{\text{COM}}}$:

COMMIT PHASE

$\mathcal{S} \leftrightarrow \mathcal{F}_{\widetilde{\text{COM}}}$: Arbitrary protocol
$\mathcal{F}_{\widetilde{\text{COM}}} \to \mathcal{K}, \mathcal{S}$: commit

REVEAL PHASE

$\mathcal{S} \leftrightarrow \mathcal{F}_{\widetilde{\text{COM}}}$: Arbitrary Protocol
$\mathcal{F}_{\widetilde{\text{COM}}} \to \mathcal{K}, \mathcal{S}$: (reveal, b)
b determined
arbitrarily by $\mathcal{F}_{\widetilde{\text{COM}}}$

Completeness: If $\mathcal{S}$ is honest, $\mathcal{F}_{\widetilde{\text{COM}}}$ never aborts.
Monitor: When $\mathcal{F}_{\widetilde{\text{COM}}}$ sends the message commit, the monitor must record one value b^* as the committed value. If $\mathcal{S}$ is honest this value must be the intended value for commitment. Later if $\mathcal{F}_{\widetilde{\text{COM}}}$ sends (reveal, b) but $b \neq b^*$ then the monitor notifies the receiver by raising an alarm.

(a) Commitment

Zero Knowledge Proof: $\langle \mathcal{F}_{\widetilde{\text{ZK}}} \rangle$
Parametrized by a polynomial time relation $R: \{0,1\}^{\ell_1} \times \{0,1\}^{\ell_2} \to \{0,1\}$. Common input to $\mathcal{S}$ and $\mathcal{K}$: $x \in \{0,1\}^{\ell_1}$.
Semi-Functionality:

$\mathcal{S} \leftrightarrow \mathcal{F}_{\widetilde{\text{ZK}}}$: Arbitrary protocol
$\mathcal{F}_{\widetilde{\text{ZK}}} \to \mathcal{K}, \mathcal{S}$: (proven, x)

Completeness: If $\mathcal{S}$ is honest, $\mathcal{F}_{\widetilde{\text{ZK}}}$ sends the proven message to $\mathcal{K}$.
Monitor: If $\mathcal{F}_{\widetilde{\text{ZK}}}$ sends the message (proven, x) to the verifier, *and* there exists no $y \in \{0,1\}^{\ell_2}$ such that $R(x,y) = 1$ then the monitor notifies the verifier by raising an alarm.

(b) Zero Knowledge Proof

Commit and Prove: $\langle \mathcal{F}_{\widetilde{\text{CAP}}} \rangle$
Parametrized by a polynomial time relation $R: \{0,1\}^{\ell_1} \times \{0,1\}^{\ell_2} \times \{0,1\}^{\ell_3} \to \{0,1\}$.
Semi-Functionality:

COMMIT PHASE:

$\mathcal{S} \leftrightarrow \mathcal{F}_{\widetilde{\text{CAP}}}$: Arbitrary protocol
$\mathcal{F}_{\widetilde{\text{CAP}}} \to \mathcal{K}, \mathcal{S}$: commit

PROOF PHASE (CAN BE MULTIPLE TIMES):

$\mathcal{S} \to \mathcal{F}_{\widetilde{\text{CAP}}} \to \mathcal{K}, \mathcal{S}$: x
$\mathcal{S} \leftrightarrow \mathcal{F}_{\widetilde{\text{CAP}}}$: Arbitrary Protocol
$\mathcal{F}_{\widetilde{\text{CAP}}} \to \mathcal{K}, \mathcal{S}$: (proven, x)

Completeness: Interacting with an honest prover $\mathcal{S}$, $\mathcal{F}_{\widetilde{\text{CAP}}}$ never aborts.
Monitor: At the end of the commit phase, the monitor internally records a value w. If the $\mathcal{F}_{\widetilde{\text{CAP}}}$ sends (proven, x) later in the proof phase then the monitor checks if there is a value y such that $R(w, x, y) = 1$. If not it raises an alarm.

(c) Commit and Prove

Client-Server Computation: $\langle \mathcal{F}_{\widetilde{\text{CSC}}} \rangle$
Parametrized by a function F.
Semi-Functionality:

$\mathcal{S} \leftrightarrow \mathcal{F}_{\widetilde{\text{CSC}}}$: Arbitrary Protocol
$\mathcal{F}_{\widetilde{\text{CSC}}} \to \mathcal{K}, \mathcal{S}$: commit
$\mathcal{K} \to \mathcal{F}_{\widetilde{\text{CSC}}} \to \mathcal{S}, \mathcal{S}$: $x_{\mathcal{K}}$
$\mathcal{S} \leftrightarrow \mathcal{F}_{\widetilde{\text{CSC}}}$: Arbitrary Protocol
$\mathcal{F}_{\widetilde{\text{CSC}}} \to \mathcal{K}, \mathcal{S}$: z (z determined
arbitrarily by $\mathcal{F}_{\widetilde{\text{CSC}}}$)

Completeness: Interacting with an honest prover $\mathcal{S}$ with input $x_\mathcal{S}$, if $\mathcal{K}$ sends it input $x_{\mathcal{K}}$, $\mathcal{F}_{\widetilde{\text{CSC}}}$ sends $x_{\mathcal{K}}$ to $\mathcal{S}$ and $z = F(x_\mathcal{S}, x_{\mathcal{K}})$ to $\mathcal{K}$.
Monitor: At the end of the first step, the monitor internally records a value x^*. If $\mathcal{F}_{\widetilde{\text{CSC}}}$ sends $x_{\mathcal{K}}$ to $\mathcal{S}$ and z to $\mathcal{K}$, the monitor checks if $z = F(x^*, x_{\mathcal{K}})$. If not it raises an alarm.

(d) 2-Party Client-Server Computation

Fig. 1. Monitored Functionalities $\langle \mathcal{F}_{\widetilde{\text{COM}}} \rangle$, $\langle \mathcal{F}_{\widetilde{\text{ZK}}} \rangle$, $\langle \mathcal{F}_{\widetilde{\text{CAP}}} \rangle$ and $\langle \mathcal{F}_{\widetilde{\text{CSC}}} \rangle$

A monitored functionality (e.g., $\langle \mathcal{F}_{\widetilde{\text{COM}}} \rangle$ described in Figure 1(a)) consists of a semi-functionality ($\mathcal{F}_{\widetilde{\text{COM}}}$ in Figure 1(a)) and some conditions on the semi-functionality. The semi-functionality is syntactically just a functionality, but it is not "ideal" enough. It is typically defined based on an arbitrary protocol. For instance the specification of $\mathcal{F}_{\widetilde{\text{COM}}}$ consists of arbitrary interaction between the server and $\mathcal{F}_{\widetilde{\text{COM}}}$ (which is unspecified in Figure 1(a), but will be later specified in such a way that binding property can be argued separately). Note that the arbitrary protocol is carried out *between the semi-functionality and a party, and not between the two parties*. This is why the semi-functionality guarantees secrecy – in the case of $\mathcal{F}_{\widetilde{\text{COM}}}$, the only message it sends to the receiver and the adversary before the reveal phase is commit. To complete the specification of the ideal functionality we need to also give a guarantee that the semi-functionality is *functional* (i.e., it can be used by the server to make commitments) and *correct* (i.e., it is binding on the server). These requirements are specified separately as properties that the semi-functionality needs to satisfy. It is all these three requirements together that make up the specification of the ideal commitment functionality. We shall call such a collection of requirements a *monitored functionality*.

IDEAL *world of Monitored Functionality.* The Monitored Functionality is proposed as an IDEAL functionality, which captures all the security properties of a given task. In Figure 1 we show the four Monitored Functionalities used in this work. We point out some of the important features of this new formalism.

In Figure 1, the semi-functionalities are not fully specified, but allows arbitrary interaction between the server and the semi-functionalities. Once a protocol is chosen, the semi-functionality will be specialized to suit that protocol. That is, the semi-functionality will carry out the client's part in the protocol. Note that the view of the server is unchanged if we replace the interaction with the semi-functionality by the protocol with the client. The important thing here is that irrespective of the protocol that is used, these semi-functionalities are designed to capture the secrecy requirements of the respective tasks. For instance, in commitment, the only messages sent to the client are "commit" and "(reveal, b)." Indeed, in all four semi-functionalities the messages reaching the client and the IDEAL world adversary are exactly those messages that a corresponding IDEAL functionality in the ES/UC model would specify. The name semi-functionality is to emphasize that they provide only the secrecy guarantee, and correctness needs to be ensured separately. But otherwise there is nothing "semi" about them – technically these are full-fledged functionalities in the ES/UC model.

Next, we draw the reader's attention to the way the correctness requirement is specified. For convenience and concreteness, we employ a new notion, called *monitors*. A monitor is a conceptual device used to specify the security requirements of a functionality. If the security requirement is violated we want the monitor to alert the parties by "raising an alarm." Each session of the functionality has its own monitor. A monitor is a (computationally unbounded) function which can inspect the entire system including all parties and functionalities (except any other monitors) and maintain its own internal state. This is in contrast to the PPT functionalities. There is only one way a monitor can affect the system, namely, by raising an alarm.

Securely Realizing a Monitored Functionality in REAL *world.* In the REAL world we would like to have protocols which can replace Monitored Functionalities. That is, if we replace the IDEAL monitored functionality (i.e., the semi-functionality and monitor) by a protocol, no environment should be able to detect the difference (we are allowed also to replace the REAL adversary $\mathcal{A}$, by an IDEAL adversary $\mathcal{S}$). This involves two things: first the protocol should securely realize the semi-functionality (in the conventional sense of [3]). But in addition, it should be able to mimic being monitored. But clearly there are no monitors in the REAL world. So we require that even in the IDEAL world having the monitor should not be detectable to the environment. Note that this is a requirement on the functionality, and not on the protocol. However, it depends on the protocol in that the functionality is fully specified depending on the protocol.

Definition 1. *We say a protocol securely realizes a monitored functionality if*
1. *for every adversary there exists a simulator such that no environment can tell between interacting with the protocol in the* REAL *world (running the protocol) and interacting with the semi-functionality of the monitored functionality in the* IDEAL *world (this is the condition for the UC theorem to hold), and*
2. *there exists a monitor satisfying the specified requirements, such that for any environment and adversary, the probability that the monitor raises an alarm is negligible, even when there are other protocols, functionalities and monitors in the system.*

Note that in the above definition, the first condition is stated for a stand-alone setting, as the UC theorem [3, 19] ensures that it holds in a composed setting also. But the second condition needs to be met for the composed setting, as we do not have a composition theorem for (computationally unbounded) monitors. (i.e., a monitor may behave entirely differently when, in some part of the system, a REAL protocol is replaced by an IDEAL functionality). Hence we need to show the existence of a monitor for the composed setting- i.e., after all REAL protocols have been transformed to IDEAL functionalities or semi-functionalities. Further there may be other monitors in the system. But the monitors are independent of each other and the only way a monitor interferes with the system is by raising an alarm. Hence other monitors can be ignored for analysing the monitor of a particular session.

Expiring Monitors. For concreteness in our analysis, we shall consider monitors $\mathfrak{M}_\tau$ which *expire* after time τ from the start of the protocol. The guarantee given by the monitor holds only till it expires. But for any τ polynomial in the security parameter, we shall show that the monitor raises an alarm with negligible probability. Thus for any τ polynomially large in the security parameter, the guarantee would hold.

Locked States. For our guarantees to be useful, we would often require that the monitor cannot inspect some parts of the system. In $\mathcal{F}_{\widetilde{\text{CSC}}}$, for example, we would like the monitor to record the server's input independent of the client's input. We cannot make such a guarantee if the client has released its input into the system earlier (either explicitly, or by giving out a commitment, even if the commitment is perfectly hiding). However, if the client's input is kept "locked" and unused until after the sever makes its commitment step, then we should provide the above guarantee. We do this in Section 6.2.

2.3 Generalized Environmental Security

We assume that the reader is somewhat familiar with the ES/UC framework [3]. An IDEAL *functionality* is specified to define a task as well as the security requirements on it. A REAL world protocol is said to *securely realize* the IDEAL functionality, if replacing access to the IDEAL functionality by execution of the REAL protocol does not let the adversaries take advantage of this in any environment. That is, for every REAL world adversary $\mathcal{A}$, there is an IDEAL world adversary $\mathcal{S}$, such that no environment will behave differently when in the REAL world (running the protocol), instead of the IDEAL world (with access to the functionality). All the parties, the adversary, the environment and the IDEAL functionalities are probabilistic polynomial time (PPT) machines.

However for most of the interesting cryptographic tasks, there are no protocols which are secure under the ES/UC model, in the standard model [3,4,6]. The only protocols for these tasks, which are ES/UC secure require some strong trust assumptions in the model (eg. honest majority, or a trusted common random string) In [19] this difficulty was overcome by altering the definition of security, to obtain a new framework called the generalized Environmental Security (gES) framework.[2] There the IDEAL world adversary is given extra computational power via oracle access to an exponential time machine (referred to as the "Imaginary Angel," or simply Angel). When an imaginary angel Γ is used, the resulting model will be called the Γ-ES model. A protocol for a functionality is said to Γ-ES-realize the functionality against PPT adversaries, if for every PPT adversary $\mathcal{A}$, we can demonstrate a PPT simulator $\mathcal{S}$ with oracle access to Γ.

As in [19], here the information provided by the imaginary angel will be about a hash function. Suitable assumptions of hardness related to this hash function will be made in Section 2. The specification of the imaginary angel is given in Section 2. Though our assumptions on the hash function and our imaginary angel are similar to those in [19], they are somewhat simpler in our case. In fact, we avoid a strong "non-malleability flavored" assumption. (Correspondingly, however, we restrict ourselves to the client-server model introduced in Section 2.1.)

[19] proves the composition theorem below for the generalized setting, for any imaginary angel, which when queried, returns a (probabilistic) function of the set of corrupted parties and the query. For further details, we refer the reader to [19].

Theorem 1. (Extended Universal Composition Theorem- Informal Statement) [19]
Let $\mathcal{C}$ be a class of real-world adversaries and $\mathcal{F}$ be an ideal functionality. Let ρ be an n-party protocol that Γ-ES-realizes $\mathcal{F}$ against adversaries of class $\mathcal{C}$. Also, suppose π is an n-party protocol in the $\mathcal{F}$-hybrid model which Γ-ES-realizes a functionality $\mathcal{F}'$ against adversaries of class $\mathcal{C}$. Then the protocol π^ρ (obtained from π by replacing invocations of $\mathcal{F}$ by invocations of the protocol ρ) Γ-ES-realizes $\mathcal{F}'$ against adversaries of class $\mathcal{C}$.

Hash Function. As in [19], our computational assumptions have to do with a "hash function." However our assumptions are weaker than those there. We assume a hash function $\mathcal{H} : \{0,1\}^{k_1} \times \{0,1\}^{k_2} \times \{0,1\} \to \{0,1\}^\ell$, with the following properties:

---

[2] In [19] it was called generalized *Environmental* Security.

A1 (Difficult to find collisions with same prefix): For all PPT circuits M, for a random $r \leftarrow \{0,1\}^{k_1}$, probability that $M(r)$ outputs (x,y) such that $\mathcal{H}(r,x,0) = \mathcal{H}(r,y,1)$ is negligible.

A2 (Collisions and Indistinguishability): For every $r \in \{0,1\}^{k_1}$, there is a distribution $\mathcal{D}_r$ over the set $\{(x,y,z)|\mathcal{H}(r,x,0) = \mathcal{H}(r,y,1) = z\} \neq \phi$, such that

$$\{(x,z)|(x,y,z) \leftarrow \mathcal{D}_r\} \approx \{(x,z)|x \leftarrow \{0,1\}^{k_2}, z = \mathcal{H}(r,x,0)\}$$

$$\{(y,z)|(x,y,z) \leftarrow \mathcal{D}_r\} \approx \{(y,z)|y \leftarrow \{0,1\}^{k_2}, z = \mathcal{H}(r,y,1)\}$$

Further, given sampling access to $\mathcal{D}_r$ to a distinguisher, these distributions still remain indistinguishable.

The last condition essentially says that the hash function is "equivocable": i.e., for every r it is possible (but computationally infeasible) to give a z such that z can be explained as a hash of 0 ($\mathcal{H}(r,x,0)$ for some x) as well as a hash of 1 ($\mathcal{H}(r,y,1)$ for some y), and both explanations look as if it came from a uniform choice of x or y. Note that for random oracles all these conditions trivially hold (where k_1, k_2, ℓ are polynomially related).

These assumptions suffice for achieving concurrent Zero Knowledge proofs and commit-and-prove functionality. To securely realize "client-server computation," we make one more assumption (a stronger variety of trapdoor permutations) in Section 5.1. All these assumptions were used in [19] as well (where, in fact, the assumptions used are stronger than the ones here).

Imaginary Angel Γ. We specify the imaginary angel Γ that we use through out this work. Γ first checks if the server $\mathcal{S}$ is corrupted or not (recall that we do not allow $\mathcal{S}$ to be adaptively corrupted). If it is corrupted Γ functions as a *null-angel*, i.e., it returns $\bot$ on any query. But if $\mathcal{S}$ is not corrupted, then when queried with a string r Γ draws a sample from $\mathcal{D}_r$ described above and returns it. This is very similar to the imagnary angel used in [19], but slightly simpler.

3 Monitored Commitment and Zero Knowledge Proof

The semi-functionalities $\mathcal{F}_{\widetilde{\text{COM}}}$ and $\mathcal{F}_{\widetilde{\text{ZK}}}$ were introduced in [19] where protocols were given for these semi-functionalities. Further, lemmas were proved there which showed binding and soundness properties of these functionalities. Our protocols in this section are very similar to (but slightly simpler than) the corresponding ones in [19]. The proofs are similar too, except that the binding and soundness properties are now proven in terms of the probability with which a monitor raises alarm.

3.1 Monitored Commitment

The monitored functionality for Commitment $\langle \mathcal{F}_{\widetilde{\text{COM}}} \rangle$ was described in Figure 1(a), as composed of the Commitment semi-functionality $\mathcal{F}_{\widetilde{\text{COM}}}$ and a monitor to ensure binding. Now we give a protocol which realizes this functionality, in Figure 2.

Commitment Protocol COM

The committing party is the server $\mathcal{S}$, and the receiving party is some client $\mathcal{K}$.

COMMIT PHASE

$$\mathcal{K} \to \mathcal{S} : r \leftarrow \{0,1\}^{k_1}$$
$$\mathcal{S} \to \mathcal{K} : c = \mathcal{H}(r, r', b) \text{ where } r' \leftarrow \{0,1\}^{k_2}$$

REVEAL PHASE

$$\mathcal{S} \to \mathcal{K} : (b, r')$$
$$\mathcal{K} : \text{if } \mathcal{H}(r, r', b) = c \text{ then accept } b \text{ as revealed}$$

Fig. 2. Commitment Protocol COM

(Given the protocol, we can go back to the specification of $\langle \mathcal{F}_{\widetilde{\text{COM}}} \rangle$ and complete the semi-functionality specification, by replacing the "Arbitrary protocol" steps with the corresponding steps from the protocol.)

As mentioned earlier, we call the protocol secure if it achieves the semi-functionality and there exists a monitor as specified by the functionality, which will raise an alarm with negligible probability. The following lemmas which assure us of this.

Lemma 1. *For any polynomial (in the security parameter k) τ, under assumption A1, there is a monitor satisfying the requirements specified by $\langle \mathcal{F}_{\widetilde{\text{COM}}} \rangle$, such that the probability of the monitor raising an alarm within time τ is negligible.*

Proof. Firstly, if the adversary does not corrupt the sender, then the monitor reads the committed value from the honest sender's input to the protocol and sets b^* to that value. It is easy to verify that this monitor meets all the requirements and never raises an alarm.

So suppose the adversary corrupts the sender. In this case the imaginary oracle functions as a null-oracle. Thus the entire system of all the parties and the environment is probabilistic polynomial time. (We need not consider other monitors, as explained above.) The monitor $\mathfrak{M}_\tau$ chooses b^* as follows: examine the state of the entire system and determine the probability p_0 of the sender (legally) revealing this commitment as 0 within time τ, and the probability p_1 of the sender revealing it as 1 within that time. Choose $b^* = 0$ if $p_0 \geq p_1$; else choose $b^* = 1$.

We shall demonstrate a (non-uniform) PPT machine M which accepts $r \leftarrow \{0,1\}^k$ and outputs (x, y) such that $\mathcal{H}(r, x, 0) = \mathcal{H}(r, y, 1)$, with a probability polynomially related to the probability of the monitor raising an alarm.

M simulates the system internally, starting at the point the session is initiated. Recall that this session is to be run with access to the IDEAL semi-functionality. But instead, M will play the role of the semi-functionality for this session. It sends the input it received r as the first message to the sender. Then the sender responds with a string c. At this point M makes two copies of the system, and runs them with independent randomness, for time τ each. If the sender eventually reveals the commitment as $(x, 0)$ in one run and as $(y, 1)$ in the other run, then M outputs (x, y).

Define random variable p_0 (respectively, p_1) as the probability that after sending c the sender reveals the commitment as 0 (respectively, 1) within time τ. The probability that the monitor raises an alarm is at most

$$\mathbf{E}[\min\{p_0, p_1\}] \leq \mathbf{E}[\sqrt{p_0 p_1}] \leq \sqrt{\mathbf{E}[p_0 p_1]} = \sqrt{\frac{1}{2} \mathbf{Pr}\left[M \text{ succeeds}\right]}$$

because after forking two copies of the system, M succeeds (i.e., it manages to output (x, y) such that $\mathcal{H}(r, x, 0) = \mathcal{H}(r, y, 1)$) when in one of the runs the event with probability p_0 occurs and in the other the event with probability p_1.

Since the probability that M succeeds is negligible by assumption on $\mathcal{H}$, so is the probability that the monitor $\mathfrak{M}_\tau$ raises an alarm. Clearly this holds for any τ polynomial in the security parameter.

Lemma 2. COM Γ-*ES-realizes* $\mathcal{F}_{\widetilde{\text{COM}}}$ *against static adversaries, under assumption A2.*

Proof (sketch): For every PPT adversary $\mathcal{A}$ we demonstrate a PPT simulator $\mathcal{S}$ such that no PPT environment $\mathcal{Z}$ can distinguish between interacting with the parties and $\mathcal{A}$ in the REAL world, and interacting with the parties and $\mathcal{S}$ in the IDEAL world. We do this in the presence of the imaginary oracle Γ.

Corrupt Server. Note that the semi-functionality is designed such that choosing the simulator $\mathcal{S}$ to be identical to $\mathcal{A}$ (except that it sends the messages to $\mathcal{F}_{\widetilde{\text{COM}}}$ instead of to $\mathcal{K}$) works.

Honest Server. During simulation $\mathcal{S}$ runs $\mathcal{A}$ internally. When $\mathcal{A}$ starts the commitment protocol, $\mathcal{S}$ initiates a session with the IDEAL functionality. When $\mathcal{A}$ sends out the first message in the protocol r, $\mathcal{S}$ forwards this to the oracle Γ and receives $(x, y, z) \leftarrow \mathcal{D}_r$. Then, when $\mathcal{F}_{\widetilde{\text{COM}}}$ gives the commit message, $\mathcal{S}$ provides $\mathcal{A}$ with z as the message from $\mathcal{S}$ to $\mathcal{K}$ (whether $\mathcal{K}$ is corrupted or not). Later if $\mathcal{F}_{\widetilde{\text{COM}}}$ gives the message (reveal, 0), $\mathcal{S}$ provides $\mathcal{A}$ with $(0, x)$ as the REAL message, and if $\mathcal{F}_{\widetilde{\text{COM}}}$ gives the message (reveal, 1), $\mathcal{S}$ provides $(1, y)$ to $\mathcal{A}$. Under the assumption on $\mathcal{D}_r$, it can be shown that this is a good simulation. ∎

3.2 Monitored Zero Knowledge Proof

In Figure 3 we show a simple protocol ZK in the $\mathcal{F}_{\widetilde{\text{COM}}}$-hybrid model which Γ-ES-realizes $\langle \mathcal{F}_{\widetilde{\text{ZK}}} \rangle$ against static adversaries. The particular relation R used in $\langle \mathcal{F}_{\widetilde{\text{ZK}}} \rangle$ is of Hamiltonicity (ie, given a graph, whether it contains a Hamiltonian cycle or not). The protocol is a simple adaptation of the well-known zero knowledge protocol for this relation. However that protocol (as well as its previous variants) is not known to be secure in a concurrent setting.

Multi-bit Commitment. Multiple bits can be committed to by running independent copies of the protocol in Section 3.1 in parallel. (Better efficiency can be achieved by making suitable assumptions on the hash function $\mathcal{H}$. But in this work we do not address this aspect of efficiency.) For convenience, we denote this collection of sessions of $\mathcal{F}_{\widetilde{\text{COM}}}$ by $\mathcal{F}^*_{\widetilde{\text{COM}}}$. For simplifying the description, we shall use the notation $\mathcal{S} \leftrightarrow \mathcal{F}^*_{\widetilde{\text{COM}}} \rightarrow \mathcal{K}$: COM(M) to denote a step where $\mathcal{S}$ sends a commitment to the bits of M, to $\mathcal{K}$ through the semi-functionality $\mathcal{F}^*_{\widetilde{\text{COM}}}$, and $\mathcal{S} \leftrightarrow \mathcal{F}^*_{\widetilde{\text{COM}}} \to \mathcal{K}$: REV(M') will denote a reveal to $M' \subset M$ later using (the same copy of) $\mathcal{F}^*_{\widetilde{\text{COM}}}$.

Note that we are providing the protocol ZK in the $\mathcal{F}^*_{\widetilde{\text{COM}}}$-hybrid model. So in the semi-functionality $\mathcal{F}_{\widetilde{\text{ZK}}}$, the "arbitrary protocol" will involve interaction of $\mathcal{F}_{\widetilde{\text{ZK}}}$ (and $\mathcal{S}$) with $\mathcal{F}^*_{\widetilde{\text{COM}}}$. This simply means that $\mathcal{F}^*_{\widetilde{\text{COM}}}$ is internally run by $\mathcal{F}_{\widetilde{\text{ZK}}}$, when interacting with $\mathcal{S}$.

The prover receives a Hamiltonian cycle $H \subset G$ as witness. First it verifies that the H is indeed a valid Hamiltonian cycle in G. (Else it aborts the protocol.) We use the above notation for commitments and reveals of $n \times n$ matrices. The adjacency matrix of a graph is naturally represented as an $n \times n$ bit-matrix. For convenience we let a permutation ϕ of $[n]$ also to be represented by an $n \times n$ bit-matrix Φ defined as $\Phi_{ij} = 1$ iff $\phi(i) = j$ (else $\Phi_{ij} = 0$).

The idea is that the prover has to commit to the pair of $n \times n$ matrices (M_1, M_2) where the verifier expects $M_1 = \phi(G)$ and $M_2 = \Phi$ for some permutation ϕ with the representation Φ. In response to sending $b = 0$ the verifier expects the prover to reveal all of M_1 and M_2, where as for $b = 1$ it expects the prover to reveal the bits in M_1 corresponding to a Hamiltonian cycle in $\phi(G)$. An edge is represented by an index (i, j) into this matrix. Given a set of edges ζ, we use $M|_\zeta$ to denote the entries in the matrix M given by the edges in ζ.

Lemma 3. *For any polynomial (in the security parameter k) τ, there is a monitor satisfying the requirements specified by $\langle \mathcal{F}_{\widetilde{\text{ZK}}} \rangle$, such that the probability of the monitor raising an alarm within time τ is negligible.*

Proof. Our monitor does exactly what the specification requires: it checks if G is Hamiltonian. If not *and* if $\mathcal{F}_{\widetilde{\text{ZK}}}$ sends the message HAMILTONIAN to $\mathcal{K}$, then the monitor raises an alarm.

We shall use the result that $\mathcal{F}_{\widetilde{\text{COM}}}$ has a monitor, to argue that the probability this monitor raises an alarm is negligible. For each of the n parallel sessions, consider the behaviour of the monitors for $\mathcal{F}_{\widetilde{\text{COM}}}$ for the $2n^2$ sessions of $\mathcal{F}_{\widetilde{\text{COM}}}$. These monitors record values b^*_{ij}, $(i, j) \in [n] \times [2n]$ internally.

For convenience, we define the following events: ALARM is the above event that the monitor raises an alarm; BADCOM is the event that some $\mathcal{F}_{\widetilde{\text{COM}}}$ monitor raises an alarm; ALLGOODQUERIES is the event that in each of the n sessions, for the bit selected by $\mathcal{K}$ the bits recorded by commitment monitors define a valid answer (i.e., for $b = 0$ the monitors have recorded $M_1 = \phi(G), M_2 = \Phi$ and for $b = 1$ the monitors have recorded an M_1 with Hamiltonian cycle. If any pair (M_1, M_2) recorded by the monitors defines a valid answer for both $b = 0$ and $b = 1$, it implies that the graph is Hamiltonian; else we call the pair "bad." Let ALLBADPAIRS be the event that in all the n sessions, bits recorded by the commitment monitors give bad pairs of matrices. Then it is easy to see (as shown in the soundness proof for the corresponding protocol, in [19]) that

$$\Pr[\text{ALARM}] \leq \Pr[\text{ALLGOODQUERIES}|\text{ALLBADPAIRS}] + \Pr[\text{BADCOM}].$$

If a pair is bad it can define a valid answer for at most one of the two possible queries. That is, with probability at most $\frac{1}{2}$, $\mathcal{K}$ makes a good query on that pair. So,

$$\Pr[\text{ALLGOODQUERIES}|\text{ALLBADPAIRS}] \leq 2^{-n}$$

Zero Knowledge Proof Protocol: ZK

Common input to $\mathcal{S}$ and $\mathcal{K}$: a graph G.

REPEAT IN PARALLEL n TIMES:

$\mathcal{S} \leftrightarrow \mathcal{F}^*_{\widetilde{\text{COM}}} \to \mathcal{K} : \text{COM}(M_1 = \phi(G), M_2 = \Phi)$, where ϕ (represented by the matrix Φ) is a randomly chosen permutation of $[n]$.

$\quad\mathcal{K} \to \mathcal{S} : b \leftarrow \{0, 1\}$

$\quad\quad$IF $b = 0$

$\mathcal{S} \leftrightarrow \mathcal{F}^*_{\widetilde{\text{COM}}} \to \mathcal{K} : \text{REV}(M_1, M_2)$

$\quad\quad$ELSE

$\mathcal{S} \leftrightarrow \mathcal{F}^*_{\widetilde{\text{COM}}} \to \mathcal{K} : \text{REV}(M_1|_\zeta)$

$\quad\quad\quad$where ζ corresponds to the edges of the cycle $\phi(H)$

$\quad\quad$ENDIF

END REPEAT

$\quad\quad\mathcal{K}$: if in all parallel repetitions

$\quad\quad\quad b = 0 \implies M_2$ represents a permutation ϕ, and M_1 represents a graph

$\quad\quad\quad\quad$ such that $\phi(G) = M_1$

$\quad\quad\quad b = 1 \implies \zeta$ corresponds to the edges of a Hamiltonian cycle,

$\quad\quad\quad\quad$ and $\forall (i, j) \in \zeta, M_{1\,ij} = 1$

$\quad\quad$then ACCEPT

Fig. 3. Protocol for the Monitored Functionality for ZK Proof

Since $\mathbf{Pr}\,[\text{BADCOM}]$ is also negligible, we conclude that $\mathbf{Pr}\,[\text{ALARM}]$ is negligible.

Lemma 4. ZK Γ-ES-realizes $\mathcal{F}_{\widetilde{\text{ZK}}}$ against static adversaries in the $\mathcal{F}_{\widetilde{\text{COM}}}$-hybrid model.

Proof. Corrupt Server. Just as in the case of $\mathcal{F}_{\widetilde{\text{COM}}}$, if the server is corrupt, a trivial simulator in the IDEAL world, which acts transparently between an internal copy of $\mathcal{A}$ and the semi-functionality $\mathcal{F}_{\widetilde{\text{ZK}}}$ perfectly simulates the protocol between the corrupt server and an honest client.

Server not Corrupt. Recall that the protocol is in the $\mathcal{F}_{\widetilde{\text{COM}}}$-hybrid model. If the server is not corrupt, the only protocol messages that $\mathcal{A}$ can see are the statement to be proven, the length of the messages from $\mathcal{S}$ to $\mathcal{F}_{\widetilde{\text{COM}}}$, the commit messages from $\mathcal{F}_{\widetilde{\text{COM}}}$, the bit b sent by $\mathcal{K}$ ($\mathcal{K}$ may be corrupt or honest), and the final proven message. All these are available to simulator $\mathcal{S}$ in the IDEAL execution too. Note that if $\mathcal{K}$ is not corrupted, the bit b can be chosen uniformly at random during simulation. On the other hand, if $\mathcal{K}$

is corrupted (before it sends out b), then this bit is indeed produced by the copy of $\mathcal{A}$ that $\mathcal{S}$ runs internally. Then it is easily verified that $\mathcal{S}$ can indeed simulate in this case *perfectly*.

The above two lemmas can be summarized as follows.

Lemma 5. *Protocol* ZK *in* $\mathcal{F}_{\widetilde{\text{COM}}}$-*hybrid model* Γ-*ES-realizes* $\langle \mathcal{F}_{\widetilde{\text{ZK}}} \rangle$ *against static adversaries.*

Using the composition theorem Theorem 1 and Lemma 2 we get a protocol ZK^{COM} in the REAL world which Γ-ES-realizes $\langle \mathcal{F}_{\widetilde{\text{ZK}}} \rangle$ against static adversaries. Note that ZK^{COM} is a 4-round protocol. In the language of Zero-Knowledge proofs, we can state this result as follows.

Theorem 2. *There is a 4-round concurrent Zero Knowledge argument for Hamiltonicity when the simulator (as well as corrupt verifiers) has sampling access to $\mathcal{D}_r$ for all $r \in \{0,1\}^{k_1}$.*

4 Monitored Commit and Prove

Somewhat surprisingly, our model of security allows very simple protocols for the commit and prove (monitored) functionality (Figure 1(c)) as well. Below is the semi-functionality $\mathcal{F}_{\widetilde{\text{CAP}}}$, with respect to a relation R, and monitor for it.

For the commit phase, we use a straight-forward extension of the bit commitment protocol COM to multiple bits (see Section 3.2). A transcript of the commitment phase consists of two messages (r, c), where $c = \overline{\mathcal{H}}(r, r', w)$, where r' is a random string privately chosen by $\mathcal{S}$ and w is the string committed to. $\overline{\mathcal{H}} : \{0,1\}^{k_1 t} \times \{0,1\}^{k_2 t} \times \{0,1\}^t \to \{0,1\}^{\ell t}$ is a multi-bit version of $\mathcal{H}$: $\overline{\mathcal{H}}((r_1 \cdots r_t), (r'_1 \cdots r'_t), (w_1 \cdots w_t)) = (\mathcal{H}(r_1, r'_1, w_1), \cdots, \mathcal{H}(r_t, r'_t, w_t))$.

We introduce some more notation to conveniently describe the protocol. Let $\mathcal{S} \leftrightarrow \mathcal{F}_{\widetilde{\text{ZK}}} \to \mathcal{K}$: ZKP$_R(x; r, c)$ denote the following specification: first, parties $\mathcal{S}$ and $\mathcal{K}$ reduce the problem "$\exists w, y, r'$ such that $R'(w, x, r, r', c, y) = 1$" to a Hamiltonicity problem instance $G(x, r, c)$, where $R'(w, x, r, r', c, y) = 1$ if and only if $R(w, x, y) = 1$ and $\overline{\mathcal{H}}(r, r', w) = c$. This reduction is carried out in such a way that given a Hamiltonian cycle in G, it is possible to recover (w, r', y) as above. Then $\mathcal{S}$ uses the semi-functionality $\mathcal{F}_{\widetilde{\text{ZK}}}$ to prove to $\mathcal{K}$ that G is Hamiltonian.

The protocol is given in Figure 4. We shall prove the following:

Lemma 6. *Protocol* CAP Γ-*ES-realizes monitored functionality* $\langle \mathcal{F}_{\widetilde{\text{CAP}}} \rangle$ *against static adversaries, under assumptions A1 and A2.*

Proof. 1. CAP Γ-*ES-realizes* $\mathcal{F}_{\widetilde{\text{CSC}}}$ *against static adversaries in the* $\mathcal{F}_{\widetilde{\text{ZK}}}$-*hybrid model.*

Corrupt Server. Just as in the case of $\mathcal{F}_{\widetilde{\text{COM}}}$ and $\mathcal{F}_{\widetilde{\text{ZK}}}$, a trivial simulator which acts transparently between an internal copy of $\mathcal{A}$ and the semi-functionality $\mathcal{F}_{\widetilde{\text{CAP}}}$ perfectly simulates the protocol between the corrupt server and an honest client.

Server not Corrupt. The protocol is in the $\mathcal{F}_{\widetilde{\text{ZK}}}$-hybrid model, and hence so is the semi-functionality $\mathcal{F}_{\widetilde{\text{CAP}}}$. When the server is not corrupted, the only protocols messages $\mathcal{A}$

Commit and Prove Protocol: CAP

COMMIT PHASE:

$$\mathcal{K} \to \mathcal{S} : r \leftarrow \{0,1\}^{k_1}$$
$$\mathcal{S} \to \mathcal{K} : c = \overline{\mathcal{H}}(r, r', w) \text{ where } r' \leftarrow \{0,1\}^{k_2}$$

PROOF PHASE (CAN BE MULTIPLE TIMES):

$$\mathcal{S} \to \mathcal{K} : x$$
$$\mathcal{S} \leftrightarrow \mathcal{F}_{\widetilde{\text{ZK}}} \to \mathcal{K} : \text{ZKP}_R(x; c)$$
$$\mathcal{K} : \text{Accept if accepted in the above protocol}$$

Fig. 4. Protocol for the Monitored Functionality for Commit and Prove

can see are the initial commitment messages r and c, lengths of the messages from $\mathcal{S}$ to $\mathcal{F}_{\widetilde{\text{ZK}}}$, and the $(\text{proven}, G(x, r, c))$ message from $\mathcal{F}_{\widetilde{\text{ZK}}}$ at the end of each proof phase. The only non-trivial task for the simulator is to produce the commitment text c. Since c will never be revealed (because the server is honest and the adversary cannot adaptively corrupt $\mathcal{S}$), $\mathcal{S}$ can simply use a commitment to a random text using r to produce a purported commitment of w.

Note that assumption A2 implies that the distributions of commitments to 0 and to 1 are indistinguishable (even with access to $\mathcal{D}_r$): that is, for all $r \in \{0,1\}^{k_1}$

$$\{z | x \leftarrow \{0,1\}^{k_2}, z = \mathcal{H}(r, x, 0)\} \approx \{z | y \leftarrow \{0,1\}^{k_2}, z = \mathcal{H}(r, y, 1)\},$$

because both the distributions are indistinguishable from $\{z | (x, y, z) \leftarrow \mathcal{D}_r\}$. From this, it is a routine exercise to show that no PPT environment (with access to the imaginary oracle Γ) can distinguish between the simulation in the IDEAL world and the execution in the $\mathcal{F}_{\widetilde{\text{ZK}}}$-hybrid model.

2. *For any polynomial (in the security parameter k) τ, there is a monitor satisfying the requirements specified by $\langle \mathcal{F}_{\widetilde{\text{CAP}}} \rangle$, such that the probability of the monitor raising an alarm within time τ is negligible.*

We restrict ourselves to the case when $\langle \mathcal{F}_{\widetilde{\text{CAP}}} \rangle$ allows only one proof phase per session. It is possible to extend it to multiple proofs, but the details become lengthy and tedious.

First we describe how a value w^* is recorded by $\mathfrak{M}_\tau$. Consider an "extractor" PPT machine M_τ which simulates the entire (composed IDEAL) system internally, starting at the point where the session of interest running our Commit-and-Prove protocol starts (this start state is given to M_τ as non-uniform advice), for at most τ time-steps. M_τ runs the system until the proof phase of started, and the prover makes the commitment step. At this point M_τ clones the system and runs the two copies independent of each other. If in both the copies the proof is accepted by the verifier, M_τ checks if the n-bit queries made by the verifier in $\text{ZKP}_R(x; r, c)$ are identical or not. If they are not identical this lets M_τ extract a Hamiltonian cycle for G (assuming the monitors for the $\mathcal{F}_{\widetilde{\text{COM}}}$s do not raise any alarm). Then M_τ derives a witness (w, r', y) from this Hamiltonian cycle, and outputs it. Else M_τ outputs $\bot$.

Now we use M_τ to describe the monitor $\mathfrak{M}_\tau$. When $\mathcal{F}_{\widetilde{\mathrm{CAP}}}$ sends commit to $\mathcal{K}$, for each w $\mathfrak{M}_\tau$ checks the probability of M_τ outputting w, and records the one with the highest such probability, say w^*. Later if $\mathcal{F}_{\widetilde{\mathrm{CAP}}}$ sends (proven, x) for some x such that for no y $R(w^*, x, y)$ holds, then it raises an alarm. Also, for purposes of analysis, when the prover executes the commitment protocol (semi-functionality) as part of the zero-knowledge proof protocols, $\mathfrak{M}_\tau$ starts the monitors for $\mathcal{F}_{\widetilde{\mathrm{COM}}}$ as sub-monitors. The monitors will also be run when the extractor M_τ runs. If any of these sub-monitors raises an alarm, then too $\mathfrak{M}_\tau$ will raise an alarm.

Clearly $\mathfrak{M}_\tau$ satisfies the requirements of the functionality (up to the time bound τ). We go on to prove that the probability that $\mathfrak{M}_\tau$ raises an alarm (which event we denote by ALARM) is negligible. In the rest of the proof, we condition on the event that none of these sub-monitors raise an alarm. Since we have already shown that this is an event of negligible probability (and only polynomially many such sub-monitors are run), this will not change our conclusions.

Now, consider the point at which M_τ forks the system. Let p_w be the probability that M_τ outputs w starting at (conditioned on) this point, within τ time-steps. Let q be the probability that $\mathcal{K}$ accepts the proof $\mathrm{ZKP}_R(x; c)$ within τ time-steps, but $\not\exists (y, r') R'(w^*, x, r, r', c, y) = 1$. Note that $\mathbf{Pr}\,[\mathrm{ALARM}] = \mathbf{E}[q]$, where the expectation is over the distribution on the state of the system at the point at which M_τ forks.

Since we assume that the sub-monitors do not raise alarm, M_τ outputs some w if the two copies it runs both accept the proof, and in the second copy the verifier sends a query different from the one in the first copy. So, $\sum_{w \neq w^*} p_w \geq q(q - 2^{-n})$. Then,

$$\mathbf{Pr}\,[M_\tau \text{ outputs } w \neq w^*] \geq \mathbf{E}[q(q - 2^{-n}]$$
$$\geq \mathbf{E}[q]^2 - 2^{-n}\mathbf{E}[q]$$
$$\geq \frac{1}{2}\mathbf{E}[q]^2 \qquad \text{if } \mathbf{E}[q] \geq 2^{-n+1}$$

If the assumption in the last line above does not hold, we would be done, because $\mathbf{Pr}\,[\mathrm{ALARM}] = \mathbf{E}[q]$. So we make that assumption and proceed.

Now we shall demonstrate a (non-uniform) PPT machine M'_τ which accepts $r \leftarrow \{0, 1\}^k$ and outputs (x, y) such that $\mathcal{H}(r, x, 0) = \mathcal{H}(r, y, 1)$, with a probability polynomially related to the probability of the monitor raising an alarm. $M'_\tau(r)$ starts M_τ and runs the commit phase by sending r. It forks M_τ after the commitment from P arrives. Then it runs the two independent copies of M_τ (which involves forking the system again), and checks if they output different values (w_1, r'_1) and (w_2, r'_2), with $w_1 \neq w_2$. If so, M'_τ derives a collision to the hash function from some bit at which w_1 and w_2 differ, and outputs the corresponding portions of r'_1, r'_2. We say that M'_τ succeeds if it gets (w_1, w_2), such that $w_1 \neq w_2$ from two runs of M_τ.

Then,
$$\Pr[M'_\tau \text{ succeeds}] \geq \sum_{w'} p_{w'} \sum_{w \neq w'} p_w$$
$$\geq \sum_{w'} p_{w'} \sum_{w \neq w^*} p_w \qquad \text{because for all } w', p_{w^*} \geq p_{w'}$$
$$= (\sum_{w'} p_{w'})(\sum_{w \neq w^*} p_w) \geq (\sum_{w \neq w^*} p_w)^2$$
$$\geq \frac{1}{4}\mathbf{E}[q]^4 = \frac{1}{4}\Pr[\text{ALARM}]^4$$

Putting it all together we have that $\Pr[\text{ALARM}] \leq (4\Pr[M'_\tau \text{ finds a collision}])^{\frac{1}{4}}$, which is negligible by assumption on $\mathcal{H}$.

5 Applications of the New Framework

As we have shown above, theoretically interesting cryptographic tools like commitment and zero-knowledge proofs can be securely realized in the new framework, relatively efficiently (compared to those in previous Environmental Security models). The reason for this is that our security requirements are much more relaxed. However this raises the question if these weakened versions of the above tools are useful to achieve security for practically interesting tasks. In this section we make some progress towards making the new framework usable for multi-party computation problems. We restrict ourselves to 2-party computations of a very specific kind, as described below.

5.1 Client-Server Computation

A 2-party Client-Server Computation functionality $\langle \mathcal{F}_{\widetilde{\text{CSC}}} \rangle$ is given earlier in Figure 1(d). Note that the client does not keep any secrets from the server $\mathcal{S}$. But the server must *commit* to its inputs (and the monitor shall record the committed input) before the client sends its inputs. First, we shall give a protocol for this Monitored functionality, before discussing some of its limitations.

Secret Commit and Prove. In order to give a protocol for $\langle \mathcal{F}_{\widetilde{\text{CSC}}} \rangle$, we need to modify the Commit-and-Prove functionality, so that if both the server and the client are honest, the adversary is not given the statements that the server proves. (This is because the adversary should not learn the client's input.) Such a functionality $\mathcal{F}_{\text{SECRET-CAP}}$ can be securely realized in the $\mathcal{F}_{\text{ENC}}$-hybrid model, where $\mathcal{F}_{\text{ENC}}$ is the encryption functionality. For this $\mathcal{F}_{\widetilde{\text{ZK}}}$ and $\mathcal{F}_{\widetilde{\text{COM}}}$ are modified to the SECRET versions, which do not send the statement proven ($\mathcal{F}_{\text{SECRET-}\widetilde{\text{ZK}}}$) or the bit revealed ($\mathcal{F}_{\text{SECRET-}\widetilde{\text{COM}}}$) to the adversary. $\mathcal{F}_{\text{SECRET-}\widetilde{\text{COM}}}$ can be securely realized by the protocol COM modified to encrypt the reveal step, using the functionality $\mathcal{F}_{\text{ENC}}$. In the static case $\mathcal{F}_{\text{ENC}}$ is known to be easy to implement, using CCA2-secure public-key encryption with new keys each time (see for instance [3]), which in turn can be implemented assuming a family of trapdoor permutations (using the construction in [21], for instance). But since we are in the Γ-ES-model, we need to revisit the assumptions used to securely realize $\mathcal{F}_{\text{ENC}}$, namely the existence of trapdoor permutations.

Client-Server Computation Protocol: CSC

The protocol is parametrized by a function F.

$$S \leftrightarrow \mathcal{F}_{\text{SECRET-}\widetilde{\text{CAP}}} \to \mathcal{K} : \text{COMMIT-PHASE}(x_S)$$
$$\mathcal{K} \to \mathcal{F}_{\text{ENC}} \to S : x_\mathcal{K}$$
$$S \to \mathcal{F}_{\text{ENC}} \to \mathcal{K} : z = F(x_S, x_\mathcal{K})$$
$$S \leftrightarrow \mathcal{F}_{\text{SECRET-}\widetilde{\text{CAP}}} \to \mathcal{K} : \text{PROOF-PHASE}(z == F(x_S, x_\mathcal{K}))$$

Fig. 5. Protocol for the Monitored Functionality for Client-Server Computation

A3 There exists a family of trapdoor permutations secure against *non-uniform PPT adversaries* which are given sampling access to $\mathcal{D}_r$ for all r.

This is also an assumption made in [19]. With this assumption in place, we get the following result.

Lemma 7. *There are protocols which Γ-ES-realize $\mathcal{F}_{\text{ENC}}$ and $\mathcal{F}_{\text{SECRET-}\widetilde{\text{CAP}}}$ against static adversaries, under assumptions A1, A2 and A3.*

5.2 The Protocol

Theorem 3. *The protocol* CSC *Γ-ES-realizes the monitored functionality $\langle \mathcal{F}_{\widetilde{\text{CSC}}} \rangle$ against static adversaries in the $\mathcal{F}_{\text{SECRET-}\widetilde{\text{CAP}}}, \mathcal{F}_{\text{ENC}}$-hybrid model.*

Proof. 1. For any polynomial (in the security parameter k) τ, there is a monitor satisfying the requirements specified by $\langle \mathcal{F}_{\widetilde{\text{CSC}}} \rangle$, such that the probability of the monitor raising an alarm within time τ is negligible.

We can build a monitor $\mathfrak{M}_\tau$ for $\langle \mathcal{F}_{\widetilde{\text{CSC}}} \rangle$ using the monitor for $\langle \mathcal{F}_{\widetilde{\text{CAP}}} \rangle$. $\mathfrak{M}_\tau$ starts the monitor for $\langle \mathcal{F}_{\widetilde{\text{CAP}}} \rangle$, and if the protocol proceeds beyond the first step, it would record a value x^* internally as the committed value. $\mathfrak{M}_\tau$ will copy that value and record it as the input of S. Later if the monitor for $\langle \mathcal{F}_{\widetilde{\text{CAP}}} \rangle$ raises an alarm, $\mathfrak{M}_\tau$ will raise an alarm. If $\mathcal{F}_{\widetilde{\text{CSC}}}$ sends the value z to $\mathcal{K}$, then $\mathcal{F}_{\widetilde{\text{CAP}}}$ must return (proven, $z = F(x^*, x_\mathcal{K})$). So if the monitor for $\langle \mathcal{F}_{\widetilde{\text{CAP}}} \rangle$ does not raise an alarm, it means indeed $z = F(x^*, x_\mathcal{K})$ and $\mathfrak{M}_\tau$ need not raise any alarm either. Thus $\mathfrak{M}_\tau$ does satisfy the reuirements specified by $\langle \mathcal{F}_{\widetilde{\text{CSC}}} \rangle$. Further the probability that $\mathfrak{M}_\tau$ raises an alarm is the same as that the monitor for $\langle \mathcal{F}_{\widetilde{\text{CAP}}} \rangle$ raises an alarm. By earlier analysis, this is indeed negligible.

2. CSC Γ-ES-realizes $\mathcal{F}_{\widetilde{\text{CSC}}}$ against static adversaries in the $\mathcal{F}_{\text{SECRET-}\widetilde{\text{CAP}}}, \mathcal{F}_{\text{ENC}}$-hybrid model.

For every PPT adversary $\mathcal{A}$ we demonstrate a PPT simulator $\mathcal{S}$ such that no PPT environment $\mathcal{Z}$ can distinguish between interacting with the parties and $\mathcal{A}$ in the REAL world, and interacting with the parties and $\mathcal{S}$ in the IDEAL world.

As usual $\mathcal{S}$ internally runs $\mathcal{A}$ (which expects to work in the $\mathcal{F}_{\text{SECRET-}\widetilde{\text{CAP}}}, \mathcal{F}_{\text{ENC}}$-hybrid with the parties running the CSC protocol), and works as an interface between $\mathcal{A}$ and the parties. When $\mathcal{A}$ starts the CSC protocol, $\mathcal{S}$ initiates a session with the IDEAL functionality $\mathcal{F}_{\widetilde{\text{CSC}}}$.

Corrupt Server. Again, as in the case of all the monitored functionalities introduced in this work, thanks to the way the semi-functionality is designed, a trivial simulator in the IDEAL world, which acts transparently between an internal copy of $\mathcal{A}$ and the semi-functionality $\mathcal{F}_{\widetilde{ZK}}$ perfectly simulates the protocol.

Server not Corrupt. The client $\mathcal{K}$ may or may not be corrupt. We analyse the two cases separately:

- *Honest $\mathcal{K}$:* In this case all that $\mathcal{A}$ can see are the lengths of the messages $x_\mathcal{S}, x_\mathcal{K}$ and $F(x_\mathcal{S}, x_\mathcal{K})$, given to it by $\mathcal{F}_{\text{SECRET-}\widetilde{\text{CAP}}}$ and $\mathcal{F}_{\text{ENC}}$. These are known to $\mathcal{S}$ (because F is publicly known), and it can send them to $\mathcal{A}$.
- *Corrupt $\mathcal{K}$:* In this case $\mathcal{S}$ gets $x_\mathcal{K}$ and $F(x_\mathcal{S}, x_\mathcal{K})$. In addition $\mathcal{A}$ expects to see the messages commit and proven from $\mathcal{F}_{\text{SECRET-}\widetilde{\text{CAP}}}$ (in the first and last steps of the protocol). These are easily provided by the simulator.

It is easy to see that in all the cases, the simulation is perfect.

From this theorem, using Lemma 7 and the composition theorem Theorem 1, we get the following corollary.

Corollary 4. *There is a protocol which Γ-ES-realizes monitored functionality $\langle \mathcal{F}_{\widetilde{\text{CSC}}} \rangle$ against static adversaries, under assumptions A1, A2 and A3.*

5.3 Extensions to Adaptive Adversaries

Above we analyzed security in the presence of static adversaries, for the sake of simplicity. Here we mention how the tools developed here can be extended to the case of adaptive adversaries. Firstly, if we expand the adversary class to allow the adaptive corruption of only the clients, it is easy to see that the analyses still hold. The only modification required is that (in the case of the "SECRET" versions of the functionalities), the encryption protocols used will need to be secure against adaptive adversaries as well.

However extending to full-fledged adaptive corruption (i.e., adaptive corruption of the server as well) requires more modifications. Note that the Imaginary Angel Γ functions as a null-angel when the server $\mathcal{S}$ is corrupted, but otherwise gives access to the distribution $\mathcal{D}_r$. If $\mathcal{S}$ is initially uncorrupted and corrupted later on, removing access to $\mathcal{D}_r$ is not enough; having had access to $\mathcal{D}_r$ in the past gives the adversary an advantage. To fix this, we can use the hash function used in [19] for commitment, which takes one more parameter, namely, the ID of the receiving party. The assumptions used and the Imaginary Angel will then be the same as in [19]. The difference with [19] is that the "basic commitment" and "basic ZK proof" protocols there cannot be directly used to satisfy the security requirements there, where as the final protocols developed are secure only against static adversaries. In our case these basic protocols can be directly used to securely realize monitored functionalities. Note however that there is no significant advantage in using the client-server model anymore if we use the same assumptions as in [19]. Indeed, the resulting protocols securely realize the monitored functionalities in the unrestricted environmental setting (without the restrictions of the client-server model), against adaptive adversaries.

6 Limitations and Challenges

6.1 Problem with $\langle \mathcal{F}_{\widetilde{\text{CSC}}} \rangle$

Though we have successfully applied our tools in the new framework to obtain a 2-party computation protocol, there are some serious limitations to this functionality. Clearly, the set of functions that are computed are limited (namely, only client-server computations). But more seriously, the guarantee given by the monitor is not satisfactory. In particular, there is no guarantee of *"independence" of inputs*. Though the monitor records a value for the server's input prior to the client sending out its input, the value recorded is allowed to be *dependent on the entire system*, and in particular on the input of the client![3]

6.2 The Solution: Restricting the Monitors

In ongoing work, we suggest ways to address this problem. There we show that if the clients keep their private inputs totally unused until the point of commitment (but may use them immediately afterwards), then the monitor can be required to record a value independent of their private inputs. As it turns out, the protocols are not altered, but some restrictions are imposed on the monitor, and some parts of the proof become significantly more involved.

We allow that some part of the state of the system can be kept "locked." This part, which we shall call the *locked state*, cannot be used in the system (until it is unlocked). The requirement on the monitor is that it does not have access to the part of the system state if it is locked at the point the monitor is required to record a value; it will have to record a value based on the rest of the system, which we shall call the *open state*.

Technically, the locked state corresponding to a protocol execution is defined at the beginning of that execution: it is the maximal part of the system state, not including any of the adversary's state, such that the *distribution* of the rest of the system state at the recording point is *independent* of it. Note that the independence requirement implies in particular that the probability of unlocking the state before the monitor finishes recording, is zero (unless the locked state is completely predictable *a priori* from the open state).

We do allow the locked state to evolve, as long as the independence is maintained (in particular, no information should pass between the locked state and the open state). Further, for full generality, we allow the locked state to be randomized: i.e., its value is a random variable. However, we shall require that this random variable is efficiently sampleable (which is implied by the assumption that the non-adverserial part of the system

---

[3] The monitor's recorded value is independent of as yet unsampled randomness in the system. So if the client's input is only a freshly sampled random value, as is the case in a ZK proof or coin-tossing protocol, this issue does not arise.

is PPT). In particular all the "future" randomness, i.e., randomness which is sampled after the monitor finishes recording, can be considered part of the locked state.[4]

As indicated earlier, the reason we allow the notion of a locked state in our framework has to do with the meaningfulness of the two-party computation scenario. With the modification sketched above in place, we can allow the client to keep its input locked, and then even the monitor does not get to see it, before recording the other party's input.

However, note that to keep an input locked, it can never be used in the system at all (until it is unlocked).[5] This is because the monitor is computationally unbounded. Note that this is related to the problem of malleability: if it was used in the system previously, somehow that can be mauled and used to make a commitment related to it. (It is an interesting problem to relax this information theoretic locking constraint to a computational equivalent.) However, interestingly we do avoid the problem of malleability while *opening* a commitment: the locked state is allowed to be unlocked *before the commitment is opened*. Indeed, if the locked states are to be kept locked until after the protocol terminates completely, restricting the monitor to the rest of the system state is automatic. But to be useful, we need to allow locked states which can be opened after the monitor records its value, but before the protocol terminates.

In work under progress we show how to prove that in all the monitored functionalities we use, the monitors can be required not to inspect the locked state of the system. Surprisingly, this complicates the construction of the monitor and the proofs considerably. Below we sketch the changes in the proof in the case of $\mathcal{F}_{\widetilde{\text{COM}}}$.

Lemma 8. *For any polynomials (in the security parameter k) τ and Π, under assumption A1, there is a monitor satisfying the requirements specified by $\langle \mathcal{F}_{\widetilde{\text{COM}}} \rangle$ which does not inspect the locked state of the system, such that the probability of the monitor raising an alarm within time τ is less than $1/\Pi$.*

Proof (sketch): When $\mathcal{F}_{\widetilde{\text{COM}}}$ sends the commit message the monitor $\mathfrak{M}_{\tau,\Pi}$ must record a bit b^* internally. First, we sketch how $\mathfrak{M}_{\tau,\Pi}$ does this. As before, the basic idea is for the monitor to look ahead in the system, and record the more likely bit that the sender will ever reveal; if the sender can reveal to both bits with significant probability, a reduction can be used to obtain a circuit for finding collisions in the hash function. But note that here $\mathfrak{M}_{\tau,\Pi}$ does not know the value of the locked state, and so it cannot calculate the bit as above. However, we can show that for no two values for the locked state, can the sender feasibly reveal the commitment in different ways. Intuitively then, the monitor can use an arbitrary value for the locked state and use it to carry out the calculation. However, there are a couple of problems with this. Firstly, revealing can

---

[4] Incidentally, in the use of semi-functionalities in [19], the only locked state is future randomness. However this is an especially simple special case, taken care of by the original proof there. [19] does not introduce or require a generalization. As it turns out generalizing to other locked states complicates our arguments considerably.

[5] In other words, the inputs are for *one time* use only. After that if it is used as a client input in a server-client computation protocol, there is no guarantee that the server's input will be independent of that input. This is a significant limitation. However note that a client's input for a "server-client" computation, with a corrupt server is the *last time* it can be used secretly, as the computation gives the client's input to the server.

depend not only on the open state of the system at the end of commitment, but also on the locked state, as it might be unlocked after the commit phase is over. In particular, for certain values of the locked state (and open state), the sender might never complete the reveal phase. So using a single value of the locked state will not suffice. The second problem is that while $\mathfrak{M}_{\tau,\Pi}$ is computationally unbounded, the reduction to finding collision should use a polynomial sized circuit. This circuit will need to be given the value(s) of the locked state with which it will emulate the system. Further, the circuit will obtain as input the random challenge in the commitment. Thus, the value(s) of the locked state that it obtains should be defined prior to seeing the random challenge.

Nevertheless, we show how to define *polynomially many values for the locked state of the system, based only on the open state of the system*, and obtain a bit b^* using just these values. To show that the probability of $\mathfrak{M}_{\tau,\Pi}$ raising an alarm within time τ is less than $1/\Pi$, we show that otherwise we can give a polynomial sized circuit (with the above mentioned values of the locked states builtin) which can find a collision in our hash function for a random challenge with significant possibility. ∎

The construction of the monitor for $\mathcal{F}_{\widetilde{\text{CAP}}}$ is also changed in a similar fashion. However, since the monitor in this case is defined based on an extractor, and the extractor itself will need to be modified to take polynomially many values of the locked state, the proof is much more involved. The monitors for the $\mathcal{F}_{\widetilde{\text{ZK}}}$ and $\mathcal{F}_{\widetilde{\text{CSC}}}$ need to be modified too. However since their description and proof is based on those of $\mathcal{F}_{\widetilde{\text{COM}}}$ and $\mathcal{F}_{\widetilde{\text{CAP}}}$ respectively they do not involve much change.

7 Conclusion

We introduced a framework of Monitored Functionalities, which provides a way to define and prove relaxed (but ES) security guarantees for (relatively simple) protocols. We also introduced a restricted model called the Client-Server model, which allows simpler protocols to be secure, and potentially under simpler computational assumptions. Both these relaxations, we believe, would help in further exploring Environmental Security (Network-Aware Security).

However, the applicability of the security guarantees from this work are somewhat limited. It is an open problem to work around these limitations, while still maintaining the relaxed nature of the security requirement so that simple protocols are possible. We suggest restricting the computational powers of the monitors (but still giving them more power than the players) as a useful direction.

There are many other ways in which this line of research can be furthered. It is a challenge to try and base these results on more conventional computational assumptions, without setups. On the other hand it should be relatively simpler to allow setups and replace the use of gES model here, by the ES/UC model. A general direction to pursue is to use Monitored Functionalities or other similar notions to give some security guarantee to many simple, efficient and intuitively secure protocols currently used in practice.

References

1. B. Barak. How to Go Beyond the Black-Box Simulation Barrier. FOCS 2001: 106-115.
2. Boaz Barak. Constant-Round Coin-Tossing with a Man in the Middle or Realizing the Shared Random String Model. FOCS 2002: 345-355.
3. R. Canetti. Universally composable security: A new paradigm for cryptographic protocols. FOCS 2001: 136-145.
4. R. Canetti and M. Fischlin. Universally composable commitments. Crypto 2001: 19-40.
5. Ran Canetti and Hugo Krawczyk. Universally Composable Notions of Key Exchange and Secure Channels. EuroCrypt 2002: 337-351.
6. R. Canetti, E. Kushilevitz, and Y. Lindell. On the limitations of universally composable two-party computation without set-up assumptions. EuroCrypt 2003: 68-86.
7. R. Canetti, Y. Lindell, R. Ostrovsky, and A. Sahai. Universally composable two-party and multi-party secure computation. STOC 2002: 494-503.
8. Danny Dolev, Cynthia Dwork, Moni Naor. Nonmalleable Cryptography. SIAM J. Comput. 30(2) 2000: 391-437.
9. Cynthia Dwork, Moni Naor, Amit Sahai. Concurrent Zero-Knowledge. STOC 1998: 409-418.
10. S. Goldwasser, Y. Lindell. Secure Computation without Agreement. DISC 2002: 17-32.
11. Joe Kilian, Erez Petrank. Concurrent and resettable zero-knowledge in poly-loalgorithm rounds. STOC 2001: 560-569.
12. Yehuda Lindell. Bounded-concurrent secure two-party computation without setup assumptions. STOC 2003: 683-692.
13. Tatsuaki Okamoto. An Extension of Zero-Knowledge Proofs and Its Applications. AsiaCrypt 1991: 368-381.
14. Rafael Pass. Simulation in Quasi-Polynomial Time, and Its Application to Protocol Composition. EuroCrypt 2003: 160-176.
15. Rafael Pass. Bounded-Concurrent Secure Multi-Party Computation with a Dishonest Majority. STOC 2004: 232-241.
16. Rafael Pass, Alon Rosen. Bounded-Concurrent Secure Two-Party Computation in a Constant Number of Rounds. FOCS 2003: 404-413.
17. Birgit Pfitzmann, Michael Waidner. Composition and integrity preservation of secure reactive systems. ACM Conference on Computer and Communications Security 2000: 245-254.
18. Manoj Prabhakaran, Alon Rosen, Amit Sahai. Concurrent Zero Knowledge with Logarithmic Round-Complexity. FOCS 2002: 366-375.
19. Manoj Prabhakaran, Amit Sahai. New Notions of Security: Achieving Universal Composability without Trusted Setup. STOC 2004: 242-251.
20. Ransom Richardson, Joe Kilian. On the Concurrent Composition of Zero-Knowledge Proofs. EuroCrypt 1999: 415-431.
21. Amit Sahai. Non-malleable Non-interactive Zero Knowledge and Adaptive Chosen Ciphertext Security. FOCS 1999: 543-553.

Handling Expected Polynomial-Time Strategies in Simulation-Based Security Proofs

Jonathan Katz[1] and Yehuda Lindell[2],[*]

[1] Department of Computer Science, University of Maryland, USA
jkatz@cs.umd.edu
[2] Department of Computer Science, Bar-Ilan University, Israel
lindell@cs.biu.ac.il

Abstract. The standard class of adversaries considered in cryptography is that of *strict* polynomial-time probabilistic machines (or circuits). However, *expected* polynomial-time machines are often also considered. For example, there are many zero-knowledge protocols for which the only simulation techniques known run in expected (and not strict) polynomial-time. In addition, it has been shown that expected polynomial-time simulation is *essential* for achieving constant-round black-box zero-knowledge protocols. This reliance on expected polynomial-time simulation introduces a number of conceptual and technical difficulties. In this paper, we develop techniques for dealing with expected polynomial-time adversaries in the context of simulation-based security proofs.

1 Introduction

Informally speaking, the simulation paradigm (introduced in [15]) states that a protocol is secure if the adversary's view in a real protocol execution can be generated solely from the information that it legitimately possesses (i.e., its input and output). The implication of this statement is that the adversary learns nothing from the protocol *execution*, since everything that the adversary sees in such an execution could be generated by the adversary itself. This paradigm can be instantiated in a number of different ways, where the differences that we refer to here relate to the complexity of the real adversary and the complexity of the simulator that generates the adversary's view.

The most straightforward way of instantiating the simulation paradigm is to require that for every strict polynomial-time adversary there exists a strict polynomial-time simulator that generates the required view. However, in many cases it is not known how to construct such simulators; rather, it is shown that for every *strict* polynomial-time adversary there exists an *expected* polynomial-time simulator that generates the required view. Essentially, this instantiation of the simulation paradigm has become the default one (at least for zero-knowledge). This reliance on expected polynomial-time simulation is problematic for the following reasons:

---

[*] Most of this work was carried out while the author was at IBM T.J. Watson.

1. **Aesthetic Considerations:** The intuition behind the simulation paradigm is that anything an adversary can learn from its interaction in a real protocol execution, it could also learn given only the input and the output. This follows because the adversary can run the simulator itself and thus obtain a view that is essentially the same as its view in a real execution. However, if the adversary is only allowed to run in strict polynomial-time while the simulator may run in expected polynomial-time, then the adversary *cannot* run the simulator (because it doesn't have enough time). One immediate solution to this problem is to allow the adversary to run in expected polynomial-time as well. However, as we will see in Section 1.1 below, this turns out to be problematic for technical reasons.

2. **Technical Considerations (Composition):** Consider the case that a secure protocol π calls a secure subprotocol ρ. Furthermore, both π and ρ are proven secure for strict polynomial-time adversaries using expected polynomial-time simulation. (Here, this means that π is proven secure under the assumption that ρ is replaced by some ideal function evaluation.) Now, the typical way of proving that π is secure when it calls the real subprotocol ρ is to first replace ρ with a simulated version, and then prove the security of π. However, this strategy will fail since it yields an *expected polynomial-time* adversary for π (because the adversary for π actually runs an internal expected polynomial-time simulation of ρ); yet π is proven secure only for strict polynomial-time adversaries.

In order to stress the implications of this difficulty, consider the following natural protocol. The parties first run a coin-tossing protocol (that uses expected polynomial-time simulation) in order to generate a common random string. Following this, the parties run a protocol that is secure in the common random string model (in this model, some trusted party provides both parties with the same uniformly distributed string). If the protocol that is designed for the common random string model is proven secure with respect to strict polynomial-time adversaries (which is usually the case), then the security of the coin-tossing protocol does not imply that the larger protocol is secure. The reason for this "gap" is the fact that simulation of the coin-tossing protocol yields an expected polynomial-time adversary, in the presence of which the protocol in the common random string model may not be secure. We remark that – seemingly due, at least in part, to these difficulties – all simulation-based composition theorems of which we are aware (e.g., [14, 4, 5]) deal only with the case of protocols proven secure via *strict* polynomial-time simulation.

In conclusion, expected polynomial-time simulation is currently a fact of life when it comes to proving the security of many cryptographic protocols. However, this causes difficulties especially when a protocol proven secure using expected polynomial-time simulation is used as a subprotocol.

1.1 Potential Ways of Resolving the Difficulties

There are at least two possible ways of dealing with the difficulties raised above:

1. Require Simulators to be "as Powerful" as Adversaries: One way of resolving the above difficulties is to require simulators and adversaries to lie in the same complexity class. Here, there are two natural choices: **(a)** require both the adversary and the simulator to run in STRICT polynomial-time, or **(b)** allow both the adversary and the simulator to run in EXPECTED polynomial-time.

Limitations of the first choice (requiring STRICT polynomial-time for both adversary and simulator) were demonstrated in [3], who show that there do not exist *constant-round* zero-knowledge protocols with black-box simulators running in strict polynomial time. We note that non black-box simulation strategies running in strict polynomial-time are known to exist [1, 2]. However, all known "highly efficient" protocols are black-box. Thus, given our current knowledge, strict polynomial-time simulation techniques still pose a limitation on efficiency.

Before considering the second choice, where both simulators and adversaries run in EXPECTED polynomial-time, we briefly address the issue of defining expected polynomial-time adversaries. Loosely speaking, Feige [7] defined that an adversary $\mathcal{A}$ attacking a protocol π runs in expected polynomial-time if it runs in expected polynomial-time when interacting with the *honest parties running* π. Here, $\mathcal{A}$ may run for an unbounded amount of time when interacting with other machines (for example, an adversarial verifier for zero-knowledge needs only run in expected polynomial-time when interacting with the honest prover). The justification for such a definition is that the goal of an adversary is to attack an honest party. Therefore, any strategy that is "efficient" when interacting with an honest party is "feasible". We call this notion **expected polynomial-time with respect to the protocol** π. A more stringent definition, advocated by Goldreich [9], requires the adversary to run in expected polynomial-time when interacting with *any interactive machine*. We call this notion **expected polynomial-time in any interaction**. Clearly, any machine that is expected polynomial-time in any interaction is also expected polynomial-time with respect to any protocol π; it is also not hard to see that the converse is not true. Thus, the second notion defines a strictly smaller set of adversaries than the first.

We are now ready to discuss the implementation of the simulation paradigm in which both the adversary and the simulator run in expected polynomial-time. Feige [7] showed that the known simulation strategies for *computational* zero-knowledge all fail when considering adversaries that run in expected polynomial-time *with respect to the protocol*. In contrast, it was shown by [16–Appendix A.1] that the Feige-Shamir zero-knowledge argument system [7, 8] remains both zero-knowledge and an argument of knowledge even when the adversarial party runs in expected polynomial-time *in any interaction*. (We stress that the result of [16] does *not* hold for adversaries that run in expected polynomial-time with respect to the protocol.) It was further demonstrated by [16–Appendix A.2] that the known simulator for the Goldreich-Kahan zero-knowledge *proof*[1] system [11] does *not* remain zero-knowledge for adversaries that run in expected

---

[1] Recall that in a proof system soundness holds even for all-powerful provers, whereas in an argument system it holds only for polynomial-time provers.

polynomial-time in any interaction (and so likewise for expected polynomial-time with respect to the protocol). Furthermore, there is no *computational proof system* that is known to remain zero-knowledge for adversaries that run in expected polynomial-time (under any definition). We therefore conclude that allowing both the adversary and the simulator to run in EXPECTED polynomial-time is problematic *because we simply don't know how to construct simulators for such adversaries*. This is in contrast to the case when both the adversary and the simulator run in strict polynomial-time which, as we have mentioned, suffers from limitations which are inherent.

We remark that requiring simulators to be "as powerful" as adversaries addresses not only the aesthetic difficulty raised above, but also the issue of composition. This is due to the fact that once the simulator lies in the same class as the adversary, the general strategy for proving secure composition (as sketched above) is a viable one.

2. Prove a Direct Composition Theorem: A second and incomparable approach addresses the technical issue of protocol composition, but does not deal with the above-mentioned aesthetic considerations. (Arguably, we can live more easily without aesthetics than without protocol composition.) In this approach, a composition theorem of the following type is proven: If two protocols π and ρ are both proven secure for strict polynomial-time adversaries while using expected polynomial-time simulation, then the composition of π with ρ is also secure for strict polynomial-time adversaries while using expected polynomial-time simulation. Such an approach may be pursued independently of the previous approach, and is worthwhile since many known protocols only satisfy the "strict/expected" notion of security. Namely, even if it is possible to construct protocols that are secure when both the adversary and the simulator run in expected polynomial-time, one may still want to use existing protocols that have been proven secure only for adversaries that run in strict polynomial-time (while using expected polynomial-time simulation).

1.2 Our Results

The main focus of this paper is to develop techniques for working with expected polynomial-time adversaries and simulation. We take the *first steps* in this direction and present two incomparable results, corresponding to the two approaches discussed in the previous section.

1. Simulation for Expected Polynomial-Time Adversaries. Our first result focuses on achieving expected polynomial-time simulation for expected polynomial-time adversaries. Before describing the result, we discuss one of the central technical problems that arises when dealing with expected polynomial-time adversaries: expected polynomial-time machines are not closed under "oracle composition". In more detail, let A be an oracle machine belonging to a class $\mathcal{C}$ and let B be any machine that also belongs to class $\mathcal{C}$. Then, we say the class $\mathcal{C}$ is closed under oracle composition if the machine A^B also belongs to $\mathcal{C}$ (when counting the steps of both A and B in their executions). This property of closure under oracle composition is important for black-box simulations (where machine A is the simulator and machine B is the adversary), and holds for the class of strict polynomial-time machines. However, the class of expected polynomial-time machines is *not* closed under oracle composition. To see this, consider the following two machines:

1. Machine A queries its oracle with the message 0 and receives back a message x. Next, A queries its oracle with x and halts.
2. Machine B receives an input q. If q equals its random tape r (where $|q| = |r| = k$, the security parameter), then B runs for 2^k steps and halts. Otherwise, it replies with r and halts.

Machine A runs in strict (and thus expected) polynomial-time. Likewise, machine B runs in expected polynomial-time because the probability (over choice of random tapes) that $q = r$ is 2^{-k} (and thus B runs for 2^k steps with probability 2^{-k}). However, the composed machine A^B always runs for more than 2^k steps. We therefore conclude that the composition of an expected polynomial-time simulator with an expected polynomial-time adversary may not yield an expected polynomial-time simulation. We stress that this problem is not just hypothetical. Rather, as we have mentioned earlier, many concrete protocols and expected polynomial-time simulators suffer from this problem [7, 16]. Furthermore, simple solutions, like truncating the execution after some polynomial number of steps, do not work; see [3] for some discussion.

Ideally, we would like to present conditions under which closure under oracle composition can be achieved for expected polynomial-time machines. This would allow us to construct an expected polynomial-time simulator that fulfills the conditions, and immediately derive simulation even when the adversary runs in expected polynomial-time. Toward this goal, we prove a theorem that shows how to automatically convert a class of simulators (characterized by a certain property) so that they remain expected polynomial-time even if the adversary runs in expected polynomial-time. More precisely, let S be a black-box simulator with the following two properties:

1. S runs in expected polynomial-time when given *any* oracle $\mathcal{A}$ (even if $\mathcal{A}$ is all-powerful). We stress that here we do not include $\mathcal{A}$'s running time in the complexity of S. We also remark that most known black-box simulators have this property.
2. Every oracle query that S makes to its oracle $\mathcal{A}$ during its simulation is "strongly indistinguishable" to $\mathcal{A}$ from some partial view of a real protocol execution. By "strongly indistinguishable", we mean that the oracle query is computationally indistinguishable for circuits of size $\alpha(k)$, for some superpolynomial function $\alpha(k) = k^{\omega(1)}$. We remark that by making an appropriate $\alpha(k)$-hardness assumption, most known black-box simulators can be easily modified so that they fulfill this property.

Let $\mathcal{A}$ be an expected polynomial-time adversary and let S be a simulator that fulfills the above properties. We show that by truncating $S^{\mathcal{A}}$ at $\alpha(k)$ steps, the resulting machine is a "good" simulator that runs in expected polynomial-time. We thus obtain a type of closure under oracle composition, as desired.

An important corollary of this theorem is a proof that, under mildly superpolynomial hardness assumptions, there exist computational zero-knowledge *proofs* for all $\mathcal{NP}$ that remain zero-knowledge even if the adversarial verifier runs in expected polynomial-time. As we have mentioned above, prior to this work no such proof system was known to exist. We note that our corollary has the following caveat: Our simulator for the zero-knowledge proof runs in expected polynomial-time only when given a statement x that is in the language L; see Section 3.3 for more details.[2]

We note that the above result does not achieve closure under oracle composition in its utmost generality, because it holds only for the above-described class of simulators. Nevertheless, many (if not most) known simulators can be modified so that they belong to this class. Furthermore, it is impossible to prove closure for *all* simulators, because closure under oracle composition for expected polynomial-time machines simply does not hold. Of course, it may still be possible to widen the class of simulators for which closure holds, and to remove the superpolynomial hardness assumptions.

2. A Composition Theorem. The above theorem holds for a restricted class of simulators, but achieves generality with respect to closure under oracle composition. Our second result is the opposite in that it holds for *all* black-box simulators, but relates only to a specific type of composition. Specifically, under a superpolynomial hardness assumption, we prove an analogue of the modular sequential composition theorem of Canetti [4] for protocols that are proven secure for strict polynomial-time adversaries using expected polynomial-time simulation. Loosely speaking, the modular sequential composition theorem of [4] states that if a secure protocol π contains *sequential* ideal calls to some functionalities, then it remains secure even when these ideal calls are replaced by *sequential* executions of subprotocols that securely realize the functionalities. The original result of [4] was previously known to hold only for protocols proven secure via *strict* polynomial-time simulation (in fact, in the full version we show that the proof of [4] fails in general for protocols proven secure via expected polynomial-time simulation). In contrast, our analogous result holds even if these protocols are proven secure using *expected* polynomial-time simulation (and only for *strict* polynomial-time adversaries). However, we also note that the proof of [4] requires no hardness assumptions, in contrast to ours which requires a superpolynomial hardness assumption.

We remark that both our results hold even for the larger class of adversaries running in expected polynomial-time with respect to the protocol under consideration [7].

Related Work. The problem of simulation in expected polynomial-time was first posed by [7]; here we provide the first (partial) answers to some of the

---

[2] Standard definitions require a simulator to generate a distribution that is indistinguishable from the view of the verifier *only* when it receives a statement $x \in L$. However, polynomial-time machines are typically required to run in polynomial-time for *all* inputs (i.e., even for $x \notin L$).

open questions posed there. The existence of constant-round zero-knowledge arguments with strict polynomial-time (non black-box) simulation was demonstrated in [1, 2]. The feasibility of obtaining constant-round arguments *of knowledge* with strict polynomial-time extraction was then shown in [3]. They also showed that such protocols do not exist when the simulator or extractor is black-box. Thus, the protocols of [1, 2, 3] provide an alternative to expected polynomial-time simulation. In this work, we take a different approach and develop techniques for working with expected polynomial-time simulation. This has the advantage of not ruling out the many protocols (including most of the highly efficient protocols) that rely on expected polynomial-time simulation.

2   Definitions and Preliminaries

The security parameter is denoted by k; for conciseness, we equate the security parameter with the input length. (We therefore consider security for "sufficiently long inputs".) We denote by $A(x, z, r)$ the output of machine A on input x, auxiliary input z, and random coins r. The running time of A is measured in terms of the length of its first input x (where $|x| = k$), and the exact running time of the deterministic computation $A(x, z, r)$ is denoted by $\text{time}_A(A(x, z, r))$. A runs in strict polynomial time if there is a polynomial $p(\cdot)$ such that for all x, z, and *all* r, it holds that $\text{time}_A(A(x, z, r)) \leq p(|x|)$. A runs in expected polynomial time if there is a polynomial $p(\cdot)$ such that for all x and z, it holds that $\mathbf{Exp}_r[\text{time}_A(A(x, z, r))] \leq p(|x|)$.

Running Time for ITMs. If A is an interactive Turing machine (ITM), we let $A(x, z, r; \cdot)$ denote the "next message function" of A on inputs x, z, and random coins r. The ITM A runs in strict polynomial time if there is a polynomial $p(\cdot)$ such that for all x, z, r, and any sequence of messages $\overline{m}$, it holds that $\text{time}_A(A(x, z, r; \overline{m})) \leq p(|x|)$.

Defining expected polynomial-time ITMs is more complicated, and at least two such definitions have been considered. We first present the definition of Feige [7]. As mentioned in the Introduction, the idea behind this definition is that any adversarial strategy that is efficient when run against the specified target is feasible. Thus, the running-time of an adversary when interacting with an arbitrary ITM (that is not the honest party under attack) is irrelevant. Informally, an ITM A is therefore said to run in expected polynomial-time with respect to a particular protocol π if there exists a polynomial $p(\cdot)$ such that for all inputs, the expected running time of A when interacting with *honest parties running π* is at most $p(|x|)$. (The expectation here is taken over the random coins of both A and the honest parties.) More formally, let $\text{time}_A(\langle A(x, z_A, r), B(y, z_B, s) \rangle)$ denote the running time of A with input x, auxiliary input z_A, and random coins r, when interacting with B having input y, auxiliary input z_B, and random coins s. Then:

Definition 1. An ITM A runs in expected polynomial-time with respect to an ITM B if there exists a polynomial $p(\cdot)$ such that for all x, y with $|x| = |y|$ and all auxiliary inputs $z_A, z_B \in \{0,1\}^*$, the following holds:

$$\mathbf{Exp}_{r,s}\left[\text{time}_A(\langle A(x, z_A, r), B(y, z_B, s)\rangle)\right] \leq p(|x|).$$

Let $\pi = (P_1, P_2)$ be a two-party protocol. Then an adversary $\mathcal{A}$ runs in expected polynomial-time with respect to π if it runs in expected polynomial-time with respect to P_1 and in expected polynomial-time with respect to P_2.

The above definition relates to the case of two-party protocols. The extension to the multiparty case is obtained by considering the expected running-time of $\mathcal{A}$ when interacting (simultaneously) with every subset of honest parties.

As we have mentioned above, the fact that an adversary $\mathcal{A}$ runs in expected polynomial-time with respect to a protocol π means nothing about its running time when it interacts with other machines. A definition of the above sort makes sense in a cryptographic context, but is arguably a somewhat strange way of defining a "complexity class". An alternative approach advocated by Goldreich [9] therefore states that an ITM runs in expected polynomial time if there exists a polynomial $p(\cdot)$ such that for all inputs, the expected running time of A when interacting with *any* (even all powerful) ITM is at most $p(|x|)$. Here, the expectation is taken over the random coins of A only. In such a case, we say that A runs in expected polynomial-time in any interaction. More formally:

Definition 2. An ITM $\mathcal{A}$ runs in expected polynomial-time in any interaction if for every ITM B it holds that A runs in expected polynomial-time with respect to B (as defined in Definition 1).

It is immediate that if an ITM A runs in expected polynomial-time in any interaction, then A also runs in expected polynomial-time with respect to any protocol π. Furthermore, it is not difficult to show that for many protocols π, the class of adversaries running in expected polynomial-time with respect to π is *strictly larger* than the class of adversaries running in expected polynomial-time in any interaction. Since all our results hold even with respect to the stronger definition, and we view it as preferable in the cryptographic context, we adopt Definition 1 in this paper.

Expected Polynomial-Time Oracle Machines. Let A be an oracle machine that receives oracle access to an ITM B. In the execution of A with B, denoted by $A^{B(y,z_B,s;\cdot)}(x, z_A, r)$, machine A receives input x, auxiliary-input z_A and random tape r, and provides queries of the form $\overline{m}$ to its oracle which are answered as $B(y, z_B, s; \overline{m})$. We distinguish between two notions of running time for an oracle machine A^B:

1. $\text{time}_A(A^{B(y,z_B,s;\cdot)}(x, z_A, r))$ denotes the exact running time of A on input x, auxiliary-input z_A, and random tape r when interacting with the oracle $B(y, z_B, s; \cdot)$, *counting calls to B as a single step* (i.e., we only "look" at the steps taken by A).

2. $\text{time}_{A+B}(A^{B(y,z_B,s;\cdot)}(x, z_A, r))$ denotes the total running time of *both* A and B in the analogous execution. *Here, the steps taken by B to answer A's queries are also counted.*

Given the above, we can define expected polynomial-time oracle machines. An oracle machine A is said to run in expected polynomial-time if there exists a polynomial $p(\cdot)$ such that for every (even all powerful) machine B, all sufficiently-long inputs x, and every auxiliary input z, $\mathbf{Exp}_r[\text{time}_A(A^B(x, z, r))] \leq p(|x|)$. Likewise, the composed machine A^B is said to run in expected polynomial-time if there exists a polynomial $p(\cdot)$ such that for all sufficiently-long inputs x and y with $|x| = |y|$, and all auxiliary inputs z_A and z_B, it holds that $\mathbf{Exp}_{r,s}[\text{time}_{A+B}(A^{B(y,z_B,s)}(x, z_A, r))] \leq p(|x|)$. Note that for any strict polynomial-time B, if A runs in expected polynomial-time (not counting the steps of B) then so does A^B (where B's steps are counted). We stress, however, that this *does not necessarily hold* when B runs in expected polynomial time (under either definition considered earlier).

Requiring an expected polynomial-time oracle machine to run in the same (expected) amount of time when interacting with any machine B, even one which is computationally unbounded, seems to be overly stringent. However, all black-box simulators that we are aware of fulfill this condition. This extra condition is also *needed* for our results. We also remark that our definition of expected polynomial-time oracle and composed machines is *asymptotic*. That is, the machine is only required to run in (expected) time $p(|x|)$ for all long enough x's. As long as all machines considered halt on all inputs (and all random tapes), this is equivalent to the standard notion. (Indeed, we will assume this "halting condition" for all machines.)

3   Simulation for Expected Polynomial-Time Adversaries

In this section, we show how protocols proven secure against *strict* poly-time adversaries using a certain class of black-box simulation can in fact be proven secure against *expected* poly-time adversaries as well.

3.1   Preliminaries

As we have mentioned, the results of this section hold for a certain class of black-box simulators. We begin with a high-level description of secure computation, and then define the class of simulators. For the sake of simplicity, we present the results here for the case of two-party protocols. The extension to multiparty protocols is straightforward.

Secure Two-Party Computation. We provide a very brief and informal overview of the definition of security for two-party computation. For more details, see [4, 10]. In the setting of two-party computation, two parties wish to jointly compute a (possibly probabilistic) functionality $f : \{0,1\}^* \times \{0,1\}^* \to \{0,1\}^* \times \{0,1\}^*$, where $f = (f_1, f_2)$. That is, upon respective inputs x and y, the parties wish to compute $f(x,y)$ so that party P_1 receives $f_1(x,y)$ and party P_2 receives $f_2(x,y)$. Furthermore, the parties wish to ensure that nothing more than the output is revealed and that the function is correctly computed, even if one of the parties behaves adversarially. These requirements (and others) are formalized by comparing a real protocol execution to an ideal execution involving a trusted party. In an ideal execution with f, the parties send their inputs x and y to a trusted party who computes $f(x,y)$ and sends $f_1(x,y)$ to P_1 and $f_2(x,y)$ to P_2. Of course, the adversary who controls one of the parties can choose to send any input it wishes to the trusted party.[3] In contrast, in a real execution the parties P_1 and P_2 run a protocol π, where one of the parties may be corrupted and thus under the complete control of the adversary $\mathcal{A}$. Informally, we say a protocol π is secure if for every real-model adversary $\mathcal{A}$ interacting with an honest party running π, there exists an ideal-model adversary $\mathcal{S}$ interacting with a trusted party computing f, such that the output of $\mathcal{A}$ and the honest party in the real model is computationally indistinguishable from the output of $\mathcal{S}$ and the honest party in the ideal model. We note that in this work we consider *static adversaries* who corrupt one of the parties before the protocol execution begins.

Notation. Let $\pi = (P_1, P_2)$ be a two-party protocol and let f be a two-party functionality. We denote by $\text{REAL}_{\pi,\mathcal{A}}(x,y,z)$ the output of a *real execution of* π where party P_1 has input x, party P_2 has input y, and the adversary $\mathcal{A}$ has input z. Likewise, we denote by $\text{IDEAL}_{f,\mathcal{S}}(x,y,z)$ the output of an *ideal execution with* f where the respective inputs are as above. Since we are interested in black-box simulation, we present the definition for a black-box simulator $\mathcal{S}$:

Definition 3. *(secure computation with black-box simulation):* Let f and π be as above. Protocol π is said to black-box securely compute f *(in the malicious model)* if there exists a non-uniform probabilistic expected polynomial-time oracle machine (ideal adversary/simulator) $\mathcal{S}$ such that for every non-uniform probabilistic polynomial-time real-model adversary $\mathcal{A}$, every non-uniform polynomial-time distinguisher D, every polynomial $p(\cdot)$, all sufficiently-long inputs x and y such that $|x| = |y|$, and all $z \in \{0,1\}^{\text{poly}(|x|)}$,

$$\left| \Pr[D(\text{IDEAL}_{f,\mathcal{S}^{\mathcal{A}(z)}}(x,y,\lambda)) = 1] - \Pr[D(\text{REAL}_{\pi,\mathcal{A}}(x,y,z)) = 1] \right| < \frac{1}{p(|x|)}.$$

We note that $\mathcal{S}$ is an expected polynomial-time oracle machine as defined earlier. That is, for every $\mathcal{A}$ the expected value of $\text{time}_\mathcal{S}(\mathcal{S}^\mathcal{A})$ is polynomial (even if $\mathcal{A}$ is computationally unbounded). To be more exact, however, the running-time of $\mathcal{S}$ may also depend on the messages it receives from the trusted party (and in particular, the random coins used by the trusted party to compute the functionality). We therefore denote by $\text{time}_\mathcal{S}(\text{IDEAL}_{f,\mathcal{S}^{\mathcal{A}(z)}}(x,y,\lambda))$ the running-time of $\mathcal{S}^\mathcal{A}$ here. Adapting the earlier notation, we denote the expected running-time of $\mathcal{S}^\mathcal{A}$ *not counting* $\mathcal{A}$'s steps by $\mathbf{Exp}_s[\text{time}_\mathcal{S}(\text{IDEAL}_{f,\mathcal{S}^{\mathcal{A}(z,r)}(1^{|z|},s)}(x,y,\lambda))]$, and its expected time *counting* $\mathcal{A}$'s steps by $\mathbf{Exp}_{r,s}[\text{time}_{\mathcal{S}+\mathcal{A}}(\text{IDEAL}_{f,\mathcal{S}^{\mathcal{A}(z,r)}(1^{|z|},s)}(x,y,\lambda))]$.

---

[3] The adversary also has control over the delivery of the output from the trusted party to the honest party. Therefore, fairness and output delivery are not guaranteed.

(The expectations above are actually also over the random-coins of the functionality. In this extended abstract, we ignore this issue.)

We now define a stronger notion of simulation which, informally, requires not only that the final output of $\text{IDEAL}_{f,\mathcal{S}^{\mathcal{A}}}$ be indistinguishable from $\text{REAL}_{\pi,\mathcal{A}}$, but also that each *partial* transcript generated during the simulation is indistinguishable from the (corresponding) partial transcript of a real execution of the protocol. Furthermore, we require that indistinguishability holds in a "strong" sense even against algorithms running in some slightly superpolynomial time. We begin by defining the following distributions:

1. $\text{SIM}_{f,\mathcal{S}^{\mathcal{A}}}(x,y,z,r,i)$ is defined by the following experiment: choose a random-tape $s \in_R \{0,1\}^*$ and run $\mathcal{S}^{\mathcal{A}(z,r;\cdot)}(1^{|z|},s)$ in the ideal model with f. Let query_i be the ith oracle query made by $\mathcal{S}$ to $\mathcal{A}$; if no such query is made, then set $\text{query}_i = \bot$. Output query_i.
2. $\text{REAL}_{\pi,\mathcal{A}}(x,y,z,r,i)$ is defined by the following experiment: choose $s \in_R \{0,1\}^*$ and run a real execution where $\mathcal{A}$ has random-tape r and the honest party has random-tape s. Let $\overline{T}$ be the vector of messages sent by the honest party to $\mathcal{A}$ in this execution, and let $\overline{T}_j$ denote the first j messages in $\overline{T}$. Next, run the experiment $\text{SIM}_{f,\mathcal{S}^{\mathcal{A}}}(x,y,z,r,i)$ above (with an independent choice of s) and obtain query_i. If $\text{query}_i = \bot$, then output $\bot$. Otherwise, let j denote the number of messages in query_i, and output $\overline{T}_j$.

We note that the reason for running SIM in the second distribution is just to decide the length of the partial transcript to output. That is, we wish to compare the distribution of query_i to the partial transcript of a real execution *of the appropriate length*. We are now ready for the formal definition.

Definition 4. (α-strong black-box simulation): Let π be a two-party protocol that is secure under black-box simulation, and let $\mathcal{S}$ be a black-box simulator for π. We say that $\mathcal{S}$ is an α-strong black-box simulator for π (and say that π is secure under α-strong black-box simulation), if for every strict polynomial-time adversary $\mathcal{A}$, every non-uniform algorithm D running in time at most $\alpha(k)$, all $i \in \mathbf{N}$, all sufficiently large x and y, and all $z, r \in \{0,1\}^*$,

$$\left| \Pr[D(\text{SIM}_{f,\mathcal{S}^{\mathcal{A}}}(x,y,z,r,i)) = 1] - \Pr[D(\text{REAL}_{\pi,\mathcal{A}}(x,y,z,r,i)) = 1] \right| < \frac{1}{\alpha(k)}.$$

If the above holds for adversaries $\mathcal{A}$ that are expected polynomial-time with respect to π, then we say that π is secure under α-strong black-box simulation for expected polynomial-time adversaries.

Extended Black-Box Simulation. Finally, we introduce a generalization of black-box simulation in which the black-box simulator is allowed to *truncate* its oracle after it exceeds some (poly-time computable) number of steps $\alpha(\cdot)$. We call such a simulator extended black-box. We argue that this generalization is natural in the sense that the simulator still does not "look" at the internal workings of its oracle. We remark that when computing $\text{time}_A(A^B)$, oracle calls are still considered a single step (even if A truncates B after some number of steps).

Of course, $\text{time}_{A+B}(A^B)$ also remains unchanged. We note that by requiring $\alpha(\cdot)$ to be polynomial-time computable, we ensure that any extended black-box simulator can be implemented by a non black-box simulator.

3.2 Simulation for Expected Polynomial-Time Adversaries

Theorem 5. Let $\alpha(k) = k^{\omega(1)}$ be a superpolynomial function that is poly-time computable, and let π be a protocol that is secure under α-strong (extended) black-box simulation for **strict** polynomial-time adversaries. Then there exists a superpolynomial function $\alpha'(k)$ such that π is secure under α'-strong extended black-box simulation for **expected** polynomial-time adversaries.

Proof: The idea behind the proof of this theorem is as follows. Since each query made by the α-strong simulator $\mathcal{S}$ to the real adversary $\mathcal{A}$ is indistinguishable from a partial real transcript *even for circuits of size $\alpha(k)$*, it follows that as long as $\mathcal{A}$ does not exceed $\alpha(k)$ steps, it cannot behave in a noticeably different way when receiving an oracle query or a real partial transcript. In particular, it cannot run longer when it receives an oracle query than it would run when interacting in a real protocol execution, and we know that it runs in expected polynomial-time in the latter case. We therefore construct a new simulator $\tilde{\mathcal{S}}$ that works in the same way as $\mathcal{S}$, except that it halts if $\mathcal{A}$ ever exceeds $O(\alpha(k))$ steps when answering a query. This enables us to prevent $\mathcal{A}$ from ever running for a very long time (something which can cause its expected running-time to be superpolynomial). Furthermore, by what we have claimed above, $\mathcal{A}$ will behave in almost the same way as before, because it can exceed $\alpha(k)$ steps only with probability that is inversely proportional to $\alpha(k)$. This will suffice for us to show that the new simulator is expected polynomial-time even if $\mathcal{A}$ is expected polynomial-time. Of course, we must also prove that the new simulation is no different than the old one. This follows again from the fact that $\mathcal{A}$ must behave in the "same way" as in a real execution, as long as $\alpha(k)$ steps are not exceeded. We now proceed with the actual proof.

Throughout the proof, we let k denote the length of x. Let $\mathcal{S}$ be the α-strong black-box simulator for π that is assumed to exist, and define $\hat{\mathcal{A}}$ as the algorithm that behaves exactly as $\mathcal{A}$ except that it outputs $\bot$ if it ever exceeds $\alpha(k)/2$ steps. Then, we construct a new simulator $\hat{\mathcal{S}}$ that receives oracle access to $\mathcal{A}$ and emulates a simulation of $\mathcal{S}$ with $\hat{\mathcal{A}}$. That is, $\hat{\mathcal{S}}$ chooses a random tape $s \in \{0,1\}^*$ and invokes $\mathcal{S}$ with random-tape s. Then, all oracle queries from $\mathcal{S}$ are forwarded by $\hat{\mathcal{S}}$ to its own oracle $\mathcal{A}$ and the oracle replies are returned to $\mathcal{S}$ unless the oracle exceeds $\alpha(k)/2$ steps while answering the query, in which case $\hat{\mathcal{S}}$ returns $\bot$ (thereby emulating $\hat{\mathcal{A}}$). Furthermore, all communication between $\mathcal{S}$ and the trusted party computing f is forwarded unmodified by $\hat{\mathcal{S}}$. We remark that $\hat{\mathcal{S}}$ is an *extended black-box simulator* because it truncates its oracle. (It makes no difference whether $\mathcal{S}$ was extended black-box or not.) We first show that $\hat{\mathcal{S}}$ runs in expected polynomial time, even when $\mathcal{A}$ runs in expected polynomial-time with respect to π.

Claim 6. For every expected polynomial-time adversary $\mathcal{A}$, the composed machine $\hat{\mathcal{S}}^{\mathcal{A}}$ runs in expected polynomial time. That is, for every $\mathcal{A}$ there exists a polynomial $p(\cdot)$ such that all sufficiently large x and y, and all $z \in \{0,1\}^*$, it holds that $\mathbf{Exp}_{r,s}[\text{time}_{\hat{\mathcal{S}}+\mathcal{A}}(\text{IDEAL}_{f,\hat{\mathcal{S}}^{\hat{\mathcal{A}}(z,r)}(1^{|z|},s)}(x,y,\lambda)] \leq p(k)$.

Proof: To prove the claim, first note that the running time of $\hat{\mathcal{S}}$ consists of two components: the steps taken by $\mathcal{S}$ and the steps taken by $\hat{\mathcal{A}}$ in answering all of the oracle queries of $\mathcal{S}$. By the linearity of expectations, it suffices to show that the expectation of each of these components is polynomial. Since $\mathcal{S}$ is an expected polynomial-time oracle machine, its expected running time is polynomial when interacting with *any* oracle (see the end of Section 2). It therefore remains to bound the total number of steps taken by $\hat{\mathcal{A}}$. This is equal to $\mathbf{Exp}_{r,s}[\sum_{i=1}^{\tau} \text{time}^{\mathcal{S}}_{\hat{\mathcal{A}}(z,r)}(i)]$, where τ is a random variable denoting the number of oracle queries made by $\mathcal{S}$, and $\text{time}^{\mathcal{S}}_{\hat{\mathcal{A}}(z,r)}(i)$ is a random variable denoting the running time of $\hat{\mathcal{A}}(z,r)$ in answering the ith query from $\mathcal{S}$. (Note that these random variables may depend on both r and s, and also on the honest party's inputs.) The expected value of τ is polynomial because $\mathcal{S}$ is an expected polynomial-time oracle machine. We now show that the expected value of $\text{time}^{\mathcal{S}}_{\hat{\mathcal{A}}(z,r)}(i)$ is also polynomial for any i. Applying Wald's inequality (see Appendix A) then completes the proof that the expected total number of steps taken by $\hat{\mathcal{A}}$ is polynomial.

For any i, it holds that $\mathbf{Exp}_{r,s}[\text{time}^{\mathcal{S}}_{\hat{\mathcal{A}}(z,r)}(i)] = \mathbf{Exp}_r[\mathbf{Exp}_s[\text{time}^{\mathcal{S}}_{\hat{\mathcal{A}}(z,r)}(i)]]$. Furthermore, since $\hat{\mathcal{A}}$ halts after $\alpha(k)/2$ steps, it follows that for any fixed r,

$$\mathbf{Exp}_s\left[\text{time}^{\mathcal{S}}_{\hat{\mathcal{A}}(z,r)}(i)\right] = \sum_{t=1}^{\alpha(k)/2} t \cdot \Pr_s\left[\text{time}^{\mathcal{S}}_{\hat{\mathcal{A}}(z,r)}(i) = t\right] = \sum_{t=1}^{\alpha(k)/2} \Pr_s\left[\text{time}^{\mathcal{S}}_{\hat{\mathcal{A}}(z,r)}(i) \geq t\right].$$

Notice that the distribution on the message sequence input to $\hat{\mathcal{A}}$ here (namely, the ith query from $\mathcal{S}$) is exactly that given by $\text{SIM}_{f,\mathcal{S}^{\hat{\mathcal{A}}}}(x,y,z,r,i)$. Now, let $\text{time}_{\hat{\mathcal{A}}(z,r)}(i)$ be a random variable denoting the running time of $\hat{\mathcal{A}}(z,r)$ when run on input distributed according to $\text{REAL}_{\pi,\hat{\mathcal{A}}}(x,y,z,r,i)$. (Recall that this is a message of the same length as query_i, that $\hat{\mathcal{A}}$ receives in a real execution.) We first claim that, for large enough x and y, for any z,r,i, and for $t \leq \alpha(k)/2$,

$$\left|\Pr_s[\text{time}_{\hat{\mathcal{A}}(z,r)}(i) \geq t] - \Pr_s[\text{time}^{\mathcal{S}}_{\hat{\mathcal{A}}(z,r)}(i) \geq t]\right| < \frac{1}{\alpha(k)}. \tag{1}$$

This follows because otherwise we obtain a non-uniform distinguisher, in contradiction to the fact that $\mathcal{S}$ is an α-strong black-box simulator. In more detail, given an auxiliary input $z' = (z,r,t)$ with $t \leq \alpha(k)/2$, and a sequence of j messages $\overline{T}_j$ we simply run $\hat{\mathcal{A}}(z,r)$ on message sequence $\overline{T}_j$, and output 1 iff $\hat{\mathcal{A}}$ exceeds t steps. For large enough k, the total running time of this distinguishing algorithm (including the overhead for maintaining a counter and running $\hat{\mathcal{A}}$) is at most $\alpha(k)$. Therefore, by Definition 4, it follows that Eq. (1) holds. We remark that the non-uniformity of Definition 4 is essential here. We thus have that:

$$\sum_{t=1}^{\alpha(k)/2} \Pr_s\left[\text{time}^{\mathcal{S}}_{\hat{\mathcal{A}}(z,r)}(i) \geq t\right] \leq \sum_{t=1}^{\alpha(k)/2} \left(\Pr_s[\text{time}_{\hat{\mathcal{A}}(z,r)}(i) \geq t] + \frac{1}{\alpha(k)}\right)$$

$$= \frac{1}{2} + \sum_{t=1}^{\alpha(k)/2} \Pr_s[\text{time}_{\hat{\mathcal{A}}(z,r)}(i) \geq t], \quad (2)$$

and therefore the expected value of $\text{time}^{\mathcal{S}}_{\hat{\mathcal{A}}(z,r)}(i)$ is bounded by the expression in Eq. (2). Using the simple observations that: **(1)** $\text{time}_{\hat{\mathcal{A}}(z,r)}(i) \leq \text{time}_{\hat{\mathcal{A}}(z,r)}$ (where the latter expression refers to the total running time of $\hat{\mathcal{A}}(z,r)$ in a real execution), and **(2)** $\text{time}_{\hat{\mathcal{A}}(z,r)} \leq \text{time}_{\mathcal{A}(z,r)}$ (because $\hat{\mathcal{A}}$ is truncated whereas $\mathcal{A}$ is not), we see that the expected value of $\text{time}^{\mathcal{S}}_{\hat{\mathcal{A}}(z,r)}(i)$ is bounded by:

$$\frac{1}{2} + \sum_{t=1}^{\alpha(k)/2} \Pr_s[\text{time}_{\mathcal{A}(z,r)} \geq t] \;\leq\; \frac{1}{2} + \mathbf{Exp}_s[\text{time}_{\mathcal{A}(z,r)}]$$

where $\mathbf{Exp}_s[\text{time}_{\mathcal{A}(z,r)}]$ is simply the expected running time of $\mathcal{A}$ in a real protocol execution with the *honest parties*. The fact that $\mathcal{A}$ runs in expected polynomial-time *with respect to* π therefore implies that the expected value of $\text{time}^{\mathcal{S}}_{\hat{\mathcal{A}}(z,r)}(i)$ is polynomial, completing the proof of Claim 6. □

Until now, we have shown that $\hat{\mathcal{S}}$ runs in expected polynomial-time. It remains to show that it is an α'-strong (extended black-box) simulator for expected polynomial-time adversaries, for some superpolynomial function $\alpha'(k)$. First, $\hat{\mathcal{S}}$ is an expected polynomial-time oracle machine because it inherits this from $\mathcal{S}$. Next, we claim that for every expected polynomial-time $\mathcal{A}$, every non-uniform algorithm D running in time at most $\alpha(k)$, all $i \in \mathbf{N}$, all sufficiently large x and y, and all $z, r \in \{0,1\}^*$,

$$\left|\Pr[D(\text{SIM}_{f,\hat{\mathcal{S}}^{\mathcal{A}}}(x,y,z,r,i)) = 1] - \Pr[D(\text{SIM}_{f,\mathcal{S}^{\mathcal{A}}}(x,y,z,r,i)) = 1]\right| < \frac{1}{\alpha''(k)}$$

for some superpolynomial function $\alpha''(k)$. This follows from the facts that **(1)** the composed machine $\hat{\mathcal{S}}^{\mathcal{A}}$ runs in expected polynomial-time, and **(2)** the only time that $\hat{\mathcal{S}}^{\mathcal{A}}$ and $\mathcal{S}^{\mathcal{A}}$ differ is if $\mathcal{A}$ exceeds $\alpha(k)/2$ steps. That is, let $p(k)$ be the expected running time of the composed machine $\hat{\mathcal{S}}^{\mathcal{A}}$. Then, by Markov's inequality, the probability that $\hat{\mathcal{S}}^{\mathcal{A}}$ will exceed $\alpha(k)/2$ steps is at most $2p(k)/\alpha(k)$. Therefore, the statistical difference between $\text{SIM}_{f,\hat{\mathcal{S}}^{\mathcal{A}}}(x,y,z,r,i)$ and $\text{SIM}_{f,\mathcal{S}^{\mathcal{A}}}(x,y,z,r,i)$ is at most $\alpha''(k) \stackrel{\text{def}}{=} 2p(k)/\alpha(k)$. Combining this with the assumption that $\mathcal{S}$ is an α-strong simulator and so $\text{SIM}_{f,\mathcal{S}^{\mathcal{A}}}(x,y,z,r,i)$ can be distinguished from $\text{REAL}_{\pi,\mathcal{A}}(x,y,z,r,i)$ with probability at most $1/\alpha(k)$, we conclude that

$$\left|\Pr[D(\text{SIM}_{f,\hat{\mathcal{S}}^{\mathcal{A}}}(x,y,z,r,i)) = 1] - \Pr[D(\text{REAL}_{\pi,\mathcal{A}}(x,y,z,r,i)) = 1]\right| < \frac{1}{\alpha'(k)}$$

where $\alpha'(k) \stackrel{\text{def}}{=} (1/\alpha(k) + 1/\alpha''(k))^{-1}$. We conclude that $\hat{\mathcal{S}}$ is an α'-strong extended black-box simulator, as required. ■

3.3 Zero-Knowledge Proofs – A Corollary

Consider now the zero-knowledge functionality for an NP-language L. This function is defined by $f(x,x) = (\lambda, \chi_L(x))$, where $\chi_L(x) = 1$ if and only if $x \in L$. A zero-knowledge protocol is a protocol π that securely realizes f for strict polynomial-time adversaries. Now, for the sake of concreteness, consider the zero-knowledge protocol of Goldreich, Micali, and Wigderson [13]. Assuming the existence of commitment schemes that are hiding for circuits of size $\alpha(k)$, it is easy to verify that the black-box simulator provided by [13] is α-strong for strict polynomial-time adversaries. Therefore, by applying Theorem 5, we obtain that the protocol of [13] is also black-box secure for adversaries that run in expected polynomial-time with respect to the protocol. The soundness condition is unaffected by the above. We therefore obtain the *first* computational zero-knowledge *proof system* that remains zero-knowledge for expected polynomial-time adversaries (with respect to either of the definitions in Section 2).[4] Thus, as a corollary of Theorem 5, we partially resolve the open questions from [7,16] discussed in the Introduction. (The result is only "partial" because we need superpolynomial hardness assumptions, and due to the caveat below.)

We remark that there is a subtle, yet important, caveat to the above. The simulator is only α-strong in the case that the input is a statement $x \in L$. This is due to the fact that when $x \notin L$, it may be possible for a distinguisher D to distinguish partial transcripts of the simulator from partial transcripts of a real execution just by checking if the statement is in the language (unless distinguishing $x \in L$ from $x \notin L$ is also assumed to be hard for circuits of size $\alpha(k)$). On the one hand, this is fine because simulators are only required to generate indistinguishable distributions in the case that $x \in L$. On the other hand, this is a problem because our simulator is not even guaranteed to *run* in expected polynomial-time for $x \notin L$. Thus, within a proof of security, one cannot invoke the zero-knowledge simulator on a statement x that may or may not be in the language, unless it is assumed that it is hard to distinguish $x \in L$ from $x \notin L$ in time $\alpha(k)$. In the full version of this paper, we discuss the ramifications of this caveat in greater detail.

3.4 Protocol Composition and Other Scenarios

We note that our result above has been stated for the stand-alone setting of secure computation. However, it actually holds for *any* setting, as long as the black-box simulator is α-strong for that setting. In particular, the result holds also for the setting of protocol composition where many protocol executions are run (and thus the simulator interacts with the trusted party many times).

---

[4] In fact, computational zero-knowledge arguments were also not known to exist for adversaries that are expected polynomial-time with respect to the protocol.

4 A Modular Composition Theorem

Our goal in this section is to prove a modular composition theorem for secure multi-party computation which is analogous to the result of Canetti [4], but which holds even for protocols proven secure against *strict* polynomial-time adversaries while using *expected* polynomial-time simulation. As in Section 3, the results of this section are stated for the two-party case; the extension to the multiparty case is straightforward.

The sequential composition theorem of [4] can be informally described as follows. Let π be a two-party protocol computing a function g, designed in an (idealized) model in which the parties have access to a trusted party who evaluates functions $f_1, \ldots, f_m$; furthermore, assume that at most one ideal function call is made during any round of π. This model is called the $(f_1, \ldots, f_m)$-hybrid model, denoted $\text{HYBRID}^{f_1, \ldots, f_m}$, because parties send real messages from the protocol π and also interact with a trusted party computing functions $f_1, \ldots, f_m$. Let $\rho_1, \ldots, \rho_m$ be a sequence of two-party protocols such that ρ_i securely computes f_i (as in Definition 3), and let $\pi^{\rho_1, \ldots, \rho_m}$ denote the "composed protocol" in which each ideal call to f_i is replaced by an invocation of ρ_i (we stress that each executed protocol ρ_i is run to completion before continuing the execution of π). The composition theorem then states that if π securely computes g in the hybrid model, and if each ρ_i securely computes f_i, then the composed real protocol $\pi^{\rho_1, \ldots, \rho_m}$ securely computes g. An important point to note is that the proof of [4] only considers the case that each of the component protocols ρ_i is proven secure via *strict* polynomial-time simulation. In fact, the *proof* of [4] demonstrably fails (in general) for the case of protocols proven secure via expected polynomial-time simulation; a counterexample is provided in the full version of this paper. In this section, we show that a suitable modification of the approach of [4] can be used to prove an analogous modular composition theorem even when each of the component protocols is proven secure via expected polynomial-time simulation.

We view this result as important both for conceptual reasons as well as for reasons of efficiency and practicality. Conceptually, there seems to be no fundamental reason that a composition theorem of this sort should not hold for the case of expected polynomial-time simulation; a number of technical barriers, however, make proving such a result difficult. From a practical point of view, many existing protocols – and, in particular, efficient ones – seem to require a proof of security via expected polynomial-time simulation. The composition theorem proven here enables protocol designers to enjoy the benefits of modular design and analysis, while ultimately allowing (more) efficient sub-protocols to be "plugged-in" for each of the components.

Preliminaries. We assume that the reader is familiar with [4], and so we borrow notation to the extent possible. In our proof, we use pseudorandom function families that are indistinguishable from random even for circuits of size $\alpha(k)$, for some superpolynomial function α. We call these α-secure pseudorandom functions.

The Composition Theorem. The composition theorem we prove is analogous to the one shown in [4] for the case of *strict* polynomial-time simulation. The only differences are that on the one hand, our proof holds also for the case of *expected* polynomial-time simulation, and on the other hand, we require *black-box* simulation and the existence of α-secure pseudorandom functions (the proof of [4] holds for any type of simulation and requires no hardness assumptions). We stress that, unlike in Section 3, here we consider the case that the real adversary runs in strict polynomial-time. Our proof of Theorem 7 is rather informal; a full and rigorous proof appears in the full version.

Theorem 7. Assume the existence of $\alpha(k)$-secure pseudorandom functions for some $\alpha(k) = k^{\omega(1)}$. Let $f_1, \ldots, f_m$ and g be two-party functions, let π be an two-party protocol that black-box securely computes g in the $(f_1, \ldots, f_m)$-hybrid model where no more than one ideal evaluation call is made at each round, and let $\rho_1, \ldots, \rho_m$ be two-party protocols such that each ρ_i securely computes f_i. Then protocol $\pi^{\rho_1, \ldots, \rho_m}$ securely computes g.

Proof: We follow the structure and notation of the proofs of [4–Theorems 5, 15] and [4–Corollaries 7, 17] as closely as possible. We focus on the case $m = 1$; the general case follows easily using the techniques described here (and is omitted due to lack of space). We begin with a high-level overview of our proof, stressing where it diverges from [4]: Let $f = f_1$ be a two-party function, π a protocol in the f-hybrid model, ρ a protocol that securely computes f, and π^ρ the composed protocol. Given a strict polynomial-time adversary $\mathcal{A}$ in the real world (who interacts with parties running π^ρ), our goal is to construct an expected polynomial-time ideal-world adversary $\mathcal{S}$ (interacting with a trusted party who evaluates g) such that $\text{IDEAL}_{g,\mathcal{S}} \stackrel{c}{\equiv} \text{REAL}_{\pi^\rho,\mathcal{A}}$. We proceed in the following steps:

- As in [4], we first construct from $\mathcal{A}$ a (natural) real-world adversary $\mathcal{A}_\rho$ who interacts with parties running ρ as a stand-alone protocol. The security of ρ implies the existence of an expected polynomial-time simulator $\mathcal{S}_\rho$, who interacts with a trusted party evaluating f, such that $\text{IDEAL}_{f,\mathcal{S}_\rho} \stackrel{c}{\equiv} \text{REAL}_{\rho,\mathcal{A}_\rho}$.
- As in [4], using $\mathcal{A}$ and $\mathcal{S}_\rho$ we construct an adversary $\mathcal{A}_\pi$ interacting with parties running π in the f-hybrid model and satisfying $\text{HYBRID}^f_{\pi,\mathcal{A}_\pi} \stackrel{c}{\equiv} \text{REAL}_{\pi^\rho,\mathcal{A}}$. Contrary to [4], we *cannot* at this point claim the existence of an expected polynomial-time ideal-world adversary $\mathcal{S}$, who interacts with a trusted party evaluating g, such that $\text{IDEAL}_{g,\mathcal{S}} \stackrel{c}{\equiv} \text{HYBRID}^f_{\pi,\mathcal{A}_\pi}$ (such a claim, if true, would complete the proof). We cannot make such a claim because $\mathcal{A}_\pi$ runs in *expected* polynomial-time but the security of π only guarantees the existence of a "simulator" for *strict* polynomial-time adversaries.
- Instead, we first construct a modified adversary $\mathcal{A}'_\pi$ (still interacting with parties running π in the f-hybrid model) that runs in expected polynomial time and for which $\text{HYBRID}^f_{\pi,\mathcal{A}'_\pi} \stackrel{c}{\equiv} \text{HYBRID}^f_{\pi,\mathcal{A}_\pi}$ under the assumption that α-secure pseudorandom functions exist. This forms the crux of our proof, and further details are given below.

– Let $\mathcal{S}_\pi$ denote a black-box simulator for π (as in Definition 3). We define an ideal-world adversary $\mathcal{S}$ by running a slightly modified version of $\mathcal{S}_\pi$ with oracle access to $\mathcal{A}'_\pi$. We then prove that (1) $\text{IDEAL}_{g,\mathcal{S}} \stackrel{c}{\equiv} \text{HYBRID}^f_{\pi,\mathcal{A}'_\pi}$; and (2) that $\mathcal{S}$ runs in expected polynomial time (even when taking the running time of $\mathcal{A}'_\pi$ into account). The proof of the second claim relies on the existence of α-secure pseudorandom functions. We stress that we do not claim the above is true when $\mathcal{S}_\pi$ is run with oracle access to an *arbitrary* expected polynomial time machine (indeed, the claims may not be true if $\mathcal{S}_\pi$ is run with oracle access to the original $\mathcal{A}_\pi$), but rather we only make these claims with regard to the *specific* $\mathcal{A}'_\pi$ that we construct.

We now proceed with the proof. Since the first steps of our proof – namely, the construction of $\mathcal{A}_\rho$, $\mathcal{S}_\rho$, and $\mathcal{A}_\pi$ – are exactly as in [4], we omit the details here but instead provide only a high-level description of the adversary $\mathcal{A}_\pi$ which runs in the f-hybrid model. Loosely speaking, $\mathcal{A}_\pi$ runs $\mathcal{A}$ until the protocol ρ is supposed to begin. At this point, $\mathcal{A}$ expects to run ρ, whereas $\mathcal{A}_\pi$ should use an ideal call to f. Therefore, $\mathcal{A}_\pi$ invokes $\mathcal{S}_\rho$ giving it the current internal state z^ρ of $\mathcal{A}$ as its auxiliary input, and forwarding the messages between $\mathcal{S}_\rho$ and the trusted party computing f. The output of $\mathcal{S}_\rho$ is an internal state of $\mathcal{A}$ at the end of the execution of ρ; adversary $\mathcal{A}_\pi$ continues by invoking $\mathcal{A}$ on this state and running $\mathcal{A}$ until the conclusion of π. We remark that $\mathcal{A}_\pi$'s random-tape is parsed into r and r^*, and $\mathcal{A}_\pi$ invokes $\mathcal{A}$ with random-tape r and $\mathcal{S}_\rho$ with random-tape r^*. This concludes the (informal) description of $\mathcal{A}_\pi$. As in [4], it holds that $\text{HYBRID}^f_{\pi,\mathcal{A}_\pi} \stackrel{c}{\equiv} \text{REAL}_{\pi^\rho,\mathcal{A}}$ In this case, however, $\mathcal{A}_\pi$ is an *expected* polynomial-time adversary.

Sidetrack – Motivation for the Proof. At this point, it is possible to provide the key idea behind the proof of the theorem. Let $\mathcal{S}_\pi$ be the simulator that is guaranteed to exist by the fact that π black-box securely computes g in the f-hybrid model. Then, the main problem that arises in the proof of [4] is that the expected running-time of $\mathcal{S}_\pi$ when given access to the oracle $\mathcal{A}_\pi$ may not be polynomial. Consider the case that the strategy of $\mathcal{S}_\pi$ involves "rewinding" $\mathcal{A}_\pi$. Then, it is possible that $\mathcal{A}_\pi$ will invoke $\mathcal{S}_\rho$ *a number of times with the same random-tape r^**. This introduces dependence between the executions, and may cause $\mathcal{S}_\rho$ to always run for a very long time. (The composition of the machines A and B described in the Introduction yielded an exponential-time machine exactly due to the fact that A invoked B with the same random-tape twice.) The first solution that comes to mind would be to have $\mathcal{A}_\pi$ choose an independent random-tape every time that it invokes $\mathcal{S}_\rho$. However, $\mathcal{S}_\pi$ works when given an oracle $\mathcal{A}_\pi$ with a *fixed* random-tape, and therefore this solution does not work. Our solution is to instead modify $\mathcal{A}_\pi$ so that it invokes $\mathcal{S}_\rho$ with a new pseudorandom tape each time (in a way reminiscent of a similar technique used in [6]). By using α-strong pseudorandom functions, we ensure that the pseudorandom tapes "look random" throughout the entire simulation by $\mathcal{S}_\pi$.

Back to the Proof. As described above, we modify $\mathcal{A}_\pi$ to an adversary $\mathcal{A}'_\pi$, using a family $\mathcal{F}$ of α-secure pseudorandom functions for $\alpha(k) = k^{\omega(1)}$. The random tape of $\mathcal{A}'_\pi$ is parsed as r, s, where r is used exactly as above (i.e., $\mathcal{A}'_\pi$ invokes $\mathcal{A}$ with random-tape r), and s is used as a key to an α-secure pseudorandom function. Then $\mathcal{A}'_\pi$ sets the random-tape r^* for $\mathcal{S}_\rho$ to $r^* = F_s(z^\rho)$, where z_ρ is the current internal state of $\mathcal{A}$ when $\mathcal{S}_\rho$ is invoked (instead of choosing it randomly like $\mathcal{A}_\pi$). In addition, $\mathcal{A}'_\pi$ halts with output $\bot$ if it ever exceeds $\alpha(k)/2$ steps overall (not including steps used in computing F_s).[5] Apart from the above, $\mathcal{A}'_\pi$ works in exactly the same way as $\mathcal{A}_\pi$. We now prove the following claims:

Claim 8. *Assuming that $\mathcal{F}$ is an α-secure family of pseudorandom functions, $\mathcal{A}'_\pi$ runs in expected polynomial time.*

Proof (sketch): Consider a modified simulator $\hat{\mathcal{A}}_\pi$ who chooses a truly random-tape r^* for $\mathcal{S}_\rho$ instead of a pseudorandom one. (In particular, the only difference between $\hat{\mathcal{A}}_\pi$ and $\mathcal{A}_\pi$ is that $\hat{\mathcal{A}}_\pi$ outputs $\bot$ if it ever exceeds $\alpha(k)/2$ steps.) Then, the expected running time of $\hat{\mathcal{A}}_\pi$ on any set of global inputs global (which includes both the inputs explicitly given to $\hat{\mathcal{A}}_\pi$ as well as the inputs and random coins of the honest parties and the random coins of the trusted party) is at most:

$$\sum_{t=1}^{\alpha(k)/2} \Pr\nolimits_{r,r^*}[\text{time}_{\hat{\mathcal{A}}_\pi}(\text{global}) \geq t] \leq \sum_{t=1}^{\alpha(k)/2} \Pr\nolimits_{r,r^*}[\text{time}_{\mathcal{A}_\pi}(\text{global}) \geq t]$$

$$\leq \sum_{t=1}^{\infty} \Pr\nolimits_{r,r^*}[\text{time}_{\mathcal{A}_\pi}(\text{global}) \geq t] \leq p_{\mathcal{A}_\pi}(k)$$

where $p_{\mathcal{A}_\pi}(\cdot)$ is the polynomial upper-bound on the expected running-time of $\mathcal{A}_\pi$, and where we ignore the time required to maintain a counter for the number of steps (since this only affects the expected running time by a multiplicative polynomial factor). Now, since r^* is actually chosen pseudorandomly by $\mathcal{A}'_\pi$, we have that for large enough k, every value of global and all $t \leq \alpha(k)/2$:

$$\left| \Pr\nolimits_{r,s}[\text{time}_{\mathcal{A}'_\pi}(\text{global}) \geq t] - \Pr\nolimits_{r,r^*}[\text{time}_{\hat{\mathcal{A}}_\pi}(\text{global}) \geq t] \right| \leq \frac{1}{\alpha(k)}. \quad (3)$$

(Eq. (3) ignores the time spent by $\mathcal{A}'_\pi$ in computing F_s because, as above, it only affects the expected running-time by a multiplicative polynomial factor). Otherwise, we can construct a distinguisher D for $\mathcal{F}$ as in the proof of Claim 6 (details appear in the full version). We conclude that the expected running-time of $\mathcal{A}'_\pi$ on global inputs global and large enough k equals

---

[5] There is an additional subtlety here, in that $\mathcal{S}_\rho$ may require a superpolynomial number of coins while the output of F_s is polynomial. However, this can be easily resolved: by construction, we never require more than $\alpha(k)$ coins for $\mathcal{S}_\rho$. Coins for $\mathcal{S}_\rho$ can thus be generated as needed by letting the ith coin required by $\mathcal{S}_\rho$ be given by $F_s(z^\rho | \langle i \rangle)$ where $\langle i \rangle$ is the $\log(\alpha(k))$-bit representation of i.

$$\sum_{t=1}^{\alpha(k)/2} \Pr\nolimits_{r,s}[\text{time}_{\mathcal{A}'_\pi}(\text{global}) \geq t] \leq \sum_{t=1}^{\alpha(k)/2} \left(\Pr\nolimits_{r,r^*}[\text{time}_{\hat{\mathcal{A}}_\pi}(\text{global}) \geq t] + \frac{1}{\alpha(k)} \right)$$

which equals at most $p_{\mathcal{A}_\pi}(k) + 1$, and so is polynomial. □

Claim 9. *Assuming that $\mathcal{F}$ is an α-secure family of pseudorandom functions, it holds that* $\text{HYBRID}^f_{\pi,\mathcal{A}'_\pi} \stackrel{c}{\equiv} \text{HYBRID}^f_{\pi,\mathcal{A}_\pi}$.

Proof (sketch): Let $\hat{\mathcal{A}}_\pi$ be the same as in Claim 8. Since the expected running-time of $\mathcal{A}_\pi$ on security parameter 1^k is polynomial (for any set of global inputs), the probability that $\mathcal{A}_\pi$ exceeds $\alpha(k)/2$ steps is negligible. Hence $\text{HYBRID}^f_{\pi,\hat{\mathcal{A}}_\pi}$ is *statistically* close to $\text{HYBRID}^f_{\pi,\mathcal{A}_\pi}$. Now, $\mathcal{A}'_\pi$ is identical to $\hat{\mathcal{A}}_\pi$ except that it uses a pseudorandom r^* while $\hat{\mathcal{A}}_\pi$ uses a truly random r^*. Since $\mathcal{A}'_\pi$ and $\hat{\mathcal{A}}_\pi$ both run in at most $\alpha(k)/2$ steps (for the case of $\mathcal{A}'_\pi$, not counting the time required to compute F_s), the assumption that $\mathcal{F}$ is an α-secure family of pseudorandom functions immediately implies that $\text{HYBRID}^f_{\pi,\mathcal{A}'_\pi}$ is computationally indistinguishable from $\text{HYBRID}^f_{\pi,\hat{\mathcal{A}}_\pi}$ (details omitted), completing the proof. □

Defining the Simulator $\mathcal{S}$. Since π *black-box* securely computes g, there exists an oracle machine $\mathcal{S}_\pi$ satisfying the conditions of Definition 3 (with appropriate modifications for comparing the f-hybrid and ideal models). Our simulator $\mathcal{S}$ works by simply invoking $\mathcal{S}_\pi$ with oracle $\mathcal{A}'_\pi$, with the limitation that it halts with output $\bot$ if it ever exceeds $\alpha(k)/2$ steps (including the running time of $\mathcal{A}'_\pi$ but, again, not including time spent computing F_s). Our aim is to show that (1) $\mathcal{S}$ runs in expected polynomial-time (even when taking the running time of $\mathcal{A}'_\pi$ into account), and (2) $\text{HYBRID}^f_{\pi,\mathcal{A}'_\pi} \stackrel{c}{\equiv} \text{IDEAL}_{g,\mathcal{S}}$. We stress that neither of these claims are immediate since $\mathcal{A}'_\pi$ is an *expected* polynomial-time adversary, and the simulator $\mathcal{S}_\pi$ has only been proven for the case that it is given a *strict* polynomial-time oracle.

Claim 10. *Assuming that $\mathcal{F}$ is an α-secure family of pseudorandom functions, $\mathcal{S}$ runs in expected polynomial time.*

Proof (sketch): We use the same general technique as in the proof of Claim 8, but the proof here is slightly more complicated. First imagine an adversary $\widetilde{\mathcal{S}}$ that differs from $\mathcal{S}$ in the following way: whenever $\mathcal{S}_\rho$ is called from within $\mathcal{A}'_\pi$, $\widetilde{\mathcal{S}}$ monitors the value of z^ρ at that point. Let z_i^ρ denote the ith value of z^ρ in the execution of $\widetilde{\mathcal{S}}$. Then instead of setting $r_i^* = F_s(z_i^\rho)$, $\widetilde{\mathcal{S}}$ instead chooses r_i^* as follows: if $z_i^\rho = z_j^\rho$ for some $j < i$, then set $r_i^* = r_j^*$. Otherwise, choose r_i^* uniformly at random (the technicalities raised in footnote 5 can be handled in the obvious way). We first show that $\widetilde{\mathcal{S}}$ runs in expected polynomial-time, and then claim (as in the proof of Claim 8) that the expected running-times of $\widetilde{\mathcal{S}}$ and $\mathcal{S}$ cannot differ "too much".

The running time of $\widetilde{\mathcal{S}}$ is the sum of three components: $\text{time}_{\mathcal{S}_\pi}$, the running time of $\mathcal{S}_\pi$ when counting its oracle calls to $\mathcal{A}'_\pi$ as a single step; $\text{time}_{\mathcal{A}'_\pi}$, the running time of $\mathcal{A}'_\pi$ (when answering oracle calls of $\mathcal{S}_\pi$) but excluding time spent running $\mathcal{S}_\rho$; and time$_{\mathcal{S}_\rho}$, and the running time of $\mathcal{S}_\rho$ when called by $\mathcal{A}'_\pi$ (each time $\mathcal{A}'_\pi$ is run).[6] By linearity of expectation, we can analyze each of these individually. The expected value of time$_{\mathcal{S}_\pi}$ is polynomial since $\mathcal{S}_\pi$ is an expected polynomial-time oracle machine (as defined in Section 2). Furthermore, since $\mathcal{A}'_\pi$ runs in *strict* polynomial time when excluding the steps of $\mathcal{S}_\rho$, and since $\mathcal{S}_\pi$ makes an expected polynomial number of calls to $\mathcal{A}'_\pi$, the expected value of time$_{\mathcal{A}'_\pi}$ is polynomial as well. It remains to analyze time$_{\mathcal{S}_\rho}$. This variable is equal to $\sum_{i=1}^{\text{time}_{\mathcal{S}_\pi}} \text{time}_{\mathcal{S}_\rho}(i)$, where time$_{\mathcal{S}_\rho}(i)$ represents the running time of $\mathcal{S}_\rho$ in its ith execution. Since the random coins r_i^* used in the ith execution of $\mathcal{S}_\rho$ are chosen at random, the expectation of time$_{\mathcal{S}_\rho}(i)$ is polynomial for all i. Wald's inequality (cf. Appendix A) thus implies that the expected value of time$_{\mathcal{S}_\rho}$ is polynomial.

Exactly as in the proof of Claim 8, the fact that $\mathcal{F}$ is α-secure can be used to show that $\mathcal{S}$ runs in expected polynomial time as well. We omit the details (which are identical) here. □

To complete the proof of the main theorem, we need to prove that $\text{IDEAL}_{g,\mathcal{S}} \stackrel{c}{\equiv} \text{HYBRID}^f_{\pi,\mathcal{A}'_\pi}$. The proof of this is largely similar to the end of the proof of Theorem 5 and appears in the full version of this paper. ■

References

1. B. Barak. How to Go Beyond the Black-Box Simulation Barrier. In *42nd FOCS*, pages 106–115, 2001.
2. B. Barak and O. Goldreich. Universal Arguments and their Applications. *17th IEEE Conference on Computational Complexity*, pages 194–203, 2002.
3. B. Barak and Y. Lindell. Strict Polynomial-Time in Simulation and Extraction. *SIAM Journal on Computing*, 33(4):783–818, 2004.
4. R. Canetti. Security and Composition of Multiparty Cryptographic Protocols. *Journal of Cryptology*, 13(1):143–202, 2000.
5. R. Canetti. Universally Composable Security: A New Paradigm for Cryptographic Protocols. In *42nd FOCS*, pages 136–145, 2001.
6. R. Canetti, O. Goldreich, S. Goldwasser, and S. Micali. Resettable Zero-Knowledge. STOC 2000.
7. U. Feige. *Alternative Models for Zero Knowledge Interactive Proofs*. Ph.D. Thesis, Weizmann Institute, 1990.
8. U. Feige and A. Shamir. Zero-Knowledge Proofs of Knowledge in Two Rounds. In *CRYPTO'89*, Springer-Verlag (LNCS 435), pages 526–544, 1989.
9. O. Goldreich. *Foundations of Cryptography: Volume 1 – Basic Tools*. Cambridge University Press, 2001.
10. O. Goldreich. *Foundations of Cryptography: Volume 2 – Basic Applications*. Cambridge University Press, 2004.
11. O. Goldreich and A. Kahan. How To Construct Constant-Round Zero-Knowledge Proof Systems for NP. *Journal of Cryptology*, 9(3):167–190, 1996.

---

[6] As discussed earlier, we again ignore time spent computing F_s.

12. O. Goldreich and H. Krawczyk. On the Composition of Zero-Knowledge Proof Systems. *SIAM Journal on Computing* 25(1):169–192, 1996.
13. O. Goldreich, S. Micali, and A. Wigderson. Proofs that Yield Nothing but Their Validity or All Languages in NP Have Zero-Knowledge Proof Systems. *Journal of the ACM* 38(1):691–729, 1991.
14. O. Goldreich and Y. Oren. Definitions and Properties of Zero-Knowledge Proof Systems. *Journal of Cryptology* 7(1):1–32, 1994.
15. S. Goldwasser, S. Micali, and C. Rackoff. The Knowledge Complexity of Interactive Proof Systems. *SIAM Journal on Computing,* 18(1):186–208, 1989.
16. Y. Lindell. Parallel Coin-Tossing and Constant-Round Secure Two-Party Computation. *Journal of Cryptology,* 16(3):143–184, 2003.

A Wald's Inequality

We state a (slightly modified version of) Wald's inequality here. The proof is provided in the full version.

Lemma 11. *Let $Y_1, Y_2, \ldots$ be an infinite sequence of non-negative random variables such that $\mathbf{Exp}\,[Y_i] \leq N$ for all i. Let τ be a non-negative integer random variable for which, for all i, $\Pr[\tau = i]$ depends only on $Y_1, \ldots, Y_i$. Define $\overline{Y} \stackrel{\text{def}}{=} \sum_{i=1}^{\tau} Y_i$ (with the sum defined as 0 in case $\tau = 0$). Then $\mathbf{Exp}\,[\overline{Y}] \leq N \cdot \mathbf{Exp}\,[\tau]$.*

Adaptively-Secure, Non-interactive Public-Key Encryption*

Ran Canetti[1], Shai Halevi[1], and Jonathan Katz[2]

[1] IBM T.J. Watson Research Center, NY, USA
[2] Department of Computer Science, University of Maryland

Abstract. Adaptively-secure encryption schemes ensure secrecy even in the presence of an adversary who can corrupt parties in an adaptive manner based on public keys, ciphertexts, and secret data of already-corrupted parties. Ideally, an adaptively-secure encryption scheme should, like standard public-key encryption, allow arbitrarily-many parties to use a single encryption key to securely encrypt arbitrarily-many messages to a given receiver who maintains only a single short decryption key. However, it is known that these requirements are impossible to achieve: no non-interactive encryption scheme that supports encryption of an unbounded number of messages and uses a single, unchanging decryption key can be adaptively secure. Impossibility holds even if secure data erasure is possible.

We show that this limitation can be overcome by *updating the decryption key over time* and making some mild assumptions about the frequency of communication between parties. Using this approach, we construct adaptively-secure, completely non-interactive encryption schemes supporting secure encryption of arbitrarily-many messages from arbitrarily-many senders. Our schemes additionally provide forward security and security against chosen-ciphertext attacks.

1 Introduction

Imagine a band of political dissidents who need to go into hiding from an oppressive regime. While in hiding, the only form of communication with the outside world is via the public media. Before going into hiding, each individual wants to publish a key that will allow *anyone* (even parties not currently known to this individual) to publish encrypted messages that only this individual can decipher. Since it is not known in advance how long these members will need to be in hiding, reasonably short public keys must suffice for encrypting an unbounded number of messages. Furthermore, messages encrypted to each dissident must remain secret even if other dissidents are caught and their secrets are extracted from them. Do encryption schemes satisfying these requirements exist?

---

* This work was supported by NSF Trusted Computing Grant #0310751 and CyberTrust Grant #0430450.

At first glance, a standard public-key encryption scheme seems to suffice. Indeed, a public-key encryption schemes allows a receiver to publish a key that can then be used by anyone to send encrypted messages to the receiver. The public key is *short* (i.e., of fixed polynomial length) and can be used by arbitrary senders (potentially unknown to the receiver at the time the key is published) to securely send arbitrarily-many messages to the receiver *without further interaction*. Furthermore, senders need not maintain any state other than the receiver's public key, and the receiver similarly need not maintain any state except for his secret key.

However, standard public-key encryption schemes do not provide the desired level of security. Standard definitions of security, including semantic security against passive attacks [GM84] as well as various notions of security against active attacks [NY90, RS91, DDN00, BDPR98], only consider the case where the adversary never learns any secret key. However, when an adversary can compromise players and learn their internal states in an *adaptive* manner, possibly depending on previously-observed ciphertexts and information learned during previous corruptions, the standard notions no longer apply. In particular, in the adaptive setting *encrypting with a CCA-secure encryption scheme is not known to provide secure communication*.

To obtain provable security against adaptive adversaries, one must ensure that the information gathered by the adversary when compromising parties (namely, their secret keys) does not give the adversary any additional advantage toward compromising the security of the yet-uncorrupted parties. The standard way of formulating this is by requiring the existence of a *simulator* that can generate "dummy ciphertexts" which can be later "opened" (i.e., by revealing an appropriate secret key) as encryptions of any message; see, e.g., [CFGN96]. A scheme satisfying this additional condition is said to be *adaptively secure*.

Several methods are known for achieving adaptively-secure encrypted communication, but none can be used in the basic setting exemplified by the above toy problem. Beaver and Haber [BH92] propose an adaptively secure encryption protocol in which the sender and receiver must interact before they can securely communicate for the first time. Furthermore, the parties must maintain a shared secret key *per connection*. This key must be continually updated, with the old key being erased, as more messages are encrypted. *Non-committing encryption schemes* [CFGN96, B97, DN00] more closely mimic the functionality of standard public-key encryption, and in particular do not require maintenance of per-connection state. (In addition, these solutions also remove the need for secure data erasure.) In these schemes, however, both the public and secret keys are at least as long as the overall number of bits to be encrypted. In fact, as noted by Nielsen [N02], any adaptively-secure scheme with non-interactive encryption *must* have a decryption key which is at least as long as the number of bits to be decrypted under this key. In a nutshell, this is because the simulator must "open" the "dummy ciphertexts" as encryptions of any given sequence of messages by presenting an appropriate secret key; therefore, the number of possible secret keys must be at least the number of possible message-sequences. The unfortunate conclusion is that a public-key encryption scheme that can encrypt an unbounded number of messages with short and unchanging keys cannot be adaptively secure. This holds even if secure data erasures are possible, and even in a weaker setting where only receivers can be corrupted.

We also comment that previous work on adaptively-secure encryption did not address resistance to chosen-ciphertext attacks.

Our Contributions. This work demonstrates that we can circumvent Nielsen's negative result if the secret decryption key is allowed to periodically change, and some mild assumptions about the frequency of communication between parties are made. That is, under standard hardness assumptions, there exist adaptively-secure, non-interactive public-key encryption schemes with short keys that can handle arbitrarily-many messages and senders. In particular, our schemes solve the toy example from above in a way that is essentially the best possible under the given constraints.

This is done by considering *key-evolving* encryption schemes [CHK03] in which the secret key is locally updated by the receiver according to a globally-known schedule (say, at the end of every day), while the public key remains fixed. The secret key for the previous period is securely *erased* once it is no longer needed. Using this approach, we construct adaptively-secure, non-interactive encryption schemes that can be used to encrypt arbitrarily-many bits as long as the number of encrypted bits (for any particular key) is bounded *per time period*. As discussed above, an assumption of this sort is essential to circumvent Nielsen's negative results. Also, this assumption is reasonable in many cases: for instance, one may easily posit some known upper bound on the number of incoming e-mails processed per day.

In addition to being adaptively secure, our schemes also provide both forward security [A97, CHK03] and security against chosen-ciphertext attacks. (We comment that although forward security is reminiscent of adaptive security, neither security property implies the other.) Accordingly, we refer to schemes satisfying our security requirements as adaptively- and forward-secure encryption (AFSE) schemes. We formalize the requirements for AFSE schemes within the UC framework [C01]. That is, we present an functionality $\mathcal{F}_{\mathrm{AFSE}}$ that captures the desired properties of AFSE schemes. This functionality is a natural adaptation of the "standard" public-key encryption functionality of [C01, CKN03] to the context of key-evolving encryption. As in the non-adaptive case, $\mathcal{F}_{\mathrm{AFSE}}$ guarantees security against active adversaries, which in particular implies security against chosen-ciphertext attacks. Using the composability properties of the UC framework, our constructions are guaranteed to remain secure in any protocol environment. Indeed, the formulation of $\mathcal{F}_{\mathrm{AFSE}}$, which blends together the notions of forward security, chosen-ciphertext security, and adaptive security of public-key encryption schemes, is another contribution of this work.

Techniques and Constructions. We first note that dealing with corruption of senders is easy, since a sender can simply erase its local state upon completing the encryption algorithm. We thus concentrate on the more difficult case of receiver corruption. We then show that it suffices to consider AFSE for the case when only a *single* message is encrypted per time period, since any such construction can be extended in a generic manner to give a scheme which can be used to encrypt any bounded number of messages per time period. With this in mind, our first construction uses the paradigm of Naor-Yung and Sahai [NY90, S99] to construct an AFSE scheme based on any forward-secure encryption (FSE) scheme and any simulation-sound non-interactive zero-knowledge (NIZK) proof system [DDOPS01]. Recall that, under the Naor-Yung/Sahai paradigm, the sender encrypts messages by essentially using two independent copies of a semantically-secure encryption scheme together with an NIZK proof of consistency. To decrypt, the receiver verifies the proof and then decrypts either one of the component ciphertexts. Naor and Yung prove that this provides security against "lunch-time" (i.e., non-adaptive) chosen-ciphertext attacks when an arbitrary NIZK proof system is used, and Sahai later showed that this technique achieves full (i.e., adaptive) CCA-security if a one-time simulation-sound NIZK proof system is used. We show that if a semantically-secure FSE scheme is used as the underlying encryption scheme, and the NIZK proof system is "fully" simulation sound (as defined in [DDOPS01]), the resulting construction is also an AFSE scheme. This approach can be extended to encrypt a polynomial number of bits per ciphertext using only a single NIZK proof. (We remark that, as opposed to the case of standard CCA-secure encryption [S99], here it is not enough that the underlying NIZK is *one-time* simulation sound.)

While the above approach is conceptually simple, it is highly impractical due to the inefficiency of known NIZKs. We thus propose an alternate approach that leads to more efficient solutions based on specific, number theoretic assumptions. As part of this approach, we first define and construct "standard" (i.e., non key-evolving) encryption schemes which are secure against lunch-time chosen-ciphertext attacks and are adaptively-secure for encryption of a *single* message (in total). We call such schemes *receiver non-committing encryption* (RNCE) schemes.[1] Our construction of an AFSE scheme proceeds by first encrypting the message using any RNCE scheme, and then encrypting the resulting ciphertext using any CCA-secure FSE scheme. Informally, this construction achieves adaptive security for an *unbounded* number of messages (as long as only one message is encrypted per time period) because the secret key of the outer FSE scheme is updated after every period and so the simulator only needs to "open" one ciphertext (i.e., the one corresponding to the current time period) as an arbitrary message. It can accomplish the latter using the "inner" RNCE scheme.

Obtaining an efficient scheme using this approach requires efficient instantiation of both components. Relatively efficient CCA-secure FSE schemes (in particular, schemes which avoid the need for NIZK proofs) are already known [CHK04, BB04]. Therefore, we focus on constructing efficient RNCE schemes

---

[1] Indeed, this is a relaxation of the notion of non-committing encryption from [CFGN96]. It is similar to the relaxation studied by Jarecki and Lysyanskaya [JL00], except that we also require security against lunch-time chosen-ciphertext attacks.

based on specific number-theoretic assumptions. Our first RNCE scheme is based on the Cramer-Shoup encryption scheme [CS98] (and adapts techniques of [JL00]) and its security is predicated on the decisional Diffie-Hellman (DDH) assumption. However, this scheme allows encryption of only a logarithmic number of bits per ciphertext. We also show a second RNCE scheme based on the schemes of [GL03, CS03] (which, in turn, build on [CS02]), whose security relies on the decisional composite residuosity assumption introduced by Paillier [P99] and which can be used to encrypt a polynomial number of bits per ciphertext.

Organization. The AFSE functionality is defined and motivated in Section 2. Our construction of AFSE using the Naor-Yung/Sahai paradigm is described in Section 3. In Section 4, we present definitions for RNCE and show two constructions of RNCE schemes based on specific number-theoretic assumptions. Finally, in Section 5 we construct an AFSE scheme from any RNCE scheme and any CCA-secure FSE scheme. In Appendix A, we include definitions of key-evolving and forward-secure encryption, while a brief overview of the UC framework and its application to secure encryption is provided in Appendix B. In this abstract we omit all proofs due to lack of space. The proofs can be found in the full version of this paper [CHK05].

2 Definition of AFSE

We define AFSE by specifying an appropriate ideal functionality in the UC security framework (cf. Appendix B). This functionality, denoted $\mathcal{F}_{\mathrm{AFSE}}$ and presented in Figure 1, is obtained by appropriately modifying the "standard" public-key encryption functionality $\mathcal{F}_{\mathrm{PKE}}$ [C01, CKN03] which is reviewed in Appendix B.1.

Intuitively, $\mathcal{F}_{\mathrm{AFSE}}$ captures the same security notions as $\mathcal{F}_{\mathrm{PKE}}$ except that it also provides a mechanism by which the receiver can "update" its secret key; $\mathcal{F}_{\mathrm{AFSE}}$ guarantees security only as long as a bounded number of messages are encrypted between key updates. In fact, for simplicity, the functionality as defined only guarantees security when a *single* ciphertext is encrypted between key updates. Say a ciphertext encrypted with respect to a particular time period t is **outstanding** until the receiver has updated its secret key a total of $t+1$ times. Then, if more than one outstanding ciphertext is requested, the functionality guarantees no security whatsoever for this ciphertext. (Formally, this is captured by handing the corresponding plaintext to the adversary.) Section 2.1 discusses how $\mathcal{F}_{\mathrm{AFSE}}$ can be extended to allow any bounded number of outstanding ciphertexts, which corresponds to ensuring security as long as at most this many messages are encrypted between key updates. It also presents a generic transformation from protocols secure for a single outstanding ciphertext to protocols secure for the general case.

For convenience, we highlight some differences between $\mathcal{F}_{\mathrm{AFSE}}$ and $\mathcal{F}_{\mathrm{PKE}}$. First, an additional parameter — a time period t — is introduced. An encryption request now additionally specifies a time period for the encryption called the

---

Functionality $\mathcal{F}_{\text{AFSE}}$

$\mathcal{F}_{\text{AFSE}}$ proceeds as follows, when parameterized by message domain ensemble $\mathcal{D} = \{D_k\}_{k \in \mathbb{N}}$ and security parameter k.

Key Generation: Upon receiving a request (KeyGen, sid) from party R^*, do: Verify that $sid = (sid', R^*)$ (i.e., that the identity R^* is encoded in the session ID). If not, then ignore this input. If yes:
  1. Hand (KeyGen, sid) to the adversary.
  2. Receive a value pk* from the adversary, and hand pk* to R^*. Initialize $t^* \leftarrow 0$ and messages-outstanding $\leftarrow 0$.

Encryption: Upon receiving from some party P a tuple (Encrypt, sid, pk, t, m) proceed as follows:
  1. If $m \in D_k$, pk = pk*, and either $t < t^*$ or messages-outstanding = 0, then send (Encrypt, sid, pk, t, P) to the adversary. In all other cases, send (Dummy-Encrypt, sid, pk, t, m, P) to the adversary (i.e., reveal the plaintext to the adversary).
  2. Receive a reply c from the adversary and send (ciphertext, c) to P. In addition, if $m \in D_k$, pk = pk*, and $t \geq t^*$, then do:
     (a) If messages-outstanding = 0, set messages-outstanding $\leftarrow 1$ and flag $\leftarrow$ outstanding. Else, set flag $\leftarrow$ dummy.
     (b) record the tuple (m, t, c, flag) in the list of ciphertexts.

Decryption: Upon receiving a tuple (Decrypt, sid, c) from player P, if $P \neq R^*$ then ignore this input. Otherwise:
  1. If the list of ciphertexts contains a tuple $(m, t, c, \star)$ with the given ciphertext c and $t = t^*$, then return m to R^*.
  2. Otherwise send a message (Decrypt, sid, t^*, c) to the adversary, receive a reply m, and forward m to R^*.

Update: Upon receiving (Update, sid) from player P, if $P = R^*$ do:
  1. Send a message (Update, sid) to the adversary.
  2. Remove from the list of ciphertexts all the tuples (m, t^*, c, flag) with the current time t^*. If any of these tuple has flag = outstanding, then reset messages-outstanding $\leftarrow 0$.
  3. Set $t^* \leftarrow t^* + 1$.

Corruptions: Upon corruption of party P, if $P = R^*$ then send to the adversary all tuples $(m, t, c, \star)$ in the list of ciphertexts with $t \geq t^*$. (If $P \neq R^*$ then do nothing.)

Fig. 1. The AFSE functionality, $\mathcal{F}_{\text{AFSE}}$

"sender time", and the functionality maintains a variable t^* called the "receiver time". The receiver time is initialized to 0, and is incremented by the receiver R^* using an Update request. A ciphertext generated for sender time t is only decrypted by $\mathcal{F}_{\text{AFSE}}$ (upon request of the appropriate receiver) when the current receiver time is $t^* = t$.

Second, $\mathcal{F}_{\text{AFSE}}$ limits the information gained by the adversary upon corruption of parties in the system. When corrupting parties other than R^*, the adversary learns nothing. When corrupting R^* at some "receiver time" t^*, the adversary does not learn any information about messages that were encrypted at "sender times" $t < t^*$. (This is akin to the level of security provided by forward-secure encryption schemes, and in fact strengthens the usual notion of adaptive security which potentially allows an adversary to learn all past messages upon corruption of a party.) In addition, *adaptive* security is guaranteed for a single message encrypted at some sender time $t \geq t^*$ (i.e., a single outstanding message).

The fact that security is guaranteed only for a single outstanding message is captured via the variable messages-outstanding, which is initialized to 0 and is set to 1 when a message is encrypted for time period t with $t \geq t^*$. When the receiver's time unit t^* advances beyond the time unit t of the outstanding ciphertext, the variable messages-outstanding is reset to 0. If another encryption request arrives with time period $t \geq t^*$ while messages-outstanding is equal to 1, then $\mathcal{F}_{\text{AFSE}}$ *discloses the entire plaintext to the adversary* (and thus does not ensure any secrecy in this case).

We remark that $\mathcal{F}_{\text{AFSE}}$ can be used in a natural way to realize a variant of the "secure message transmission functionality" [C01, AF04] in synchronous networks with respect to adaptive adversaries. We omit further details.

2.1 Handling Multiple Outstanding Ciphertexts

While the functionality $\mathcal{F}_{\text{AFSE}}$ and all the constructions in this work are described assuming a bound of at most one outstanding ciphertext, both the functionality and the constructions can be generalized to the case of any *bounded* number of outstanding ciphertexts (corresponding to a bounded number of messages encrypted per time period). Generalizing the functionality is straightforward, so we do not describe it here. As for constructions, any AFSE scheme which is secure for the case of a single outstanding ciphertext can be extended generically so as to be secure for any bounded number ℓ of outstanding ciphertext in the following way: The public key of the new scheme consists of ℓ independent keys $\mathsf{pk}_1, \ldots, \mathsf{pk}_\ell$ generated using the original scheme. To encrypt a message m, the sender computes the "nested encryption" $E_{\mathsf{pk}_1}(E_{\mathsf{pk}_2}(\cdots E_{\mathsf{pk}_\ell}(m) \cdots))$ and sends the resulting ciphertext to the receiver. One can show that this indeed realizes $\mathcal{F}_{\text{AFSE}}$ for at most ℓ outstanding ciphertexts. The formal proof, however, is more involved and is omitted.

2.2 Realizing $\mathcal{F}_{\text{AFSE}}$ Using Key-Evolving Encryption Schemes

We present our constructions as key-evolving encryption *schemes* (i.e., as a collection of algorithms) rather than as protocols (as technically required by the UC framework). For completeness, we describe the (obvious) transformation from key-evolving encryption schemes to protocols geared toward realizing $\mathcal{F}_{\text{AFSE}}$.

Recall that a key-evolving encryption scheme consists of four algorithms (Gen, Upd, Enc, Dec), where (Gen, Enc, Dec) are the key generation, encryption, and decryption routines (as in a standard encryption scheme, except that the encryption and decryption routines also take as input a time period t), and Upd is the secret-key update algorithm that takes as input the current secret key and time unit, and outputs the secret key for the next time unit. The definition is reviewed in Appendix A.

Given a key evolving encryption scheme $S = (\mathsf{Gen}, \mathsf{Upd}, \mathsf{Enc}, \mathsf{Dec})$, one may construct the protocol π_S as follows: An activation of π_S with input message KeyGen, Update, Encrypt, or Decrypt is implemented via calls to the algorithms Gen, Upd, Enc, or Dec, respectively. The only state maintained by π_S between activations is the secret key that was generated by Gen (and that is modified in each activation of Update), and the current time period. Any other local variables that are temporarily used by any of the algorithms are erased as soon as the activation completes. With this transformation we can now define an AFSE scheme:

Definition 1. *A key-evolving encryption scheme S is an adaptively- and forward-secure encryption (AFSE) scheme if the protocol π_S resulting from the transformation above securely realizes $\mathcal{F}_{\mathrm{AFSE}}$ with respect to adaptive adversaries.*

3 AFSE Based on Forward-Secure Encryption

In this section we show how to construct an AFSE scheme from any FSE scheme secure against chosen-plaintext attacks along with any simulation-sound NIZK proof system. (See Appendix A for definitions of key-evolving encryption and forward security, both against chosen-plaintext and chosen-ciphertext attacks.) We describe in detail a construction that allows encryption of only a single bit per ciphertext and then discuss how this may be generalized to allow for encryption of any polynomial number of bits per ciphertext. Our construction uses a simple twist of the Naor-Yung/Sahai transformation [NY90, S99]; when applied to two FSE schemes, the resulting scheme yields not only CCA security but also security against adaptive corruptions. We comment that, as opposed to the case of non-adaptive CCA security, "one-time" simulation sound NIZK proofs are *not sufficient* to achieve security against adaptive corruptions; instead, we require NIZK proofs satisfying the stronger notion of unbounded simulation soundness [DDOPS01].

The Construction. Let $\mathcal{E}' = (G', U', E', D')$ be a key-evolving encryption scheme, and let $\mathcal{P} = (\ell, P, V)$ be an NIZK proof system (where $\ell(k)$ is the length of the common random string for security parameter k) for the following NP language

$$L_{\mathcal{E}'} \stackrel{\text{def}}{=} \{(t, \mathsf{pk}'_0, c'_0, \mathsf{pk}'_1, c'_1) :$$
$$\exists\, m, r_0, r_1 \text{ s.t. } c'_0 = E'(\mathsf{pk}'_0, t; m; r_0),\ c'_1 = E'(\mathsf{pk}'_1, t; m; r_1)\}.$$

We construct a new key-evolving encryption scheme $\mathcal{E} = (G, U, E, D)$ as follows:

Key Generation, G. On security parameter 1^k, run two independent copies of the key generation algorithm of $\mathcal{E}'$ to obtain $(\mathsf{pk}'_0, \mathsf{sk}'_0) \leftarrow G'(1^k)$ and

$(\mathsf{pk}'_1, \mathsf{sk}'_1) \leftarrow G'(1^k)$. Choose a random bit $b \in \{0, 1\}$ and a random $\ell(k)$-bit string $\mathsf{crs} \in \{0, 1\}^{\ell(k)}$. The public key is the triple $(\mathsf{pk}'_0, \mathsf{pk}'_1, \mathsf{crs})$, and the secret key is (b, sk'_b). Erase the other key $\mathsf{sk}'_{\bar{b}}$.

Key Update, U. Key update is unchanged, namely $U(t, (b, \mathsf{sk}')) = (b, U'(t, \mathsf{sk}'))$.

Encryption, E. To encrypt a bit $m \in \{0, 1\}$ at time t, first pick two independent random strings r_0, r_1 as needed for the encryption algorithm E' and compute $c'_0 \leftarrow E'(\mathsf{pk}'_0, t; m; r_0)$, $c'_1 \leftarrow E'(\mathsf{pk}_1, t; m; r_1)$, and a proof that $(t, \mathsf{pk}'_0, c'_0, \mathsf{pk}'_1, c'_1) \in L_{\mathcal{E}'}$; namely $\pi \leftarrow P(\mathsf{crs}; t, \mathsf{pk}'_0, c'_0, \mathsf{pk}'_1, c'_1; m, r_0, r_1)$. The ciphertext is the triple $c = (c'_0, c'_1, \pi)$.

Decryption, D. To decrypt a ciphertext $c = (c'_0, c'_1, \pi)$ at time t, first run the verifier $V(\mathsf{crs}; t, \mathsf{pk}'_0, c'_0, \mathsf{pk}'_1, c'_1)$. If V rejects, the output is $\bot$. Otherwise, the recipient uses (b, sk'_b) to recover $m \leftarrow D'(\mathsf{sk}'_b; c'_b)$.

We claim the following theorem:

Theorem 1. *If $\mathcal{E}'$ is forward-secure against chosen-plaintext attacks (fs-CPA, cf. Definition 4) and if (P, V) is an unbounded simulation-sound NIZK proof system [DDOPS01–Def. 6], then $\mathcal{E}$ is an AFSE scheme.*

The proof appears in the full version, but we provide some intuition here. Underlying our analysis is the observation that a simulator (who can generate proofs for false assertions) can come up with a valid-looking "dummy ciphertext" whose component ciphertexts encrypt *different* messages (i.e., both 0 and 1). The simulator, who also knows both underlying decryption keys, can thus open the dummy ciphertext as an encryption of either 0 or 1, depending on which decryption key is presented to an adversary. (Note further that the adversary will be unable to generate dummy ciphertexts of this form due to the simulation soundness of the NIZK proof system.) The above argument demonstrates adaptive security for a single encrypted bit. Adaptive security for an unbounded number of bits (as long as only one ciphertext is outstanding) holds since the secret keys of the underlying FSE schemes evolve after each encryption. We remark that one-time simulation soundness for (P, V) would not be sufficient here, since the simulator must generate *multiple* "fake ciphertexts" and the hybrid argument that works in the non-adaptive case (see [S99]) does not work here.

AFSE for Longer messages. To obtain a construction of an AFSE scheme for n-bit messages, one can simply use n pairs of public keys generated using $\mathcal{E}'$ (the receiver now chooses at random one secret key from each pair to store, while the other is erased). The rest is an obvious extension of the proof intuition from above, with the only subtle point being that the resulting ciphertext contains a single NIZK proof computed over the entire vector of n ciphertext pairs (with the language being defined appropriately).

4 Receiver Non-committing Encryption

This section defines and constructs receiver non-committing encryption (RNCE) that is secure against "lunch-time attacks" (aka CCA1-secure). We note that

RNCE was considered in [JL00] for the more basic case of chosen-plaintext attacks. Section 5 shows how to combine any RNCE scheme with any FSE scheme secure against chosen-ciphertext attacks to obtain a secure AFSE scheme. Since our proposed constructions of RNCE schemes are quite efficient (and since relatively-efficient constructions of FSE schemes secure against chosen-ciphertext attacks are known [CHK03, CHK04, BB04]), we obtain (relatively) efficient AFSE schemes.

On a high level, a receiver non-committing encryption scheme is one in which a simulator can generate a single "fake ciphertext" and later "open" this ciphertext (by showing an appropriate secret key) as any given message. These "fake ciphertexts" should be indistinguishable from real ciphertexts, even when an adversary is given access to a decryption oracle before the fake ciphertext is known.

4.1 Definition of RNCE

Formally, a receiver non-committing encryption (RNCE) scheme consists of five PPT algorithms $(G, E, D, \tilde{F}, \tilde{R})$ such that:

- $G, E,$ and D are the key-generation, encryption, and decryption algorithms. These are defined just as for a standard encryption scheme, except that the key generation algorithm also outputs some auxiliary information z in addition to the public and secret keys pk and sk.
- The fake encryption algorithm $\tilde{F}$ takes as input $(\text{pk}, \text{sk}, z)$ and outputs a "fake ciphertext" $\tilde{c}$.
- The reveal algorithm $\tilde{R}$ takes as input $(\text{pk}, \text{sk}, z)$, a "fake ciphertext" $\tilde{c}$, and a message $m \in \mathcal{D}$. It outputs a "secret key" $\tilde{\text{sk}}$. (Intuitively, $\tilde{\text{sk}}$ is a secret key for which $\tilde{c}$ decrypts to m.)

We make the standard correctness requirement; namely, for any pk, sk, z output by G and any $m \in \mathcal{D}$, we have $D(\text{sk}; E(\text{pk}; m)) = m$.

Our definition of security requires, informally, that for any message m an adversary cannot distinguish whether it has been given a "real" encryption of m along with a "real" secret key, or a "fake" ciphertext along with a "fake" secret key under which the ciphertext decrypts to m. This should hold even when the adversary has non-adaptive access to a decryption oracle. We now give the formal definition.

Definition 2 (RNC-security). *Let $\mathcal{E} = (G, E, D, \tilde{F}, \tilde{R})$ be an RNCE scheme. We say that $\mathcal{E}$ is RNC-secure (or simply "secure") if the advantage of any* PPT *algorithm A in the game below is negligible in the security parameter k.*

1. *The key generation algorithm $G(1^k)$ is run to get $(\text{pk}, \text{sk}, z)$.*
2. *The algorithm A is given 1^k and pk as input, and is also given access to a decryption oracle $D(\text{sk}; \cdot)$. It then outputs a challenge message $m \in \mathcal{D}$.*
3. *A bit b is chosen at random. If $b = 1$ then a ciphertext $c \leftarrow E(\text{pk}; m)$ is computed, and A receives (c, sk). Otherwise, a "fake" ciphertext $\tilde{c} \leftarrow \tilde{F}(\text{pk}, \text{sk}, z)$*

and a "fake" secret key $\tilde{\sf sk} \leftarrow \tilde{R}({\sf pk},{\sf sk},z;\tilde{c},m)$ are computed, and A receives $(\tilde{c},\tilde{\sf sk})$. (After this point, A can no longer query its decryption oracle.) A outputs a bit b'.

The advantage of A is defined as $2 \cdot \left|\Pr[b' = b] - \frac{1}{2}\right|$.

It is easy to see that the RNC-security of $(G, E, D, \tilde{F}, \tilde{R})$ according to Definition 2 implies in particular that the underlying scheme (G, E, D) is secure against non-adaptive chosen-ciphertext attacks. It is possible to augment Definition 2 so as to grant the adversary access to the decryption oracle even after the ciphertext is known, but we do not need this stronger definition for our intended application (Section 5). We also comment that the Naor-Yung construction [NY90] is RNC-secure for 1-bit messages (if the secret key is chosen at random from the two underlying secret keys); a proof can be derived from [NY90] as well as our proof of Theorem 1.

4.2 A Secure RNCE Scheme for Polynomial-Size Message Spaces

Here, we show that the Cramer-Shoup cryptosystem [CS98] can be modified to give a secure RNCE scheme for *polynomial-size* message spaces. Interestingly, because our definition of security only involves non-adaptive chosen-ciphertext attacks, we can base our construction on the simpler and more efficient "Cramer-Shoup lite" scheme. In fact, the only difference is that we encode a message m by the group element g^m, rather than encoding it directly as the element m. (This encoding is essential for the reveal algorithm $\tilde{R}$.[2])

In what follows, we let $\mathcal{G} = \{\mathbb{G}_k\}_{k \in \mathbb{N}}$ be a family of finite, cyclic groups (written multiplicatively), where each group $\mathbb{G}_k$ has (known) prime order q_k and $|q_k| = k$. For simplicity, we describe our RNCE scheme for the message space $\{0, 1\}$; however, we will comment briefly afterward how the scheme can be extended for any polynomial-size message space.

Key Generation, G. Given the security parameter 1^k, let $\mathbb{G}$ denote $\mathbb{G}_k$ and q denote q_k. Choose at random $g_1 \leftarrow \mathbb{G} \setminus \{1\}$, and also choose random $\alpha, x_1, x_2, y_1, y_2 \leftarrow \mathbb{Z}_q$. Set $g_2 = g_1^\alpha$; $h = g_1^{x_1} g_2^{x_2}$; and $d = g_1^{y_1} g_2^{y_2}$. The public key is ${\sf pk} = (g_1, g_2, h, d)$, the secret key is ${\sf sk} = (x_1, x_2, y_1, y_2)$, and the auxiliary information is $z = \alpha$.

Encryption, E. Given a public key ${\sf pk} = (g_1, g_2, h, d)$ and a message $m \in \{0, 1\}$, choose a random $r \in \mathbb{Z}_q$, compute $u_1 = g_1^r$ $u_2 = g_2^r$, $e = g_1^m h^r$ and $v = d^r$. The ciphertext is $\langle u_1, u_2, e, v \rangle$.

Decryption, D. Given a ciphertext $\langle u_1, u_2, e, v \rangle$ and secret key ${\sf sk} = (x_1, x_2, y_1, y_2)$, proceed as follows: First check whether $u_1^{y_1} u_2^{y_2} = v$. If not, then output $\bot$. Otherwise, compute $w = e/u_1^{x_1} u_2^{x_2}$. If $w = 1$ (i.e., the group identity), output 0; if $w = g_1$, output 1. (If $w \notin \{1, g_1\}$ then output $\bot$.)

---

[2] Looking ahead, it is for this reason that the present construction only handles *polynomial-size* message spaces: the receiver only directly recovers g^m, and must search through the message space to find the corresponding message m.

Fake Encryption, $\tilde{F}$. Given $\mathsf{pk} = (g_1, g_2, h, d)$ and $\mathsf{sk} = (x_1, x_2, y_1, y_2)$, choose at random $r \in \mathbb{Z}_q$. Then compute $\tilde{u}_1 = g_1^r$, $\tilde{u}_2 = g_1 g_2^r$, $\tilde{e} = g_1^{x_2} h^r$ and $\tilde{v} = \tilde{u}_1^{y_1} \tilde{u}_2^{y_2}$, and output the "fake" ciphertext $\tilde{c} = \langle \tilde{u}_1, \tilde{u}_2, \tilde{e}, \tilde{v} \rangle$.

Reveal Algorithm, $\tilde{R}$. Given $\mathsf{pk} = (g_1, g_2, h, d)$, $\mathsf{sk} = (x_1, x_2, y_1, y_2)$, $z = \alpha$, a "fake" ciphertext $\langle \tilde{u}_1, \tilde{u}_2, \tilde{e}, \tilde{v} \rangle$, and a message $m \in \{0, 1\}$, set $x_2' = x_2 - m$ and $x_1' = x_1 + m\alpha$ (both in $\mathbb{Z}_q$) and output the "fake" secret key $\tilde{\mathsf{sk}} = (x_1', x_2', y_1, y_2)$.

One can check that the secret key $\tilde{\mathsf{sk}}$ matches the public key pk, since

$$g_1^{x_1'} g_2^{x_2'} = g_1^{x_1 + m\alpha} g_2^{x_2 - m} = (g_1^{x_1} g_2^m) g_2^{x_2 - m} = g_1^{x_1} g_2^{x_2} = h;$$

moreover, $\tilde{\mathsf{sk}}$ decrypts the "fake" ciphertext $\langle \tilde{u}_1, \tilde{u}_2, \tilde{e}, \tilde{v} \rangle$ to m, since

$$\frac{e}{\tilde{u}_1^{x_1'} \tilde{u}_2^{x_2'}} = \frac{g_1^{x_2} (g_1^{x_1'} g_2^{x_2'})^r}{(g_1^r)^{x_1'} (g_1 g_2^r)^{x_2'}} = \frac{g_1^{x_2 + r x_1'} g_2^{r x_2'}}{g_1^{r x_1' + x_2'} g_2^{r x_2'}} = g_1^{x_2 - x_2'} = g_1^m.$$

The above scheme can be immediately extended to support any polynomial-size message space: encryption, fake encryption, and reveal would be exactly the same, and decryption would involve computation of w, as above, followed by an exhaustive search through the message space to determine $m \stackrel{\text{def}}{=} \log_{g_1} w$. A proof of the following appears in the full version:

Theorem 2. *If the DDH assumption holds for $\mathcal{G}$, then the above scheme is RNC-secure.*

4.3 A Secure RNCE Scheme for Exponential-Size Message Spaces

The RNCE scheme in the previous section can be used only for message spaces of size polynomial in the security parameter, as the decryption algorithm works in time linear in the size of the message space. We now show a scheme that supports message spaces of size exponential in the security parameter. Just as in the previous section, we construct our scheme by appropriately modifying a (standard) cryptosystem secure against chosen-ciphertext attacks. Here, we base our construction on schemes developed independently by Gennaro and Lindell [GL03] and Camenisch and Shoup [CS03], building on earlier work by Cramer and Shoup [CS02]. Security of our scheme, as in these earlier schemes, is predicated on the decisional composite residuosity (DCR) assumption [P99].

Let p, q, p', q' be distinct primes with $p = 2p' + 1$ and $q = 2q' + 1$ (i.e., p, q are *strong* primes). Let $n = pq$ and $n' = p'q'$, and observe that the group $\mathbb{Z}_{n^2}^*$ can be decomposed as the direct product $\mathbb{G}_n \cdot \mathbb{G}_{n'} \cdot \mathbb{G}_2 \cdot \mathbf{T}$, where each $\mathbb{G}_i$ is a cyclic group of order i and $\mathbf{T}$ is the order-2 subgroup of $\mathbb{Z}_{n^2}^*$ generated by $(-1 \bmod n^2)$. This implies that there exist homomorphisms $\phi_n, \phi_{n'}, \phi_2, \phi_T$ from $\mathbb{Z}_{n^2}^*$ onto $\mathbb{G}_n$, $\mathbb{G}_{n'}$, $\mathbb{G}_2$, and $\mathbf{T}$, respectively, and every $x \in \mathbb{Z}_{n^2}^*$ is uniquely represented by the 4-tuple $(\phi_n(x), \phi_{n'}(x), \phi_2(x), \phi_T(x))$. We use also the fact that the element $\gamma \stackrel{\text{def}}{=} (1 + n) \bmod n^2$ has order n in $\mathbb{Z}_{n^2}^*$ (i.e., it generates a group isomorphic to $\mathbb{G}_n$) and furthermore $\gamma^a \bmod n^2 = 1 + an$, for any $0 \leq a < n$.

Let $\mathbf{P}_n \stackrel{\text{def}}{=} \{x^n \bmod n^2 : x \in \mathbb{Z}_{n^2}^*\}$ denote the subgroup of $\mathbb{Z}_{n^2}^*$ consisting of all n^{th} powers; note that $\mathbf{P}_n$ is isomorphic to the direct product $\mathbb{G}_{n'} \cdot \mathbb{G}_2 \cdot \mathbf{T}$. The DCR assumption (informally) is that, given n, it is hard to distinguish a random element of $\mathbf{P}_n$ from a random element of $\mathbb{Z}_{n^2}^*$.

Our RNCE scheme is defined below. In this description, we let $\mathcal{G}$ be an algorithm that on input 1^k randomly chooses two primes p', q' as above with $|p'| = |q'| = k$. Also, for a positive real number r we denote by $[r]$ the set $\{0, \ldots, \lfloor r \rfloor - 1\}$.

Key Generation, G. Given the security parameter 1^k, use $\mathcal{G}(1^k)$ to select two random k-bit primes p', q' for which $p = 2p'+1$ and $q = 2q'+1$ are also prime, and set $n = pq$ and $n' = p'q'$. Choose random $x, y \in [n^2/4]$ and a random $g' \in \mathbb{Z}_{n^2}^*$, and compute $g = (g')^{2n}$, $h = g^x$, and $d = g^y$. The public key is $\mathsf{pk} = (n, g, h, d)$, the secret key is $\mathsf{sk} = (x, y)$, and the auxiliary information is $z = n'$.

Encryption, E. Given a public key as above and a message $m \in [n]$, choose random $r \in [n/4]$, compute $u = g^r$, $e = \gamma^m h^r$, and $v = d^r$ (all in $\mathbb{Z}_{n^2}^*$), and output the ciphertext $c = \langle u, e, v \rangle$.

Decryption, D. Given a ciphertext $\langle u, e, v \rangle$ and secret key (x, y), check whether $u^{2y} = v^2$; if not, output $\bot$. Then, set $\hat{m} = (e/u^x)^{n+1}$. If $\hat{m} = 1 + mn$ for some $m \in [n]$, then output m; otherwise, output $\bot$.

Correctness follows, since for a valid ciphertext $\langle u, e, v \rangle$ we have $u^{2y} = (g^r)^{2y} = d^{2r} = v^2$, and also $(e/u^x)^{n+1} = (\gamma^m h^r/g^{rx})^{n+1} = (\gamma^m)^{n+1} = \gamma^m = 1 + mn$ (using for the third equality the fact that the order of γ is n).

Fake Encryption, $\tilde{F}$. Given $\mathsf{pk} = (n, g, h, d)$ and $\mathsf{sk} = (x, y)$, choose at random $r \in [n/4]$, compute $\tilde{u} = \gamma \cdot g^r$, $\tilde{e} = \tilde{u}^x$, and $\tilde{v} = \tilde{u}^y$ (all in $\mathbb{Z}_{n^2}^*$), and output the "fake" ciphertext $\tilde{c} = \langle \tilde{u}, \tilde{e}, \tilde{v} \rangle$.

Reveal Algorithm, $\tilde{R}$. Given $\mathsf{pk} = (n, g, h, d)$, $\mathsf{sk} = (x, y)$, $z = n'$, a "fake" ciphertext $\langle \tilde{u}, \tilde{e}, \tilde{v} \rangle$ as above, and a message $m \in [n]$, proceed as follows: Using the Chinese Remainder Theorem and the fact that $\gcd(n, n') = 1$, find the unique $x' \in [nn']$ satisfying $x' = x \bmod n'$, and $x' = x - m \bmod n$, and output the secret key $\tilde{\mathsf{sk}} = (x', y)$.

It can be verified that the secret key $\tilde{\mathsf{sk}}$ matches the public key pk and also decrypts the "fake" ciphertext to the required message m: For the second component y this is immediate and so we focus on the first component x'. First, the order of g divides n' and so $g^x = g^{x' \bmod n'} = g^{x \bmod n'} = g^x = h$. Furthermore, using also the fact that the order of γ in $\mathbb{Z}_{n^2}^*$ is n, we have

$$\left(\frac{\tilde{e}}{\tilde{u}^{x'}}\right)^{n+1} = \left(\frac{\gamma^x g^{rx}}{\gamma^{x'} g^{rx'}}\right)^{n+1} = \left(\gamma^{x-x' \bmod n}\right)^{n+1} = \gamma^m.$$

In the full version we define the decisional composite residuosity assumption (DCR) with respect to $\mathcal{G}$ (cf. [P99]), and show:

Theorem 3. *If the DCR assumption holds for $\mathcal{G}$, then the above scheme is RNC-secure.*

5 AFSE Based on Receiver Non-committing Encryption

We describe a construction of an AFSE scheme based on any secure RNCE scheme and any FSE scheme secure against chosen-ciphertext attacks. Let $\mathcal{E}' = (G', E', D', \tilde{F}, \tilde{R})$ be an RNCE scheme, and let $\mathcal{E}'' = (G'', U'', E'', D'')$ be a key-evolving encryption scheme. The message space of $\mathcal{E}'$ is $\mathcal{D}$, and we assume that ciphertexts of $\mathcal{E}'$ belong to the message space of $\mathcal{E}''$. We construct a new key-evolving encryption scheme $\mathcal{E} = (G, U, E, D)$ with message space $\mathcal{D}$ as follows:

Key Generation, G. On security parameter 1^k, run the key-generation algorithms of both schemes, setting $(\mathsf{pk}', \mathsf{sk}', z) \leftarrow G'(1^k)$ and $(\mathsf{pk}'', \mathsf{sk}''_0) \leftarrow G''(1^k)$. The public key is $(\mathsf{pk}', \mathsf{pk}'')$ and the initial secret key is $(\mathsf{sk}', \mathsf{sk}''_0)$. (The extra information z is ignored.)

Key update, U. The key-update operation is derived as one would expect from $\mathcal{E}''$; namely: $U(t; \mathsf{sk}', \mathsf{sk}''_t) = (\mathsf{sk}', U''(t; \mathsf{sk}''_t))$.

Encryption, E. To encrypt a message $m \in \mathcal{D}$ at time t, first compute $c' \leftarrow E'(\mathsf{pk}'; m)$ and then $c \leftarrow E''(\mathsf{pk}'', t; c')$. The resulting ciphertext is just c.

Decryption, D. To decrypt a ciphertext c, set $c' \leftarrow D''(\mathsf{sk}''_t; c)$ and then compute $m \leftarrow D'(\mathsf{sk}'; c')$.

Theorem 4. *If $\mathcal{E}'$ is RNC-secure, and if $\mathcal{E}''$ is forward-secure against chosen-ciphertext attacks, then the combined scheme given above is an AFSE scheme.*

We provide some informal intuition behind the proof of the above theorem. The most interesting scenario to consider is what happens upon player corruption, when the adversary obtains the secret key for the current time period t^*. We may immediately note that messages encrypted for prior time periods $t < t^*$ remain secret; this follows from the FSE encryption applied at the "outer" layer. Next, consider adaptive security for the (at most one) outstanding ciphertext which was encrypted for some time period $t \geq t^*$. Even though the adversary can "strip off" the outer later of the encryption (because the adversary now has the secret key for time period t^*), RNC security of the inner layer ensures that a simulator can open the inner ciphertext to any desired message. The main point here is that the simulator only needs to "fake" the opening of *one* inner ciphertext, and thus RNC security suffices. (Still, since the simulator does not know in advance what ciphertext it will need to open, it actually "fakes" *all* inner ciphertexts.) Chosen-ciphertext attacks are dealt with using the chosen-ciphertext security of the outer layer, as well as the definition of RNC security (where "lunch-time security" at the inner layer is sufficient). Also, we note that reversing the order of encryptions does not work: namely, using $RNCE(FSE(m))$ does not yield adaptive security, even if the RNCE scheme is fully CCA secure.

References

[AF04] M. Abe and S. Fehr. Adaptively Secure Feldman VSS and Applications to Universally-Composable Threshold Cryptography. *Crypto 2004*, LNCS vol. 3152, pp. 317–334, 2004. Full version available at eprint.iacr.org/2004/119.

[A97] R. Anderson. Two Remarks on Public Key Cryptology. Invited lecture, given at *ACM CCCS '97*. Available at http://www.cl.cam.ac.uk/ftp/users/rja14/forwardsecure.pdf.

[B97] D. Beaver. Plug and Play Encryption. *Crypto 1997*, LNCS vol. 1294, pp. 75–89, 1997.

[BH92] D. Beaver and S. Haber. Cryptographic Protocols Provably Secure Against Dynamic Adversaries. *Eurocrypt 1992*, LNCS vol. 658, pp. 307–323, 1992.

[BDPR98] M. Bellare, A. Desai, D. Pointcheval, and P. Rogaway. Relations among Notions of Security for Public-Key Encryption Schemes. *Crypto 1998*, LNCS vol. 1462, pp. 26–45, 1998.

[BB04] D. Boneh and X. Boyen. Efficient Selective-ID Secure Identity Based Encryption Without Random Oracles. *Eurocrypt 2004*, LNCS vol. 3027, pp. 223–238, 2004.

[CS03] J. Camenisch and V. Shoup. Practical Verifiable Encryption and Decryption of Discrete Logarithms. *Crypto 2003*, LNCS vol. 2729, pp. 126–144, 2003.

[C01] R. Canetti. Universally Composable Security: A New Paradigm for Cryptographic Protocols. *42nd IEEE Symposium on Foundations of Computer Science (FOCS)*, pp. 136–145, 2001. Also available as ECCC TR 01-16, or from http://eprint.iacr.org/2000/067.

[CFGN96] R. Canetti, U. Feige, O. Goldreich, and M. Naor. Adaptively Secure Computation. *28th ACM Symposium on Theory of Computing (STOC)*, pp. 639–648, 1996. Full version in MIT-LCS-TR #682, 1996.

[CHK03] R. Canetti, S. Halevi, and J. Katz. A Forward-Secure Public-Key Encryption Scheme. *Eurocrypt 2003*, LNCS vol. 2656, pp. 255–271, 2003. Full version available at http://eprint.iacr.org/2003/083.

[CHK04] R. Canetti, S. Halevi, and J. Katz. Chosen-Ciphertext Security from Identity-Based Encryption. *Eurocrypt 2004*, LNCS vol. 3027, pp. 207–222, 2004. Full version available at http://eprint.iacr.org/2003/182.

[CHK05] R. Canetti, S. Halevi, and J. Katz. Adaptively-Secure, Non-Interactive Public-Key Encryption. Full version available at http://eprint.iacr.org/2004/317.

[CKN03] R. Canetti, H. Krawczyk, and J.B. Nielsen. Relaxing Chosen Ciphertext Security. *Crypto 2003*, LNCS vol. 2729, pp. 565–582, 2003. Full version available at http://eprint.iacr.org/2003/174.

[CS98] R. Cramer and V. Shoup. A Practical Public Key Cryptosystem Provably Secure Against Chosen Ciphertext Attack. *Crypto 1998*, LNCS vol. 1462, pp. 13–25, 1998.

[CS02] R. Cramer and V. Shoup. Universal Hash Proofs and a Paradigm for Adaptive Chosen Ciphertext Secure Public-Key Encryption. *Eurocrypt 2001*, LNCS vol. 2332, pp. 45–63, 2001.

[DN00] I. Damgård and J. B. Nielsen. Improved Non-Committing Encryption Schemes Based on General Complexity Assumptions. *Crypto 2000*, LNCS vol. 1880, pp. 432–450, 2000.

[DDOPS01] A. De Santis, G. Di Crescenzo, R. Ostrovsky, G. Persiano, and A. Sahai. Robust Non-Interactive Zero Knowledge. *Crypto 2001*, LNCS vol. 2139, pp. 566–598, 2001.
[DDN00] D. Dolev, C. Dwork, and M. Naor. Non-Malleable Cryptography. *SIAM. J. Computing* 30(2): 391-437, 2000.
[GL03] R. Gennaro and Y. Lindell. A Framework for Password-Based Authenticated Key Exchange. *Eurocrypt 2003*, LNCS vol. 2656, pp. 524–543, 2003. Full version available at http://eprint.iacr.org/2003/032.
[GM84] S. Goldwasser and S. Micali. Probabilistic Encryption. *J. Computer System Sciences* 28(2): 270-299, 1984.
[HMS03] Dennis Hofheinz, Joern Mueller-Quade, and Rainer Steinwandt. On Modeling IND-CCA Security in Cryptographic Protocols. Available at http://eprint.iacr.org/2003/024.
[JL00] S. Jarecki and A. Lysyanskaya. Adaptively Secure Threshold Cryptography: Introducing Concurrency, Removing Erasures. *Eurocrypt 2000*, LNCS vol. 1807, pp. 221–242, 2000.
[NY90] M. Naor and M. Yung. Public-Key Cryptosystems Provably-Secure against Chosen-Ciphertext Attacks. *22nd ACM Symposium on Theory of Computing (STOC)*, pp. 427–437, 1990.
[N02] J.B. Nielsen. Separating Random Oracle Proofs from Complexity Theoretic Proofs: The Non-Committing Encryption Case. *Crypto 2002*, LNCS vol. 2442, pp. 111–126, 2002.
[P99] P. Paillier. Public-Key Cryptosystems Based on Composite Degree Residuosity Classes. *Eurocrypt 1999*, LNCS vol. 1592, pp. 223–238, 1999.
[RS91] C. Rackoff and D. Simon. Non-Interactive Zero-Knowledge Proof of Knowledge and Chosen Ciphertext Attack. *Crypto 1991*, LNCS vol. 576, pp. 433–444, 1991.
[S99] A. Sahai. Non-Malleable Non-Interactive Zero Knowledge and Adaptive Chosen-Ciphertext Security. *40th IEEE Symposium on Foundations of Computer Science (FOCS)*, pp. 543–553, 1999.

A Key-Evolving and Forward-Secure Encryption

We review the definitions of key-evolving and forward-secure encryption schemes from [CHK03].

Definition 3. *A (public-key) key-evolving encryption (*ke-PKE*) scheme is a 4-tuple of* PPT *algorithms* (Gen, Upd, Enc, Dec) *such that:*

- *The key generation algorithm* Gen *takes as input a security parameter* 1^k *and the total number of time periods* N. *It returns a public key* pk *and an initial secret key* sk_0.
- *The key update algorithm* Upd *takes as input* pk, *an index* $t < N$ *of the current time period, and the associated secret key* sk_t. *It returns the secret key* sk_{t+1} *for the following time period.*
- *The encryption algorithm* Enc *takes as input* pk, *an index* $t \leq N$ *of a time period, and a message* M. *It returns a ciphertext* C.

- The decryption algorithm Dec *takes as input* pk, *an index* $t \leq N$ *of the current time period, the associated secret key* sk_t, *and a ciphertext* C. *It returns a message* M.

We require that $\mathsf{Dec}(\mathsf{sk}_t; t; \mathsf{Enc}(\mathsf{pk}_t, t, M)) = M$ *holds for all* $(\mathsf{pk}, \mathsf{sk}_0)$ *output by* Gen, *all time periods* $t \leq N$, *all correctly generated* sk_t *for this* t, *and all messages* M.

Definition 4. *A ke-PKE scheme is* forward-secure against chosen plaintext attacks *(fs-CPA) if for all polynomially-bounded functions* $N(\cdot)$, *the advantage of any* PPT *adversary in the following game is negligible in the security parameter:*

Setup: $\mathsf{Gen}(1^k, N(k))$ *outputs* (PK, SK_0). *The adversary is given* PK.

Attack: *The adversary issues one* breakin(i) *query and one* challenge(j, M_0, M_1) *query, in either order, subject to* $0 \leq j < i < N$. *These queries are answered as follows:*

- *On query* breakin(i), *key* SK_i *is computed via* $\mathsf{Upd}(PK, i-1, \cdots \mathsf{Upd}(PK, 0, SK_0)\cdots)$. *This key is then given to the adversary.*
- *On query* challenge(j, M_0, M_1), *a random bit* b *is selected and the adversary is given* $C^* = \mathsf{Enc}(PK, j, M_b)$.

Guess: *The adversary outputs a guess* $b' \in \{0, 1\}$; *it succeeds if* $b' = b$. *The adversary's* advantage *is the absolute value of the difference between its success probability and* $1/2$.

Forward security against (adaptive) chosen-ciphertext attacks (fs-CCA security) is defined by the natural extension of the above definition in which the adversary is given decryption oracle access during both the "Attack" and "Guess" stages.

B The UC Framework, Abridged

We provide a brief review of the universally composable security framework [C01]. The framework allows for defining the security properties of cryptographic tasks so that security is maintained under general composition with an unbounded number of instances of arbitrary protocols running concurrently. Definitions of security in this framework are called universally composable (UC).

In the UC framework, the security requirements of a given task (i.e., the functionality expected from a protocol that carries out the task) are captured via a set of instructions for a "trusted party" that obtains the inputs of the participants and provides them with the desired outputs (in one or more iterations). Informally, a protocol securely carries out a given task if running the protocol with a realistic adversary amounts to "emulating" an ideal process where the parties hand their inputs to a trusted party with the appropriate functionality and obtain their outputs from it, without any other interaction.

The notion of emulation in the UC framework is considerably stronger than that considered in previous models. Traditionally, the model of computation includes the parties running the protocol and an adversary $\mathcal{A}$ that controls the communication channels and potentially corrupts parties. "Emulating an ideal process" means that for any adversary $\mathcal{A}$ there should exist an "ideal process adversary" (or simulator) $\mathcal{S}$ that causes the *outputs* of the parties in the ideal process to have similar distribution to the outputs of the parties in an execution of the protocol. In the UC framework the requirement on $\mathcal{S}$ is more stringent. Specifically, an additional entity, called the environment $\mathcal{Z}$, is introduced. The environment generates the inputs to all parties, reads all outputs, and in addition interacts with the adversary in an arbitrary way throughout the computation. A protocol is said to securely realize functionality $\mathcal{F}$ if for any "real-life" adversary $\mathcal{A}$ that interacts with the protocol and the environment there exists an "ideal-process adversary" $\mathcal{S}$, such that *no environment* $\mathcal{Z}$ can tell whether it is interacting with $\mathcal{A}$ and parties running the protocol, or with $\mathcal{S}$ and parties that interact with $\mathcal{F}$ in the ideal process. In a sense, $\mathcal{Z}$ serves as an "interactive distinguisher" between a run of the protocol and the ideal process with access to $\mathcal{F}$.

The following *universal composition theorem* is proven in [C01]. Consider a protocol π that operates in the $\mathcal{F}$-hybrid model, where parties can communicate as usual and in addition have ideal access to an unbounded number of *copies* of the functionality $\mathcal{F}$. Let ρ be a protocol that securely realizes $\mathcal{F}$ as sketched above, and let π^ρ be identical to π with the exception that the interaction with *each copy* of $\mathcal{F}$ is replaced with an interaction with a *separate instance* of ρ. Then, π and π^ρ have essentially the same input/output behavior. In particular, if π securely realizes some functionality $\mathcal{I}$ in the $\mathcal{F}$-hybrid model then π^ρ securely realizes $\mathcal{I}$ in the standard model (i.e., without access to any functionality).

B.1 The Public-Key Encryption Functionality $\mathcal{F}_{\text{PKE}}$

(This section is taken almost verbatim from [CKN03].) Within the UC framework, public-key encryption is defined via the public-key encryption functionality, denoted $\mathcal{F}_{\text{PKE}}$ and presented in Figure 2. Functionality $\mathcal{F}_{\text{PKE}}$ is intended to capture the functionality of public-key encryption and, in particular, is written in a way that allows realizations consisting of three non-interactive algorithms without any communication. (The three algorithms correspond to the key generation, encryption, and decryption algorithms in traditional definitions.)

Referring to Figure 2, we note that *sid* serves as a unique identifier for an instance of functionality $\mathcal{F}_{\text{PKE}}$ (this is needed in a general protocol setting when this functionality can be composed with other components, or even with other instances of $\mathcal{F}_{\text{PKE}}$). It also encodes the identity of the decryptor for this instance. The "public key value" pk has no particular meaning in the ideal scenario beyond serving as an identifier for the public key related to this instance of the functionality, and this value can be chosen arbitrarily by the attacker. Also, in the ideal setting ciphertexts serve as identifiers or tags with no particular relation to the encrypted messages (and as such are also chosen by the adversary without knowledge of the plaintext). Still, rule 1 of the decryption operation guarantees that "legitimate ciphertexts" (i.e., those produced and recorded by the functionality under an Encrypt request) are decrypted correctly, while the resultant plaintexts

---

Functionality $\mathcal{F}_{\text{PKE}}$

$\mathcal{F}_{\text{PKE}}$ proceeds as follows, when parameterized by message domain ensemble $\mathcal{D} = \{D_k\}_{k \in \mathbf{N}}$ and security parameter k.

Key Generation: Upon receiving a value (KeyGen, sid) from some party R^*, verify that $sid = (sid', R^*)$. If not, then ignore the input. Otherwise:
 1. Hand (KeyGen, sid) to the adversary.
 2. Receive a value pk* from the adversary, and hand pk* to R^*.
 3. If this is the first KeyGen request, record R^* and pk*.

Encryption: Upon receiving from some party P a value (Encrypt, sid, pk, m) proceed as follows:
 1. If $m \notin D_k$ then return an error message to P.
 2. If $m \in D_k$ then hand (Encrypt, sid, pk, P) to the adversary. (If pk $\neq$ pk* or pk* is not yet defined then hand also the entire value m to the adversary.)
 3. Receive a "ciphertext" c from the adversary, record the pair (c, m), and send (ciphertext, c) to P. (If pk $\neq$ pk* or pk* is not yet defined then do *not* record the pair (c, m).)

Decryption: Upon receiving a value (Decrypt, sid, c) from R^* (and R^* only), proceed as follows:
 1. If there is a recorded pair (c, m) then hand m to R^*. (If there is more than one such pair then use the first one.)
 2. Otherwise, hand the value (Decrypt, sid, c) to the adversary. When receiving a value m' from the adversary, hand m' to R^*.

Fig. 2. The public-key encryption functionality, $\mathcal{F}_{\text{PKE}}$

remain unknown to the adversary. In contrast, ciphertexts that were not legitimately generated can be decrypted in any way chosen by the ideal-process adversary. (Since the attacker obtains no information about legitimately-encrypted messages, we are guaranteed that illegitimate ciphertexts will be decrypted to values that are independent from these messages.) Note that the same illegitimate ciphertext can be decrypted to different values in different activations. This provision allows the decryption algorithm to be non-deterministic with respect to ciphertexts that were not legitimately generated.

Another characteristic of $\mathcal{F}_{\text{PKE}}$ is that, when activated with a KeyGen request, it always responds with an (adversarially-chosen) encryption key pk$'$. Still, only the first key to be generated is recorded, and only messages that are encrypted with that key are guaranteed to remain secret. Messages encrypted with other keys are disclosed to the adversary in full. This modeling represents the fact that a single copy of the functionality captures the security requirements of only a single instance of a public-key encryption scheme (i.e., a single pair of encryption and decryption keys). Other keys may provide correct encryption and decryption, but do not guarantee any security (see [CKN03] for further discussion about possible alternative formulations of the functionality).

Adaptive Security of Symbolic Encryption*

Daniele Micciancio and Saurabh Panjwani

University of California, San Diego, La Jolla, CA 92093, USA
{daniele, panjwani}@cs.ucsd.edu

Abstract. We prove a computational soundness theorem for the symbolic analysis of cryptographic protocols which extends an analogous theorem of Abadi and Rogaway (J. of Cryptology 15(2):103–127, 2002) to a scenario where the adversary gets to see the encryption of a sequence of adaptively chosen symbolic expressions. The extension of the theorem of Abadi and Rogaway to such an adaptive scenario is nontrivial, and raises issues related to the classic problem of selective decommitment, which do not appear in the original formulation of the theorem.

Although the theorem of Abadi and Rogaway applies only to passive adversaries, our extension to adaptive attacks makes it substantially stronger, and powerful enough to analyze the security of cryptographic protocols of practical interest. We exemplify the use of our soundness theorem in the analysis of group key distribution protocols like those that arise in multicast and broadcast applications. Specifically, we provide cryptographic definitions of security for multicast key distribution protocols both in the symbolic as well as the computational framework and use our theorem to prove soundness of the symbolic definition.

Keywords: Symbolic encryption, adaptive adversaries, soundness theorem, formal methods for security protocols.

1 Introduction

Traditionally, security protocols have been designed and analyzed using two competing approaches: the symbolic one, and the computational one. The symbolic approach is characterized by an abstract (adversarial) execution model, where cryptographic operations and objects are treated as an abstract data type, not only when used by honest protocol participants, but also when used by malicious players attacking the system. This allows for simple proofs of security, typically based on syntactic properties of the messages exchanged during the execution of the protocol. The computational approach is based on a more detailed execution model that accounts for a much wider class of adversaries attacking the system, namely arbitrary probabilistic polynomial time bounded adversaries that

---

* Research supported in part by NSF grants 0313241 and 0430595. Any opinions, findings, and conclusions or recommendations expressed in this material are those of the author(s) and do not necessarily reflect the views of the National Science Foundation.

do not necessarily respect the cryptographic abstractions used by the protocol. The stronger security guarantees offered by the computational approach come at a substantial price in complexity: proofs of security in this framework typically involve subtle probabilistic arguments, complicated running time analysis, and the ubiquitous use of computational assumptions, like the intractability of factoring large integers.

Recently, there has been a lot of interest in combining the two approaches, with the generic goal of coming up with abstract models that allow computationally sound symbolic security analysis, i.e., a method to translate symbolic security proofs into precise computational statements about the security of concrete protocol executions in the computational framework.

Our work follows a line of research initiated by Abadi and Rogaway in [2], where a simple language of encrypted expressions is defined, together with a computationally sound symbolic semantics. Technically, [2] introduces a mapping from expressions to *patterns* that characterize the information leaked by the expressions when evaluated using a computationally secure encryption scheme. The structure of the soundness result of [2] is rather simple: an adversary attacking the system produces a symbolic expression and subsequently receives the computational evaluation of either the expression or its pattern. The soundness theorem of [2] states that if the expression satisfy certain syntactic restrictions (namely, it does not contain encryption cycles) then the adversary cannot efficiently determine if it received the evaluation of the expression or its pattern.

The result of [2] is an interesting first step demonstrating the feasibility of computationally sound symbolic security analysis. The class of encrypted expressions considered in [2] is fairly general, and allows to describe the messages transmitted in many practical protocols. However, the result itself is too simple for direct application to security analysis of protocols. Intuitively, the scenario considered in [2] involves a party sending a single message to another party over an authenticated channel, and a passive adversary monitoring the channel. In practice, security protocols involve the exchange of several messages, among two or more parties, and in different directions. Moreover, messages may depend on the computational interpretation of previously chosen messages and/or external inputs that are not known at the beginning of the protocol.

Our Results. In this paper, we consider an extension of [2] wherein the adversary produces a *sequence* of expressions, which are subsequently evaluated according to a common key assignment. If all the expressions in the sequence were specified at the same time, then this wouldn't be any different from the original soundness theorem of [2], as all the expressions in the sequence could be concatenated into a single expression. What makes our extension interesting and nontrivial is the fact that the sequence is adversarially specified in an *adaptive* way, so that each expression in the sequence may depend on the computational evaluation of the previous ones. The ability to specify the expressions adaptively allows the adversary to generate probability distributions that do not correspond to any fixed sequence of expressions, and immediately raises issues related to the classic problem of selective decommitment [9] and adaptive corruption [5]. (See

Section 3 for details.) In order to avoid these problems we introduce some syntactical restrictions on the expressions, beside the acyclicity condition already considered by [2]. Informally, the syntactic restrictions postulate that each key is used in two stages: a key distribution stage during which the key can be used as a message, and a key deployment stage during which the key is used to encrypt other messages. Using these syntactic restrictions, we are able to bypass the selective decommitment problems, and prove a soundness result for symbolic encryption with adaptively chosen expressions.

Our soundness result allows to analyze a generic class of protocols that involve communication between multiple parties over an authenticated network.[1] The execution model for these protocols involves an adversary that observes all messages sent over the network and can adaptively change the execution flow of the protocol (e.g., through interaction with the execution environment), but is not allowed to modify or delete any of the messages sent or received by the legitimate parties.

Our soundness result for adaptively chosen encrypted expressions substantially increases the expressive power of the soundness theorem of [2], making it powerful enough to analyze practical cryptographic protocols. We exemplify the use of our soundness theorem in the analysis of group key distribution protocols, like those used in multicast applications [23, 6, 22]. In the multicast key distribution problem, a data source wants to broadcast information to a dynamically changing group of parties, in such a way that at any given point in time only current group members can decipher the transmitted messages. The problem is typically solved by establishing a secret key, known to all and only the "current" group members. Each time a user joins or leaves the group, a group controller broadcasts some messages which are used by the new set of members to update the group key.

We give formal definitions of security for multicast key distribution protocols, in both the computational framework and the symbolic framework, and show that if a protocol is secure in the symbolic setting, then the (implementation of the) protocol is also secure in the computational setting (provided the messages used in the protocol conform with the syntactic restrictions of our soundness theorem). Most multicast key distribution protocols we are aware of (e.g., [23, 6]) satisfy these restrictions and can thus be proven secure (against powerful computational adversaries) using the symbolic definition. To the best of our knowledge, formal definitions for security of these protocols have not been discussed in the literature prior to this work nor has there been any attempt to relate security analysis of these protocols in the symbolic framework (as is done implicitly in many papers) to computational security guarantees on their implementations. Our soundness theorem is an important building block in that direction.

---

[1] Authenticated channels are a widely used model in cryptographic protocol design, and can be implemented on top of non-authenticated networks using standard techniques, like message authentication codes and digital signatures.

Related Work. Several improvements and refinements have followed the original work of Abadi and Rogaway [2], but they are mostly orthogonal to our results. In [19], Micciancio and Warinschi prove a converse of the soundness theorem, showing that if a sufficiently strong encryption scheme is used, then a computationally bounded adversary can recover all and only the information captured by the symbolic patterns. The result is further refined by Gligor and Horvitz [10], who give an exact characterization of the computational requirements on the encryption scheme under which the completeness theorem of [19] holds true.

An extension of the soundness result of [2] to multiple message/player settings, is presented by Abadi and Jürjens in [1]. This result considers an arbitrary set of parties exchanging several messages over an authenticated network over time. However, the protocol specification language used in [1] only allows to describe protocols in which all messages transmitted during the protocol execution can be uniquely determined before the execution of the protocol begins. In other words, the result of [1] does not account for scenarios where the messages are chosen adaptively, and from a technical point of view, it is much closer to [2] than to our work.

The papers [20, 14] present two different extensions of the framework of [2] that allow for active attacks, i.e., adversaries that have a total control of the communication network and may drop, alter, or inject messages in the network. Both works are based on encryption schemes satisfying the stronger notion of security against chosen ciphertext attack (CCA [21, 8]). Our results hold for any encryption scheme secure against chosen plaintext attack (CPA [11]). Moreover, the results of [20, 14] have a qualitatively different and somehow more complex formulation than the results presented in this paper. In [20] Micciancio and Warinschi consider trace properties[2] of both symbolic and computational executions of cryptographic protocols and relate the two models by proving that (if a CCA secure encryption scheme is used) any protocol that satisfies a trace property in all its symbolic executions, also satisfies the corresponding trace property in computational executions with overwhelming probability. We remark that the results in [20] are incomparable with those presented in this paper as trace properties do not allow to readily model indistinguishability properties as considered in this paper. Laud's result [14] is quite different from the other soundness results considered so far. Rather than considering a computational and a symbolic execution models, and relating the two, [14] only considers a computational execution model and a set of symbolic program transformations, and proves that the symbolic transformations are computationally sound in the sense that they preserve computational secrecy properties when both the original and transformed program are executed in the computational model.

Other approaches to the problem of computationally sound symbolic analysis are exemplified by [4, 3, 15, 13]. In [4, 3] Backes, Pfitzmann and Waidner present

---

[2] These are a class of properties, extensively used in the formal verification of distributed protocols, which can be represented as the allowable sequences of internal states (or external actions) performed by the honest protocol participants.

an implementation of Dolev-Yao style terms achieving a simulation based security definition within a general computational framework. It is important to notice that while [4, 3] allow to formulate and prove computational security properties of protocols built using their library, their results do not apply to protocols that make direct use of encryption schemes satisfying standard security notions, like CPA or CCA indistinguishability. We remark that [3] relies on syntactic restrictions on the use of symmetric encryption similar to those used in this paper. Our work shows that the difficulties encountered in [3] in trying to lift these restrictions are not specific to their universally composable security framework, but arise already in much simpler scenarios as those considered in this paper. In [15] Lincoln, Mitchell, Mitchell and Scedrov present a probabilistic process calculus that can be used to analyze computational security properties of cryptographic protocols. All these works are both substantially more complex and powerful than (though, technically incomparable to) the line of work initiated by [2], as they allow to describe arbitrary probabilistic polynomial time computations. The work of Impagliazzo and Kapron [13] approaches the problem of computationally sound symbolic analysis from still another side. They present an axiomatic system with limited forms of recursion that can be used to carry out proofs of the type used in the analysis of basic cryptographic constructions without the explicit use of nested quantifiers and asymptotic notation. An interesting question is whether the soundness theorem proved in this paper can be proved within the logic of [13].

Organization. After giving some basic definitions in Section 2, we present our soundness theorem in Section 3. The proof of the soundness theorem is given in Section 5, after describing an application to multicast key distribution in Section 4. Section 6 concludes with a discussion of future work and open problems.

2 Preliminaries

Let **Keys** and **Const** be two sets of symbols called *keys* and *constants* respectively. We can assume that both sets are finite, and have size bounded by a polynomial in the security parameter. For a given value of the security parameter, let **Keys** $:= \{K_1, \cdots, K_n\}$ be the set of keys. We define a language, **Exp**, of expressions, called *basic expressions*, that is generated using the following syntactic rules:

$$M \to (M, M) | \{M\}_{K_1} | \{M\}_{K_2} | \cdots | \{M\}_{K_n}$$
$$M \to \text{Each symbol in } \mathbf{Keys} \cup \mathbf{Const}$$

The rule $M \to (M, M)$ symbolizes a pairing operation while $M \to \{M\}_{K_i}$ for any K_i symbolizes encryption under K_i. Sequences of expressions can be converted into a single expression using the paring operation in the obvious way, e.g., the sequence $(M[1], \ldots, M[q])$ can be represented by the expression $(M[1], (M[2], \ldots, (M[q-1], M[q]) \ldots))$. For any sequence $(M[1], \ldots, M[q])$ and indexes $1 \leq i \leq j \leq q$, we use notation $M[i..j]$ to denote the subsequence $(M[i], M[i+1], \ldots, M[j])$.

For any $M \in \mathbf{Exp}$, a key that occurs as a plaintext in some sub-expression of M is referred to as a *message key* while one that is used to encrypt some sub-expression is called an *encryption key*. We denote the set of all message (resp. encryption) keys of M by $\mathbf{MsgKeys}(M)$ (resp. $\mathbf{EncKeys(M)}$). We say that a key K_i encrypts K_j (or K_j is *encrypted under* K_i) in M, denoted $K_i \rightarrow_M K_j$, if M contains a sub-expression $\{M'\}_{K_i}$ such that $K_j \in \mathbf{MsgKeys}(M')$. As in [2], we call a key *recoverable* in M if it is in $\mathbf{MsgKeys}(M)$ and occurs in it either unencrypted or encrypted under keys that are, in turn, recoverable in it. The set of all recoverable keys of M is denoted $\mathbf{RecKeys}(M)$. The set of all unrecoverable encryption keys in M, i.e. the set $\mathbf{EncKeys}(M) \setminus \mathbf{RecKeys}(M)$, is denoted $\mathbf{UEncKeys}(M)$. As an example, if $M = ((K_1, \{K_2\}_{K_1}), \{K_4\}_{K_3})$, then $\mathbf{EncKeys}(M) = \{K_1, K_3\}$; $\mathbf{MsgKeys}(M) = \{K_1, K_2, K_4\}$; $\mathbf{RecKeys}(M) = \{K_1, K_2\}$ and $\mathbf{UEncKeys}(M) = \{K_3\}$.

Formal Semantics. The information that can be extracted from an expression using known keys and the decryption algorithm can be represented by a syntactic object, called *pattern*. We use a definition of patterns recently proposed in [16] which characterizes encryption schemes satisfying the standard notion of semantic security under chosen plaintext attack [11]. (These patterns are slightly different from those used in [2, 19, 10], which correspond to encryption schemes satisfying a variant of semantic security. A definition similar to ours was also used in [12].) We define the *structure* of an expression $M \in \mathbf{Exp}$, denoted $\texttt{struct}(M)$, as the expression obtained by substituting all message keys in M by a symbol K', all encryption keys in it by K and all constants by a symbol $\texttt{c}$, where $K, K' \notin \mathbf{Keys}$ and $\texttt{c} \notin \mathbf{Const}$ are all fresh symbols. For example, $\texttt{struct}(\{\mathbf{0}\}_{K_2}, \{(K_2, \{K_1\}_{K_2})\}_{K_3}) = (\{\texttt{c}\}_K, \{(K', \{K'\}_K)\}_K)$.

Definition 1. *For any $M \in \mathbf{Exp}$, the pattern of M given a set of keys T, denoted $\texttt{pat}(M, T)$, is an expression defined recursively as follows:*

- *If $M \in \mathbf{Keys} \cup \mathbf{Const}$, then $\texttt{pat}(M, T) = M$.*
- *If $M = (M_1, M_2)$, then $\texttt{pat}(M, T) = (\texttt{pat}(M_1, T), \texttt{pat}(M_2, T))$.*
- *If $M = \{M'\}_{K_i}$ and $K_i \in T$, then $\texttt{pat}(M, T) = \{\texttt{pat}(M', T)\}_{K_i}$.*
- *If $M = \{M'\}_{K_i}$ and $K_i \notin T$, then $\texttt{pat}(M, T) = \{\texttt{struct}(M')\}_{K_i}$.*

The pattern of M, denoted $\texttt{pattern}(M)$, is defined as $\texttt{pat}(M, \mathbf{RecKeys}(M))$.

This definition of patterns captures the intuitive idea that given a bitstring interpretation of any expression encrypted under say K_i, an adversary can learn everything about the expression if he knows K_i, but can learn nothing more than its structure if he does not know K_i.

Just as in [2], we say that two expressions $M_1, M_2 \in \mathbf{Exp}$ are *equivalent*, denoted $M_1 \simeq M_2$, if their patterns, when viewed as strings of symbols, are identical up to renaming of their keys. That is, M_1 and M_2 are equivalent if there exists an injective map, ϕ, from the keys in M_1 to the keys in M_2 such that when every key K_i (other than a structure key) in $\texttt{pattern}(M_1)$ is substituted with $\phi(K_i)$, the resulting expression is identical to pattern(M_2). For example, if we consider the expressions

$$M_1 = ((K_1, \{\mathbf{0}\}_{K_2}), \{K_2, \{K_1\}_{K_2}\}_{K_3})$$
$$M_2 = ((K_1, \{\mathbf{1}\}_{K_6}), \{K_5, \{K_6\}_{K_1}\}_{K_2})$$
$$M_3 = ((K_1, \{K_5\}_{K_2}), \{K_5, \{K_6\}_{K_1}\}_{K_2})$$

we have $M_1 \simeq M_2$ but $M_1 \not\simeq M_3$.

Computational Semantics. We define computational semantics of all expressions, including all basic expressions and their respective patterns, using a single procedure. (We denote the set of all such expressions by **Exp'** below). For any symmetric encryption scheme, $\Pi = \{\mathcal{K}, \mathcal{E}, \mathcal{D}\}$, let $\mathcal{K}^m(\eta)$ denote the random variable corresponding to a vector of m keys sampled independently using the key generation algorithm $\mathcal{K}$, giving it security parameter η as input. The procedure for defining computational semantics of expressions takes the security parameter η as input and works in two steps.

1. We generate a key vector τ from the distribution $\mathcal{K}^{n+2}(\eta)$ (where $n = |\mathbf{Keys}|$) and map all keys in $\mathbf{Keys} \cup \{K', K\}$ to elements in this vector. Specifically, for all $i \in \{1, \cdots, n\}$, $\tau[i]$ corresponds to K_i, $\tau[n+1]$ to K' and $\tau[n+2]$ to K.
2. In the second step, we look at expressions in **Exp'** and for each expression M we define the *bitstring interpretation of M given τ* as a random variable, $[\![M]\!]_{\Pi,\tau}$, in the following recursive manner:
   – If $M \in \mathbf{Const} \cup \{\mathbf{c}\}$, then $[\![M]\!]_{\Pi,\tau}$ is the bitstring representation of M, using some standard encoding.
   – If $M = K_i \in \mathbf{Keys}$, then $[\![M]\!]_{\Pi,\tau} \equiv \tau[i]$. If $M = K'$, then $[\![M]\!]_{\Pi,\tau} \equiv \tau[n+1]$.
   – If $M = (M_1, M_2)$ for some $M_1, M_2 \in \mathbf{Exp'}$, then $[\![M]\!]_{\Pi,\tau}$ is the random variable corresponding to $([\![M_1]\!]_{\Pi,\tau}, [\![M_2]\!]_{\Pi,\tau})$ (we use some standard efficiently computable and invertible encoding for the pairing operation).
   – If $M = \{M'\}_{K_i}$ for some $M' \in \mathbf{Exp'}$ and $K_i \in \mathbf{Keys}$, then $[\![M]\!]_{\Pi,\tau}$ is the random variable corresponding to $\mathcal{E}_{\tau[i]}([\![M']\!]_{\Pi,\tau})$. If $M = \{M'\}_K$ for some $M' \in \mathbf{Exp'}$, then $[\![M]\!]_{\Pi,\tau}$ is the random variable corresponding to $\mathcal{E}_{\tau[n+2]}([\![M']\!]_{\Pi,\tau})$.

Security of Encryption. We consider encryption schemes that are semantically secure against chosen plaintext attacks. For any symmetric encryption scheme $\Pi = \{\mathcal{K}, \mathcal{E}, \mathcal{D}\}$, a left-right oracle, $\mathcal{LR}_{\Pi,b}$ for Π is a program that first generates a key k using the key generating algorithm $\mathcal{K}$ and then for each query, (m_0, m_1) (m_0 and m_1 being bitstrings of equal length), given to it replies with the ciphertext $\mathcal{E}_k(m_b)$. Π is called ind-cpa secure if for any probabilistic polynomial-time distinguisher D the following quantity

$$\mathbf{Adv}^{ind-cpa}_{\Pi}(D, \eta) = |\Pr[D^{\mathcal{LR}_{\Pi,0}(\eta)}(\eta) = 1] - \Pr[D^{\mathcal{LR}_{\Pi,1}(\eta)}(\eta) = 1]|$$

is a negligible function of η, i.e., it is less than $1/\eta^c$ for any $c > 0$ and sufficiently large η.

3 Soundness

We consider a setting in which an adversary gets to see the computational evaluation of a sequence of (adaptively chosen) expressions. We want to model the fact that the adversary does not learn anything about the expressions, beside whatever information can be deduced from their patterns.

We formalize the problem using a cryptographic experiment as follows. Fix a symmetric encryption scheme $\Pi = \{\mathcal{K}, \mathcal{E}, \mathcal{D}\}$. Let A be any probabilistic polynomial-time machine that issues queries consisting of pairs[3] of basic expressions, the ith query of A being denoted by $(M_0[i], M_1[i])$. The experiment runs in one of two worlds, decided based on a bit b sampled uniformly at random in the beginning. After selecting b, the adversary is executed (given some security parameter as input) and the queries of the adversary are answered using an oracle, $\mathcal{O}_{\Pi,b}$, parameterized by the encryption scheme Π and the bit b. This oracle first selects a random key vector τ using the key generation algorithm of Π, $\mathcal{K}$, and for each query $(M_0[j], M_1[j])$, replies with a sample from the distribution $[\![M_b[j]]\!]_{\Pi,\tau}$, i.e. the bitstring interpretation of the bth expression in the query (with respect to Π and τ). A concise description of the oracle and the experiment appears below.

Oracle $\mathcal{O}_{\Pi,b}(\eta)$
Let $\tau \xleftarrow{\$} \mathcal{K}^{n+2}(\eta)$
For the jth query received, $(M_0[j], M_1[j])$, reply with a sample from $[\![M_b[j]]\!]_{\Pi,\tau}$

Expt$_{\Pi}^{adpt}(A)$
Let $b \xleftarrow{\$} \{0, 1\}$. Fix η.
Run $A^{\mathcal{O}_{\Pi,b}(\eta)}(\eta)$

The goal of the adversary is to guess the value of b with probability better than random, and under the constraint that the two sequences of expressions queried to the oracle have the same pattern. More specifically, let q denote the number of queries made by A in any execution of the experiment, and let $M_b = M_b[1..q]$ be the sequence of expressions encrypted by the oracle $\mathcal{O}_{\Pi,b}(\eta)$. (Without loss of generality, q can be assumed to be a fixed polynomial in the security parameter, e.g., a polynomial upper bound on the running time of A.) We require that $M_0 \simeq M_1$. For technical reasons, in order to prove our soundness theorem we need to introduce some additional restrictions on the syntax of M_0 and M_1.

Definition 2. *A sequence of basic expressions, $M_b[1], M_b[2], \cdots, M_b[q]$, is called legal if it satisfies the following two properties:*

1. *The expressions M_0 and M_1 contain no encryption cycles.[4]*
2. *No unrecoverable encryption key in $M_b[1..i]$ occurs as a message key in $M_b[j]$ for any $j > i$. That is, for all $i < j \leq q$*

---

[3] As standard in cryptography, we use distinguishability of pairs of messages to model leakage of partial information about those messages.
[4] An expression M is said to contain an encryption cycle if there exist keys $K_{i_1}, K_{i_2}, \cdots K_{i_m}$ such that $K_{i_1} \to_M K_{i_2} \to_M \cdots K_{i_{m-1}} \to_M K_{i_m} \to_M K_{i_1}$. Examples of such expressions are $\{K_1\}_{K_1}$ and $(\{K_1\}_{K_2}, \{(K_3, \{K_2\}_{K_3})\}_{K_1})$.

$$\mathbf{UEncKeys}(M_b[1..i]) \cap \mathbf{MsgKeys}(M_b[j]) = \emptyset$$

For example, if $M_b[1] = \{K_i\}_{K_j}$ then it is illegal to have $M_b[2] = \{K_j\}_{K_l}$ or even $M_b[2] = K_j$.

The first requirement is standard in cryptography, and was already used in [2]. The second requirement is also very natural and it informally states that each key is used in two stages: a key distribution stage where the key is used as a message, and a subsequent deployment stage where the key is used to encrypt other messages and keys. This is the way keys are used in many cryptographic protocols, e.g., the key distribution protocols of [23,6]. Intuitively, the reason we introduce this requirement is that if a key is first used to encrypt messages, and then revealed, the symbolic patterns of previously received messages may change and, consequently, our proof breaks down. At a more technical level, in the absence of the second requirement, an adversary can play the following game: first issue the expressions $\{M_1\}_{K_1}, \ldots, \{M_l\}_{K_l}$, and then, after getting the corresponding ciphertexts, ask for a randomly chosen set of keys $\{k_{i_1}, \ldots, k_{i_m}\}$ ($i_1, \cdots, i_m \in \{1, \cdots, l\}$) by issuing the expression $(K_{i_1}, \ldots, K_{i_m})$. The question of whether security of the ciphertexts (other than those that can be decrypted using the revealed keys) is ensured in this game is the classic problem of selective decryption for which no answer is known to date [9].

An adversary in $\mathbf{Expt}_\Pi^{adpt}$ is called *legal* if the queries issued by it are such that both $M_0[1], \cdots, M_0[q]$ and $M_1[1], \cdots, M_1[q]$ are legal sequences and $M_0 \simeq M_1$. The advantage of A in the experiment, denoted $\mathbf{Adv}_\Pi^{adpt}(A, \eta)$, is defined as the following quantity:

$$\mathbf{Adv}_\Pi^{adpt}(A, \eta) = |\mathbf{P}[A^{\mathcal{O}_{\Pi,0}(\eta)}(\eta) = 1] - \mathbf{P}[A^{\mathcal{O}_{\Pi,1}(\eta)}(\eta) = 1]|$$

where the probabilities are taken based on the randomness used by A and $\mathcal{O}_{\Pi,b}$.

We now state our soundness theorem:

Theorem 1. *If Π is an* ind-cpa *secure encryption scheme, then for any legal adversary A, $\mathbf{Adv}_\Pi^{adpt}(A, \eta)$ is a negligible function of η.*

We provide an overview of the proof of the soundness theorem in Section 5 but before doing that, we discuss an application to multicast key distribution.

4 Application to Secure Multicast

In this section, we present an example to illustrate how our soundness theorem can be used in the analysis of real cryptographic protocols. Our example is the *multicast key distribution problem* in which a large set of users communicates using a multicast (or broadcast) channel and at any time some of these users, called "group members", share a secret key which is known only to them and not to the rest of the users. The group members change dynamically and in order to maintain the secrecy property of the group key over time, a central authority, called the *group center*, broadcasts messages to enable the members to update the key whenever a new member joins or an old member leaves the group. In other words, the center "rekeys" the group whenever its composition changes. The goal is to ensure that at any point in time, the non-members are unable to compute the group key, even if several of them collude together and share all their information in an attempt to do so.

This problem arises in many practical scenarios and has been studied extensively by the cryptography as well as computer networks communities. (See for example [23, 6, 7, 22, 18].) However, there seems to have been very little attempt towards formulating a sound cryptographic model for the problem and proving security of any of the proposed solutions using standard cryptographic techniques. Although some works implicitly use a Dolev-Yao like framework in arguing for security of multicast key distribution protocols, it is not clear how such analysis relates to actual security of the protocols. Our soundness theorem provides a useful tool in relating proofs of security for these protocols in the formal framework to security proofs in the standard computational framework used in cryptography.

4.1 Security in the Computational Framework

We model a multicast key distribution protocol as a set of three programs $\Gamma = \{\mathcal{I}, \mathcal{C}, \mathcal{U}\}$ where $\mathcal{I}$ is an initialization program, $\mathcal{C}$ is the group center's program used to compute the rekey messages and $\mathcal{U}$ is the program run by the group members $U = \{u_1, \cdots, u_N\}$.

These programs work as follows. $\mathcal{I}$ takes the security parameter η as input and outputs the initial state of the center, $s_0^\mathcal{C}$, the initial states of all users $s_0^1, s_0^2, \cdots, s_0^N$ and the initial group membership $G_0 \subset U$. (Typically, $G_0 = \emptyset$.) The center's program, $\mathcal{C}$, takes as input η, the current state $s_t^\mathcal{C}$ and a command com_t and returns a message m_t (the rekey message at time t) and the updated state of the center $s_{t+1}^\mathcal{C}$. Each command com_t given to the center is either of the form $add(u_i)$ (which adds a new user to the group) or of the form $del(u_i)$ (which removes an existing member from the group). The users' program, $\mathcal{U}$, takes as input η, a user index i ($\leq N$), the previous state s_{t-1}^i of user u_i, and the current rekey message m_t, and outputs a string k_t^i and the updated state s_t^i of u_i. For correctness, we require that at every time instant $t \geq 0$, k_t^i be identical for every member u_i in the current group G_t. This value is called the group key at time t and is denoted k_t.

Security Definition: The security of multicast key distribution is modelled using an adversary that controls a subset of corrupted users and adaptively issues commands to change membership of the group. The adversary's goal is to gain information about the group key when none of the corrupted users are part of the group. Formally, for any protocol $\Gamma = \{\mathcal{I}, \mathcal{C}, \mathcal{U}\}$, we consider the following experiment, which we denote by $\textbf{Expt}_\Gamma^{gkd}$. First, $\mathcal{I}$ is used to generate the initial states of the group center and all users, and the adversary $\mathcal{A}$ is given a set of corrupted users $B \subset U$, together with their initial states $\{s_0^i | u_i \in B\}$. The adversary then issues a sequence of t commands $\text{com}_1, \cdots, \text{com}_t$ and for each command $\text{com}_{t'}$, it is given the corresponding rekey message $m_{t'}$, computed according to program $\mathcal{C}$ and the group center's initial state produced by $\mathcal{I}$. (At any point in time, the users in B may or may not be in the group). Let $k_1, \cdots, k_t$ be the group keys at times $1, \cdots, t$ as computed by the honest group members. Let also $T \subseteq \{1, \cdots, t\}$ be the set of time instants when none of the corrupted users are in the group, and let $\bar{k}_T = \{k_i : i \in T\}$ be the corresponding keys. The security requirement is that the keys in $\bar{k}_T$ are pseudorandom. More precisely, let $\bar{k}'_T$ be a set of $|T|$ uniformly and independently chosen keys, and let b be a random bit. At the end of the experiment, the adversary is given either $\bar{k}_T$ or $\bar{k}'_T$ (depending on whether $b = 0$ or $b = 1$, respectively) and her goal is to correctly guess the value of b.[5]

Let $p_{\mathcal{A}}(B, b)$ be the probability that $\mathcal{A}$ outputs 1 in $\mathbf{Expt}_{\Gamma}^{gkd}$ when the corrupted set of users is B (here probabilities are taken based on the random choices of $\mathcal{A}, \mathcal{I}$ and $\mathcal{C}$). The advantage function of $\mathcal{A}$ in the experiment is defined as:

$$\mathbf{Adv}_{\Gamma}^{gkd}(\mathcal{A}, B, \eta) = |p_{\mathcal{A}}(B, 0) - p_{\mathcal{A}}(B, 1)|$$

Definition 3. *A multicast key distribution protocol Γ is secure if for all probabilistic polynomial-time adversaries $\mathcal{A}$ and all sets $B \subseteq U$, $\mathbf{Adv}_{\Gamma}^{gkd}(\mathcal{A}, B, \eta)$ is a negligible function of η.*

We remark that the definition above allows the adversary to change the group membership in an adaptive way, but does not permit *adaptive corruption* of the users, i.e. the set of corrupted users must be chosen before the protocol starts executing.

4.2 Computationally Sound Security in the Dolev-Yao Model

We now define security of multicast key distribution in the Dolev-Yao framework and for this we consider a special class of key distribution protocols that encompasses most of the protocols used in practical applications. Let $\Gamma_F = \{\mathcal{I}_F, \mathcal{C}_F, \mathcal{U}_F\}$ denote a multicast key distribution protocol in the Dolev-Yao framework. The program $\mathcal{I}_F$ works just as $\mathcal{I}$ in the previous definition except that it initializes the state of every user u_i as a fixed symbolic key K_i that is unique to that user and the state of the center as the set of all the unique keys $K_1, K_2, \cdots, K_N$, where N is a bound on the number of users.[6] The program $\mathcal{C}_F$ takes commands of the form $add(u_i)$ and $del(u_i)$ as before but for each command com_t, returns an *expression* M_t (denoting the rekey message for time t). The internal state of $\mathcal{C}_F$ at time t consists of all unique keys, all rekey messages sent till

---

[5] We remark that this definition can be made stronger by giving to the adversary either the key $k_{t'}$ (if $b = 0$) or a random key $k'_{t'}$ (if $b = 1$) at every time instant t' for which $B \cap G_{t'} = \emptyset$ (instead of giving the set of these keys, $\bar{k}_t$ or $\bar{k}'_t$, at the end of the experiment as is done above). This strengthening does not affect our result in any way and we use the above definition only for the sake of simplicity.

[6] In practice, the group center can store a compact representation of all these keys using a pseudorandom function.

time t and the group composition at time t, G_t. $\mathcal{U}_F$ takes a user index i as input and returns a key, K_t^i, that can be obtained by applying the Dolev-Yao rules on all the rekey messages received till the current time, given the knowledge of the key K_i. K_t^i should also be such that it is not used as an encryption key in any of the rekey messages sent by the group center at any time[7]. Again, for correctness we require that for all time instants t, K_t^i be identical for each group member u_i at time t. We let $\bar{M}_t$ denote the expression $(M_1, M_2, \ldots, M_t)$ and for any set $B \subseteq U$, we let K_B denote the set of unique keys of all users in the set $B \subseteq U$.

Definition 4. *A multicast key distribution protocol Γ_F is secure in the Dolev-Yao framework if for every sequence of commands, $\mathrm{com}_1, \mathrm{com}_2, \cdots, \mathrm{com}_t$ and for every subset $B \subseteq U$, the following holds: Let $K_{t'}$ and $G_{t'}$ be the group key and group member set at time $t' \leq t$, and let T be the set of all t' such that $B \cap G_{t'} = \emptyset$. Then, $((\bar{M}_t, K_T), K_B) \simeq ((\bar{M}_t, K'_T), K_B)$, where $K_T = \{K_{t'} : t' \in T\}$ and K'_T is a set of $|T|$ fresh keys.*

For any protocol Γ_F in the Dolev-Yao framework, the translation of Γ_F in the computational framework with respect to a symmetric encryption scheme Π, is the protocol Γ_F^Π which behaves identically to Γ_F with the difference that a key assignment τ is generated for the set of all keys ever used in the protocol execution (using the key generation algorithm for Π) and each symbolic expressions M (a key or a rekey message) used in Γ_F is replaced with the bitstring interpretation of M, $[\![M]\!]_{\Pi,\tau}$. Using our soundness theorem, we can now show the following connection between the above two definitions.

Theorem 2. *Let Γ_F be a multicast key distribution protocol in the Dolev-Yao framework with the property that for any sequence of commands $\mathrm{com}_1, \cdots, \mathrm{com}_t$, the sequence of rekey messages, $M_1, M_2, \cdots, M_t$, returned by the center's program $\mathcal{C}_F$ is a legal sequence. Let Π be any $\mathrm{ind\text{-}cpa}$ secure symmetric encryption scheme. If Γ_F is secure in the Dolev-Yao framework (Definition 4), then Γ_F^Π is secure in the computational framework (Definition 3).*

Proof (Sketch): Suppose, towards contradiction, that Γ_F satisfies Definition 4, but Γ_F^Π does not satisfy Definition 3. Let $\mathcal{A}$ be a computational adversary and $B \subset U$ a set of initially corrupted users such that $\mathbf{Adv}_{\Gamma_F^\Pi}^{gkd}(\mathcal{A}, B, \eta)$ is non-negligible (in η). Given any such choice of $\mathcal{A}$ and B, we can build an adversary $\mathcal{A}'$ that uses $\mathcal{A}$ as a black-box and has non-negligible advantage in the experiment $\mathbf{Expt}_\Pi^{adpt}$ defined in our soundness theorem. $\mathcal{A}'$ first queries its oracle on the unique keys of all users in B and invokes $\mathcal{A}$ on input B and the corresponding keys. For any query $\mathrm{com}_{t'}$ of $\mathcal{A}$, $\mathcal{A}'$ uses the program $\mathcal{C}_F$ to determine the rekey message $M_{t'}$ and uses its oracle to determine the computational interpretation of the same (which it then returns to $\mathcal{A}$). Finally, $\mathcal{A}'$ queries its oracle on the pair

---

[7] The reason we introduce this requirement is that if a key is used to encrypt a message, then the key is no necessarily pseudorandom anymore, as the encryption scheme may leak partial information about the key.

(K_T, K'_T) where $T = \{t' : B \cap G_{t'} = \emptyset\}$ and K'_T is a set of $|T|$ fresh (symbolic) keys and the reply is passed on to $\mathcal{A}$. $\mathcal{A}'$ outputs whatever $\mathcal{A}$ outputs.

Given that the sequence of rekey messages generated in any run of Γ_F is a legal sequence and the fact that no key in K_T is ever used as an encryption key in any of the messages, it follows that the adversary $\mathcal{A}'$ constructed above is a legal adversary. It is easy to see that the advantage of $\mathcal{A}'$ in $\mathbf{Expt}_{\Pi}^{adpt}$ is exactly the same as that of $\mathcal{A}$ in $\mathbf{Expt}_{\Gamma_F^\Pi}^{gkd}$ (which means that if the latter is a non-negligible quantity, so is the former). This leads us to a contradiction of our soundness theorem. ∎

We remark that many practical group key distribution schemes (e.g., [23]) satisfy the precondition of Theorem 2 (that requires all sequences of rekey messages generated by the protocol to be legal sequences). Moreover, these protocols can be easily proved secure in the symbolic framework. It follows that their natural implementation is also secure in the computational framework.

5 Proof of the Soundness Theorem

This section provides an overview of the proof of our soundness theorem. More details appear in the full version of the paper [17].

5.1 Defining Orders for Legal Sequences of Expressions

For any acyclic expression $M \in \mathbf{Exp}$, the "encrypts" relation defines a partial order on the keys in M and we consider the restriction of this partial order on just its unrecoverable encryption keys. Any total order on the set of unrecoverable encryption keys of M that is consistent with such a partial order is called a *good order* for M. For example, in the expression

$$M := (((K_1, \{K_2\}_{K_3}), \{(K_6, \{K_1\}_{K_3})\}_{K_2}), \{K_4\}_{K_5})$$

the unrecoverable encryption keys are K_2, K_3 and K_5, and we have $K_3 \to_M K_2$. This gives us a partial order $K_3 \leq K_2$ on $\mathbf{UEncKeys}(M)$ and so the good orders for M are $K_3 \leq K_2 \leq K_5$, $K_3 \leq K_5 \leq K_2$ and $K_5 \leq K_3 \leq K_2$.

We now re-interpret the definition of legal sequences of expressions given in Section 3. Recall that for any such sequence, $M[1], M[2], \cdots, M[q]$, the expression $M = M[1..q]$ is acyclic and for any $i < j \leq q$, no unrecoverable encryption key in $M[1..i]$ is a message key in $M[j]$. The latter condition implies that for any $i \in \{1, \cdots, q-1\}$ no unrecoverable encryption key in $M[1..i]$ can be recoverable in $M[1..i+1]$, and therefore the sets

$\mathbf{UEncKeys}(M[1\ldots 1]) \subseteq \mathbf{UEncKeys}(M[1\ldots 2]) \subseteq \cdots \subseteq \mathbf{UEncKeys}(M[1..q])$

form a monotonically nondecreasing sequences.

This relation enables us to partition the unrecoverable encryption keys of M into q sets such that the ith set in the partition, say $\mathbf{UEnc}_i$, contains the keys that are used as encryption keys in $M[i]$ but in none of $M[1], M[2], \cdots, M[i-1]$, i.e. $\mathbf{UEnc}_i := \mathbf{UEncKeys}(M[1..i]) \setminus \mathbf{UEncKeys}(M[1..i-1])$. The definition of legal sequences implies that for any $i < j \leq q$, no key from $\mathbf{UEnc}_j$ can encrypt any key from $\mathbf{UEnc}_i$ in M. Now, using the fact that M is acyclic, we can find a good order $\leq$ for every $M[1..i]$ such that for any $1 \leq i_1 < j_1 \leq i$, the keys from $\mathbf{UEnc}_{i_1}$ "precede" all the keys from $\mathbf{UEnc}_{j_1}$ in $\leq$ i.e. for all $K_{i'} \in \mathbf{UEnc}_{i_1}$ and $K_{j'} \in \mathbf{UEnc}_{j_1}$, $K_{i'} \leq K_{j'}$. We select the lexicographically first order among all orders having this property and denote it by $\leq_{M[1..i]}$. The order $\leq_{M[1..1]}$ is just the lexicographically first good order defined on $M[1]$ and for each $i < q$, the ordering produced by $\leq_{M[1..i-1]}$ is a prefix of that produced by $\leq_{M[1..i]}$.

As an example, consider the following sequence of expressions:

$$M[1] = (K_1, \{K_2\}_{K_3})$$
$$M[2] = \{(K_6, \{K_1\}_{K_3})\}_{K_2}$$
$$M[3] = \{K_4\}_{K_5}$$

Observe that this sequence is consistent with our definition of legal sequences. The expressions $M[1..2]$ and $M[1..3]$ are

$$M[1..2] = ((K_1, \{K_2\}_{K_3}), \{K_6, \{K_1\}_{K_3}\}_{K_2})$$
$$M[1..3] = (((K_1, \{K_2\}_{K_3}), \{K_6, \{K_1\}_{K_3}\}_{K_2}), \{K_4\}_{K_5})$$

We have $\mathbf{UEncKeys}(M[1..1]) = \{K_3\}$; $\mathbf{UEncKeys}(M[1..2]) = \{K_2, K_3\}$ and $\mathbf{UEncKeys}(M[1..3]) = \{K_2, K_3, K_5\}$ and so $\mathbf{UEnc}_1 = \{K_3\}$; $\mathbf{UEnc}_2 = \{K_2\}$ and $\mathbf{UEnc}_3 = \{K_5\}$. The lexicographically first good orders for $M[1..2]$ and $M[1..3]$ are given by $K_3 \leq K_2$ and $K_3 \leq K_2 \leq K_5$. These relations are denoted $\leq_{M[1..2]}$ and $\leq_{M[1..3]}$ respectively.

5.2 Defining Hybrid Oracles

The proof of the soundness theorem uses a hybrid technique. We define a set of $2n + 2$ hybrid oracles (where $n = |\mathbf{Keys}|$)[8] and relate the success probability of any legal adversary in distinguishing between any neighboring pair of these oracles to its success probability in distinguishing between the instances of the oracle $\mathcal{O}_{\Pi,b}$ (viz. $\mathcal{O}_{\Pi,0}$ and $\mathcal{O}_{\Pi,1}$) used in experiment $\mathbf{Expt}_\Pi^{adpt}$. We then use this relation to show how any legal adversary with a non-negligible advantage in $\mathbf{Expt}_\Pi^{adpt}$ (i.e. a non-negligible success probability in distinguishing between $\mathcal{O}_{\Pi,0}$ and $\mathcal{O}_{\Pi,1}$) can be used to mount a successful attack against the ind-cpa security of the underlying encryption scheme Π.

We denote the hybrid oracles by $\mathcal{O}_{\Pi,0}^0, \mathcal{O}_{\Pi,0}^1, \cdots \mathcal{O}_{\Pi,0}^n, \mathcal{O}_{\Pi,1}^n, \mathcal{O}_{\Pi,1}^{n-1}, \cdots, \mathcal{O}_{\Pi,1}^0$. The extreme oracles, $\mathcal{O}_{\Pi,0}^0$ and $\mathcal{O}_{\Pi,1}^0$, correspond to the instantiations of $\mathcal{O}_{\Pi,b}$ with $b = 0$ and $b = 1$ respectively. The behavior of oracle $\mathcal{O}_{\Pi,0}^0$ is close to that of $\mathcal{O}_{\Pi,0}^1$, the behavior of $\mathcal{O}_{\Pi,0}^1$ is close to that of $\mathcal{O}_{\Pi,0}^2$ and so on up to $\mathcal{O}_{\Pi,0}^n$. Simi-

---

[8] Without loss of generality, the number of key symbols that can potentially be used by the adversary in generating queries can be assumed to be a fixed polynomial in the security parameter.

larly, $\mathcal{O}_{\Pi,1}^n$'s behavior is similar to that of $\mathcal{O}_{\Pi,1}^{n-1}$'s, $\mathcal{O}_{\Pi,1}^{n-1}$'s close to $\mathcal{O}_{\Pi,1}^{n-2}$'s and so on up to $\mathcal{O}_{\Pi,1}^0$. For each $i \in \{1, 2, \cdots, n\}$, the oracle $\mathcal{O}_{\Pi,0}^i$ is defined as follows:

Oracle $\mathcal{O}_{\Pi,0}^i(\eta)$

1. Let $\tau \xleftarrow{\$} \mathcal{K}^{n+2}(\eta)$
2. For the jth query received, $(M_0[j], M_1[j])$, do the following:
   (a) Compute the order $\leq_{M_0[1..j]}$ and let S be the set of those keys in $M_0[j]$ that are among the smallest i keys of $\leq_{M_0[1..j]}$.
   (b) Let M^{new} be the expression obtained by substituting, for all $K_l \in S$, all sub-expressions in $M_0[j]$ of the form $\{M'\}_{K_l}$ with $\{\texttt{struct}(M')\}_{K_l}$.
   (c) Return $[\![M^{new}]\!]_{\Pi,\tau}$.

The oracles $\mathcal{O}_{\Pi,1}^i$ (for $i \in [n]$) are defined analogously with the difference that in steps 2(a) and 2(b), $M_0[j]$ gets replaced by $M_1[j]$ and $\leq_{M_0[1..j]}$ by $\leq_{M_1[1..j]}$. The following fact about oracles $\mathcal{O}_{\Pi,0}^n$ and $\mathcal{O}_{\Pi,1}^n$ is easy to deduce:

Lemma 1. *Whenever the queries received by the oracles come from a legal adversary, the two oracles $\mathcal{O}_{\Pi,0}^n$ and $\mathcal{O}_{\Pi,1}^n$ have identical behavior i.e. for any sequence of queries, the distribution of the replies given by one of them is exactly the same as that of the replies given by the other.* ∎

For any $i \in [n]$ and $b \in \{0,1\}$, we define the advantage of A in distinguishing between oracles $\mathcal{O}_{\Pi,b}^i$ and $\mathcal{O}_{\Pi,b}^{i-1}$, $\mathbf{Adv}_{\Pi,i,b}^{adpt}$, as the following quantity:

$$\mathbf{Adv}_{\Pi,i,b}^{adpt}(A, \eta) = \left| \mathbf{P}[A^{\mathcal{O}_{\Pi,b}^i(\eta)}(\eta) = 1] - \mathbf{P}[A^{\mathcal{O}_{\Pi,b}^{i-1}(\eta)}(\eta) = 1] \right|$$

The following lemma relates these advantages to the advantage of A in $\mathbf{Expt}_\Pi^{adpt}$.

Lemma 2. $\sum_{i=1}^n \sum_{b \in \{0,1\}} \mathbf{Adv}_{\Pi,i,b}^{adpt}(A, \eta) \geq \mathbf{Adv}_\Pi^{adpt}(A, \eta)$ ∎

5.3 The Reduction

Given any legal adversary A in experiment $\mathbf{Expt}_\Pi^{adpt}$, we construct a distinguisher D attacking the ind-cpa security of Π such that the advantage of D in performing an ind-cpa attack on Π is related (by a polynomial multiplicative factor) to A's advantage in $\mathbf{Expt}_\Pi^{adpt}$. This essentially implies that any successful attack in $\mathbf{Expt}_\Pi^{adpt}$ can be effectively translated into an attack on the underlying encryption scheme itself.

Our construction of D will be such that the advantage of D in $\mathbf{Expt}_\Pi^{ind-cpa}$ will be $1/poly$ times the *expected* advantage of A in distinguishing between oracles $\mathcal{O}_{\Pi,b}^i$ and $\mathcal{O}_{\Pi,b}^{i-1}$, where i and b are treated as random variables sampled uniformly from $\{1, \cdots, n\}$ and $\{0, 1\}$. More precisely, the construction will be such that

$$\mathbf{Adv}_\Pi^{ind-cpa}(D, \eta) = \frac{1}{n} \mathbf{E}_{i \xleftarrow{\$} [n]; b \xleftarrow{\$} \{0,1\}} \left(\mathbf{Adv}_{\Pi,i,b}^{adpt}(A, \eta) \right)$$

$$= \frac{1}{n} \sum_{i=1}^n \sum_{b \in \{0,1\}} \left(\frac{1}{n} \cdot \frac{1}{2} \cdot \mathbf{Adv}_{\Pi,i,b}^{adpt}(A, \eta) \right)$$

Now, applying Lemma 2 we get

$$\mathbf{Adv}_{\Pi}^{ind-cpa}(D,\eta) \geq \frac{1}{2n^2} \mathbf{Adv}_{\Pi,i,b}^{adpt}(A,\eta)$$

Thus, a non-negligible advantage of A in experiment $\mathbf{Expt}_{\Pi}^{adpt}$ would imply a non-negligible advantage of D in $\mathbf{Expt}_{\Pi}^{ind-cpa}$ and the theorem would follow immediately from this.

The Construction. The distinguisher D works as follows: it first selects a random number i' in the range $\{1, \cdots, n\}$ and a random bit $b' \in \{0,1\}$ and then tries to simulate the behavior of the oracle pair $\{\mathcal{O}_{\Pi,b'}^{i'}, \mathcal{O}_{\Pi,b'}^{i'-1}\}$ using its own oracle $\mathcal{LR}_{\Pi,b}$. D runs A inside it and answers A's queries using its simulated setup. To carry out the simulation, it guesses a value i randomly from $\{1, \cdots, n\}$ and hopes that each query, $(M_0[j], M_1[j])$, issued by A would be such that the i'th key in the order $\leq_{M_{b'}[1..j]}$ is K_i (or else the number of unrecoverable encryption keys in $M_{b'}[1..j]$ is smaller than i' and K_i is neither a recoverable key nor an unrecoverable encryption key in $M_{b'}[1..j]$). If D fails in its guess, it gives up and outputs 0. Else, it treats the key used by $\mathcal{LR}_{\Pi,b}$ as corresponding to K_i and answers A's queries in such a way that the behavior of $\mathcal{LR}_{\Pi,b}$ with $b = 0$ (resp. $b = 1$) corresponds to the simulation of the oracle $\mathcal{O}_{\Pi,b'}^{i'-1}$ (resp. $\mathcal{O}_{\Pi,b'}^{i'}$).

Adversary $D^{\mathcal{LR}_{\Pi,b}(\eta)}(\eta)$

1. Let $i' \xleftarrow{\$} \{1,2,\cdots,n\}$ and $b' \xleftarrow{\$} \{0,1\}$.
2. Guess a value $i \xleftarrow{\$} \{1,2,\cdots,n\}$.
3. Generate a key vector τ as follows: $(\tau[1], \ldots, \tau[i-1], \tau[i+1], \ldots, \tau[n+2])$ $\xleftarrow{\$} \mathcal{K}^{n+1}(\eta)$ (the ith entry in τ is empty and the rest are random keys).
4. Run $A(\eta)$.
5. When A issues the jth query $(M_0[j], M_1[j])$, do the following:
   (a) Compute the order $\leq_{M_{b'}[1..j]}$. Check if either K_i is the i'th key in $\leq_{M_{b'}[1..j]}$; OR $|\mathbf{UEncKeys}(M_{b'}[1..j])| < i'$ and $K_i \notin \mathbf{UEncKeys}(M_{b'}[1..j]) \cup \mathbf{RecKeys}(M_{b'}[1..j])$
   (b) If so, do the following:
      i. Let S be the set of those keys in $M_{b'}[j]$ that are among the *smallest* $(i'-1)$ keys of $\leq_{M_{b'}[1..j]}$.
      ii. Let M^{new} be the expression obtained by substituting, for all $K_l \in S$, all sub-expressions in $M_{b'}[j]$ of the form $\{M'\}_{K_l}$ with $\{\mathbf{struct}(M')\}_{K_l}$.
      iii. Return $Sample^{\mathcal{LR}_{\Pi,b}}(M^{new}, K_i, \tau)$ to A.
   (c) Else, output 0 and halt.
6. Output whatever A outputs.

The crux of the code lies in the subroutine *Sample* (invoked at the end of step 5(b)) which is given below

Procedure $Sample^{\mathcal{LR}_{\Pi,b}}(M,K_i,\tau)$

1. If M is a constant or **c**, return the corresponding bitstring.
2. If $M = K_l$ (for some $l \in [n]$), return $\tau[l]$. If $M = K'$, return $\tau[n+1]$.
3. If $M = (M_1, M_2)$, return $(Sample(M_1, K_i, \tau), Sample(M_2, K_i, \tau))$.
4. If $M = \{M'\}_{K_l}$ and $l \neq i$, return $\mathcal{E}_{\tau[l]}(Sample(M', K_i, \tau))$. If $M = \{M'\}_{K'}$, return $\mathcal{E}_{\tau[n+2]}(Sample(M', K_i, \tau))$.
5. If $M = \{M'\}_{K_l}$ and $l = i$, return $\mathcal{LR}_{\Pi,b}(Sample(M', K_j, \tau), [\![\texttt{struct}(M')]\!]_{\Pi,\tau})$.

The proofs of the two lemmas and the analysis of the distinguisher can be found in the full version of the paper[17].

6 Future Work

We have proved a generalization of the soundness theorem of [2] in which the adversary can issue a sequence of adaptively chosen expressions, rather than a single expression, and demonstrated the usefulness of the theorem in an application to secure multicast key distribution. For simplicity, in this paper we considered a language of expressions that make use of only symmetric encryption operations, but most of the techniques can be easily extended to other cryptographic primitives whose security can be expressed as an indistinguishability property. Examples of such primitives are public key encryption, in which two different keys are used to encrypt and decrypt messages, and pseudorandom number generators, that can be used to expand a key into a sequence of multiple seemingly independent keys. Some of these extensions (e.g., the use of pseudorandom generators) are especially interesting in the context of multicast security protocols as the protocol of [6] (which was shown to be optimal in the Dolev-Yao model in [18]) makes use of these operations.

The proof of our soundness theorem introduces a syntactic restriction (besides the acyclicity condition already used in [2]) about the order in which each key is used as a message or as an encryption key. An interesting question is whether either restriction can be lifted, possibly using a special encryption scheme with additional security properties (a good candidate might be non-committing encryption introduced by Canetti, Feige, Goldreich and Naor in [5]). Although most practical protocols satisfy the syntactic restrictions in our soundness theorem, removing the ordering restriction would allow to model attack scenarios with adaptive corruption of users, where, when the adversary wants to corrupt user i (holding a secret key k_i as its internal state) it can simply issue the expression K_i to learn the value of the key. Currently this is allowed only if K_i has not already been used to encrypt other messages. Designing protocols that are secure against adaptive corruption raises issues similar to the selective decommitment problem discussed in Section 3, and is not easily addressed using the techniques developed in this paper. We leave the investigation of multicast key distribution protocols secure under adaptive corruption of the users to future work.

References

1. M. Abadi and J. Jürjens. Formal eavesdropping and its computational interpretation. In N. Kobayashi and B. Pierce, editors, *Proceedings of the 4th International Symposium on Theoretical Aspects of Computer Software - TACS 2001*, volume 2215 of *Lecture Notes in Computer Science*, pages 82–94, Sendai, Japan, Oct. 2001. Springer-Verlag.
2. M. Abadi and P. Rogaway. Reconciling two views of cryptography (the computational soundness of formal encryption). *Journal of Cryptology*, 15(2):103–127, 2002.
3. M. Backes and B. Pfitzmann. Symmetric encryption in a simulatable Dolev-Yao style cryptographic library. In *Proceedings of the 17th IEEE computer security foundations Workshop*, pages 204–218, Pacific Grove, CA, USA, June 2004. IEEE Computer Society.
4. M. Backes, B. Pfitzmann, and M. Waidner. A Composable Cryptographic Library with Nested Operations. In *Proceedings of the 10th ACM conference on computer and communications security - CCS 2003*, pages 220–230, Washington, DC, USA, Oct. 2003. ACM.
5. R. Canetti, U. Feige, O. Goldreich, and M. Naor. Adaptively Secure Multiparty Computation. In *Proceedings of the twenty-eighth annual ACM symposium on the theory of computing - STOC '96*, pages 639–648, Philadelphia, Pennsylvania, USA, May 1996. ACM.
6. R. Canetti, J. Garay, G. Itkis, D. Micciancio, M. Naor, and B. Pinkas. Multicast security: A taxonomy and some efficient constructions. In *INFOCOM 1999. Proceedings of the Eighteenth Annual Joint conference of the IEEE computer and communications societies*, volume 2, pages 708–716, New York, NY, Mar. 1999. IEEE.
7. R. Canetti, T. Malkin, and K. Nissim. Efficient communication-storage tradeoffs for multicast encryption. In J. Stern, editor, *Advances in Cryptology - EUROCRYPT '99, Proceedings of the International Conference on the Theory and Application of Cryptographic Techniques*, volume 1592 of *Lecture Notes in Computer Science*, Prague, Czech Republic, May 1999. Springer-Verlag.
8. D. Dolev, C. Dwork, and M. Naor. Nonmalleable Cyptography. *SIAM Journal on Computing*, 30(2):391–437, 2000. Preliminary version in STOC 1991.
9. C. Dwork, M. Naor, O. Reingold, and L. Stockmeyer. Magic Functions. *Journal of the ACM*, 50(6):852–921, Nov. 2003.
10. V. Gligor and D. O. Horvitz. Weak Key Authenticity and the Computational Completeness of Formal Encryption. In D. Boneh, editor, *Advances in cryptology - CRYPTO 2003, proceedings of the 23rd annual international cryptology conference*, volume 2729 of *Lecture Notes in Computer Science*, pages 530–547, Santa Barbara, California, USA, Aug. 2003. Springer-Verlag.
11. S. Goldwasser and S. Micali. Probabilistic encryption. *Journal of Computer and System Sience*, 28(2):270–299, 1984. Preliminary version in Proc. of STOC 1982.
12. J. C. Herzog. *Computational Soundness for Standard Assumptions of Formal Cryptography*. PhD thesis, Massachusetts Institute of Technology, Boston, USA, 2004.
13. R. Impagliazzo and B. Kapron. Logics for Reasoning about Cryptographic Constructions. In *Proceedings of the 44rd annual symposium on foundations of computer science - FOCS 2003*, pages 372–383, Cambridge, MA, USA, Nov. 2003. IEEE.

14. P. Laud. Symmetric Encryption in Automatic Analyses for Confidentiality against Active Adversaries. In *IEEE symposium on security and Privacy*, pages 71–85, Berkeley, CA, USA, May 2004. IEEE Computer Society.
15. P. D. Lincoln, J. C. Mitchell, M. Mitchell, and A. Scedrov. A probabilistic polytime framework for protocol analysis. In *Proceedings of the fifth ACM conference on computer and communications security - CCS '98*, pages 112–121, San Francisco, California, USA, Nov. 1998. ACM.
16. D. Micciancio. Towards Computationally Sound Symbolic Security Analysis, June 2004. Tutorial. Slides available at http://dimacs.rutgers.edu/Workshops/Protocols/slides/micciancio.pdf.
17. D. Micciancio and S. Panjwani. Adaptive Security of Symbolic Encryption, Nov. 2004. Full version of this paper. Available from http://www-cse.ucsd.edu/users/spanjwan/papers.html.
18. D. Micciancio and S. Panjwani. Optimal communication complexity of generic multicast key distributio. In C. Cachin and J. Camenisch, editors, *Advances in cryptology - EUROCRYPT 2004, proceedings of the internarional conference on the theory and application of cryptographic techniques*, volume 3027 of *Lecture Notes in Computer Science*, Interlaken, Switzerland, May 2004. Springer-Verlag.
19. D. Micciancio and B. Warinschi. Completeness theorems for the abadi-rogaway logic of encrypted expressions. *Journal of Computer Security*, 12(1):99–129, 2004. Preliminary version in WITS 2002.
20. D. Micciancio and B. Warinschi. Soundness of Formal Encryption in the presence of Active Adversaries. In M. Naor, editor, *Theory of cryptography conference - Proceedings of TCC 2004*, volume 2951 of *Lecture Notes in Computer Science*, pages 133–151, Cambridge, MA, USA, Feb. 2004. Springer-Verlag.
21. C. Rackoff and D. R. Simon. Non-interactive zero-knowledge proof of knowledge and chosen ciphertext attack. In J. Feigenbaum, editor, *Advances in Cryptology - CRYPTO '91, Proceedings*, volume 576 of *Lecture Notes in Computer Science*, Santa Barbara, California, USA, Aug. 1991. Springer-Verlag.
22. S. Rafaeli and D. Hutchinson. A survey of key management for secure group communication. *ACM Computing Surveys*, 35(3):309–329, Sept. 2003.
23. C. K. Wong, M. Gouda, and S. S. Lam. Secure group communications using key graphs. *IEEE/ACM Transactions on Networking*, 8(1):16–30, Feb. 2000. Preliminary version in SIGCOMM 1998.

Chosen-Ciphertext Security of Multiple Encryption

Yevgeniy Dodis[1,*] and Jonathan Katz[2,**]

[1] Dept. of Computer Science, New York University
[2] Dept. of Computer Science, University of Maryland

Abstract. Encryption of data using multiple, independent encryption schemes ("multiple encryption") has been suggested in a variety of contexts, and can be used, for example, to protect against partial key exposure or cryptanalysis, or to enforce threshold access to data. Most prior work on this subject has focused on the security of multiple encryption against *chosen-plaintext attacks*, and has shown constructions secure in this sense based on the chosen-plaintext security of the component schemes. Subsequent work has sometimes assumed that these solutions are also secure against *chosen-ciphertext attacks* when component schemes with stronger security properties are used. Unfortunately, this intuition is false for all existing multiple encryption schemes.

Here, in addition to formalizing the problem of chosen-ciphertext security for multiple encryption, we give simple, efficient, and generic constructions of multiple encryption schemes secure against chosen-ciphertext attacks (based on *any* component schemes secure against such attacks) in the standard model. We also give a more efficient construction from any (hierarchical) identity-based encryption scheme secure against selective-identity *chosen plaintext* attacks. Finally, we discuss a wide range of applications for our proposed schemes.

1 Introduction

Encrypting data using multiple, independent instantiations of a basic encryption scheme (or schemes) is a simple — yet powerful — approach which can be used both to improve security as well as to provide additional functionality not present in any of the underlying schemes. The security implications of *multiple encryption* (as we refer to it here) were noted as early as Shannon [38], who proposed using "product ciphers" to enhance the security of symmetric-key primitives. This idea was further explored and rigorously formalized in a number of subsequent works (e.g., [32, 21, 31]) analyzing the security of *cascade ciphers* (in which a message m is encrypted via $\mathcal{E}_{k_1}(\mathcal{E}'_{k_2}(m))$, where k_1, k_2 are

---

[*] This work was supported by the NSF under CAREER Award CCR-0133806 and Trusted Computing Grant CCR-0311095.
[**] This work was supported by NSF Trusted Computing Grant CCR-0310751.

two independent keys and $\mathcal{E}, \mathcal{E}'$ are symmetric-key encryption schemes) in the symmetric-key setting. The approach can be applied to public-key encryption as well; for example, "cascaded encryption" (which we will call *sequential encryption*) was advocated as part of the NESSIE recommendation [34]: "[f]or very high level security we note that double encryption... gives a good range of security".

Multiple encryption, of which sequential encryption is but one example, offers at least two potential security advantages: First, the resulting scheme may be secure as long as *any one of* the component schemes is secure (indeed, sequential encryption is secure against chosen-plaintext attacks as long as either of the component schemes are). Thus, multiple encryption offers a way to "hedge one's bets" about the security of any particular scheme (see also the recent work of Herzberg [28]). This is especially important when the security of different schemes depends upon different, and incomparable, cryptographic assumptions. A second potential advantage of multiple encryption is that the resulting encryption scheme may in fact be *more secure* than any of the component schemes; this is the rationale, for example, behind using triple-DES (see also [1]).

Beyond the security-oriented advantages listed above, multiple encryption schemes potentially offer *functionality* not present in any of the component schemes. We briefly highlight two applications of multiple encryption, and defer a more detailed discussion of these and other applications to Section 6:

Threshold Encryption. In a threshold encryption scheme [16], the data is encrypted in such a way that only particular sets of users can recover it; typically, a scheme requires any t-out-of-n users in order to decrypt, but more general access structures can also be considered. Multiple encryption gives *generic* constructions of threshold encryption in either the private- or public-key settings. For example, to enforce n-out-of-n decryption in the private-key setting, one may provide each user i with an independent key k_i and encrypt a message M via $\mathcal{E}^1_{k_1}(M_1), \ldots, \mathcal{E}^i_{k_i}(M_i)$, where the M_i are chosen at random subject to $\oplus_{i=1}^n M_i = M$ (and the $\mathcal{E}^i$ may, in general, be different schemes). Let J (with $|J| < n$) represent the set of corrupted players; i.e., if $j \in J$ then the adversary has the key k_j. The above scheme, which we will refer to as *parallel encryption*, informally satisfies the following level of security against chosen-plaintext attacks: as long as *any* encryption scheme $\mathcal{E}^i$ with $i \notin J$ is secure, the message remains secret. Thus, in addition to enabling threshold access to the data, this scheme also allows one again to "hedge one's bets" about the security of any particular scheme (as in the case of sequential encryption, discussed earlier).

(Strong) Key-Insulated Encryption. Multiple encryption has also been used to give a generic construction of a key-insulated public-key encryption scheme secure against chosen-plaintext attacks [20]. Without going into the full details — and omitting some details unimportant for the present discussion — in this case a message M is encrypted by first splitting the message into *shares* $M_1, \ldots, M_i$ and then encrypting each share M_i with respect to a particular public key PK_i. (This general technique is similar to the parallel encryption discussed above; indeed, parallel encryption is obtained if the shares constitute an n-out-of-n sharing of M.) If the message is "split" a second time (before the sharing described above), and one of these shares is encrypted with a public key whose secret key is known only to the user, it is possible to obtain a generic construction of *strong key-insulated encryption* [20].

Other Applications. We remark that multiple encryption is applicable to many other domains as well, including anonymous routing [14, 26], broadcast encryption [22], proxy encryption (see [19]), and certificate-based encryption [24]. We defer a more detailed discussion to Section 6.

1.1 Motivation for Our Work

Chosen-ciphertext security ("CCA security") is as much of a concern in each of the above settings as it is in the case of standard encryption. One might hope to achieve CCA security for any of the above settings by simply "plugging in" an appropriate CCA-secure multiple encryption scheme. However (with one recent exception; see below), *we are unaware of any previous work which considers chosen-ciphertext security for multiple encryption*. To be clear: there has been much work aimed at giving solutions for *specific* applications using *specific* number-theoretic assumptions: for example, in the context of CCA-secure threshold encryption [40, 13, 30], broadcast encryption [20], and key-insulated encryption [18]. However, this type of approach suffers from at least two drawbacks: first, it does not provide *generic* solutions, but instead only provides solutions based on very specific assumptions. Second, the derived solutions are application-dependent, and must be constantly "re-invented" and modified each time one wants to apply the techniques to a new domain. Although solutions based on specific assumptions are often more efficient than generic solutions, it is important to at least be aware that a generic solution exists so that its efficiency can be directly compared with a solution based on specific assumptions. Indeed, we argue in Section 6 that for some applications, a generic solution may be roughly as efficient as (or may offer reasonable efficiency tradeoffs as compared to) the best currently-known solutions based on specific assumptions.

Making the problem even more acute is that currently-known schemes for multiple encryption are demonstrably *insecure* against chosen-ciphertext attacks (this holds even with respect to the weakest definition considered here; see Section 3.1). Zhang, et al. [41] have also recently noticed this problem, and appear to be the first to have considered chosen-ciphertext security for multiple encryption. We compare our work to theirs in the following section.

1.2 Our Contributions

Our results may be summarized as follows:

Definitions of Security. We provide formal definitions of chosen-ciphertext security for multiple encryption. Interestingly, multiple definitions make sense in this context, and we introduce three such definitions and briefly comment on the relationships between them. We also which of these definitions is the "right" one for a number of different applications.

CCA-Secure Multiple Encryption. We show two constructions of CCA-secure multiple encryption schemes which are *generic* (i.e., they may be constructed based on *any* CCA-secure standard encryption scheme) and are proven secure *in the standard model*. Our first construction achieves a "basic" level of security which suffices for many (but not all!) applications of multiple encryption. Our second construction satisfies the strongest notion of security proposed here, and suffices for all applications we consider. We also show a more efficient construction based on any (hierarchical) identity-based encryption scheme secure against selective-identity *chosen plaintext* attacks.

Applications. As mentioned earlier, our work was motivated by the applications of CCA-secure multiple encryption to a variety of settings; we therefore conclude the paper by sketching a number of applications of the constructions given here. Our resulting schemes are, for most cases, the first known *generic* constructions achieving CCA security in the given setting. Furthermore, in some cases the solutions we give are roughly as efficient as (or even more efficient than) previous solutions which were based on very specific assumptions. As one example, we show a CCA-secure threshold encryption scheme with completely *non-interactive* decryption (and a proof of security in the standard model); for the two-party case, our solution is roughly as efficient as the only previous solution [30].

Comparison to Previous Work. Our definitions differ from those given by Zhang, et al. [41], and the definitions given in their work are weaker than those given here. In fact, the best construction given by Zhang, et al. only satisfies the weakest of our definitions; therefore, their constructions are not sufficient for certain applications such as threshold encryption. (Indeed, they concentrate primarily on the application to key-insulated encryption, while we consider a much wider range of applications.) Finally, their constructions require the random oracle model whereas our results all hold in the standard model.

2 Preliminaries

We begin by introducing some notation. A (standard) public-key encryption scheme $\mathcal{E} = (\mathsf{Gen}, \mathsf{Enc}, \mathsf{Dec})$ consists of three PPT algorithms: the key-generation algorithm Gen takes as input security parameter 1^k and outputs a encryption key EK and a decryption key DK. The randomized encryption algorithm Enc takes as input EK, a label ℓ, and a message m, and outputs a ciphertext C; for brevity, we sometimes omit EK and write this as $C \leftarrow \mathsf{Enc}^\ell(m)$. The decryption algorithm Dec takes as input DK, a ciphertext C, and a label ℓ; it outputs a message m, or $\bot$ if C is "invalid". We write this as $m \leftarrow \mathsf{Dec}^\ell(C)$ (where we again sometimes omit DK). We assume $\mathsf{Dec}^\ell(\mathsf{Enc}^\ell(m)) = m$ for any message m and label ℓ. Security for encryption is defined following [3, 39]. In particular, we use "CPA-secure" to refer to what is called IND-CPA security in [3], and "CCA-secure" to refer to what is called IND-CCA2 in [3] (modified to take labels into account as in [39]).

A signature scheme $\Sigma =$ (Sig-Gen, Sig, Ver) consists of three PPT algorithms: the key-generation algorithm Sig-Gen takes as input a security parameter 1^k and outputs a signing key SK and a verification key VK. The signing algorithm Sig takes as input SK and a message m, and outputs a signature σ; we will sometimes omit SK and write $\sigma \leftarrow \text{Sig}(m)$. The verification algorithm Ver takes as input VK, a message m, and a signature σ; it outputs 1 iff the signature is valid. We write this as $a \leftarrow \text{Ver}(m, \sigma)$ (again, sometimes omitting VK). We require that $\text{Ver}(m, \text{Sig}(m)) = 1$, for all m.

Unless specified otherwise, the notion of security we consider for signature schemes is that of strong unforgeability under adaptive chosen-message attacks, following [27, 4]. We also use the notion of *one-time signature schemes* which satisfy an analogous definition of security except that an adversary is only allowed to request a signature on a single message.

Secret Sharing Schemes. A secret sharing scheme is a pair of transformations $SSS =$ (Share, Rec).[1] Share$(\cdot)$ is a probabilistic transformation which takes a message M and outputs n secret shares $s_1, \ldots, s_n$ and possibly one public share pub. Rec is a deterministic transformation which takes n shares $s'_1, \ldots, s'_n$ (some of which might be $\bot$) and (if present) the public share pub, and outputs some message M' (possibly $\bot$). The basic correctness property states that $\text{Rec}(\text{Share}(M)) = M$. Security may be quantified by the following thresholds:

- t_p — the *privacy* threshold. Determines the maximum number of shares which (together with pub) reveal "no information" about the message.
- t_f — the *fault-tolerance* threshold. Determines the minimum number of correct shares which (together with pub) suffice to recover the message, when the other shares are *missing*.
- t_r — the *robustness* threshold. Determines the minimum number of correct shares which (together with pub) suffice to recover the message, when the other shares are *adversarially set*.
- t_s — the *soundness* threshold. Determines the minimum number of correct shares which (together with pub) ensure that it is impossible to recover an *incorrect* message $M' \notin \{M, \bot\}$, when the other shares are *adversarially set*.

The above must satisfy $t_p + 1 \leq t_f \leq t_r \leq n$ and $t_s \leq t_r$. The security properties corresponding to the thresholds above can all be formalized in a straightforward way, so we omit them. In a basic secret sharing scheme, only privacy and fault-tolerance are addressed. This is useful when all the parties holding the corresponding shares are trustworthy, but some shares may have been leaked to an adversary and/or some parties may be (temporarily) unavailable. Shamir's scheme [37] is the classical example; this scheme achieves information-theoretic privacy, has no public share, and achieves $t_f = t_p + 1$ and $|M| = |s_i|$. Generalizing this idea [23], one can achieve arbitrary $t_f > t_p$. Krawczyk [29] extended Shamir's scheme to the computational setting by using the scheme to share a

---

[1] Sometimes, we may also have a setup procedure which prepares public parameters. For simplicity, we omit this from our description.

short symmetric key k, and then encrypting the message M using k. The resulting ciphertext can either be stored publicly, or shared among the servers using an *information dispersal scheme* [35] (i.e., a secret sharing scheme which achieves fault-tolerance and/or robustness, but has $t_p = 0$). In fact, this approach can be applied to any information-theoretic secret sharing scheme to obtain a computational scheme with share size proportional to the security parameter and public part proportional to the message length. When fault-tolerance is not needed, one can also use computational all-or-nothing transforms (AONTs) [36, 11] to achieve extremely short shares.

Sometimes, basic secret sharing schemes already enjoy certain robustness properties. For example, Shamir's scheme achieves $t_r = (n + t_f)/2$. Moreover, there are several simple methods to transform any (t_p, t_f, n)-secret sharing scheme into a *robust* (t_p, t_f, t_r, t_s, n)-secret sharing scheme (in a computational sense), achieving optimal values $t_s = 0$ and $t_r = t_f$. We describe two such methods now. In both methods, the dealer first computes the regular sharing $(s_1, \ldots, s_n, \mathsf{pub})$ of M. In the first method, the dealer then generates signing/verification keys $(\mathsf{SK}, \mathsf{VK})$ for a signature scheme, and sets $s'_i = (s_i, \mathsf{Sig}_{\mathsf{SK}}(i, s_i))$, $\mathsf{pub}' = (\mathsf{pub}, \mathsf{VK})$. To reconstruct, users apply the original reconstruction algorithm only to shares whose signatures are correct. In the second method, the dealer uses a commitment scheme to commit to (i, s_i); let c_i (resp., d_i) be the corresponding commitment (resp., decommitment). The dealer then sets $s'_i = (s_i, d_i)$, $\mathsf{pub}' = (\mathsf{pub}, c_1, \ldots, c_n)$. As before, users will only use those shares whose proper decommitment is revealed. In this second method the size of the public information is $O(n)$, but using, e.g., Merkle trees this storage can be reduced considerably at the expense of slightly increasing the share size.

3 Multiple Encryption

We now define a multiple encryption scheme.

Definition 1. *A (non-interactive) public-key multiple encryption scheme is a tuple of* PPT *algorithms* $\mathcal{TE} = (\mathsf{TGen}, \mathsf{TEnc}, \mathsf{Split}, \mathsf{TDec}, \mathsf{Combine})$ *such that:*

- TGen, *the* key generation algorithm, *is a probabilistic algorithm which takes as input a security parameter* 1^k *and outputs a public key* TEK *along with n secret keys* $\mathbf{TDK} = (\mathsf{TDK}_1, \ldots, \mathsf{TDK}_n)$.
- TEnc, *the* encryption algorithm, *is a probabilistic algorithm which takes as input a public key* TEK, *a message M, and a label L. It outputs a ciphertext* $C \leftarrow \mathsf{TEnc}^L(M)$.
- Split, *the* splitting algorithm, *is a deterministic algorithm which takes as input a public key* TEK, *a ciphertext C and a label L. It either outputs $\bot$, or n ciphertext shares* $\mathbf{C} = (C_1, \ldots, C_n)$ *and some auxiliary info* aux.
- TDec, *the* partial decryption algorithm, *takes as input* $i \in \{1, \ldots, n\}$, *a secret key* TDK_i, *and a ciphertext share* C_i; *it outputs the message share* M_i

or the distinguished symbol $\perp$. We denote the output of this algorithm by $\mathsf{TDec}^i(C_i)$. We also let $\mathbf{DEC}(\mathbf{C},\mathsf{aux}) \stackrel{\text{def}}{=} (\mathsf{TDec}^1(C_1), \ldots, \mathsf{TDec}^n(C_n), \mathsf{aux})$.
- Combine, *the combining algorithm, takes as input shares* $\mathbf{M} = (M_1, \ldots, M_n)$ *and the auxiliary info* aux, *and outputs a message* M *or* $\perp$.

Correctness (refined later) requires that for all TEK, **TDK** *output by* TGen, *all messages* M *and labels* L, *we have:* $\mathsf{Combine}(\mathbf{DEC}(\mathsf{Split}^L(\mathsf{TEnc}^L(M)))) = M$.

Before discussing security, a few remarks are in place. It is important to recognize that multiple encryption might be used in a number of different scenarios. In one scenario, the set of decryption keys **TDK** (or some subset of these keys) are co-located, so a single user receiving a ciphertext C would perform the splitting, partial decryption, and combining by itself. In another scenario, there are a set of n servers and server i stores TDK_i. Here, a user receiving a ciphertext C would perform the splitting himself to obtain $\mathbf{C}$, aux, would keep aux, and would send the ciphertext share C_i to server i for decryption. Server i would respond with M_i and the various message shares would be combined by the user to recover the original message. These different ways of thinking about the decryption process are each appropriate for different applications of multiple encryption.

When decryption keys TDK_i are stored at different locations (i.e., on different servers), the above definition implies that servers do not communicate with each other and do not keep any intermediate state. Also, we remark that we could have ignored the splitting algorithm altogether and simply have TDK_i operate on the entire ciphertext C (performing any splitting itself, as necessary). The reason for not doing so is that C might contain information which is not "relevant" to server i, and thus sending the entire ciphertext to each server might be wasteful. In fact, our solutions achieve $\sum |C_i| = O(|C|)$, so the total communication between the user and *all* servers is proportional to the size of the original ciphertext.

In either of the above scenarios (i.e., whether the decryption keys are co-located or stored at different servers), it is possible for some of the decryption keys to be compromised by an adversary. This raises the first security issue, which is that of *message privacy*. When keys are stored on separate servers, there is also the possibility that some servers may be compromised in their entirety; this raises the additional issue of *decryption robustness*. Since the security issues in the latter case are stronger than those in the former case, for the remainder of this section we will speak in terms of a central user running the splitting/combining algorithm and n servers performing the partial decryption of each share.

Message Privacy. We assume that the adversary may learn $t_p < n$ decryption keys, where t_p is the *privacy threshold*. Formally, given a set $I = \{i_1, \ldots, i_{t_p}\}$, an adversary is given a randomly-generated public key TEK, the set of secret keys $\mathsf{TDK}_I = \{\mathsf{TDK}_{i_1}, \ldots, \mathsf{TDK}_{i_{t_p}}\}$, and oracle access to some oracle $\mathcal{O}$ whose meaning will be clarified shortly. $\mathcal{B}$ outputs two messages M_0, M_1 (along with some label L), and receives a challenge ciphertext $C \leftarrow \mathsf{TEnc}(M_b)$ for a randomly-chosen b. The adversary succeeds if it correctly guesses b, and the adversary's advantage is defined as the absolute value of the difference between its success probability and $1/2$. If the oracle $\mathcal{O}$ is "empty", we say that $\mathcal{B}$ is performing a *(multiple)* *chosen-plaintext attack* (MCPA). As for the *(multiple) chosen-ciphertext attack*, there are several meaningful flavors described below in the order of increasing adversarial power.

In the weakest such attack, denoted wMCCA ("weak MCCA"), we have $\mathcal{O} = \mathsf{Combine}(\mathbf{DEC}(\mathsf{Split}^{(\cdot)}(\cdot)))$ (where the adversary is prohibited from submitting (C, L) to this oracle). Namely, $\mathcal{B}$ only gets access to the entire decryption process without seeing any partial decryption results and without being able to ask questions to the decryption servers directly. While this notion already suffices for some applications, it assumes that the adversary can never see the intermediate decryption shares. In a (regular) MCCA attack, we let $\mathcal{O} = \mathbf{DEC}(\mathsf{Split}^{(\cdot)}(\cdot))$ (as before, we forbid the adversary from submitting (C, L) to this oracle). Namely, we still assume that the ciphertext gets passed through a proper splitting procedure but $\mathcal{B}$ also learns the intermediate decryption results $M_1, \ldots, M_n$. As we shall see, this attack is sufficient for most applications of multiple encryption.

However, sometimes we need to consider an even stronger attack denoted sMCCA (for "strong MCCA"), where we have $\mathcal{O} = \mathsf{TDec}^{(\cdot)}(\cdot)$. Namely, we allow $\mathcal{B}$ to ask arbitrary and questions to the individual decryption servers. Of course, to make sense of this attack, we need to add some restrictions. First and most obvious, for a challenge ciphertext C (with label L) we disallow questions (i, C_i), where C_i is the ciphertext share for server i that results from "splitting" C using label L. Second and less obvious, we assume (for all i) that the mapping Split_i from (C, L) to C_i is *weakly collision-resistant*. This means that no PPT adversary $\mathcal{A}$ can succeed with non-negligible probability in the following game: $\mathcal{A}(\mathbf{TDK})$ supplies some pair (M, L) to the encryption oracle, and gets back a ciphertext $C \leftarrow \mathsf{TEnc}^L(M)$. $\mathcal{A}$ succeeds if it can output a pair $(C', L') \neq (C, L)$ and an index i such that $\mathsf{Split}_i(C, L) = \mathsf{Split}_i(C', L')$. Indeed, without this latter condition it seems unnecessarily restrictive to prohibit the adversary $\mathcal{B}$ in the sMCCA game from asking questions $(i, C_i = \mathsf{Split}_i(C, L))$. This is because there is a chance such a question might have "legally" come from a different ciphertext $(C', L') \neq (C, L)$. We further observe that when the Split procedure *does* satisfy this condition, the sMCCA attack is at least as strong as the MCCA attack,[2] and it is easy to see that this conclusion does *not* hold without weak collision resistance. Therefore, we will insist on weak collision-resistance when talking about sMCCA attacks.

Definition 2. *Let $X \in \{\mathsf{MCPA}, \mathsf{wMCCA}, \mathsf{MCCA}, \mathsf{sMCCA}\}$. We say multiple encryption scheme $\mathcal{TE}$ is X-secure with privacy threshold t_p, if the advantage of any PPT adversary $\mathcal{B}$ performing attack X with any set I of size t_p is negligible.*

Decryption Robustness. The correctness property of Definition 1 only ensures correct decryption when all algorithms are honestly and correctly executed. Just as in the case of secret sharing, however, one may often desire fault-tolerance,

---

[2] This is so since one can simulate (with all but negligible probability) any "allowed" call to $\mathbf{DEC}(\mathsf{Split}^{(\cdot)}(\cdot))$ by n "allowed" calls to $\mathsf{TDec}^{(\cdot)}(\cdot)$.

robustness, and/or soundness. As in the case of secret sharing, these are parameterized by thresholds t_f, t_r, t_s, whose meaning is completely analogous to their meaning in the case of secret sharing (described earlier). Our solutions can achieve optimal $t_s = 0$, $t_r = t_f$, and any $t_p < t_f$.

3.1 Insecurity of Known Multiple Encryption Schemes

It is instructive to note that known constructions of multiple encryption schemes (even when instantiated with a CCA-secure standard encryption scheme) are insecure under the weakest definition of chosen-ciphertext security considered above. We briefly illustrate this for the simplest case of $n = 2$.

In sequential encryption, M is encrypted via $C \leftarrow \mathsf{Enc}_{\mathsf{EK}_1}(\mathsf{Enc}_{\mathsf{EK}_2}(M))$. An adversary, when given the decryption key DK_1 and a challenge ciphertext C, can break the encryption scheme as follows: decrypt C using DK_1 to obtain $C' \in \mathsf{Enc}_{\mathsf{EK}_2}(M)$ and then *re-encrypt* C' using EK_1; this results in a second, different ciphertext $\tilde{C}$. Now, by submitting $\tilde{C}$ to its decryption oracle, the adversary will receive in return the original message M.

Attacks are also possible for the case of parallel encryption. Here, a message M is encrypted as $C = \langle C_1, C_2 \rangle$, where $C_1 \leftarrow \mathsf{Enc}_{\mathsf{EK}_1}(s_1)$, $C_2 \leftarrow \mathsf{Enc}_{\mathsf{EK}_2}(s_2)$, and s_1 and s_2 are chosen at random subject to $s_1 \oplus s_2 = M$. Now, even without being given any decryption keys, an adversary given a challenge ciphertext C can compute $\tilde{C}_1 \leftarrow \mathsf{Enc}_{\mathsf{EK}_1}(0)$ and $\tilde{C}_2 \leftarrow \mathsf{Enc}_{\mathsf{EK}_2}(0)$, and then submit the ciphertexts $\langle \tilde{C}_1, C_2 \rangle$ and $\langle C_1, \tilde{C}_2 \rangle$. Note that the adversary thus obtains both s_1 and s_2 separately, from which it can recover the original message $M = s_1 \oplus s_2$.

4 Generic Constructions

In this section we describe how to build MCCA- and sMCCA-secure multiple encryption schemes from any (standard) CCA-secure encryption scheme $\mathcal{E}$. In our schemes, the decryption keys will simply be decryption keys DK_i independently-generated by $\mathcal{E}$, and partial decryption will essentially require only a single decryption with this key. Our results achieve: (1) ciphertext length linear in the length of the plaintext message; (2) communication with each server *independent* of the number of servers and the length of the message. We also stress that when the decryption keys are held by several servers, no interaction between servers is required. A drawback is that the ciphertext and public-key lengths in our solutions are proportional to the number of decryption keys n (of course, for small n, such as the important case of $n = 2$, this is not a problem). We believe that this dependence in unavoidable if we are not willing to assume any algebraic structure on $\mathcal{E}$. Indeed, in the following section we show how this dependence can be avoided when starting from (hierarchical) identity-based encryption.

For the remainder of this section, let $\mathcal{SSS} = (\mathsf{Share}, \mathsf{Rec})$ be a (t_p, t_f, t_r, t_s, n)-secret sharing scheme. All multiple encryption schemes we construct will inherent the same thresholds t_p, t_f, t_r, t_s, which elegantly allows us to push all the privacy and robustness constraints onto the much simpler secret sharing primitive.

4.1 Achieving Chosen-Ciphertext Security

Recall that in parallel encryption the message M is first shared using $\mathcal{SSS}$, and then each share is separately encrypted using an independent key. As noted earlier, this folklore scheme is not secure against chosen-ciphertext attacks (even against a weak MCCA attack and with no corrupted keys). We show a simple and elegant way to extend parallel encryption so as to solve this problem, without introducing much extra complexity. In brief, we use a secure one-time signature scheme $\Sigma = (\text{Sig-Gen}, \text{Sig}, \text{Ver})$ to bind all the local ciphertexts to each other (and to the label L). The main twist which makes this work is that we also bind the verification key of Σ to each of the ciphertexts.

Before giving the formal description of our solution, we illustrate our construction for the case $n = 2$ (with $t_f = t_r = t_s = 2, t_p = 1$, and no labels). The public key consists of two independently-generated keys $\mathsf{EK}_1, \mathsf{EK}_2$, and the secret key contains the corresponding decryption keys $\mathsf{DK}_1, \mathsf{DK}_2$. Let $\mathsf{Enc}_1 \stackrel{\text{def}}{=} \mathsf{Enc}_{\mathsf{EK}_1}$ and similarly for Enc_2. To encrypt M, a sender first "splits" M by choosing random s_1 and setting $s_2 = M \oplus s_1$. The sender then generates a key pair $(\mathsf{VK}, \mathsf{SK})$ for a one-time signature scheme, and computes $C_1 \leftarrow \mathsf{Enc}_1^{\mathsf{VK}}(s_1)$ and $C_2 \leftarrow \mathsf{Enc}_2^{\mathsf{VK}}(s_2)$. Finally, the sender computes $\sigma = \mathsf{Sig}_{\mathsf{SK}}(C_1, C_2)$; the complete ciphertext is $\langle \mathsf{VK}, C_1, C_2, \sigma \rangle$. Decryption is done in the obvious way: if σ is not a valid signature on C_1, C_2 with respect to VK, the ciphertext is invalid. Otherwise, DK_1 and DK_2 are used to obtain s_1 and s_2 from which the original message $M = s_1 \oplus s_2$ can be recovered.

We now generalize this solution to arbitrary n and using an arbitrary secret sharing scheme $\mathcal{SSS} = (\mathsf{Share}, \mathsf{Rec})$.

- $\mathsf{TGen}(1^k)$: For $i = 1, \ldots, n$, let $(\mathsf{EK}_i, \mathsf{DK}_i) \leftarrow \mathsf{Gen}(1^k)$ and set $\mathbf{TEK} = (\mathsf{EK}_1, \ldots, \mathsf{EK}_n)$, $\mathsf{TDK}_i = \mathsf{DK}_i$, so that $\mathbf{TDK} = (\mathsf{DK}_1, \ldots, \mathsf{DK}_n)$. Below, let $\mathsf{Enc}_i \stackrel{\text{def}}{=} \mathsf{Enc}_{\mathsf{EK}_i}$ and $\mathsf{Dec}_i \stackrel{\text{def}}{=} \mathsf{Dec}_{\mathsf{DK}_i}$.
- $\mathsf{TEnc}^L(M)$: Let $(s_1, \ldots, s_n, \mathsf{pub}) \leftarrow \mathsf{Share}(M)$, and $(\mathsf{VK}, \mathsf{SK}) \leftarrow \mathsf{Sig\text{-}Gen}(1^k)$. Set $C_i = \mathsf{Enc}_i^{\mathsf{VK}}(s_i)$ (for $i = 1, \ldots, n$) and then compute the signature $\sigma = \mathsf{Sig}_{\mathsf{SK}}(C_1, \ldots, C_n, \mathsf{pub}, L)$. Output $C = (C_1, \ldots, C_n, \mathsf{pub}, \mathsf{VK}, \sigma)$.
- $\mathsf{Split}^L(C)$: Parse C as $(C_1, \ldots, C_n, \mathsf{pub}, \mathsf{VK}, \sigma)$, and reject if verification fails; i.e., if $\mathsf{Ver}_{\mathsf{VK}}((C_1, \ldots, C_n, \mathsf{pub}, L), \sigma) = 0$. Otherwise, set ciphertext share $\hat{C}_i = (C_i, \mathsf{VK})$ and $\mathsf{aux} = \mathsf{pub}$.
- $\mathsf{TDec}^i(C_i, \mathsf{VK})$: Output $s'_i = \mathsf{Dec}_i^{\mathsf{VK}}(C_i)$.
- $\mathsf{Combine}(s'_1, \ldots, s'_n, \mathsf{pub})$: Output $\mathsf{Rec}(s'_1, \ldots s'_n, \mathsf{pub})$.

As with the folklore scheme, each decryption server simply performs a single regular (now CCA-secure) decryption, but here using a label which is the verification key of a one-time signature scheme (and which is used to bind all the ciphertexts together). We claim:

Theorem 1. *If $\mathcal{E}$ is CCA-secure, $\mathcal{SSS}$ is a (t_p, t_f, t_r, t_s, n)-secret sharing scheme, and Σ is a secure one-time signature scheme, then $\mathcal{TE}$ is MCCA-secure with thresholds t_p, t_f, t_r, t_s.*

Proof. Robustness thresholds t_f, t_r, t_s follow immediately from those of the secret sharing scheme, due to the definition of Combine = Rec. We now argue message privacy.

Assume there exists some PPT adversary $\mathcal{B}$ attacking MCCA-security who has some non-negligible advantage. Recall, this $\mathcal{B}$ has oracle access to $\mathcal{O}(\cdot, \cdot) = \mathbf{DEC}(\mathsf{Split}^{(\cdot)}(\cdot))$, chooses some messages M_0, M_1 and a label L, gets an unknown ciphertext $C = (C_1, \ldots, C_n, \mathsf{pub}, \mathsf{VK}, \sigma)$, and tries to guess whether this corresponds to the encryption of M_0 or M_1 (with label L). Let X denote the event that $\mathcal{B}$ asks $\mathcal{O}$ a query $(C', L') \neq (C, L)$, where C' includes the same verification key $\mathsf{VK}' = \mathsf{VK}$ as the challenge, but σ' is a new, valid signature (with respect to VK) of the corresponding "message" $(C_1', \ldots, C_n', \mathsf{pub}', L')$. It is immediate that $\Pr[X] = \mathsf{negl}(k)$, or else an easy argument (omitted) shows that we can use $\mathcal{B}$ to construct a PPT adversary breaking the security of the one-time signature scheme Σ with non-negligible advantage.

We can therefore construct an adversary $\mathcal{B}'$ who never makes a query to $\mathcal{O}$ using the same verification key as in the challenge ciphertext, yet whose advantage is negligibly close to the advantage of $\mathcal{B}$. Let ε_0 denote the advantage of $\mathcal{B}'$, and assume w.l.o.g. that $\mathcal{B}'$ corrupts servers $\{n - t_p + 1, \ldots, n\}$. We refer to this game involving $\mathcal{B}'$ as G_0, and now gradually change this game into games $G_1, \ldots, G_{n-t_p}$. In general, G_i is identical to G_0, except for one step in the computation of the challenge ciphertext C. Recall, in G_0 we have $C_j \leftarrow \mathsf{Enc}_j^{\mathsf{VK}}(s_j)$, where s_j is the j-th share of the secret sharing scheme. In game G_i we instead do this only for $j > i$, but set $C_i \leftarrow \mathsf{Enc}_i^L(0)$ for $j \leq i$ (where 0 is some arbitrary fixed message in our space). In other words, G_{i-1} and G_i are identical except G_{i-1} sets $C_i \leftarrow \mathsf{Enc}_i^{\mathsf{VK}}(s_i)$, while G_i sets $C_i \leftarrow \mathsf{Enc}_i^{\mathsf{VK}}(0)$ Denote by ε_i the advantage of $\mathcal{B}'$ in predicting the challenge bit b in game G_i. We claim that for every $1 \leq i \leq n - t_p$ we have $|\varepsilon_i - \varepsilon_{i-1}| = \mathsf{negl}(k)$.

To show the claim, using $\mathcal{B}'$ we construct an adversary $\mathcal{A}_i$ who succeeds in breaking CCA-security of $\mathcal{E}$ with advantage $\delta_i = \frac{1}{2}|\varepsilon_{i-1} - \varepsilon_i|$. Since $\mathcal{E}$ is assumed to be CCA-secure, the claim follows. $\mathcal{A}_i$ gets an encryption key EK for $\mathcal{E}$, sets $\mathsf{EK}_i = \mathsf{EK}$, and generates the remaining $(n-1)$ public/secret keys by himself. These public keys, as well as the last t_p secret keys, are given to $\mathcal{B}'$. Adversary $\mathcal{A}_i$ then honestly simulates the run of G_{i-1}/G_i until $\mathcal{B}'$ submits the challenge (M_0, M_1, L). At this point, $\mathcal{A}_i$ chooses a random bit b, generates (SK, VK), computes the shares $(s_1, \ldots, s_n, \mathsf{pub}) \leftarrow \mathsf{Share}(M_b)$, and prepares C_j for $j \neq i$ just as in G_{i-1} and G_i. Furthermore, $\mathcal{A}'$ outputs the challenge $(s_i, 0, \mathsf{VK})$ in its own CCA game. Upon receiving the challenge ciphertext C, it sets $C_i = C$, signs whatever is needed, and passes the resulting challenge ciphertext to $\mathcal{B}'$. It only remains to specify how $\mathcal{A}_i$ deals with oracle queries of $\mathcal{B}'$. Notice that $\mathcal{A}_i$ can decrypt all ciphertexts C_j' for $j \neq i$ by himself, since the appropriate decryption keys are known. As for C_i', since (by construction) $\mathcal{B}'$ does not reuse the challenge value VK, this means that $\mathcal{A}_i$ can always submit C_i' to its decryption oracle using the label $\mathsf{VK}' \neq \mathsf{VK}$. Finally, $\mathcal{A}_i$ outputs 1 iff $\mathcal{B}'$ correctly predicts b. This completes the description of $\mathcal{A}_i$, and it is not hard to see that $\mathcal{A}_i$ gives a perfect simulation of either game G_{i-1} or G_i depending on which of s_i or 0 was encrypted. The claim regarding $|\varepsilon_{i-1} - \varepsilon_i|$ follows easily.

Now, since $(n - t_p)$ is polynomial in k and ε_0 is assumed to be non-negligible, we get that ε_{n-t_p} is non-negligible as well. But let us now examine the game G_{n-t_p} more closely. When encrypting the challenge M_b, only $t = t_p$ shares (and the value pub) are used in creating the ciphertext. But then the privacy of the secret sharing scheme implies that ε_{n-t_p} must be negligible, a contradiction. (This is not hard to see, and we omit the obvious details.)

Replacing Signatures with MACs. At the expense of settling for (weaker) wMCCA-security, we can use the recent technique of Boneh and Katz [10] to replace the one-time signature scheme by the more efficient combination of a message authentication code (MAC) and a weak form of commitment. The idea is to commit to a MAC key τ, then encrypt both the message M and the decommitment d using the secret sharing technique above, but with the public verification key VK replaced by the commitment c. Finally, τ is used to compute a message authentication code on the entire resulting ciphertext. In brief, the reason this only yields wMCCA-security is that the message authentication code computed over the ciphertext (as opposed to the one-time signature computed above) can only be verified *after* all the shares are collected. More details are given in Appendix A.

4.2 Achieving Strong Chosen-Ciphertext Security

The scheme above does not enjoy sMCCA-security since, in particular, the mapping Split_i from (C, L) to C_i is not weakly collision-resistant; indeed, it ignores all ciphertexts other than C_i. A natural first thought is to simply append a hash α of the entire ciphertext C to each of the local decryption shares C_i (and let each server simply ignore α). While this may make each Split_i weakly collision-resistant, it will *not* achieve sMCCA-security: Since the servers ignore α anyway, the adversary can simply replace α by "garbage" while keeping the rest of the C_i the same; this will result in a "valid" decryption request to each server, but will result in a proper decryption of C_i to the adversary.

A natural way to fix this is to let each server check the hash α by sending to the server the entire ciphertext C. In fact, if we are willing to send the entire ciphertext to each server, we no longer need α: each server can just perform the corresponding splitting procedure on C by itself. In fact, doing so will trivially give sMCCA-security. However, sending all of C (and having each server perform the splitting procedure) may be wasteful in some scenarios; it therefore remains interesting to explore improved solutions with lower user-server communication and in which more of the work is shifted to the user rather than the servers.

For the case of the *particular* MCCA-secure scheme $\mathcal{TE}_{cca}$ of the previous section, sMCCA-security can be achieved at a very small additional cost. Let $\mathcal{H} = \{H\}$ be a family of collision-resistant hash functions. We now describe the modified scheme $\mathcal{TE}_{scca}$.

- TGen(1^k). Sample $H \leftarrow \mathcal{H}$ and for $i = 1, \ldots, n$, let $(\mathsf{EK}_i, \mathsf{DK}_i) \leftarrow \mathsf{Gen}(1^k)$. Set $\overline{\mathsf{TEK}} = (\mathsf{EK}_1, \ldots, \mathsf{EK}_n, H)$, $\overline{\mathsf{TDK}}_i = \mathsf{DK}_i$. Below, denote $\mathsf{Enc}_i = \mathsf{Enc}_{\mathsf{EK}_i}$, $\mathsf{Dec}_i = \mathsf{Dec}_{\mathsf{DK}_i}$.
- $\mathsf{TEnc}^L(M)$. Let $(s_1, \ldots, s_n, \mathsf{pub}) \leftarrow \mathsf{Share}(M)$, and $(\mathsf{VK}, \mathsf{SK}) \leftarrow \mathsf{Sig\text{-}Gen}(1^k)$. Set $C_i = \mathsf{Enc}_i^{\mathsf{VK}}(s_i)$ for $i = 1, \ldots, n$; then compute $\alpha = H(C_1, \ldots, C_n, \mathsf{pub}, L)$ and $\sigma = \mathsf{Sig}_{\mathsf{SK}}(\alpha)$. Output $C = (C_1, \ldots, C_n, \mathsf{pub}, \mathsf{VK}, \sigma)$.
- $\mathsf{Split}^L(C)$. Parse $C = (C_1, \ldots, C_n, \mathsf{pub}, \mathsf{VK}, \sigma)$, set $\alpha = H(C_1, \ldots, C_n, \mathsf{pub}, L)$, and reject if $\mathsf{Ver}_{\mathsf{VK}}(\alpha, \sigma) = 0$. Otherwise, set the ciphertext share to be $\hat{C}_i = (C_i, \mathsf{VK}, \alpha, \sigma)$ and set $\mathsf{aux} = \mathsf{pub}$.
- $\mathsf{TDec}^i(C_i, \mathsf{VK}, \alpha, \sigma)$. Output $\mathsf{Dec}_i^{\mathsf{VK}}(C_i)$ if $\mathsf{Ver}_{\mathsf{VK}}(\alpha, \sigma) = 1$, and $\perp$ otherwise.
- $\mathsf{Combine}(s'_1, \ldots, s'_n, \mathsf{pub})$. Output $\mathsf{Rec}(s'_1, \ldots s'_n, \mathsf{pub})$.

Thus, the only effective change is to force each server to verify a signature (of a one-time signature scheme) before performing the decryption. The cost of this will typically be small compared to the cost of decryption.

We now consider the security of the above. On an intuitive level, when an adversary makes a decryption query, either: (1) the adversary reuses a previous VK, which implies that it uses a previous α (due to unforgeability of the signature scheme), which in turn implies that the query is illegal (since H is collision-resistant); or (2) the adversary uses a new VK, in which case the chosen-ciphertext security of the underlying encryption schemes (which use VK as a label) implies that the resulting ciphertexts are unrelated to the challenge. Notice, there is no need for the server to check that α is the correct hash; having a valid signature of α implicitly assures the server that either this decryption query is unrelated to the challenge, or α is indeed correct due to the unforgeability of the one-time signature scheme. Notice also that once again the communication between the user and each server is independent of n. The above intuition can in fact be used to prove the following theorem:

Theorem 2. *If $\mathcal{E}$ is CCA-secure, $\mathcal{SSS}$ is a (t_p, t_f, t_r, t_s, n)-secret sharing scheme, Σ is a secure one-time signature scheme, and $\mathcal{H}$ is collision-resistant, then $\mathcal{TE}_{scca}$ is sMCCA-secure with thresholds t_p, t_f, t_r, t_s.*

Proof. As before, robustness thresholds t_f, t_r, t_s follow immediately from those of the secret sharing scheme since $\mathsf{Combine} = \mathsf{Rec}$. We now argue message privacy. Here we need to argue two things: indistinguishability of the scheme against sMCCA attack and weak collision resistance of the splitting procedure.

We start with the second part. Take any adversary $\mathcal{A}$ attacking weak collision resistance of $\mathcal{TE}_{scca}$. $\mathcal{A}$ gets the entire secret key **TDK**, produces a pair (M, L), gets $C \leftarrow \mathsf{TEnc}^L(M)$, and outputs $(C', L') \neq (C, L)$ and an index i. If it is the case that $\mathsf{Split}_i(C, L) = \mathsf{Split}_i(C', L')$ then (by definition of Split) this means that $(C_i, \mathsf{VK}, \alpha, \sigma) = (C'_i, \mathsf{VK}', \alpha', \sigma')$. But then $H(C_1 \ldots C_n, \mathsf{pub}, L) = \alpha = \alpha' = H(C'_1 \ldots C'_n, \mathsf{pub}', L')$ and this violates collision-resistance of $\mathcal{H}$.

Next, we show security against sMCCA attack. Assume there exists some adversary $\mathcal{B}$ attacking sMCCA-security who has some non-negligible advantage. Recall, $\mathcal{B}$ has oracle access to $\mathcal{O}(\cdot, \cdot) = \mathsf{TDec}^{(\cdot)}(\cdot)$, chooses some messages M_0, M_1 and a label L, gets a challenge ciphertext $C = (C_1, \ldots, C_n, \mathsf{pub}, \mathsf{VK}, \sigma)$, and tries to predict whether this ciphertext corresponds to an encryption of M_0 or of M_1 (with label L). Let X denote the event that $\mathcal{B}$ asks $\mathcal{O}$ a query $(i, (C'_i, \mathsf{VK}, \alpha', \sigma'))$, where VK is the same verification key as the one used in the challenge ciphertext but σ' is a *new, valid* signature with respect to VK of the corresponding message α'. Namely, σ' is a valid signature of α', but $(\alpha', \sigma') \neq (\alpha, \sigma)$. Clearly, $\Pr(X) = \mathsf{negl}(k)$ or otherwise $\mathcal{B}$ can be used to break the security of the one-time signature scheme Σ.

We thus assume that X never occurs in the run of $\mathcal{B}$, yet $\mathcal{B}$ still has non-negligible advantage. Since $\mathcal{B}$ is forbidden to ask any challenge query of the form $(i, (C_i, \mathsf{VK}, \alpha, \sigma))$, this means that every query $(i, (C'_i, \mathsf{VK}', \alpha', \sigma'))$ that $\mathcal{B}$ makes satisfies one of three conditions: (1) $\mathsf{Ver}_{\mathsf{VK}'}(\alpha', \sigma) = 0$, in which case the response is automatically $\perp$ (and so we can assume that $\mathcal{B}$ never makes such a query); (2) $(\mathsf{VK}', \alpha', \sigma') = (\mathsf{VK}, \alpha, \sigma)$, but $C'_i \neq C_i$ (recall, we proved that $\mathsf{VK}' = \mathsf{VK}$ implies $(\alpha', \sigma') = (\alpha, \sigma)$, so the only way for this query to be legal while keeping $\mathsf{VK}' = \mathsf{VK}$ is to have $C'_i \neq C_i$); (3) $\mathsf{VK}' \neq \mathsf{VK}$. Since we excluded queries of type (1), we combine cases (2) and (3) to conclude that every query of $\mathcal{B}$ must have $(C'_i, \mathsf{VK}') \neq (C_i, \mathsf{VK})$.

Given this observation, the rest of the proof is almost identical to the proof of Theorem 1 (with obvious syntactic modifications). Namely, we create hybrid games in which encryptions of the shares of M_b are gradually replaced by encryptions of 0. As in the proof of the previous theorem, we show that any such change cannot be noticed by $\mathcal{B}$ since the corresponding encryption scheme $\mathcal{E}_i$ is CCA-secure. The only new aspect of this proof is the description of how $\mathcal{A}_i$ handles $\mathcal{B}$'s queries $(j, (C'_j, \mathsf{VK}', \alpha', \sigma'))$. When $j \neq i$, then $\mathcal{A}_i$ can simply decrypt by itself, as before. For $j = i$, $\mathcal{A}_i$ first checks the validity of the signature, and then asks its own decryption oracle to decrypt (C'_i, VK'). So all we need to argue is that this query is different from $\mathcal{A}_i$'s own challenge (C_i, VK) (which $\mathcal{A}_i$ is forbidden to ask). But this is precisely what we argued about $\mathcal{B}$'s behavior in the previous paragraph.

Remark 1. The existence of collision-resistant hash functions does not seem to follow from the existence of CCA-secure encryption schemes. However, by slightly sacrificing the efficiency of our construction, we can rely on universal one-way hash functions (UOWHFs) (which *are* implied by the existence of CCA-secure encryption) thus making our construction completely generic. Briefly, instead of using a single $H \in \mathcal{H}$ in the public key, the sender will choose a new $H \leftarrow \mathcal{H}$ for every encryption. The description of H will then be included as part of the ciphertext, signed together with α, and be included as part of each server's share. Since one can achieve $|H| \sim \log n$ [6], this still keeps the user-server communication very low.

5 Direct Constructions from Selective Identity IBE/HIBE Schemes

We assume the reader is familiar with the basic terminology of identity-based encryption (IBE) and hierarchical identity-based encryption (HIBE); see [8, 25].

Recently, Canetti et al. [12] gave a simple and elegant construction transforming a "weak" (so called selective-identity-secure) IBE scheme secure against CPA attacks into a CCA-secure (standard) public-key encryption scheme. Their transformation uses a secure one-time signature scheme, by first encrypting the message M with the identity VK (for newly chosen keys (SK, VK)), and then signing the resulting ciphertext with SK. The receiver, who stores the master secret key for the IBE scheme, can then decrypt the ciphertext if the signature is valid.

We could then use the resulting CCA-secure encryption schemes in our transformations to get CCA-secure multiple encryption schemes, where each server would store a master key for an independent IBE scheme. However, this will result in generating $(n+1)$ one-time keys and signatures per ciphertext, which is wasteful. Instead, we notice that the same verification key VK can be used as the identity for all n IBE schemes, and then used to sign the concatenation of n ciphertexts (or its hash). This gives a much more efficient direct construction with only a single one-time signature per ciphertext.

However, just like our original scheme, the public key of the resulting multiple encryption is still proportional to the number of parties n. We now show that using a *two-level hierarchical* IBE scheme (secure against selective-identity CPA-attack), we can make the first relatively generic multiple encryption scheme whose public key is *independent* of the number of players (although the ciphertext size still is). Specifically, the public key pk is simply the mater public key of the two-level HIBE. The i-th decryption key TDK_i consists of level-1 identity-based secret key corresponding to identity i. To encrypt a message M, the sender (as before) generates a key pair $(SK, VK) \leftarrow \text{Sig-Gen}(1^k)$ and applies a secret sharing scheme to the message M resulting in shares $s_1 \ldots s_n$ (and pub). Now, however, the sender encrypts s_i "to" the level-2 identity (i, VK), and then signs the resulting ciphertexts (or their hash) using SK. Each server i can still decrypt its share since it knows the level-1 secret key for the parent identity i, while the collusion-resistance of the HIBE easily implies that no other coalition of servers can get any information from this share. We omit a formal proof in this abstract.

We remark that Boneh and Boyen [7] have recently constructed simple and efficient selective-identity IBE/HIBE schemes, which immediately give rise to simple and efficient multiple encryption schemes using our paradigm. In particular, using their HIBE scheme we get an efficient multiple encryption scheme with a constant-size public key. We also notice that the technique of replacing signatures by MACs [10] also applies here to obtain more efficient wMCCA-secure multiple encryption.

6 Applications

We outline in brief a number of applications of multiple encryption.

CCA-Secure Threshold Encryption. In the generally-considered model for threshold encryption, there is a *combiner* who receives a ciphertext and sends some information to various servers who may then potentially interact, either with each other or with the combiner. The information sent to the servers is typically assumed to be the ciphertext itself, but in general (and in our case in particular) it is possible to transmit a smaller amount of information to each server. In either case, the servers then send decryption shares back to the combiner, who uses these to recover the original message. In a chosen-ciphertext attack on a threshold encryption scheme (see, e.g., [40, 13, 30]), an adversary can expose the decryption shares stored at some number of servers, observe a ciphertext C, and send ciphertexts $C' \neq C$ to the combiner. When it does so, in addition to receiving the decryption of C', it is also typically assumed that the adversary can observe all communication in the network, both between the servers and the combiner as well as between the servers themselves.

It is not hard to see that the adversarial model thus described corresponds exactly to a MCCA-attack. Moreover, if the combiner itself is untrusted (and can send what it likes to the servers), we effectively have a sMCCA-attack. Thus, any MCCA/sMCCA-secure multiple encryption scheme with privacy threshold t_p immediately gives a threshold encryption scheme with the same privacy threshold. Furthermore, a MCCA/sMCCA-secure multiple encryption scheme with robustness threshold t_r immediately gives a threshold encryption scheme in which the ciphertext can be correctly decrypted as long as t_r servers remain uncorrupted. Thresholds t_f and t_s can be interpreted similarly.

Our techniques thus give the first *generic* construction for CCA-secure threshold encryption (note that no previous generic solution existed even in the random oracle model). We remark further that for small values of n, our schemes are competitive with previous threshold schemes. For example, when $n = 2$ and we use the Cramer-Shoup [15] encryption scheme as our building block, we obtain a CCA-secure two-party public-key encryption scheme (in the standard model) which has more efficient decryption than the scheme recently proposed by MacKenzie [30]. In fact, although this construction increases the encryption time and ciphertext size by (roughly) a factor of two as compared to [30], the time required for decryption (by each server) is actually a factor of 10 *more* efficient; furthermore decryption in our case is completely non-interactive. As another example, if we use RSA-OAEP as our building block we obtain a very efficient solution for CCA-secure, RSA-based threshold encryption with completely non-interactive decryption (in the random oracle model).

CCA-Secure Key-Insulated and Strong Key-Insulated Encryption. We assume the reader is somewhat familiar with the key-insulated model, as well as with the generic constructions of [20] (which achieve only CPA security). In a key-insulated public-key encryption scheme there is a *server* and a *user*; at the beginning of each time period, the user communicates with the server to update the user's secret key. Ciphertexts sent during any time period can be decrypted by the user alone, without any further communication with the server. The main property of such schemes is that exposing the secret information stored by the user during many time periods leaves *all* non-exposed periods secure.

At a high level, in the generic solution of [20] the server stores n secret keys for a standard encryption scheme (and the n corresponding public keys constitute the public key of the key-insulated scheme). At the beginning of each time period, some ℓ of these secret keys are given to the user. To encrypt a message during a particular time period, the sender first splits the message into ℓ shares using a secret-sharing scheme, and then encrypts each of these shares using one of the ℓ keys associated with the current time period. The keys are chosen in such a way so that multiple exposures of the user do not compromise "enough" of the ℓ keys associated with any other time periods. (In [20], it is shown how to "tune" n and ℓ to achieve the desired level of security in a reasonably-efficient way.)

It is immediately apparent that the above encryption process (namely, splitting the message and then encrypting each share with an independent key) corresponds exactly to multiple encryption. For this particular application, a single user stores all ℓ keys that are used to decrypt during a given time period; therefore, a chosen-ciphertext attack against a key-insulated cryptosystem is equivalent to a wMCCA attack on a multiple encryption scheme (that is, an adversary does not get to see the individual shares output by each partial decryption algorithm). Thus, any wMCCA-secure multiple encryption scheme can be used to achieve CCA-secure key-insulated encryption. We remark that robustness is not needed for this particular application since all keys are stored by a single entity (namely, the user).

Dodis, et al. [20] also show a generic conversion from any CPA-secure key-insulated scheme to a CPA-secure *strong* key-insulated scheme (where in a strong key-insulated scheme, encrypted messages are kept confidential even from the server itself). In their conversion, they split the plaintext message into two shares, encrypt one share using any "basic" key-insulated scheme, and encrypt the second share using a key that is stored (at all times) only by the user. Again, it can be seen that this solution corresponds to "double" encryption; thus, the techniques outlined in this paper suffice to construct generic CCA-secure *strong* key-insulated schemes from any CCA-secure key insulated scheme (thereby answering a question left open by [5]).

CCA-Secure Certificate-Based Encryption. The notion of certificate-based encryption (CBE) was recently introduced by Gentry [24]. In this model, a certificate — or, more generally, a signature — acts not only as a "certification" of the public key of a particular entity, but serves also as a decryption key. In particular, to decrypt a message a key-holder needs *both* its secret key and an up-to-date certificate from its certification authority (CA). Certificate-based encryption combines the aspects of identity-based encryption (IBE) and public-key encryption (PKE). Specifically, the sender of the message does not need to check whether the user is properly certified before sending the message, and the user can decrypt the message *only if* he has been certified (this is called *implicit certification*, a feature of IBE but not of PKE). Additionally, (1) the certificates from the CA can be sent to the user in the clear (as in PKE but unlike IBE), and (2) the CA cannot decrypt messages sent to to the user since he does not know the user's private key (i.e., there is no *escrow*, again like PKE but unlike IBE).

From the above description, one would expect that it should be possible to construct a CBE scheme using a simple combination of any IBE and regular PKE. In fact, this was the intuitive description of CBE as presented by Gentry [24], and this approach achieves security against chosen-plaintext attacks. Unfortunately, this does not suffice to achieve security against chosen-ciphertext attacks. As a result, [24] only constructed a CCA-secure CBE scheme based on specific assumptions, and left open the problem of designing a *generic* CCA-secure CBE scheme. Using the techniques from this paper with $n = 2$, but applying them to an IBE and a PKE (instead of two PKEs), we can easily resolve this open question. Note that ones only needs a wMCCA-secure multiple encryption scheme with no robustness in this case, since the user holds both keys and never reveals any intermediate results.

Our technique also applies to most of the CBE extensions presented by Gentry, such as hierarchical CBE (which combines CCA-secure hierarchical IBE and PKE) and the general technique (based on subset covers) to reduce CA computation in a multi-user environment.

CCA-Secure Broadcast Encryption. A *broadcast encryption* scheme allows the sender to securely distribute data to a dynamically changing set of users over an insecure channel, with the possibility of "revoking" users when they are no longer "qualified". One of the most challenging settings for this problem is that of *stateless receivers*, where each user is given a fixed set of keys which cannot be updated for the lifetime of the system. This setting was considered by Naor, Naor, and Lotspiech [33], who also present a general "subset cover framework" for this problem. Although originally used in the symmetric-key setting, Dodis and Fazio [17] extended the subset cover framework to the public-key setting, where anybody can encrypt the data using a single public key of the system.

Without getting into technical details, each user (more or less) stores a certain, user-specific subset of secret keys, while all the public keys are freely available to everybody (specifically, are efficiently derived from a single "global public key"; in the case of [17] this is done by using an appropriate identity-based mechanism whose details are not important for the present discussion). When one wants to revoke a certain subset of users, one cleverly chooses a small subset P of public keys satisfying the following two properties: (1) every non-revoked user possesses at least one secret key corresponding to some key in P; but (2) every revoked user possesses no secret keys in P. Once this is done, a message is simply encrypted in parallel using every key in P.

Clearly, the above corresponds exactly to a multiple encryption scheme with $t_p = 0$ and $t_f = 1$. However, as acknowledged in [33, 17], the resulting broadcast encryption scheme is at best secure against "lunch-time" chosen-ciphertext attacks even if the underlying encryption scheme being used is CCA-secure. Using the techniques of this paper, we can resolve this problem and extend the subset-cover framework to achieve CCA-security (provided, of course, that the corresponding basic encryption schemes are CCA-secure). This results in the first generic CCA-secure broadcast encryption scheme. When instantiated with any of the two subset cover methods given in [33, 17], we obtain two "semi-generic" constructions of CCA-secure broadcast encryption: from any regular (e.g. [8]) or any hierarchical (e.g. [25]) identity-based encryption scheme, respectively. Each of these schemes, when properly instantiated, will offer several advantages over the only previously known CCA-secure broadcast encryption scheme [18] (which was based on specific assumptions), including a fixed public-key size, an unbounded number of revocations, and qualitatively stronger traitor-tracing capabilities.

We remark that although wMCCA-security is already enough for this application, a more communication-efficient solution can be achieved using our sMCCA-secure scheme (since each user can then simply "ignore" the majority of the ciphertext which is "not relevant" to him).

Cryptanalysis-Tolerant CCA-Secure Encryption. As discussed in the Introduction, a multiple encryption scheme may be viewed as achieving "cryptanalysis-tolerance" for public-key encryption: namely, a message can be encrypted with respect to multiple encryption schemes (using independent keys) such that the message remains confidential as long as any *one* of these schemes remains secure (see [28] for further discussion of this concept). Herzberg [28] shows constructions of cryptanalysis-tolerant CPA-secure encryption schemes; the techniques outlined here resolve the question of constructing cryptanalysis-tolerant CCA-secure encryption schemes.

CCA-Secure Proxy Encryption. Proxy encryption [19] may be viewed as non-interactive, two-party, threshold encryption, where one server is the end-user and the other server is called the *proxy*. The proxy receives the ciphertext C, partially decrypts it into some ciphertext C', and forwards C' to the end-user. The user stores the second part of the decryption key and can now recover the message M from C'. In [19], the authors give a formal treatment of proxy encryption but left open the question of constructing a generic, CCA-secure scheme. The generic 2-party multiple encryption scheme presented in this paper resolves this open question in the natural way. We remark that we require MCCA-security for this application, since the attacker (who is one of the servers) has full oracle access to the other server.

Other Applications. We believe that multiple encryption schemes will find even more uses; we highlight two. One interesting direction is to apply multiple encryption to the construction of "anonymous channels" [14] using, e.g., "onion routing" [26]. It would be interesting to see if our methods can be extended to give CCA-secure constructions in this setting. For the second application, we mention recent work of Boneh, et al. [9] on searchable public-key encryption. Here, one wants to design an encryption scheme for which one can encrypt some keyword W as a ciphertext C such that that: (1) given some trapdoor T_W one can test whether C is an encryption of W; (2) without such trapdoor, one gets no information about W, even when given many other trapdoors T_X for $X \neq W$ (except that W is not one of these X's). It is not hard to see that this concept is also related to anonymous IBE, where the ciphertext should not reveal anything about the identity of the recipient of the message. Alternately, it is also related to key-insulated encryption in which the ciphertext does not reveal the time period for which the ciphertext was encrypted. In all these cases, one can adapt the generic construction of key-insulated encryption from [20], discussed earlier in this section, to obtain a CPA-secure version of the corresponding primitive, provided that the regular encryption $\mathcal{E}$ is *key-indistinguishable* [2]. Indeed, one of the constructions in [9] exactly follows this route. Using the techniques in this paper, we can obtain generic CCA-secure searchable encryption, recipient-anonymous IBE, or time-anonymous key-insulated encryption, provided one uses a CCA-secure, key-indistinguishable encryption scheme (such as the Cramer-Shoup encryption scheme [15], shown to be key-indistinguishable by [2]).

References

1. B. Aiello, M. Bellare, G. Di Crescenzo, and R. Venkatesan. Security Amplification by Composition: the Case of Doubly-Iterated, Ideal Ciphers. Crypto '98.
2. M. Bellare, A. Boldyreva, A. Desai, and D. Pointcheval. Key-Privacy in Public-Key Encryption. Asiacrypt 2001.
3. M. Bellare, A. Desai, D. Pointcheval, and P. Rogaway. Relations among Notions of Security for Public-Key Encryption Schemes. Crypto '98.
4. M. Bellare and C. Namprempre. Authenticated Encryption: Relations Among Notions and Analysis of the Generic Composition Paradigm. Asiacrypt 2000.
5. M. Bellare and A. Palacio. Protecting against Key Exposure: Strongly Key-Insulated Encryption with Optimal Threshold. Available at http://eprint.iacr.org/2002/064.
6. M. Bellare and P. Rogaway. Collision-Resistant Hashing: Towards Making UOWHFs Practical. Crypto '97.
7. D. Boneh and X. Boyen. Efficient Selective-ID Secure Identity Based Encryption Without Random Oracles. Eurocrypt 2004.
8. D. Boneh and M. Franklin. Identity-Based Encryption From the Weil Pairing. Crypto 2001.
9. D. Boneh, G. Di Crescenzo, R. Ostrovsky, and G. Persiano. Searchable Public Key Encryption. Eurocrypt 2004.
10. D. Boneh and J. Katz. Improved Efficiency for CCA-Secure Cryptosystems Built Using Identity Based Encryption. RSA — Cryptographers' Track 2005, to appear.
11. R. Canetti, Y. Dodis, S. Halevi, E. Kushilevitz, and A. Sahai. Exposure-Resilient Functions and All-or-Nothing Transforms. Eurocrypt 2000.
12. R. Canetti, S. Halevi, and J. Katz. Chosen-Ciphertext Security from Identity-Based Encryption. Eurocrypt 2004.
13. R. Canetti and S. Goldwasser. An Efficient Threshold Public-Key Cryptosystem Secure Against Adaptive Chosen-Ciphertext Attack. Eurocrypt '99.
14. D. Chaum. Untraceable Electronic Mail, Return Addresses, and Digital Pseudonyms. *Comm. ACM* 24(2): 84–88 (1981).
15. R. Cramer and V. Shoup. A Practical Public Key Cryptosystem Provably Secure Against Chosen Ciphertext Attack. Crypto '98.
16. Y. Desmedt. Society and Group-Oriented Cryptography: a New Concept. Crypto '87.
17. Y. Dodis and N. Fazio. Public Key Broadcast Encryption for Stateless Receivers. *ACM Workshop on Digital Rights Management*, 2002.

18. Y. Dodis and N. Fazio. Public Key Broadcast Encryption Secure Against Adaptive Chosen Ciphertext Attack. PKC 2003.
19. Y. Dodis and A. Ivan. Proxy Cryptography Revisited. NDSS 2003.
20. Y. Dodis, J. Katz, S. Xu, and M. Yung. Key-Insulated Public-Key Cryptosystems. Eurocrypt 2002.
21. S. Even and O. Goldreich. On the Power of Cascade Ciphers. *ACM Trans. Comp. Systems* 3: 108–116 (1985).
22. A. Fiat and M. Naor. Broadcast Encryption. Crypto '93.
23. M. Franklin and M. Yung. Communication Complexity of Secure Computation. STOC '92.
24. C. Gentry. Certificate-Based Encryption and the Certificate Revocation Problem. Eurocrypt 2003.
25. C. Gentry and A. Silverberg. Hierarchical Id-Based Cryptography. Asiacrypt 2002.
26. D. Goldschlag, M. Reed, and P. Syverson. Onion Routing. *Comm. ACM* 42(2): 39–41 (1999).
27. S. Goldwasser, S. Micali, and R. Rivest. A Digital Signature Scheme Secure Against Adaptive Chosen-Message Attacks. *SIAM J. Computing* 17(2): 281–308, 1988.
28. A. Herzberg. On Tolerant Cryptographic Constructions. Available at http://eprint.iacr.org/2002/135/.
29. H. Krawczyk. Secret Sharing Made Short. Crypto '93.
30. P. MacKenzie. An Efficient Two-Party Public Key Cryptosystem Secure Against Adaptive Chosen Ciphertext Attack. PKC 2003.
31. U. Maurer and J. Massey. Cascade Ciphers: the Importance of Being First. *J. Crypto* 6(1): 55–61 (1993).
32. R. Merkle and M. Hellman. On the Security of Multiple Encryption. *Comm. ACM* 24(7): 465–467 (1981).
33. D. Naor, M. Naor, and J. Lotspiech. Revocation and Tracing Schemes for Stateless Receivers. Crypto 2001.
34. NESSIE consortium. Portfolio of Recommended Cryptographic Primitives. Manuscript, Feb. 2003. Available at http://www.cosic.esat.kuleuven.ac.be/nessie/deliverables/decision-final.pdf.
35. M. Rabin. Efficient Dispersal of Information for Security, Load Balancing, and Fault Tolerance. *J. ACM* 36(2): 335–348 (1989).
36. R. Rivest. All-or-Nothing Encryption and the Package Transform. FSE '97.
37. A. Shamir. How to Share a Secret. *Comm. ACM* 22(11): 612–613 (1979).
38. C. Shannon. Communication Theory of Secrecy Systems. *Bell System Technical Journal*, vol. 28, Oct. 1949.
39. V. Shoup. A Proposal for an ISO Standard for Public-Key Encryption, version 2.1. Available at http://eprint.iacr.org/2001/112/
40. V. Shoup and R. Gennaro. Securing Threshold Cryptosystems Against Chosen Ciphertext Attack. *J. Crypto* 15(2): 75–96 (2002).
41. R. Zhang, G. Hanaoka, J. Shikata, and H. Imai. On the Security of Multiple Encryption, or CCA-security+CCA-security=CCA-security? Public Key Cryptography (PKC) 2004. Also available at http://eprint.iacr.org/2003/181.

A   Replacing Signatures by MACs

Recall, a message authentication code (MAC) is given by a deterministic algorithm Tag which outputs an "existentially unforgeable" tag $T = \mathsf{Tag}_\tau(M)$ for a given message M using a secret key τ. In fact, a "one-time" message authentication code (defined analogously to a one-time signature scheme) is sufficient for our purposes. We define a relaxed commitment scheme $\mathcal{C} = (\mathsf{Setup}, \mathsf{Commit}, \mathsf{Open})$ (termed *encapsulation* in [10]) as follows: $\mathsf{Setup}(1^k)$ outputs the public commitment key CK, which is always input to both Commit and Open and is omitted for brevity. Commit takes no inputs and produces a triple of values (τ, c, d), where τ is a (random) key, c is the commitment to this key, and d is the corresponding decommitment. $\mathsf{Open}(c, d)$ should produce τ under normal circumstances. The hiding property states that τ "looks random" given c (i.e., one cannot efficiently distinguish (CK, c, τ) from (CK, c, r) for random r). The relaxed binding property states that given a random triple (τ, c, d) output by Commit, it is infeasible to produce $d' \neq d$ such that $\mathsf{Open}(c, d') \notin \{\tau, \bot\}$. It is easy to construct simple and efficient MACs and relaxed commitment schemes (see [10]).

Given the above, we construct $\mathcal{TE}_{wcca}$ as follows:

- $\mathsf{TGen}(1^k)$. Let $\mathsf{CK} \leftarrow \mathsf{Setup}(1^k)$, and for $i = 1 \ldots n$, let $(\mathsf{EK}_i, \mathsf{DK}_i) \leftarrow \mathsf{Gen}(1^k)$. Set $\mathbf{TEK} = (\mathsf{EK}_1 \ldots \mathsf{EK}_n, \mathsf{CK})$, $\mathsf{TDK}_i = \mathsf{DK}_i$, so that $\mathbf{TDK} = (\mathsf{DK}_1 \ldots \mathsf{DK}_n)$. Below, denote $\mathsf{Enc}_i = \mathsf{Enc}_{\mathsf{EK}_i}$, $\mathsf{Dec}_i = \mathsf{Dec}_{\mathsf{DK}_i}$.
- $\mathsf{TEnc}^L(M)$. Let $(\tau, c, d) \leftarrow \mathsf{Commit}(1^k)$ and $(s_1, \ldots, s_n, \mathsf{pub}) \leftarrow \mathsf{Share}(M, d)$. Set $C_i = \mathsf{Enc}_i^c(s_i)$ ($i = 1 \ldots n$) and compute $\sigma = \mathsf{Tag}_\tau(C_1, \ldots, C_n, \mathsf{pub}, L)$. Output $C = (C_1, \ldots, C_n, \mathsf{pub}, c, \sigma)$.
- $\mathsf{Split}^L(C)$. Parse $C = (C_1, \ldots, C_n, \mathsf{pub}, c, \sigma)$, and let ciphertext share $\hat{C}_i = (C_i, c)$, and $\mathsf{aux} = (\mathsf{pub}, c, L)$.
- $\mathsf{TDec}^i(C_i, c)$. Output $s'_i = \mathsf{Dec}_i^c(C_i)$.
- $\mathsf{Combine}(s'_1, \ldots, s'_n, (\mathsf{pub}, c, L))$. Let $(M, d) = \mathsf{Rec}(s'_1, \ldots s'_n, \mathsf{pub})$ (if invalid, reject). Let $\tau = \mathsf{Open}(c, d)$. Reject if $\sigma \neq \mathsf{Tag}_\tau(C_1, \ldots, C_n, \mathsf{pub}, L)$. Otherwise, output M.

Theorem 3. $\mathcal{TE}_{wcca}$ *is wMCCA-secure with thresholds t_p, t_f, t_r, t_s, provided $\mathcal{E}$ is CCA-secure, $\mathcal{SSS}$ is (t_p, t_f, t_r, t_s, n)-robust, $\mathcal{C}$ is a relaxed commitment scheme, and MAC is a one-time message authentication code.*

We give the complete proof in the full version, here only briefly sketching our argument (which is based on [10]). The problem is the apparent circularity in the usage of the MAC as Tag is applied to data which depends on the MAC key τ. Intuitively, what saves us here is the relaxed binding property which holds even when the adversary *knows* d. This means that when the attacker is given the challenge ciphertext C, it has to either (1) try to use new value c (which does not help due to the CCA-security of the underlying encryption scheme which uses c as a label); or (2) reuse the same c and cause an invalid $d' \neq d$ to be recovered (which leads to rejection anyway); or (3) reuse the same pair (c, d), which results in the same τ and then also to rejection due to the one-time security of the MAC. The latter argument is the most delicate, and its proof in fact requires several sub-arguments. See [10] for further details.

Public-Key Steganography with Active Attacks

Michael Backes and Christian Cachin

IBM Zurich Research Laboratory,
CH-8803 Rüschlikon, Switzerland
{mbc, cca}@zurich.ibm.com

Abstract. A complexity-theoretic model for public-key steganography with active attacks is introduced. The notion of *steganographic security against adaptive chosen-covertext attacks (SS-CCA)* and a relaxation called *steganographic security against publicly-detectable replayable adaptive chosen-covertext attacks (SS-PDR-CCA)* are formalized. These notions are closely related to *CCA-security* and *PDR-CCA-security* for public-key cryptosystems. In particular, it is shown that any SS-(PDR-)CCA stegosystem is a (PDR-)CCA-secure public-key cryptosystem and that an SS-PDR-CCA stegosystem for any covertext distribution with sufficiently large min-entropy can be realized from any PDR-CCA-secure public-key cryptosystem with pseudorandom ciphertexts.

1 Introduction

Steganography is the art and science of hiding information by embedding messages within other, seemingly harmless messages. As the goal of steganography is to hide the *presence* of a message, it can be seen as the complement of cryptography, whose goal is to hide the *content* of a message.

Consider two parties linked by a public communications channel which is under the control of an adversary. The parties are allowed to exchange messages as long as they are not adding a hidden meaning to their conversation. A genuine communication message is called *covertext*; but if the sender of a message has embedded hidden information in a message, it is called *stegotext*. The adversary, who also knows the distribution of the covertext, tries to detect whether a given message is covertext or stegotext.

Steganography has a long history as surveyed by Anderson and Petitcolas [2], but formal models for steganography have only recently been introduced. Several information-theoretic formalizations [4, 24, 15] and one complexity-theoretic model [12] have addressed *private-key* steganography, where the participants share a common secret key. These models are all limited to a passive adversary, however, who can only read messages on the channel.

Von Ahn and Hopper [22] have recently formalized *public-key* steganography with a passive adversary and, in a restricted model, also with an active adversary. Their notion offers security against "attacker-specific" chosen-stegotext attacks, where the recipient must know the identity of the sender, however; this is a limitation of the model compared to the bare public-key scenario.

In this paper, we introduce a complexity-theoretic model for public-key steganography with active attacks, where the participants a priori do not need shared secret information and the adversary may write to the channel and mount a so-called *adaptive chosen-covertext attack*. This attack seems to be the most general attack conceivable against a public-key stegosystem. It allows the adversary to send an arbitrary sequence of adaptively chosen covertext messages to a receiver and to learn the interpretation of every message, i.e., if the receiver considers a message to be covertext or stegotext, plus the decoding of the embedded message in the latter case. (Note that here and in the sequel, a message on the channel is sometimes also called a "covertext" when we do not want to distinguish between stegotext and covertext in the proper sense.)

We do not address denial-of-service attacks in this work, where the adversary tries to disrupt the hidden communication among the participants. Although they also qualify as "active" attacks and are very important in practice, we think that protection against them can be addressed orthogonally to the methods presented here.

Our model is based on the intuition that a public-key stegosystem essentially is a public-key cryptosystem with the additional requirement that its output conforms to a given covertext distribution. As in previous formalizations of steganography [4, 12, 9, 22], the covertext distribution is publicly known in the sense that it is accessible through an oracle that samples the distribution. We introduce the notions of *steganographic security against adaptive chosen-covertext attacks (SS-CCA)* and *steganographic security against publicly-detectable replayable adaptive chosen-covertext attacks (SS-PDR-CCA)* and show that they are closely linked to the analogous notions for public-key cryptosystems, called *security against adaptive chosen-ciphertext attacks* (or *CCA-security*) [16] and *security against publicly-detectable replayable adaptive chosen-ciphertext attacks* [5] (or *PDR-CCA-security*), respectively. (PDR-CCA-security is the same as *benign malleability* [19] and *generalized CCA-security* [1].)

In particular, we show that stegosystems are related to public-key cryptosystems in the following ways:

Theorem 1 (informal statement). *Any SS-(PDR-)CCA stegosystem is a (PDR-)CCA-secure public-key cryptosystem.*

Theorem 2 (informal statement). *An SS-PDR-CCA stegosystem for covertext distributions with sufficiently large min-entropy can be constructed from any PDR-CCA-secure public-key cryptosystem whose ciphertexts are pseudorandom (i.e., computationally indistinguishable from a random bit string).*

A corollary of Theorem 2 is that SS-PDR-CCA stegosystems exist in the standard model under the Decisional Diffie-Hellman (DDH) assumption and in the random oracle model under the assumption of trapdoor one-way permutations. The stegosystem constructed in the proof of Theorem 2 uses the "rejection sampler" construction found in essentially all previous work in the area [12, 9, 22], which is already described by Anderson and Petitcolas [2]. However, our system embeds more hidden bits per stegotext than any previous system. This follows from an improved analysis of the rejection sampler. It is not known if a result analogous to Theorem 2 holds for CCA-security; finding an SS-CCA stegosystem that works for an arbitrary covertext distribution with sufficiently large min-entropy remains an interesting open problem.

Our model for public-key steganography is introduced in Section 2, where also the relation to previous models for steganography is discussed in detail. Section 3 recalls the definitions of CCA- and PDR-CCA-security for public-key cryptosystems, states our results formally, and presents the proof of Theorem 1. Section 4 gives the construction of an SS-PDR-CCA stegosystem and proves Theorem 2.

2 Definitions

2.1 Notation

A function $f: \mathbb{N} \to \mathbb{R}_{\geq 0}$ is called *negligible* if for every constant $c \geq 0$ there exists $k_c \in \mathbb{N}$ such that $f(k) < \frac{1}{k^c}$ for all $k > k_c$. Given some set S, a subset of *almost all* elements contains all but a negligible fraction of elements from S. A (randomized) algorithm is called *efficient* if its running time is bounded by a polynomial except with negligible probability (over the coin tosses of the algorithm).

Let $x \leftarrow y$ denote the algorithm that assigns a value y to x. If $\mathsf{A}(\cdot)$ is a (randomized) algorithm, the notation $x \leftarrow \mathsf{A}(y)$ denotes the algorithm that assigns to x a randomly selected value according to the probability distribution induced by $\mathsf{A}(\cdot)$ with input y over the set of its outputs.

If S is a probability distribution, then the notation $x \stackrel{R}{\leftarrow} S$ denotes any algorithm which assigns to x an element randomly selected according to S. If S is a finite set, then the notation $x \stackrel{R}{\leftarrow} S$ denotes the algorithm which assigns to x an element selected at random from S with uniform distribution over S.

If $p(\cdot, \cdot, \cdots)$ is a predicate, the notation

$$\Pr[x \stackrel{R}{\leftarrow} S; y \stackrel{R}{\leftarrow} T; \cdots : p(x, y, \cdots)]$$

denotes the probability that $p(x, y, \cdots)$ will be true after the ordered execution of the algorithms $x \stackrel{R}{\leftarrow} S, y \stackrel{R}{\leftarrow} T, \cdots$. If X is a (randomized) algorithm, a distribution, or a set, then $\Pr_X[x]$ is short for $\Pr_{x \stackrel{R}{\leftarrow} X}[x]$, which is short for $\Pr[s \stackrel{R}{\leftarrow} X : s = x]$.

The *statistical distance* between two distributions $\mathcal{X}$ and $\mathcal{Y}$ over the same set X is defined as $\|\mathcal{X} - \mathcal{Y}\| = \max_{X_0 \subseteq X} |\sum_{x \in X_0} \Pr_\mathcal{X}(x) - \Pr_\mathcal{Y}(x)|$. The *min-entropy* of a distribution $\mathcal{X}$ over an alphabet X is defined as $H_\infty(\mathcal{X}) = -\log \max_{x \in X} \Pr_\mathcal{X}[x]$. (All logarithms are to the base 2.)

2.2 Public-Key Stegosystems

We define a public-key stegosystem as a triple of algorithms for key generation, message encoding, and message decoding, respectively. The notion corresponds to a public-key cryptosystem in which the ciphertext should conform to a target covertext distribution.

For the scope of this work, the covertext is modeled by a distribution $\mathcal{C}$ over a given set C. The distribution is only available via an oracle; it samples $\mathcal{C}$ upon request, with each sample being independent. In other words, it outputs a sequence of independent and identically distributed covertexts. W.l.o.g., $\Pr_\mathcal{C}[c] > 0$ for all $c \in C$.

The restriction to independent repetitions is made here only to simplify the notation and to focus on the contribution of this work. All our definitions and results can be extended in the canonical way to the very general model of a covertext *channel* as introduced by Hopper et al. [12]. They model a channel as an unbounded sequence of values drawn from a set C whose distribution may depend in arbitrary ways on past outputs; access to the channel is given only by an oracle that samples from the channel.

Such a channel underlies only one restriction: The sampling oracle must allow random access to the channel distribution, i.e., the oracle can be queried with an arbitrary prefix of a possible channel output and will return the next symbol according to the channel distribution. In other words, the channel sampler cannot only be rewound to an earlier state of its execution but also restarted from a given state. (Hence it may be difficult to use an email conversation among humans for a covertext channel since that cannot easily be restarted.)

The sampling oracle for the covertext distribution is available to all users and to the adversary. In order to avoid technical complications, assume w.l.o.g. that the sampling oracle is implemented by a probabilistic polynomial-time algorithm and therefore does not help an adversary beyond its own capabilities (for example, with solving a computationally hard problem).

Definition 1. *[Public-Key Stegosystem] Let $\mathcal{C}$ be a distribution on a set C of covertexts. A public-key stegosystem is a triple of probabilistic polynomial-time algorithms (SK, SE, SD) with the following properties.*

- *The key generation algorithm SK takes as input the security parameter k and outputs a pair of bit strings (spk, ssk), called the [stego] public key and the [stego] secret key. W.l.o.g. SK induces the uniform distribution over the set of possible key pairs for security parameter k.*
- *The steganographic encoding algorithm SE takes as inputs the security parameter k, a public key spk and a message $m \in \{0,1\}^{l(k)}$, where $l(k)$ is an arbitrary polynomial, and outputs a covertext $c \in C$. The plaintext m is often called the embedded message.*
- *The steganographic decoding algorithm SD takes as inputs the security parameter k, a secret key ssk, and a covertext $c \in C$ and outputs either a message $m \in \{0,1\}^{l(k)}$ or a special symbol $\perp$. An output value of $\perp$ indicates a decoding error, for example, when SD has determined that no message is embedded in c.*

We require that for almost all (spk, ssk) output by $\mathsf{SK}(1^k)$ and all $m \in \{0,1\}^{l(k)}$, the probability that $\mathsf{SD}(1^k, ssk, \mathsf{SE}(1^k, spk, m)) \neq m$ is negligible in k.

Note that except for the presence of the covertext distribution, this definition is equivalent to that of a public-key cryptosystem. Although all algorithms have oracle access to $\mathcal{C}$, only SE needs it in the stegosystems considered in this paper. For ease of notation, the security parameter will be omitted henceforth.

The probability that the decoding algorithm outputs the correct embedded message is referred to as the *reliability* of the stegosystem. Although one might also allow a non-negligible decoding error in the definition of a stegosystem (as done in previous work [12]), we require that the decoding error probability is negligible in order to maintain the analogy between a stegosystem and a cryptosystem.

Security definition. Coming up with the "right" security definition for a cryptographic primitive has always been a challenging task because the sufficiency of a security property cannot be demonstrated by running the cryptosystem. Only its insufficiency can be shown by pointing out a specific attack, but finding an attack is usually hard. Often, security definitions had to be strengthened when a primitive was used as part of a larger system. Probably the most typical example is the security of public-key cryptosystems: the original notion of semantic security [11], which considers only a passive or eavesdropping adversary, was later augmented to security against adaptive chosen-ciphertext attacks or non-malleability, which allows also for active attacks [16, 10, 3].

We introduce here the notion of *steganographic security against adaptive chosen-covertext attacks*, abbreviated *SS-CCA*, and its slightly relaxed variant *steganographic security against publicly-detectable replayable chosen-covertext attacks*, abbreviated *SS-PDR-CCA*. Both notions are based on the intuition that a stegosystem is essentially a cryptosystem with a prescribed ciphertext distribution. We first recall the definition of *compatible [publicly computable] relations*, adopted from public-key cryptosystem to stegosystems, on which the definition of SS-PDR-CCA is based.

Definition 2. *[Compatible Relation [19]] Let* $\Sigma = (\mathsf{SK}, \mathsf{SE}, \mathsf{SD})$ *be a stegosystem. A family of binary relations* $\equiv_{spk}$ *(indexed by the public keys of* Σ*) on covertext pairs is called a* compatible *relation family for* Σ *if for almost all key pairs* (spk, ssk) *we have:*

- *For any two covertexts c and c', if $c \equiv_{spk} c'$ then* $\mathsf{SD}(ssk, c) = \mathsf{SD}(ssk, c')$, *except with negligible probability over the random choices of the algorithm* SD.
- *For any two covertexts c and c', it can be determined except with negligible probability whether $c \equiv_{spk} c'$ using a probabilistic polynomial-time algorithm taking inputs spk, c, and c'.*

SS-CCA and SS-PDR-CCA are defined by the following experiment. Let an arbitrary distribution $\mathcal{C}$ on a set C be given and consider a (stego-)adversary, defined by two arbitrary probabilistic polynomial-time algorithms SA_1 and SA_2. For the SS-PDR-CCA experiment, let also an arbitrary compatible relation family $\equiv_{spk}$ be given. The experiment consists of five stages, where both notions only differ in the fourth stage.

Key Generation: A key pair (spk, ssk) is generated by the key generation algorithm SK.

First Decoding Stage: Algorithm SA_1 is run with the public key spk as input and has access to the sampling oracle for $\mathcal{C}$ and to a decoding oracle SO_1. The decoding oracle knows the secret key ssk. Whenever it receives a covertext c, it runs $\mathsf{SD}(ssk, c)$ and returns the result to SA_1.

When SA_1 finishes its execution, it outputs a tuple (m^*, s), where $m^* \in \{0, 1\}^l$ is a message and s is some additional information which the algorithm wants to preserve.

Challenge: A bit b is chosen at random and a *challenge covertext* c^* is determined depending on it: If $b = 0$ then $c^* \leftarrow \mathsf{SE}(spk, m^*)$ else $c^* \xleftarrow{R} \mathcal{C}$. c^* is given to algorithm SA_2, who should guess the value of b, i.e., determine whether the message m^* has been embedded in c^* or whether c^* has simply been chosen according to $\mathcal{C}$.

Second Decoding Stage: SA_2 is run on input c^*, and s, i.e., it knows the challenge covertext and the state provided by SA_1.

For SS-CCA, SA_2 may access a decoding oracle SO_2^{cca}, which is analogous to SO_1 except that upon receiving query c^*, oracle SO_2^{cca} returns $\perp$.

For SS-PDR-CCA, SA_2 has access to a decoding oracle $SO_2^{pdr\text{-}cca,\equiv_{spk}}$, which is identical to SO_2^{cca} except that it does not allow any query that is equivalent to c^* under $\equiv_{spk}$. In particular, upon receiving query c, $SO_2^{pdr\text{-}cca,\equiv_{spk}}$ returns $\perp$ if $c \equiv_{spk} c^*$; otherwise, it returns $\mathsf{SD}(ssk, c)$.

Guessing Stage: When SA_2 finishes its execution, it outputs a bit b'.

The stego-adversary succeeds in distinguishing stegotext from covertext if $b' = b$ in the above experiment. We require that for a secure stegosystem, no efficient adversary can distinguish stegotext from covertext except with negligible probability over random guessing.

Definition 3. *[Steganographic Security against Active Attacks] Let C be a distribution on a covertext set C and let $\Sigma = (\mathsf{SK}, \mathsf{SE}, \mathsf{SD})$ be a stegosystem. We say that Σ is steganographically secure against adaptive chosen-covertext attacks (SS-CCA) with respect to C if for all probabilistic polynomial-time adversaries (SA_1, SA_2), there exists a negligible function ϵ such that*

$$\Pr\Big[(spk, ssk) \leftarrow \mathsf{SK};\ (m^*, s) \leftarrow SA_1^{SO_1}(spk);\ b \stackrel{R}{\leftarrow} \{0,1\};$$

$$\textit{if } b = 0 \textit{ then } c^* \leftarrow \mathsf{SE}(spk, m^*) \textit{ else } c^* \stackrel{R}{\leftarrow} C :$$

$$SA_2^{SO_2^{cca}}(spk, m^*, c^*, s) = b\Big] = \frac{1}{2} + \epsilon(k).$$

Similarly, we say that Σ is steganographically secure against publicly-detectable replayable adaptive chosen-covertext attacks (SS-PDR-CCA) with respect to C if there exists a compatible relation family $\equiv_{spk}$ such that for all probabilistic polynomial-time adversaries (SA_1, SA_2), there exists a negligible function ϵ such that the above equation holds with SO_2^{cca} replaced by $SO_2^{pdr\text{-}cca,\equiv_{spk}}$.

Note that this leaves the adversary free to query the decoding oracle with any element of the covertext space *before* the challenge is issued. By definition, an SS-CCA stegosystem is also SS-PDR-CCA.

2.3 Discussion

The relation to public-key cryptosystems. A stegosystem should enable two parties to communicate over a public channel in such a way that the presence of a message in the conversation cannot be detected by an adversary. It seems natural to conclude from this that the adversary must not learn any useful information about an embedded message, should there be one. The latter property is the subject of cryptography: hiding the content of a message transmitted over a public channel. This motivates the approach of von Ahn and Hopper [22] and of this paper that models a public-key stegosystem after a public-key cryptosystem in which the ciphertext conforms to a particular covertext distribution.

The most widely accepted formal notion of a public-key cryptosystem secure against an active adversary is *indistinguishability of encryptions against an adaptive chosen-ciphertext attack* (CCA-security) [16] and is equivalent to *non-malleability of ciphertexts* in the same attack model [10, 3]. CCA-security is defined by an experiment with almost the same stages as above, except that the first part of the adversary outputs *two* messages m_0 and m_1, of which one is chosen at random and then encrypted. The resulting value c^*, also called the *target ciphertext*, is returned to the adversary and the adversary has to guess what has been encrypted. In the second query stage, the adversary is allowed to obtain decryptions of *any* ciphertext except for c^*.

This appears to be the minimal requirement to make the definition of a cryptosystem meaningful, but it has turned out to be overly restrictive in some cases. For example, consider a CCA-secure cryptosystem where a useless bit is appended to each ciphertext during encryption and that is ignored during decryption. Although this clearly does not affect the security of the cryptosystem, the modified scheme is no longer CCA-secure.

Several authors have relaxed CCA-security to allow for such "benign" modifications [19, 1, 5]. The corresponding relaxed security notion has been called *publicly-detectable replayable CCA-security* or *PDR-CCA-security* by Canetti et al. [5] because the modifications are apparent without knowledge of the secret key. The difference to CCA-security is that in the second query stage, the adversary is more restricted and does not allow any query that is equivalent to the target ciphertext under some compatible relation that can be derived from the public key. The intuition is that such a cryptosystem allows anyone to modify a ciphertext into an equivalent one if this is apparent from the public key, and therefore to "replay" the target ciphertext.

Our notion of an SS-CCA stegosystem is analogous to a CCA-secure cryptosystem, in that it only excludes the target covertext from the queries to the second decoding oracle. Likewise, our notion of an SS-PDR-CCA stegosystem contains a restriction that is reminiscent of a PDR-CCA-secure cryptosystem, by not allowing queries that are publicly identifiable transformations of the challenge covertext. These similarities are no coincidence: We show in Section 3 that any SS-CCA stegosystem is a CCA-secure public-key cryptosystem, and similarly for their replayable counterparts.

Canetti et al. [5] also propose a further relaxation of CCA-security called *replayable CCA-security* (or *R-CCA-security*), where anyone can generate new ciphertexts that decrypt to the same value as a given ciphertext, but the equivalence may not be publicly detectable. We note that it is possible to formulate the corresponding notion of *steganographic security against replayable chosen-ciphertext attacks* (*SS-R-CCA*) by suitably modifying Definition 3. Our results of Sections 3 and 4 can be adapted analogously.

Related work on steganography. The first published model of a steganographic system is the "Prisoners' Problem" by Simmons [21]. This work addresses the particular situation of message authentication among two communicating parties, where a so-called *subliminal channel* might be used to transport a hidden message in the view of an adversary who tries to detect the presence of a hidden message. Although a subliminal channel in that sense is only made possible by the existence of message authentication in the model, it can be seen as the first formulation of a general model for steganography.

Cachin [4] presented an information-theoretic model for steganography, which was the first to explicitly require that the stegotext distribution is indistinguishable from the covertext distribution to an adversary. Since the model is unconditional, a statistical information measure is used.

Hopper et al. [12] give the first complexity-theoretic model for private-key steganography with passive attacks; they point out that a stegosystem is similar to a cryptosystem whose ciphertext is indistinguishable from a given covertext. In Section 3 we establish such an equivalence formally for public-key systems with active attacks.

Dedić et al. [9] study the efficiency of stegosystems that have black-box access to the covertext distribution and provide lower bounds on their efficiency.

Recently, von Ahn and Hopper [22] have formalized public-key steganography with a passive adversary, i.e., one who can mount a chosen-message attack. The resulting notion is the analogue of a cryptosystem with security against chosen-plaintext attacks (i.e., a cryptosystem with semantic security). They also formalize the notion of a stegosystem that offers security against "attacker-specific" chosen-stegotext attacks; this means that the decoder must know the identity of the encoder, however, and restricts the usefulness of their notion compared to SS-CCA and SS-PDR-CCA.

No satisfying formal model for public-key steganography with active attacks has been published so far, although the subject was discussed by several authors, and some systems with heuristic security have been proposed [8, 2]. A crucial element that seems to make our formalizations useful is the restriction of the stage-two decoding oracle depending on the challenge covertext.

3 Results

This section investigates the relation between SS-(PDR-)CCA stegosystems and (PDR-)CCA-secure public-key cryptosystems. Two results are presented:

1. Any SS-CCA stegosystem is a CCA-secure public-key cryptosystem and, similarly, any SS-PDR-CCA stegosystem is a PDR-CCA-secure public-key cryptosystem.
2. An SS-PDR-CCA stegosystem for covertext distributions with sufficiently large min-entropy can be constructed from any PDR-CCA-secure public-key cryptosystem whose ciphertexts are pseudorandom.

We first recall the formal definitions for public-key encryption with CCA- and PDR-CCA-security, respectively. A *public-key cryptosystem* is a triple $(\mathsf{K}, \mathsf{E}, \mathsf{D})$ of probabilistic polynomial-time algorithms. Algorithm K, on input the security parameter k, generates a pair of keys (pk, sk). The encryption and decryption algorithms, E and D, have the property that for almost all pairs (pk, sk) generated by K and for any plaintext message $m \in \{0,1\}^{l(k)}$ where l is an arbitrary polynomial in k, the probability that $\mathsf{D}(1^k, sk, \mathsf{E}(1^k, pk, m)) \neq m$ is negligible in k. (The security parameter is omitted henceforth.)

CCA-security and PDR-CCA-security for a public-key encryption scheme are defined by the following experiment. Consider an adversary defined by two arbitrary polynomial-time algorithms A_1 and A_2. First, a key pair (pk, sk) is generated by K. Next, A_1 is run on input the public key pk and may access a decryption oracle O_1. Oracle O_1 knows the secret key sk, and whenever it receives a ciphertext c, it applies D with key sk to c and returns the result to A_1. When A_1 finishes its execution, it outputs a triple (m_0, m_1, s), where $m_0, m_1 \in \{0,1\}^l$ are two arbitrary messages and s is some additional state information. Now a bit b is chosen at random and m_b is encrypted using E under key pk, resulting in a ciphertext c^*. Algorithm A_2 is given m_0 and m_1, ciphertext c^*, and state s, and has to guess the value of b, i.e., whether m_0 or m_1 has been encrypted. For CCA-security, A_2 may access a decryption oracle O_2^{cca}, which is analogous to O_1 and knows sk, but returns $\bot$ upon receiving query c^*. For PDR-CCA-security, the cryptosystem also specifies a compatible relation family $\equiv_{pk}$ according to Definition 2 with the stegosystem being replaced by the cryptosystem. A_2 may access a decryption oracle $O_2^{pdr\text{-}cca,\equiv_{pk}}$, which is identical to O_1^{cca} except that it answers $\bot$ for any query c with $c \equiv_{pk} c^*$. Finally, A_2 outputs a bit b' as its guess for b.

A secure cryptosystem requires that no efficient adversary can distinguish an encryption of m_0 from an encryption of m_1 except with negligible probability.

Definition 4. *[(PDR-)CCA-Security for Public-Key Cryptosystems [3,5]] Let $\Omega = (K, E, D)$ be a public-key cryptosystem. We say that Ω is CCA-secure if for all probabilistic polynomial-time adversaries $A = (A_1, A_2)$, there exists a negligible function ϵ such that*

$$\Pr\left[(pk, sk) \leftarrow K;\; (m_0, m_1, s) \leftarrow A_1^{O_1}(pk);\; b \xleftarrow{R} \{0,1\};\right.$$
$$\left. c^* \leftarrow E(pk, m_b);\; A_2^{O_2^{cca}}(pk, m_0, m_1, c^*, s) = b\right] = \frac{1}{2} + \epsilon(k).$$

We say that Ω is PDR-CCA-secure if there exists a compatible relation family $\equiv_{pk}$ such that the above condition holds with O_2^{cca} replaced by $O_2^{pdr\text{-}cca,\equiv_{pk}}$.

The following is our first main result.

Theorem 1. *Let $\Sigma = (SK, SE, SD)$ be a public-key stegosystem. If Σ is SS-CCA (SS-PDR-CCA) with respect to some distribution C, then Σ is a CCA-secure (PDR-CCA-secure) public-key cryptosystem.*

Proof. Note first that Σ satisfies the definition of a public-key cryptosystem. We prove that Σ is (PDR-)CCA-secure by a reduction argument. Assume that Σ is not a (PDR-)CCA-secure cryptosystem and hence there exists an (encryption-)adversary (A_1, A_2) that breaks the (PDR-)CCA-security of Σ, i.e., it wins in the experiment of Definition 4 with probability $\frac{1}{2} + \delta(k)$ for some non-negligible function δ. Let $\equiv_{pk}$ denote a compatible relation family for Σ in the case of PDR-CCA security. We construct a (stego-)adversary (SA_1, SA_2) against Σ as a stegosystem with respect to C that has black-box access to (A_1, A_2) as follows.

Key Generation: When SA_1 receives a public-key pk, it invokes A_1 with this key.

First Decoding Stage: Whenever A_1 queries its decryption oracle O_1 with a ciphertext c, SA_1 passes c on to its decoding oracle SO_1, waits for the response and forwards the response to A_1.

When A_1 halts and outputs (m_0, m_1, s), the stego-adversary SA_1 chooses a random bit b', and outputs $(m_{b'}, (m_0, m_1, b', s))$.

Challenge: A challenge covertext c^* is computed according to the definition of a stegosystem and given to SA_2.

Second Decoding Stage: SA_2 receives inputs $m_{b'}, c^*$, and (m_0, m_1, b', s) and invokes A_2 on inputs m_0, m_1, c^*, and s. Otherwise, SA_2 behaves in the same way as SA_1 during the first decoding stage, forwarding the decryption requests that A_2 makes to

O_2 to the respective decoding oracle SO_2^{cca} or $SO_2^{pdr\text{-}cca,\equiv_{pk}}$ and the responses back to A_2. If the distinction between SO_2^{cca} and $SO_2^{pdr\text{-}cca,\equiv_{pk}}$ is irrelevant, we simply write SO_2, similarly for the decryption oracle O_2.

Guessing Stage: When A_2 outputs a bit b^*, the stego-adversary SA_2 tests if $b^* = b'$ and outputs 0 if true, and 1 otherwise.

We now analyze the environment simulated by the stego-adversary (SA_1, SA_2) to the encryption-adversary (A_1, A_2), and the probability that the stego-adversary can distinguish stegotext from covertext.

Clearly, key generation and the first decoding stage perfectly simulate the decryption oracle to adversary A_1. During the challenge, a random bit b is chosen and a challenge covertext is computed as $c^* \leftarrow \mathsf{SE}(pk, m_{b'})$ in case $b = 0$ and as $c^* \xleftarrow{R} \mathcal{C}$ in case $b = 1$.

Note that when $b = 1$, algorithm A_2 and its final output b^* are independent of b'. Hence, we have $\Pr[b' = b^* | b = 1] = \frac{1}{2}$ and the stego-adversary has no advantage over randomly guessing b' in that case. When $b = 0$, we show that during the second decoding phase, SA_2 correctly simulates the decryption oracle O_2 to A_2. For SS-CCA, correct simulation for queries $c \neq c^*$ is clear by definition. For a query $c = c^*$, the decoding oracle SO_2^{cca} will output $\bot$, and so will the decryption oracle O_2^{cca}, which gives a correct simulation again. For SS-PDR-CCA, correct simulation for queries $c \not\equiv_{pk} c^*$ is again clear by definition. For queries c with $c \equiv_{pk} c^*$, the decoding oracle $SO_2^{pdr\text{-}cca,\equiv_{pk}}$ will output $\bot$, and so will the decryption oracle $O_2^{pdr\text{-}cca,\equiv_{pk}}$.

Since the encryption-adversary A_2 by assumption breaks the (PDR-)CCA-security of the cryptosystem, and A_2 is independent of b' when $b = 1$ as argued above, it obtains all its advantage in the case $b = 0$ and we have $\Pr[b' = b^* | b = 0] = \frac{1}{2} + \delta(k)$. By the definition of SA_2, this is also the probability that the stego-adversary guesses b correctly when $b = 0$. Hence, the overall probability that SA_2 guesses b correctly is $\frac{1}{2} + \frac{\delta(k)}{2}$, which exceeds $\frac{1}{2}$ by a non-negligible quantity and shows that Σ is not SS-(PDR-)CCA with respect to any $\mathcal{C}$.

Theorem 1 shows that an SS-CCA stegosystem is a special case of a CCA-secure public-key cryptosystem, and similarly for their replayable variants. In the converse direction, we show now that some PDR-CCA-secure public-key cryptosystems, namely those with "pseudorandom ciphertexts," can also be used to construct SS-PDR-CCA stegosystems. Constructing an SS-CCA stegosystem from a CCA-secure public-key cryptosystem — or from other assumptions, for that matter — for an arbitrary covertext distribution with sufficiently large min-entropy remains an open problem.

In a cryptosystem with pseudorandom ciphertexts, the encryption algorithm outputs a bit string that is indistinguishable from a random string of the same length for any efficient distinguisher that has knowledge of the public key. We make the assumption that the encryption of a plaintext of length $l(k)$ always results in a ciphertext of length $n(k)$, for some polynomial n in k.

Definition 5. *[Public-key Cryptosystem with Pseudorandom Ciphertexts [22]] A public-key cryptosystem $(\mathsf{K}, \mathsf{E}, \mathsf{D})$ is said to have pseudorandom ciphertexts if for all probabilistic polynomial-time adversaries $A = (A_1, A_2)$, there exists a negligible function ϵ such that*

$$\Pr\Big[(pk, sk) \leftarrow K;\ (m, s) \leftarrow A_1(pk);\ c_0 \leftarrow E(pk, m);\ c_1 \stackrel{R}{\leftarrow} \{0,1\}^{n(k)};$$
$$b \stackrel{R}{\leftarrow} \{0,1\};\ A_2(pk, m, c_b, s) = b\Big] = \frac{1}{2} + \epsilon(k).$$

It seems difficult to construct SS-(PDR-)CCA stegosystems for *any* covertext distribution. We show that it is possible for covertexts whose distribution conforms to a sequence of independently repeated experiments and has sufficiently large min-entropy. (According to the remark in Section 2.2, this result generalizes to an arbitrary covertext channel.) Given a covertext distribution $\mathcal{C}$ and positive t, let $\mathcal{C}^t$ denote the probability distribution consisting of a sequence of t independent repetitions of $\mathcal{C}$.

The next theorem is our second main result. Its proof is the subject of Section 4.

Theorem 2. *SS-PDR-CCA stegosystems with respect to a covertext distribution $\mathcal{C}^t$ for any $\mathcal{C}$ with sufficiently large min-entropy can be efficiently constructed from any PDR-CCA-secure cryptosystem with pseudorandom ciphertexts.*

Theorem 2 leaves us with the task of finding a PDR-CCA-secure cryptosystem with pseudorandom ciphertexts. Such cryptosystems exist under a variety of standard assumptions if one asks for security against a *passive* adversary only, i.e., security against *chosen-plaintext attacks (CPA)*. For example, von Ahn and Hopper [22] demonstrate a scheme that is as secure as RSA and one that is secure under the Decisional Diffie-Hellman (DDH) assumption. It is also straightforward to verify that the generic method of encrypting a single bit by xoring it with the hard-core predicate of a trapdoor one-way permutation has pseudorandom ciphertexts.

But any PDR-CCA-secure cryptosystem can be turned into one with pseudorandom ciphertexts using the following method, suggested by Lindell [13]: Take the ciphertext output by the PDR-CCA-secure encryption algorithm and encrypt it again, using a second cryptosystem with pseudorandom ciphertexts, which is secure against chosen-plaintext attacks. Decryption proceeds analogously, by first applying the decryption operation of the second cryptosystem and then the decryption operation of the PDR-CCA-secure cryptosystem. It can be verified that the composed cryptosystem retains PDR-CCA-security because the stage-two decryption oracle knows both secret keys. This method yields SS-PDR-CCA stegosystems in three different models as follows.

By applying the above generic CPA-secure cryptosystem with pseudorandom ciphertexts to a generic non-malleable cryptosystem [10, 18], we obtain an SS-PDR-CCA stegosystem under general assumptions.

Corollary 1. *Provided that trapdoor one-way permutations exist, there is an SS-PDR-CCA stegosystem in the common random string model.*

Using the above DDH-based cryptosystem with pseudorandom ciphertexts combined with the Cramer-Shoup cryptosystem [7], we obtain also an efficient SS-PDR-CCA stegosystem in the standard model.

Corollary 2. *Under the Decisional Diffie-Hellman assumption, there is an SS-PDR-CCA stegosystem.*

A more practical cryptosystem with pseudorandom ciphertexts exists also in the random oracle model: the OAEP+ scheme of Shoup [20]. OAEP+ is a CCA-secure cryptosystem based on an arbitrary trapdoor one-way permutation.

Corollary 3. *Provided that trapdoor one-way permutations exist, there is an SS-PDR-CCA stegosystem in the random oracle model.*

4 An SS-PDR-CCA Stegosystem

In this section, we propose a stegosystem that is steganographically secure against publicly-detectable replayable adaptive chosen-covertext attacks.

This stegosystem works for any covertext distribution that consists of a sequence of independent repetitions of a base-covertext distribution. Deviating from the notation of Section 2, we denote the base-covertext distribution by $\mathcal{C}$ and the covertext distribution used by the stegosystem by $\mathcal{C}^t = \Pi_{i=1}^t \mathcal{C}$. As noted in Section 2.2, through the introduction of a history, our construction also generalizes to arbitrary covertext channels.

Let (K, E, D) be a PDR-CCA-secure public-key cryptosystem with pseudorandom ciphertexts and compatible relation $\equiv_{pk}$. Suppose its cleartexts are l-bit strings and its ciphertexts are n-bit strings.

A class G of functions $X \to Y$ is called *strongly 2-universal* [23] if, for all distinct $x_1, x_2 \in X$ and all (not necessarily distinct) $y_1, y_2 \in Y$, exactly $|G|/|Y|^2$ functions from G take x_1 to y_1 and x_2 to y_2. Such a function family is sometimes simply called a *strongly 2-universal hash function* for brevity.

4.1 Description

The SS-PDR-CCA stegosystem consists of a triple of algorithms (keygen, encode, decode). The idea behind it is to encrypt a message using the public-key cryptosystem first and to embed the resulting ciphertext into a covertext sequence, as shown in Figure 1.

Fig. 1. The encoding process of the stegosystem: a message is first encrypted and then embedded using Algorithm sample. The decoding process works analogously in the reverse direction

---

Algorithm sample$^\mathcal{C}$

Input: security parameter k, a function $g : C \to \{0,1\}^f$, and a value $b \in \{0,1\}^f$
Output: a covertext x
1: $j \leftarrow 0$
2: **repeat**
3: $\quad x \xleftarrow{R} \mathcal{C}$
4: $\quad j \leftarrow j + 1$
5: **until** $g(x) = b$ **or** $j = k$
6: **return** x

The encoding method is based on the following algorithm sample, which has oracle access to $\mathcal{C}$ and samples a base-covertext according to $\mathcal{C}$ such that a given f-bit string b is embedded in it. This algorithm is the well-known rejection sampler [2, 12, 17, 9], generalized to embed multi-bit messages instead of only single-bit messages.

Intuitively, algorithm sample returns a covertext chosen from distribution $\mathcal{C}$, but restricted to that subset of $\mathcal{C}$ which is mapped to the given b by g. sample may also fail and return a covertext c with $g(c) \neq b$, but this happens only with negligible probability in k. As will be shown in Section 4.2, when b is a random f-bit string, g is chosen randomly from a 2-universal hash function, and $\mathcal{C}$ has sufficient min-entropy, then the output distribution of sample is statistically close to $\mathcal{C}$.

We now turn to the description of the stegosystem. Let $f \leq \gamma \log k$ for a positive constant $\gamma < 1$ and let $G : \mathcal{C} \to \{0,1\}^f$ denote a strongly 2-universal hash function.

Algorithm keygen chooses a random $g \xleftarrow{R} G$ and computes a tuple $(pk, sk) \leftarrow \mathsf{K}$, by running the key generation algorithm of the cryptosystem. The output of keygen is the tuple $(spk, ssk) = ((pk, g), sk)$.

Algorithm encode first encrypts an input message m using the given encryption algorithm E, which outputs in a ciphertext y. Assuming w.l.o.g. that y is an n-bit string (bounded by a polynomial in k) and $n = tf$, encode then repeatedly invokes sample to embed y in pieces of f bits a time into a sequence of t covertext symbols. Formally:

Algorithm encode

Input: security parameter k, a public key $spk = (pk, g)$, and a message $m \in \{0,1\}^l$ to encode

Output: a covertext $(c_1, \ldots, c_t)$

1: $y \leftarrow \mathsf{E}(pk, m)$
2: parse y as $y_1 \| \cdots \| y_t$, where $y_i \in \{0,1\}^f$
3: **for** $i = 1$ to t **do**
4: $\quad c_i \leftarrow \mathsf{sample}^{\mathcal{C}}(k, g, y_i)$
5: return $(c_1, \ldots, c_t)$

Algorithm decode proceeds analogously. From each of the t symbols in the covertext, a string of f bits is extracted by g; then the concatenation of these bit strings is decrypted by D, and the resulting value is returned (this is either an l-bit message or the symbol $\bot$):

Algorithm decode

Input: security parameter k, a secret key $ssk = (sk, g)$, and a covertext $(c_1, \ldots, c_t) \in \mathcal{C}^t$ to decode

Output: a decoded l-bit message or $\bot$

1: **for** $i = 1$ to t **do**
2: $\quad y_i \leftarrow g(c_i)$
3: $y \leftarrow y_1 \| \cdots \| y_t$
4: $x \leftarrow \mathsf{D}(sk, y)$
5: return x

The compatible relation $\equiv_{spk}$ of the stegosystem is computed as follows: given a pair of covertexts $(c_1, \ldots, c_t)$ and $(c'_1, \ldots, c'_t)$, they are first mapped to a pair of ciphertexts y and y', respectively, by running lines 1–3 of Algorithm decode. Then the relation is determined according to $y \equiv_{pk} y'$.

4.2 Analysis

This section is devoted to an analysis of the above stegosystem. Theorems 3 and 4 below together imply Theorem 2.

Theorem 3. *(keygen, encode, decode) is a valid stegosystem for covertext distributions with sufficiently large min-entropy.*

Proof (Sketch). According to Definition 1, the only non-trivial steps are to show that the algorithms are efficient and that the stegosystem is reliable, i.e., that

$$\mathsf{decode}(ssk, \mathsf{encode}(spk, m)) = m$$

for almost all pairs (spk, ssk) and all $m \in \{0,1\}^l$ except with negligible probability.

Efficiency follows immediately from the construction, the assumption $f \leq \gamma \log k$, and the efficiency of the public-key cryptosystem.

For reliability, it suffices to analyze the output of encode because the decoding operation is deterministic.

Consider iteration i in Algorithm encode, in which Algorithm sample tries to find a covertext x that is mapped to y_i by g. Because g is chosen from a strongly 2-universal class of hash functions, the entropy smoothing theorem [14] implies that over the random choices of g and $c \xleftarrow{R} \mathcal{C}$, the random variable $(g, g(c))$ is exponentially close to the uniform distribution over f-bit strings, provided $\mathcal{C}$ has enough min-entropy. Hence, there exists a negligible quantity $\epsilon(k) \ll 2^{-f}$ such that for almost all g, the distance of $g(c)$ from the uniform distribution is at most $\epsilon(k)$ over the choice $c \xleftarrow{R} \mathcal{C}$. Thus, the probability that in any particular iteration of sample, an x is chosen with $g(x) \neq y_i$, is at most $1 - 2^{-f} + \epsilon(k)$.

For any such g, since the k iterations and choices of $\mathcal{C}$ in sample are independent, the algorithm returns c with $g(c) \neq y_i$ only with some negligible probability $\epsilon'(k)$ for $f \leq \gamma \log k$. Hence, by the union bound, the probability that any iteration of Algorithm encode fails to embed the correct value is at most $t\epsilon'(k)$, which is negligible.

The proof of security is based on the following result. It shows that the joint distribution of the output from Algorithm sample and $\mathcal{G}$ is statistically close to the joint distribution of $\mathcal{C}$ and $\mathcal{G}$, where $\mathcal{G}$ denotes the distribution of choosing g uniformly from G, and where sample is run with a uniformly chosen b. The proof of Proposition 1 is given in the full version of the paper.

Proposition 1. *If the min-entropy of the covertext distribution $\mathcal{C}$ is large enough compared to f, then the statistical distance between $(\mathcal{S}(k), \mathcal{G})$ and $(\mathcal{C}, \mathcal{G})$ is negligible.*

Theorem 4. *For a covertext distribution $\mathcal{C}^t$ such that $\mathcal{C}$ has sufficiently large min-entropy and provided that $(\mathsf{K}, \mathsf{E}, \mathsf{D})$ is a PDR-CCA-secure public-key cryptosystem with pseudorandom ciphertexts, the stegosystem (keygen, encode, decode) is SS-PDR-CCA.*

Proof (Sketch). We prove that the stegosystem (keygen, encode, decode) is SS-PDR-CCA by a reduction argument. Assume that it is not SS-PDR-CCA and and hence there exists a (stego-)adversary (SA_1, SA_2) that succeeds in the experiment of Definition 3 with probability $\frac{1}{2} + \delta(k)$ for some non-negligible function δ. We construct an (encryption-)adversary (A_1, A_2) that has black-box access to (SA_1, SA_2) and breaks the PDR-CCA-security of (K, E, D) as follows.

Key Generation: When A_1 receives a public-key pk generated by K, it chooses $g \xleftarrow{R} G$, computes $spk \leftarrow (pk, g)$, and invokes SA_1 with spk.

First Decryption Stage: When SA_1 sends a query $(c_1, \ldots, c_t)$ to its decoding oracle SO_1, then A_1 computes $y \leftarrow y_1 \| \cdots \| y_t$ for $y_i \leftarrow g(c_i)$, gives y to its decryption oracle O_1, waits for the response and forwards the response to SA_1.

Challenge: When SA_1 halts and outputs (m^*, s), the encryption-adversary A_1 chooses an arbitrary plaintext message $m' \in \{0,1\}^l$, different from m^*, and outputs a triple (m^*, m', g). According to the definition of a public-key cryptosystem, a challenge ciphertext y^* is computed. Now A_2 is invoked with inputs pk, m^*, m', y^*, and g. It parses y^* as a sequence $y_1^* \| \cdots \| y_t^*$ of f-bit strings, computes $c_i^* \leftarrow \mathsf{sample}^\mathcal{C}(k, g, y_i^*)$ for $i = 1, \ldots, t$, and invokes SA_2 with inputs (pk, g), m^*, $(c_1^*, \ldots, c_t^*)$, and s.

Second Decryption Stage: A_2 behaves in the same way as A_1 during first decryption stage: It computes a ciphertext y from any decoding request that SA_2 makes as above, submits y to the decryption oracle O_2, and returns the answer to SA_2.

Guessing Stage: When SA_2 outputs a bit b^*, indicating its guess as to whether message m^* is contained in the challenge covertext $(c_1^*, \ldots, c_t^*)$, the encryption-adversary A_2 returns b^* as its own guess of whether m^* or m' is encrypted in y^*.

We now analyze the environment simulated by the encryption-adversary (A_1, A_2) to the stego-adversary (SA_1, SA_2) and the probability that the encryption-adversary can distinguish the encrypted messages.

Clearly, during key generation and the first decoding stage, the simulation for the stego-adversary SA_1 is perfect. During the encoding stage, a random bit b is chosen according to Definition 4 and the challenge ciphertext is computed as $y^* \leftarrow \mathsf{E}(pk, m^*)$ if $b = 0$ and $y^* \leftarrow \mathsf{E}(pk, m')$ if $b = 1$.

When $b = 0$, then, according to the definition of A_1, the challenge covertext c^* is computed in the same way as expected by the stego-adversary in the experiment of Definition 3 and the simulation is perfect.

When $b = 1$, however, SA_2 expects $(c_1^*, \ldots, c_t^*)$ to be a random covertext drawn according to $\mathcal{C}^t$, but receives $c_i^* = \mathsf{sample}^\mathcal{C}(k, g, y_i^*)$ for $i = 1, \ldots, t$ instead, where the concatenation of the y_i^* is an encryption of m' under key pk with E.

Proposition 1 implies that for every $i \in \{1, \ldots, t\}$, the statistical distance between $\mathcal{C}$ and the distribution of c_i^* as computed by Algorithm sample when run with input a uniformly chosen f-bit string is bounded by a negligible quantity $\epsilon_1^*(k)$. Furthermore, since the cryptosystem (K, E, D) has pseudorandom ciphertexts, for every distinguisher SA_2 there exists a negligible quantity $\epsilon_2^*(k)$ such that its advantage (over guessing randomly) in distinguishing between y^* as used by A_2 and the uniform distribution on n-bit strings is at most $\epsilon_2^*(k)$.

By combining these two facts, it follows that the behavior of the stego-adversary SA_2 who observes $(c_1^*, \ldots, c_t^*)$ in the simulation when $b = 1$ does not differ from its behavior in experiment of Definition 3, where it observes covertext $\mathcal{C}^t$, with more than probability $\epsilon^*(k) = t\epsilon_1^*(k) + \epsilon_2^*(k)$.

By definition, the output of the encryption-adversary A_2 is the same as that of the stego-adversary SA_2. Since SA_2 succeeds with probability $\frac{1}{2} + \delta(k)$ in attacking the stegosystem and since the simulated view of SA_2 is correct except with probability $\epsilon^*(k)$ when $b = 1$, the probability that SA_2 breaks PDR-CCA-security is $\frac{1}{2} + \delta(k) - \frac{\epsilon^*(k)}{2}$, which exceeds $\frac{1}{2}$ by a non-negligible quantity and establishes the theorem.

References

1. J. H. An, Y. Dodis, and T. Rabin, "On the security of joint signatures and encryption," in *Advances in Cryptology: EUROCRYPT 2002* (L. Knudsen, ed.), vol. 2332 of *Lecture Notes in Computer Science*, Springer, 2002.
2. R. J. Anderson and F. A. Petitcolas, "On the limits of steganography," *IEEE Journal on Selected Areas in Communications*, vol. 16, May 1998.
3. M. Bellare, A. Desai, D. Pointcheval, and P. Rogaway, "Relations among notions of security for public-key encryption schemes," in *Advances in Cryptology: CRYPTO '98* (H. Krawczyk, ed.), vol. 1462 of *Lecture Notes in Computer Science*, Springer, 1998.
4. C. Cachin, "An information-theoretic model for steganography," *Information and Computation*, vol. 192, pp. 41–56, July 2004. Parts of this paper appeared in Proc. 2nd Workshop on Information Hiding, Springer, 1998.
5. R. Canetti, H. Krawczyk, and J. Nielsen, "Relaxing chosen-ciphertext security," in *Advances in Cryptology: CRYPTO 2003* (D. Boneh, ed.), vol. 2729 of *Lecture Notes in Computer Science*, Springer, 2003.
6. T. M. Cover and J. A. Thomas, *Elements of Information Theory*. Wiley, 1991.
7. R. Cramer and V. Shoup, "A practical public-key cryptosystem provably secure against adaptive chosen-ciphertext attack," in *Advances in Cryptology: CRYPTO '98* (H. Krawczyk, ed.), vol. 1462 of *Lecture Notes in Computer Science*, Springer, 1998.
8. S. Craver, "On public-key steganography in the presence of an active warden," in *Information Hiding, 2nd International Workshop* (D. Aucsmith, ed.), vol. 1525 of *Lecture Notes in Computer Science*, pp. 355–368, Springer, 1998.
9. N. Dedić, G. Itkis, L. Reyzin, and S. Russell, "Upper and lower bounds on black-box steganography," in *Proc. 2nd Theory of Cryptography Conference (TCC)* (J. Kilian, ed.), Lecture Notes in Computer Science, Springer, 2005.
10. D. Dolev, C. Dwork, and M. Naor, "Non-malleable cryptography," *SIAM Journal on Computing*, vol. 30, no. 2, pp. 391–437, 2000.
11. S. Goldwasser and S. Micali, "Probabilistic encryption," *Journal of Computer and System Sciences*, vol. 28, pp. 270–299, 1984.
12. N. J. Hopper, J. Langford, and L. von Ahn, "Provably secure steganography," in *Advances in Cryptology: CRYPTO 2002* (M. Yung, ed.), vol. 2442 of *Lecture Notes in Computer Science*, Springer, 2002.
13. Y. Lindell. Personal communication, Jan. 2004.
14. M. Luby, *Pseudorandomness and Cryptographic Applications*. Princeton University Press, 1996.
15. T. Mittelholzer, "An information-theoretic approach to steganography and watermarking," in *Information Hiding, 3rd International Workshop, IH'99* (A. Pfitzmann, ed.), vol. 1768 of *Lecture Notes in Computer Science*, pp. 1–16, Springer, 1999.

16. C. Rackoff and D. R. Simon, "Non-interactive zero-knowledge proof of knowledge and chosen ciphertext attack," in *Advances in Cryptology: CRYPTO '91* (J. Feigenbaum, ed.), vol. 576 of *Lecture Notes in Computer Science*, pp. 433–444, Springer, 1992.
17. L. Reyzin and S. Russell, "Simple stateless steganography." Cryptology ePrint Archive, Report 2003/093, 2003. http://eprint.iacr.org/.
18. A. Sahai, "Non-malleable non-interactive zero knowledge and adaptive chosen-ciphertext security," in *Proc. 40th IEEE Symposium on Foundations of Computer Science (FOCS)*, pp. 543–553, 1999.
19. V. Shoup, "A proposal for an ISO standard for public key encryption." Cryptology ePrint Archive, Report 2001/112, 2001. http://eprint.iacr.org/.
20. V. Shoup, "OAEP reconsidered," *Journal of Cryptology*, vol. 15, no. 4, pp. 223–249, 2002.
21. G. J. Simmons, "The prisoners' problem and the subliminal channel," in *Advances in Cryptology: Proceedings of Crypto 83* (D. Chaum, ed.), pp. 51–67, Plenum Press, 1984.
22. L. von Ahn and N. J. Hopper, "Public-key steganography," in *Advances in Cryptology: Eurocrypt 2004* (C. Cachin and J. Camenisch, eds.), vol. 3027 of *Lecture Notes in Computer Science*, pp. 322–339, Springer, 2004.
23. M. N. Wegman and J. L. Carter, "New hash functions and their use in authentication and set equality," *Journal of Computer and System Sciences*, vol. 22, pp. 265–279, 1981.
24. J. Zöllner, H. Federrath, H. Klimant, A. Pfitzmann, R. Piotraschke, A. Westfeld, G. Wicke, and G. Wolf, "Modeling the security of steganographic systems," in *Information Hiding, 2nd International Workshop* (D. Aucsmith, ed.), vol. 1525 of *Lecture Notes in Computer Science*, pp. 344–354, Springer, 1998.

Upper and Lower Bounds on Black-Box Steganography

Extended Abstract

Nenad Dedić, Gene Itkis, Leonid Reyzin, and Scott Russell

Boston University Computer Science,
111 Cummington St.
Boston MA 02215 USA
{nenad, itkis, reyzin, srussell}@cs.bu.edu

Abstract. We study the limitations of steganography when the sender is not using any properties of the underlying channel beyond its entropy and the ability to sample from it. On the negative side, we show that the number of samples the sender must obtain from the channel is exponential in the rate of the stegosystem. On the positive side, we present the first secret-key stegosystem that essentially matches this lower bound regardless of the entropy of the underlying channel. Furthermore, for high-entropy channels, we present the first secret-key stegosystem that matches this lower bound *statelessly* (i.e., without requiring synchronized state between sender and receiver).

1 Introduction

Steganography's goal is to conceal the presence of a secret message within an innocuous-looking communication. In other words, steganography consists of hiding a secret *hiddentext* message within a public *covertext* to obtain a *stegotext* in such a way that any observer (except, of course, the intended recipient) is unable to distinguish between a covertext *with* a hiddentext and one *without*.

The first rigorous complexity-theoretic formulation of secret-key steganography was provided by Hopper, Langford and von Ahn [HLvA02]. In this formulation, *steganographic secrecy* of a stegosystem is defined as the inability of a polynomial-time adversary to distinguish between observed distributions of unaltered covertexts and stegotexts. (This is in contrast with many previous works, which tended to be information-theoretic in perspective; see, e.g., [Cac98] and other references in [HLvA02, Cac98].)

The model of [HLvA02], which we adopt with slight changes, assumes that the two communicating parties have some underlying channel $\mathcal{C}$ of covertext documents that the adversary expects to see. They also share a secret key (public-key steganography is addressed in [vAH04, BC04]). The sender is allowed to draw documents from $\mathcal{C}$; the game for the sender is to alter $\mathcal{C}$ imperceptibly for the adversary, while transmitting a meaningful hiddentext message to the recipient.

Conversely, the game for the (passive) adversary is to distinguish the distribution of transmitted messages from $\mathcal{C}$.

1.1 Desirable Characteristics of a Stegosystem

Black-Box. In order to obtain a stegosystem of broad applicability, one would like to make as few assumptions as possible about the understanding of the underlying channel. Indeed, as Hopper et al. [HLvA02] point out, the channel (such as human email traffic or images of various scenes) may well be very complex and not easily described. For example, if the parties are using photographs of city scenes as covertexts, it is reasonable to assume that the sender can obtain such photographs, but unreasonable to expect the sender and the recipient to know a polynomial-time algorithm that can construct such photographs from uniformly distributed random strings. In this work, we therefore concentrate on the study of *black-box* steganography. Namely, the sender and the recipient need not know anything about the underlying channel distribution (beyond a lower bound on its min-entropy). The sender's only access to the channel is via an oracle that draws a random sample from the channel distribution. The recipient need not access the channel at all.

Efficient and Secure. Stegosystems have several performance characteristics. First, of course, it is desirable that the encoding algorithm of sender and the decoding algorithm of the receiver be efficient. A particularly important characteristic of the efficiency of the sender is the number of samples that the sender is required to draw from $\mathcal{C}$. In fact, in all proposed black-box stegosystems, sender computation is proportional to the number of samples drawn, with actual computation per sample being quite minimal. Because most real-life channels are quite complex, the drawing of the samples is likely to dominate the running time of an actual implementation.

Another important performance measure is the transmission rate of the stegosystem, which is the number of hiddentext bits transmitted per single stegotext document sent (a document is the value returned by a single request to the channel sampling oracle—e.g., a photograph). Transmission rate is tied to reliability, which is the probability of successful decoding of an encoded message (correspondingly, unreliability is one minus reliability). The goal is to construct stegosystems that are reliable and transmit at a high rate (it is, of course, easier to transmit at a high rate if reliability is low and the recipient will not understand much of what is transmitted).

Finally, even a most efficient stegosystem is useless if not secure. Quantitatively, insecurity is defined as the adversary's advantage in distinguishing stegotext from $\mathcal{C}$ (and security as one minus insecurity). Naturally, we are interested in stegosystems with insecurity as close to 0 as possible.

The efficiency and security of a stegosystem, even if it is black-box, may depend on the channel distribution. In particular, we will be interested in the dependence on the channel min-entropy h. Ideally, a stegosystem would work well even for low-min-entropy channels.

Stateless. It is desirable to construct *stateless* stegosystems, so that the sender and the recipient need not maintain synchronized state in order to communicate long messages. Indeed, the need for synchrony may present a particular problem in steganography in case messages between sender and recipient are dropped or arrive out of order. Unlike in counter-mode symmetric encryption, where the counter value can be sent along with the ciphertext in the clear, here this is not possible: the counter itself would also have to be steganographically encoded to avoid detection, which brings us back to the original problem of steganographically encoding multibit messages.

1.2 Our Contributions

We study the optimal efficiency achievable by black-box steganography, and present secret-key stegosystems that are nearly optimal. Specifically, we demonstrate the following results:

- A lower bound, which states that a secure and reliable black-box stegosystem with rate of w bits per document sent requires the encoder to take at least $c2^w$ samples from the channel per w bits sent, for some constant c. The value of c depends on security and reliability, and tends to $1/(2e)$ as security and reliability approach 1. This lower bound applies to secret-key as well as public-key stegosystems.
- A stateful black-box secret-key stegosystem STF that transmits w bits per document sent, takes 2^w samples per w bits, has unreliability of 2^{-h+w} per document, and negligible insecurity, which is independent of the channel. (A very similar construction was independently discovered by Hopper [Hop04–Construction 6.10].)
- A stateless black-box secret-key stegosystem STL that transmits w bits per document sent, takes 2^w samples per w bits, has unreliability $2^{-\Theta(2^h)}$, and insecurity negligibly close to $l^2 2^{-h+2w}$ for lw bits sent.

Note that for both stegosystems, the rate vs. number of samples tradeoff is very close to the lower bound—in fact, for channels with sufficient entropy, the optimal rate allowed by the lower bound and the achieved rate differ by $\log_2 2e < 2.5$ bits (and some of that seems due to slack in the bound). Thus, our bound is quite tight, and our stegosystems quite efficient. The proof of the lowerbound involves a surprising application of the huge random objects of [GGN03], specifically of the truthful implementation of a boolean function with interval-sum queries. The lowerbound demonstrates that significant improvements in stegosystem performance must come from assumptions about the channel.

The stateless stegosystem STL can be used whenever the underlying channel distribution has sufficient min-entropy h for the insecurity to be acceptably low. It is extremely simple, requiring just evaluations of a pseudorandom function for encoding and decoding, and very reliable.

If the underlying channel does not have sufficient min-entropy, then the stateful stegosystem STF can be used, because its insecurity is independent of the channel. While it requires shared synchronized state between sender and receiver, the state information is only a counter of the number of documents sent so far. If min-entropy of the channel is so low that the error probability of 2^{-h+w} is too high for the application, reliability of this stegosystem can be improved through the use of error-correcting codes over the 2^w-ary alphabet (applied to the hiddentext before stegoencoding), because failure to decode correctly is independent for each w-bit block. Error-correcting codes can increase reliability to be negligibly close to 1 at the expense of reducing the asymptotic rate from w to $w - (h+2)2^{-h+w}$. Finally, of course, the min-entropy of any channel can be improved from h to nh by viewing n consecutive samples as a single draw from the channel; if h is extremely small to begin with, this will be more efficient than using error-correcting codes (this improvement requires both parties to be synchronized modulo n, which is not a problem in the stateful case).

This stateful stegosystem STF also admits a few variants. First, the logarithmic amount of shared state can be eliminated at the expense of adding a linear amount of private state to the sender and reducing reliability slightly (as further described in 4.1), thus removing the need for synchronization between the sender and the recipient. Second, under additional assumptions about the channel (e.g., if each document includes time sent, or has a sequence number), STF can be made completely stateless. The remarks of this paragraph and the previous one can be equally applied to [Hop04–Construction 6.10].

1.3 Related Work

The bibliography on the subject of steganography is extensive; we do not review it all here, but rather recommend references in [HLvA02].

Constructions. In addition to introducing the complexity-theoretic model for steganography, [HLvA02] proposed two constructions of black-box[1] secret-key stegosystems, called Construction 1 and Construction 2.

Construction 1 is stateful and, like our stateful construction STF, boasts negligible insecurity regardless of the channel. However, it can transmit only 1 bit per document, and its reliability is limited by $1/2 + 1/4(1 - 2^{-h})$ per document sent, which means that, regardless of the channel, each hiddentext bit has probability at least $1/4$ of arriving incorrectly (thus, to achieve high reliability, error-correcting codes with expansion factor of at least $1/(1 - H_2(1/4)) \approx 5$ are needed). In contrast, STF has reliability that is exponentially (in the min-entropy) close to 1, and thus works well for any channel with sufficient entropy. Furthermore, it can transmit at rate w for any $w < h$, provided the encoder has sufficient time for the 2^w samples required. It can be seen as a generalization of Construction 1.

---

[1] Construction 2, which, strictly speaking, is not presented as a black-box construction in [HLvA02], can be made black-box through the use of extractors (such as universal hash functions) in place of unbiased functions, as shown in [vAH04].

Construction 2 of [HLvA02] is stateless. Like the security of our stateless construction STL, its security depends on the min-entropy of the underlying channel. While no exact analysis is provided in [HLvA02], the insecurity of Construction 2 seems to be roughly $\sqrt{l}2^{(-h+w)/2}$ (due to the fact that the adversary sees l samples either from $\mathcal{C}$ or from a known distribution with bias roughly $2^{(-h+w)/2}$ caused by a public extractor; see Appendix A), which is higher than the insecurity of STL (unless l and w are so high that $h < 3w + 3\log l$, in which case both constructions are essentially insecure, because insecurity is higher than the inverse of the encoder's running time $l2^w$). Reliability of Construction 2, while not analyzed in [HLvA02], seems close to the reliability of STL. The rate of Construction 2 is lower (if other parameters are kept the same), due to the need for randomized encryption of the hiddentext, which necessarily expands the number of bits sent.

It is important to note that the novelty of STL is not the construction itself, but rather its analysis. Specifically, its stateful variant appeared as Construction 1 in the Extended Abstract of [HLvA02], but the analysis of the Extended Abstract was later found to be flawed by [KMR02]. Thus, the full version of [HLvA02] included a different Construction 1. We simply revive this old construction, make it stateless, generalize it to w bits per document, and, most importantly, provide a new analysis for it.

In addition to the two constructions of [HLvA02] described above, and independently of our work, Hopper in [Hop04] proposed two more constructions: Constructions 6.10 ("MultiBlock") and 3.15 ("NoState"). As already mentioned, MultiBlock is essentially the same as our STF. NoState is an interesting variation of Construction 1 of [HLvA02], that addresses the problem of maintaining shared state at the expense of lowering the rate even further.

Bounds on the Rate and Efficiency. Hopper in [Hop04–Section 6.2] establishes a bound on the rate vs. efficiency tradeoff. Though quantitatively similar to ours (in fact, tighter by the constant of $2e$), this bound applies only to a restricted class of black-box stegosystems: essentially, stegosystems that encode and decode one block at a time and sample a fixed number of documents per block. The bound presented in this paper applies to any black-box stegosystem, as long as it works for a certain reasonable class of channels, and thus can be seen as a generalization of the bound of [Hop04]. Our proof techniques are quite different than those of [Hop04], and we hope they may be of independent interest. We refer the reader to Section 3.3 for an elaboration. Finally it should be noted that non-black-box stegosystems can be much more efficient—see [HLvA02, vAH04, Le03, LK03].

2 Definitions

2.1 Steganography

The definitions here are essentially those of [HLvA02]. We modify them in three ways. First, we view the channel as producing documents (symbols in some, possibly very large, alphabet) rather than bits. This simplifies notation and makes min-entropy of the channel more explicit. Second, we consider stegosystem reliability as a parameter rather than a fixed value. Third, we make the length of the adversary's description (and the adversary's dependence on the channel) more explicit in the definition.

The Channel. Let Σ be an alphabet; we call the elements of Σ documents. A channel $\mathcal{C}$ is a map that takes a history $\mathcal{H} \in \Sigma^*$ as input and produces a probability distribution $D_{\mathcal{H}} \in \Sigma$. A history $\mathcal{H} = s_1 s_2 ... s_n$ is *legal* if each subsequent symbol is obtainable given the previous ones, i.e., $Pr_{D_{s_1 s_2 ... s_{i-1}}}[s_i] > 0$. Min-entropy of a distribution D is defined as $H_\infty(D) = \min_{s \in D}\{-\log_2 Pr_D[s]\}$. Min-entropy of $\mathcal{C}$ is the $\min_{\mathcal{H}} H_\infty(D_{\mathcal{H}})$, where the minimum is taken over legal histories $\mathcal{H}$.

Our stegosystems will make use of a channel sampling oracle M, which, on input $\mathcal{H}$, outputs a symbol s according to $D_{\mathcal{H}}$.

Definition 1. *A* black-box secret-key stegosystem *is a pair of probabilistic polynomial time algorithms* $S = (SE, SD)$ *such that, for a security parameter* κ,

1. *SE has access to a channel sampling oracle M for a channel $\mathcal{C}$ and takes as input a randomly chosen key $K \in \{0,1\}^\kappa$, a string $m \in \{0,1\}^*$ (called the* hiddentext*), and the channel history $\mathcal{H}$. It returns a string of symbols $s_1 s_2 \ldots s_l \in \Sigma^*$ (called the* stegotext*)*
2. *SD takes as input a key $K \in \{0,1\}^\kappa$, a stegotext $s_1 s_2 \ldots s_l \in \Sigma^*$ and a channel history $\mathcal{H}$, and returns a hiddentext $m \in \{0,1\}^*$.*

We further assume that the length l of the stegotext output by SE depends only on the length of hiddentext m but not on its contents.

Stegosystem Reliability. The *reliability* of a stegosystem S with security parameter κ for a channel $\mathcal{C}$ and messages of length l is defined as

$$\mathbf{Rel}_{S(\kappa), \mathcal{C}, l} = \min_{m \in \{0,1\}^l, \mathcal{H}} \{ \Pr_{K \in \{0,1\}^\kappa}[SD(K, SE^M(K, m, \mathcal{H}), \mathcal{H}) = m] \}.$$

Unreliability (as a parallel to insecurity) is defined as $\mathbf{UnRel}_{S(\kappa), \mathcal{C}, l} = 1 - \mathbf{Rel}_{S(\kappa), \mathcal{C}, l}$.

The Adversary. We consider only passive adversaries who mount a chosen hiddentext attack on S (stronger adversarial models for steganography have also been considered, see e.g. [HLvA02, vAH04, BC04]). The goal of such an adversary is to distinguish whether it is seeing encodings of the hiddentext it supplied to the encoder, or simply random draws from the channel. To this end, define an oracle $O(\cdot, \mathcal{H})$ that produces random draws from the channel starting with history $\mathcal{H}$ as follows: on input $m \in \{0,1\}^*$, O computes the length l of the stegotext that $SE^M(K, m)$ would have output, and outputs $s_1 s_2 \ldots s_l$ where each s_i is drawn according to $D_{\mathcal{H} \circ s_1 s_2 \ldots s_{i-1}}$.

Definition 2. W^2 is a (t,d,q,l) passive adversary for stegosystem S if

1. W runs in expected time t (including the running time needed by the stegoencoder to answer its queries) and has description of length d (in some canonical language).
2. W has access to $\mathcal{C}$ via the sampling oracle $M(\cdot)$.
3. W can make an expected number of q queries of combined length l bits to an oracle which is either $SE^M(K,\cdot,\cdot)$ or $O(\cdot,\cdot)$.
4. W outputs a bit indicating whether it was interacting with SE or with O.

Stegosystem Security. The *advantage* $\mathbf{Adv}^{SS}$ (here SS stands for "Steganographic Secrecy") of W against S with security parameter κ for a channel $\mathcal{C}$ is defined as

$$\mathbf{Adv}^{SS}_{S(\kappa),\mathcal{C}}(W) = \left| \Pr_{K \leftarrow \{0,1\}^\kappa}[W^{M,SE^M(K,\cdot,\cdot)} = 1] - \Pr[W^{M,O(\cdot,\cdot)} = 1] \right|.$$

For a given (t,d,q,l), the *insecurity* of a stegosystem S with respect to channel $\mathcal{C}$ is defined as

$$\mathbf{InSec}^{SS}_{S(\kappa),\mathcal{C}}(t,d,q,l) = \max_{(t,d,q,l) \text{ adversary } W} \{\mathbf{Adv}^{SS}_{S(\kappa),\mathcal{C}}(W)\},$$

and security **Sec** as $1 - \mathbf{InSec}$.

Note that the adversary's algorithm can depend on the channel $\mathcal{C}$, subject to the restriction on the algorithm's total length d. In other words, the adversary can possess some description of the channel in addition to the black-box access provided by the channel oracle. This is a meaningful strengthening of the adversary: indeed, it seems imprudent to assume that the adversary's knowledge of the channel is limited to whatever is obtainable by black-box queries (for instance, the adversary has some idea of a reasonable email message or photograph should look like). It does not contradict our focus on black-box steganography: it is prudent for the honest parties to avoid relying on particular properties of the channel, while it is perfectly sensible for the adversary, in trying to break the stegosystem, to take advantage of whatever information about the channel is available.

2.2 Pseudorandom Functions

We use pseudorandom functions [GGM86] as a tool. Because the adversary in our setting has access to the channel, any cryptographic tool used must be secure even given the information provided by the channel. Thus, our underlying assumption is the existence of pseudorandom functions that are secure given the channel oracle, which is equivalent [HILL99] to the existence of one-way functions

---

[2] The adversary in the context of steganography is sometimes referred to as the "warden." The idea of the adversary as a warden and the use of W to designate it is a consequence of original problem formulation in [Sim83].

that are secure given the channel oracle. Thus is the minimal assumption needed for steganography [HLvA02].

Let $\mathcal{F} = \{F_{\text{seed}}\}_{\text{seed} \in \{0,1\}^*}$ be a family of functions, all with the same domain and range. For a probabilistic adversary A, and channel $\mathcal{C}$ with sampling oracle M, the *PRF-advantage of A over $\mathcal{F}$* is defined as

$$\mathbf{Adv}^{\text{PRF}}_{\mathcal{F}(n),\mathcal{C}}(A) = \left| \Pr_{\text{seed} \leftarrow \{0,1\}^n}[A^{M, F_{\text{seed}}(\cdot)} = 1] - \Pr_g[A^{M, g(\cdot)} = 1] \right|,$$

where g is a random function with the same domain and range. For a given (t, d, q), the *insecurity* of a pseudorandom function family $\mathcal{F}$ with respect to channel $\mathcal{C}$ is defined as

$$\mathbf{InSec}^{\text{PRF}}_{\mathcal{F}(n),\mathcal{C}}(t, d, q, l) = \max_{(t,d,q,l) \text{ adversary } A} \{\mathbf{Adv}^{\text{SS}}_{\mathcal{F}(n),\mathcal{C}}(A)\},$$

where the maximum is taken over all adversaries that run in expected time t, whose description size is at most d, and that make an expected number of q queries to their oracles.

3 The Lower Bound

Recall that we define the rate of a stegosystem as the *average number of hiddentext bits per document sent* (this should not be confused with the average number of hiddentext bits per *bit* sent; note also that this is the sender's rate, not the rate of information actually decoded by the recipient, which is lower due to unreliability). We set out to prove that a reliable stegosystem with black-box access to the channel with rate w, must make roughly $l2^w$ queries to the channel to send a message of length lw. Intuitively, this should be true because each document carries w bits of information on average, but since the encoder knows nothing about the channel, it must keep on sampling until it gets the encoding of those w bits, which amounts to 2^w samples on average.

In particular, it suffices for the purposes of this lower bound to consider a restricted class of channels: the distribution of the sample depends only on the length of the history (not on its contents). We will write $D_1, D_2, ..., D_i, ...$, instead of $D_\mathcal{H}$, where i is the length of the history $\mathcal{H}$. Furthermore, it will suffice for us to consider only distributions D_i that are uniform on a subset of Σ. We will identify the distribution with the subset (as is often done for uniform distributions).

Let $|D_i| = H = 2^h$ and $|\Sigma| = S$. Because the encoder receives the min-entropy h of the channel as input, if $H = S$, then encoder knows the channel completely (it's simply uniform on Σ), and our lower bounds do not hold, because no sampling from the channel is necessary. Thus, we require that h be smaller than $\log_2 S$. Let $R = 1/(1 - H/S)$.

Our proof proceeds in two parts. First, we consider a stegoencoder SE that does not output anything that it did not receive as a response from the channel-sampling oracle. To be reliable, such an encoder has to make many queries, as shown in Lemma 1. Second, we show that to be secure, a black-box SE cannot output anything it did not receive from the channel-sampling oracle.

The second half of the proof is somewhat complicated by the fact that we want to assume security only against bounded adversaries: namely, ones whose description size and running time are polynomial in the description size and running time of the encoder (in particular, polynomial in $\log S$ rather than S). This requires us to come up with pseudorandom subsets D_i of Σ that have concise descriptions and high min-entropy, and whose membership is impossible for the stegoencoder to predict. In order to do that, we utilize techniques from the truthful implementation of a boolean function with interval-sum queries of [GGN03] (truthfulness is important because min-entropy has to be high unconditionally).

3.1 Lower Bound When Only Query Results Are Output

We consider the following channel: if $D_1, D_2, \ldots$ are subsets of Σ, we write $\boldsymbol{D} = D_1 \times D_2 \times \ldots$ to denote the channel that, on history length i, outputs an uniformly random element of D_i; if $|D_1| = |D_2| = \ldots = 2^h$ then we say that $\boldsymbol{D}$ is a *flat h-channel*. Normally, one would think of the channel sampling oracle for $\boldsymbol{D}$ as making a fresh random choice from D_i when queried on history length i. Instead, we will think of the oracle as having made all its choices in advance. Imagine that the oracle already took "enough" samples:

$s_{1,1}, s_{1,2}, \ldots, s_{1,j}, \ldots$ from D_1,
$s_{2,1}, s_{2,2}, \ldots, s_{2,j}, \ldots$ from D_2,
$\ldots,$
$s_{i,1}, s_{i,2}, \ldots, s_{i,j}, \ldots$ from D_i
$\ldots.$

We will denote the string containing all these samples by $\mathcal{S}$, and refer to it as a *draw-sequence* from the channel. We will give our stegoencoder access to an oracle (also denoted by $\mathcal{S}$) that, each time it's queried with i, returns the next symbol from the sequence $s_{i,1}, s_{i,2}, \ldots, s_{i,j}, \ldots$. Choosing $\mathcal{S} \in \Sigma^{**}$ at random and giving the stegoencoder access to it is equivalent to giving the encoder access to the usual channel-sampling oracle M for our channel $\boldsymbol{D}$.

Assume $SE^{\mathcal{S}}(K, m, \mathcal{H}) = t = t_1 t_2 \ldots t_l$, where $t_i \in \Sigma$. Note that t_i is an element of the sequence $s_{i,1}, s_{i,2}, \ldots, s_{i,j}, \ldots$. If t_i is the j-th element of this sequence, then it took j queries to produce it. We will denote by *weight of t with respect to* $\mathcal{S}$, the number of queries it took to produce t: $W(t, \mathcal{S}) = \sum_{i=1}^{k} \min\{j \mid s_{i,j} = y_i\}$. In the next lemma, we prove (by looking at the *decoder*) that for any $\mathcal{S}$, most messages have high weight.

Lemma 1. *Let $F : \Sigma^* \to \{0,1\}^*$ be an arbitrary (possibly unbounded) deterministic stegodecoder that takes a sequence $t \in \Sigma^l$ and outputs a message m of length lw bits.*

Then the probability that a random lw-bit message has an encoding of weight significantly less than $(1/e)l2^w$, is small. More precisely, for any $\mathcal{S} \in \Sigma^{**}$ and any $N \in \mathbb{N}$:

$$\Pr_{m\in\{0,1\}^{lw}}[(\exists t \in \Sigma^l)(F(t) = m \land W(t,\mathcal{S}) \le N)] \le \frac{\binom{N}{l}}{2^{lw}} < \left(\frac{Ne}{l2^w}\right)^l.$$

Proof. Simple combinatorics show that the number of different sequences t that have weight less than N (and hence the number of messages that have encodings of weight less than N) is at most $\binom{N}{l}$: indeed, it is simply the number of positive integer solutions to $x_1 + \ldots + x_l \le N$, which is the number of ways to put l bars among $N - l$ stars (the number of stars to the right of the i-th bar corresponds to $x_i - 1$), or, equivalently, the number of ways choose l positions out of N. The total number of messages is 2^{lw}. The last inequality follows from $\binom{N}{l} < \left(\frac{Ne}{l}\right)^l$. □

Observe that taking the probability over a random lw-bit message, as we do above, is meaningful. Indeed, if the distribution of messages encoded is not uniform, then compression could reduce their size and thus improve the efficiency of the stegosystem, rendering our bound pointless. Our lower bound applies when the designer of the stegosystem assumes that the messages are distributed uniformly. (For any other distribution, data compression should be applied before stegoencoding.)

3.2 Secure Stegosystems Almost Always Output Query Answers

The next step is to prove that the encoder of a secure black-box stegosystem must output only what it gets from the oracle, with high probability. Assume D is a flat h-channel chosen uniformly at random. Then it is easy to demonstrate that, if the encoder outputs in position i a symbol $s_i \in \Sigma$ that it did not receive as a response to a query to D_i, the chances that s_i is in the support of D_i are H/S. It can then be shown that, if the stegoencoder has insecurity ϵ, then it cannot output something it did not receive as response to a query with probability higher than $\epsilon/(1 - H/S)$.

The problem with the above argument is the following: it assumes that the adversary can test whether s_i the support of D_i. This is not possible if we assume D_i is completely random and the adversary's description is small compared to $S = |\Sigma|$. However, it does serve as a useful warm-up, and leads to the following theorem when combined with the results of the previous section.

Theorem 1. *Let (SE, SD) be a black-box stegosystem with insecurity ϵ against an adversary who has an oracle for testing membership in the support of $\mathcal{C}$, unreliability ρ and rate w for an alphabet Σ of size S. Then there exists a channel with min-entropy $h = \log_2 H$ such that the probability that the encoder makes at most N queries to send a random message of length lw, is upper bounded by*

$$\left(\frac{Ne}{l2^w}\right)^l + \rho + \epsilon R,$$

and the expected number of queries per stegotext symbol is therefore at least

$$\frac{2^w}{e}\left(\frac{1}{2} - \rho - \epsilon R\right),$$

where $R = 1/(1 - H/S)$.

Proof. See the full version [DIRR04]. □

3.3 Lower Bound for Computationally Bounded Parties

We now want to establish the same lower bound without making such a strong assumption about the security of the stegosystem. Namely, we do not want to assume that the insecurity ϵ is low unless the adversary's description size and running time are small ("small," when made rigorous, will mean some fixed polynomials in the description size and running time, respectively, of the stegoencoder, and a security parameter for a function that is pseudorandom against the stegoencoder). Recall that our definitions allow the adversary to depend on the channel; thus, our goal is to construct channels that have short descriptions for the adversary but look like random flat h-channels to the black-box stegoencoder. In other words, we wish to replace a random flat h-channel with a pseudorandom one.

We note that the channel is pseudorandom only in the sense that it has a short description, so as to allow the adversary to be computationally bounded. The min-entropy guarantee, however, can not be replaced with a "pseudo-guarantee": else the encoder is being lied to, and our lower bound is no longer meaningful. Thus, a simpleminded approach, such as using a pseudorandom predicate with bias H/S applied to each symbol and history length to determine whether the symbol is in the support of the channel, will not work here: because S is constant, eventually (for some history length) the channel will have lower than guaranteed min-entropy (moreover, we do not wish to assume that S is large in order to demonstrate that this is unlikely to happen; our lower bound should work for any alphabet). Rather, we need the pseudorandom implementation of the channel to be truthful[3] in the sense of [GGN03], and so rely on the techniques developed therein.

The result is the following theorem.

Theorem 2. *There exist polynomials p, q and constants c_1, c_2 with the following property. Let $S(\kappa)$ be a black-box stegosystem with description size δ, insecurity $\mathbf{InSec}^{SS}_{S(\kappa),\mathcal{C}}(t, d, q, l)$, unreliability ρ, rate w and running time τ for an alphabet Σ of size S. Assume there exists a pseudorandom function family $\mathcal{F}(n)$ with insecurity $\mathbf{InSec}^{PRF}_{\mathcal{F}(n)}(t, d, q)$. Then there exists a channel $\mathcal{C}$ with min-entropy $h = \log_2 H$ such that the probability that the encoder makes at most N queries to send a random message of length lw, is upper bounded by*

$$\left(\frac{Ne}{l2^w}\right)^l + \rho + R\mathbf{InSec}^{SS}_{S(\kappa),\mathcal{C}}(q(\tau), n + c_1, 1, lw) + $$
$$(R+1)\left(\mathbf{InSec}^{PRF}_{\mathcal{F}(n)}(p(\tau), \delta + c, p(\tau)) + 2^{-n}\right),$$

and the expected number of queries per stegotext symbol is therefore at least

---

[3] In this case, truthfulness implies that for each history length, the support of the channel has exactly H elements.

$$\frac{2^w}{e}\left(\frac{1}{2}-\rho-\text{RInSec}^{\text{SS}}_{S(\kappa),\mathcal{C}}(q(\tau),n+c_1,1,lw)\right)-$$
$$\frac{2^w}{e}(R+1)\left(\text{InSec}^{\text{PRF}}_{\mathcal{F}(n)}(p(\tau),\delta+c,p(\tau))+2^{-n}t\right),$$

where $R = 1/(1 - H/S)$.

Proof. See the full version [DIRR04]. □

Discussion. The proof of Theorem 2 relies fundamentally on Theorem 1. In other words, to prove a lower bound in the computationally bounded setting, we use the corresponding lower bound in the information-theoretic setting. To do so, we replace an object of an exponentially large size (the channel) with one that can be succinctly described. This replacement substitutes *some* information-theoretic properties with their computational counterparts. However, for a lower bound to remain "honest" (i.e., not restricted to uninteresting channels), some global properties must remain information-theoretic. This is where the truthfulness of huge random objects of [GGN03] comes to the rescue. We hope that other interesting impossibility results can be proved in a similar fashion, by adapting an information-theoretic result using the paradigm of [GGN03]. We think truthfulness of the objects will be important in such adaptations for the same reason it was important here.

Note that the gap in the capabilities of the adversary and encoder/decoder is different in the two settings: in the information-theoretic case the adversary is given unrestricted computational power, while in the computationally bounded case it is assumed to run in polynomial time, but is given the secret channel seed. However, in the information-theoretic case we may remove the gap altogether, by providing both the adversary and the encoder/decoder with a channel membership oracle, and still obtain a lower bound analogous[4] to that of Theorem 2. We see no such opportunity to remove the gap in the computationally bounded case (e.g., equipping the encoder/decoder with the channel seed seems to break our proof). Removing this asymmetry in the computationally bounded case seems challenging and worth pursuing.

4 The Stateful Construction STF

The construction STF relies on a pseudorandom function family $\mathcal{F}$. In addition to the security parameter κ (the length of the PRF key K), it depends on the rate parameter w. Because it is stateful, both encoder and decoder take a counter *ctr* as input.

---

[4] A lower bound on the number of samples per document sent, becomes trivially zero if the encoder is given as much time as it pleases, in addition to the membership oracle of the flat channel. Yet it should not be difficult to prove that it must then run for $O(2^w)$ steps per document sent.

Our encoder is similar to the rejection-sampler-based encoder of [HLvA02] generalized to w bits: it simply samples elements from the channel until the pseudorandom function evaluated on the element produces the w-bit symbol being encoded. The crucial difference of our construction is the following: to avoid introducing bias into the channel, if the same element is sampled twice, the encoder simply flips a random coin to decide whether to output that element with probability 2^{-w}. Hopper in [Hop04–Construction 6.10] independently proposes a similar construction, except instead of flipping a fresh random coin, the encoder evaluates the pseudorandom function on a new counter value (there is a separate counter associated to each sampled document, indicating how many times the document has been sampled), thus conserving randomness.

Observe that, assuming $\mathcal{F}$ is truly random rather than pseudorandom, each sample from the channel has probability 2^{-w} of being output, independent of anything else, because each time fresh randomness is being used. Of course, this introduces unreliability, which is related to the probability of drawing the same element from $D_\mathcal{H}$ twice.

Procedure STF.$SE(K, w, m, \mathcal{H}, ctr)$:
  Let $m = m_1 \ldots m_l$, where $|m_i| = w$
  for $i \leftarrow 1$ to l:
    $j \leftarrow 0;\; f \leftarrow 0;\; ctr \leftarrow ctr + 1$
    repeat :
      $j \leftarrow j + 1$
      $s_{i,j} \leftarrow M(\mathcal{H})$
      if $\exists j' < j$ s.t. $s_{i,j} = s_{i,j'}$
        let $c \in_R \{0,1\}^w$
        if $c = m_i$ then $f \leftarrow 1$
      else if $F_K(ctr, s_{i,j}) = m_i$
        then $f \leftarrow 1$
    until $f = 1$
    $s_i \leftarrow s_{i,j};\; \mathcal{H} \leftarrow \mathcal{H}\|s_i$
  output $s = s_1 s_2 \ldots s_l$

Procedure STF.$SD(K, w, s, ctr)$:
  Let $s = s_1 \ldots s_l$, where $s_i \in \Sigma$
  for $i = 1$ to l
    $ctr \leftarrow ctr + 1$
    $m_i \leftarrow F_K(ctr, s_i)$
  output $m = m_1 m_2 \cdots m_l$

Theorem 3. *The stegosystem STF has insecurity* $\mathbf{InSec}^{SS}_{STF(\kappa,w)}(t, d, l, lw) = \mathbf{InSec}^{PRF}_{\mathcal{F}(\kappa)}(t + O(1), d + O(1), l2^w)$. *For each i, the probability that s_i is decoded incorrectly is* $2^{-h+w} + \mathbf{InSec}^{PRF}_{\mathcal{F}(\kappa)}(2^w, O(1), 2^w)$, *and unreliability is at most* $l(2^{-h+w} + \mathbf{InSec}^{PRF}_{\mathcal{F}(\kappa)}(2^w, O(1), 2^w))$.

Proof. Insecurity bound is apparent from the fact that if $\mathcal{F}$ were truly random, then the system would be perfectly secure, because its output is distributed identically to $\mathcal{C}$ (simply because the encoder samples from the channel, and independently at random decides which sample to output, because the random function is never applied more than once to the same input). Hence, any adversary for the stegosystem would distinguish $\mathcal{F}$ from random.

The reliability bound per symbol can be demonstrated as follows. Assuming $\mathcal{F}$ is random, the probability that $s_i = s_{i,j}$ is $(1-2^{-w})^{j-1}2^{-w}$. If that happens, the probability that $\exists j' < j$ such that $s_{i,j} = s_{i,j'}$ is at most $(j-1)2^{-h}$. Summing up and using standard formulas for geometric series, we get

$$\sum_{j=1}^{\infty}(j-1)2^{-h}\left(1-2^{-w}\right)^{j-1}2^{-w} =$$

$$= 2^{-h-w}\sum_{j=1}^{\infty}\left((1-2^{-w})^j\left(\sum_{k=0}^{\infty}(1-2^{-w})^k\right)\right) < 2^{w-h}. \qquad \square$$

Note that errors are independent for each symbol, and hence error-correcting codes over alphabet of size 2^w can be used to increase reliability: one simply encodes m before feeding it to SE. Observe that, for a truly random $\mathcal{F}$, if an error occurs in position i, the symbol decoded is uniformly distributed among all elements of $\{0,1\}^w - \{m_i\}$. Therefore, the stegosystem creates a 2^w-ary symmetric channel with error probability $2^{w-h}(1-2^{-w}) = 2^{-h}(2^w-1)$ (this comes from more careful summation in the above proof). Its capacity is $w - H[1 - 2^{-h}(2^w - 1), 2^{-h}, 2^{-h}, \ldots, 2^{-h}]$ (where H is Shannon entropy of a distribution) [McE02–p. 58]. This is equal to $w+(2^w-1)2^{-h}\log 2^{-h}+(1-2^{-h}(2^w-1))\log(1-2^{-h}(2^w-1))$. Assuming error probability $2^{-h}(2^w - 1) \leq 1/2$ and using $\log(1-x) \geq -2x$ for $0 \leq x \leq 1/2$, we get that the capacity of the channel created by the encoder is at least $w + 2^{-h}(2^w - 1)(-h-2) \geq w - (h+2)2^{-h+w}$. Thus, as l grows, we can achieve rates close to $w - (h+2)2^{-h+w}$ with near perfect security and reliability (independent of h).

4.1 Stateless Variants of STF

Our stegosystem STF is stateful because we need F to take ctr as input, to make sure we never apply the pseudorandom function more than once to the same input. This will happen automatically, without the need for ctr, if the channel $\mathcal{C}$ has the following property: for any histories $\mathcal{H}$ and $\mathcal{H}'$ such that $\mathcal{H}$ is the prefix of $\mathcal{H}'$, the supports of $D_{\mathcal{H}}$ and $D_{\mathcal{H}'}$ do not intersect. For instance, when documents have monotonically increasing sequence numbers or timestamps, no shared state is needed.

To remove the need for shared state for all channels, we can do the following. We remove ctr as an input to F, and instead provide STF.SE with the set Q of all values received so far as answers from M. We replace the line "if $\exists j' < j$ s.t. $s_{i,j} = s_{i,j'}$" with "if $s_{i,j} \in Q$" and add the line "$Q \leftarrow Q \cup \{s_{i,j}\}$" before the end of the inner loop. Now shared state is no longer needed for security, because we again get fresh coins on each draw from the channel, even if it collides with a draw made for a previous hiddentext symbol. However, reliability suffers, because the larger l is, the more likely a collision will happen. A careful analysis, omitted here, shows that unreliability is $l^2 2^{-h+w}$ (plus the insecurity of the PRF).

Unfortunately, this variant requires the encoder to store the set Q of all the symbols ever sampled from $\mathcal{C}$. Thus, while it removes shared state, it requires a lot of private state. This storage can be reduced somewhat by use of Bloom filters [Blo70] at the expense of introducing potential false collisions and thus further decreasing reliability. An analysis utilizing the bounds of [BM02] (omitted here) shows that using a Bloom filter with $(h - w - \log l)/\ln 2$ bits per entry will increase unreliability by only a factor of 2, while potentially reducing storage significantly (because the symbols of Σ require at least h bits to store, and possibly more if the $D_\mathcal{H}$ is sparse).

5 The Stateless Construction STL

The stateless construction STL is simply STF without the counter and collision detection (and is a generalization to rate w of the construction that appeared in the extended abstract of [HLvA02]). Again, we emphasize that the novelty is not in the construction but in the analysis. The construction requires a reliability parameter k, to make sure that expected running time of the encoder does not become infinite due to a low-probability event of infinite running time.

Procedure STL.$SE(K, w, k, m, \mathcal{H})$:
  Let $m = m_1 \ldots m_l$, where $|m_i| = w$
  for $i \leftarrow 1$ to l:
    $j \leftarrow 0$
    repeat :
      $j \leftarrow j + 1$
      $s_{i,j} \leftarrow M(\mathcal{H})$
    until $F_K(s_{i,j}) = m_i$ or $j = k$
    $s_i \leftarrow s_{i,j}$; $\mathcal{H} \leftarrow \mathcal{H}\|s_i$
  output $s = s_1 s_2 \ldots s_l$

Procedure STL.$SD(K, w, s)$:
  Let $s = s_1 \ldots s_l$, where $s_i \in \Sigma$
  for $i = 1$ to l
    $m_i \leftarrow F_K(s_i)$
  output $m = m_1 m_2 \cdots m_l$

Theorem 4. *The stegosystem* STL *has insecurity*

$$\mathbf{InSec}^{SS}_{STL(\kappa,w,k),\mathcal{C}}(t,d,l,lw) \in$$
$$O(2^{-h+2w}l^2 + le^{-k/2^w}) + \mathbf{InSec}^{PRF}_{\mathcal{F}(\kappa)}(t+O(1), d+O(1), l2^w).$$

More precisely,

$$\mathbf{InSec}^{SS}_{STL(\kappa,w,k),\mathcal{C}}(t,d,l,lw) < 2^{-h}\left(l(l+1)2^{2w} - l(l+3)2^w + 2l\right)$$
$$+ 2l\left(1 - \frac{1}{2^w}\right)^k$$
$$+ \mathbf{InSec}^{PRF}_{\mathcal{F}(\kappa)}(t+1, d+O(1), l2^w).$$

Proof. The proof of Theorem 4 consists of a hybrid argument. The first step in the hybrid argument is replace the stegoencoder SE with SE_1, which is the same as SE except that it uses a truly random G instead of pseudorandom F, which accounts for the term $\mathbf{InSec}^{PRF}_{\mathcal{F}(\kappa)}(t+O(1), d+O(1), l2^w)$. Then, rather than consider directly the statistical difference between $\mathcal{C}$ and the output of SE_1 on an

lw-bit message, we bound it via a series of steps involving related stegoencoders (these are not encoders in the sense defined in Section 2, as they do not have corresponding decoders; they are simply related procedures that help in the proof).

We now describe these encoders SE_2, SE_3, and SE_4. SE_2 is the same as SE_1, except that it maintains a set Q of all answers received from M so far. After receiving an answer $s_{i,j} \leftarrow M(\mathcal{H})$, it checks if $s_{i,j} \in Q$; if so, it aborts and outputs "Fail"; else, it adds $s_{i,j}$ to Q. It also aborts and outputs "Fail" if j ever reaches k during an execution of the inner loop. SE_3 is the same as SE_2, except that instead of thinking of random function G as being fixed before hand, it creates G "on the fly" by repeatedly flipping coins to decide the w-bit value assigned to $s_{i,j}$. Since, like SE_2, it aborts whenever a collision between strings of covertexts occurs, the function will remain consistent. Finally, SE_4 is the same as SE_3, except that it never aborts with failure.

In a sequence of lemmas, we bound the statistical difference between the outputs of SE_1 and SE_2; show that it is the same as the statistical difference between the outputs of SE_3 and SE_4; and show that the outputs of SE_2 and SE_3 are distributed identically. Finally, observe that SE_4 does nothing more than sample from the channel and then randomly and obliviously to the sample keep or discard it. Hence, its output is distributed identically to the channel. The details of the proof are contained in the full version [DIRR04]. □

Theorem 5. *The stegosystem* STL *has unreliability*

$$\mathbf{UnRel}^{SS}_{\mathrm{STL}(\kappa,w,k),\mathcal{C},l} \leq$$
$$l\left(2^w \exp\left[-2^{h-2w-1}\right] + \exp\left[-2^{-w-1}k\right]\right) + \mathbf{InSec}^{PRF}_{\mathcal{F}(\kappa)}(t, d, l2^w),$$

where t and d are the expected running time and description size, respectively, of the stegoencoder and the stegodecoder combined.

Proof. As usual, we consider unreliability if the encoder is using a truly random G; then, for a pseudorandom F, the encoder and decoder will act as a distinguisher for F (because whether something was encoded correctly can be easily tested by the decoder), which accounts for the $\mathbf{InSec}^{PRF}$ term.

Now, fix channel history $\mathcal{H}$ and w-bit message m, and consider the probability that $G(D_\mathcal{H})$ is so skewed that the weight of $G^{-1}(m)$ in $D_\mathcal{H}$ is less $c2^{-w}$ for some constant $c < 1$ (note that the expected weight is 2^{-w}). Let $\Sigma = \{s_1 \ldots s_n\}$ be the alphabet, and let $\Pr_{D_\mathcal{H}}[s_i] = p_i$. Define random variable X_i as $X_i = 0$ if $G(s_i) = m$ and $X_i = p_i$ otherwise. Then the weight of $G^{-1}(m)$ equals $1 - \sum_{i=1}^n X_i$. Note that the expected value of $\sum_{i=1}^n X_i = 1 - 2^{-w}$. Using Hoeffding's inequality (Theorem 2 of [Hoe63]), we obtain

$$\Pr[1 - \sum_{i=1}^n X_i \leq cR] \leq \exp\left[-2(1-c)^2 2^{-2w} / \sum_{i=1}^n p_i^2\right]$$
$$\leq \exp\left[-2(1-c)^2 2^{-2w} / 2^{-h} / \sum_{i=1}^n p_i\right]$$
$$= \exp\left[-2(1-c)^2 2^{h-2w}\right],$$

where the second to last step follows from $p_i \leq 2^{-h}$ and the last step follows from $\sum_{i=1}^{n} p_i = 1$. If we now set $c = 1/2$ and take the union bound over all message $m \in \{0,1\}^w$, we get $2^w \exp\left[-2^{h-2w-1}\right]$.

Assuming $G(D_\mathcal{H})$ is not so skewed, the probability of failure is

$$(1 - c2^{-w})^k \leq \exp\left[-c2^{-w}k\right] .$$

The result follows from the union bound over l. □

Acknowledgements

We are grateful to Nick Hopper for clarifying related work.

The authors were supported in part by the National Science Foundation under Grant No. CCR-0311485. Scott Russell's work was also facilitated in part by a National Physical Science Consortium Fellowship and by stipend support from the National Security Agency.

References

[BC04]   Michael Backes and Christian Cachin. Public-key steganography with active attacks. Technical Report 2003/231, Cryptology e-print archive, http://eprint.iacr.org, 2004.

[Blo70]   B. Bloom. Space/time tradeoffs in hash coding with allowable errors. *Communications of the ACM*, 13(7):422–426, July 1970.

[BM02]   A. Broder and M. Mitzenmacher. Network applications of bloom filters: A survey. In *Proceedings of the Fortieth Annual Allerton Conference on Communication, Control and Computing*, 2002.

[Cac98]   C. Cachin. An information-theoretic model for steganography. In *Second Internation Workshop on Information Hiding*, volume 1525 of *Lecture Notes in Computer Science*, pages 306–316, 1998.

[DIRR04] Nenad Dedić, Gene Itkis, Leonid Reyzin, and Scott Russell. Upper and lower bounds on black-box steganography. Technical Report 2004/246, Cryptology e-print archive, http://eprint.iacr.org, 2004.

[GGM86] Oded Goldreich, Shafi Goldwasser, and Silvio Micali. How to construct random functions. *Journal of the ACM*, 33(4):792–807, October 1986.

[GGN03] Oded Goldreich, Shafi Goldwasser, and Asaf Nussboim. On the implementation of huge random objects. In *44th Annual Symposium on Foundations of Computer Science*, pages 68–79, Cambridge, Massachusetts, October 2003.

[HILL99] J. Håstad, R. Impagliazzo, L.A. Levin, and M. Luby. Construction of pseudorandom generator from any one-way function. *SIAM Journal on Computing*, 28(4):1364–1396, 1999.

[HLvA02] N. Hopper, J. Langford, and L. von Ahn. Provably secure steganography. Technical Report 2002/137, Cryptology e-print archive, http://eprint.iacr.org, 2002. Preliminary version in Crypto 2002.

[Hoe63]   W. Hoeffding. Probability inequalities for sums of bounded random variables. *Journal of the American Statistical Association*, 58(301):13–30, March 1963.

[Hop04] Nicholas J. Hopper. *Toward a Theory of Steganography*. PhD thesis, Carnegie Mellon University, Pittsburgh, PA, USA, July 2004. Available as Technical Report CMU-CS-04-157.

[KMR02] Lea Kissner, Tal Malkin, and Omer Reingold. Private communication to N. Hopper, J. Langford, L. von Ahn, 2002.

[Le03] Tri Van Le. Efficient provably secure public key steganography. Technical Report 2003/156, Cryptology e-print archive, http://eprint.iacr.org, 2003.

[LK03] Tri Van Le and Kaoru Kurosawa. Efficient public key steganography secure against adaptively chosen stegotext attacks. Technical Report 2003/244, Cryptology e-print archive, http://eprint.iacr.org, 2003.

[McE02] Robert J. McEliece. *The Theory of Information and Coding*. Camridge University Press, second edition, 2002.

[Rey04] Leonid Reyzin. A Note On the Statistical Difference of Small Direct Products. Technical Report BUCS-TR-2004-032, CS Department, Boston University, September 21 2004. Available from http://www.cs.bu.edu/techreports/.

[Sim83] G. J. Simmons. The prisoners' problem and the subliminal channel. In David Chaum, editor, *Advances in Cryptology: Proceedings of Crypto 83*, pages 51–67. Plenum Press, New York and London, 1984, 22–24 August 1983.

[vAH04] Luis von Ahn and Nicholas J. Hopper. Public-key steganography. In Christian Cachin and Jan Camenisch, editors, *Advances in Cryptology— EUROCRYPT 2004*, volume 3027 of *Lecture Notes in Computer Science*. Springer-Verlag, 2004.

A On Using Public ε-Biased Functions

Many stegosystems [HLvA02, vAH04, BC04] (particularly public-key ones) use the following approach: they encrypt the plaintext using encryption that is indistinguishable from random, and then use rejection sampling with a public function $f : \Sigma \to \{0,1\}^w$ to stegoencode the plaintext.

For security, f should have small bias on $D_\mathcal{H}$: i.e., for every $c \in \{0,1\}^w$, $\Pr_{s \in D_\mathcal{H}}[s \in f^{-1}(c)]$ should be close to 2^{-w}. It is commonly suggested that a universal hash function with a published seed (e.g., as part of the public key) be used for f.

Assume the stegosystem has to work with a memoryless channel $\mathcal{C}$, i.e., one for which the distribution D is the same regardless of history. Let E be the distribution induced on Σ by the following process: choose a random $c \in \{0,1\}^w$ and then keep choosing $s \in D$ until $f(s) = c$. Note that the statistical difference between D and E is exactly the bias ε of f. We are interested in the statistical difference between D^l and E^l.

For a universal hash function f that maps a distribution of min-entropy h to $\{0,1\}^w$, the bias is roughly $\varepsilon = 2^{(-h+w)/2}$. As shown in [Rey04], if $l < 1/\varepsilon$ (which is reasonable to assume here), statistical difference between D^l and E^l is roughly at least $\sqrt{l}\varepsilon$.

Hence, the approach based on public hash functions results in statistical insecurity of about $\sqrt{l}2^{(-h+w)/2}$.

Fair-Zero Knowledge

Matt Lepinski, Silvio Micali, and Abhi Shelat

Massachusetts Institute of Technology, Cambridge MA 02114, USA
{lepinski, silvio, abhi}@csail.mit.edu
http://crypto.csail.mit.edu/~abhi

Abstract. We introduce *Fair Zero-Knowledge*, a multi-verifier ZK system where every proof is guaranteed to be "zero-knowledge for all verifiers." That is, if an honest verifier accepts a fair zero-knowledge proof, then he is assured that all other verifiers also learn nothing more than the verity of the statement in question, even if they maliciously collude with a cheating prover.

We construct Fair Zero-Knowledge systems based on standard complexity assumptions (specifically, the quadratic residuosity assumption) and an initial, one-time use of a physically secure communication channel (specifically, each verifier sends the prover a private message in an envelope). All other communication occurs (and must occur) on a broadcast channel.

The main technical challenge of our construction consists of provably removing any possibility of using *steganography* in a ZK proof. To overcome this technical difficulty, we introduce tools —such as Unique Zero Knowledge— that may be of independent interest.

1 Introduction

A New Worry. A traditional zero-knowledge proof enjoys two crucial properties, *soundness* and *zero knowledge*, each guarding the interests of mutually cautious parties. Soundness protects the verifier: a malicious prover has practically no chance to convince the verifier of a false statement. Zero knowledge protects the prover: a malicious verifier has practically no chance of learning anything about the statement in question beyond the fact that it is indeed true.

A new threat emerges, however, when there are *multiple* verifiers. In such a situation, a malicious prover may *collude* with some of the verifiers by generating proofs that convey additional information to them while remaining zero-knowledge to all others. Indeed, an honest verifier that accepts a ZK proof of a given theorem learns nothing more that the verity of the theorem statement in question, but *can he be sure that the same holds for his "colleagues?"*

Notice that the traditional definition of a zero-knowledge proof is *orthogonal* to the above concern. Let us illustrate this point by constructing the following (somewhat artificial) NIZK proof system, (P', V') —which uses as a subroutine (P, V), the original NIZK proof system of [BSMP91].

> P' initially chooses (PK, SK), the public and secret key of a uniquely decryptable public-key cryptosystem. Later on, whenever it receives as an input

a member x of a NP-language L together with a witness w for $x \in L$, P' first computes w', an encryption of w relative to PK, and then outputs a proof string π' which consists of (1) PK, (2) w', and (3) a NIZK proof — according to (P, V) and some common reference string σ — of the statement "there is a decryption key corresponding to PK (i.e., SK) such that after decrypting w' with said key, one obtains a witness for $x \in L$."

Clearly each such π' is accepted by (and is zero-knowledge for) all honest verifiers. But, a malicious prover P' may, without notice, ensure that it is "much more informative" for some colluding verifiers. There are very subtle ways for him to accomplish this, but the simplest one consists of having P' provide each colluding verifier with SK, so that each subsequent π' reveals the corresponding witness w in its entirety to each colluding verifier!

A New Goal. We wish to build a multi-verifier zero-knowledge system that is provably *fair*. That is, we wish to guarantee that whenever an honest verifier accepts a fair ZK proof, then he is assured that all other verifiers too (whether honest or colluding with a malicious prover) learn nothing more than the verity of the statement in question. In other words, we wish to *extend the zero-knowledgeness property of a ZK system to protect also the Verifier(s) and not just the Prover!*

A Motivating Example. In repeated auctions of similar items,[1] it may be desirable that all bids in an individual auction (including the winning bid) remain secret in subsequent ones. This goal appears to be a golden opportunity for encryption and zero-knowledge proofs, but special care must be taken. The following example illustrates.

A closely-watched auctioneer possessing a public encryption key PK sells a series of n lithographs from the same etching by repeating the following two-step process, once for each lithograph. First, the bidders publicly announce their individual bids encrypted with PK; second, the auctioneer proves in zero knowledge who the winner of the current lithograph is. At the very end, the auctioneer privately collects all right amounts from the winners.

Using zero-knowledge proofs in Step 2 aims at providing the minimum amount of knowledge enabling the bidders to decide whether they should continue bidding in subsequent auctions. At a superficial level, this aim seems to be achieved: First, if the auctioneer is honest, then *standard* zero-knowledge proofs guarantee that no additional bid information is leaked prematurely. Second, even if the auctioneer were dishonest, by virtue of being closely-watched he could not use any "side channels" to divulge additional bid information to a selected subset of the bidders. A better analysis, however, shows that our new worry naturally arises in this setting: no matter how closely watched, a dishonest auctioneer might use

---

[1] A well-studied problem in economics.

the ZK proof *itself* as a mechanism to leak bid information to a colluding bidder (thus giving him an advantage in later rounds).

In sum, a *standard* ZK proof of who is winner of an individual auction is not enough here: what is really needed is a *Fair* ZK proof!

Fair Zero Knowledge. Syntactically, Fair ZK is a two-phase process. The first phase consists of a *preprocessing* protocol, where various quantities (e.g., public keys) are established, and private channels are (seemingly necessarily) used. The second phase consists of a *proving* protocol, where it is imperative that the Prover be restricted to communicate via broadcast only. (Were a malicious prover connected to some colluding verifiers via private channels during the proving phase, it would be impossible to prevent the selective dissemination of witness information!) In a sense, Prover and Verifiers might execute the preprocessing phase on Earth, but for the proving phase, Prover is sent to the moon from where anything he says is heard by all Verifiers.

Semantically, Fair Zero Knowledge guarantees that, for any NP-theorem that (1) has a single witness and (2) is chosen after preprocessing ends, the prover cannot undetectably communicate anything more than the truthfulness of the theorem in question to any verifier, no matter what arrangements they might have made beforehand.

Postponing for a moment a discussion of our "unique-witness constraint," notice that Fair Zero Knowledge does not provide any guarantees for theorems whose witnesses are known beforehand to the prover. In this case, the (to be) prover could have already divulged witness information to anyone he wanted, and protocol designers have no responsibility for what happens before the protocol starts![2]

A bit more precisely, assume that, for a unique-witness NP-language L and for all $i = 1, 2, \ldots$, (a) $x_i \in L$ is chosen on-line and its statement made public, and (b) w_i, the witness of x_i, is privately given to the Prover. Then, Fair ZK enables the Prover to prove to a set of verifiers that every x_i, *individually* and *in order*, indeed belongs to L, so that

1. *When the Prover is honest, no set of malicious verifiers can learn anything more than the mere sequence of statements "$x_i \in L$ and it is the ith theorem;" and*
2. *No dishonest prover can convey any knowledge other than $x_i \in L$ to any colluding verifier —no matter how much information they secretly share up to the last round of preprocessing— without being detected by all honest verifiers.*

---

[2] In this extended abstract we wish to keep the notion of Fair ZK as simple as possible. In the final version, however, we shall allow colluding verifiers to have prior partial information y about a proven theorem x, and extend Fair ZK to ensure that they cannot get any knowledge from the proofs beyond y and x's truthfulness. The difficulty of doing this right lies in the fact that a malicious prover himself may know some other information z about x beforehand, including what information y some verifiers know.

We are actually able to construct a non-interactive version of Fair ZK under physical-channel and standard complexity assumptions. Namely, under the quadratic residuosity assumption, there exists a Fair ZK proof system whose prover —after a preprocessing protocol in which he receives an envelope from each verifier— proves every theorem x_i non-interactively by broadcasting a proof string π_i.

Technical Challenges. There are two obstacles in constructing Fair ZK.

The first, and main obstacle is *preventing steganography*. The prover of any zero knowledge proof must be probabilistic, and this very probabilism ushers in the ability for a malicious prover to communicate with colluding verifiers over "subliminal channels." That is, a malicious prover can use the probabilism of a ZK proof to convey steganographically, to a subset of verifiers, witness information about any NP theorem —even one chosen after he has been "sent to the moon." For example, in the above repeated-auction process, all theorems are of the type "the winner of the current auction is bidder X", and thus are generated after the prover (i.e., the auctioneer) starts being closely watched. Yet, a malicious auctioneer may agree beforehand with a subset of colluding bidders on a way to encode the actual amount of the highest bid in the bits he broadcasts in the corresponding ZK proof.[3] In fact, as early as 1987, Desmedt, Goutier and Bengio [DGB87] illuminated this problem by showing that a particular zero-knowledge protocol due to Fiat and Shamir can easily be used as a perfect subliminal channel. More generally, Langford, Hopper, and von Ahn [NHvA02] show that whenever there is entropy, steganographic communication —provably undetectable by honest parties— always exists.

Perhaps surprisingly, in light of their result, we show how to *provably prevent steganography in our context*. In our approach, we construct a novel type of ZK system, uniZK, in which the prover's probabilism is confined to a preprocessing phase after which not only is he made totally deterministic, but *his determinism is actually made universally verifiable*. Very roughly, in a uniZK system the prover first establishes a suitable public key, so that, for any NP-theorem x having a single witness, it is universally verifiable that there exists a single ZK way to prove x that is acceptable by an honest verifier. Such "unique provability" therefore *provably bans steganography from* uniZK *proofs*.

Note that in our application we need *verifiable determinism*, and not just determinism. Naively, one might consider constructing a uniZK system by replacing the probabilistic prover of any NIZK system with one who chooses a short random seed for a pseudo-random function [GGM86] and acts deterministically ever after. However, while this would be conceptually simple to do, it would also be impossible for an efficient verifier to *check* that the prover indeed behaved in such a fashion instead of flipping new coins for each proof. Thus, an honest verifier may not be convinced that a malicious prover is not steganographically

---

[3] For example, a naive approach is to make the first 20 bits of the proof the same as those of the winning bid.

conveying additional witness information to colluding verifiers. In sum, prover determinism might be easy, but verifiable prover determinism is not!

After so overcoming steganography, a second obstacle remains in building Fair Zero Knowledge. While, in a uniZK system, knowledge of the prover's public key ensures that there is only one acceptable proof, knowledge of the corresponding secret key may enable anyone to read off the entire witness from such a proof! Thus, we must ensure that the prover is the only one possessing knowledge of his secret key. If the generation of his public-secret key pair were totally up to him, however, this would be impossible, because the prover and his accomplices may agree on which public-secret keys he will choose. Instead, we show how to generate and distribute the prover's keys by a protocol involving all verifiers so that, as long as there is one honest verifier, then only the prover will know the resulting secret key. It is in this subprotocol that we make a single use of a physically secure communication channel: namely every verifier sends to the prover a single message in *an envelope*. After this, all communication (in the prover-key-generation subprotocol and in all subsequent uniZK proofs) is via broadcast.

More on Preprocessing. Our protocol and even our definition of Fair ZK includes preprocessing. The reason for this is that Zero Knowledge, as any secure protocol, requires randomness and (as discussed above) any amount of entropy enables undetectable steganography, which defeats fairness. We prove, however, that we can confine the necessary entropy "somewhere" where it is actually "innocuous". Such a place is our preprocessing phase. Though steganography may be rampant during preprocessing, it is also useless since the theorems to be proved in zero-knowledge have not yet been selected – and thus no information about their proofs can be conveyed. Nor can such information be conveyed afterward preprocessing, since all communication after preprocessing is via broadcast and verifiably unique!

More On Envelopes. Usage of a physically secure channel is crucial to our preprocessing. In our application, it is unclear how to simulate these channels by an "encrypt-and-broadcast" process, since such methodology must start with the prover choosing a suitable encryption key, and he could always choose a key whose corresponding secret key is already known to his accomplices. In such a case, any message sent encrypted to the prover by an honest verifier will be understood by a dishonest one, defeating the very reason for encrypting it. By delivering a message to the prover in an envelope, however, honest verifiers are guaranteed that the message will indeed remain secret to any malicious verifier, thus "dividing the state of knowledge of the prover from that of the verifier" at a specific moment of the protocol. (The protocol must then ensure —e.g., via "steganography-free broadcasting"— that these divided states of knowledge will indeed continue to remain so!)

But if physically secure channels must be used, why envelopes rather than traditional private channels? The point is that traditional private channels are "bidirectional." We instead need to prevent a malicious prover, after receiving a private message M from an honest verifier along a physically secure channel, from forwarding M to a colluding verifier along another, similar channel. Thus, we require mono-directional channels from the verifiers to the prover. Envelopes in our protocol are just good (and well known!) examples of mono-directional physically secure channels. Let us remark that, since envelopes may be more inconvenient than broadcasting in many a setting, it is a feature of our protocol that envelope communication is confined to a single round!

More on Witness Uniqueness. Let us now explain why we define Fair ZK for NP-languages whose members have a unique witness. We allow a Fair ZK proof to depend (via an underlying uniZK proof) on the given input witness w. Thus, if the prover knows two or more witnesses for $x \in L$, he can have "a multiplicity of Fair ZK proofs to choose from," which would again enable steganographic communication. For instance, the NP-complete language of 3-colorability appears to be unsuitable for Fair ZK proofs as we define them, because from any coloring of a graph one can immediately compute 5 more colorings by just permuting the three colors!

Note, however, that our unique NP-witness requirement is often automatically satisfied in cryptographic applications. This is so because underlying complexity problems (e.g., integer factorization, discrete logarithm, etc.) often have unique solutions, and appropriate NP reductions can be used so as to "preserve such uniqueness." For example, the desired ZK proofs of our motivating example are for unique-witness languages, because all bids are encrypted by means of a uniquely decryptable cryptosystems.

More generally, Fair ZK actually applies to *computationally unique-witness languages*, that is, to languages for which it is hard for the prover to generate a second witness from a first one. (For example, this encompasses statements which refer to most computationally binding commitment schemes.) In sum therefore, this enlarged constraint is very mild in a cryptographic setting, making Fair ZK widely applicable.

Notice, that while Fair ZK is quite meaningful when applied to NP-languages having computationally unique witnesses, uniZK can be meaningfully defined for all NP-languages.[4] Thus, in the next section we define uniZK for all NP-languages, and then, in Section 3, we define Fair ZK only for those languages having computationally unique witnesses.

2 Notation

We shall follow, verbatim, [BSMP91] and [GMR88]. A function $\mu(\cdot)$ from non-negative integers to reals is called *negligible* if for every constant $c > 0$ and all sufficiently large n, $\mu(n) < n^{-c}$. An *efficient* algorithm is a probabilistic algorithm running in expected polynomial time. If S is a probability space,

---
[4] Essentially, as we shall see, " any witness efficiently maps to a uniZK proof, and vice versa."

then "$x \leftarrow S$" denotes the probabilistic algorithm consisting of choosing an element x at random according to S and returning x. If p is a predicate, then the notation "$x \leftarrow S|p(x)$" denotes the assignment consisting of choosing an element x at random according to S, and returning the first x such that $p(x)$ is true. Let $S_1, S_2, \ldots$ be probability spaces, then the notation $\Pr[x_1 \leftarrow S_1;\ x_2 \leftarrow S_2;\ \ldots\ :\ p(x_1, x_2, \ldots)]$ denotes the probability that the predicate $p(x_1, x_2, \ldots)$ is true after the ordered execution of the assignments $x_1 \leftarrow S_1;\ x_2 \leftarrow S_1;\ \ldots$ If $S, T, \ldots$ are probability spaces, the notation $\{x \leftarrow S; y \leftarrow T; \cdots\ :\ (x, y, \cdots)\}$ denotes the new probability space over $\{(x, y, \cdots)\}$ generated by the ordered execution of the assignments $x \leftarrow S,\ y \leftarrow T, \cdots$.

3 Unique Non-interactive Zero-Knowledge

We define Unique Non-interactive Zero-Knowledge (uniZK) proofs as special types of NIZK proofs. Thus, we begin this section with a review of NIZK, and then proceed to give a precise formalization and construction of uniZK.

NIZK, in a Nut Shell. An NIZK proof system [BFM88] [BSMP91] for a NP-language L consists of a pair of efficient algorithms, a prover P and a verifier V, and a public, random string, σ, called the *reference string*. When proving that the statement "x is a member of L", it is assumed that P is also privately given a witness w for $x \in L$. The proof process is extremely simple: P computes a single string π (for proof), $\pi = P(x, w, \sigma)$, and sends it to V. The verifier, on inputs x, σ and proof string π accepts or rejects, without having to reply to P (hence, non-interactively). This process can be repeated, with the same reference string, for an unbounded number of theorems (i.e., members of L).

Semantically, an NIZK satisfies the usual ZK properties of *Completeness*, *Soundness* and *Zero Knowledgeness*. In this non-interactive setting, completeness means that, for every reference string and every genuine member of L, the verifier accepts all honestly generated proofs. soundness means that, for most reference strings, no acceptable "proof" π^* exists for any $x^* \notin L$. Zero-Knowledgeness means that there exists an efficient simulator S that first generates a reference string σ' and then, for any sequence of theorems, $x_i, x_2, \ldots$, (and without any witness information) generates strings $\pi'_1, \pi'_2, \ldots$, such that the sequence $\sigma', \pi'_1, \pi'_2, \ldots$ is indistinguishable from the sequence consisting of a random reference string followed by the proofs that an honest prover would generate for the same theorem sequence —with the proper witness information!

The construction of [BSMP91] actually satisfies (but does not claim) a stronger notion of Zero Knowledgeness, that was put forward in [FLS90]. Namely, the simulator S (rather than being given the sequence of theorems $x_1, x_2, \ldots$ up-front) must produce each string π_i knowing theorem x_i but not future ones. (Thus, although we adopt this stronger notion of zero knowledgeness for uniZK, we can base our uniZK construction on the NIZK system of [BSMP91].)

Adding Verifiably Unique Provability to NIZK. As anticipated in the Introduction, we wish to define uniZK for all NP-languages (rather than for those having computationally unique witnesses). We do so by demanding that, for any $x \in L$, any prover —honest or malicious— "may produce a single uniZK proof for every witness he knows." How can this be formalized?

The easiest way would be demanding that, every $x \in L$, no matter how many witnesses it may have, has a single uniZK proof. Unfortunately, no such uniZK system may exist. (We certainly do not know how to construct one.)

A second way might be demanding the existence of a unique uniZK proof for each NP-witness. Unfortunately, relative to our steganography-free goals, such a definition may not be sufficiently meaningful, because it leaves open the possibility for a malicious prover to choose from a multiplicity of uniZK proofs by "rewriting" then. Assume that an efficient, malicious prover P' were given a witness w of a theorem x belonging to an NP-language L with computationally unique witnesses. Then, w would be the only witness of $x \in L$ known to P', and by Completeness, P' could certainly produce one uniZK proof, π_w. But now, if from π_w one could also compute additional uniZK proofs for $x \in L$, P' could compute a multiplicity of uniZK proofs for $x \in L$ from a single witness!

We thus formalize uniZK by demanding that (for most reference strings σ and public keys PK) the honest algorithm P forms an *easy-to-invert bijection* between the witness set of $x \in L$ (denoted W_x) and the set of acceptable uniZK proofs (denoted $\Pi_{PK}(x, \sigma)$). This captures the notion that any prover "can only produce a single uniZK proof for any witness he knows:" his ability to produce multiple uniZK proofs from a single witness can solely originate from his ability of producing multiple witnesses from a single one.

To complete our formalization, we must handle the case of a cheating prover who posts an invalid public key PK^*; that is, a key that does not pass a proper inspection of a honest verifier. In this case, it is reasonable for the verifier to reject any subsequent proof: after all, he knows for certain that the prover is malicious! Therefore, our definition requires that either the set of acceptable proofs $\Pi_{PK^*}(x, \sigma)$ is empty, or else there exists a secret key SK^* such that $P(x, \cdot, \sigma, SK^*)$ forms an efficient bijection from W_x to $\Pi_{PK^*}(x, \sigma)$. For this to be meaningful, however, such SK^* should be unique, that is, there must be a function sk (possibly hard to compute) mapping any "reasonable looking" public key PK^* to the right SK^*.

In sum, our definition states that unless $\Pi_{PK^*}(x, \sigma)$ is empty, $P(x, \cdot, \sigma, sk(PK^*))$ forms an efficient bijection from W_x to $\Pi_{PK^*}(x, \sigma)$.

3.1 Formal Definition

Let L be an NP language, and R_L be its corresponding, polynomial-time relation. We say that a sequence of pairs of strings, $(x_1, w_1), (x_2, w_2), \ldots$, is a *theorem-witness sequence for L* if each $x_i \in L$ and $w_i \in R_L(x_i)$.

Definition 1. *A triple of efficient algorithms, (G, P, V), where P is deterministic, is a* unique non-interactive zero-knowledge (uniZK) proof system *for an NP-language L if there exists a positive constant c and a negligible function μ such that the following properties are satisfied:*

Completeness: $\forall$ theorem-witness sequences $(x_1, w_1), (x_2, w_2), \ldots$ for L, and for all $k > 2$

$$\Pr\left[\begin{array}{l}(PK,SK) \leftarrow G(1^k);\, \sigma \leftarrow \{0,1\}^{k^c};\, \pi_1 = P(x_1, w_1, \sigma, SK, 1); \\ \pi_2 = P(x_2, w_2, \sigma, SK, 2) \ldots : \bigwedge_i V(x_i, \sigma, PK, \pi_i, i) = 1\end{array}\right] = 1$$

Soundness: $\forall k > 2$ and $\forall$ algorithms P^*

$$\Pr\left[\sigma \leftarrow \{0,1\}^{k^c}; (x^*, PK^*, \pi^*, i) \leftarrow P^*(\sigma) : x^* \notin L \wedge V(\sigma, x^*, PK^*, \pi^*, i) = 1\right] < \mu(k)$$

Zero-Knowledgeness: $\exists$ an efficient algorithm S such that $\forall$ theorem-witness sequences $(x_1, w_1), (x_2, w_2), \ldots$ for L, the following two ensembles are computationally indistinguishable:

$$\left\{\begin{array}{l}(PK,SK) \leftarrow G(1^k);\, \sigma \leftarrow \{0,1\}^{k^c};\, \pi_1 = P(x_1, w_1, \sigma, SK, 1); \\ \pi_2 = P(x_2, w_2, \sigma, SK, 2) \ldots : (\sigma, PK, \pi_1, \pi_2, \ldots)\end{array}\right\}_k$$

$$\left\{\begin{array}{l}(PK', SK', \sigma') \leftarrow S(1^k);\, \pi_1' \leftarrow S(SK', x_1, 1); \\ \pi_2' \leftarrow S(SK', x_2, 2), \ldots : (\sigma', PK', \pi_1', \pi_2', \ldots)\end{array}\right\}_k$$

Uniqueness: $\exists$ a deterministic function $sk(\cdot)$ and an efficient deterministic algorithm P^{-1} such that $\forall x \in L$, $\forall i > 0$, and $\forall PK^* \in \{0,1\}^*$,

$$\Pr\left[\begin{array}{l}\sigma \leftarrow \{0,1\}^{k^c};\, (|\Pi^i_{PK^*}(x, \sigma)| > 0) \Rightarrow \\ \quad P(\sigma, x, \cdot, sk(PK^*), i) : W_x \xrightarrow{1-1} \Pi^i_{PK^*}(x, \sigma) \wedge \\ \quad P^{-1}(\sigma, x, \cdot, sk(PK^*), i) : \Pi^i_{PK^*}(x, \sigma) \xrightarrow{1-1} W_x\end{array}\right] > 1 - \mu(k)$$

where $W_x = \{w : w \in R_L(x)\}$ and $\Pi^i_{PK^*}(x, \sigma) = \{\pi : V(x, \sigma, PK', \pi, i) = 1\}$.

3.2 Constructing uniZK

We can construct a uniZK system based on the hardness of the quadratic residuosity problem[GM84], for Blum integers, by modifying the protocol of Blum, De Santis, Micali and Persiano [BSMP91]. [5]

Theorem 1. *If quadratic residuosity is hard, then there exist* uniZK *systems for 3SAT.*

Proof Sketch: The key generator, $G(1^k)$, produces a public key consisting of a randomly selected k-bit Blum integer, x, and a quadratic non-residue, y mod x. We denote the tuple (x, y) as a *proving pair*. The secret key consists of the factorization of x.

Let $(a_1, \ldots, a_m)$ be a tuple of k-bit integers that have Jacobi symbol 1 mod x. If $(b_1, \ldots, b_m)$ is tuple of bits then we say that $(a_1, \ldots, a_m)$ has type

---

[5] We can also make a uniZK system for CIRCUIT-SAT by combining the single-theorem protocol of Damgård [DAM92] with the multi-theorem techniques of Blum, De Santis, Micali and Persiano.

$(b_1, \ldots, b_m)$ if each a_i is a square mod x if and only if b_i is 0. If $(c_1, \ldots, c_m)$ is a tuple of k-bit integers then we say that $(a_1, \ldots, a_m)$ and $(c_1, \ldots, c_m)$ have the *same type* if a_i is a square mod x if and only if c_i is a square mod x.

A prover who knows the factorization of x can prove that the tuple $(a_1, \ldots, a_m)$ has type $(b_1, \ldots, b_m)$ by providing, for each i, a square root of $a_i y^{b_i}$ mod x. Similarly, a prover can prove that $(a_1, \ldots, a_m)$ and $(c_1, \ldots, c_m)$ have the same type by providing, for each i, a square root of $a_i c_i$ mod x. To make these proofs unique, whenever the prover provides a square root, he provides the Jacobi-symbol 1 square root which is less than $n/2$. (Since x is a Blum integer, there is exactly one such root for every quadratic residue.) The verifier rejects any proof in which a different square root is provided.

Following [BSMP91], we first present a proving algorithm, P, for the single theorem case. Let 3-SAT be the language of satisfiable boolean 3-CNF formulas. Let $\phi \in 3-SAT$ be a theorem with m clauses and variables $v_1, \ldots, v_n$ and let w be a satisfying assignment for ϕ.

1. Break the reference string into two parts, ρ and τ where $|\rho| = 16k^3$ and $|\tau| = 64k^2 n + 48k^3 m$.
2. Parse ρ into k-bit integers; skip any values that are greater than x or have Jacobi symbol -1.
3. Prove that each of the remaining k-bit integers in ρ has either type 0 or type 1 by giving a square root mod x or a square root of it times y mod x. As in [BSMP91], this proves that (x, y) is a properly-formed *proving pair*, that is, that x is a Blum Integer and y is a quadratic non-residue mod x.
4. Parse τ into k-bit integers as in Step 2.
5. Acquire n pairs of k-bit integers such that each pair is either of type $(1, 0)$ or type $(0, 1)$. To do this, parse a section of τ as $8kn$ pairs. Then for each pair (s, t) (in order) either give a square root of st mod x and discard the pair or give a square root of sty mod x and select the pair. Once n pairs have been selected, discard any remaining pairs.
6. Now define a value u_i corresponding to each variable v_i in ϕ as follows: let u_i be the quadratic residue in the ith pair acquired in Step 5 if v_i is false in w, and to the non-residue in the pair otherwise.
7. Let v_d, v_e and v_f be the three variables that appear in clause j of ϕ. For each clause j of ϕ, form a triple (a_j, b_j, c_j) where a_j is equal to u_d if v_d appears non-negated in the clause or to the product of u_d and y mod x otherwise. The values b_j and c_j are analogously defined.
8. Parse the remaining portion of τ as $8k^2 m$ triples of k-bit integers. Among the jth set of $8k^2$ triples, select 8 triples that all have different types as follows: within a set of $8k^2$ triples, inspect each triple in order and either *select it* or provide a proof that it is of the same type as a previously selected triple. If at the end, 8 triples have been selected, then either all 8 triples are of different types, or one type did not occur within the set at all. In the former case, prove that one of the selected triples has type $(0, 0, 0)$ and discard it. Denote the remaining 7 *selected* triples as $((\alpha_j^1, \beta_j^1, \gamma_j^1), \ldots, (\alpha_j^7, \beta_j^7, \gamma_j^7))$.

9. Finally, for each j, show that for some $1 \leq t \leq 7$, (a_j, b_j, c_j) is of the same type as $(\alpha_j^t, \beta_j^t, \gamma_j^t)$. Note, this proves that the clause is satisfied since the identified triple $(\alpha_j, \beta_j, \gamma_j)$ is not of type $(0, 0, 0)$.

In the following, we refer to the portion of π generated by step I in the honest prover algorithm as π_I. The single-theorem verifier algorithm, V, proceeds as follows.

1. Run the honest-prover algorithm as per step 1, 2, 4 and 7 to generate π_2, π_4, π_7 and verify that the corresponding proof string parts are equivalent. Also verify that every root given in the proof string has Jacobi symbol 1 and is less than $n/2$. Reject if not.
2. As per [BSMP91], verify π_3, which is the proof that (x, y) is well-formed.
3. Verify π_5 by making sure that each pair is handled, and that the proof string contains a proper root of the pair.
4. Verify π_8 by checking that for each set of triples, the prover has handled the pairs in order, and that each of the proofs given between triples is sound. Finally, verify that the opened pair is of type $(0, 0, 0)$
5. For each clause, verify the proof that it is associated with one of it's remaining seven selected triples.

As in [BSMP91], we now transform the single theorem system to a multiple theorem one by breaking the random string into three pieces, ρ, τ_1 and τ_2. We use ρ to prove that (x_0, y_0) in a proper proving pair[6]. This is done exactly as in Step 3. At this point, x_0 and y_0 can be used with τ_2 to prove the first theorem as in the single theorem case (starting from Step 4 since the correctness of (x_0, y_0) has already been established).

At this point, our construction diverges from [BSMP91]. Originally, for the second theorem, the prover in [BSMP91] randomly selects completely new proving pairs (x_{00}, y_{00}) and (x_{01}, y_{01}) and then uses (x_0, y_0) and τ_1 along with the single theorem system to prove the auxiliary theorem, "(x_{00}, y_{00}) and (x_{01}, y_{01}) are properly formed proving pairs."[7] This approach, however, does not work in our setting because selecting new random values after posting the public key compromises the Uniqueness property.

To circumvent this difficulty, we add a seed, s, for a pseudo-random function f [GGM86] to the prover's secret key, and a perfectly binding commitment to s to the prover's public key. Now whenever the prover in [BSMP91] is instructed to prove that

"$(x_{0b_1...b_i 0}, y_{0b_1...b_i 0})$ and $(x_{0b_1...b_i 1}, y_{0b_1...b_i 1})$ are properly formed proving pairs"

our prover instead proves that

---

[6] We have changed notation from (x, y) above to (x_0, y_0) in order to match the notation from [BSMP91]

[7] In general, [BSMP91] describes a tree structure in which $(x_{0b_1...b_i}, y_{0b_1...b_i})$ is used to certify $(x_{0b_1...b_i 0}, y_{0b_1...b_i 0})$ and $(x_{0b_1...b_i 1}, y_{0b_1...b_i 1})$ which are then used to prove the $b_1...b_i 0^{\text{th}}$ and $b_1...b_i 1^{\text{th}}$ theorems.

"$(x_{0b_1...b_i0}, y_{0b_1...b_i0})$ and $(x_{0b_1...b_i1}, y_{0b_1...b_i1})$ are generated using the BDMP honest prover algorithm with coins $f_s(0b_1...b_i)$"

Observe that this auxiliary theorem is an NP-statement whose length is a fixed polynomial in k and can therefore be proven using the single theorem uniZK system with a sufficiently long τ_1. This assures both that $(x_{0b_1...b_i0}, y_{0b_1...b_i0})$ and $(x_{0b_1...b_i1}, y_{0b_1...b_i1})$ have the necessary properties and also that the prover had no choice in selecting these values (given his public key).[8]

We can also extend our system to work for theorems of arbitrary size by using techniques similar to those in [BSMP91]. Let ϕ be an arbitrarily long formula and let $(\hat{x}, \hat{y})$ be the next proving pair in the tree construction described above. First, use $(\hat{x}, \hat{y})$ to complete steps 4 through 7. Observe that we cannot continue with step 8 because τ_2 is not long enough to accommodate all of the clauses of ϕ. Instead, for each clause, we form the NP-statement

In clause j of ϕ, the triple (a_j, b_j, c_j) contains one non-residue mod $\hat{x}$.

Note that the length of this statement is fixed and independent of the size of ϕ. Therefore, by making τ_2 sufficiently long, we can prove each of these statement as separate theorems using the successor pairs of $(\hat{x}, \hat{y})$ as per the multi-theorem construction. Note that the prover has no choices to make since the form of the statement and the order in which they are proven are fixed by the statement ϕ.

Security Properties. The proof that this scheme is complete, sound, and zero-knowledge closely follows the corresponding proofs in [BSMP91]. Therefore, we will only sketch a proof that our construction satisfies Uniqueness.

First, we consider Uniqueness in the single-theorem case. Define the secret key extraction function, $sk()$, to take in a proving pair $PK = (x, y)$ and return the factorization of x. We now observe that if PK is not properly constructed, then with overwhelmingly high probability over the choice of random string, the verifier will reject any proof (because of soundness in Step 3), and therefore $\Pi_{PK}(\sigma, \phi)$ will be empty and uniqueness is automatic.

Therefore, we restrict attention to the case when PK is properly formed. First we observe that P (with auxiliary inputs σ, ϕ and the factorization of x) is a deterministic function and that by completeness it maps W_x into $\Pi_{PK}(\sigma, \phi)$. We then put forward an efficient algorithm P^{-1} (with the same auxiliary inputs) and show that it is the inverse of P. Finally, we show P and P^{-1} are bijections by proving that P^{-1} is an injection.

In the following we refer to the portion of π generated by step I in the honest prover algorithm as π_I. Let P^{-1} on input $\pi \in \Pi_{PK}(\sigma, \phi)$ inspect the portion π_6, use the factorization of x to determine the quadratic character (mod x) of $u_1, \ldots, u_n$, and output the corresponding assignment w. Note by inspection of

---

[8] Note here that we need to use a commitment scheme with only a single valid decommit message (to assure that the prove does not have a choice in selecting the witness for the auxiliary theorem).

step 6 P^{-1} returns the exact assignment that was used to generate π, so P^{-1} is the inverse of P.

All that remains to be shown is that P^{-1} is injective. We do this by showing that if $\pi^* \neq \pi = P(\sigma, \phi, w, sk(PK))$ and yet $P^{-1}(\sigma, \phi, \pi^*, sk(PK)) = w$ then $\pi^* \notin \Pi_{PK}(\sigma, \phi)$. We establish this using case analysis. Suppose the first point at which π and π^* differ is portion π_I. For all cases, except for $I = 8$, the proof is straightforward based on the Verifier's algorithm.

For case $I = 8$, we first argue that the sub-proof used to show that two triples are of the same type is sound. This follows directly from the fact that (x, y) is properly formed.

We next show that π^* cannot select two triples of the same type. If π^* selects two triples of the same type, then some type, is not selected. With high probability, this unselected type appears in the set of $8k^2$ triples. Therefore, the Verifier rejects π^* since π^* cannot prove that the unselected type is similar to a previously selected triple. Hence, π^* must select all 8 types.

If π and π^* select the same 8 triples, then the fact that π^* is rejected follows from the fact that each quadratic residue has exactly one Jacobi symbol 1 root less than $x/2$.

Assume π and π^* select different triples. If π selects a triple that π^* does not, then π^* must give a false proof that this triple was the same as a previously selected one, and we already know that the Verifier rejects such proofs. Alternatively, if π^* selects a triple not selected by π, then π^* cannot contain 8 different types, and we know that the Verifier rejects in this case as well.

This completes our proof of uniqueness in the single-theorem case. The only difference in the multi-theorem case is that π and π^* might use different pairs $(x, y) \neq (x^*, y^*)$ to prove theorem i. This means that (x^*, y^*) is not the output of the honest prover algorithm with coins specified by the committed seed in the prover's public key. In this case, by the soundness of the single-theorem proof system, the verifier will reject any auxiliary proof certifying (x^*, y^*). □

Remark: Choosing The Right NP-Complete Problem. We deliberately choose 3SAT (over, say, 3-Colorability) because, in order to satisfy the Uniqueness property, our multi-theorem construction requires a reduction from general NP-statements to 3-SAT formula which preserves the number of witnesses (in our case, one to one). Notice that even parsimonious reductions for 3-colorability map one witness to six possible colorings.

Remark: Choosing The Right Complexity Assumption. There are several NIZK systems based on the more general assumption that trap-door permutations exist (e.g., [FLS90] and [KP98]). Adapting such systems to admit Unique proofs, however, seems to require substantially new techniques.

4 Fair Zero-Knowledge Proofs

Informally, the goal of Fair ZK is to be a ZK proof system which remains secure even when the prover maliciously colludes with some subset of the verifiers.

This goal is embodied by the four properties of completeness, soundness, zero-knowledgeness, and fairness. Completeness states that if the prover and all verifiers are honest, than all true theorems are provable. Soundness states that even if a (computationally unbounded) dishonest prover collaborates with malicious verifiers during the set-up stage, no honest verifier will accept a false theorem. Zero knowledgeness states that even if all verifiers are malicious, they are unable to extract from the prover any extra information except that x_i is true and it is the ith theorem. Zero knowledgeness is formalized by the existence of an efficient simulator S that generates the same view that the malicious verifiers would have seen had they interacted with the *honest* prover about the same sequence of theorems (without seeing the corresponding witnesses). Importantly, S succeeds even if it is given each theorem one at a time (without knowing what future theorems might be). Fairness states that, *as long as an honest verifier accepts all of the theorems*, then, no matter how a dishonest prover might collude with a set of malicious verifiers, no verifier learns anything other than "x_i is true and it is the ith theorem." This is again formalized via a second simulator S^* that generates the same views that the malicious verifiers would have seen if they interacted with the *dishonest* prover. Again, S^* succeeds even though it is given the sequence of theorems one at at time. As far as we know, this is the first use of the simulator paradigm to protect the secrets of one dishonest party from another dishonest party.

Remarks

1. The primary difficulty with simulating a dishonest prover is that the prover has a witness, and the simulator does not! Clearly, if the prover decides to cheat and output the witness (or some partial information about it) in lieu of a valid proof, there is no hope for a simulator to produce indistinguishable transcripts. Thus, the best one can hope for is to require that simulated proofs are indistinguishable from real proofs *conditioned* on the event that an honest verifier accepts *all* the real proofs.
2. It is crucial to the applicability of Fair ZK that it applies to an unbounded sequence of theorems. And it is this feature that prevents us (at least for now) from relying on general cryptographic assumption. In particular, "single-theorem" Fair ZK can be achieved without number-theoretic assumptions by suitable modifying [DMP91].
3. In order to guarantee that no verifier gets additional knowledge about theorem x_i, an honest verifier must monitor all "utterances" of the prover as soon as he hands him an envelope in the preprocessing phase. In particular, the honest verifier must also monitor the first $i-1$ proofs : If all honest verifiers are "out to lunch", a dishonest prover may send sk to her accomplices!
4. The order in which a sequence of theorems is proven must be fixed. Giving the prover freedom to choose this order provides yet another opportunity for steganography. (Achieving Fair ZK requires us to run a tight ship!) However, the prover may receive all theorems and witnesses, if available, immediately after completing the setup protocol successfully.

4.1 Formal Definition

A *setup protocol* is a protocol, $(\mathcal{P}, \mathcal{V}_1, \ldots, \mathcal{V}_n)$, with a distinguished ITM $\mathcal{P}$, the prover (referred to as player 0), and n ITMs, $\mathcal{V}_1, \ldots, \mathcal{V}_n$, the verifiers (respectively referred to as players 1 through n). All players in this protocol exchange message via broadcast; in addition the verifiers may also send messages in envelopes and the prover also receives messages in envelopes. Each execution e of the setup protocol produces a common public output $pk \in \{0,1\}^* \cup \{\bot\}$ and a secret output sk for the prover.

In an execution e of this protocol with security parameter 1^k, we denote by $\text{VIEW}_i(e)$ the triple $(1^k, \rho_i, M_i)$, where ρ_i is the random tape for player i and M_i is the set of messages received by player i during the execution. If $T = (a, b, \ldots)$ is a sequence of players, then denote by $\text{VIEW}_T(e)$ the sequence of views $(\text{VIEW}_a(e), \text{VIEW}_b(e), \ldots)$.

We denote by $(pk, sk), e \leftarrow \langle \mathcal{P} \xleftarrow{1^k} \mathcal{V}_1, \ldots, \mathcal{V}_n \rangle$ the random variable obtained by uniformly and independently selecting a random tape ρ_i for each player i, executing the setup protocol with security parameter 1^k and random tapes ρ_i's, and outputting the so generated execution e, with its corresponding outputs pk and sk.

Definition 1 (Fair Zero Knowledge). Let L be a (computationally) unique-witness language[9] A *Fair zero-knowledge proof system* for L consists of (1) a setup protocol, $(\mathcal{P}, \mathcal{V}_1, \ldots, \mathcal{V}_m)$; (2) an efficient deterministic *proving algorithm* P, (3) an efficient verification algorithm V, and a negligible function, μ, such that the following properties are satisfied:

Completeness. $\forall$ theorem-witness sequences $(x_1, w_1), (x_2, w_2), \ldots$ for L and $\forall k \in \mathbb{Z}^+$,

$$\Pr\left[\begin{array}{l}(pk, sk), e \leftarrow \langle P \xleftarrow{1^k} V_1, \ldots, V_m \rangle; \\ \pi_1 \leftarrow P(x_1, w_1, sk, 1);\ \pi_2 \leftarrow P(x_2, w_2, sk, 2);\ \ldots \\ :\ \bigwedge_i V(x_i, \pi_i, pk, i) = 1 \end{array}\right] > 1 - \mu(k)$$

Soundness. $\forall\ P^*, V_1^*, \ldots, V_{i-1}^*, V_{i+1}^*, \ldots, V_n^*$, $\forall$ sufficiently large $k \in \mathbb{Z}^+$,

$$\Pr\left[\begin{array}{l}(pk^*, sk^*), e \leftarrow \langle P^* \xleftarrow{1^k} V_1^*, \ldots, V_{i-1}^*, V_i, V_{i+1}^*, \ldots, V_n^* \rangle; \\ (x^*, \pi^*, i) \leftarrow P^*(\text{VIEW}_0(e)) \\ :\ x^* \notin L \wedge V(x^*, \pi^*, pk^*, i) = 1 \end{array}\right] < \mu(k)$$

Zero-Knowledgeness. $\forall$ efficient ITMs $V_1^*, \ldots, V_n^*$, $\exists$ an efficient algorithm S such that $\forall$ theorem-witness sequences $(x_1, w_1), (x_2, w_2), \ldots$ for L,

---

[9] A *computationally unique-witness* language is one in which, given a witness w for a statement $x \in L$, it is hard to produce a new witness for the same statement.

$$\left\{\begin{array}{l}(pk,\alpha,\text{VIEW})\leftarrow S(1^k);\\ \quad \pi_1 \leftarrow S(x_1,\alpha,1);\\ \quad \pi_2 \leftarrow S(x_2,\alpha,2);\ldots\\ pk,\text{VIEW},x_1,\pi_1,x_2,\pi_2\ldots\end{array}\right\}_k \stackrel{c}{\approx} \left\{\begin{array}{l}(pk,sk),e \leftarrow \langle P \xleftrightarrow{1^k} V_1^*,\ldots,V_n^*\rangle;\\ \quad \pi_1 \leftarrow P(x_1,w_1,sk,1);\\ \quad \pi_2 \leftarrow P(x_2,w_2,sk,2);\ldots\\ pk,\text{VIEW}_{1,\ldots,n}(e),x_1,\pi_1,x_2,\pi_2\ldots\end{array}\right\}_k$$

Fairness. $\forall$ efficient $P^*, V_1^*,\ldots,V_{i-1}^*, V_{i+1}^*,\ldots,V_n^*$, $\exists$ an efficient S^* such that $\forall$ theorem-witness sequences $(x_1,w_1),(x_2,w_2),\ldots$ for L, the following two ensembles are computationally indistinguishable:

$$\left\{\begin{array}{l}(pk^*,\alpha,\text{VIEW})\leftarrow S^*(1^k);\\ \quad \pi_1^* \leftarrow S^*(x_1,\alpha,1);\ \pi_2^* \leftarrow S^*(x_2,\alpha,2);\ldots\\ \quad : pk^*,\text{VIEW},x_1,\pi_1^*,x_2,\pi_2^*\ldots\end{array}\right\}_k$$

$$\left\{\begin{array}{l}(pk^*,sk^*),e \leftarrow \langle P^* \xleftrightarrow{1^k} V_1^*,\ldots,V_{i-1}^*,V_i,V_{i+1}^*,\ldots,V_n^*\rangle;\\ \pi_1^* \leftarrow P^*(x_1,w_1,sk^*,1);\ \pi_2^* \leftarrow P^*(x_2,w_2,sk^*,2);\ \ldots\\ : pk^*,\text{VIEW}_{1,\ldots,i-1,i+1,\ldots,n}(e),x_1,\pi_1^*,x_2,\pi_2^*,\ldots\end{array}\right|\bigwedge_i V(x_i,\pi_i,pk^*,i)=1\right\}_k$$

4.2 Constructing Fair ZK

Our goal is to defeat steganographic attacks by using uniZK. However, we cannot allow the prover to pick his own secret key (since he might share it with a verifier beforehand). Therefore, we need to incorporate randomness from *all* of the verifiers during the selection of a prover secret key. We show that if we allow the prover to receive a *single envelope* from each verifier during the preprocessing, that we can transform any UniZK system into a Fair zero-knowledge proof system.

Theorem 2. *The existence of a* uniZK *proof system and a family of trap-door one-way permutations implies a fair zero knowledge proof system in which, during the preprocessing phase, each verifier sends the prover a single envelope.*

Proof Sketch: We first present our fair zero-knowledge protocol and then sketch a proof that the protocol satisfies the security properties specified in our definition of fair zero-knowledge.

– Preprocessing Phase:
  1. The players engage in a simulatable coin-flipping protocol which is secure against an unbounded Prover to generate a reference string, σ. For example, in order to generate a single bit, the prover uses a perfectly binding commitment scheme to commit to a random bit. Then all the verifiers broadcast (in turn) a commitment of a randomly chosen bit, then decommit the bits in the opposite order, and finally the prover decommits her bit. The output is defined as the xor of all opened bits. This can be repeated sequentially to generate longer reference strings.
  2. The players partially execute the secure function evaluation protocol from [GMW87] with privacy threshold set to $n-1$ in order to compute the following n-valued function:

$$F(\varepsilon,\ldots,\varepsilon) = \{ \overset{\text{Prover's output}}{SK_{uzk}}, \overset{\text{Verifiers' outputs}}{PK_{uzk},\ldots,PK_{uzk}}\}$$

That is, the function produces a private output for the prover consisting of a uniZK secret key, and produces the corresponding uniZK public key as the output for all of the verifiers.

The GMW protocol is executed until all of the players have shares of each of the output values, but have not yet sent each other these shares.

3. All of the shares for the verifiers' outputs are broadcast to all parties. Note that there is no need to encrypt these shares as all of the verifiers have the same output values. As in the original GMW protocol, all parties use interactive zero-knowledge proofs in order to prove to all other parties that the share they have broadcast is correctly computed.

4. In the final round all verifiers send their shares of the prover's output as well as all random coins that they used during preprocessing to the prover using an envelope channel.

5. The prover runs the honest verifier algorithm to verify that the shares sent by the verifiers were computed correctly. It then computes its private output, namely the SK_{uzk}, by combining the shares. At this point, the prover has unique knowledge of a uniZK secret key, and all parties have a corresponding uniZK public key and a reference string, σ.

- Proof Phase:
  In the proof phase, the prover can prove any number of theorems by using the uniZK prover algorithm with secret key SK_{uzk} and reference string, σ. The verifiers use the corresponding uniZK verifier algorithm with public key PK_{uzk} and random string σ to verify each proof. As soon as a single proof fails to verify, the verifiers are instructed to reject all subsequent proofs.

Security Properties. The completeness of this system is straightforward. Therefore, we will only sketch proofs that our system satisfies Soundness, Zero-Knowledgeness and Fairness.

Soundness. Here we must show that any prover who manages to cheat against a set of verifiers either breaks the correctness property of the coin flipping protocol, or breaks the soundness property of the uniZK system, both of which are unconditionally secure. Assume that the output from the coin-flipping protocol is truly random. In this case, the prover's algorithm for cheating can be used without modification to break the soundness of the uniZK system. Note, even if the Prover breaks the correctness of the SFE, thereby generating the uniZK keys of her choice, this does not allow the Prover to break soundness since the uniZK system is sound, even when the prover chooses his key after seeing the reference string.

Zero-knowledgeness. Our simulator, S, works as follows:

1. First run the uniZK simulator in order to generate a reference string σ^* as well as a public and private key, PK_{uniZK} and SK_{uniZK}.
   The goal is to now manipulate the coin-flipping protocol and the secure function evaluation in order to produce σ^* and PK_{uniZK}.

2. Use the simulator for the coin-flipping protocol in order to generate a transcript with output σ^*.
3. Begin running the secure function evaluation protocol. At any point during which S is required to send a message on behalf of any party, write the message to the transcript as an honest party would.
4. During the last step when each party broadcasts its share of the public output and proves that it was formed correctly, S uses it ability to rewind the malicious parties in order to do two things. First, it learns the shares of each of the malicious parties by proceeding honestly. It then rewinds the malicious parties, and broadcasts shares on behalf of the honest parties to force the public output to be PK_{uniZK}. Finally, by rewinding, it simulates the zero-knowledge proofs that the broadcast shares are correct.
5. The envelopes that are sent from the malicious parties are opened and the random coins inside are used for verification. Upon failure, S aborts.
6. S now uses the uniZK simulator in order to generate the proofs for the sequence of theorems that arrive using its key SK_{uniZK} and σ^*.

In order to prove that the transcripts produced by this Simulator are indistinguishable from those of a real execution, we first note that the transcript for the coin-flipping protocol is generated by a simulator and thus indistinguishable. During the SFE portion of the protocol, all of the steps are identical until Step 4. During the last two steps, S is using another simulator to generate indistinguishable transcripts for a zero-knowledge proof. Therefore, any distinguisher of the Fair ZK protocol's transcripts can be trivially used to break the zero-knowledge property of the proof used in this step. Since the envelope traffic is not part of the view of the verifiers, it does not matter what is sent in them. Therefore, the verifiers have no information about the SK_{uniZK} since they have no information about the Prover's share. Therefore, any distinguisher between the key produced by S and the key produced in a real execution can be used to break the uniZK simulator.

Similarly, the proof strings are henceforth produced by the uniZK simulator and therefore any distinguisher can also be used (in a straightforward reduction) to break the uniZK simulator.

Fairness. The same simulator used to prove the zero-knowledgeness property is also used to prove the Fairness property. The only difference is that the simulator must use the cheating prover algorithm in order to generate all of the prover messages during the preprocessing phase. Note that during preprocessing the simulator is able to directly run the malicious prover algorithm because until a witness is given to the prover, the prover has no secrets which the simulator does not know. This step ensures that any secret agreements between the cheating prover and any malicious set of verifiers reflect themselves during the simulated transcripts (and therefore maintain indistinguishability with real executions).

Once the envelopes are sent to the prover, the uniqueness property of the uniZK system guarantees that for each theorem either the prover gives the single acceptable uniZK proof (which can be simulated) or she sends any other string in which case the honest verifier algorithm rejects. In the former case, fairness is guaranteed by the indistinguishability of uniZK. In the later case, fairness is vacuous because an honest verifier rejects. □

Corollary 1. *Under the Quadratic Residuosity assumption, there exists a fair zero knowledge proof system consisting of a preprocessing phase during which each verifier sends the prover a single envelope.*

This result follows directly from Theorem 2 and Theorem 1.

References

[BFM88] Manuel Blum, Paul Feldman, and Silvio Micali. Non-interactive zero-knowledge and its applications (extended abstract). In *STOC 1988*, pages 103–112, 1988.

[BSMP91] Manuel Blum, Alfredo De Santis, Silvio Micali, and Giuseppe Persiano. Noninteractive zero-knowledge. *SIAM J. Computing*, 20(6):1084–1118, 1991.

[DAM92] I. Damgard. Non-interactive circuit based proofs and non-interactive perfect zeroknowledge with preprocessing. In *EUROCRYPT '92*. Springer-Verlag, 1992.

[DGB87] Y. Desmedt, C. Goutier, and S. Bengio. Special uses and abuses of the Fiat-Shamir passport protocol. In *CRYPTO '87*. Springer-Verlag, 1987.

[DMP91] Alfredo DeSantis, Silvio Micali, and Giuseppe Persiano. Non-interactive zero-knowledge with preprocessing. In *CRYPTO 1988*. Springer-Verlag, 1991.

[FLS90] Uriel Feige, Dror Lapidot, and Adi Shamir. Multiple non-interactive zero knowledge proofs based on a single random string. In *Proc. 31th FOCS*, pages 308–317, 1990.

[GMW87] O. Goldreich, S. Micali, and A. Wigderson. How to play any mental game. In *Proc of STOC '87*, pages 218–229. ACM, 1987.

[GGM86] Oded Goldreich, Shafi Goldwasser, and Silvio Micali. How to construct random functions. *Journal of the ACM*, 33(4):792–807, October 1986.

[GMR89] S. Goldwasser, S. Micali, and C. Rackoff. The knowledge complexity of interactive proof-systems. *SIAM. J. Computing*, 18(1):186–208, February 1989.

[GM84] Shafi Goldwasser and Silvio Micali. Probabilistic encryption. *Journal of Computer and System Science*, 28(2), 1984.

[GMR88] Shafi Goldwasser, Silvio Micali, and Ronald L. Rivest. A digital signature scheme secure against adaptive chosen-message attacks. *SIAM J. Computing*, 17(2):281–308, April 1988.

[KP98] Joe Kilian and Erez Petrank. An efficient noninteractive zero-knowledge proof system for np with general assumptions. *J. Cryptology*, 11(1):1–27, 1998.

[NHvA02] John Langford Nicholas Hopper and Luis von Ahn. Provably secure steganography. In *CRYPTO 2002*. Springer-Verlag, 2002.

How to Securely Outsource Cryptographic Computations

Susan Hohenberger[1,*] and Anna Lysyanskaya[2,**]

[1] CSAIL, Massachusetts Institute of Technology,
Cambridge, MA 02139 USA
srhohen@mit.edu

[2] Computer Science Department, Brown University,
Providence, RI 02912 USA
anna@cs.brown.edu

Abstract. We address the problem of using untrusted (potentially malicious) cryptographic helpers. We provide a formal security definition for *securely outsourcing* computations from a computationally limited device to an untrusted helper. In our model, the adversarial environment writes the software for the helper, but then does not have direct communication with it once the device starts relying on it. In addition to security, we also provide a framework for quantifying the *efficiency* and *checkability* of an outsourcing implementation. We present two practical outsource-secure schemes. Specifically, we show how to securely outsource modular exponentiation, which presents the computational bottleneck in most public-key cryptography on computationally limited devices. Without outsourcing, a device would need $O(n)$ modular multiplications to carry out modular exponentiation for n-bit exponents. The load reduces to $O(\log^2 n)$ for any exponentiation-based scheme where the honest device may use two untrusted exponentiation programs; we highlight the Cramer-Shoup cryptosystem [13] and Schnorr signatures [28] as examples. With a relaxed notion of security, we achieve the same load reduction for a new CCA2-secure encryption scheme using only one untrusted Cramer-Shoup encryption program.

1 Introduction

Modern computation has become pervasive: pretty much any device these days, from pacemakers to employee ID badges, is expected to be networked with other components of its environment. This includes devices, such as RFID tags, that are not designed to carry out expensive computations. In fact, RFID tags do not even have a power source. This becomes a serious concern when we want to guarantee that these devices are integrated into the network securely: if a device

---

* Supported by an NDSEG Fellowship.
** Supported by NSF Career grant CNS-0347661.

is computationally incapable of carrying out cryptographic algorithms, how can we give it secure and authenticated communication channels?

In this paper, we study the question of how a computationally limited device may *outsource* its computation to another, potentially malicious, but much more computationally powerful device. In addition to powering up from an external power source, an RFID tag would have some external helper entity do the bulk of the computation that the RFID tag needs done in order to securely and authentically communicate with the outside world. The non-triviality here is that, although this external helper will be carrying out most of the computation, it can, potentially, be operated by a malicious adversary. Thus, we need to ensure that it does not learn anything about what it is actually computing; and we also need to, when possible, detect any failures.

There are two adversarial behaviors that the helper software might engage in: *intelligent* and *unintelligent* failures. Intelligent failures occur any time that the helper chooses to deviate from its advertised functionality based on knowledge it gained of the inputs to the computation it is aiding. For example, the helper might refuse to *securely* encrypt any message once it sees the public key of a competing software vendor; it might pass any signature with its manufacturer's public key without checking it; it might even choose to broadcast the honest device's secret key to the world. The first goal of any outsourcing algorithm should be to hide as much information as possible about the actual computation from the helper, thus removing its ability to bias outputs or expose secrets. Obviously, software may also unintelligently fail. For example, the helper might contain a malicious bug that causes it to fail on every 1,000th invocation regardless of who is using it. Thus, we face a real challenge: get helper software to do *most* of the computations for an honest device, without telling it anything about what it is actually doing, and then check its output!

In this paper, we give the definition of security for outsourced computation, including notions of efficiency and checkability. We also provide two practical outsource-secure schemes.

In Section 3, we show how to securely outsource variable-exponent, variable-base modular exponentiation. Modular exponentiation has been considered prohibitively expensive for embedded devices. Since it is required by virtually any public-key algorithm, it was believed that public-key cryptography for devices such as RFID tags is impossible to achieve. Our results show that outsourced computation makes it possible for such devices to carry out public-key cryptography. Without outsourcing, a device would need $O(n)$ modular multiplications to carry out a modular exponentiation for an n-bit exponent. Using two untrusted programs that purportedly compute exponentiations (and with the restriction that at most one of them will deviate from its advertised functionality on a non-negligible fraction of inputs), we show that an honest device can get away with doing only $O(\log^2 n)$ modular multiplications itself – while able to catch an error with probability $\frac{1}{2}$. This result leads to a dramatic reduction in the burden placed on the device to support Cramer-Shoup encryption [13] and Schnorr signatures [28] with error rates of $\frac{1}{8}$ and $\frac{1}{4}$ respectively. (Consider that after a small number of uses, malfunctioning software is likely to be caught.)

In Section 4, we show how to securely outsource a CCA2-secure variant of Cramer-Shoup encryption, using only one Cramer-Shoup encryption program as an untrusted helper. Since this is a randomized functionality, its output cannot generally be checked for correctness. However, suppose we can assume that the untrusted helper malfunctions on only a negligible fraction of adversarially chosen inputs; for example, suppose it is encryption software that works properly except when asked to encrypt a message under a certain competitor's public key. Normally, software that fails on only a negligible fraction of *randomly-chosen* inputs can be tolerated, but in the context of secure outsourcing we cannot tolerate any *intelligent* failures (i.e., failures based on the actual public key and message that the user wishes to encrypt). That is, secure outsourcing requires that the final solution, comprised of trusted and untrusted components, works with high probability for all inputs. Consider that Alice may have unwittingly purchased helper software for the sole purpose of encrypting messages under one of the few public keys for which the software is programmed to fail. Thus, in this scenario, we provide a solution for Alice to securely encrypt *any* message under *any* public key with high probability (where the probability is no longer taken over her choice of message and key). One can easily imagine how to hide the message and/or public key for RSA or El Gamal based encryption schemes; however, our second result is non-trivial because we show how to do this for the *non-malleable* Cramer-Shoup encryption scheme, while achieving the same asymptotic speed-up as before.

Related Work. Chaum and Pedersen [11] previously introduced "wallets with observers" where a third party, such as a bank, is allowed to install a piece of hardware on a user's computer. Each transaction between the bank and the user is designed to use this hardware, which the bank trusts, but the user may not. This can be viewed as a special case of our model.

This work shares some similarities with the TPM (Trusted Platform Module) [29], which is currently receiving attention from many computer manufacturers. Like the TPM, our model separates software into two categories: trusted and untrusted. Our common goal is to minimize the necessary trusted resources. Our model differs from TPM in that we have the trusted component controlling all the input/output for the system, whereas TPM allows some inputs/outputs to travel directly between the environment and untrusted components.

In the 1980s, Ben-Or et al. used multiple provers as a way of removing intractability assumptions in interactive proofs [4], which led to a series of results on hiding the input, and yet obtaining the desired output, from an honest-but-curious oracle [1, 2, 3]. Research in program checking merged into this area when Blum, Luby, and Rubinfeld [5, 7, 6] considered checking *adaptive, malicious* programs (i.e., oracles capable of intelligently failing).

The need for a formal security definition of outsourcing is apparent from previous research on using untrusted servers for RSA computations, such as the work of Matsumoto et al. [22] which was subsequently broken by Nguyen and Shparlinski [24]. We incorporate many previous notions including: the idea of an untrusted helper [19], confining untrusted applications and yet allowing a sanitized space for trusted applications to operate [30], and oracle-based checking of untrusted software [23]. Our techniques in Section 4 also offer novel approaches to the area of message and key *blinding* protocols [10, 18, 31].

Secure outsourcing of exponentiations is a popular topic [27, 28, 17, 8, 25, 22, 1, 2, 3, 12], but past approaches either focus on fixed-base (or fixed-exponent) exponentiation or meet a weaker notion of security.

2 Definition of Security

Suppose that we have a cryptographic algorithm *Alg*. Our goal is to split *Alg* up into two components: (1) a trusted component T that *sees* the input to *Alg* but is not very computationally intensive; (2) luckily T can make oracle queries to the second component, U, which is an untrusted component (or possibly components) that can carry out computation-intensive tasks.

Informally, we say that T *securely outsources* some work to U, and that (T, U) thereby form an *outsource-secure* implementation of a cryptographic algorithm *Alg* if (1) together, they implement *Alg*, i.e., $Alg = T^U$ and (2) suppose that, instead of U, T is given oracle access to a malicious U' that records all of its computation over time and, every time it is invoked, tries to act maliciously – e.g., not work on some adversarially selected inputs; we do not want such a malicious U', despite carrying out most of the computation for $T^{U'}(x)$, to learn anything *interesting* about the input x. For example, we do not want a malicious U' to trick T into rejecting a valid signature because U' sees the verification key of a competing software vendor or a message it does not like.

To define outsource-security more formally, we first ask ourselves how much security can be guaranteed. The least that U' can learn is that T actually received some input. In some cases, for a cryptographic algorithm $Alg = T^U$ that takes as input a secret key SK, and an additional input x, we may limit ourselves to hiding SK but not worry about hiding x. For example, we might be willing to give a ciphertext to the untrusted component U', but not our secret key. At other times, we may want to hide everything meaningful from U'. Thus, the inputs to *Alg* can be separated into two logical groups: (1) inputs that should remain hidden from the untrusted software U' at all times (for example, keys and messages), and (2) inputs that U' is entitled to know if it is to be of any help in running *Alg* (for example, if *Alg* is a time-stamping scheme, then U' may need to know the current time). Let us denote these two types of input as *protected* and *unprotected*.

Similarly, *Alg* has protected and unprotected outputs: those that U' is entitled to find out, and those that it is not. For example, if $Alg = T^U$ is an encryption program it may ask U' to help it compute a part of the ciphertext, but then wish to conceal other parts of the ciphertext from U'.

However, U' is not the only malicious party interacting with *Alg*. We model the adversary A as consisting of two parts: (1) the adversarial environment E

that submits adversarially chosen inputs to Alg; (2) the adversarial software U' operating in place of oracle U. One of the fundamental assumptions of this model is that E and U' may first develop a joint strategy, but once they begin interacting with an honest party T, they no longer have a direct communication channel. Now, E may get to see some of the protected inputs to Alg that U' does not. For example, E gets to see *all* of its own adversarial inputs to Alg, although T might hide some of these from U'. Consider that if U' was able to see some values chosen by E, then E and U' can agree on a joint strategy causing U' to stop working upon receiving some predefined message from E. Thus, there are going to be some inputs that are known to E, but hidden from U', so we ought to formalize how different their views need to be.

We have three logical divisions of inputs to Alg: (1) secret – information only available to T (e.g., a secret key or a plaintext); (2) protected – information only available to T and E (e.g., a public key or a ciphertext); (3) unprotected – information available to T, E, and U' (e.g., the current time). These divisions are further categorized based on whether the inputs were generated *honestly* or *adversarially*, with the exception that there is no *adversarial, secret* input – since by definition it would need to be both generated by and kept secret from E. Similarly, Alg has secret, protected, and unprotected outputs. Thus, let us write that Alg takes five inputs and produces three outputs. This is simplified notation since these inputs may be related to each other in some way. For example, the secret key is related to the public key.

As an example of this notation, consider a signing algorithm $sign$ such that we want to hide from the malicious software U' the secret key SK and the message m that is being signed, but not the time t at which the message is signed. The key pair was generated using a correct key generation algorithm and the time was honestly generated, while the message may have been chosen adversarially. Also, we do not want the malicious U' to find out anything about the signature that is output by the algorithm. Then we write $sign(SK, \varepsilon, t, m, \varepsilon) \to (\varepsilon, \sigma, \varepsilon)$ to denote that the signature σ is the protected output, there are no secret or unprotected outputs, SK is the honest, secret input, t is the honest, unprotected input, m is the adversarial, protected input, and there are no other inputs. This situation grows more complex when we consider Alg operating in a compositional setting where the protected outputs of the last invocation might become the adversarial, unprotected inputs of the next; we will further discuss this subtlety in Remark 2.

Let us capture an algorithm with this input/output behavior in a formal definition:

Definition 1 (Algorithm with outsource-IO). *An algorithm Alg obeys the outsource input/output specification if it takes five inputs, and produces three outputs. The first three inputs are generated by an honest party, and are classified by how much the adversary $A = (E, U')$ knows about them. The first input is called the* honest, secret *input, which is unknown to both E and U; the second is called the* honest, protected *input, which may be known by E, but is protected from U; and the third is called the* honest, unprotected *input, which may be known by both E and U. In addition, there are two adversarially-chosen inputs* *generated by the environment E: the* adversarial, protected *input, which is known to E, but protected from U; and the the* adversarial, unprotected *input, which may be known by E and U. Similarly, the first output called* secret *is unknown to both E and U; the second is* protected*, which may be known to E, but not U; and the third is* unprotected*, which may be known by both parts of A.*

At this point, this is just input/output notation, we have not said anything about actual security properties. We now discuss the definition of security.

The two adversaries E, U' can only communicate with each other by passing messages through T, the honest party. In the real world, a malicious manufacturer E might program its software U' to behave in an adversarial fashion; but once U' is installed behind T's firewall, manufacturer E should no longer be able to directly send instructions to it. Rather, E may try to establish an *indirect* communication channel with U' via the unprotected inputs and outputs of *Alg*. For example, if E knows that the first element in a signature tuple is unprotected (meaning, T always passes the first part of a signature tuple, unchanged, to U'), it might encode a message in that element instructing U' to "just tell T the signature is valid" – even though it may not be. Alternatively, an indirect communication channel might be realized by U' smuggling secrets about the computation it helped T with, through the unprotected outputs, back to E. For example, if, in the course of helping T with decryption, U' learned the secret key, it might append that key to the next unprotected output it creates for T. Obviously, T must use U' with great care, or he will be completely duped.

Our definition of outsource-security requires that anything secret or protected that a malicious U' can learn about the inputs to T^U from being T's oracle instead of U, it can also learn without that. Namely, there exists a simulator S_2 that, when told that $T^U(x)$ was invoked, simulates the view of U' without access to the secret or protected inputs of x. This property ensures that U' cannot intelligently choose to fail.

Similarly, our definition of outsource-security must also prevent the malicious environment E from gaining any knowledge of the secret inputs and outputs of T^U, even when T is using malicious software U' written by E. Again, there exists a simulator S_1 that, when told that $T^{U'}(x)$ was invoked, simulates the view of E without access to the secret inputs of x.

Definition 2 (Outsource-security). *Let $Alg(\cdot, \cdot, \cdot, \cdot, \cdot)$ be an algorithm with outsource-IO. A pair of algorithms (T, U) is said to be an outsource-secure implementation of an algorithm Alg if:*
Correctness. *T^U is a correct implementation of Alg.*
Security. *For all probabilistic polynomial-time adversaries $A = (E, U')$, there exist probabilistic expected polynomial-time simulators (S_1, S_2) such that the following pairs of random variables are computationally indistinguishable. Let us say that the honestly-generated inputs are chosen by a process I.*
Pair One: $EVIEW_{real} \sim EVIEW_{ideal}$ *(The external adversary, E, learns nothing.):*

– The view that the adversarial environment E obtains by participating in the following REAL process:

$$EVIEW^i_{real} = \{(istate^i, x^i_{hs}, x^i_{hp}, x^i_{hu}) \leftarrow I(1^k, istate^{i-1});$$
$$(estate^i, j^i, x^i_{ap}, x^i_{au}, stop^i) \leftarrow E(1^k, EVIEW^{i-1}_{real}, x^i_{hp}, x^i_{hu});$$
$$(tstate^i, ustate^i, y^i_s, y^i_p, y^i_u) \leftarrow T^{U'(ustate^{i-1})}(tstate^{i-1}, x^{j^i}_{hs}, x^{j^i}_{hp}, x^{j^i}_{hu}, x^i_{ap}, x^i_{au}):$$
$$(estate^i, y^i_p, y^i_u)\}$$

$EVIEW_{real} = EVIEW^i_{real}$ if $stop^i = TRUE$.

The real process proceeds in rounds. In round i, the honest (secret, protected, and unprotected) inputs $(x^i_{hs}, x^i_{hp}, x^i_{hu})$ are picked using an honest, stateful process I to which the environment does not have access. Then the environment, based on its view from the last round, chooses (0) the value of its $estate_i$ variable as a way of remembering what it did next time it is invoked; (1) which previously generated honest inputs $(x^{j^i}_{hs}, x^{j^i}_{hp}, x^{j^i}_{hu})$ to give to $T^{U'}$ (note that the environment can specify the index j^i of these inputs, but not their values); (2) the adversarial, protected input x^i_{ap}; (3) the adversarial, unprotected input x^i_{au}; (4) the Boolean variable $stop^i$ that determines whether round i is the last round in this process. Next, the algorithm $T^{U'}$ is run on the inputs $(tstate^{i-1}, x^{j^i}_{hs}, x^{j^i}_{hp}, x^{j^i}_{hu}, x^i_{ap}, x^i_{au})$, where $tstate^{i-1}$ is T's previously saved state, and produces a new state $tstate^i$ for T, as well as the secret y^i_s, protected y^i_p and unprotected y^i_u outputs. The oracle U' is given its previously saved state, $ustate^{i-1}$, as input, and the current state of U' is saved in the variable $ustate^i$. The view of the real process in round i consists of $estate^i$, and the values y^i_p and y^i_u. The overall view of the environment in the real process is just its view in the last round (i.e., i for which $stop^i = TRUE$).

– The IDEAL process:

$$EVIEW^i_{ideal} = \{(istate^i, x^i_{hs}, x^i_{hp}, x^i_{hu}) \leftarrow I(1^k, istate^{i-1});$$
$$(estate^i, j^i, x^i_{ap}, x^i_{au}, stop^i) \leftarrow E(1^k, EVIEW^{i-1}_{ideal}, x^i_{hp}, x^i_{hu});$$
$$(astate^i, y^i_s, y^i_p, y^i_u) \leftarrow Alg(astate^{i-1}, x^{j^i}_{hs}, x^{j^i}_{hp}, x^{j^i}_{hu}, x^i_{ap}, x^i_{au});$$
$$(sstate^i, ustate^i, Y^i_p, Y^i_u, replace^i) \leftarrow S^{U'(ustate^{i-1})}_1(sstate^{i-1}, \ldots$$
$$\ldots x^{j^i}_{hp}, x^{j^i}_{hu}, x^i_{ap}, x^i_{au}, y^i_p, y^i_u);$$
$$(z^i_p, z^i_u) = replace^i(Y^i_p, Y^i_u) + (1 - replace^i)(y^i_p, y^i_u):$$
$$(estate^i, z^i_p, z^i_u)\}$$

$EVIEW_{ideal} = EVIEW^i_{ideal}$ if $stop^i = TRUE$.

The ideal process also proceeds in rounds. In the ideal process, we have a stateful simulator S_1 who, shielded from the secret input x^i_{hs}, but given the non-secret outputs that Alg produces when run all the inputs for round i, decides to either output the values (y_p^i, y_u^i) generated by Alg, or replace them with some other values (Y_p^i, Y_u^i). (Notationally, this is captured by having the indicator variable $replace^i$ be a bit that determines whether y_p^i will be replaced with Y_p^i.) In doing so, it is allowed to query the oracle U'; moreover, U' saves its state as in the real experiment.

Pair Two: $UVIEW_{real} \sim UVIEW_{ideal}$ *(The untrusted software, U', learns nothing.)*:

- The view that the untrusted software U' obtains by participating in the REAL process described in Pair One. $UVIEW_{real} = ustate^i$ if $stop^i = TRUE$.
- The IDEAL process:

$$UVIEW_{ideal}^i = \{(istate^i, x_{hs}^i, x_{hp}^i, x_{hu}^i) \leftarrow I(1^k, istate^{i-1});$$
$$(estate^i, j^i, x_{ap}^i, x_{au}^i, stop^i) \leftarrow E(1^k, estate^{i-1}, x_{hp}^i, x_{hu}^i, y_p^{i-1}, y_u^{i-1});$$
$$(astate^i, y_s^i, y_p^i, y_u^i) \leftarrow Alg(astate^{i-1}, x_{hs}^{j^i}, x_{hp}^{j^i}, x_{hu}^{j^i}, x_{ap}^i, x_{au}^i);$$
$$(sstate^i, ustate^i) \leftarrow S_2^{U'(ustate^{i-1})}(sstate^{i-1}, x_{hu}^{j^i}, x_{au}^i) :$$
$$(ustate^i)\}$$

$UVIEW_{ideal} = UVIEW_{ideal}^i$ if $stop^i = TRUE$.
In the ideal process, we have a stateful simulator S_2 who, equipped with only the unprotected inputs (x_{hu}^i, x_{au}^i), queries U'. As before, U' may maintain state.

There are several interesting observations to make about this security definition.

Remark 1. The states of all algorithms, i.e., I, E, U', T, S_1, S_2, in the security experiments above are initialized to $\emptyset$. Any joint strategy that E and U' agree on prior to acting in the experiments must be embedded in their respective codes. Notice the intentional asymmetry in the access to the untrusted software U' given to environment E and the trusted component T. The environment E is allowed non-black-box access to the software U', since E may have written code for U'; whereas U' will appear as a black-box to T, since one cannot assume that a malicious software manufacturer will (accurately) publish its code. Or, consider the example of an RFID tag outsourcing its computation to a more powerful helper device in its environment. In this case we cannot expect that, in the event that it is controlled by an adversary, such a helper will run software that is available for the purposes of the proof of security.

Remark 2. For any outsource-secure implementation, the adversarial, unprotected input x_{au} must be empty. If x_{au} contains even a single bit, then a covert channel is created from E to U', in which k bits of information can be transfered after k rounds. In such a case, E and U' could jointly agree on a secret value beforehand, and then E could slowly smuggle in that k-bit secret to U'. Thus,

$UVIEW_{real}$ would be distinguishable from $UVIEW_{ideal}$, since E may detect that it is interacting with Alg instead of $T^{U'}$ (since Alg's outputs (y_p^i, y_u^i) are always correct), and communicate this fact to U' through the covert channel. A non-empty x_{au} poses a real security threat, since it would theoretically allow a software manufacturer to covertly reprogram its software *after* it was installed behind T's firewall and *without* his consent.

Remark 3. No security guarantee is implied in the event that the environment E and the software U' are able to communicate without passing messages through T. For example, in the event that E captures all of T's network traffic and then steals T's hard-drive (containing the memory of U') – all bets are off!

RFID tags and other low-resource devices require that a large portion of their cryptographic computations be outsourced to better equipped computers. When a cryptographic algorithm Alg is divided into a pair of algorithms (T, U), in addition to its security, we also want to know how much work T saves by using U. We want to compare the work that T must do to safely use U to the work required for the fastest known implementation of the functionality T^U.

Definition 3 (α-efficient, secure outsourcing). *A pair of algorithms (T, U) are an α-efficient implementation of an algorithm Alg if (1) they are an outsource-secure implementation of Alg, and (2) $\forall$ inputs x, the running time of T is $\leq$ an α-multiplicative factor of the running time of $Alg(x)$.*

For example, say U relieves T of at least half its computational work; we would call such an implementation $\frac{1}{2}$-efficient. The notion above considers only T's computational load compared to that of Alg. One might also choose to formally consider U's computational burden or the amount of precomputation that T can do in his idle cycles versus his on-demand load. We will not be formally considering these factors.

The above definition of outsource-security does not prevent U' from deviating from its advertised functionality, rather it prevents U' from intelligently choosing her moments for failure based on any secret or protected inputs to Alg (e.g., a public key or the contents of a message). Since this does not rule out unintelligent failures, it is desirable that T have some mechanism for discovering that his software is unsound. Thus, we introduce another characteristic of an outsourcing implementation.

Definition 4 (β-checkable, secure outsourcing). *A pair of algorithms (T, U) are a β-checkable implementation of an algorithm Alg if (1) they are an outsource-secure implementation of Alg, and (2) $\forall$ inputs x, if U' deviates from its advertised functionality during the execution of $T^{U'}(x)$, T will detect the error with probability $\geq \beta$.*

Recall that the reason T purchased U in the first place was to *get out of doing work*, so any testing procedure should be *far more* efficient than computing the function itself; i.e., the overall scheme, including the testing procedure, should remain α-efficient. We combine these characteristics into one final notion.

Definition 5 ((α, β)-outsource-security). *A pair of algorithms (T, U) are an (α, β)-outsource-secure implementation of an algorithm Alg if they are both α-efficient and β-checkable.*

3 Outsource-Secure Exponentiation Using Two Untrusted Programs

Since computing exponentiations modulo a prime is, by far, the most expensive operation in many discrete-log based cryptographic protocols, much research has been done on how to reduce this work-load. We present a method to securely outsource most of the work needed to compute a variable-exponent, variable-base exponentiation modulo a prime, by combining two previous approaches to this problem: (1) using preprocessing tricks to speed-up *offline* exponentiations [27, 28, 17, 8, 25] and (2) *untrusted* server-aided computation [22, 1, 2, 3].

The preprocessing techniques (introduced by Schnorr [27, 28], broken by de Rooij [14, 15, 17], and subsequently fixed by others [16, 9, 21, 8, 25]) seek to optimize the production of random $(k, g^k \bmod p)$ pairs used in signature generation (e.g., El Gamal, Schnorr, DSA) and encryption (e.g., El Gamal, Cramer-Shoup). By *offline*, we mean the randomization factors that are independent of a key or message; that is, exponentiations for a fixed base g, where the user requires nothing more of the exponent k than that it appear random. We leverage these algorithms to speed-up *online* exponentiations as well; that is, given random values $x \in \mathbb{Z}_{ord(G)}$ and $h \in \mathbb{Z}_p^*$, compute $h^x \bmod p$. Generally speaking, given *any* oracle that provides T with random pairs $(x, g^x \bmod p)$ (we discuss the exact implementation of this oracle in Section 3.2), we give a technique for efficiently computing *any* exponentiation modulo p. To do this, we use *untrusted* server-aided (or program-aided) computation.

Blum, Luby, and Rubinfeld gave a general technique for computing and checking the result of a modular exponentiation using four untrusted exponentiation programs – that cannot communicate with each other after deciding on an initial strategy [6]. Their algorithm *leaks* only the size of the inputs (i.e., $|x|, |g|$ for known p) to the programs and runs in time $O(n \log^2 n)$ for an n-bit exponent (this includes the running time of each program). The output of the Blum et al. algorithm is always guaranteed to be correct.

We provide a technique for computing and checking the result of a modular exponentiation using two untrusted exponentiation boxes $U' = (U_1', U_2')$ – that again, cannot communicate with each other after deciding on an initial strategy. In this strategy, at most one of them can deviate from its advertised functionality on a non-negligible fraction of the inputs. Our algorithm reveals no more information than the size of the input and the running time is reduced to $O(\log^2 n)$ multiplications for an n-bit exponent.[1] More importantly, we focus on minimizing the computations done by T to compute an exponentiation, which

---

[1] Disclaimer: these running times assume certain security properties about the EBPV generator [25] which we discuss in detail in Section 3.2.

is $O(\log^2 n)$. This is an asymptotic improvement over the $1.5n$ multiplications needed to compute an exponentiation using square-and-multiply. We gain some of this efficiency by only requiring that an error in the output be detected with probability $\frac{1}{2}$. The rationale is that software malfunctioning on a non-negligible amount of random inputs will not be on the market long.

Our $(O(\frac{\log^2 n}{n}), \frac{1}{2})$-outsource-secure exponentiation implementation, combined with previously known preprocessing tricks, yields a technique for using two untrusted programs U_1, U_2 to securely do most of the resource-intensive work in discrete log based protocols. By way of example, we highlight an asymptotic speed-up in the running time of an honest user T (from $O(n)$ to $O(\log^2 n)$) for the Cramer-Shoup cryptosystem [13] and Schnorr signature verification [27, 28] when using U_1, U_2. Let's lay out the assumptions for using the two untrusted programs more carefully.

3.1 The Two Untrusted Program Model

In the *two untrusted program model*, E writes the code for two (potentially different) programs U_1', U_2'. E then gives this software to T, advertising a functionality that U_1' and U_2' may or may not accurately compute, and T installs this software in a manner such that all subsequent communication between any two of E, U_1' and U_2' must pass through T. The new adversary attacking T (i.e., trying to read T's messages or forge T's signatures) is now $A = (E, U_1', U_2')$.

The *one-malicious* version of this model assumes that at most one the programs U_1', U_2' deviates from its advertised functionality on a non-negligible fraction of the inputs; but we do not know which one and security means that there is a simulator for both. This is the equivalent of buying the "same" advertised software from two different vendors and achieving security as long as one of them is honest without knowing which one.

The concept of an honest party gaining information from two (or more) possibly dishonest, but physically separated parties was first used by Ben-Or, Goldwasser, Kilian, and Wigderson [4] as a method for obtaining interactive proofs without intractability assumptions. Blum et. al. expanded this notion to allow an honest party to check the output of a function rather than just the validity of a proof [7, 6]. Our work, within this model, demonstrates, rather surprisingly, that an honest party T can leverage adversarial software to do the vast majority of its cryptographic computations!

Our $(O(\frac{\log^2 n}{n}), \frac{1}{2})$-outsource-secure implementation of Exp, exponentiation modulo a prime function, appears in Figure 1 and Section 3.3. Figure 1 also demonstrates how to achieve an asymptotic speed-up in T's running time, using Exp and other known preprocessing techniques [25], for the Cramer-Shoup cryptosystem [13] and Schnorr signature verification [27, 28] (this speed-up was already known for Schnorr signature generation [25]).

3.2 *Rand1, Rand2*: Algorithms for Computing $(b, g^b \bmod p)$ Pairs

The subroutine *Rand1* in Figure 1 is initialized by a prime p, a base $g_3 \in \mathbb{Z}_p^*$, and possibly some other values, and then, on each invocation must produce a

---

Outsource-secure Encryption and Signatures: $Alg = (T, U_1, U_2)$
(in the two untrusted program model)

Global Setup (denoted gp as honest, unprotected inputs)
  ◦ Security parameter: 1^k.
  ◦ Global Encryption parameters: a group G of prime order q with generators g_1, g_2, a (can be weakly) collision-resistant hash function $H : \{0,1\} \to \mathbb{Z}_q$.
  ◦ Global Signature parameters: a k-bit prime q, $p = 2q+1$, a generator g_3 for $\mathbb{Z}_p^*$, and a collision-resistant hash function $H : \{0,1\}^* \to \mathbb{Z}_q$.

Advertised Functionality of U_1 and U_2
  ◦ $U_1(b, g) \to g^b$
  ◦ $U_2(b, g) \to g^b$

Subroutines Executed by T with access to U_1, U_2
  ◦ $Rand1 \to (b, g_3^b)$. T computes alone as in Section 3.2.
  ◦ $Rand2 \to (b, g_1^b, g_2^b)$. T computes alone as in Section 3.2.
  ◦ $Exp(a, u) \to u^a$. T uses U_1 and U_2 to compute u^a as in Section 3.3.

Functionality of $Alg = (T, U_1, U_2)$

Outsource-Secure Cramer-Shoup Cryptosystem [13]
Key Generation: Generated by an honest process on input 1^k:
  $PK = (B = g_1^{x_1} g_2^{x_2}, C = g_1^{y_1} g_2^{y_2}, D = g_1^z), SK = (x_1, x_2, y_1, y_2, z)$.
Encryption: $Alg.Enc(m, (PK, t), gp, \varepsilon, \varepsilon) \to (\varepsilon, \tau, \varepsilon)$.
  On input $PK = (B, C, D), m \in G$, and $t \in \{0,1\}^*$,
  1. T computes $Rand2 \to (r, u_1 = g_1^r, u_2 = g_2^r)$.
  2. T computes $Exp(r, D) \to D^r, e = D^r m, \kappa = H(u_1, u_2, e, t)$.
  3. T computes $Exp(r, B) \to B^r, Exp(r\kappa, C) \to C^{r\kappa}, v = B^r C^{r\kappa}$.
  4. T outputs the ciphertext $\tau = (u_1, u_2, e, v, t)$.
Decryption: $Alg.Dec(SK, \varepsilon, gp, \tau, \varepsilon) \to (m, \varepsilon, \varepsilon)$.
  (If E generates ciphertext, $Alg.Dec(SK, \varepsilon, gp, \tau, \varepsilon) \to (\varepsilon, m, \varepsilon)$.)
  On input $SK = (x_1, x_2, y_1, y_2, z)$ and $\tau = (u_1, u_2, e, v, t)$,
  1. T computes $\kappa = H(u_1, u_2, e, t)$.
  2. T computes $Exp(x_1 + \kappa y_1, u_1) \to \alpha, Exp(x_2 + \kappa y_2, u_2) \to \beta$.
  3. T checks if $\alpha\beta = v$; if not, it outputs "invalid".
  4. Otherwise, T computes $Exp(z, u_1) \to \delta$ and outputs $m = e/\delta$.

Outsource-Secure Schnorr Signatures [27, 28]
Key Generation: Generated by an honest process: $SVK = g^a, SSK = a$.
Signature Generation: $Alg.Sign(SSK, m, gp, \varepsilon, \varepsilon) \to (\varepsilon, \sigma, \varepsilon)$.
  On input $SSK = a$ and $m \in \{0,1\}^*$,
  1. T computes $Rand1 \to (k, r = g_3^k)$.
  2. T computes $e = H(r||m)$ and $s = ae + k \mod q$.
  3. T outputs the signature $\sigma = (r, s)$.
Signature Verification: $Alg.Vf(\varepsilon, SVK, gp, (m, \sigma), \varepsilon) \to (\varepsilon, \{0,1\}, \varepsilon)$.
  On input $SVK = y, m \in \{0,1\}^*$, and $\sigma = (r, s)$,
  1. T checks that $1 \le r \le p-1$, if not, it outputs 0.
  2. T computes $e = H(r||m), Exp(s, g_3) \to \alpha$ and $Exp(e, y) \to \beta$.
  3. T checks that $\alpha = \beta r$. If so, T outputs 1; otherwise it outputs 0.

Fig. 1. An honest user T, given untrusted exponentiation boxes U_1, U_2, achieves outsource-secure encryption and signatures

random, independent pair of the form $(b, g_3^b \bmod p)$, where $b \in \mathbb{Z}_q$. The subroutine $Rand2$ is the natural extension of $Rand1$ initialized by two bases g_1, g_2 and producing triplets $(b, g_1^b \bmod p, g_2^b \bmod p)$. Given a, perhaps expensive, initialization procedure, we want to see how expeditiously these subroutines can be executed by T.

One naive approach is for a *trusted* server to compute a table of random, independent pairs and triplets in advance and load it into T's memory. Then on each invocation of $Rand1$ or $Rand2$, T simply retrieves the next value in the table. (We will see that this table, plus access to the untrusted servers, allows an honest device to compute any exponentiation by doing only 9 multiplications regardless of exponent size!)

For devices that are willing to do a little more work, in exchange for requiring less storage, we apply well-known preprocessing techniques for this exact functionality. Schnorr first proposed an algorithm which, takes as input a small set of truly random (k, g^k) pairs and then, produces a long series of "nearly random" (r, g^r) pairs as a means of speeding-up signature generation in smartcards [27]. However, the output of Schnorr's algorithm is too dependent, and de Rooij found a series of equations that allow the recovery of a signer's secret key [14]. A subsequent fix by Schnorr [28] was also broken by de Rooij [15, 17]. Since then several new preprocessing algorithms were proposed [16, 9, 21, 8, 25]. Among the most promising is the EBPV generator by Nguyen, Shparlinski, and Stern [25], which adds a feedback extension (i.e., reuse of the output pairs) to the BPV generator proposed by Boyko, Peinado, and Venkatesan [8], which works by taking a subset of truly random (k, g^k) pairs and combing them with a random walk on expanders on Cayley graphs to reduce the dependency of the pairs in the output sequence. The EBPV generator, secure against adaptive adversaries, runs in time $O(\log^2 n)$ for an n-bit exponent. (This holds for the addition of a second base in $Rand2$ as well.)

A critical property that we will shortly need from $Rand1$ and $Rand2$ is that their output sequences be computationally indistinguishable from a truly random output sequence. It is conjectured that with sufficient parameters (i.e., number of initial (k, g^k) pairs, etc.) the output distribution of the EBPV generator is statistically-close to the uniform distribution [25]. We make this working assumption throughout our paper. In the event that this assumption is false, our recourse is to use the naive approach above and, thus, further reduce our running time, in exchange for additional memory.

3.3 *Exp*: Outsource-Secure Exponentiation Modulo a Prime

Our main contribution for Section 3 lies in the subroutine *Exp* from Figure 1. In *Exp*, T out-sources its exponentiation computations, while maintaining its privacy, by invoking U_1 and U_2 on a series of (exponent, base) pairs that appear random in the limited view of the software.

The *Exp* Algorithm. Let primes p, q be the global parameters, where $\mathbb{Z}_p^*$ has order q. *Exp* takes as input $a \in \mathbb{Z}_q$ and $u \in \mathbb{Z}_p^*$, and outputs $u^a \bmod p$. As used in Figure 1, *Exp*'s input a may be secret or (honest/adversarial) protected; its input u may be (honest/adversarial) protected; and its output is always secret or protected. Exp also receives the (honest, unprotected) global parameters gp; there are no adversarial, unprotected inputs. All (secret/protected) inputs are computationally blinded before being sent to U_1 or U_2.

To implement this functionality using (U_1, U_2), T runs $Rand1$ twice to create two *blinding pairs* (α, g^α) and (β, g^β). We denote

$$v = g^\alpha \text{ and } v^b = g^\beta, \text{ where } b = \beta/\alpha.$$

Our goal is to logically break u and a into random looking pieces that can then be computed by U_1 and U_2. Our first logical divisions are

$$u^a = (vw)^a = v^a w^a = v^b v^c w^a, \text{ where } w = u/v \text{ and } c = a - b.$$

As a result of this step, u is hidden, and the desired value u^a is expressed in terms of random v and w. Next, T must hide the exponent a. To that end, it selects two *blinding elements* $d \in \mathbb{Z}_q$ and $f \in G$ at random. Our second logical divisions are

$$v^b v^c w^a = v^b (fh)^c w^{d+e} = v^b f^c h^c w^d w^e, \text{ where } h = v/f \text{ and } e = a - d.$$

Next, T fixes two *test* queries per program by running $Rand1$ to obtain (t_1, g^{t_1}), (t_2, g^{t_2}), (r_1, g^{r_1}) and (r_2, g^{r_2}). T queries U_1 (in random order) as

$$U_1(d, w) \to w^d, \ U_1(c, f) \to f^c, \ U_1(t_1/r_1, g^{r_1}) \to g^{t_1}, \ U_1(t_2/r_2, g^{r_2}) \to g^{t_2},$$

and then queries U_2 (in random order) as

$$U_2(e, w) \to w^e, \ U_2(c, h) \to h^c, \ U_2(t_1/r_1, g^{r_1}) \to g^{t_1}, \ U_2(t_2/r_2, g^{r_2}) \to g^{t_2}.$$

(Notice that *all* queries to U_1 can take place before any queries to U_2 must be made.) Finally, T checks that the test queries to U_1 and U_2 both produce the correct outputs (i.e., g^{t_1} and g^{t_2}). If not, T outputs "error"; otherwise, it multiplies the real outputs of U_1, U_2 with v^b to compute u^a as

$$v^b f^c h^c w^d w^e = v^{b+c} w^{d+e} = v^a w^a = (vw)^a = u^a.$$

We point out that this exponentiation outsourcing only needs the $Rand1$ functionality; the $Rand2$ functionality discussed in Figure 1 is used for the Cramer-Shoup outsourcing.

Theorem 1. *In the one-malicious model, the above algorithms $(T, (U_1, U_2))$ are an outsource-secure implementation of Exp, where the input (a, u) may be honest, secret, or honest, protected, or adversarial, protected.*

Proof of Theorem 1 is in the full version of the paper. The correctness property is fairly straight-forward. To show security, both simulators S_1 and S_2 send random (exponent, base) pairs to the untrusted components U_1' and U_2'. One can see that such an S_2 simulates a view that is computationally indistinguishable from the real world view for the untrusted helper. However, it is our construction of S_1 which must simulate a view for the environment that requires the one-malicious model. Consider what might happen in the real world if both U_1' and U_2' deviate from their advertised functionalities. While the event that U_1' misbehaves is independent of the input (a, u), and the same is true for the event that U_2' misbehaves, the event that *both* of them misbehave is *not independent* of the input (a, u).

Lemma 1. *In the one-malicious model, the above algorithms $(T, (U_1, U_2))$ are an $O(\frac{\log^2 n}{n})$-efficient implementation of Exp.*

Proof. Raising an arbitrary base to an arbitrary power by the square-and-multiply method takes roughly $1.5n$ modular multiplications (MMs) for an n-bit exponent. *Exp* makes six calls to *Rand*1 plus 9 other MMs (additions are negligible by comparison). *Exp* takes $O(\log^2 n)$ MMs using the EBPV generator [25] for *Rand*1 and $O(1)$ MMs when using a table-lookup for *Rand*1. □

Lemma 2. *In the one-malicious model, the above algorithms $(T, (U_1, U_2))$ are a $\frac{1}{2}$-checkable implementation of Exp.*

Proof. By Theorem 1, U_1 (resp,. U_2) cannot distinguish the two test queries from the two real queries T makes. If U_1 (resp., U_2) fails during any execution of *Exp*, it will be detected with probability $\frac{1}{2}$. □

We combine Theorem 1, Lemmas 1 and 2, and known preprocessing techniques [25] to arrive at the following result. (Schemes differ in β-checkability depending on the number of *Exp* calls they make.)

Theorem 2. *In the one-malicious model, the algorithms $(T, (U_1, U_2))$ in Figure 1 are (1) an $(O(\frac{\log^2 n}{n}), \frac{1}{2})$-outsource-secure implementation of Exp, (2) an $(O(\frac{\log^2 n}{n}), \frac{7}{8})$-outsource-secure implementation of the Cramer-Shoup cryptosystem [13], and (3) an $(O(\frac{\log^2 n}{n}), \frac{3}{4})$-outsource-secure implementation of Schnorr Signatures [27, 28].*

4 Outsource-Secure Encryption Using One Untrusted Program

Suppose Alice is given an encryption program that is guaranteed to work correctly on all but a negligible fraction of adversarially-chosen public keys and messages. She wants to trick this software into efficiently helping her encrypt *any* message for *any* intended recipient – even those in the (unknown) set for which her software adversarially fails. This is a non-trivial exercise when one wants to hide both the public key and the message from the software – and even more so, when one wants to achieve CCA2-secure encryption, as we do.

Section 3 covered an $O(\frac{\log^2 n}{n})$-efficient outsource-secure CCA2 encryption scheme using *two* untrusted programs. Here, using only *one* untrusted program, we remain $O(\frac{\log^2 n}{n})$-efficient and CCA2-secure. To efficiently use only one program, one must assume that the software behaves honestly on random inputs with high-probability. After all, it isn't hard to imagine a commercial product that works most of the time, but has a few surprises programmed into it such that on a few inputs it malfunctions. Moreover, some assumption about the correctness of a probabilistic program is necessary since there will be no means of checking its output. (To see this, consider that there is no way for T to know if the "randomness" U' used during the probabilistic encryption was genuinely random or a value known by E.) In Figure 2, present an outsource-secure implementation for CCA2-secure encryption only. We leave open the problem of efficiently outsourcing the decryption of *these* ciphertexts, as well as *any* signature verification algorithm, using only one untrusted program.

The One Untrusted Program Model. This model is analogous to the two untrusted program model in Section 3.1, where only one of U_1, U_2 is available to T and the advertised functionality of U is *tagged* Cramer-Shoup encryption [13]. (Recall that ciphertexts in tagged CS encryption include a public, non-malleable string called a tag.)

4.1 *Com*: Efficient, Statistically-Hiding Commitments

We use Halevi and Micali's commitment scheme based on collision-free hash families [20]. Let $HF : \{0,1\}^{O(k)} \to \{0,1\}^k$ be a family of universal hash functions and let $MD : \{0,1\}^* \to \{0,1\}^k$ be a collision-free hash function. Given any value $m \in \{0,1\}^*$ and security parameter k, generate a statistically-hiding commitment scheme as follows: (1) compute $s = MD(m)$, (2) pick $h \in HF$ and $x \in \{0,1\}^{O(k)}$ at random, so that $h(x) = s$ and (3) compute $y = MD(x)$. (One can construct h by randomly selecting A and computing $b = s - Ax$ modulo a prime set in HF.) The commitment is $\phi_C = (y, h)$. The decommitment is $\phi_D = (x, m)$. Here, we denote the commitment scheme as *Com* and the decommitment scheme as *Decom*.

4.2 CCA2 and Outsource-Security of T^U Encryption

First, we observe that the Cramer-Shoup variant in Figure 2 is CCA2-secure [26]. Here, we only need to look at the *honest* algorithm T^U.

Theorem 3. *The cryptosystem T^U is secure against adaptive chosen-ciphertext attack (CCA2) assuming the CCA2-security of Cramer-Shoup encryption [13] and the security of the Halevi-Micali commitment scheme [20].*

The full proof of Theorem 3 is included in the full version of this paper. It follows a fairly standard reduction from tagged Cramer-Shoup.

Using *Exp*, we achieve the same asymptotic speed-up as in Section 3. Checking the output of this probabilistic functionality is theoretically impossible. Thus, we summarize the properties of this scheme as:

Theorem 4. *The algorithms $(T, U).Enc$ in Figure 2 are an $O(\frac{\log^2 n}{n})$-efficient, outsource-secure implementation of CCA2-secure encryption.*

Outsource-secure Encryption: $Alg = (T, U)$
(in the one untrusted program model)

Global Setup (denoted gp as honest, unprotected inputs)
○ Security parameter: 1^k.
○ Global Encryption parameters: a group G of prime order q with generators g_1, g_2, a weakly collision-resistant hash function $H : \{0,1\} \to \mathbb{Z}_q$, and a statistically-hiding commitment scheme $Com : \{0,1\}^* \to \phi_C$ with a decommitment of ϕ_D.

Advertised Functionality of U: CCA2-Secure Cramer-Shoup Encryption [13]
○ $U(pk, m, t) \to \tau = (u_1, u_2, e, v, t)$. (See Figure 1 for details.)

Subroutines Executed by T
○ $Rand1 \to (b, g_1^b)$. (See Section 3.2 for details.)

Functionality of $Alg = (T, U)$: CCA2 and Outsource-Secure Cryptosystem
Key Generation: Generated by an honest process on input 1^k: $PK = (B = g_1^{x_1} g_2^{x_2}, C = g_1^{y_1} g_2^{y_2}, D = g_1^z)$, $SK = (x_1, x_2, y_1, y_2, z)$.
Encryption: $Alg.Enc(m, (PK, t), gp, \varepsilon, \varepsilon) \to (\varepsilon, \phi_D, \tau)$.
On input $PK = (B, C, D)$, $m \in G$, and $t \in \{0,1\}^*$,
1. T computes $Rand1 \to (x'_1, g_1^{x'_1})$, $Rand1 \to (y'_1, g_1^{y'_1})$, $Rand1 \to (z', g_1^{z'})$.
2. T computes $PK' = (Bg_1^{x'_1}, Cg_1^{y'_1}, Dg_1^{z'})$.
3. T selects a random $w \in G$ and computes $\beta = wm$.
4. T computes $(\phi_C, \phi_D) = Com(\beta||t||x'_1||y'_1||z')$.
5. T calls $U(PK', w, \phi_C) \to \tau$, where $\tau = (u_1, u_2, e, v, \phi_C)$.
6. T outputs the ciphertext (τ, ϕ_D).
Decryption: $Alg.Dec(SK, (\tau, \phi_D), gp, \varepsilon, \varepsilon) \to (m, \varepsilon, \varepsilon)$.
(If E generates ciphertext, $Alg.Dec(SK, \varepsilon, gp, (\tau, \phi_D), \varepsilon) \to (\varepsilon, m, \varepsilon)$.)
On input $SK = (x_1, x_2, y_1, y_2, z)$ and $(\tau = (u_1, u_2, e, v, \phi_C), \phi_D)$,
1. T computes $(\beta||t||x'_1||y'_1||z') = Decom(\phi_C, \phi_D)$.
2. T computes $\hat{x}_1 = x_1 + x'_1$, $\hat{y}_1 = y_1 + y'_1$, $\hat{z} = z + z'$.
3. T computes $\kappa = H(u_1, u_2, e, \phi_C)$, $\alpha = u_1^{\hat{x}_1 + \kappa \hat{y}_1}$, and $\pi = u_2^{x_2 + \kappa y_2}$.
4. T checks if $\alpha \pi = v$; if not, it outputs "invalid".
5. Otherwise, T computes $w = e/u_1^{\hat{z}}$ and outputs $m = \beta/w$ with tag t.

Fig. 2. An honest user T, given untrusted Cramer-Shoup encryption software U, achieves outsource-secure encryption. Note that the speed-up is for encryption only, not decryption

Proof sketch. All inputs to U, besides the (honest, unprotected) global parameters, are computationally blinded by T. The public key is re-randomized using $Rand1$; a random message $w \in G$ is selected; and the tag ϕ_C, that binds these new values to the old key and message, is a statistically-hiding commitment. Thus, both S_1 and S_2 query U' on random triplets of the form $(PK \in (\mathbb{Z}_p^*)^3, w \in G, t \in \{0,1\}^{|\phi_C|})$. In pair one, S_1 always sets $replace^i = 0$, since the output of $T^{U'}$ in the real experiment is wrong with negligible probability. For efficiency, observe that the commitment scheme Com can be implemented with only a constant number of modular multiplications.

Acknowledgments. We are grateful to Ron Rivest for many useful discussions and for suggesting the use of the generators in Section 3.2. We also thank Srini Devadas, Shafi Goldwasser, Matt Lepinski, Alon Rosen, and the anonymous referees for comments on earlier drafts.

References

1. M. Abadi, J. Feigenbaum, and J. Kilian. On hiding information from an oracle. *Journal of Comput. Syst. Sci.*, 39(1):21–50, 1989.
2. D. Beaver and J. Feigenbaum. Hiding instances in multioracle queries. In *Proceedings of STAC '90*, pages 37–48, 1990.
3. D. Beaver, J. Feigenbaum, J. Kilian, and P. Rogaway. Locally random reductions: Improvements and applications. *Journal of Cryptology*, 10(1):17–36, 1997.
4. M. Ben-Or, S. Goldwasser, J. Kilian, and A. Wigderson. Multi-prover interactive proofs: How to remove intractability assumptions. In *Proceedings of STOC*, pages 113–131, 1988.
5. M. Blum and S. Kannan. Designing programs that check their work. *Journal of the ACM*, pages 269–291, 1995.
6. M. Blum, M. Luby, and R. Rubinfeld. Program result checking against adaptive programs and in cryptographic settings. *DIMACS Series in Discrete Mathematics and Theoretical Computer Science*, pages 107–118, 1991.
7. M. Blum, M. Luby, and R. Rubinfeld. Self-testing/correcting with applications to numerical problems. *Journal of Computer and System Science*, pages 549–595, 1993.
8. V. Boyko, M. Peinado, and R. Venkatesan. Speeding up discrete log and factoring based schemes via precomputations. In *Proceedings of Eurocrypt '98*, volume 1403 of LNCS, pages 221–232, 1998.
9. E. F. Brickell, D. M. Gordon, K. S. McCurley, and D. B. Wilson. Fast exponentiation with precomputation. In *Proceedings of Eurocrypt '92*, volume 658 of LNCS, pages 200–207, 1992.
10. D. Chaum. Security without identification: transaction systems to make big brother obsolete. *Communications of the ACM*, 28(10):1030–1044, 1985.
11. D. Chaum and T. P. Pedersen. Wallet Databases with Observers. In *Proceedings of Crypto '92*, volume 740 of LNCS, pages 89–105, 1992.
12. D. Clarke, S. Devadas, M. van Dijk, B. Gassend, and G. E. Suh. Speeding up Exponentiation using an Untrusted Computational Resource. Technical Report Memo 469, MIT CSAIL Computation Structures Group, August 2003.
13. R. Cramer and V. Shoup. Design and analysis of practical public-key encryption schemes secure against adaptive chosen ciphertext attack. *SIAM Journal of Computing*, 2003. To appear. Available at http://www.shoup.net/papers.
14. P. de Rooij. On the security of the Schnorr scheme using preprocessing. In *Proceedings of Eurocrypt '91*, volume 547 of LNCS, pages 71–80, 1991.
15. P. de Rooij. On Schnorr's preprocessing for digital signature schemes. In *Proceedings of Eurocrypt '93*, volume 765 of LNCS, pages 435–439, 1993.
16. P. de Rooij. Efficient exponentiation using precomputation and vector addition chains. In *Proceedings of Eurocrypt 1994*, volume 950 of LNCS, pages 389–399, 1994.
17. P. de Rooij. On Schnorr's preprocessing for digital signature schemes. *Journal of Cryptology*, 10(1):1–16, 1997.

18. M. Franklin and M. Yung. The blinding of weak signatures (extended abstract). In *Proceedings of Eurocrypt '95*, volume 950 of LNCS, pages 67–76, 1995.
19. I. Goldberg, D. Wagner, R. Thomas, and E. A. Brewer. A secure environment for untrusted helper applications. In *Proceedings of the 6th Usenix Security Symposium*, 1996.
20. S. Halevi and S. Micali. Practical and provably-secure commitment schemes from collision-free hashing. In *Proceedings of Crypto '96*, volume 1109 of LNCS, pages 201–212, 1996.
21. C. H. Lim and P. J. Lee. More flexible exponentiation with precomputation. In *Proceedings of Crypto '94*, volume 839 of LNCS, pages 95–107, 1994.
22. T. Matsumoto, K. Kato, and H. Imai. Speeding up secret computations with insecure auxiliary devices. In *Proceedings of Crypto '88*, volume 403 of LNCS, pages 497–506, 1988.
23. G. C. Necula and S. P. Rahul. Oracle-based checking of untrusted software. *ACM SIGPLAN Notices*, 36(3):142–154, 2001.
24. P. Q. Nguyen and I. Shparlinski. On the insecurity of a server-aided RSA protocol. In *Proceedings of Asiacrypt 2001*, volume 2248 of LNCS, pages 21–35, 2001.
25. P. Q. Nguyen, I. E. Shparlinski, and J. Stern. Distribution of modular sums and the security of server aided exponentiation. In *Proceedings of the Workshop on Comp. Number Theory and Crypt.*, pages 1–16, 1999.
26. C. Rackoff and D. Simon. Noninterative zero-knowledge proof of knowledge and chosen ciphertext attack. In *Proceedings of Crypto '91*, volume 576 of LNCS, pages 433–444, 1991.
27. C.-P. Schnorr. Efficient identification and signatures for smart cards. In *Proceedings of Crypto '89*, volume 435 of LNCS, 1989.
28. C.-P. Schnorr. Efficient signature generation by smart cards. *Journal of Cryptography*, 4:161–174, 1991.
29. Trusted Computing Group. Trusted computing platform alliance, main specification version 1.1b, 2004. Date of Access: February 10, 2004.
30. D. A. Wagner. Janus: an approach for confinement of untrusted applications. Technical Report CSD-99-1056, UC Berkeley, 12, 1999.
31. B. R. Waters, E. W. Felten, and A. Sahai. Receiver anonymity via incomparable public keys. In *Proceedings of the 10th ACM CCS Conference*, pages 112–121, 2003.

Secure Computation of the Mean and Related Statistics

Eike Kiltz[1,3], Gregor Leander[1], and John Malone-Lee[2]

[1] Fakultät für Mathematik,
Ruhr-Universität Bochum, 44780 Bochum, Germany
gregor.leander@ruhr-uni-bochum.de
[2] University of Bristol, Department of Computer Science,
Woodland Road, Bristol, BS8 1UB, UK
malone@cs.bris.ac.uk
[3] University of Southern California at San Diego,
9500 Gilman Drive, La Jolla, CA 92093-0114, USA
ekiltz@cs.ucsd.edu

Abstract. In recent years there has been massive progress in the development of technologies for storing and processing of data. If statistical analysis could be applied to such data when it is distributed between several organisations, there could be huge benefits. Unfortunately, in many cases, for legal or commercial reasons, this is not possible.

The idea of using the theory of multi-party computation to analyse efficient algorithms for *privacy preserving data-mining* was proposed by Pinkas and Lindell. The point is that algorithms developed in this way can be used to overcome the apparent impasse described above: the owners of data can, in effect, pool their data while ensuring that privacy is maintained.

Motivated by this, we describe how to securely compute the mean of an attribute value in a database that is shared between two parties. We also demonstrate that existing solutions in the literature that could be used to do this leak information, therefore underlining the importance of applying rigorous theoretical analysis rather than settling for ad hoc techniques.

1 Introduction

In recent years there has been massive progress in the development of technologies for networking, storage and data processing. Such progress has allowed the creation of enormous databases storing unprecedented quantities of information. This possibility to store and process huge quantities of information throws up the question of privacy. The need for privacy may be a legal requirement, in the UK for example there are strict rules for any party that holds information about individuals [24]. It may also be motivated by commercial interests: a pharmaceutical company does not want the results of its trials to become available while products are still being developed based on these results. On the other hand, if it were not for such privacy considerations, there could be significant mutual benefit in pooling data for research purposes, whether it be scientific, economic or market research. This apparent impasse is the motivating factor for our work.

We consider a situation in which there are two parties, each owning a database. Suppose that there is some attribute present in both databases. We propose a protocol which the two parties can use to evaluate the mean of this attribute value for the union of their databases. This is done in such a way that, at the end of the protocol, the two parties learn the mean and nothing else. No trusted party is required.

Related Work. The problem that we have described above is a case of *secure two-party computation*. This notion was first investigated by Yao who proposed a general solution [28]. The two-party case was subsequently generalised to the multi-party case [3, 6, 18]. Although these solutions are general, they may not be terribly efficient when used with huge inputs and complex algorithms. We are therefore interested in a tailor-made protocol for the problem in question.

In [20, 21] Pinkas and Lindell analysed an algorithm for data-mining in the model for secure two-party computation. This work has stimulated research in the cryptography community into tools for working securely with large, distributed databases [1, 15].

Several algorithms that could be used for two parties to compute the mean of their combined data have already been proposed [9, 11, 12]. None of these solutions have been analysed in the formal model of security for two-party computation; moreover, in the appendices of this paper we demonstrate that they leak information. Similar weaknesses are also found in related protocols proposed elsewhere [7, 9, 10, 11, 12, 13, 25, 26].

Outline. The paper proceeds as follows. In Section 2 we discuss the notion of secure two-party computation that we will be working with. We describe how our protocol works in Section 3 and Section 4: in Section 3 we assume the existence of oracles to compute the low-level functions required by our protocol and in Section 4 we give secure implementations of these oracles. We conclude Section 4 by comparing of our protocol with Yao's general solution for two-party computation applied to computing the mean. In the appendices we discuss other solutions that have been proposed and show why they are insecure.

2 Secure Two-Party Computation: Definitions and Results

We define secure two-party computation following Goldreich [17]. An equivalent model and results may be found in [4]. Henceforth, all two-party protocols will involve the two parties P1 and P2.

We will consider *semi-honest* adversaries. A semi-honest adversary is an adversary that follows the instructions defined by the protocol; however, it might try to use the information that it obtains during the execution of the protocol to learn something about the input of the other party. Using techniques such as the GMW compiler of Canetti et al. [5], a protocol that is secure against semi-honest adversaries can be made secure against adversaries that attempt to deviate from the protocol.

2.1 Definitions

Using the notation of Pinkas and Lindell [20, 21], let $f : \{0,1\}^* \times \{0,1\}^* \to \{0,1\}^* \times \{0,1\}^*$ be a function. Denote the first element of $f(x_1, x_2)$ by $f_1(x_1, x_2)$ and the second by $f_2(x_1, x_2)$. Let π be a two-party protocol for computing f. The views of P1 and P2 during an execution of $\pi(x_1, x_2)$, denoted $\text{view}_1^\pi(x_1, x_2)$ and $\text{view}_2^\pi(x_1, x_2)$ respectively, are

$$\text{view}_1^\pi(x_1, x_2) := (x_1, r_1, m_{1,1}, \ldots, m_{1,t}) \text{ and}$$
$$\text{view}_2^\pi(x_1, x_2) := (x_2, r_2, m_{2,1}, \ldots, m_{2,t})$$

where r_i denotes Pi's random input, and $m_{i,j}$ denotes the j-th message received by Pi. The outputs P1 and P2 during an execution of $\pi(x_1, x_2)$ are denoted $\text{output}_1^\pi(x_1, x_2)$ and $\text{output}_2^\pi(x_1, x_2)$ respectively. We define

$$\text{output}^\pi(x_1, x_2) := (\text{output}_1^\pi(x_1, x_2), \text{output}_2^\pi(x_1, x_2)).$$

Definition 1 (Privacy w.r.t. semi-honest behaviour). *We say that π privately computes a function f if there exist probabilistic, polynomial-time algorithms S_1 and S_2 such that*

$$\{S_1(x_1, f_1(x_1, x_2)), f(x_1, x_2)\} \equiv \{\text{view}_1^\pi(x_1, x_2), \text{output}^\pi(x_1, x_2)\}, \text{ and} \quad (1)$$
$$\{S_2(x_2, f_2(x_1, x_2)), f(x_1, x_2)\} \equiv \{\text{view}_2^\pi(x_1, x_2), \text{output}^\pi(x_1, x_2)\} \quad (2)$$

where $\equiv$ denotes computational indistinguishability.

Equations (1) and (2) state that the view of the parties can be simulated given access to the party's input and output only. Recall that the adversary here is semi-honest and therefore the view is exactly according to the protocol definition. Note that it is not sufficient for simulator S_i to generate a string indistinguishable from $\text{view}_i(x_1, x_2)$: the joint distribution of the simulator's output and the functionality output $f(x_1, x_2)$ must be indistinguishable from $\{\text{view}_i^\pi(x_1, x_2), \text{output}^\pi(x_1, x_2)\}$. This is necessary for probabilistic functionalities [17].

A Simpler Formulation for Deterministic Functionalities. In the case that the functionality f is deterministic, it suffices to require that simulator S_i generates the view of party Pi, without considering the joint distribution with the output. That is, we can require that there exist S_1 and S_2 such that

$$\{S_1(x_1, f_1(x_1, x_2))\} \equiv \{\text{view}_1^\pi(x_1, x_2)\}, \text{ and}$$
$$\{S_2(x_2, f_2(x_1, x_2))\} \equiv \{\text{view}_2^\pi(x_1, x_2)\}.$$

The reason that this suffices is that when f is deterministic, $\text{output}^\pi(x_1, x_2)$ must equal $f(x_1, x_2)$. See [17] for a more complete discussion.

Private Approximations. If we privately compute an approximation of a function, we may reveal more information than we would by computing the function itself. To capture this we use the framework of Feigenbaum et al. [14] for private approximations. We restrict ourself to the case of deterministic functions f.

We say that $\hat{f}$ is an ε-approximation of f if, for all inputs (x_1, x_2),

$$|f(x_1, x_2) - \hat{f}(x_1, x_2)| < \varepsilon.$$

Definition 2. *We say that $\hat{f}$ is* functionally private with respect to f *if there exist a probabilistic, polynomial-time algorithm S such that*

$$S(f(x_1, x_2)) \equiv \hat{f}(x_1, x_2)$$

where $\equiv$ denotes computational indistinguishability.

Definition 3. *Let f be a deterministic function. We say that π* privately computes an ε-approximation of function f *if π privately computes a (possibly randomised) function $\hat{f}$ such that $\hat{f}$ is functionally private with respect to f and $\hat{f}$ is an ε-approximation of f.*

2.2 Secure Composition of Two-Party Protocols

Before stating the composition theorem that we will use, we first need to define two notions: *oracle-aided protocols* and *reducibility of protocols*.

Definition 4 (Oracle-aided protocols). *An* oracle-aided protocol *is a protocol augmented by two things: (1) pairs of oracle-tapes, one for each party; and (2) oracle-call steps. An oracle-call step proceeds by one party sending a special oracle-request message to the other party. Such a message is typically sent after the party has written a string, its* query, *to its write-only oracle-tape. In response the other party writes its own query to its write-only oracle-tape and responds to the requesting party with an* oracle-call *message. At this point the oracle is invoked and the result is that a string is written onto the read-only oracle-tape of each party. Note that these strings may not be the same. This pair of strings is the* oracle answer.

In an oracle-aided protocol, oracle-call steps are only ever made sequentially, never in parallel.

Definition 5 (Reducibility of protocols)

- *An oracle-aided protocol* uses the oracle-functionality f *if its oracle answers according to f. That is, when the oracle is invoked with requesting party query x_1 and responding party query x_2, the oracle answers $f_1(x_1, x_2)$ to the requesting party and $f_2(x_1, x_2)$ to the responding party.*
- *An oracle-aided protocol using the oracle functionality f is said to privately compute a function g if there exist polynomial-time algorithms S_1 and S_2 that satisfy (1) and (2) (from Definition 1) respectively, where the corresponding views of the oracle-aided protocol g are defined in the natural manner.*

- An oracle-aided protocol privately reduces g to f if it privately computes g when using the oracle-functionality f. If this is so, we say that g is privately reducible to f.

Theorem 1 (Composition theorem [17]). *Suppose that g is privately reducible to f and that there exists a protocol for privately computing f. Then, the protocol g' derived from g by replacing oracle calls to f with the protocol for computing f privately computes g.*

Theorem 1 above will greatly simplify our analysis. It allows us to first describe and analyse an oracle-aided protocol in Section 3 before separately discussing how to implement the low-level details in Section 4.

3 An Oracle-Aided Protocol for Computing the Mean

In this section we describe oracle-aided protocol for computing an approximation of the mean. We first describe the oracles that we will use in Section 3.1 before discussing the actual protocol in Section 3.2. The notation that we will be using in the remainder of the paper is described below.

Notation. Let a be a real number. We denote by $\lfloor a \rfloor$ the largest integer $b \leq a$, by $\lceil a \rceil$ the smallest integer $b \geq a$, and by $\lceil a \rfloor$ the largest integer $b \leq a + 1/2$. We denote by $\mathsf{trunc}(a)$ the integer b such that $b = \lceil a \rceil$ if $a < 0$ and $b = \lfloor a \rfloor$ if $a \geq 0$; that is, $\mathsf{trunc}(a)$ rounds a towards 0.

Let p be a positive integer. All arithmetic modulo p is done centred around 0; that is $c \bmod p = c - \lceil c/p \rfloor p$.

3.1 The Oracles

Much of our computation is going to occur in a suitably large finite field. Henceforth we denote this field $\mathbb{F}_p$. We will elaborate on what "suitably large" means in Section 4 when we discuss how to implement the oracles used by our protocol.

We will make use of various oracles for two-out-of-two sharing. These are described below. In all the definitions we assume that P1 and P2 input y_1 and y_2 to the oracle respectively, and we let f be a function of (y_1, y_2).

Oracle for additive shares of $f(y_1, y_2)$ **over** $\mathbb{F}_p$: The oracle chooses s_1 at random from $\mathbb{F}_p$, sends s_1 to P1, and sends $s_2 = f(y_1, y_2) - s_1$ to P2. Players now hold s_1 and s_2 such that $s_1 + s_2 = f(y_1, y_2)$.

Oracle for multiplicative shares of $f(y_1, y_2)(\neq 0)$ **over** $\mathbb{F}_p^*$: The oracle chooses s_1 at random from $\mathbb{F}_p^*$, sends s_1 to P1, and sends $s_2 = f(y_1, y_2)/s_1$ to P2. Players now hold s_1 and s_2 such that $s_1 s_2 = f(y_1, y_2)$.

Oracle for additive shares of $f(y_1, y_2)$ **over the integers:** This definition requires a little more care. First we assume that $f(y_1, y_2) \in [-A, A]$ for some A and we let ρ be a security parameter. Now, the oracle chooses s_1 at random from $[-A2^\rho, A2^\rho]$, sends s_1 to P1, and sends $s_2 = f(y_1, y_2) - s_1$ to P2. Players now hold s_1 and s_2 such that $s_1 + s_2 = f(y_1, y_2)$.

Note 1. By defining an appropriate f, the oracles above can be used to convert shares of one type to shares of another type.

The primitive that we will make the most use of is *oblivious polynomial evaluation* (OPE).

Oracle for OPE: One of the players takes the role of the *sender* and inputs a polynomial P of (public) degree l over $\mathbb{F}_p$ to the oracle. The second player, the *receiver*, inputs $z \in \mathbb{F}_p$ to the oracle. The oracle responds to the receiver with $P(z)$. The sender requires no response.

The idea of OPE was first considered in [22] where an efficient solution requiring $O(l)$ exponentiations and with a communication cost of $O(l \log p)$ was proposed. This method uses a 1-out-of-N oblivious transfer (OT) from [23]. This OT protocol requires $O(1)$ exponentiations. We note that all exponentiations can be carried out over a different – potentially smaller – finite field.

3.2 The Protocol

Suppose that P1 has n_1 entries in its databases and P2 has n_2. Denote these $\{x_{1,1}, x_{1,2}, \ldots, x_{1,n_1}\}$ and $\{x_{2,1}, x_{2,2}, \ldots, x_{2,n_2}\}$ respectively. Let

$$x_1 = \sum_{i=1}^{n_1} x_{1,i} \text{ and } x_2 = \sum_{i=1}^{n_2} x_{2,i}.$$

Without loss of generality we will assume that x_1 and x_2 are integers; appropriate scaling can be applied otherwise. Our problem becomes computing

$$M = \frac{x_1 + x_2}{n_1 + n_2}$$

where P1 knows (x_1, n_1) and P2 knows (x_2, n_2). We will assume that there are some publicly known values N_1 and N_2 such that

$$-2^{N_1} \leq x_1 + x_2 \leq 2^{N_1}, \ 0 < n_1 + n_2 < 2^{N_2}.$$

We describe an oracle-aided protocol for privately computing an approximation

$$\hat{M} \approx \frac{x_1 + x_2}{n_1 + n_2}.$$

By adding random noise we show how our protocol can be modified to privately approximate the mean M in the framework of Feigenbaum et al. [14].

Let m be the closest power of 2 to $n_1 + n_2$, that is

$$2^{m-1} + 2^{m-2} \leq n_1 + n_2 < 2^m + 2^{m-1} \text{ and } m \leq N_2.$$

Let m_1 and m_2 be defined analogously for n_1 and n_2 respectively. Let

$$k = \max\{m_1, m_2\} + 1 \in \{m, m+1\}.$$

With k thus defined we have

$$n_1 + n_2 = 2^k(1-\varepsilon) \text{ where } -\frac{1}{2} < \varepsilon \le \frac{5}{8}.$$

We can express

$$\frac{1}{n_1+n_2} = \frac{1}{2^k(1-\varepsilon)} = \frac{1}{2^k}\left(\sum_{i=0}^\infty \varepsilon^i\right) = \frac{1}{2^k}\left(\sum_{i=0}^d \varepsilon^i\right) + \frac{1}{2^k}R_d$$

where $|R_d| < \frac{8}{3}(\frac{5}{8})^{d+1} < 2^{-\frac{2}{3}d+1}$.
It follows that

$$\frac{2^{N_2 d+k}}{n_1+n_2} = \sum_{i=0}^d (2^{N_2}\varepsilon)^i 2^{N_2(d-i)} + 2^{N_2 d}R_d. \qquad (3)$$

Let

$$Z = \sum_{i=0}^d (2^{N_2}\varepsilon)^i 2^{N_2(d-i)}. \qquad (4)$$

We are almost ready to describe our protocol, first we define some polynomials that it will use. For $i = 2, \ldots, N_2 + 1$, let $P_i(X)$ be a degree $N_2 - 1$ polynomial such that, for $X \in \{2, \ldots, N_2 + 1\}$

$$P_i(X) = \begin{cases} 1 & X = i \\ 0 & \text{otherwise.} \end{cases}$$

With the definitions above we can now describe the protocol. In the description, when we say that P1 or P2 "inputs" something we mean that it inputs it to an oracle that computes the required functionality. Without loss of generality we can assume $x_1 + x_2 \ne 0$ (we do this to allow field multiplication); the case $x_1 + x_2 = 0$ can be handled as a special case.

Note 2. We assume that $\mathbb{F}_p$ is sufficiently large that whenever a conversion from the integers to $\mathbb{F}_p$ is necessary, for an oracle or a player, it can be done in the obvious way without having to do any modular reduction. We will see what this means for the value of p in Section 4.

Protocol 1. Oracle-aided protocol to compute a private 2^{-t}-approximation $\hat{M}$ of $M = \frac{x_1+x_2}{n_1+n_2}$

Set $d = \lceil \frac{3}{2}(t + N_1 + 2) \rceil$.
1. P1 and P2 input n_1 and n_2 respectively. Oracle returns additive shares a_1^F, a_2^F of Z over $\mathbb{F}_p$.

2. P1 and P2 input a_1^F and a_2^F respectively. Oracle returns additive shares a_1^I, a_2^I of Z over $\mathbb{Z}$.
3. For $j = 2, \ldots, N_2 + 1$:
   - P1 computes $b_{1,j} = \lfloor a_1^I / 2^j \rfloor$
   - P2 computes $b_{2,j} = \lfloor a_2^I / 2^j \rfloor$
4. Parties locally convert their shares back into additive $\mathbb{F}_p$ shares $c_{i,j}$, where share $c_{i,j}$ is the $\mathbb{F}_p$ equivalent of integer share $b_{i,j}$.
5. P1 and P2 input m_1 and m_2 respectively. Oracle returns additive shares d_1, d_2 of k over $\mathbb{F}_p$.
6. P1 chooses e_1 at random from $\mathbb{F}_p$ and defines the polynomial

$$R_1(X) = \sum_{i=2}^{N_2+1} c_{1,i} P_i(d_1 + X) - e_1.$$

P1 runs an OPE protocol with P2 so that P2 learns $e_2 = R_1(d_2)$ and P1 learns nothing.

7. P2 chooses f_2 at random from $\mathbb{F}_p$ and defines the polynomial

$$R_2(X) = \sum_{i=2}^{N_2+1} c_{2,i} P_i(d_2 + X) - f_2.$$

P2 runs an OPE protocol with P1 so that P1 learns $f_1 = R_2(d_1)$ and P2 learns nothing.

8. P1 and P2 input $e_1 + f_1$ and $e_2 + f_2$ respectively. Oracle returns multiplicative shares g_1, g_2 of $(e_1 + f_1) + (e_2 + f_2)$ over $\mathbb{F}_p$.
9. P1 inputs x_1 and P2 inputs x_2. Oracle returns multiplicative shares h_1, h_2 of $x_1 + x_2$ over $\mathbb{F}_p$.
10. P1 computes $\hat{M}_1 = g_1 h_1 \cdot 2^{-N_2 d}$ and sends it to P2.
11. P2 computes $\hat{M}_2 = g_2 h_2 \cdot 2^{-N_2 d}$ and sends it to P1.
12. P1 computes and outputs $\hat{M} = \hat{M}_1 \hat{M}_2$.
13. P2 computes and outputs $\hat{M} = \hat{M}_1 \hat{M}_2$.

Theorem 2. *Protocol 1 correctly computes an 2^{-t}-approximation of $M = \frac{x_1 + x_2}{n_1 + n_2}$.*

Proof. By (3), (4) and by definition of the oracles used in steps 1 and 2, after step 2 of the protocol, P1 and P2 hold a_1^I and a_2^I respectively such that

$$\left| \frac{2^{N_2 d + k}}{n_1 + n_2} - (a_1^I + a_2^I) \right| \leq 2^{N_2 d} R_d.$$

Once the local computation takes place in step 3, P1 and P2 hold $b_{1,k}$ and $b_{2,k}$ such that

$$\left| \frac{2^{N_2 d}}{n_1 + n_2} - (b_{1,k} + b_{2,k}) \right| \leq 2^{N_2 d - k} R_d + 2 \leq 2^{N_2 d} R_d + 2.$$

Using the bound on the error term R_d we get

$$\left|\frac{2^{N_2 d}}{n_1 + n_2} - (b_{1,k} + b_{2,k})\right| \leq 2^{d(N_2 - \frac{2}{3}) + 2}. \tag{5}$$

However, the players cannot identify k.

By definition of the oracle invoked at step 5 and the construction of the polynomials R_1 and R_2, after step 7 P1 and P2 hold (e_1, f_1) and (e_2, f_2) respectively such that

$$e_1 + e_2 = c_{1,k} \text{ and } f_1 + f_2 = c_{2,k}.$$

Moreover, by (5) and the properties of the conversion used at step 4, when e_1, e_2, f_1 and f_2 are considered as integers we have

$$\left|\frac{2^{N_2 d}}{n_1 + n_2} - ((e_1 + f_1) + (e_2 + f_2))\right| \leq 2^{d(N_2 - \frac{2}{3}) + 2}. \tag{6}$$

Once the final steps have been executed, by (6) and by definition of N_1, when $\hat{M}$ is treated as an integer we have

$$\left|\left(\frac{x_1 + x_2}{n_1 + n_2}\right) 2^{N_2 d} - \hat{M} 2^{N_2 d}\right| \leq 2^{d(N_2 - \frac{2}{3}) + N_1 + 2}.$$

Therefore, once the factor of $2^{N_2 d}$ is removed from $\hat{M}$, we obtain an approximation such that

$$\left|\left(\frac{x_1 + x_2}{n_1 + n_2}\right) - \hat{M}\right| \leq 2^{-\frac{2}{3} d + N_1 + 2} \leq 2^{-t}. \tag{7}$$

□

Theorem 3. *Protocol 1 is private.*

Proof. To prove privacy we must define simulators S_1 and S_2 as in Definition 1. We describe simulator S_1 as a list of steps 1-13 to output what comes into view of P1 at the appropriate step of Protocol 1. The description of S_2 is similar.

$S_1(x_1, \hat{M})$
1. Choose a_1^F at random from $\mathbb{F}_p$.
2. Choose a_1^I at random from $[-A2^\rho, A2^\rho]$.
3. Do nothing - local computation going on.
4. Do nothing - local computation going on.
5. Choose d_1 at random from $\mathbb{F}_p$.
6. Do nothing - P1 learns nothing from an oracle for OPE.
7. Choose f_1 at random from $\mathbb{F}_p$ - P2 would choose f_2 at random and so f_1 has the correct distribution.
8. Choose g_1 at random from $\mathbb{F}_p^*$.
9. Choose h_1 at random from $\mathbb{F}_p^*$.
10. Do nothing - local computation going on.
11. Compute $M_2 = \hat{M}/(g_1 h_1) \mod p$.
12. Do nothing - local computation going on.
13. Output $\hat{M}$.

□

Achieving Functional Privacy. We sketch how our protocol can be modified to achieve functionally privacy with respect to the mean function. This is done by adding random noise before outputting the approximation (see also [14]).

Assume that Protocol 2 is being used to compute a 2^{-2t}-approximation of the mean M. The modified protocol proceeds as before as far as step 9. In the new step 10, P1 inputs $g_1 h_1$ and P2 inputs $g_2 h_2$ to an oracle that returns additive shares of $g_1 h_1 g_2 h_2$ over $\mathbb{Z}$. Each player performs division of the resulting shares by $2^{N_2 d}$ locally. The players now have additive shares of $\hat{M}$. Let us denote these $\hat{M}'_1$ and $\hat{M}'_2$. If the players output their shares at this point, the result would be identical to that for Protocol 1; however, before P1 and P2 output $\hat{M}'_1$ and $\hat{M}'_2$ respectively, the players individually add uniform random noise in the range $[-2^{-t}, 2^{-t}]$ to their shares. Only once this random noise is added do the players individually output their shares with precision 2^{-2t}. Adding these shares gives an approximation $\hat{M}'$ of the mean M.

It is easy to verify that the modified protocol computes a $2^{-2t} + 2^{-t+1}$-approximation $\hat{M}'$ of M. It remains to show that the $\hat{M}'$ computed in the modified protocol is functionally private with respect to M: we require a simulator as described in Definition 2. Suppose that a simulator given $M = \frac{x_1 + x_2}{n_1 + n_2}$ adds uniform random noise R_1 and R_2 in the range $[-2^{-t}, 2^{-t}]$ and outputs $S(M) = M + R_1 + R_2$ with precision 2^{-2t}. It can be readily checked that the statistical difference between $S(M)$ and $\hat{M}$ is about 2^{-t}. This implies that the function computed by the modified protocol is functionally private with respect to the mean. The properties of the modified protocol give us the following theorem.

Theorem 4. *There exists a protocol that privately computes an approximation of the mean.*

3.3 The Variance and Standard Deviation

Using the notation of Section 3.2, let

$$\tilde{x}_1 = \sum_{i=1}^{n_1} x_{1,i}^2, \quad \tilde{x}_2 = \sum_{i=1}^{n_2} x_{2,i}^2 \text{ and } \tilde{M} = \frac{\tilde{x}_1 + \tilde{x}_2}{n_1 + n_2}.$$

It is easy to use the techniques of Protocol 1 to compute the sample variance

$$\sigma^2 = \frac{1}{n_1 + n_2} \left(\sum_{i=1}^{n_1} (x_{1,i} - M)^2 + \sum_{i=1}^{n_2} (x_{2,i} - M)^2 \right) = \frac{\tilde{x}_1 + \tilde{x}_2}{n_1 + n_2} - M^2$$

as follows. One first computes multiplicative shares of M by following Protocol 1 until P1 and P2 hold $\hat{M}_1$ and $\hat{M}_2$ respectively. These shares are squared locally to give P1 and P2 multiplicative shares of M^2 which are then then converted to additive shares a_1 and a_2. The next step is to apply Protocol 1 replacing x_1 and x_2 with $\tilde{x}_1$ and $\tilde{x}_2$ respectively until P1 and P2 hold multiplicative shares of $\tilde{M}$ which are then converted to additive shares b_1 and b_2. Now, $b_1 - a_1$ and $b_2 - a_2$ is an additive sharing of the variance as required. The standard deviation is obtained by taking the non-negative square root.

4 Implementing the Oracles

Here we describe how to implement the various oracles used by Protocol 1. The security of the resulting construction then follows from Theorem 1. These protocols all use an oblivious polynomial evaluation protocol OPE, for example that proposed in [22] could be used.

Conversion Protocols. Protocol 2 (ATM) can be used for converting additive to multiplicative shares over $\mathbb{F}_p$ as required in steps 8 and 9 of Protocol 1.

Protocol 2. $\mathsf{ATM}(a_1, a_2)$ where $a_1 + a_2 = x (\neq 0)$

---

1. P1 chooses m_1 at random from $\mathbb{F}_p^*$ and constructs the polynomial $P(X) = m_1^{-1} a_1 + m_1^{-1} X$.
2. P1 runs OPE with P2 so that P2 learns $m_2 = P(a_2) = m_1^{-1} a_1 + m_1^{-1} a_2 = m_1^{-1} x$.

At the end the parties hold multiplicative shares m_1, m_2 of x.

Protocol 3 (MTA) can be used for converting multiplicative to additive shares over $\mathbb{F}_p$. It will be necessary for Protocol 5 that we describe shortly.

Protocol 3. $\mathsf{MTA}(m_1, m_2)$ where $m_1 m_2 = x$

---

1. P1 chooses a_1 at random from $\mathbb{F}_p^*$ and constructs the polynomial $P(X) = -a_1 + m_1 X$.
2. P1 runs OPE with P2 so that P2 learns $a_2 = P(m_2) = -a_1 + m_1 m_2 = x - a_1$.

At the end the parties hold additive shares a_1, a_2 of x.

Step 2 of Protocol 1 require a protocol to convert additive shares from $\mathbb{F}_p$ into additive shares from the integers. This is not as straightforward as it sounds. Suppose that $z = z_1^F + z_2^F$ over $\mathbb{F}_p$. The corresponding equation over the integers is $z = z_1^F + z_2^F - lp$ for some l. We therefore need to know l in order to make the conversion.

Suppose that our parties have shares (z_1^F, z_2^F) over $\mathbb{F}_p$ where

$$-2^{n-1} < z = z_1^F + z_2^F \bmod p < 2^{n-1}$$

for some n. If $p > 2^{\rho+n+4}$, where ρ is a security parameter (see Section 4.1), the parties can use Protocol 4 to compute additive shares z_1^I, z_2^I of z over the integers. Protocol 4 is taken from [2] and specialised to the two-party case.

Protocol 4. $\mathsf{FTI}(z_1^F, z_2^F)$ where $z = z_1^F + z_2^F \bmod p$

Let $t = \rho + n + 2$. P1 and P2 execute the following steps.
1. P2 reveals $a_2 = \mathrm{trunc}\left(\frac{z_2^F}{2^t}\right)$ to P1.
2. P1 computes $l = \lceil \frac{z_1^F + 2^t a_2}{p} \rceil$.
3. P1 chooses an integer b_1 at random from $[-p 2^\rho, p 2^\rho]$ and reveals $b_2 = 0 - b_1$ to P2.
4. P1 sets $z_1^I = z_1^F + b_1 - lp$.
5. P2 sets $z_2^I = z_2^F + b_2$.

At the end the parties hold additive shares z_1^I and z_2^I of z over the integers.

Other Protocols. Step 1 of Protocol 1 requires a protocol for sharing Z additively over $\mathbb{F}_p$. Before describing a protocol to do this we give a protocol, Protocol 5, for sharing $2^{N_2}\varepsilon$.

Protocol 5. Sharing of $2^{N_2}\varepsilon$

---

1. The parties run $\mathsf{ATM}(n_1, n_2)$ to obtain multiplicative shares a_1, a_2 of $n_1 + n_2$.
2. The parties run Protocol 7 followed by ATM to obtain multiplicative shares b_1, b_2 of 2^k.
3. P1 computes $c_1 = 2^{N_2} a_1 b_1^{-1} \bmod p$, P2 computes $c_2 = a_2 b_2^{-1} \bmod p$.
4. The parties run $\mathsf{MTA}(c_1, c_2)$ to obtain additive shares d_1, d_2 of $c_1 c_2 \bmod p$.
5. P1 computes $e_1 = d_1 - 2^{N_2 - 1} \bmod p$ and P2 computes $e_2 = d_2 - 2^{N_2 - 1} \bmod p$.

At the end the parties hold additive shares e_1, e_2 of

$$2^{N_2}\varepsilon = (n_1 + n_2) 2^{N_2 - k} - 2^{N_2}.$$

We now have Protocol 6 for computing additive shares of Z over $\mathbb{F}_p$.

Protocol 6. Sharing of Z

1. The parties run Protocol 5 to obtain additive shares a_1, a_2 of $2^{N_2}\varepsilon$.
2. P1 chooses b_1 at random and defines the polynomial
$$P(X) = \sum_{i=0}^{d}(a_1 + X)^i 2^{N_2(d-i)} - b_1.$$
3. P1 runs OPE with P2 so that P2 learns $b_2 = P(a_2)$ and P1 learns nothing.

At the end the Parties hold additive shares b_1, b_2 of Z.

At Step 5 of Protocol 1 we require a protocol for obtaining an additive sharing of k over $\mathbb{F}_p$. Protocol 7 below does this; it requires a polynomials Q_a and Q_b that we define first.

We have $m_1, m_2 \in \{1, \ldots, N_2\}$ and so $m_1 - m_2 \in \{-(N_2-1), \ldots, (N_2-1)\} = S$; there are $2N_2 - 1$ possibilities. Let $Q_a(X)$ and $Q_b(X)$ be the polynomials of degree $|S| - 1 = 2N_2 - 2$ such that for $s \in S$,

$$Q_a(s) = \begin{cases} 0 & \text{if } s < 0 \\ 1 & \text{if } s \geq 0 \end{cases} \quad \text{and} \quad Q_b(s) = \begin{cases} 0 & \text{if } s \leq 0 \\ 1 & \text{if } s > 0 \end{cases} \tag{8}$$

Protocol 7. Sharing of k and 2^k

1. P1 chooses a_1 at random from $\mathbb{F}_p$ and defines the polynomial
$$P_1(X) = Q_a(m_1 - X)2^{m_1+1} - a_1.$$
2. P1 runs OPE with P2 so that P2 learns $a_2 = P_1(m_2)$ and P1 learns nothing.
3. P2 chooses b_2 at random from $\mathbb{F}_p$ and defines the polynomial
$$P_2(X) = Q_b(m_2 - X)2^{m_2+1} - b_2.$$
4. P2 runs OPE with P1 so that P1 learns $b_1 = P_2(m_1)$ and P2 learns nothing.

At the end the parties hold additive shares a_1+b_1, a_2+b_2 of 2^k. By replacing 2^{m_1+1} and 2^{m_2+1} with m_1+1 and m_2+1 respectively, the same technique can be used for sharing k.

It is straightforward to prove the correctness and the privacy of the protocols in this section in the same manner as Theorem 2 and Theorem 3 for Protocol 1. The security of the protocol derived from Protocol 1 by replacing each oracle call with the appropriate protocol from this section then follows from Theorem 1.

4.1 Complexity

All that remains is to analyse the complexity of the final protocol and discuss the security parameters.

Our protocol clearly runs in a constant number of communication rounds between the two parties. The complexity of our protocol depends chiefly on the accuracy of the result; this corresponds to the length d of the Taylor expansion. After execution of Protocol 1, by (7) both parties end up with a real number $\hat{M}$ such that

$$\left| \frac{x_1 + x_2}{n_1 + n_2} - \hat{M} \right| \leq 2^{-\frac{2}{3}d + N_1 + 2} \leq 2^{-t}. \tag{9}$$

Let us consider the size of the finite field $\mathbb{F}_p$. For Protocol 1, we have to choose p to be sufficiently large so that no unwanted wrap-around (modulo p) can occur. The value Z that is computed in the first step satisfies the bound

$$0 < Z \leq \sum_{i=0}^{d} \left((5/8)2^{N_2}\right)^i 2^{N_2(d-i)} = 2^{N_2 d} \sum_{i=0}^{d} (5/8)^i \leq 2^{N_2 d + 2}.$$

Consequently for Protocol 4 (FTI protocol) to work we need a prime $p > 2^{N_2 d + \rho + 8}$, where ρ is a security parameter, typically chosen as $\rho = 80$.

In steps 10 and 11 we have to ensure that the value of $g_1 h_1 g_2 h_2 \approx 2^{N_2 d}(x_1 + x_2)/(n_1 + n_2)$ does not exceed p. Now, we have the bound $|g_1 h_1 g_2 h_2| \leq 2^{N_2 d + N_1}$. From these bounds on Z and $g_1 h_1 g_2 h_2$, we conclude that the prime p must satisfy $\log p > N_2 d + \max\{\rho + 8, N_1\}$. By an improved FTI protocol (using an implicit representation of l in terms of shares) we can improve the requirement made to the prime p to $\log p > N_2 d + N_1$. This does not affect the asymptotic running time of the FTI protocol.

The complexity of the protocol is clearly dominated by two operations: (1) Protocol 6 to compute shares of Z using OPE with a polynomial of degree d; and (2) step 5 (implemented by Protocol 7), and steps 6 and 7 of Protocol 1 using OPE with polynomials of degree $2N_2 - 2$ and $N_2 - 1$ respectively.

Assume $N = N_1 + N_2$. Using the OPE protocol from [22] this makes a computation cost of $O(N_2 + d) = O(N + t)$ exponentiations and a communication cost of $O((N_2 + d) \log p) = O((N + t)^2 N)$.

4.2 Comparison with the Generic Solution

In [28] Yao presents a constant-round protocol for privately computing any probabilistic polynomial-time functionality. We compare the complexity of our solution with the one obtained by using Yao's generic solution.

Assume the given probabilistic polynomial-time functionality is given as a binary circuit with N inputs and G gates. Without going into details, Yao's protocol requires a communication of $O(G)$ times the length of the output of a pseudo-random function (which we denote by β and is typically 80 bits long). The main computational overhead of the protocol is the computation of the N oblivious transfers plus the application of G pseudorandom functions.

In our case set $N = N_1 + N_2$. A circuit computing an 2^{-t}-approximation of $M = \frac{x_1+x_2}{n_1+n_2}$ should compute the Taylor series, namely $d = O(t + N)$ multiplications in $\mathbb{F}_p$. Assuming multiplication requires circuits of quadratic size [1] we estimate the number of gates as $G = O((t + N)N^2)$. This results in a communication cost of $O(\beta(t + N)\log^2 p) = O((t + N)^3 N^2 \beta)$ which is larger than the cost of our protocol by a factor of $(t + N)N\beta$. On the other hand, the number of oblivious transfers (and so the number of exponentiations) for the generic protocol is $O(N)$. If we also take into account the $O((N + t)N^2)$ applications of the pseudorandom generator, the computation cost remains much the same in both cases.

For comparison, note that there is a computation–communication tradeoff for oblivious transfer suggested in [23]. This can reduce the number of exponentiations by a factor of c at the price of increasing the communication by a factor of 2^c.

We also note that, using our techniques, all application of Yao's circuits in the log protocol from [20, 21] can be abandoned. This essentially leads to a slight improvement – a multiplicative factor β – in the communication cost of [20, 21].

Acknowledgement. We thank the anonymous referees for the helpful comments. The first author was partially supported by a DAAD postdoc fellowship.

References

1. G. Aggarwal, N. Mishra, and B. Pinkas. Secure computation of the kth-ranked element. In *Advances in Cryptology - EUROCRYPT 2004*, volume 3027 of *Lecture Notes in Computer Science*, pages 40–55. Springer-Verlan, 2004.
2. J. Algesheimer, J. Camenisch, and V. Shoup. Efficient computation modulo a shared secret with application to the generation of shared safe-prime products. In *Advances in Cryptology - CRYPTO 2002*, volume 2442 of *Lecture Notes in Computer Science*, pages 417–432. Springer-Verlag, 2002.
3. M. Ben-Or, S. Goldwasser, and A. Wigderson. Completeness theorems for non-cryptographic fault-tolerant distributed computation. In 20th *ACM Symposium on Theory of Computing*, pages 1–10. ACM Press, 1988.
4. R. Canetti. Security and composition of multiparty cryptographic protocols. *Journal of Cryptology*, 13(1):143–202, 2000.
5. R. Canetti, Y. Lindell, R. Ostrovsky, and A. Sahai. Universally composable two-party computation. In 34th *ACM Symposium on Theory of Computing*, pages 494–503. ACM Press, 2002.
6. D. Chanm, C. Crépeau, and I. Damgård. Multiparty unconditionally secure protocols. In 20th *ACM Symposium on Theory of Computing*, pages 11–19. ACM Press, 1988.

---

[1] There exist circuits of size $O(\log p \log^2 \log p)$ for multiplication; however, assuming quadratic circuits seems reasonable for realistic values of $\log p$. We note that this convention is also made in [2, 20, 21].

7. W. Du. *A Study of Several Specific Secure Two-party Computation Problems.* PhD thesis, Department of Computer Science, Purdue University, 2001.
8. W. Du and J. Atallah. Privacy-preserving cooperative scientific computations. In *14th IEEE Computer Security Foundations Workshop*, pages 273–282, 2001.
9. W. Du and M. J. Atallah. Privacy-preserving cooperative statistical analysis. In *2001 ACSAC: Annual Computer Security Applications Conference*, pages 102–110, 2001.
10. W. Du and M. J. Atallah. Secure multi-party computation problems and their applications: A review and open problems. In *New Security Paradigms Workshop*, pages 11–20, 2001.
11. W. Du, Y. S. Han, and S. Chen. Privacy-preserving multivariate statistical analysis: Linear regression and classification. In *4th SIAM International Conference on Data Mining*, 2004.
12. W. Du and Z. Zahn. Building decision tree classifier on private data. In *Workshop on Privacy, Security, and Data Mining at The 2002 IEEE International Conference on Data Mining (ICDM)*, pages 1–8, 2002.
13. W. Du and Z. Zhan. A practical approach to solve secure multi-party computation problems. In *New Security Paradigms Workshop*, pages 127–135, 2002.
14. J. Feigenbaum, Y. Ishai, T. Malkin, K. Nissim, M. Strauss, and R. N. Wright. Secure multiparty computation of approximations. In *ICALP*, volume 2076 of *Lecture Notes in Computer Science*, pages 927–938. Springer-Verlan, 2001. Full version on Cryptology ePrint Archive, Report 2001/024.
15. M. J. Freedman, K. Nissim, and B. Pinkas. Efficient private matching and set intersection. In *Advances in Cryptology - EUROCRYPT 2004*, volume 3027 of *Lecture Notes in Computer Science*, pages 1–19. Springer-Verlan, 2004.
16. J. von zur Gathen and J. Gerhard. *Modern computer algebra.* Cambridge University Press, New York, 2 edition, 2003.
17. O. Goldreich. *Foundations of Cryptography, Volume 2, Basic Applications.* Cambridge University Press, 2004.
18. O. Goldreich, S. Micali, and A. Wigderson. How to play any mental game: A completeness theorem for protocols with honest majority. In *19th ACM Symposium on Theory of Computing*, pages 218–229. ACM Press, 1997.
19. O. Goldreich and R. Vainish. How to solve any protocol problem - an efficiency improvement. In *Advances in Cryptology - CRYPTO '87*, volume 293 of *Lecture Notes in Computer Science*, pages 73–86. Springer-Verlag, 1987.
20. Y. Lindell and B. Pinkas. Privacy preserving data mining. In *Advances in Cryptology - CRYPTO 2000*, volume 1800 of *Lecture Notes in Computer Science*, pages 35–24. Springer-Verlag, 2000.
21. Y. Lindell and B. Pinkas. Privacy preserving data mining. *Journal of Cryptology*, 15(3):117–206, 2002.
22. M. Naor and B. Pinkas. Oblivious transfer and polynomial evaluation. In *31st ACM Symposium on Theory of Computing*, pages 245–254. ACM Press, 1999. Full version available at http://www.wisdom.weizmann.ac.il/\%7Enaor/onpub.html.
23. M. Naor and B. Pinkas. Efficient oblivious transfer protocols. In *12th Annual ACM-SIAM Symposium on Discrete Algorithms (SODA)*, pages 448–457, 2001.
24. UK Government. Data protection act 1998. Available at http://www.hmso.gov.uk/acts/acts1998/19980029.htm.
25. J. Vaidya and C. Clifton. Privacy preserving naive bayes classifier for vertically partitioned data. In *4th SIAM International Conference on Data Mining*, 2004.
26. J. S. Vaidya. *Privacy Preserving Data Mining over Vertically Partitioned Data.* PhD thesis, Department of Computer Science, Purdue University, 2004.

27. X. Wang and V. Y. Pan. Acceleration of Euclidean algorithm and rational number reconstruction. *SIAM Journal on Computing*, 32(2):548–556, 2003.
28. A. C. Yao. How to generate and exchange secrets. In 27th *Annual Symposium on Foundations of Computer Science*, pages 162–167. IEEE Computer Science Press, 1986.

A Flaws in Existing Protocols

In this Section we review some protocols in the area of private information retrieval appearing or used in [7, 9, 10, 11, 12, 13, 25, 26]. We identify two general design techniques in protocols that lead to insecure solutions. First, to hide an (integer) number, multiplying by a random integer and publishing the resulting product is a bad idea as we outline below. This kind of "hiding by multiplication" technique only makes sense over a finite field where multiplication by a random field element results in a random field element.

Second, as a tradeoff between security and efficiency, it could be tempting to design protocols that leak some secret information. However, when revealing some information about a secret value, one has to take extreme care not to apply the protocol to another dependent instance (sequential composition) that may result in leaking more information than initially intended. This is discussed further below.

The Mean Protocol of Du and Atallah [9]. In this subsection we review the mean protocol from [9] (see also [7]) and show that it leaks information.

Protocol 8. Protocol to compute the mean $M = \frac{x_1 + x_2}{n_1 + n_2}$.

---

1. P1 generates two random integer numbers r and s (chosen from a sufficiently large interval).
2. P1 runs OPE with P2 twice. The first time P2 learns $a = r(x_1 + x_2)$ the second time it learns $b = s(n_1 + n_2)$. P1 learns nothing.
3. P1 sends $t = s/r$ to P2.
4. P2 computes $t \cdot \frac{a}{b} = \frac{x_1 + x_2}{n_1 + n_2}$ and sends it to P1.

First note that, after the second step of the protocol, P2 learns something about x_1: P2 knows all the divisors of x_2 and so if $a | x_2$ and $a \nmid r(x_1 + x_2)$ then it knows that $a \nmid x_1$. We note that protocols 6 and 7 from [13] and the log protocol from Du and Zhan (Section 5.2 in [12]) suffer from the same problem as the above protocol and hence leak secret information.

Now we show that in some cases P1 can completely determine x_2 and n_2. Suppose that the random numbers r and s in the first step are integers chosen from the interval $[0, 2^k - 1]$. (The distribution of r and s is not specified in [9]. To properly hide $x_1 + x_2$ and $n_1 + n_2$, r and s have to be chosen uniformly at random from a large enough interval. We can assume that r and s are integer values, otherwise appropriate scaling can be applied.) To properly hide $x_1 + x_2$ and $n_1 + n_2$ in Step 2, we assume that $k \approx 2N$ where $N = \max\{N_1, N_2\}$. Suppose now that $r > s$. The problem comes in Step 3 where the rational number $t = s/r$ must be transfered to P2. Using the well known technique of rational number approximation, we show that under some choice of parameters, t leaks r and s – a complete break of the protocol.

Transferring a rational number t typically is done by using fixed point arithmetic. Let t' be the fixed point representation of t, so t' equals t up to a (public) number d of digits: $t' = u'/2^d$ for an integer $0 \le u' < 2^d$. To guarantee P2 a sufficiently good approximation $t' \cdot \frac{a}{b}$ of $t \cdot \frac{a}{b} = \frac{x_1+x_2}{n_1+n_2}$ in the last step, we need $d > k + 2N + 1$. Now, P2 runs an algorithm to compute a rational number approximation of t', that is to find (if it exists) the unique pair of coprime integers $r' < s'$ in the interval $[0, 2^k - 1]$ such that

$$|u'/2^d - s'/r'| < 1/2^{2k+1}. \qquad (10)$$

There are algorithms for this with complexity $O(d \log^3 d)$. See [16, 27] for example.

Assume that r and s are coprime (by a theorem of Dirichlet this happens with asymptotic probability $6/\pi^2$). Now, $|u'/2^d - s/r| < 1/2^d \le 1/2^{2k+1}$. Hence there must exist two integers r' and s' satisfying (10). Since such integers are unique we must have $r' = r$ and $s' = s$. Of course, given r and s, P2 can compute x_1 and n_1 – a complete break of the protocol.

We note that the mean protocol from Abdallah and Du is used in several places [10, 25, 26] ([25, 26] also discuss alternative solutions).

The Matrix Multiplication Protocol of Du, Han, and Chen [11].

Here we review the matrix multiplication protocol from [11] and indicate why certain applications of it leak information. Such an application is used in [11].

We start explaining a multiplication protocol (Protocol 2 of [11]). On input of two $n \times n$ matrices $A = (A_{ij})$ and $B = (B_{ij})$ (for what follows we do not need to further specify where the elements of A and B are chosen from) it outputs shares V_1 and V_2 such that $V_1 + V_2 = AB$.

We vertically divide the $n \times n$ matrix M into two equal-sized sub-matrices M_l and M_r of size $n \times n/2$ (we assume n is even); we horizontally divide $M^I = M^{-1}$ into two equal-sized sub-matrices M_t^I and M_b^I of size $n/2 \times n$.

Protocol 9. MUL(A,B)

---

1. Both players generate a random, public, invertible $n \times n$ matrix M.
2. P1 computes $A_1 = AM_l$, $A_2 = AM_r$ and sends A_2 to P2.
3. P2 computes $B_1 = M_t^I B$, $B_2 = M_b^I B$ and sends B_2 to P1.

4. P1 computes $V_1 = A_2B_2$ and P2 computes $V_2 = A_1B_1$.
   At the end of the protocol P1 holds V_1, P2 holds V_2 such that $V_1+V_2 = AB$.

It is easy to see that the indicated protocol correctly computes shares V_1 and V_2 of the matrix product AB. However, as also noted in [11], the protocol leaks information to both players: P2, for example, learns the $n/2 \times n$ matrix $A_2 = AM_r$ in Step 2, where M_r is a public matrix it knows. Intuitively (if M is properly chosen) this provides some information about the secret matrix A. To be more precise, since M_r is a $n \times n/2$ matrix, for each fixed column j of the matrix A, P2 gets a system of $n/2$ linear equations with n unknown variables A_{ij}, $1 \leq i \leq n$. In total P2 gets $n^2/2$ linear equations with n^2 unknown variables A_{ij}, $1 \leq i,j \leq n$. We note that in [11] the matrices M in the first step are chosen according to a more complex distribution. However, our simplification does not affect what follows.

If only applying the protocol once this may not be a problem. However, when using this protocol more than once one has to be extremely careful: Any composition of this protocol applied to *dependent instances* may lead to a complete break of the protocol.

This problem occurs in Section 4.5 of [11] when computing the inverse of a matrix. Protocol MUL is applied to two dependent matrices: to a (random) matrix Q *and* to its inverse Q^{-1}. Unfortunately this inversion protocol is a crucial building block for all main results in [11].

We now describe the inversion protocol. The setting is that P1 holds matrix A, P2 holds matrix B, and both want to securely compute shares V_1 and V_2 such that $V_1 + V_2 = (A+B)^{-1}$. This is done in two main steps. In the first step P2 generates two random, invertible matrices, P and Q. The two players use MUL so that P1 learns $C = P(A+B)Q$ but P2 learns nothing about C. In a second step P1 locally computes $C^{-1} = Q^{-1}(A+B)^{-1}P^{-1}$ and both players run a protocol to compute matrices V_1 and V_2 such that $V_1 + V_2 = QC^{-1}P = (A+B)^{-1}$. We will show that after the execution of the protocol, P1 can learn P and Q.

In the first step P2 must reveal some information about Q, namely in the MUL protocol P1 learns $M_b^I Q$ for some public $n/2 \times n$ matrix M_b^I. In the second step (as an intermediate step), P2 has to create shares of $Q = Q_1 + Q_2$ and send Q_1 to P1. P1 has to create shares of $C^{-1} = C_1 + C_2$ and send C_2 to P2. Now both run a multiplication protocol to get shares W_1 and W_2 such that $W_1 + W_2 = C^{-1}Q = (C_1 + C_2)(Q_1 + Q_2) = C_1Q_1 + C_1Q_2 + C_2Q_2 + C_2Q_1$. To compute the shares W_1 and W_2, the players have to run the protocol MUL twice: on inputs (C_1, Q_2) and on inputs (C_2, Q_2). But at the point where the players run the multiplication protocol MUL on inputs (C_1, Q_2), P1 learns $Q_2\hat{M}_r = (Q^{-1} - Q_1)\hat{M}_r$, where the $n \times n/2$ matrix $\hat{M}_r$ and the $n \times n$ matrix Q_1 are known to P1.

So far in this inversion protocol P1 has learnt

$$S := M_b^I Q \text{ and} \tag{11}$$

$$T := (Q^{-1} - Q_1)\hat{M}_r + Q_1\hat{M}_r = Q^{-1}\hat{M}_r \tag{12}$$

for a known $n \times n/2$ matrix M_b^I, and for a known $n/2 \times n$ matrix $\hat{M}_r$.

Let $Q = (Q_{ij})$ and $Q^{-1} = (Q_{ij}^{-1})$. Equation (11) provides $n^2/2$ linear equations in the unknown Q_{ij}, (12) provides $n^2/2$ linear equations in the unknown Q_{ij}^{-1}. Since Q^{-1} depends on Q, combining (11) and (12) we get n^2 (not necessarily linear) equations in the unknown Q_{ij}. If M and $\hat{M}$ where chosen at random, with high probability the equations are independent and hence knowledge of S and T provides enough information for P1 to compute Q. For small n, the matrix Q can be computed efficiently. By a similar argument P1 also learns P enabling it to compute P2's input B and hence to completely break the protocol.

To save the MUL protocol from [11], one could argue that when the matrix $\hat{M}$ is chosen properly (depending on Q, Q^{-1} and M), then one may hope that the resulting linear equations obtained by P1 can be made dependent on the previous ones such that no new information about Q is released to P1. However, in [11] the protocols using MUL or the inversion protocol as a sub-protocol become very complex (as do the dependencies) and so this is very likely to become impractical.

We note that the main results of [11] still hold when one replaces the MUL protocol by one that is provably secure. We suspect that the protocol proposed by Atallah and Du [8] (which itself builds on an idea by Goldreich and Vainish [19]) can be proved secure.

Keyword Search and Oblivious Pseudorandom Functions

Michael J. Freedman[1], Yuval Ishai[2], Benny Pinkas[3], and Omer Reingold[4]

[1] New York University
mfreed@cs.nyu.edu
[2] Technion
yuvali@cs.technion.ac.il
[3] HP Labs, Israel
benny.pinkas@hp.com
[4] Weizmann Institute of Science
omer.reingold@weizmann.ac.il

Abstract. We study the problem of privacy-preserving access to a database. Particularly, we consider the problem of privacy-preserving keyword search (KS), where records in the database are accessed according to their associated keywords and where we care for the privacy of both the client and the server. We provide efficient solutions for various settings of KS, based either on specific assumptions or on general primitives (mainly oblivious transfer). Our general solutions rely on a new connection between KS and the oblivious evaluation of pseudorandom functions (OPRFs). We therefore study both the definition and construction of OPRFs and, as a corollary, give improved constructions of OPRFs that may be of independent interest.

Keywords: Secure keyword search, oblivious pseudorandom functions, private information retrieval, secure two-party protocols, privacy-preserving protocols.

1 Introduction

Keyword search (KS) is a fundamental database operation. It involves two main parties: a server, holding a database comprised of a set of records and their associated keywords, and a client, who may send queries consisting of keywords and receive the records associated with these keywords. A natural question in the area of secure computation is the design of protocols for efficient, privacy-preserving keyword search. These protocols enable keyword queries while providing privacy for *both* parties: namely, (1) hiding the queries from the database (client privacy) and (2) preventing the clients from learning anything but the results of the queries (server privacy).

To be more specific, the private keyword-search problem may be defined by the following functionality. The database consists of n pairs $\{(x_i, p_i)\}_{i \in [n]}$; we denote x_i as the keyword and p_i as the payload (database record). A query from a client is a searchword w, and the client obtains the result p_i if there is a value i for which $x_i = w$ and obtains a special symbol $\perp$ otherwise. Given that KS allows clients to input an arbitrary searchword, as opposed to selecting p_i by an input i, keyword search is strictly stronger than the better-studied problems of oblivious transfer (OT) and symmetrically private information retrieval (SPIR).

1.1 Contributions

Our applied and conceptual contributions can be divided into the following:

- **Specific Protocols for KS.** We construct direct instantiations of KS protocols, providing privacy for both parties, based on the use of oblivious polynomial evaluation and homomorphic encryption. The protocols have a communication complexity which is logarithmic in the size of the domain of the keywords and polylogarithmic in the number of records, and they require only one round of interaction, even in the case of malicious clients.[1] All previous fully-private KS protocols either require a linear amount of communication or multiple rounds of interaction, even in the semi-honest model.

- **KS Using Oblivious Pseudorandom Functions (OPRFs).** We describe a generic, yet very efficient, *reduction* from KS to what we call *semi-private* KS, in which only the client's privacy is maintained. Specifically, we show that any KS protocol providing (only) client privacy can be upgraded to provide server privacy as well, by using an additional *oblivious evaluation of pseudorandom functions*. This reduction is motivated by the fact that efficient semi-private KS is quite easy to obtain by combining PIR with a suitable data structure supporting keyword searches [20, 7].[2] Thus, we derive a general construction of fully-private KS protocols based on PIR, a data structure supporting keyword searches, and an OPRF.

- **New Notion of OPRF.** Motivated by the KS application and the above general reduction, we put forward a new relaxed notion of OPRF which facilitates more efficient constructions and is of independent theoretical interest.

- **Constructions of OPRF.** We show a construction of an OPRF protocol based on the DDH assumption. In addition, one of the our main contributions is a general construction of (relaxed) OPRF from OT. This construction is based on techniques from [23, 25], yet improves on these works as (1) it preserves privacy against (up to t) *adaptive* queries, (2) it is obliviously evaluated in *constant number of rounds*, and (3) it handles *exponential domain size*. These improvements are partially relevant also in the context of t-out-of-n OT, as originally studied in [23, 25]. We note that this is a *black-box*

---

[1] In the case of malicious parties, we use a slightly relaxed notion of security, following one suggested in the context of OT [1, 23] (see Section 2).
[2] In fact, if we allow a setup phase with linear communication complexity, we can obtain a semi-private KS supporting adaptive queries by simply sending the entire database to the client.

construction of t-time OPRF from OT. From a theoretical point-of-view, one of the most interesting open questions left by our work is to find an efficient black-box construction of *fully*-adaptive OPRF and KS (supporting an arbitrary number of queries) which only makes a black-box use of OT. In contrast, such a construction is easy to obtain by making a non-black-box use of OT. Thus, we have a rare example of a non-black-box construction in cryptography for which no black-box construction is known, even in the random-oracle model. In fact, we are not aware of any other such simple and natural example that fully resides in the semi-honest model.

1.2 Related Work

The work of Kushilevitz and Ostrovsky [20], which was the first to suggest a single-server PIR protocol, described how to use PIR together with a hash function for obtaining a semi-private KS protocol (we denote a KS protocol as "semi-private" if it does not ensure server privacy). Chor et al. [7] described how to implement semi-private KS using PIR and any data structure supporting keyword queries, and they added server privacy using a trie data structure and many rounds. Our reduction from KS to semi-secure KS provides a more efficient and general alternative, requiring only a small constant number of rounds.

Ogata and Kurosawa [27] show an ad-hoc solution for KS for adaptive queries, using a setup stage with linear communication. The security of their main construction is based on the random oracle assumption and on a non-standard assumption (related to the security of blind signatures). The system requires a public-key operation per item for every new query.

A problem somewhat related to KS is that of "search on encrypted data" [30, 3]. The scenario involves giving encrypted data to a third party. This party is later given a trapdoor key, enabling it to search the encrypted data for specific keywords, while hiding any other information about the data. This problem seems easier than ours since the search key is provided by the party which previously encrypted the data. Furthermore, there are protocols for "search on encrypted data" (*e.g.*, [30]) which use only symmetric-key crypto. therefore it is unlikely that they can be used for implementing KS, as KS implies OT.

Another related problem is that of "secure set intersection" [10], where two parties whose inputs consist of sets X, Y privately compute $X \cap Y$. KS is a special case of this problem with $|X| = 1$. On the other hand, set intersection can be reduced to KS by running a KS invocation for every item in X. Thus, our results can be applied to obtain efficient solutions to the set-intersection problem.

Cryptographic Primitives. We make use of several standard cryptographic primitives that can be defined as instances of private two-party computation between a server and a client, including oblivious transfer (OT) [29, 9], single-server private information retrieval (PIR) [8, 20], symmetrically-private information retrieval (SPIR) [11, 23], and oblivious polynomial evaluation (OPE) [23]. Some specific constructions for non-adaptive KS require a semantically-secure homomorphic encryption system.

1.3 Organization

The remainder of this paper is structured as follows. We provide definitions and variants of keyword search in Section 2. Section 3 describes some direct constructions of (non-adaptive) KS protocols based on OPE and homomorphic encryption. In Section 4 we introduce our new relaxed notion of OPRF and use it to obtain a reduction from fully-private KS to semi-private KS. We conclude by providing efficient implementations of OPRFs in Section 5.

2 Preliminaries

This section defines the private keyword search problem and some of its variants. We assume the reader's familiarity with standard simulation-based definitions of secure computation (cf. [5, 13]).

2.1 Private Keyword Search

The system is comprised of a server S and a client C. The server's input is a database X of n pairs (x_i, p_i), each consisting of a keyword and a payload. Keywords can be strings of an arbitrary length and payloads are padded to some fixed length. We may also assume, without loss of generality, that all x_i are distinct. The client's input is a *searchword* w. If there is a pair in the database in which the keyword is equal to the searchword, then the output is the corresponding payload. Otherwise the output is a special symbol $\perp$.

Private keyword search (KS for short) requires privacy for both the client and the server, *i.e.*, neither party learns anything more than is defined by the above transaction. The strongest way of formalizing this requirement is by appealing to general definitions of secure computation from the literature. That is, a KS protocol can be defined as a secure two-party protocol realizing the above KS functionality. However, when constructing *specific* KS protocols—rather than general reductions from KS to other primitives—efficiency considerations dictate a slight relaxation of this definition which still suffices to capture the core correctness and privacy requirements. Specifically, when simulating a malicious server, the relaxed definition only requires one to simulate the server's view alone, without considering its joint distribution with the honest client's output. (In the setting of semi-honest parties, this relaxed definition is equivalent to the original one.)

With respect to a malicious server, this relaxed definition only requires that the client's query w remains private: It does not require the server to commit to or even "know" a database to which a client's search is effectively applied. Such a relaxation is standard for related primitives such as OT (cf. [1, 23]) or PIR (cf. [20, 4]). Moreover, it seems necessary for obtaining protocols that require only a single round of interaction yet still achieve security against malicious parties. (We note, however, that our protocols can be amended to satisfy the stronger definition by adding proofs of knowledge.)

It is interesting to contrast the goals of KS and those of zero-knowledge sets [22]. While KS provides privacy for both parties but does not require the server to commit to its input, zero-knowledge sets require the server to commit to its input but provides privacy for the server yet not the client.

The requirements of a private KS protocol can be divided into *correctness*, *client privacy*, and *server privacy* components. We first define these properties independently, and then define a private KS protocol as a protocol that satisfies these definitions. (To avoid cumbersome notation, we omit the auxiliary inputs required for sequential composition.)

Definition 1 (Correctness). *If both parties are honest, then, after running the protocol on inputs (X, w), the client outputs p_i such that $w = x_i$, or $\perp$ if no such i exists.*

Definition 2 (Client's privacy: indistinguishability). *For any PPT S' executing the server's part and for any inputs X, w, w', the views that S' sees on input X, in the case that the client uses the searchword w and the case that it uses w', are computationally indistinguishable.*

For both client and server privacy, indistinguishability is parameterized by a privacy parameter k, given to both parties as a common input. Note that this definition, protecting only the privacy of the client's query w, captures the aforementioned relaxation.

In order to show that the client does not learn more or different information from the protocol than from merely obtaining its output, we compare the protocol to the *ideal implementation*. In the ideal implementation, a trusted third party gets the server's database X and the client's query w as input, and outputs the corresponding payload to the client. Privacy requires that the protocol does not leak to the client more information than in the ideal implementation.

Definition 3 (Server's privacy: comparison with the ideal model). *For every PPT machine C' substituting the client in the real protocol, there exists a PPT machine C'' that plays the client's role in the ideal implementation, such that on any inputs (X, w), the view of C' is computationally indistinguishable from the output of C''. (In the semi-honest model $C' = C$.)*

Remark 1. The protocols from Section 3, as originally described, will actually satisfy the following incomparable notion of server privacy: any computationally *unbounded* client C' can be simulated by an *unbounded* simulator C''. This can be viewed as a pure form of information-theoretic privacy. Inefficient simulation seems necessary in order to obtain 1-round KS protocols (see a discussion in [1] for the similar case of OT). However, it is easy to convert these protocols to ones that support efficient simulation, using standard zero-knowledge proofs of knowledge: Clients should prove that they know the secret key corresponding to the public key they generate. Such proofs need to be performed only once, during the system's initialization.

Definition 4 (Private KS protocol). *A two-party protocol satisfying Definitions 1 (correctness), 2 (client privacy) and 3 (server privacy).*

The above definition can be immediately applied to protocols computing any deterministic client-server functionality f. We refer to such a protocol as *private* protocol for f.

Finally, we will later use KS protocols in which the server privacy is not preserved (*i.e.*, satisfy only Definitions 1 and 2). We refer to such protocols as *semi-private* KS protocols.

2.2 Problem Variants

The default KS primitive can be extended and generalized in several ways. We first outline three orthogonal variations on the basic model, and then define the two main settings on which we focus.

- **Multiple Queries.** The default notion of KS allows the client to search for a single keyword. While this procedure can be repeated several times, one may seek more efficient solutions allowing the client to retrieve t keywords at a reduced cost. This generalized notion of t-time KS is straightforward to define and makes sense even when $t \gg n$, since the client does not necessarily have an a-priori knowledge of the keywords. (This is in contrast to the case of 1-out-of-n OT or SPIR, where there is no point in letting $t > n$, since the entire database can be learned using t queries.)

- **Allowing Setup.** By default, KS does not assume any previous interaction between the client and server. To facilitate prompt responses to future queries, the client and server may engage in a setup phase involving a polynomial amount of work. During the online phase, each keyword search may then only require a sub-linear amount of work.

- **Adaptive Queries.** In the default *non-adaptive* setting, the client may ask multiple queries, but the queries must be defined before it receives the server's first answer. In the *adaptive* setting, the client can decide on the value of each query after receiving the answers to previous queries. An adaptive t-time KS protocol allows the client to make at most t adaptive queries. The privacy definition in this case extends the above in a natural way, similarly to that of adaptive OT in [24].

The results of this work have applications to all of the above variations. However, to make the presentation more focused, we restrict our discussion to two "typical" settings for KS:

Non-Adaptive t-Time KS Without Setup. In our default notion of KS, when t is unspecified, it is taken to be 1. This setting's main goal in this setting is to obtain solutions whose total *communication complexity* is sub-linear in n. Thus, the problem can be viewed as an extension of PIR and SPIR.

Adaptive t-Time KS with Setup. In this setting, allowing t adaptive queries, the setup phase typically consists of a single message in which the server sends the database in encrypted form to the client. (This is the default setting also considered in [23, 25, 27].) In general, however, the setup may be polynomial in the database size. Ideally, each adaptive query should involve a small amount of work—sub-linear in the database size—including both communication and computation. When t is unspecified, it is taken to be an arbitrary polynomial in the database size, where this polynomial may be larger than the cost of the setup. Thus, one cannot apply solutions that separately handle each future query.

For brevity, we subsequently refer to these settings as *non-adaptive KS* and *adaptive KS*, respectively.

3 Non-adaptive KS from OPE

In this section, we construct a non-adaptive keyword search protocol using oblivious polynomial evaluation (OPE) [23]. The basic idea of the construction is to encode the database entries in $X = \{(x_1, p_1), \ldots, (x_n, p_n)\}$ as values of a polynomial, i.e., to define a polynomial Q such that $Q(x_i) = (p_i)$. Note that this design is different than previous applications of OPE, where a polynomial (of degree k) was used only as a source for $(k+1)$-wise independent values. Compared to our other constructions and to previous solutions from the literature, this construction is unique in achieving sub-linear communication overhead in a single round of communication.[3]

The following scheme uses any generic OPE to build a KS protocol. We then show a specific implementation of the OPE based on homomorphic encryption.

Protocol 1 (Generic polynomial based KS)
Input: Client: an evaluation point w; Server: $\{x_i, p_i\}_{i \in [n]}$, all x_i's are distinct
Output: Client: p_i if $w = x_i$, nothing otherwise; Server: nothing

1. *The server defines L bins and maps the n items into the L bins using a random, publicly-known hash function H with a range of size L. H is applied to the database's keywords, i.e., (x_i, p_i) is mapped to bin $H(x_i)$. Let m be a bound such that, with high probability, at most m items are mapped to any single bin. (At this point, we keep L and m as parameters.)*
2. *For every bin j, the server defines two polynomials P_j and Q_j of degree $(m-1)$. The polynomials are defined such that for every pair (x_i, p_i) mapped to bin j, it holds that $P_j(x_i) = 0$ and $Q_j(x_i) = (p_i | 0^\ell)$, where ℓ is a statistical security parameter.*
3. *For each bin j, the server picks a new random value r_j and defines the polynomial $Z_j(w) = r_j \cdot P_j(w) + Q_j(w)$.*
4. *The two parties run an OPE protocol in which the client evaluates all L polynomials at the searchword w.*
5. *The client learns the result of $Z_{H(w)}(w)$, i.e., of the polynomial associated with the bin $H(w)$. If this value if of the form $p|0^\ell$ the client outputs p, otherwise it outputs $\perp$.*

---

[3] Protocol 1 uses a public hash function H. To run it in the "plain" model, the client can pick the hash function and send it to the server in its first message.

To instantiate this generic scheme, we need to detail the following three open issues: (1) the OPE method used by the parties, (2) the number of bins L, and (3) the method by which the client receives the OPE output for the relevant bin. Additionally, one could consider using carefully-chosen hashing methods to obtain a balanced allocation of items into bins, although this approach would not yield substantial improvements.

An OPE Method. Our construction uses an OPE method based on homomorphic encryption[4] (such as Paillier's system [28]) in the following way. We first introduce this construction in terms of a single database bin.

- The server's input is a polynomial of degree m, where $P(w) = \sum_{i=0}^{m} a_i w^i$. The client's input is a value w.
- The client sends to the server homomorphic encryptions of the powers of w up to the mth power, i.e., $Enc(w), Enc(w^2), \ldots, Enc(w^m)$.
- The server uses the homomorphic properties to compute the following:

$$\prod_{i=0}^{m} Enc(a_i w^i) = Enc(\sum_{i=0}^{m} a_i w^i) = Enc(P(w))$$

The server sends this result back to the client.

In the case of semi-honest parties, it is clear that the OPE protocol is correct and private. Furthermore, the protocol can be applied in parallel to multiple polynomials, and the structure of the protocol enforces that the client evaluates all polynomials at the same point.

Now, consider that the server's input is L polynomials, one per bin. The protocol's overhead for computing all polynomials is the following. The client computes and sends m encryptions. Every polynomial P_j used by the server is of degree $d_j \leq m$ (where $d_j + 1$ items are mapped to bin j), and the server can evaluate it using $d_j + 1$ homomorphic multiplications of plaintexts. Thus, the total work of the server is $\sum_{j=0}^{L-1}(d_j + 1) = n$ exponentiations. The server returns just a single value for each of the L polynomials.

A Simple Protocol. Let the server assign the n items to L bins arbitrarily and evenly, ensuring that L items are assigned to every bin; thus, $L = \sqrt{n}$. The client need not know which items are mapped to which bin. The client's message during the OPE consists of $L = O(\sqrt{n})$ homomorphic encryptions; the server evaluates L polynomials by performing n homomorphic multiplications (exponentiations), and replies with the $L = \sqrt{n}$ results. This protocol has a communication overhead of $O(\sqrt{n})$, $O(n)$ computation overhead at the server's side, and $O(\sqrt{n})$ computation overhead at the client's side.

Reducing Communication: Receiving the OPE Output Using PIR. Note that the client does not need to learn the outputs of all polynomials but

---

[4] Other OPE constructions could be based on the hardness of noisy polynomial interpolation or on using $\log |\mathcal{F}|$ 1-out-of-2 OTs, where $\mathcal{F}$ is the underlying field [23].

rather only the value of the polynomial associated with the bin to which w could be mapped. To further lower the communication complexity, the protocol uses a public hash-function H and invokes PIR to retrieve the result of the relevant polynomial evaluation. Namely, the function H is chosen independently of the content of the database, and it is used to map items to bins. After the server evaluates the L polynomials on the client's input w, the client runs a 1-out-of-L PIR scheme to learn the result of the polynomial of bin $H(w)$.

The total communication overhead is $O(m) \approx n/L$ (client to server) plus the overhead of the PIR scheme. A good choice is to use a PIR scheme with a poly-logarithmic communication overhead, such as the scheme of Cachin et al. [4] (based on the Φ-hiding assumption) or the schemes of Chang [6] or Lipmaa [21] (based on the Paillier and Damgård-Jurik cryptosystems, respectively). In these cases, setting $L = n/\log n$ gives a total communication of $O(\text{polylog } n)$. We note that the client can combine the first message from its KS scheme with that of its PIR scheme. Thus, the round overhead of the combined protocol is the same as that of the PIR protocol alone. The computation overhead of the server is $O(n)$ plus that of a PIR scheme with L inputs; the client's overhead is $O(m)$ plus that of a PIR scheme with L inputs.

Theorem 1. *There exists a KS system for semi-honest parties with a communication overhead of $O(\text{polylog n})$ and a computation overhead of $O(\log n)$ "public-key" operations for the client and $O(n)$ for the server. The security of the KS system is based on the assumptions used for proving the security of the KS protocol's homomorphic encryption system and of the PIR system.*

Proof (sketch for semi-honest parties): Given a pair (x_i, p_i) in the server's input such that $w = x_i$, it is clear that the client outputs p_i. If $w \neq x_i$ for all i, the client outputs $\bot$ with probability at least $1 - 2^{-\ell}$. The protocol is therefore correct. Since the server receives semantically-secure homomorphic encryptions and the PIR protocol protects the privacy of the client, the protocol ensures the client's privacy: The server cannot distinguish between any two client inputs x, x'. Finally, the protocol protects the server's privacy: If a polynomial Z with fresh randomness is prepared for every query on every bin, then the result of the client's query w is random if w is not a root of P, *i.e.*, if w is not in the server's input X. A party running the client's role in the ideal model can therefore simulate the client's view in the real execution.

Handling Malicious Servers. Assume that the PIR protocol provides client privacy in the face of a malicious server (as is the case with virtually all PIR protocols from the literature). Then the protocol is secure against malicious servers (per Definition 2), as the only information that the server receives, in addition to messages of the PIR protocol, is composed of semantically-secure encryptions of powers of the client's input searchword.

Handling Malicious Clients. If the client is malicious then server privacy is not guaranteed by Protocol 1 as given. For example, a malicious client could send encryptions that do not correspond to powers of a value w. However, if the OPE protocol used in Protocol 1 is secure against malicious clients, then the overall protocol provides security against malicious clients, regardless of the security of the PIR protocol. (Note that there are no server privacy requirements on PIR; it is used merely to reduce communication complexity.)

One conceptually-simple solution therefore requires the client to prove that the encryptions it sends in the OPE protocol are well-formed, *i.e.*, correspond to encryptions of a sequence of values $w, w^2, \ldots, w^m$. Unfortunately, such a proof in the standard model requires more than a single round of messages.

A more efficient solution can be based on a known reduction of the OPE of a polynomial of degree m, to m OPEs of linear polynomials [12]. The overhead of the resulting protocol is similar to that of a direct OPE of the polynomial, and the protocol consists of only a single round (the m OPEs of the linear polynomials are done in parallel). We describe the reduction of [12] in the full version of this paper.

When the OPE protocol (based on homomorphic encryption) is applied to a linear polynomial, any encrypted value (w) sent by the client corresponds to a valid input to the polynomial, and thus the OPE of the linear polynomial computes a legitimate value of the polynomial. Therefore, if we ensure that the client sends a legitimate encryption we obtain a linear OPE (and thus a general OPE) secure against malicious clients.

When considering concrete instantiations of the OPE protocol, we note that the El Gamal cryptosystem has the required property, namely that any ciphertext can be decrypted.[5] The El Gamal cryptosystem can therefore be used for implementing a single-round OPE secure against a malicious client. Yet, the El Gamal system has a different drawback: given that it is multiplicatively homomorphic, it can only be used for an OPE in which the receiver obtains $g^{P(x)}$, rather than $P(x)$ itself. Thus, a direct use of El Gamal in KS is only useful for short payloads, as it requires encoding the payload in the exponent and asking the receiver to compute its discrete log.

We can slightly modify the KS protocol to use El Gamal yet still support payloads of arbitrary length. A detailed description appears in the full version of the paper. The main idea, however, is to have the server map the items to $n/\log n$ bins as usual, but define, for every bin j, a random polynomial Z_j of degree $m = O(\log n)$. For an item (x_i, p_i), the server encrypts $p_i|0^\ell$ using the key $g^{Z_{H(x_i)}(x_i)}$. The client sends a first message for an El Gamal-based OPE, namely encryptions of $g^w, g^{w^2}, \ldots, g^{w^m}$. The server then prepares, for every bin j, a message $\langle\, g^{Z_j(w)}, \{Enc_{Z_j(x_{j,i})}(p_{j,i}|0^\ell)\}_{i \in [m]}\,\rangle$, where the $x_{j,i}$'s are the messages mapped to bin j. The client uses PIR to learn the message of its bin of interest, and then can decrypt the payload corresponding to w if $\exists\, x_{j,i} = w$.

The only difference with this modified protocol is that the message learned during the PIR is of size $O(|p_i| \log n)$ rather than of size $O(|p_i|)$. The overall

---

[5] Unfortunately, as was observed for example in [1], the Paillier cryptosystem is not verifiable. That is, given a public key and a ciphertext, it is not known how to verify that the ciphertext is valid and can be correctly decrypted.

communication complexity does not change, however, since the PIR has polylogarithmic overhead. We obtain essentially the same overhead, including round complexity, as Protocol 1. (Note also that the security of the new protocol is proved in the model of Remark 1.)

Multiple Invocations. The privacy of the server in Protocol 1 and its variants is based on the fact that the client can evaluate each polynomial Z at most once. Therefore, fresh randomness r_i must be used in order to generate new polynomials $Z_1, \ldots, Z_L$ for every invocation of the protocol. This means that using the protocol for multiple queries must essentially be done by independent invocations of the protocol.

4 Keyword Search from OPRFs

In this section, we describe a general *reduction* of KS to semi-private KS using oblivious pseudorandom functions (OPRFs). Unlike the protocol from the previous section, this reduction can yield fully-adaptive KS protocols. We first recall the original notion of OPRFs from the literature [26] and then introduce a new natural relaxation, which arguably suffices for most applications. Finally, we describe our reduction from KS to semi-private KS using the relaxed notion of OPRF. New constructions of such OPRFs will be presented in the next section.

4.1 Oblivious Pseudorandom Functions

The strongest definition of OPRF is as a secure two-party protocol realizing the functionality $g(r, w) = (\lambda, f_r(w))$ for some pseudorandom function family f_r. (Here and in the following, the first input or output corresponds to the server $\mathcal{S}$ and the second to the client $\mathcal{C}$; by λ we denote an empty output.) As usual, the term "secure" can be interpreted in several ways. For consistency with the security definitions of Section 2 and the constructions of the next section, we interpret "secure" here as "private". We note, however, that the definitions and results of this section naturally extend the case of full security.

Definition 5 (Strongly-private OPRF (s-OPRF)). *A two-party protocol π is said to be a strongly-private OPRF (or strong OPRF for short) if there exists some PRF family f_r, such that π privately realizes the following functionality.*

- *Inputs: Client holds an evaluation point w; Server holds a key r.*
- *Outputs: Client outputs $f_r(w)$; Server outputs nothing.*

One can similarly define adaptive and non-adaptive t-time variants of strong OPRFs. Note that server privacy guarantees that a malicious client $\mathcal{C}'$ cannot learn anything about r except what follows from $f_r(w')$ for some w'. Composability of secure computation [5] implies that a 1-time s-OPRF can be invoked multiple times (with the same r and different w_i) to realize an adaptive t-time s-OPRF, where t can be an arbitrary polynomial in the security parameter.[6]

It follows from known reductions between cryptographic primitives that strong OPRF exists if OT exists [14, 19, 16]. We note, however, that the construction of s-OPRF from OT makes a non-black-box use of the OT primitive, even in the semi-honest setting: The OT-based protocol for evaluating the PRF depends on the function's circuit representation [19], which in turn depends on the representation of the OT primitive from which the PRF is derived.

A New Relaxed Type of OPRF. As noted above, a strong OPRF guarantees that the client learn no additional information about the PRF key r. As we shall see, some natural and efficient OPRF protocols do not satisfy the strong definition, yet are sufficient for the KS application. We thus turn our consideration to relaxing the definition of server privacy to the following.

Roughly speaking, we require that following the execution of the OPRF protocol, the client obtains no additional information about the *outputs* of a *random* function f_r, other than what follows from a legitimate set of queries, whose size is bounded by t in the t-time case. (Recall that the strong definition requires that no information be learned about the *key* of an *arbitrary* function f_r.) In other words, the outputs of f_r on unqueried inputs cannot be distinguished from the outputs of a *random* function, even given the client's view. Note that this does not prevent the client from learning substantial partial information about r (which does not provide information about other values of f_r).[7]

This intuitive property is relatively straightforward to formalize in the case of a semi-honest client. Specifically, one may require that following the protocol's execution, the client cannot efficiently distinguish between f_r and a random function if it only queries them on points not included in its queries $w_1, \ldots, w_t$. Obtaining a suitable definition for the case of malicious clients, however, requires more care. In particular, the inputs on which the client queries f_r in a particular execution of the protocol may not even be well-defined.

We formalize our relaxed notion of OPRF by a careful modification of the underlying functionality. The client's privacy is defined as before. However, for the purpose of defining the server's privacy, we view f_r as a *randomized* functionality (with randomness r picked by the TTP in the ideal implementation), and

---

[6] Note that our definitions of KS and OPRF do not require protecting the client against a malicious server who may choose different keys r in different invocations. On the other hand, our definition coincides with that of [5] for the case of simulating a (potentially malicious) client.

[7] As a concrete simple example, consider the following pseudo-random function based on the Naor-Reingold construction. The key r consists of two sets $x_1, \ldots, x_m$ and $y_1, \ldots, y_m$; the function is defined for inputs (i, j) such that $1 \leq i, j \leq m$, and its value is $f_r(i,j) = g^{x_i y_j}$ in a group where the DDH assumption holds and g is a generator. Consider a 1-time OPRF protocol where a client whose input is (i, j) learns x_i and y_j and uses them to compute $f_r(i, j)$. Although these values reveal part of the key r to the client, the other outputs of the function remain pseudo-random.

we allow *both* the client and the server to provide inputs to and receive outputs from this functionality.

Definition 6 (Relaxed OPRF (r-OPRF)). *A two-party protocol π is said to be a (non-adaptive, 1-time) relaxed OPRF if there exists some PRF family f_r, such that the following hold.*

CORRECTNESS AND CLIENT'S PRIVACY. *These properties remain the same as in Definition 5, i.e., using the functionality $g(r, w) = (\bot, f_r(w))$.*

SERVER'S PRIVACY. *To define server's privacy in π, we make the following mental experiment. Consider an augmented protocol $\tilde{\pi}$ in which the input of $\mathcal{S}$ consists of n evaluation points $x_1, \ldots, x_n$ (instead of a key r) and the input of $\mathcal{C}$ is an evaluation point w (as in π). Protocol $\tilde{\pi}$ proceeds as follows: (1) $\mathcal{S}$ picks a key r at random; (2) $\mathcal{S}, \mathcal{C}$ invoke π on inputs (r, w); (3) $\mathcal{S}$ outputs $(f_r(x_1), \ldots, f_r(x_n))$ and $\mathcal{C}$ outputs its output in π. We require that the augmented protocol $\tilde{\pi}$ provide server security with respect to the following randomized functionality $\tilde{g}$:*

- *Inputs: Client holds an evaluation point w; Server holds an arbitrary set of evaluation points $(x_1, \ldots, x_n)$.*
- *Outputs: Client outputs $f_r(w)$ and Server outputs $(f_r(x_1), \ldots, f_r(x_n))$, where the key r is uniformly chosen by the functionality.*[8]

Specifically, for any (efficient, malicious) client $\mathcal{C}'$ attacking $\tilde{\pi}$, there is a simulator $\mathcal{C}''$ playing the client's role in the ideal implementation of $\tilde{g}$, such that on all inputs $((x_1, \ldots, x_n), w)$, the view of $\mathcal{C}'$ concatenated with the output of $\mathcal{S}$ in π is computationally indistinguishable from the output of $\mathcal{C}''$ concatenated with that of $\mathcal{S}$ in the ideal implementation of $\tilde{g}$.

This definition applies to the non-adaptive 1-time case. In the t-time case, we replace w with $w_1, \ldots, w_t$, and $f_r(w)$ with $(f_r(w_1), \ldots, f_r(w_t))$. In the adaptive case, the protocols $\pi, \tilde{\pi}$ and the functionalities $g, \tilde{g}$ have multiple phases, where the client's input w in each phase may depend on the outputs of previous phases.

The above server's privacy requirement implies that the client's view gives no information about the server's inputs and outputs $(x_1, f_r(x_1)), \ldots, (x_n, f_r(x_n))$, other than what follows from some valid set $(w'_1, f_r(w'_1)), \ldots, (w'_t, f_r(w'_t))$. Moreover, this holds for an arbitrary choice of points x_i made by the server (including those possibly intersecting w'_i). In fact, this is precisely the requirement needed for the keyword-search application.

Finally, we note that Definition 6 is indeed a relaxation of Definition 5.

Claim. If π is an s-OPRF, then it is also an r-OPRF.

Proof: The server's privacy requirement of Definition 5 implies, in particular, that on a *uniformly-chosen* r and an arbitrary w, the view V' of a malicious

---

[8] Equivalently, f_r can be replaced here with a totally random function. We prefer the current formulation because of its closer correspondence with the notion of s-OPRF, as well as the convention that ideal functionalities are efficiently computable.

client $\mathcal{C}'$ concatenated with r is indistinguishable from the output V'' of its simulator $\mathcal{C}''$ concatenated with r. This in turn implies that $(V', \{(f_r(x_i)\}_{i \in [n]})$ is indistinguishable from $(V'', \{(f_r(x_i)\}_{i \in [n]})$, as required by Definition 6.

4.2 Reducing KS to Semi-private KS

We now present a general method of using (either variant of) OPRF to upgrade any semi-private KS protocol into fully-private KS.

Recall that a semi-private KS protocol is a KS protocol which guarantees privacy for the client but not for the server, similar to the privacy offered by PIR protocols. (The notion of semi-private KS was first considered in [7], where it was referred to as *private information retrieval by keywords*.) Semi-private KS can be simply implemented by letting the server send its input X, or (better yet) a data structure Y representing X, to the client. When the communication is required to be sublinear, semi-private KS can be implemented using PIR to probe the data structure Y, as suggested in [7].

Using the following high-level idea, we can now construct a fully-private KS protocol from a semi-private KS protocol: The server uses a PRF to assign random pseudo-identities to the original keywords x_i (as well as mask the payloads p_i), and the client uses an OPRF protocol to learn the values of the PRF on the selected searchword(s). Since the PRF values on unselected searchwords remain random from the client's point-of-view, knowledge of the original and pseudo-identity pairs of the selected searchwords does not provide any more information than does knowledge of just the set of searchwords that are in the database along with their payloads.

More formally, given a semi-private KS protocol and a (possibly relaxed) OPRF realizing f_r, the KS protocol proceeds as follows. For simplicity, we address below the case non-adaptive KS with $t=1$.

Protocol 2 (A KS protocol based on semi-private KS and r-OPRF)

1. The server picks a random key r for the PRF. For $1 \leq i \leq n$, it parses $f_r(x_i)$ as $(\hat{x}_i, \hat{p}_i)$ and constructs a pseudo-database $X' = \{(x'_i, p'_i)\}_{i \in [n]}$ with $x'_i = \hat{x}_i$ and $p'_i = p_i \oplus \hat{p}_i$. (Both X and X' must be treated as unordered sets, whose representation does not reveal the index i of each element; alternatively, one may think of X and X' as lexicographically-sorted sequences.)
2. The parties invoke the r-OPRF protocol, with server input r and client input w. As a result, the client learns $f_r(w)$ and parses it as $(\hat{w}, \hat{p})$.
3. The parties invoke the semi-private KS protocol with server input X' and client input $\hat{w}$. As a result, the client learns whether $\hat{w} \in X'$, and if so, also learns the corresponding payload p'_i. If $\hat{w} \in X'$, the client outputs $p'_i \oplus \hat{p}$; otherwise, it outputs $\perp$.

We stress that, due to the lack of server's privacy in *semi*-private KS, we should make the worst-case assumption that the client learns the *entire* pseudo-database X' in Step 3. Still, the use of an OPRF in Step 2 guarantees that the client does not learn more than it is entitled to.

Remark 2. If a setup phase with linear communication is allowed, the semi-private KS in Step 3 can be replaced by having X' (or a corresponding data structure Y') sent to the client in the clear following Step 1.

Theorem 2. *Protocol 2 is a private KS protocol.*

Proof (sketch): The protocol's correctness is easy to verify. The client's privacy follows immediately from its privacy in the OPRF and the semi-private KS.

Server's Privacy. Letting π denote the r-OPRF protocol, it is convenient to reformulate the above protocol in the following equivalent way:

- The parties invoke the augmented protocol $\tilde{\pi}$ (from Definition 6) on server input $(x_1, \ldots, x_n)$ and client input w. At the end of this protocol, $\mathcal{S}$ outputs $(f_r(x_1), \ldots, f_r(x_n))$ and $\mathcal{C}$ outputs $f_r(w)$.
- The server parses each $f_r(x_i)$ as $(\hat{x}_i, \hat{p}_i)$ and creates a pseudo-database $X' = \{(x'_i, p'_i)\}_{i \in [n]}$ with $x'_i = \hat{x}_i$ and $p'_i = p_i \oplus \hat{p}_i$, as before. Again, the client parses $f_r(w)$ as $(\hat{w}, \hat{p})$. The parties invoke the semi-private KS protocol with server input X' and client input $\hat{w}$. As a result, the client learns whether $\hat{w} \in X'$, in which case the client outputs $p'_i \oplus \hat{p}$; otherwise, it outputs $\perp$.

By Definition 6, when considering only the client's simulation, $\tilde{\pi}$ must be secure with respect to the randomized functionality $\tilde{g}$ mapping $(x_1, \ldots, x_n)$ and w to $(f_r(x_1), \ldots, f_r(x_n))$ and $f_r(w)$, respectively. Hence, using protocol composition [5], it suffices to prove the server's privacy in a simpler "hybrid" protocol, where the invocation of $\tilde{\pi}$ is replaced by a call to an oracle (or TTP) computing $\tilde{g}$. Moreover, by the pseudorandomness of f_r, we can replace the oracle $\tilde{g}$ by a similar oracle $\tilde{G}$ in which f_r is replaced by a truly random function.

The resultant hybrid protocol is in fact *perfectly* private. Given a malicious client $\mathcal{C}'$ attacking the hybrid protocol, a corresponding simulator $\mathcal{C}''$ can proceed as follows. $\mathcal{C}''$ invokes $\mathcal{C}'$ on input w. In the first step, after learning the query w' which $\mathcal{C}'$ sends to the oracle computing $\tilde{G}$, the simulator $\mathcal{C}''$ sends the query w' to the TTP computing KS. As a response, it gets p_i if $w' = x_i$ or $\perp$ if no such i exists. Now the second step can be simulated jointly with the response $(\hat{w}, \hat{p})$ of the $\tilde{G}$ oracle. First, $\mathcal{C}''$ chooses X' to be a uniformly-random pseudo-database of size n. Next, it simulates $(\hat{w}, \hat{p})$ so that they are consistent with X' and the response of KS: if a payload p was obtained from KS, then $\hat{w}$ is taken to be a random keyword from X' and $\hat{p}$ is set to the exclusive-or of the keyword's corresponding payload and p; otherwise, $\hat{w}$ and $\hat{p}$ are chosen at random from their respective domains. Finally, $\mathcal{C}''$ simulates the view of $\mathcal{C}'$ in the semi-private KS protocol by simply running the protocol on inputs (X', w').

Efficiency. The cost of the protocol is dominated by that of the semi-private KS and the OPRF. In the t-time non-adaptive model, this cost is typically dominated by that of the semi-private KS, which in turn is dominated by the cost of the underlying PIR protocol. We note that the latter cost can be amortized over t non-adaptive queries [2, 18]. In the adaptive model—more generally, in any setting allowing setup—the offline cost is dominated by linear communication in the size of the database, and the online cost by the efficiency of the underlying OPRF. We now consider efficient implementations of the OPRF primitive.

5 Constructing OPRFs

A generic implementation of an s-OPRF can be based on general secure two-party evaluation. Namely, the server has as input a key r of a PRF f_r and, whenever the client wants to evaluate f_r on x, the parties perform a secure function evaluation (SFE), during which the client learns $f_r(x)$. As noted above, this gives rise to a *non-black-box* reduction from strong OPRF to OT. In this section, we discuss two other types of constructions:

- Constructions of fully-adaptive s-OPRFs based on specific assumptions (mainly on DDH). These constructions are either given or implicit in [26, 24] and are more efficient than the generic SFE-based construction sketched above.
- General constructions of t-time adaptive r-OPRFs making a *black-box* use of OT. From a theoretical point of view, one of the most interesting open questions left by our work is to come up with any efficient black-box construction of fully-adaptive r-OPRFs. This is indeed a rare example of a non-black-box construction in cryptography for which no black-box construction is known.

For simplicity, we discuss these constructions mainly from the viewpoint of the semi-honest model.

5.1 Strong OPRFs Based on DDH or Factoring

Naor and Reingold gave two constructions of PRFs based on number-theoretic assumptions in [26]: one based on the Decisional Diffie-Hellman assumption (DDH), and the other based on the hardness of factoring. The constructions have a simple algebraic structure, and they were used to give oblivious, fully-adaptive evaluations for these functions. While more efficient than general secure function evaluation, these s-OPRFs have the disadvantage of requiring a linear number of rounds and a linear number of exponentiations. Implicit in the work of Naor and Pinkas on OT [23, 24],[9] one can find a significantly more efficient evaluation of the DDH-based PRFs of [26]. We now sketch this construction.

Initialization: Let g be a generator of a group G_g of prime order p for which the DDH assumption holds. The key $\bar{r}$ of the pseudo-random function $f_{\bar{r}} : \{0,1\}^m \mapsto G_g$ contains m values $\{r_1, \ldots, r_m\}$, sampled uniformly at random in Z_p^*. The function $f_{\bar{r}}(x)$ is defined to be $g^{\Pi_{x_i=1} r_i}$, for any m-bit $x = x_1 x_2 \ldots x_m$. (This function was shown in [26] to be pseudorandom.)

---

[9] The construction was used to generate values that mask the server's input in an adaptive OT protocol.

Secure Evaluation: The client has inputs $x = x_1 x_2 \ldots x_m$. The server selects m values $\{a_1, \ldots, a_m\}$ sampled uniformly at random in Z_p^*. For each i, the parties perform a 1-out-of-2 OT (denoted by $\binom{2}{1}$-OT), with the server using as inputs the two values a_i and $a_i \cdot r_i$. Thus, the client learns a_i if $x_i = 0$ and $a_i \cdot r_i$ otherwise. In addition, the server sends $\hat{g} = g^{1/\prod_{i=1}^m a_i}$ in the clear. Let A be the product of the values learned by the client, then $A = (\prod_{i=1}^m a_i) \cdot (\prod_{x_i=1} r_i)$. Thus, the client can compute $\hat{g}^A$ and learn the desired value $f_{\bar{r}}(x)$.

Security: This protocol's security follows from the security of the OT protocol: The distribution of the m values learned by the m OTs, combined with $g^{1/\prod_{i=1}^m a_i}$, can be easily sampled given access to $f_{\bar{r}}(x)$ alone.

Efficiency: The computational cost of the protocol (for both client and server) is m $\binom{2}{1}$-OTs and one exponentiation. The main cost in communication is that incurred by the m OTs. Given the work on batch OT of [17], the OTs performed by the oblivious evaluation protocol above can be considered, for practical purposes, to be almost as efficient as private-key operations. In particular, using these s-OPRFs in the transformation of Section 4.2 gives quite an efficient solution to KS. Unlike [27], this solution is in the standard model—rather than in the random oracle model—and only relies on standard assumptions.

5.2 Relaxed OPRFs Based on Black-Box OT

We now present a new construction of adaptive t-time r-OPRFs based on general assumptions, using the OT and PRF primitives in a black-box manner. (In fact, as discussed earlier, PRF is itself black-box implied by OT [14, 16, 15].) Our starting point is a construction of Naor and Pinkas [23] that gives PRFs—originally designed for sub-exponential domains—with some weak form of oblivious evaluation.

Consider a set of known PRFs $\{g_s\}$ over the domain $[N] = [M]^2$. Naor and Pinkas [23] construct related PRFs $\{f_{\bar{r}}\}$ over the same domain. First, let each key $\bar{r}$ be composed of two sets of M random g keys (i.e., $\bar{r}_1 = \{r_{1,1}, \ldots, r_{1,M}\}$ and $\bar{r}_2 = \{r_{2,1}, \ldots, r_{2,M}\}$). Then, define $f_{\bar{r}}(x)$ as $g_{r_{1,x_1}}(x) \oplus g_{r_{2,x_2}}(x)$ for any $x = (x_1, x_2) \in M^2$.[10]

We can now use $f_{\bar{r}}$ in place of g_s to our advantage, as there exists a somewhat oblivious way of evaluating $f_{\bar{r}}(x)$. Namely, perform two independent $\binom{M}{1}$-OTs to retrieve $r_{1,x_1} \in \bar{r}_1$ and $r_{2,x_2} \in \bar{r}_2$, and then evaluate $f_{\bar{r}}(x)$ as desired using these random keys. Of course, the client now learns r_{1,x_1} and r_{2,x_2} in addition to just $f_{\bar{r}}(x)$. Still, it is easy to argue that $f_{\bar{r}}$, when restricted to all inputs other than x, remains pseudorandom. With a small additional effort, $f_{\bar{r}}$ can be turned into a 1-time r-OPRF.

What happens if we perform an oblivious evaluation of $f_{\bar{r}}$ on t different inputs? In this case, the client learns up to t keys in both $\bar{r}_1$ and $\bar{r}_2$, allowing it to evaluate $f_{\bar{r}}$ in up to t^2 places, which is certainly undesirable. Still, $f_{\bar{r}}$ maintains a considerable amount of pseudorandomness, as its output looks random other

---

[10] This is a simple version of the construction; some useful optimizations are possible.

than at these t^2 locations. In light of this property, [23] gives a technique that can be translated into a construction of a *non-adaptive t*-time r-OPRF.

The PRF $F(\cdot)$ used in this construction is the exclusive-or of some ℓ functions $f_{\bar{r}^i}(\sigma^i(\cdot))$, where $f_{\bar{r}^i}$ is defined as before and each σ^i is a random permutation over $[N]$. All random inputs (for the sub-keys, $\bar{r}_1^i$ and $\bar{r}_2^i$, and for the permutations σ^i) are chosen independently by the server for all $1 \leq i \leq \ell$. The evaluation of $F(\cdot)$ on t inputs $x_1 \ldots x_t$ proceeds in ℓ rounds. In the ith round, σ^i is sent to the client and the parties perform t oblivious evaluations of $f_{\bar{r}^i}$, as above.

This construction's main idea is the following: In each round, the client may learn at most t^2 values of the current $f_{\bar{r}^i}(\sigma^i(\cdot))$—a $t \times t$ sub-matrix—from the total of M^2 possible values over which the PRF is defined. However, to learn the value of F for $t+1$ distinct inputs, the client must learn all intermediate values for *each one of the ℓ functions* $f_{\bar{r}^i}(\sigma^i(\cdot))$ on these $t+1$ inputs. The random permutations σ^i—each learned only during the execution of subsequent rounds— ensure that this will only happen with negligible probability. See [23] for more details. Note that for this probabilistic argument to hold, the number of rounds ℓ must depend on the security parameter.

Challenges. The above construction raises the following challenges left open by [23] and the subsequent [24]: (1) Can the construction be made secure against adaptive queries? We note that the adaptive solutions given in [24] rely either on specific assumptions or on random oracles. (2) Can one obtain oblivious evaluation in a *constant number of rounds*? Note that the number of rounds of the above protocols depends on the security parameter. (3) Can the construction handle an exponential domain size N? Various difficulties arise when naively extending the above solution to larger values of N. First, the random permutations σ^i are too large to sample and transmit. Second, one has to extend the construction to higher dimensions than two and view $[N]$ as $[M]^\ell$ for non-constant ℓ: We certainly want M to be sub-exponential, given that we are performing $\binom{M}{1}$-OTs. We can indeed perform this extension, but the natural method as used below reveals many more values of the PRFs: In t queries, the client learns t sub-keys in every dimension. Thus, it can evaluate the function at t^ℓ locations, where t^ℓ may be exponentially large (specifically, polynomially related to N). This expansion seems to complicate the analysis and, in particular, implies a larger number of rounds that also depends on t.

Our Construction. In this section, we simultaneously answer all of the above challenges: We obtain *adaptive t-time* r-OPRFs that can handle an *exponential* domain size and can be securely evaluated in a *constant number of rounds*.

The technique of [23] for turning their 1-time r-OPRF into a t-time r-OPRF is based on providing only indirect access to the functions $f_{\bar{r}^i}$. Namely, the value of the PRF F on x depends on the values $f_{\bar{r}^i}(\sigma^i(x))$, rather than on $f_{\bar{r}^i}(x)$. However, since the permutation σ_i is transmitted in its entirety to the client, this type of indirection is not very useful for obliviousness by itself. Instead, the protocol must be designed using several functions, revealing additional information (each σ_i) in synchronous stages.

Instead, we will use only one function $f_{\bar{r}}$ and therefore will need only a single permutation σ for the indirect access to $f_{\bar{r}}$. Rather than transmitting the entire permutation σ to the client, we allow the client access only to t locations of σ in some oblivious way. Since σ is now not completely known to the client, we overcome both the need for a super-constant number of rounds and the large cost of sending σ for large domain sizes. Of course, if σ is random or pseudorandom, then the oblivious evaluation of σ is exactly the problem we wanted to solve in the first place! Therefore, we relax this randomness requirement by replacing σ with $(t+2)$-wise independent functions (although, in fact, even weaker requirements suffice).[11] We proceed to the detailed construction of adaptive t-time r-OPRFs, focusing on the setting where N is exponential.

Notion of Privacy. In the above description and below, we argue that a function is an oblivious PRF if it remains pseudorandom on all inputs other than the ones retrieved by the client. This makes our discussion simpler and more intuitive. However, this type of definition seems only to make sense in the semi-honest model (as otherwise, the inputs retrieved by the client may not be well-defined). Even in the semi-honest model, this notion—though sufficiently strong for the KS application—falls short of obtaining the requirements of a r-OPRF, which are defined in terms of simulation. Nevertheless, the protocol below gives a t-time r-OPRF: All that is needed is that the basic PRFs $\{g_s\}$ used by this protocol will have the additional property that, given t inputs and t corresponding outputs, a random seed s can be sampled under the restriction that g_s is consistent with these inputs and outputs. This is easy to obtain if each g_s is an exclusive-or of a PRF and a t-wise independent function (as l-wise independent functions usually have such "interpolation" property).

Extending the 1-Time r-OPRF to Higher Dimensions. Let $\{g_s\}$ be PRFs over a domain $[N] = [M]^\ell$. Define the related PRFs $\{f_{\bar{r}}\}$ over the same domain, where each key $\bar{r}$ is composed of ℓ sets $\{\bar{r}_1, \ldots \bar{r}_\ell\}$ of M random g keys, where $\bar{r}_i = \{r_{i,1}, \ldots, r_{i,M}\}$. Thus, $\bar{r}$ defines an $\ell \times M$ matrix. For any $x = \{x_1, \ldots x_\ell\} \in M^\ell$, the value $f_{\bar{r}}(x)$ is defined to be $\bigoplus_{i=1}^{\ell} g_{i,x_i}(x)$.

The 1-time oblivious evaluation of $f_{\bar{r}}(x)$ goes as follows. First, perform ℓ independent $\binom{M}{1}$-OTs to retrieve $r_{i,x_i} \in \bar{r}_i$, for $i = 1, \ldots, \ell$. Then, the client can evaluate $f_{\bar{r}}(x)$ as desired. As mentioned above, t evaluations of $f_{\bar{r}}$ may now give information on t^ℓ values. However, $f_{\bar{r}}$ remains pseudorandom when restricted to all inputs other than x.

Oblivious Evaluation of $(t+2)$-Wise Independent Functions. The second ingredient in our construction is a family $H = \{h : [N] \mapsto [N]\}$ of $(t+2)$-wise independent functions. This definition means that, restricted to any $(t+2)$

---

[11] A different variant of the construction uses the 1-time r-OPRFs based on [23] instead of the random permutations. This construction may be more efficient in some settings of the parameters. On the other hand, it seems theoretically inferior and somewhat more complicated (e.g., it requires two levels of indirection). We therefore omit it from this version for clarity.

inputs, a function h sampled from H is completely random.[12] We also rely on H to have an oblivious evaluation (or a t-time oblivious evaluation). This problem is an easier task than that of r-OPRFs. In particular, as $(t+2)$-wise independent functions exist unconditionally, they have oblivious evaluation based on OTs in a black-box manner. Note that while this observation is based on general secure evaluation, more efficient oblivious evaluations can be designed for specific families of hash functions: for example, an OPE-based evaluation can be used for a polynomial-based $(t+2)$-wise independent hash function.

The New Adaptive t-Time r-OPRFs. We set $M = 2t$ and assume without loss of generality that ℓ is at least the security parameter.[13] The key of these adaptive t-time r-OPRFs is composed of a $(t+2)$-wise independent hash function $h \in H$ and a key $\bar{r}$ of the ℓ-dimension 1-time r-OPRF $f_{\bar{r}}(\cdot)$ defined above. The value of this function $F_{h,\bar{r}}$ on any input $x \in [N]$ is given by $F_{h,\bar{r}}(x) \stackrel{\text{def}}{=} f_{\bar{r}}(h(x))$. The oblivious evaluation of $F_{h,\bar{r}}(x)$ proceeds by first evaluating $y = h(x)$ and then evaluating $f_{\bar{r}}(y)$, using the corresponding oblivious evaluation protocols.

Security of the Construction (Sketch). We want to claim that after t evaluations of $F_{h,\bar{r}}(\cdot)$, its restriction on all other inputs is indistinguishable from a random function. Intuitively, this is true since each dimension has $2t$ keys of which the client learns at most t, and the probability that another value of the function is evaluated using only these learned keys is at most $2^{-\ell}$. Consider the hybrid function $R(h(\cdot))$, where R is a random function. It is easy to argue that $R(h(\cdot))$ is indistinguishable from random: It only can be distinguished from random by querying inputs that cause collisions of h. Since conditioned on the values of h already learned by the client, h is still pair-wise independent, collisions are encountered with negligible probability. It remains to argue that $R(h(\cdot))$ is indistinguishable from $f_{\bar{r}}(h(\cdot))$. Note that at most t^ℓ values of $f_{\bar{r}}$ are compromised by the client, and $f_{\bar{r}}$ is still pseudorandom on the rest. To distinguish $R(h(\cdot))$ from $f_{\bar{r}}(h(\cdot))$, the distinguisher needs to query with an input that causes the output of h to fall into the compromised set. As the fraction of compromised $f_{\bar{r}}$-inputs is negligible (at most $2^{-\ell}$), this happens with negligible probability.

Acknowledgements. Michael Freedman is supported by a National Defense Science and Engineering Graduate Fellowship. Yuval Ishai is supported by Israel Science Foundation grant 36/03. Omer Reingold is the incumbent of the Walter and Elise Haas Career Development Chair at the Weizmann Institute of Science and is supported by US-Israel Binational Science Foundation Grant 2002246.

---

[12] In fact, h can be only statistically close to random or even just pseudorandom.
[13] This implies r-OPRFs also for smaller values of N, although further optimizations may be possible for these cases.

References

1. Bill Aiello, Yuval Ishai, and Omer Reingold. Priced oblivious transfer: How to sell digital goods. In *EUROCRYPT*, Innsbruck, Austria, May 2001.
2. Amos Beimel, Yuval Ishai, and Tal Malkin. Reducing the servers' computation in private information retrieval: Pir with preprocessing. In *CRYPTO*, Santa Barbara, CA, August 2000.
3. Dan Boneh, Giovanni Di Crescenzo, Rafail Ostrovsky, and Giuseppe Persiano. Public key encryption with keyword search. In *EUROCRYPT*, Interlaken, Switzerland, May 2004.
4. Christian Cachin, Silvio Micali, and Markus Stadler. Computationally private information retrieval with polylogarithmic communication. In *EUROCRYPT*, Prague, Czech Republic, May 1999.
5. Ran Canetti. Security and composition of multiparty cryptographic protocols. *Journal of Cryptology*, 13(1):143–202, 2000.
6. Yan-Cheng Chang. Single database private information retrieval with logarithmic communication. In *Proc. 9th ACISP*, Sydney, Australia, July 2004.
7. Benny Chor, Niv Gilboa, and Moni Naor. Private information retrieval by keywords. Technical Report TR-CS0917, Dept. of Computer Science, Technion, 1997.
8. Benny Chor, Oded Goldreich, Eyal Kushilevitz, and Madhu Sudan. Private information retrieval. In *Proc. 36th FOCS*, Milwaukee, WI, 23–25 October 1995.
9. Shimon Even, Oded Goldreich, and Abraham Lempel. A randomized protocol for signing contracts. *Communications of the ACM*, 28(6):637–647, 1985.
10. Michael J. Freedman, Kobbi Nissim, and Benny Pinkas. Efficient private matching and set intersection. In *EUROCRYPT*, Interlaken, Switzerland, May 2004.
11. Yael Gertner, Yuval Ishai, Eyal Kushilevitz, and Tal Malkin. Protecting data privacy in private information retrieval schemes. In *Proc. 30th ACM STOC*, Dallas, TX, May 1998.
12. Niv Gilboa. *Topics in Private Information Retrieval*. PhD thesis, Technion - Israel Institute of Technology, 2000.
13. Oded Goldreich. *Foundations of Cryptography: Basic Tools*. Cambridge University Press, 2001.
14. Oded Goldreich, Shafi Goldwasser, and Silvio Micali. How to construct random functions. *Journal of the ACM*, 33(4):792–807, October 1986.
15. Johan Håstad, Russell Impagliazzo, Leonid A. Levin, and Michael Luby. Construction of pseudorandom generator from any one-way function. *SIAM Journal on Computing*, 28(4):1364–1396, 1999.
16. Russell Impagliazzo and Michael Luby. One-way functions are essential for complexity based cryptography. In *Proc. 30th FOCS*, Research Triangle Park, NC, October–November 1989.
17. Yuval Ishai, Joe Kilian, Kobbi Nissim, and Erez Petrank. Extending oblivious transfers efficiently. In *CRYPTO*, Santa Barbara, CA, August 2003.
18. Yuval Ishai, Eyal Kushilevitz, Rafail Ostrovsky, and Amit Sahai. Batch codes and their applications. In *Proc. 36th ACM STOC*, Chicago, IL, June 2004.
19. Joe Kilian. Founding cryptography on oblivious transfer. In *Proc. 20th ACM STOC*, Chicago, IL, May 1988.
20. Eyal Kushilevitz and Rafail Ostrovsky. Replication is not needed: Single database, computationally-private information retrieval. In *Proc. 38th FOCS*, Miami Beach, FL, October 1997.
21. Helger Lipmaa. An oblivious transfer protocol with log-squared communication. Crypto ePrint Archive, Report 2004/063, 2004.

22. Silvio Micali, Michael Rabin, and Joe Kilian. Zero-knowledge sets. In *Proc. 44th FOCS*, Cambridge, MA, October 2003.
23. Moni Naor and Benny Pinkas. Oblivious transfer and polynomial evaluation. In *Proc. 31st ACM STOC*, Atlanta, GA, May 1999.
24. Moni Naor and Benny Pinkas. Oblivious transfer with adaptive queries. In *CRYPTO*, Santa Barbara, CA, August 1999.
25. Moni Naor and Benny Pinkas. Efficient oblivious transfer protocols. In *Proc. 12th SIAM SODA*, Washington, DC, January 2001.
26. Moni Naor and Omer Reingold. Number-theoretic constructions of efficient pseudo-random functions. In *Proc. 38th FOCS*, Miami Beach, FL, October 1997.
27. Wakaha Ogata and Kaoru Kurosawa. Oblivious keyword search. Crypto ePrint Archive, Report 2002/182, 2002.
28. Pascal Paillier. Public-key cryptosystems based on composite degree residuosity classes. In *EUROCRYPT*, Prague, Czech Republic, May 1999.
29. Michael O. Rabin. How to exchange secrets by oblivious transfer. Technical Report TR-81, Harvard Aiken Computation Laboratory, 1981.
30. Dawn Xiaodong Song, David Wagner, and Adrian Perrig. Practical techniques for searches on encrypted data. In *Proc. IEEE Symposium on Security and Privacy*, Berkeley, CA, May 2000.

Evaluating 2-DNF Formulas on Ciphertexts

Dan Boneh[1,*], Eu-Jin Goh[1], and Kobbi Nissim[2,**]

[1] Computer Science Department, Stanford University,
Stanford CA 94305-9045, USA
{dabo, eujin}@cs.stanford.edu
[2] Department of Computer Science, Ben-Gurion University,
Beer-Sheva 84105, Israel
kobbi@cs.bgu.ac.il

Abstract. Let ψ be a 2-DNF formula on boolean variables $x_1, \ldots, x_n \in \{0,1\}$. We present a homomorphic public key encryption scheme that allows the public evaluation of ψ given an encryption of the variables $x_1, \ldots, x_n$. In other words, given the encryption of the bits $x_1, \ldots, x_n$, anyone can create the encryption of $\psi(x_1, \ldots, x_n)$. More generally, we can evaluate *quadratic* multi-variate polynomials on ciphertexts provided the resulting value falls within a small set. We present a number of applications of the system:

1. In a database of size n, the total communication in the basic step of the Kushilevitz-Ostrovsky PIR protocol is reduced from $\sqrt{n}$ to $\sqrt[3]{n}$.
2. An efficient election system based on homomorphic encryption where voters do not need to include non-interactive zero knowledge proofs that their ballots are valid. The election system is proved secure without random oracles but still efficient.
3. A protocol for universally verifiable computation.

1 Introduction

Secure computation allows several parties to compute a function of their joint inputs without revealing more than what is implied by their own inputs and the function outcome. Any polynomial time functionality can be computed by a secure protocol, requiring polynomial resources [32, 16]. These seminal results are obtained by a generic transformation that converts an insecure computation of a functionality to a secure version (often referred to as the 'garbled circuit' transformation).

Secure protocols generated from the garbled circuit transformation typically have poor efficiency. In particular, the communication complexity of the resulting protocols is proportional to the *size* of a circuit evaluating the functionality, and hence precludes sub-linear communication protocols. The result is that unless circuits are very small, the garbled circuit transformation is seldom used in protocols.

---
* Supported by NSF.
** Work done while the author was at Microsoft Research, SVC.

To avoid using the garbled circuit transformation, researchers have sought for tools that give more efficient protocols for specific functionalities. *Homomorphic encryption* enables "computing with encrypted data" and is hence a useful tool for secure protocols. Current homomorphic public key systems [17, 11, 25] have limited homomorphic properties: given two ciphertexts Encrypt($\mathcal{PK}, x$) and Encrypt($\mathcal{PK}, y$), anyone can compute either the sum Encrypt($\mathcal{PK}, x + y$), or the product Encrypt($\mathcal{PK}, xy$), but not both.[1] The problem of constructing 'doubly homomorphic' encryption schemes where one may both 'add and multiply' is a long standing open question already mentioned by Rivest et al. [29].

Homomorphic encryption schemes have many applications, such as protocols for electronic voting schemes [7, 2, 8, 9], computational private information retrieval (PIR) schemes [20], and private matching [13]. Systems with more general homomorphisms (such as both addition and multiplication) will benefit all these problems.

1.1 Our Results

A Homomorphic Encryption Scheme. We present a homomorphic public key encryption scheme based on finite groups of composite order that support a bilinear map. Using a construction along the lines of Paillier [25], we obtain a system with an additive homomorphism. In addition, the bilinear map allows for *one* multiplication on encrypted values. As a result, our system supports arbitrary additions and one multiplication (followed by arbitrary additions) on encrypted data. This property in turn allows the evaluation of multi-variate polynomials of total degree 2 on encrypted values. Our applications follow from this new capability.

The security of our scheme is based on a new hardness assumption that we put forward – the *subgroup decision problem*. Namely, given an element of a group of composite order $n = q_1 q_2$, it is infeasible to decide whether it belongs to a subgroup of order q_1.

Applications. As a direct application of the new homomorphic encryption scheme, we construct a protocol for obliviously evaluating 2-DNFs. Our protocol gives a quadratic improvement in communication complexity over garbled circuits. We show how to get a private information retrieval scheme (PIR) as a variant of the 2-DNF protocol. Our PIR scheme is based on that of Kushilevitz-Ostrovsky [20] and improves the total communication in the basic step of their PIR protocol from $\sqrt{n}$ to $\sqrt[3]{n}$ for a database of size n.

As noted above, our encryption scheme lets us evaluate quadratic multivariate polynomials on ciphertexts provided the resulting value falls within a small set; in particular, we can compute dot products on ciphertexts. We use

---

[1] An exception is the scheme by Sander et al. [30] that is doubly homomorphic over a semigroup. On the other hand, the homomorphism comes with the cost of a constant factor expansion per semigroup operation. See also its comparison with our results in Section 1.1 below.

this property to create a gadget that enables the verification that an encrypted value is one of two 'good' values. We use this gadget to construct an efficient election protocol where voters do not need to provide proofs of vote validity. Finally, we generalize the election protocol to a protocol of universally verifiable computation.

Comparison to Other Public-Key Homomorphic Systems. Most homomorphic systems provide only one homomorphism, either addition, multiplication, or xor. One exception is the system of Sander et al. [30] that provides the ability to evaluate NC^1 circuits on encrypted values. Clearly their construction also applies to 2-DNF formula. Unfortunately, the ciphertext length in their system grows exponentially in the depth of the 2-DNF formula when written using constant fan-in gates. In our system, the ciphertext size is independent of the formula size or depth; this property is essential for improving the communication complexity basic step of the Kushilevitz-Ostrovsky PIR protocol.

Organization. The rest of this paper is organized as follows. In Section 2 we review the bilinear groups underlying our construction and put forward our new hardness assumption. Section 3 details the construction of a semantically secure public key encryption scheme, its security and homomorphic properties. The basic application to 2-DNF evaluation is presented in Section 4, followed by the election and universally verifiable computation protocols in sections 5 and 6. Section 7 summarizes our results and poses some open problems.

2 Preliminaries

We briefly review the groups underlying our encryption scheme.

2.1 Bilinear Groups

Our construction makes use of certain finite groups of composite order that support a bilinear map. We use the following notation:

1. $\mathbb{G}$ and $\mathbb{G}_1$ are two (multiplicative) cyclic groups of finite order n.
2. g is a generator of $\mathbb{G}$.
3. e is a bilinear map $e : \mathbb{G} \times \mathbb{G} \to \mathbb{G}_1$. In other words, for all $u, v \in \mathbb{G}$ and $a, b \in \mathbb{Z}$, we have $e(u^a, v^b) = e(u, v)^{ab}$. We also require that $e(g, g)$ is a generator of $\mathbb{G}_1$.

We say that $\mathbb{G}$ is a bilinear group if there exists a group $\mathbb{G}_1$ and a bilinear map as above. In the next section we also add the requirement that the group action in $\mathbb{G}, \mathbb{G}_1$, and the bilinear map can be computed in polynomial time.

Constructing Bilinear Groups of a Given Order n. Let $n > 3$ be a given square-free integer that is not divisible by 3. We construct a bilinear group $\mathbb{G}$ of order n as follows:

1. Find the smallest positive integer $\ell \in \mathbb{Z}$ such that $p = \ell n - 1$ is prime and $p = 2 \bmod 3$.

2. Consider the group of points on the (super-singular) elliptic curve $y^2 = x^3 + 1$ defined over $\mathbb{F}_p$. Since $p = 2 \bmod 3$ the curve has $p + 1 = \ell n$ points in $\mathbb{F}_p$. Therefore the group of points on the curve has a subgroup of order n which we denote by $\mathbb{G}$.
3. Let $\mathbb{G}_1$ be the subgroup of $\mathbb{F}_{p^2}^*$ of order n. The modified Weil pairing on the curve [22, 19, 3, 23] gives a bilinear map $e : \mathbb{G} \times \mathbb{G} \to \mathbb{G}_1$ with the required properties.

2.2 The Subgroup Decision Problem

We define an algorithm $\mathcal{G}$ that given a security parameter $\tau \in \mathbb{Z}^+$ outputs a tuple $(q_1, q_2, \mathbb{G}, \mathbb{G}_1, e)$ where $\mathbb{G}, \mathbb{G}_1$ are groups of order $n = q_1 q_2$ and $e : \mathbb{G} \times \mathbb{G} \to \mathbb{G}_1$ is a bilinear map. On input τ, algorithm $\mathcal{G}$ works as follows:

1. Generate two random τ-bit primes q_1, q_2 and set $n = q_1 q_2 \in \mathbb{Z}$.
2. Generate a bilinear group $\mathbb{G}$ of order n as described at the end of Section 2.1. Let g be a generator of $\mathbb{G}$ and $e : \mathbb{G} \times \mathbb{G} \to \mathbb{G}_1$ be the bilinear map.
3. Output $(q_1, q_2, \mathbb{G}, \mathbb{G}_1, e)$.

We note that the group action in $\mathbb{G}, \mathbb{G}_1$ as well as the bilinear map can be computed in polynomial time in τ.

Let $\tau \in \mathbb{Z}^+$ and let $(q_1, q_2, \mathbb{G}, \mathbb{G}_1, e)$ be a tuple produced by $\mathcal{G}(\tau)$ where $n = q_1 q_2$. Consider the following problem: given $(n, \mathbb{G}, \mathbb{G}_1, e)$ and an element $x \in \mathbb{G}$, output '1' if the order of x is q_1 and output '0' otherwise; That is, without knowing the factorization of the group order n, decide if an element x is in a subgroup of $\mathbb{G}$. We refer to this problem as the *subgroup decision problem*. For an algorithm $\mathcal{A}$, the advantage of $\mathcal{A}$ in solving the subgroup decision problem SD-Adv$_\mathcal{A}(\tau)$ is defined as:

$$\text{SD-Adv}_\mathcal{A}(\tau) = \left| \Pr\left[\mathcal{A}(n, \mathbb{G}, \mathbb{G}_1, e, x) = 1 \; : \; \begin{array}{c} (q_1, q_2, \mathbb{G}, \mathbb{G}_1, e) \leftarrow \mathcal{G}(\tau), \\ n = q_1 q_2, \; x \leftarrow \mathbb{G} \end{array} \right] \right.$$
$$\left. - \Pr\left[\mathcal{A}(n, \mathbb{G}, \mathbb{G}_1, e, x^{q_2}) = 1 \; : \; \begin{array}{c} (q_1, q_2, \mathbb{G}, \mathbb{G}_1, e) \leftarrow \mathcal{G}(\tau), \\ n = q_1 q_2, \; x \leftarrow \mathbb{G} \end{array} \right] \right|.$$

Definition 1. *We say that $\mathcal{G}$ satisfies the subgroup decision assumption if for any polynomial time algorithm $\mathcal{A}$ we have that SD-Adv$_\mathcal{A}(\tau)$ is a negligible function in τ.*

Informally, the assumption states that the uniform distribution on $\mathbb{G}$ is indistinguishable from the uniform distribution on a subgroup of $\mathbb{G}$. Recall that the factorization of the order of $\mathbb{G}$ is hidden so that the order of subgroups of $\mathbb{G}$ remains unknown to a polynomial time adversary.

3 A Homomorphic Public-Key System

We can now describe our public key system. The system resembles the Paillier [25] and the Okamoto-Uchiyama [24] encryption schemes. We describe the three algorithms making up the system:

KeyGen(τ): Given a security parameter $\tau \in \mathbb{Z}^+$, run $\mathcal{G}(\tau)$ to obtain a tuple $(q_1, q_2, \mathbb{G}, \mathbb{G}_1, e)$. Let $n = q_1 q_2$. Pick two random generators $g, u \xleftarrow{R} \mathbb{G}$ and set $h = u^{q_2}$. Then h is a random generator of the subgroup of $\mathbb{G}$ of order q_1. The public key is $\mathcal{PK} = (n, \mathbb{G}, \mathbb{G}_1, e, g, h)$. The private key is $\mathcal{SK} = q_1$.

Encrypt($\mathcal{PK}, M$): We assume the message space consists of integers in the set $\{0, 1, \ldots, T\}$ with $T < q_2$. We encrypt bits in our main application, in which case $T = 1$. To encrypt a message m using public key $\mathcal{PK}$, pick a random $r \xleftarrow{R} \{0, 1, \ldots, n-1\}$ and compute

$$C = g^m h^r \in \mathbb{G}.$$

Output C as the ciphertext.

Decrypt($\mathcal{SK}, C$): To decrypt a ciphertext C using the private key $\mathcal{SK} = q_1$, observe that

$$C^{q_1} = (g^m h^r)^{q_1} = (g^{q_1})^m$$

Let $\hat{g} = g^{q_1}$. To recover m, it suffices to compute the discrete log of C^{q_1} base $\hat{g}$. Since $0 \le m \le T$ this takes expected time $\tilde{O}(\sqrt{T})$ using Pollard's lambda method [21–p.128].

Note that decryption in this system takes polynomial time in the size of the message space T. Therefore, the system as described above can only be used to encrypt short messages. Clearly one can use the system to encrypt longer messages, such as session keys, using any mode of operation that converts a cipher on a short block into a cipher on an arbitrary long block. We note that one can speed-up decryption by precomputing a (polynomial-size) table of powers of $\hat{g}$ so that decryption can occur in constant time.

3.1 Homomorphic Properties

The system is clearly additively homomorphic. Let $(n, \mathbb{G}, \mathbb{G}_1, e, g, h)$ be a public key. Given encryptions $C_1, C_2 \in G_1$ of messages $m_1, m_2 \in \{0, 1, \ldots, T\}$ respectively, anyone can create a uniformly distributed encryption of $m_1 + m_2 \bmod n$ by computing the product $C = C_1 C_2 h^r$ for a random r in $\{0, 1, \ldots, n-1\}$.

More importantly, anyone can multiply two encrypted messages *once* using the bilinear map. Set $g_1 = e(g, g)$ and $h_1 = e(g, h)$. Then g_1 is of order n and h_1 is of order q_1. Also, write $h = g^{\alpha q_2}$ for some (unknown) $\alpha \in \mathbb{Z}$. Suppose we are given two ciphertexts $C_1 = g^{m_1} h^{r_1} \in \mathbb{G}$ and $C_2 = g^{m_2} h^{r_2} \in \mathbb{G}$. To build an encryption of the product $m_1 \cdot m_2 \bmod n$ given only C_1 and C_2, do: 1) pick a random $r \in \mathbb{Z}_n$, and 2) set $C = e(C_1, C_2) h_1^r \in \mathbb{G}_1$. Then

$$C = e(C_1, C_2) h_1^r = e(g^{m_1} h^{r_1}, g^{m_2} h^{r_2}) h_1^r = g_1^{m_1 m_2} h_1^{m_1 r_2 + r_2 m_1 + \alpha q_2 r_1 r_2 + r}$$
$$= g_1^{m_1 m_2} h_1^{\tilde{r}} \in \mathbb{G}_1$$

where $\tilde{r} = m_1 r_2 + r_2 m_1 + \alpha q_2 r_1 r_2 + r$ is distributed uniformly in $\mathbb{Z}_n$ as required. Thus, C is a uniformly distributed encryption of $m_1 m_2 \bmod n$, but in the group

$\mathbb{G}_1$ rather than $\mathbb{G}$ (this is why we allow for just one multiplication). We note that the system is still additively homomorphic in $\mathbb{G}_1$.

Note. In some applications we avoid blinding with h^r, making the homomorphic computation deterministic.

Quadratic Polynomials. Let $F(x_1,\ldots,x_u)$ be a u-variate polynomial of total degree 2. The discussion above shows that given the encryptions $C_1,\ldots,C_u$ of values $x_1,\ldots,x_u$, anyone can compute the encryption of $C = F(x_1,\ldots,x_u)$. On the other hand, to decrypt C, the decryptor must already know that the result $F(x_1,\ldots,x_u)$ lies in a certain polynomial size interval.

3.2 Security

We now turn to proving semantic security of the system under the subgroup decision assumption. The proof is standard and we briefly sketch it here.

Theorem 1. *The public key system of Section 3 is semantically secure assuming $\mathcal{G}$ satisfies the subgroup decision assumption.*

Proof. Suppose a polynomial time algorithm $\mathcal{B}$ breaks the semantic security of the system with advantage $\epsilon(\tau)$. We construct an algorithm $\mathcal{A}$ that breaks the subgroup decision assumption with the same advantage. Given $(n, \mathbb{G}, \mathbb{G}_1, e, x)$ as input, algorithm $\mathcal{A}$ works as follows:

1. $\mathcal{A}$ picks a random generator $g \in \mathbb{G}$ and gives algorithm $\mathcal{B}$ the public key $(n, \mathbb{G}, \mathbb{G}_1, e, g, x)$.
2. Algorithm $\mathcal{B}$ outputs two messages $m_0, m_1 \in \{0, 1, \ldots, T\}$ to which $\mathcal{A}$ responds with the ciphertext $C = g^{m_b} x^r \in \mathbb{G}$ for a random $b \xleftarrow{R} \{0,1\}$ and random $r \xleftarrow{R} \{0, 1, \ldots, n-1\}$.
3. Algorithm $\mathcal{B}$ outputs its guess $b' \in \{0,1\}$ for b. If $b = b'$ algorithm $\mathcal{A}$ outputs 1 (meaning x is uniform in a subgroup of $\mathbb{G}$); otherwise $\mathcal{A}$ outputs 0 (meaning x is uniform in $\mathbb{G}$).

It is easy to see that when x is uniform in $\mathbb{G}$, the challenge ciphertext C is uniformly distributed in $\mathbb{G}$ and is independent of the bit b. Hence, in this case $\Pr[b = b'] = 1/2$. On the other hand, when x is uniform in the q_1-subgroup of $\mathbb{G}$, then the public key and challenge C given to $\mathcal{B}$ are as in a real semantic security game. In this case, by the definition of $\mathcal{B}$, we know that $\Pr[b = b'] > 1/2 + \epsilon(\tau)$. It now follows that $\mathcal{A}$ satisfies SD-Adv$_\mathcal{A}(\tau) > \epsilon(\tau)$ and hence $\mathcal{A}$ breaks the subgroup decision assumption with advantage $\epsilon(\tau)$ as required. □

We note that if $\mathcal{G}$ satisfies the subgroup decision assumption then semantic security also holds for ciphertexts in $\mathbb{G}_1$. These ciphertexts are the output of the multiplicative homomorphism. If semantic security did not hold in $\mathbb{G}_1$, then it would also not hold in $\mathbb{G}$ because one can always translate a ciphertext in $\mathbb{G}$ to a ciphertext in $\mathbb{G}_1$ by "multiplying" by the encryption of 1. Hence, by Theorem 1, semantic security must also hold for ciphertexts in $\mathbb{G}_1$.

4 Two Party Efficient SFE for 2-DNF

In this section we show how to use our homomorphic encryption scheme to construct efficient secure function evaluation protocols. Our basic result is a direct application of the additive and multiplicative homomorphisms of our public key encryption scheme. We consider a two-party scenario where Alice holds a Boolean formula $\phi(x_1,\ldots,x_n)$ and Bob holds an assignment $a = a_1,\ldots,a_n$. As the outcome, Bob learns $\phi(a)$. We restrict our attention to 2-DNF formulas:

Definition 2. *A 2-DNF formula over the variables $x_1,\ldots,x_n$ is of the form $\vee_{i=1}^{k}(\ell_{i,1} \wedge \ell_{i,2})$ where $\ell_{i,1}, \ell_{i,2} \in \{x_1,\ldots,x_n,\bar{x}_1,\ldots,\bar{x}_n\}$.*

We first give a protocol for the model of semi-honest parties, and then modify it to cope with a malicious Bob, capitalizing on an 'input verification' gadget.

In the semi-honest model, both parties are assumed to perform computations and send messages according to their prescribed actions in the protocol. They may also record whatever they see during the protocol (i.e. their own input and randomness, and the messages they receive). On the other hand, a malicious party may deviate arbitrarily from the protocol. We sketch the security definitions for the simple case where only one party (Bob) is allowed to learn the output. We refer readers to Goldreich's book [15] for the complete definitions.

Security in the Semi-Honest Model. The definition is straightforward since only one party (Bob) is allowed to learn the output:

- Bob's security – indistinguishability: We require that Alice cannot distinguish between the different possible inputs Bob may hold.
- Alice's security – comparison to an ideal model: Alice's security is formalized by considering an ideal trusted party that gets the inputs $\phi()$ and a, and gives $\phi(a)$ to Bob. We require in the real implementation that Bob does not get any information beyond whether a satisfies $\phi()$.

Security Against Malicious Parties. The security definition for this model captures both the privacy and correctness of the protocol and is limited to the case where only one of the parties is corrupt. Informally, the security definition is based on a comparison with an ideal trusted party model (here the corrupt party may give an arbitrary input to the trusted functionality). The security requirement is that for any strategy a corrupt party may play in a real execution of the protocol, there is an efficient strategy it could play in the ideal model with computationally indistinguishable outcomes.

4.1 The Basic Protocol

Protocol 2-DNF in Figure 1 uses our homomorphic encryption scheme for efficiently evaluating 2-DNFs with semi-honest parties. We get a three message protocol with communication complexity $O(n \cdot \tau)$ — a quadratic improvement in communication with respect to Yao's garbled-circuit protocol [32] that yields communication proportional to the potential formula length, $\Theta(n^2)$.

> INPUT: Alice holds a 2-DNF formula $\phi(x_1,\ldots,x_n) = \vee_{i=1}^k (\ell_{i,1} \wedge \ell_{i,2})$ and Bob holds an assignment $a = a_1,\ldots,a_n \in \{0,1\}^n$. Both parties' inputs include a security parameter τ.
>
> 1. Bob performs the following:
>   (a) He invokes $\mathsf{KeyGen}(\tau)$ to compute keys $\mathcal{SK}, \mathcal{PK}$, and sends $\mathcal{PK}$ to Alice.
>   (b) He computes and sends $\mathsf{Encrypt}(\mathcal{PK}, a_j)$ for $j = 1,\ldots,n$.
> 2. Alice performs the following:
>   (a) She computes an arithmetization Φ of ϕ by replacing "$\vee$" by "$+$", "$\wedge$" by "$\cdot$" and "$\bar{x}_j$" by "$(1-x_j)$". Note that Φ is a polynomial in $x_1,\ldots,x_n$ with *total* degree 2.
>   (b) Alice computes the encryption of $r \cdot \Phi(a)$ for a randomly chosen r using the encryption scheme's homomorphic properties. The result is sent to Bob.
> 3. If Bob receives an encryption of 0, he outputs 0; otherwise, he outputs 1.

Fig. 1. Protocol 2-DNF

Claim. Protocol 2-DNF is secure against semi-honest Alice and Bob.

Proof (Sketch). Alice's security follows as the distribution on Bob's output only depends on whether $\phi()$ is satisfied by a or not. Bob's security follows directly from the semantic security of the encryption scheme. □

Note. Protocol 2-DNF (as well Malicious-Bob-2-DNF below) is secure even against a computationally unlimited Bob. Interestingly, the garbled circuit protocol (where Alice garbles ϕ) has the opposite property where it can be secured against an unbounded Alice but not an unbounded Bob. (See also Cachin et al. [5] for a discussion of computing on encrypted data versus garbled circuits).

4.2 Example Application – Private Information Retrieval

A private information retrieval (PIR) scheme allows a user to retrieve information from an n-bit database without revealing any information on which bit he is interested in [6, 20]. SPIR (symmetric PIR) is a PIR scheme that also protects the database privacy – a (semi-honest) user will only learn one of the database bits [14]. In this section, we show how an immediate application of protocol 2-DNF results in a PIR/SPIR scheme. Our constructions are based on that of Kushilevitz and Ostrovsky [20].

A SPIR Scheme. We get a SPIR scheme with communication $O(\tau \cdot \sqrt{n})$ as an immediate application of protocol 2-DNF. Without loss of generality, we assume that the database size n is a perfect square and treat the database as a table D of dimensions $\sqrt{n} \times \sqrt{n}$. Using this notation, suppose Bob wants to retrieve entry (I, J) of D. Alice (the database holder) holds the 2-DNF formula ϕ over $x_1,\ldots,x_{\sqrt{n}}, y_1,\ldots,y_{\sqrt{n}}$:

$$\phi(x_1,\ldots,x_{\sqrt{n}}, y_1,\ldots,y_{\sqrt{n}}) = \vee_{D_{i,j}=1} (x_i \wedge y_j),$$

and Bob's assignment a sets x_I and y_J to 1 and all other variables to 0. Bob and Alice carry out the 2-DNF protocol with this assignment and 2-DNF formula. It is clear that $\phi(a) = D_{I,J}$ as required.

An Alternative Construction. Using the 2-DNF protocol for SPIR restricts database entries to bits. We provide an alternative construction that allows each database entry to contain up to $O(\log n)$ bits. We consider the data as a table of dimensions $\sqrt{n} \times \sqrt{n}$ as above. To retrieve entry (I, J) of D, Bob creates two polynomials $p_1(x)$ and $p_2(x)$ of degree $\sqrt{n} - 1$ such that $p_1(i)$ is zero on $0 \leq i < \sqrt{n}$ except for $p_1(I) = 1$, and similarly $p_2(j)$ is zero on $0 \leq j < \sqrt{n}$ except for $p_2(J) = 1$. Bob sends to Alice the encryption of the coefficients of $p_1(x)$ and $p_2(x)$. Alice uses the encryption scheme's homomorphic properties to compute the encryption of

$$D_{I,J} = \sum_{0 \leq i,j < \sqrt{n}} p_1(i) p_2(j) D_{i,j}.$$

We allow $D_{i,j}$ to be b-bit values where $b = O(\log n)$. Bob recovers $D_{i,j}$ in time $O(2^{b/2})$ by computing a discrete logarithm e.g. using the baby-step giant-step algorithm.

A PIR Scheme. Standard communication balancing of our SPIR scheme results in a PIR scheme where each party sends $O(\tau \cdot \sqrt[3]{n})$ bits. In particular, view the database as comprising of $n^{1/3}$ chunks, each chunk containing $n^{2/3}$ entries, where Bob is interested in retrieving entry (I, J, K) of D. Bob sends Alice the coefficients of two polynomials $p_1(x)$ and $p_2(x)$ of degree $\sqrt[3]{n} - 1$ such that $p_1(i) = p_2(i) = 0$ on $0 \leq i < \sqrt[3]{n}$ except for $p_1(I) = p_2(J) = 1$. Alice uses the encryption scheme's homomorphic properties to compute encryptions of

$$D_{I,J,k} = \sum_{0 \leq i,j < \sqrt[3]{n}} p_1(i) p_2(j) D_{i,j,k}$$

for $0 \leq k < \sqrt[3]{n}$. Alice sends the $\sqrt[3]{n}$ resulting ciphertexts to Bob who decrypts the Kth entry.

Recursively applying this balancing (as in Kushilevitz-Ostrovsky [20]) results in a protocol with communication complexity $O(\tau n^\epsilon)$ for any $\epsilon > 0$. We note that the recursion depth to reach ϵ is lower in our case compared to that of Kushilevitz-Ostrovsky [20] by a constant factor of $\log_2 3$.

4.3 Security of the 2-DNF Protocol Against a Malicious Bob

A malicious Bob may try to learn about Alice's 2-DNF formula by sending Alice an encryption of a non-boolean assignment $a_1, \ldots, a_n$. He may also let Alice evaluate ϕ for an encrypted assignment that Bob cannot decrypt himself. Both types of behaviors do not correspond to a valid run in the ideal model.

To prevent the first attack, we present a gadget that allows Alice to ensure a ciphertext she receives contains one of two 'valid' messages v_0, v_1. This gadget is applicable outside of the scope of 2-DNF as we demonstrate in sections 5 and 6. The second attack is prevented using standard methods — Alice presents Bob with a challenge that cannot be resolved unless he can decrypt. This decryption ability is then used when Bob is simulated to create valid inputs for the trusted party.[2]

A Gadget for Checking $c \in \{v_0, v_1\}$. This gadget exploits our ability to evaluate a polynomial of total degree 2 on the encryption of c. We choose a polynomial that has v_0 and v_1 as zeros as follows: given an encryption of a value c, Alice uses the homomorphic properties of the encryption scheme to compute $r \cdot (c - v_0) \cdot (c - v_1)$ for a randomly chosen r. For $c \in \{v_0, v_1\}$, this computation results in the encryption of 0. For other values of c, the result is random. In the special case of $c \in \{0, 1\}$, Alice computes $r \cdot c \cdot (c - 1)$.

The Protocol. The result is protocol Malicious-Bob-2-DNF described in Figure 2.

INPUT: as in protocol 2-DNF in Figure 1.
1. Alice and Bob engage in the following 'proof of decryption ability' protocol:
   (a) Bob invokes $\mathsf{KeyGen}(\tau)$ to compute keys $\mathcal{SK}, \mathcal{PK}$ and sends $\mathcal{PK}$ to Alice.
   (b) Alice chooses τ random bits $m_1, \ldots, m_\tau$ and sends their encryptions $\mathsf{Encrypt}(\mathcal{PK}, m_1), \ldots, \mathsf{Encrypt}(\mathcal{PK}, m_\tau)$ to Bob.
   (c) Bob replies with a decryption $m'_1, \ldots, m'_\tau$ of the received encryptions. Alice aborts if any of Bob's decryptions is incorrect.
2. Bob computes and sends $\mathsf{Encrypt}(\mathcal{PK}, a_j)$ for $j = 1, \ldots, n$.
3. Alice performs the following:
   (a) She computes an arithmetization Φ of ϕ as in protocol 2-DNF.
   (b) Using the homomorphic properties of the encryption scheme, she computes the encryption of $r \cdot \Phi(a) + \sum_{i=1}^n r_i \cdot a_i \cdot (a_i - 1)$ for randomly chosen r, r_i. She sends the result to Bob.
4. If Bob receives an encryption of 0, he outputs 0; otherwise, he outputs 1.

Fig. 2. Protocol Malicious-Bob-2-DNF

Claim. Protocol 2-DNF is secure against semi-honest Alice and malicious Bob.

Proof (Sketch). Security against semi-honest Alice follows as in protocol 2-DNF. Security against malicious Bob follows by simulation. Note that the 'proof of

---

[2] The 'standard' use of this technique is to give Bob a random message for a challenge. Bob's simulator would then use the self reducibility properties of the encryption scheme to (i) map an encrypted message $\mathsf{Encrypt}(\mathcal{PK}, m)$ to an encryption of a random message, say $\mathsf{Encrypt}(\mathcal{PK}, m + r)$, (ii) use Bob's procedure to retrieve $m' = m + r$, and (iii) retrieve $m = m' - r$. As the message space is limited in our scheme due to decryption limitations, we need a slightly modified scheme.

decryption ability' sub-protocol can be used to decrypt Bob's message in Step 2 of the protocol, hence providing the inputs to the trusted party. □

5 An Efficient Election Protocol Without Random Oracles

In this section, we describe an electronic election protocol where voters submit boolean ("yes/no") votes. Such protocols were first considered by Benaloh and Fisher [7, 2] and more recently by Cramer et al. [8, 9].

A key component of electronic election schemes is a proof, attached to each vote, of its correctness (or validity); for example, a proof that the vote really is an encryption of 0 or 1. Otherwise, voters may corrupt the tally by sending an encryption of an arbitrary value. Such proofs of validity are typically zero-knowledge (or witness indistinguishable) proofs. These interactive zero knowledge proofs of bit encryption are efficiently constructed (using zero knowledge identification protocols) for standard homomorphic encryption schemes such as ElGamal [11, 18], Pedersen [26, 8], or Paillier [25, 10]. The proof of validity is then usually made non-interactive using the Fiat-Shamir heuristic of replacing communication with an access to a random oracle [12]. In the actual instantiation, the random oracle is replaced by some 'cryptographic function' such a hash function. Security is shown hence to hold in an ideal model with access to the random oracle, and not in the standard model [27].

Our election protocol has the interesting feature that voters do not need to include proofs of validity or any other information except for their encrypted votes when casting their ballots. Instead, the election authorities can jointly verify that a vote is valid based solely on its encryption. The technique is based on the gadget we constructed in Section 2. This gadget allows us to avoid using the Fiat-Shamir heuristic and yet makes our scheme efficient. As a result, our election scheme is very efficient from the voter's point of view as it requires only a single encryption operation (two exponentiations) to create a ballot.[3]

5.1 The Election Scheme

Our scheme belongs to the class of election protocols proposed by Cramer et al. [8, 9] where votes are encrypted using a homomorphic encryption scheme.

For robustness, we use a threshold version of the encryption scheme in Section 3. For simplicity (following Shoup [31]), we assume that a trusted dealer first generates the public/private keys, shares the private keys between the election authorities, and then deletes the private key (a generic secure computation may be used to replace the trusted dealer, as this is an offline phase). With this assumption, a threshold version of our encryption scheme can be constructed using standard techniques from discrete log threshold cryptosystems [26].

---

[3] Curiously, this voting scheme is probably the most efficient for the voter, taking into account the efficiency of operating in an elliptic curve group.

Correctness of Threshold Decryption. One caveat is that threshold decryption requires a zero knowledge of correct partial decryption from each election authority that contributes a share of its private key. Since the number of election authorities is typically a small constant, the proof of correct partial decryption can be performed interactively with relative efficiency between election authorities; transcripts of such interactions are made public for verification (note that transcripts do not leak information on votes). Another possible technique is to use a trusted source of random bits (such as a beacon [28]) among the election authorities, or for the authorities to collectively generate a public source of random bits. In a typical run of our protocol, the election authorities run only a limited number of these proofs (see below), hence the usage of either technique results in a reasonably efficient protocol, and allows us to avoid using the Fiat-Shamir heuristic.

We note that these techniques can also be used in existing election protocols for verifying a voter's ballot, which avoids the Fiat-Shamir heuristic; but the resulting protocol becomes unwieldy and inefficient especially when the number of voters is large (and we expect that there is at least several orders of magnitude more voters than election authorities).

Vote Verification. Here we use the verification gadget of Section 2 in combination with threshold decryption. We let all authorities compute an encryption of $v \cdot (v - 1)$ and then jointly decrypt the result. To save on computation, we check a batch of votes at once (i.e. $\sum r_i \cdot v_i \cdot (v_i - 1)$ where the r_i's are chosen by the verifiers) and then run a binary search to identify the invalid votes [1].

The Protocol. We assume the existence of an online bulletin board where the parties participating in the protocol post messages. Our election protocol works as follows:

Setup: As discussed above, a trusted dealer first generates the public parameters and the private key for the encryption scheme of Section 3, and shares the private key between the k election authorities so that at least t out the k election authorities are needed to decrypt. Finally, the trusted dealer deletes the private key and has no further role in the protocol. The public parameters are posted on a public bulletin board.

Denote one of the k election authorities as a leader (the election authority that organizes a quorum for decryption requests). After the public parameters are posted to the public board, the leader publishes an encryption of the bit 1 and the random bits used to create that encryption. Denote this encryption of the bit 1 as E_1. With the random bits, the other $k-1$ election authorities can check that E_1 is indeed an encryption of 1.

Vote Casting: Voters cast their ballots by encrypting a bit indicating their vote, and then publishing the encrypted bit to the public bulletin board.

Vote Verification: When a ballot v has been posted, all k election authorities compute a ciphertext c corresponding to $v \cdot (v - 1)$ where E_1 is used as the encryption of "1" (hence, c is 'deterministic' given the encryption of v). The leader forms a quorum of $t-1$ other election authorities to decrypt c

(the other election authorities agree to participate only if c agrees with the ciphertext they computed). If c decrypts to something other than 0, then the vote v is invalid and is discarded.

For better efficiency in optimistic scenarios, any number of votes $v_1,\ldots,v_k$ can be verified in bulk by first computing $r_1 \cdot v_1 \cdot (v_1-1) + \ldots + r_k \cdot v_k \cdot (v_k-1)$ where the r_i's are collectively chosen by the election authorities, and then checking that the decryption of the result is 0. All invalid votes are efficiently located by binary search. We note that in general it suffices for r_i to be relatively short, as the chance of $\sum r_i \cdot v_i \cdot (v_i-1)$ being zero when some of the summed votes are invalid is exponentially small in $|r|$.

Vote Tabulation and Tally Computation: After all the votes are posted and verified, all k election authorities each add all the valid encrypted votes on the public board together (using the additive homomorphic property of the encryption scheme) to form the tallied vote V. The leader obtains a quorum of election authorities to decrypt V. Each election authority decides whether to participate in the decryption request by comparing V with her own tally.

We note that our election protocol also possesses the necessary properties of voter privacy (from semantic security of the encryption scheme), universal verifiability (from the homomorphic property of the encryption scheme and also because all votes and proof transcripts are posted to the bulletin board), and robustness (from the threshold encryption scheme). The reader is referred to [7, 2, 8, 9] for discussions of these properties.

6 Universally Verifiable Computation

We now describe a related application for the gadget of Section 2. Consider an authority performing a computation, defined by a (publicly known) circuit C over the joint private inputs $a = (a_1,\ldots,a_n)$ of the n users. The authority publishes the outcome $C(a)$ in a way that 1) lets everyone check that the computation was performed correctly, but 2) does not reveal any other information on the private inputs. Besides voting, other applications of universally verifiable computation include auctions.

To simplify our presentation, we only consider the case where $a_i \in \{0,1\}$; general inputs are treated similarly using any binary representation. We describe a single authority protocol that is easily transformed into a threshold multi-authority protocol using standard methods.

6.1 A Protocol for Verifying $C(a)$

Setup. The authority uses $\mathsf{KeyGen}(\tau)$ to generate a public-key/private-key pair $\mathcal{PK}, \mathcal{SK}$. We assume the existence of a bulletin board where each user i posts an encryption $c_i = \mathsf{Encrypt}(\mathcal{PK}, a_i)$ of her input. We also assume the existence of a random function H accessible by all parties, which implies that we prove security only in the random oracle model. As in the previous section, we can do without a random oracle in the multi-authority case (details omitted).

We first give a high level overview of the protocol. After all users post their encrypted inputs onto the bulletin board, the authority decrypts each user's input and evaluates the circuit on these inputs. In the process of evaluating the circuit, the authority computes and publishes ciphertexts for all the wire values in C. In addition, the authority also publishes an encryption of the bit 1 and the random bits used to create that encryption. Denote this encryption of the bit 1 as E_1. Finally, the authority publishes an encrypted value V and a witness-indistinguishable proof that V is an encryption of 0; for now, we defer the exact definition of V.

To convince a verifier that the circuit was computed correctly, the authority needs to prove that 1) all inputs are binary, and 2) all gate outputs are correct. We show how a verifier checks that both conditions hold using validators v that can be publicly constructed from the public encrypted wire values. We first show how to construct validators for the user inputs and gate outputs before showing how to use these validators to verify the computation.

Building Validators for User Inputs. In the process of evaluating the circuit, the authority publishes the values on every wire of the circuit. We enumerate the wires and denote the value on wire i as a_i. We also denote the encryption of a_i as c_i. Recall that each user posts $c_i = \mathsf{Encrypt}(\mathcal{PK}, a_i)$ of her input a_i on the bulletin board (the a_i's are never revealed in the clear). For each input wire i with value a_i, let $r_i = H(i, c_i)$; note that r_i can be computed by any of the parties. The validator for a_i is $v_i = r_i \cdot a_i \cdot (1 - a_i)$, and the computation occurs modulo q_2 where q_2 is one of the factors of the modulus n (recall that $\mathcal{SK} = q_1$). It is easy to see that the encryption of v_i can be computed by anyone given H, c_i, and E_1.

We note that even given q_2 and allowing a polynomial (in the security parameter τ) number of applications of H, the probability that an adversary successfully generates an invalid c_i with $v_i = 0$ is bounded by $O(\mathrm{poly}(\tau)/q_2)$, which is negligible in τ.

Building Validators for Gate Outputs. Let $g \in C$ be a binary gate for which both input wires x, y are validated. In addition, let $G(x, y)$ be the bivariate polynomial of total degree 2 that realizes the gate g. For example, an AND gate has $G_{\mathrm{AND}}(x, y) = xy$, an OR gate has $G_{\mathrm{OR}}(x, y) = x + y - xy$, and a NOT gate has $G_{\mathrm{NOT}}(x) = 1 - x$. The validator for the output wire (enumerated z) of gate g is $v_z = r_z \cdot (a_z - G(x, y))$ where $r_z = H(z, c_z)$ and c_z is the encryption of the value a_z on wire z. Again, it is easy to see that any party can compute the encryption of v_z given H, c_x, c_y, c_z, and E_1.

Verifying the Circuit Using Validators. Using the homomorphic properties of the encryption scheme, anyone can compute (by herself) an encryption of the sum of validators for the circuit. Note that if all posted encryptions are correct, then the sum of validators is zero. Otherwise, it is zero with only a negligible probability. The authority supplies its own version of the encrypted validator sum called V, together with a zero-knowledge proof that the resulting sum is zero; in this case, the encryption is of the form h^r so one can use protocols designed for the Pedersen encryption [26]. To verify the circuit computation, a verifier computes her own validator sum V', checks that $V' = V$, and then verifies the witness-indistinguishable proof that V is an encryption of 0.

7 Summary and Open Problems

We presented a homomorphic encryption scheme that supports addition and one multiplication. We require that the values being encrypted lie in a small range as is the case when encrypting bits. These homomorphic properties enable us to evaluate multi-variate polynomials of total degree 2 given the encrypted inputs. We described a number of applications of the system. Most notably, using our encryption scheme, we (i) reduced the amount of communication in the basic step of the Kushilevitz-Ostrovsky PIR, (ii) improved the efficiency of election systems based on homomorphic encryption, and (iii) implemented universally verifiable secure computation. We hope this scheme will have many other applications.

We end with a couple of open problems related to our encryption scheme:

n-**Linear Maps.** The multiplicative homomorphism was possible due to properties of bilinear maps. We note that an n-linear map would enable us to evaluate polynomials of total degree n rather than just quadratic polynomials. This provides yet another motivation for constructing cryptographic n-linear maps [4].

Message Space. Our scheme is limited in the size of message space due to the need to compute discrete logarithms during decryption. An encryption scheme that allows for a large message space would enable more applications, such as an efficient shared RSA key generation.

References

1. M. Bellare, J. Garay, and T. Rabin. Fast batch verification for modular exponentiation and digital signatures. In *Proceedings of Eurocrypt '98*, volume 1403, 1998.
2. J. Benaloh. *Verifiable Secret-Ballot Elections*. PhD thesis, Yale University, 1987.
3. D. Boneh and M. Franklin. Identity based encryption from the Weil pairing. *SIAM Journal of Computing*, 32(3):586–615, 2003. Extended abstract in *Proceedings of Crypto 2001*.
4. D. Boneh and A. Silverberg. Applications of multilinear forms to cryptography. In *Topics in Algebraic and Noncommutative Geometry*, number 324 in Contemporary Mathematics. American Mathematical Society, 2003.
5. C. Cachin, J. Camenisch, J. Kilian, and J. Müller. One-round secure computation and secure autonomous mobile agents. In *27th International Colloquium on Automata, Languages and Programming (ICALP '2000)*, volume 1853 of *Lecture Notes in Computer Science*, pages 512–523. Springer-Verlag, Berlin Germany, July 2000.

6. B. Chor, O. Goldreich, E. Kushilevitz, and M. Sudan. Private information retrieval. In *36th Annual Symposium on Foundations of Computer Science*, pages 41–50, Milwaukee, Wisconsin, 23–25 Oct. 1995. IEEE.
7. J. Cohen and M. Fischer. A robust and verifiable cryptographically secure election scheme. In *Proceedings of 26th IEEE Symposium on Foundations of Computer Science*, pages 372–382, 1985.
8. R. Cramer, M. Franklin, B. Schoenmakers, and M. Yung. Multi-authority secret-ballot elections with linear work. In U. Maurer, editor, *Proceedings of Eurocrypt 1996*, volume 1070 of *LNCS*, pages 72–83. Springer, 1996.
9. R. Cramer, R. Gennaro, and B. Schoenmakers. A secure and optimally efficient multi-authority election scheme. *European Transactions on Telecommunications*, 8(5):481–490, Sep 1997.
10. I. Damgård and M. Jurik. A generalisation, a simplification and some applications of Paillier's probabilistic public-key system. In K. Kim, editor, *Proceedings of Public Key Cryptography 2001*, volume 1992 of *LNCS*, pages 119–136. Springer, 2001.
11. T. ElGamal. A public key cryptosystem and a signature scheme based on discrete logarithms. *IEEE Transactions on Information Theory*, 31(4):469–472, Jul 1985.
12. A. Fiat and A. Shamir. How to prove yourself: Practical solutions to identification and signature problems. In A. Odlyzko, editor, *Proceedings of Crypto 1986*, volume 263 of *LNCS*, pages 186–194. Springer, 1986.
13. M. J. Freedman, K. Nissim, and B. Pinkas. Efficient private matching and set intersection. In C. Cachin and J. Camenisch, editors, *Proceedings of Eurocrypt 2004*, volume 3027 of *LNCS*, pages 1–19. Springer-Verlag, May 2004.
14. Y. Gertner, Y. Ishai, E. Kushilevitz, and T. Malkin. Protecting data privacy in private information retrieval schemes. *Journal of Computer and System Sciences*, 60(3):592–629, 2000.
15. O. Goldreich. *The Foundations of Cryptography - Volume 2*. Cambridge Univesity Press, 2004.
16. O. Goldreich, S. Micali, and A. Wigderson. Proofs that yield nothing but their validity and a methodology of cryptographic protocol design (extended abstract). In *27th Annual Symposium on Foundations of Computer Science*, pages 174–187, Toronto, Ontario, Canada, 27–29 Oct. 1986. IEEE.
17. S. Goldwasser and S. Micali. Probabilistic encryption & how to play mental poker keeping secret all partial information. In *Proceedings of the fourteenth annual ACM symposium on Theory of computing*, pages 365–377. ACM Press, 1982.
18. M. Jakobsson and A. Juels. Millimix: Mixing in small batches. Technical Report 99-33, Center for Discrete Mathematics and Theoretical Computer Science (DIMACS), Oct 1999.
19. A. Joux. A one round protocol for tripartite Diffie-Hellman. In W. Bosma, editor, *Proceedings of 4th Algorithmic Number Theory Symposium*, number 1838 in LNCS, pages 385–394. Springer, Jul 2000.
20. E. Kushilevitz and R. Ostrovsky. Replication is not needed: Single database, computationally-private information retrieval (extended abstract). In *38th Annual Symposium on Foundations of Computer Science*, pages 364–373, Miami Beach, Florida, 20–22 Oct. 1997. IEEE.
21. A. J. Menezes, P. C. Van Oorschot, and S. A. Vanstone. *Handbook of Applied Cryptography*. CRC Press, 1997.
22. V. Miller. Short programs for functions on curves. Unpublished manuscript, 1986.
23. V. Miller. The Weil pairing, and its efficient calculation. *J. of Cryptology*, 17(4), 2004.

24. T. Okamoto and S. Uchiyama. A new public-key cryptosystem as secure as factoring. In K. Nyberg, editor, *Proceedings of Eurocrypt 1998*, volume 1403 of *LNCS*, pages 308–318. Springer-Verlag, May 1998.
25. P. Pallier. Public-key cryptosystems based on composite degree residuosity classes. In J. Stern, editor, *Proceedings of Eurocrypt 1999*, volume 1592 of *LNCS*, pages 223–238. Springer-Verlag, May 1999.
26. T. P. Pedersen. A threshold cryptosystem without a trusted party. In D. Davies, editor, *Proceedings of Eurocrypt 1991*, volume 547 of *LNCS*, pages 522–526. Springer, 1991.
27. D. Pointcheval and J. Stern. Security proofs for signature schemes. In U. Maurer, editor, *Proceedings of Eurocrypt 1996*, volume 1070 of *LNCS*, pages 387–398. Springer, 1996.
28. M. Rabin. Transaction protection by beacons. *Journal of Computer and System Science*, 27(2):256–267, 1983.
29. R. Rivest, L. Adleman, and M. Dertouzos. On data banks and privacy homomorphisms. *Foundations of Secure Computation*, 1978.
30. T. Sander, A. Young, and M. Yung. Non-interactive CryptoComputing for NC^1. In *Proceedings of the 40th Symposium on Foundations of Computer Science (FOCS)*, pages 554–567, New York, NY, USA, Oct. 1999. IEEE Computer Society Press.
31. V. Shoup. Practical threshold signatures. In B. Preneel, editor, *Proceedings of Eurocrypt 2000*, volume 1807 of *LNCS*, pages 207–220. Springer, 2000.
32. A. C. Yao. Protocols for secure computations. In *Proceedings of the 23rd Symposium on Foundations of Computer Science (FOCS)*, pages 160–164. IEEE Computer Society Press, 1982.

Share Conversion, Pseudorandom Secret-Sharing and Applications to Secure Computation

Ronald Cramer[1], Ivan Damgård[2,*], and Yuval Ishai[3,**]

[1] CWI, Amsterdam and Mathematical Institute, Leiden University
cramer@cwi.nl
[2] Aarhus University
ivan@daimi.au.dk
[3] Technion, Haifa
yuvali@cs.technion.ac.il

Abstract. We present a method for converting shares of a secret into shares of the same secret in a different secret-sharing scheme using only local computation and no communication between players. In particular, shares in a replicated scheme based on a CNF representation of the access structure can be converted into shares from any linear scheme for the same structure.

We show how this can be combined with any pseudorandom function to create, from initially distributed randomness, any number of Shamir secret-sharings of (pseudo)random values without communication. We apply this technique to obtain efficient non-interactive protocols for secure computation of low-degree polynomials, which in turn give rise to other applications in secure computation and threshold cryptography. For instance, we can make the Cramer-Shoup threshold cryptosystem by Canetti and Goldwasser fully non-interactive, or construct non-interactive threshold signature schemes secure without random oracles.

The latter solutions are practical only for a relatively small number of players. However, in our main applications the number of players is typically small, and furthermore it can be argued that no solution that makes a black-box use of a pseudorandom function can be more efficient.

1 Introduction

A secret-sharing scheme enables a dealer to distribute a secret among n players, such that only some predefined qualified subsets of the players can recover the secret from their joint shares and others learn nothing about it. The collection of qualified sets that can reconstruct the secret is called an *access structure*. One

---

[*] Supported by BRICS, Basic research in Computer Science, Center of the Danish National Research Foundation and FICS, Foundations in Cryptography and Security, funded by the Danish Natural Sciences Research Council.
[**] Research supported by Israel Science Foundation grant 36/03.

useful type of secret-sharing schemes are *threshold schemes*, in which the access structure includes all sets of more than t players, for some threshold t.

Secret-sharing schemes have found numerous applications in cryptography. In most of these applications, one tries to use the "best" known scheme for the access structure at hand. Indeed, in the most popular threshold case, applications typically rely on Shamir's scheme [28], which is optimal with respect to its share size. It turns out, however, that there are contexts where it is not desirable, or even not at all possible, to use the most succinct available secret-sharing scheme. (Some examples will be provided below.) In such contexts it may be beneficial to share a secret using one secret-sharing scheme and later convert its shares to a different representation, corresponding to another secret-sharing scheme, enjoying the relative advantages of both schemes.

NON-INTERACTIVE SHARE CONVERSION. Motivated by this general scenario, as well as by the more concrete applications discussed below, we introduce and study the following notion of local conversion between secret-sharing schemes. For secret sharing schemes $\mathcal{S}, \mathcal{S}'$, we say that $\mathcal{S}$ is *locally convertible* to $\mathcal{S}'$, if any valid $\mathcal{S}$-sharing of a secret s may be converted by means of local transformations (performed by each player separately) to valid, though not necessarily random, $\mathcal{S}'$-sharing of the same secret s. Before describing our results on share conversion and their applications, we turn to describe a special class of secret-sharing schemes that play a central role in these results. REPLICATED SECRET-SHARING.

A very useful type of "inefficient" secret-sharing scheme is the so-called *replicated scheme* [23].[1] The replicated scheme for an access structure Γ proceeds as follows. First, the dealer splits the secret s into additive shares, where each additive share corresponds to some maximal *unqualified* set $T \notin \Gamma$. That is, we view s as an element of some finite field K, and write $s = \sum_{T \in \mathcal{T}} r_T$, where $\mathcal{T}$ is the collection of all maximal unqualified sets, and where the additive shares r_T are random subject to the restriction that they add up to s. Then, the dealer distributes to each player P_j all additive shares r_T such that $j \notin T$.

PSEUDORANDOM SECRET-SHARING. In the threshold case, the replicated scheme involves $\binom{n}{t}$ additive shares and is thus far worse than Shamir's scheme in terms of share size. However, it enjoys the following key property: shares of a *random* secret $s \in K$ consist of replicated instances of random and *independent* elements from K. This property comes handy in applications which require a large number of (pseudo-)random secret-shared values: viewing each replicated share r_T as an independent key to a pseudorandom function, we may get a virtually unlimited supply of independent pseudorandom secrets, each shared using the replicated scheme. Thus, we may use this method to obtain replication-shared secrets at a very low amortized cost. The main difficulty is that these shared secrets cannot be securely used in a higher level application without paying the $\binom{n}{t}$ communi-

---

[1] This scheme can also be obtained from the formula-based construction of [6] by using a CNF representation of the access structure. Hence, it is sometimes referred to in the literature as a CNF-based scheme.

cation overhead for each secret being used. The goal of share conversion, in this case, would be to locally convert replicated shares of each shared secret used by the application into an equivalent Shamir representation. This would allow to enjoy the best of both worlds, combining the share independence advantage of the replicated scheme with the succinctness advantage of Shamir's scheme.

1.1 Our Results

Our contribution goes in two directions. First, we put forward the notion of share conversion and obtain some results on the possibility and efficiency of share conversion. Second, we present various applications of our share conversion results, mainly within the domains of multiparty computation and threshold cryptography. We now provide a more detailed account of these results.

Results on Share Conversion. Our main result is that shares from the replicated scheme described above can be locally converted into shares of any *linear* scheme[2] for the same (or smaller) access structure. In particular, shares from the replicated scheme for a threshold structure can be converted into Shamir-shares for the same structure. We start by describing a simple conversion procedure for the latter special case, and then generalize it to arbitrary linear schemes and access structures. The general conversion result relies on a representation of the access structure by a *canonical span program* [24].

The share convertibility relation induces a partial order on secret-sharing schemes. Under this order, the replicated scheme is *maximal* in the class of all linear schemes realizing a given access structure. We also identify a minimal scheme in this class, and prove some negative results regarding the possibility and efficiency of share conversion. In particular, we show that the $\binom{n}{t}$ overhead cannot be avoided when converting replicated shares to Shamir shares.

Applications. As discussed above, share conversion can be combined with any pseudorandom function to securely create, from initially distributed randomness, a virtually unlimited supply of Shamir secret sharings of (pseudo)random values without further interaction. We present several applications of this idea in a setting where the cost of pre-distributing $\binom{n}{t}$ keys can be afforded.

DISTRIBUTED PRFS. We obtain a communication-efficient variant of a distributed PRF construction of Naor, Pinkas, and Reingold [27] by converting its replicated shares to Shamir shares. A natural application of distributed PRFs is to distributing a key-distribution center.

SECURE MULTIPARTY COMPUTATION. We present efficient protocols for securely evaluating low-degree polynomials, requiring only two rounds of interaction: a round of broadcast messages followed by a round of point-to-point messages.

---

[2] In a linear secret-sharing scheme the secret is taken from a finite field, and each player's share is obtained by computing some linear function of the secret and the dealer's randomness. Essentially all known useful schemes are linear.

(If no setup is allowed, then this is provably impossible [19].) Using known techniques, these results for low-degree polynomials can be extended to general functions. In the case of functions which only take a random input (e.g., a function dealing cards to poker players) the first round of broadcasts can be eliminated. Thus, we get efficient and fully non-interactive protocols for distributing a trusted dealer in a wide array of applications.

THRESHOLD CRYPTOGRAPHY. The above results on multiparty computation can be specialized to obtain non-interactive implementations of threshold cryptosystems, taking advantage of their simple algebraic structure. For instance, we show how to make the Cramer-Shoup threshold cryptosystem by Canetti and Goldwasser [11] fully non-interactive and how to construct non-interactive threshold signature schemes secure without random oracles.

Towards assessing the practicality of these solutions, we note that many of them, in particular the threshold cryptography applications, are designed for a client-server model, where we would typically have a small number of quite powerful servers and some number of (possibly less powerful) clients. Even the application to general multiparty computation can be naturally set in a client-server model, for instance when a large number of players provide input to a computation, but where it is not practical to have them all participate in the computation – this is then left to a small number of servers. Our solutions fit nicely into such a scenario, since they only require the servers to handle the $\binom{n}{t}$ overhead locally, and since here n is the number of servers which is typically small.

It is important to understand that all our applications could have been realized with the same functionality without using share conversion. Instead, the players would work directly on the replicated shares. For instance, this is exactly what was done for distributed PRF's in [27] and could be done for threshold cryptography and multiparty computation by adapting Maurer's techniques [25] in a straightforward way. However, such solutions would be much less practical. First, the $\binom{n}{t}$ overhead would now also apply to the communication, and hence also to the local computation of the clients. Second, various possibilities for optimization would be lost. For instance, the threshold cryptography applications typically require servers to use shares of fresh (pseudo)random values every time they are called by a client. Using share conversion, many sets of such (Shamir) shares can be generated off-line and stored compactly, making the on-line work and storage of servers efficient in n and t. Without share conversion, the $\binom{n}{t}$ overhead would apply to the entire storage generated off-line, destroying the on-line advantage. Finally, share conversion also yields significant savings in the local computation performed by the *servers*. Without share conversion, the servers' computation in our applications would increase (roughly) by a factor of either $\binom{n}{t}$ or $\binom{n}{t}^2$.

CONCEPTUAL CONTRIBUTION. All of the above applications are related to the use of replicated secret-sharing in conjunction with pseudorandomness. But there are also other types of applications which seem to benefit from the use of replicated secret-sharing in different, and sometimes unexpected, ways. For instance, replicated shares yield the most efficient information-theoretic private information retrieval protocols and locally decodable codes [5], the best round complexity for verifiable secret-sharing [18], and the simplest protocols for secure multiparty computation with respect to generalized adversaries [3, 25]. Our results on share conversion provide an explanation for the usefulness of the replicated scheme, suggesting that anything that can be achieved using linear secret-sharing can also be achieved (up to an $\binom{n}{t}$ overhead) using this specific scheme. This may also serve as a useful guideline for the design of cryptographic protocols (e.g., when attempting to improve [5]).

Related Work. The idea of distributing pseudorandom functions by replicating independent keys has been previously used by Micali and Sidney [26] and by Naor et al. [27]. However, without the tool of share conversion, their protocols are either very expensive in communication or lose some of their appealing features. We note that an alternative number-theoretic construction of distributed PRFs, suggested in [27], is not suitable for our applications due to the "multiplicative" representation of the output.

Most relevant to the current work is the work on compressing cryptographic resources by Gilboa and Ishai [20]. The problem considered there is that of using replicated pseudorandom sources to securely "compress" useful correlation patterns. In particular, a conversion from replicated shares of a random secret to Shamir shares of a random secret (though not necessarily the same secret) is implicit in their results. The results of [20] do not explicitly refer to the access structure associated with a given correlation pattern, and do not imply our general results on share conversion.

2 Preliminaries

We define an n-player secret sharing scheme by a tuple $\mathcal{S} = (K, (S_1, \ldots, S_n), R, D)$, where K is a finite *secret-domain* (typically a finite field), each S_j is a finite *share domain* from which P_j's share is picked (typically $S_j = K^{a_j}$ for some a_j), R is a probability distribution from which the dealer's random input is picked, and D is a share distribution function mapping a secret s and a random input r to an n-tuple of shares from $S_1 \times \cdots \times S_n$. We say that $\mathcal{S}$ *realizes* an access structure $\Gamma \subseteq 2^{[n]}$ if it satisfies the following.

- Correctness: For any qualified set $Q = \{j_1, \ldots, j_m\} \in \Gamma$ there exists a *reconstruction function* $\text{rec}_Q : S_{j_1} \times \cdots \times S_{j_m} \to K$ such that for every secret $s \in K$, $\Pr[\text{rec}_Q(D(s, R)_Q) = s] = 1$, where $D(s, R)_Q$ denotes a restriction of $D(s, R)$ to its Q-entries.
- Privacy: for any unqualified set $U \notin \Gamma$ and secrets $s, s' \in K$ the random variables $D(s, R)_U$ and $D(s', R)_U$ are identically distributed.

In a *linear* secret-sharing scheme (LSSS) the secret-domain K is a finite field, and the randomness R is a uniformly random m-tuple $(r_1, \ldots, r_m) \in K^m$. The share distribution function D is a linear function of $s, r_1, \ldots, r_m$.

In this work we will refer to the following specific LSSS:

1. SHAMIR'S SECRET-SHARING [28]. Let K be a finite field such that $|K| > n$. Each player P_j is assigned a unique non-zero element from K, which we denote j (by abuse of notation if K is not a prime field). In the *t-private Shamir scheme*, the dealer picks t random and independent field elements $r_1, \ldots, r_t$, which define the univariate polynomial $f(y) = s + r_1 y + r_2 y^2 + \ldots + r_t y^t$, and distributes to each player P_j the share $s_j = f(j)$.
2. REPLICATED SECRET-SHARING [23]. Let $\Gamma \subseteq 2^{[n]}$ be a (monotone) access structure, and let $\mathcal{T}$ include all *maximal* unqualified sets of Γ. The *replicated scheme* for Γ, denoted $\mathcal{R}_\Gamma$, proceeds as follows. To share a secret $s \in K$ the dealer first *additively* share s into $|\mathcal{T}|$ shares, each labelled by a different set from $\mathcal{T}$; that is, it lets $s = \sum_{T \in \mathcal{T}} r_T$ where the shares r_T are otherwise-random field elements. Then, the dealer distributes to each player P_j all shares r_T such that $j \notin T$; that is, P_j's share vector is $(r_T)_{T \not\ni j}$. Privacy follows from the fact that members of every maximal unqualified set $T \in \mathcal{T}$ jointly miss exactly one additive share, namely the share r_T (hence members of any unqualified set miss *at least* one share). On the other hand, since Γ is monotone, a qualified set $Q \in \Gamma$ cannot be contained in any unqualified set; hence, members of Q jointly view all shares r_T and can thus reconstruct s.
3. DNF-BASED SECRET-SHARING [23]. In the *DNF-based scheme*, the secret is additively shared between the members of each minimal qualified set, where each additive sharing uses independent randomness. This scheme can be obtained by applying the construction of [6] to the monotone DNF representation of Γ.

In the case of threshold access structures, the latter two schemes may be practical only in contexts where $\binom{n}{t}$ is not too large. Their asymptotic complexity is polynomial in n when $t = O(1)$ or $n - t = O(1)$.

3 Share Conversion

In this section we present our main results on local share conversion. We start by defining this notion, which induces a partial order on secret-sharing schemes.

Definition 1 (Share conversion). *Let $\mathcal{S}, \mathcal{S}'$ be two secret-sharing scheme over the same secret-domain K. We say that $\mathcal{S}$ is locally convertible to $\mathcal{S}'$ if there exist local conversion functions $g_1, \ldots, g_n$ such that the following holds. If $(s_1, \ldots, s_n)$ are valid shares of a secret s in $\mathcal{S}$ (i.e., $\Pr[D(s, R) = (s_1, \ldots, s_n)] > 0$), then $(g_1(s_1), \ldots, g_n(s_n))$ are valid shares of the same secret s in $\mathcal{S}'$. We denote by g the concatenation of all g_i, namely $g(s_1, \ldots, s_n) = (g_1(s_1), \ldots, g_n(s_n))$, and refer to g as a* share conversion function.

Note that the above definition does not require that random shares of a secret s in $\mathcal{S}$ will be converted into random shares of s in $\mathcal{S}'$. However, due to the locality feature of the conversion, converted shares cannot reveal more information about s than the original shares. Moreover, in typical applications of our technique the converted shares $\mathcal{S}'$ will indeed be random.

3.1 From Replicated Shares to Shamir

We first address the important special case of threshold structures. Suppose that a secret s has been shared according to the t-private replicated scheme. Thus, we may write:

$$s = \sum_{A \subseteq [n]\,:\,|A|=n-t} r_A$$

where r_A has been given to all players in A.

To locally convert these shares into shares of s according to the t-private Shamir scheme, we assign to player P_i the point i in the field. Now, for each set $A \subseteq [n]$ of cardinality $n-t$, let f_A be the (unique) degree-t polynomial such that:

1. $f_A(0) = 1$ and
2. $f_A(i) = 0$ for all $i \in [n] \setminus A$.

Each player P_j can compute a share s_j as follows:

$$s_j = \sum_{A \subseteq [n]\,:\,|A|=n-t, j \in A} r_A \cdot f_A(j).$$

We claim that this results in a set of shares from Shamir's scheme, consistent with the original secret s. To see this, define a polynomial

$$f = \sum_{A \subseteq [n]\,:\,|A|=n-t} r_A \cdot f_A.$$

Clearly, f has degree (at most) t, and it is straightforward to verify that condition 1 above on the f_A's implies $f(0) = s$ and condition 2 implies $f(j) = s_j$.

3.2 Conversion in General

We now generalize the previous conversion result to non-threshold structures. Specifically, we show that shares of the replicated scheme for an arbitrary access structure Γ can be locally converted into shares of any other LSSS for Γ (in fact, even for any $\Gamma' \subset \Gamma$).

To this end, it will be useful to rely on a representation of LSSS via *span programs*, a linear algebra based model of computation introduced by Karchmer and Wigderson [24]. A span program over the variables $\{x_1, \ldots, x_n\}$ assigns to each literal x_i or $\bar{x}_i$ some subspace of a linear space V. The span program *accepts* an assignment $z \in \{0,1\}^n$ if the n subspaces assigned to the satisfied literals span some fixed nonzero vector in V, referred to as the *target vector*. We will be interested in the monotone version of this model, formalized below.

Definition 2 (MSP). *A monotone span program (MSP) is a triple $\mathcal{M} = (K, M, \rho)$, where K is a finite field, M is an $a \times b$ matrix over K, and $\rho : [a] \to [n]$ labels the rows of M by player indices. The size of $\mathcal{M}$ is the number of rows a. For any set $A \subseteq [n]$ let M_A denote the submatrix obtained by restricting M to its rows with labels from A (and similarly for any other matrix with a rows). We* say that $\mathcal{M}$ accepts A if the rows of M_A span the all-ones vector $\mathbf{1}$. We denote by $\Gamma_{\mathcal{M}}$ the collection of all sets in $2^{[n]}$ that are accepted by $\mathcal{M}$, and by $\mathcal{T}_{\mathcal{M}}$ the collection of maximal sets not accepted by $\mathcal{M}$.

Note that for any MSP $\mathcal{M}$, the structure $\Gamma_{\mathcal{M}}$ is monotone. We also note that any nonzero vector could have been used as a target vector; however, the specific choice of $\mathbf{1}$ will be convenient in what follows. We now associate with any MSP $\mathcal{M}$ a corresponding LSSS in which the total number of field elements distributed by the dealer is equal to the size of $\mathcal{M}$.

Definition 3 (LSSS induced by MSP). Let $\mathcal{M} = (K, M, \rho)$ be an MSP, where M is an $a \times b$ matrix. The LSSS induced by $\mathcal{M}$, denoted by $\mathcal{S}_{\mathcal{M}}$, proceeds as follows. To share a secret $s \in K$:

- Additively share s into $\mathbf{r} = (r_1, \ldots, r_b)$.
- Evaluate $\mathbf{s} = M\mathbf{r}$, and distribute to each player P_j the entries $\mathbf{s}_{\{j\}}$ (i.e., those corresponding to rows labelled by j).

It is easy to verify that the induced scheme $\mathcal{S}_{\mathcal{M}}$ is indeed linear and, in fact, that any LSSS is induced by some corresponding MSP $\mathcal{M}$. Finally, the following claim from [24] establishes the expected link between the MSP semantics and the secret-sharing semantics.

Claim. [24] The scheme $\mathcal{S}_{\mathcal{M}}$ realizes the access structure $\Gamma_{\mathcal{M}}$.

Towards proving the main result of this section, it will be convenient to use the notion of *canonic span programs*, introduced in [24]. We use the following monotone version of their construction.

Definition 4 (Canonic MSP). Let $\mathcal{M} = (K, M, \rho)$ be an MSP, where M is an $a \times b$ matrix. We define a canonic MSP $\hat{\mathcal{M}} = (K, \hat{M}, \rho)$ as follows. $\hat{\mathcal{M}}$ has the same size and row labeling as $\mathcal{M}$, but possibly a different number of columns. Let $\mathcal{T} = \mathcal{T}_{\mathcal{M}}$ be the collection of maximal unqualified sets of $\Gamma_{\mathcal{M}}$. For every $T \in \mathcal{T}$, let $\mathbf{w}^T$ be a length-b column vector satisfying $M_T \cdot \mathbf{w}^T = 0$ and $\mathbf{1} \cdot \mathbf{w}^T = 1$.[3] For each maximal unqualified set $T \in \mathcal{T}$, the matrix $\hat{M}$ will include a corresponding column $\mathbf{c}^T \stackrel{def}{=} M \cdot \mathbf{w}^T$ (so that altogether $\hat{M}$ has as many columns as sets in $\mathcal{T}_{\mathcal{M}}$).

It can be shown that $\Gamma_{\hat{\mathcal{M}}} = \Gamma_{\mathcal{M}}$ [24]. (This can also be derived as a corollary of the next two lemmas.) The scheme $\mathcal{S}_{\hat{\mathcal{M}}}$ induced by the canonic program $\hat{\mathcal{M}}$ may be viewed as a randomness-inefficient implementation of $\mathcal{S}_{\mathcal{M}}$. We will use $\mathcal{S}_{\hat{\mathcal{M}}}$ as an intermediate scheme in the process of converting shares of the replicated scheme for $\Gamma_{\mathcal{M}}$ into shares of $\mathcal{S}_{\mathcal{M}}$.

Lemma 1. *The scheme $\mathcal{S}_{\hat{\mathcal{M}}}$ is locally convertible to $\mathcal{S}_{\mathcal{M}}$ via the identity function $g(\mathbf{s}) = \mathbf{s}$.*

---

[3] The existence of such $\mathbf{w}^T$ may be argued as follows: Since $\mathcal{M}$ does not accept T, the linear system $(M_T)^T \cdot \mathbf{x} = \mathbf{1}$ has no solution (where $(M_T)^T$ is the transpose of M_T); hence there must be a way to linearly combine the equations so that a contradiction of the form $\mathbf{0} \cdot \mathbf{x} = 1$ is obtained.

Proof. We need to show that any valid shares in $\mathcal{S}_{\hat{\mathcal{M}}}$ could have also been obtained in $\mathcal{S}_{\mathcal{M}}$ under the same secret s. Let $\hat{\mathbf{r}} \in K^b$ be some additive sharing of $s = \mathbf{1} \cdot \hat{\mathbf{r}}$ induced by the dealer's randomness in $\mathcal{S}_{\hat{\mathcal{M}}}$. Let $\mathbf{r} = W\hat{\mathbf{r}}$ where W is a concatenation of all column vectors $\mathbf{w}^T$ in the order used for constructing $\hat{M}$. By the construction of $\hat{M}$ we have $\hat{M} = MW$ and so $\hat{M}\hat{\mathbf{r}} = MW\hat{\mathbf{r}} = M\mathbf{r}$. Thus, $\mathbf{r}$ produces the same shares in $\mathcal{S}_{\mathcal{M}}$ as $\hat{\mathbf{r}}$ produces in $\mathcal{S}_{\hat{\mathcal{M}}}$. Finally, since every $\mathbf{w}^T$ must satisfy $\mathbf{1} \cdot \mathbf{w}^T = 1$, we have $\mathbf{1} \cdot \mathbf{r} = \mathbf{1} \cdot W\hat{\mathbf{r}} = \mathbf{1} \cdot \hat{\mathbf{r}}$, and thus $\mathbf{r}$ is consistent with the same secret s. □

Lemma 2. *Let $\mathcal{R}_\Gamma$ be the replicated scheme realizing Γ over a finite field K, $\mathcal{M}' = (K, M', \rho')$ an MSP such that $\Gamma' \stackrel{def}{=} \Gamma_{\mathcal{M}'}$ satisfies $\Gamma' \subseteq \Gamma$, and $\hat{\mathcal{M}}' = (K, \hat{M}', \rho')$ a canonic MSP of $\mathcal{M}'$. Then, $\mathcal{R}_\Gamma$ is locally convertible to $\mathcal{S}_{\hat{\mathcal{M}}'}$.*

Proof. Suppose first that $\Gamma' = \Gamma$. Let $\mathcal{T}$ be the collection of maximal unqualified sets of Γ. The $\mathcal{R}_\Gamma$-shares viewed by a player P_i are $s_i \stackrel{def}{=} (r_T)_{T \not\ni i}$, where $\mathbf{r}$ is an additive sharing of the secret s. Define the i-th local conversion function to be

$$g_i(s_i) = \sum_{T \not\ni i} r_T \cdot \mathbf{c}_{\{i\}}^T.$$

Since each column $\mathbf{c}^T$ of $\hat{M}'$ has only zeros in its T-entries, the above functions g_i jointly define the conversion function g which maps $\mathcal{R}_\Gamma$-shares $\mathbf{s}$, obtained by replicating additive shares $\mathbf{r} = (r_T)_{T \in \mathcal{T}}$, into the $\mathcal{S}_{\hat{\mathcal{M}}'}$-shares $\mathbf{s}' = \hat{M}'\mathbf{r}$. The correctness of this conversion is witnessed by letting $\mathbf{r}' = \mathbf{r}$, namely the same additive sharing of s producing $\mathbf{s}$ in $\mathcal{R}_\Gamma$ will also produce $g(\mathbf{s})$ in $\mathcal{S}_{\hat{\mathcal{M}}'}$.

The general case, where Γ' may be a proper subset of Γ, is only slightly more involved. Let $\mathcal{T}'$ denote the maximal unqualified sets in Γ', and assign to each $T \in \mathcal{T}$ some set $T' \in \mathcal{T}'$ containing it. For each $T' \in \mathcal{T}'$, define $r_{T'}$ to be the sum of all r_T such that T is assigned to T' (or 0 if there is no T assigned to T'). Then, the local conversion functions may be defined by $g_i(s_i) = \sum_{T' \not\ni i} r_{T'} \mathbf{c}_{\{i\}}^{T'}$, and the correctness of the induced conversion g is witnessed by letting $\mathbf{r}' = (r_{T'})_{T' \in \mathcal{T}'}$. □

As a direct corollary of the last two lemmas (and using the transitivity of local conversions) we get the main result of this section:

Theorem 1. *The replicated scheme $\mathcal{R}_\Gamma$, realizing Γ over a field K, is locally convertible to any LSSS over K realizing an access structure $\Gamma' \subseteq \Gamma$.*

The above proof in fact provides a *constructive* way for defining the local conversion function from $\mathcal{R}_\Gamma$ to any LSSS $\mathcal{S}$ realizing Γ (or a subset of Γ), given an MSP for $\mathcal{S}$.

Theorem 1 shows that the (CNF-based) replicated scheme $\mathcal{R}_\Gamma$ is maximal with respect to the convertibility relation among all LSSS realizing Γ. Turning to the other extreme, we now argue that the DNF-based scheme (defined in Section 2) is minimal with respect to convertibility.

Theorem 2. *Any LSSS realizing Γ is convertible to the DNF-based scheme for Γ.*

Proof sketch. Suppose that s has been shared according to some LSSS $\mathcal{S}$ for Γ. We need to show that each minimal qualified set $Q \in \Gamma$ can locally compute an additive sharing of s. This easily follows from the linearity of the reconstruction function rec_Q. □

3.3 Negative Results for Share Conversion

We now show some negative results related to the possibility and efficiency of share conversion. We start by showing that the convertibility relation is non-trivial, in the sense that not all schemes realizing the same access structure are convertible to each other. In fact, we show that Shamir shares cannot be generally converted to replicated shares.

Claim. Let $\mathcal{S}$ be the 1-private 3-player Shamir scheme over a field K ($|K| > 3$) and $\mathcal{S}'$ be the replicated scheme with the same parameters. Then $\mathcal{S}$ is not locally convertible to $\mathcal{S}'$.

Proof. By the correctness requirement, the value of g on any valid 3-tuple (s_1, s_2, s_3) of $\mathcal{S}$-shares must take the form $g(s_1, s_2, s_3) = ((r'_2, r'_3), (r'_1, r'_3), (r'_1, r'_2))$. We now use the locality requirement to show that g must be a constant function, contradicting the correctness requirement. Suppose that one of the local functions g_i is non-constant. Assume wlog that $g_1(0) \neq g_1(1)$ and that they differ in their first output r'_2. Then, either $g(0, s_2, 0)$ outputs illegal $\mathcal{S}'$-shares for all $s_2 \in K$ or $g(1, s_2, 0)$ outputs illegal $\mathcal{S}'$-shares for all $s_2 \in K$ (since in either of these cases the first share of P_1 is different from the second share of P_3). Since there exist both valid $\mathcal{S}$-shares of the form $(0, s_2, 0)$ and of the form $(1, s_2, 0)$, we obtain the desired contradiction. □

Motivated by the following applications, it is natural to ask whether one can reduce the amount of replication in the replicated scheme $\mathcal{R}_\Gamma$ and still allow to convert its shares to other useful LSSS for Γ. Specifically, let $\mathcal{S}$ be a secret-sharing scheme for Γ with the property that a qualified set of players can reconstruct not only the secret, but also the shares of all players. Note that Shamir's scheme enjoys this property. We show that the scheme $\mathcal{R}_\Gamma$ cannot be replaced by a more efficient replicated scheme which is still convertible to $\mathcal{S}$ and at the same time is private with respect to all unqualified sets of $\mathcal{S}$.

Definition 5. *A generic conversion scheme from replicated shares to $\mathcal{S}$ consists of a set of independently distributed random variables $R_1, \ldots, R_m$, an assignment of a subset B_j of these to each player P_j, and local conversion functions g_j such that if each P_j applies g_j to the variables in B_j, we obtain values $(s_1, ..., s_n)$ forming consistent $\mathcal{S}$-shares of some secret s. Furthermore, given the information accessible to any unqualified set of Γ, the uncertainty of s is non-zero.*

Note that neither $\mathcal{S}$ nor the conversion functions g_j are assumed to be linear. Also note that the convertibility requirement formulated above is weaker than our default requirement. However, we are about to show a negative result which is only made stronger this way.

Proposition 1. *For any generic conversion scheme for $\mathcal{S}$ as defined above, it holds that m is at least the number of maximal unqualified sets.*

Proof. Fix any maximal unqualified set T, and let B_T be the set of R_i's known to T. We may assume that for each $R_i \in B_j$, it is the case that $H(s_j|B_j \setminus R_i) > 0$, i.e., R_i is necessary for s_j. If there was not the case, we could remove R_i from B_j and get a more efficient scheme. For each player $P_j \notin T$, we let $C_{j,T} = B_j \setminus T$, thus representing the information available to P_j but not to T. Each such set must be non-empty, otherwise T could determine the value of s.

Now, the set $T \cup P_j$ is qualified, and hence, for any other $P_i \notin T$, it is the case that the share $s_i = g_i(B_i)$ is uniquely determined from $B_T \cup B_j$ – by assumption on $\mathcal{S}$. It follows that $B_i \subset B_T \cup B_j$ and therefore that $C_{i,T} \subset C_{j,T}$. If this was not the case, then by independence of the R_i's, s_i would not be uniquely determined from $B_T \cup B_j$. Since this argument works for any $P_j \notin T$, it follows that in fact $C_{i,T} = C_{j,T}$, so we call this set C_T for short.

Now, consider a different maximal unqualified set T'. We will be done if we show that $C_T \cap C_{T'} = \emptyset$, since this and each C_T being non-empty means that there must be as many R_i's as there are sets T.

So assume some $R_i \in C_T \cap C_{T'}$, and consider a player P_j who is in $T' \setminus T$. This means that P_j knows all variables in C_T, in particular also R_i, but this is a contradiction since R_i is also in $C_{T'}$ and $B_{T'} \cap C_{T'} = \emptyset$ by construction. □

4 Applications

The ability to convert replicated shares to Shamir shares allows to create, from initially distributed randomness, any number of Shamir secret sharings of (pseudo) random values without communication.[4] In this section we present several applications of this idea.

We begin by describing some useful sub-protocols that are common to most of these applications. The first protocol provides precisely the functionality described above: secure generation of (pseudo)random Shamir-shared secrets without communication. Recall the share conversion procedure described in Section 3.1. A secret s has been shared according to the t-private replicated scheme, namely $s = \sum_{A \subseteq [n]\, :\, |A|=n-t} r_A$ where r_A has been given to all players in A. To locally convert these shares into Shamir shares, each player P_j computes its share as $s_j = \sum_{|A|=n-t, j \in A} r_A \cdot f_A(j)$, where f_A is a degree-t polynomial determined by A.

The main observation is that when the secret s is random, all replicated shares r_A will be random and independent. Hence we may use the initially distributed r_A as keys to a PRF $\psi.(\cdot)$, and as long as players agree on a common input a to the function, all players in A can compute $\psi_{r_A}(a)$ and use it in the above construction in place of r_A. Concretely, we get the following.

---

[4] While we focus the attention on Shamir-based schemes for threshold access structures, the results of this section can be extended to linear schemes realizing general access structures.

Protocol. Pseudorandom Secret-Sharing (PRSS)
Common inputs: a value a and independent keys $\{r_A\}$ that have been predistributed as above. Each player P_j computes his share s_j as:

$$s_j = \sum_{A \subseteq [n]\,:\,|A|=n-t, j \in A} \psi_{r_A}(a) \cdot f_A(j) \qquad (1)$$

Note that if we choose K to be of characteristic 2, we can modify the PRSS protocol so that the shared value is guaranteed to be 0 or 1 by simply using a PRF that always outputs 0 or 1. We call this Binary Pseudorandom Secret-Sharing (BPRSS). Assuming that $t < n/3$ it is easy to turn this into a non-interactive verifiable secret-sharing scheme, in a model where a broadcast primitive is available: we simply arrange it such that a Dealer knows all the involved keys. This allows him to compute the pseudorandom shared value and correct it into the value he wants to share:

Protocol. Non-Interactive Verifiable Secret-Sharing (NIVSS)
Common inputs: a value a and keys $\{r_A\}$ as above. A dealer D holds all keys as well as an input value $v \in K$. Each player P_j computes a preliminary share $\tilde{s}_i$ as in Eq. 1. Using his knowledge of the keys, D computes the secret s determined by the preliminary shares. D then broadcasts $(v - s)$. Each P_j computes his share as $\tilde{s}_j + (v - s)$.

It is straightforward to verify that this creates a valid Shamir sharing of v if D is honest, and will create a valid sharing of some value no matter how D acts. Furthermore, since $t < n/3$, this value can be reconstructed using standard error correction techniques as long as at most t of the shares are wrong. Finally, for the privacy, we have the following.

Lemma 3. *Consider an adversary Adv that corrupts up to t of the players, but not D. Adv may invoke the protocol NIVSS multiple times, (adaptively) choosing a secret v_j and a distinct evaluation point a_j at each invocation. The adversary gets to see the executions of NIVSS where in the j-th invocation a_j is used as the common input and either (case 0) v_j or (case 1) a random independent value is given as input to D. Assuming the underlying PRF is secure, cases 0 and 1 are computationally indistinguishable.*

Proof. Assume that some Adv can distinguish case 0 and 1, and make the (worst case) assumption that Adv corrupts t players. This means that only one key r_A is unknown to Adv, where A consists of the $n - t$ uncorrupted players. We build an algorithm Alg that breaks the PRF. It gets oracle access to either $\psi_{r_A}()$ or a random oracle and must tell the two apart. Alg gets the inputs a_j, v_j from Adv and at each invocation simply invokes the NIVSS protocol on these inputs, except that it calls the oracle whenever it needs to compute $\psi_{r_A}()$. It is now straightforward to verify that if Alg's oracle is random, Adv will see an exact emulation of case 1: in this case the value of s computed at each invocation will be uniformly random and independent of previous values (by uniqueness of a_j) and so will $v_j - s_j$. On the other hand, if Alg talks to $\psi_{r_A}()$ we emulate exactly case 0. Thus Adv's ability to distinguish case 0 and 1 translates to breaking the PRF with the same advantage. □

It is easy to adapt the NIVSS protocol such that we are guaranteed that the shared value is 0 or 1, along the same lines as the BPRSS protocol. We will refer to this as BNIVSS. Note also that if we just want to create a shared random value known to the dealer, we can simply omit the broadcast step and use the preliminary shares as final shares. We will refer to this as *NIVSS without broadcast*.

The technique for pseudo-random secret-sharing can be generalized to create sharings of a particular value, such as zero. We explain how this is done for the same threshold t access structure as before, but for polynomials of degree $2t$, since this is what we need in the following. Generalizations to other degrees follow easily. Consider a set A of size $n - t$ and consider the set of polynomials

$$F_A = \{f| \ deg(f) \leq 2t, f(0) = 0, j \notin A \Rightarrow f(j) = 0\}.$$

If we think of the set of all degree-$(2t)$ polynomials as a vector space over K, it is easy to see that F_A is a subspace of dimension $2t + 1 - t - 1 = t$. So we choose for each A, once and for all, a basis for F_A consisting of t polynomials $f_A^1, ..., f_A^t$. Finally, we distribute initially t keys $r_A^1, ..., r_A^t$ to every player in A. This leads to the following protocol:

Protocol. Pseudorandom Zero-Sharing (PRZS)
Common input: a value a, keys $\{r_A^i | i = 1..t, |A| = n - 1\}$ that have been predistributed as above. Each player P_j computes his share s_j as:

$$s_j = \sum_{A \subseteq [n] \ : \ |A| = n-t, j \in A} \sum_{i=1}^{t} \psi_{r_A^i}(a) \cdot f_A^i(j)$$

It is straightforward to verify that this results in shares consistent with the polynomial $f_0 = \sum_{A, |A|=n-t, P_j \in A} \sum_{i=1}^{t} \psi_{s_A^i}(a) \cdot f_A^i$, that $deg(f_0) \leq 2t$ and that $f_0(0) = 0$.

The above ideas for "non-interactive random secret-sharing" are less efficient if the number of sets A is large.[5] On the other hand, the pseudo-random function is used as a black-box and hence any pseudo-random function can be used. By Proposition 1, our solution is optimal among a class of generic schemes making a black-box use of a PRF.

Distributed PRFs. An immediate application of the basic PRSS protocol described above is to the problem of distributing pseudorandom functions (or KDCs), studied by Naor, Pinkas and Reingold [27]. A distributed PRF should

---

[5] Note that when t is constant, the number of sets A is polynomial in n. Thus, share conversion allows to efficiently achieve a constant level of privacy with an arbitrarily high level of robustness.

allow a client to query its value on any chosen input a by contacting n servers, where the output should remain private from any collusion of t servers. Moreover, even if t servers are actively corrupted, the client should still learn the right output. A simple solution to this problem is to use the PRSS protocol, where the client sends a to each server P_j and receives s_j, the corresponding Shamir-share of the output, in return. A similar scheme that was suggested in [27] relies on replicated PRFs but does not use share conversion. Thus, the communication complexity of the scheme from [27] is very high, as servers are required to send all replicated shares to the client.

4.1 Applications to Secure Multiparty Computation

We show how the pseudorandom secret-sharing approach can be used to securely compute low-degree polynomials via an efficient two-round protocol. We then discuss an extension of this result to general functions.

It will be convenient to use the following model for secure computation: the input will be supplied by m *input clients* $I_1, ..., I_m$. The computation will be performed by n *servers* $P_1, .., P_n$. The outputs are to be distributed to v *output clients* $O_1, ..., O_v$. Since one player can play several of these roles, this is a generalization of the standard model, which fits well with the applications we give later. We will assume that input clients can broadcast information to the servers and we also assume secure point to point channels between servers and output clients. A typical protocol will have the following two-round structure: in Round 1 each input client broadcasts values to the servers (one value for each input) and in Round 2 each server sends (over a secure channel) a message to each output client. For some applications Round 1 will not be necessary, in which case we get fully non-interactive protocols.

We assume an adversary that can corrupt any number of clients and up to t servers. We consider both the case of a passive and an active adversary. The adversary is for now assumed to be static (non-adaptive).

We will make the following set-up assumptions: sets of keys for a pseudorandom function have been distributed to input clients and servers, such that each input client can act as the dealer in the NIVSS protocol and the servers can execute the PRSS and PRZS protocols.

Secure Computation of Low-Degree Polynomials. We show how to securely compute the value of a degree d multivariate polynomial $Q()$ in m variables, where I_j supplies the j-th variable x_j. We assume that the output value $Q(x)$ is to become known to all output clients. Generalizations to more input variables, more polynomials and different polynomials for different output clients follow easily. For a passive adversary, we assume $dt < n$, while for an active adversary we assume $(d+2)t < n$. The protocol proceeds as follows:

1. In round 1, each input client I_j acts as the dealer in the NIVSS protocol using x_j as his private input. (If the inputs x_i should be restricted to take binary values, BNIVSS is used instead of NIVSS.) Let $x_{j,i}$ be the share obtained by server P_i. We execute the PRZS protocol adapted such that we create

shares of a degree dt polynomial that evaluates to 0 on 0. Let z_i be the share obtained by P_i. Each P_i now computes $Q(x_{1,i}, ..., x_{m,i}) + z_i$.
2. In round 2, each server sends $Q(x_{1,i}, ..., x_{m,i}) + z_i$ to all output clients.
3. Each output client considers the values it receives as values of a degree dt polynomial f - where up to t values may be wrong if the adversary is active. He reconstructs the value $f(0)$ (using standard error correction in the active adversary case) and defines this to be his output.

Note that if we only wanted to compute shares of the value $Q(x_1, ..., x_m)$ we could do this by simply omitting steps 2 and 3. Using Canetti's definition of secure function evaluation from [9], we get:

Theorem 3. *The above protocol computes the function $Q(x_1, ..., x_m)$ securely against a passive, static adversary if $dt < n$ and against an active, static adversary if $(d+2)t < n$.*

Proof sketch. For some adversary Adv, the required Ideal model adversary, or simulator, works as follows: it simulates the broadcast of honest input clients by broadcasting random values. Since it knows the keys that corrupt input clients use in the NIVSS protocols, it can compute the values that these clients are sharing, and send them to the ideal functionality. Note that it also knows the keys held by corrupt servers so it can compute the share $sh_j = Q(x_{1,i}, ..., x_{m,i}) + z_i$ that each corrupt P_j holds of the result. When given the output value y, it therefore chooses a random polynomial f such that $f(0) = y$ and $f(j) = s_j$ for each corrupt P_j. And for each honest P_i, it sends $f(i)$ to corrupt output clients in round 2, to simulate the contributions of honest servers to the result.

To argue that the simulation works, we can argue along the same lines as for Lemma 3. If for some adversary Adv and some set of inputs $x_1, ..., x_m$, the output from the real process could be distinguished from that of the ideal process, we could build an algorithm Alg for breaking the pseudorandom function. Alg will have oracle access to $\psi_s()$ for all keys s not known to Adv, or to a set of random oracles. Alg will now use the given inputs $x_1, ..., x_m$ and keys for the corrupt players that it chooses itself, to execute the protocol with Adv. It emulates the honest players according to the protocol, except that it calls its oracles whenever an honest player would have used a key not known to Adv. One can now verify that if the oracles contain pseudorandom functions, we produce output distributed exactly as in the real process, whereas of they are random, we produce output according to the ideal process. It is in this last part that we need $dt < n$ $((d+2)t < n)$ since this ensures that the polynomial f reconstructed by the honest output clients will determine the correct output value y, regardless of whether f was constructed by the simulator or by Alg.

Thus the ability to distinguish between the real and the ideal process translates to breaking the pseudorandom function with the same advantage. □

Note that this result extends easily to computing polynomials where some of the inputs are to be chosen at random: we just use the PRSS protocol to create shares of these inputs.

General MPC in 2 Rounds. Using known techniques, one can reduce the secure computation of general functions to that of degree-3 polynomials. In case of functions that can be efficiently represented by (boolean or arithmetic) branching programs, such a reduction is given by constructions of randomizing polynomials from [21, 22]. A similar reduction for arbitrary (polynomial-time computable) functions is possible under standard intractability assumptions [1]. Alternatively, it is possible to modify the garbled circuit technique of Yao [31] and Beaver, Micali and Rogaway [2] to obtain 2-round protocols for arbitrary functions using our protocol for degree-3 polynomials. Using either approach, one can get 2-round general MPC protocols with security threshold $t < n/3$ ($t < n/5$) in the passive (active) case.

Distributing a Trusted Dealer. An important class of multi-party functionalities are those that distribute correlated random resources to output clients (without taking any inputs). For such functionalities, we can distribute a trusted dealer via a totally non-interactive protocol in which each server send a single message to each output client. Applications of such functionalities range from emulating a trusted poker dealer to providing players with correlated resources for general MPC protocols (e.g., [14, 16]). For the applications to threshold cryptography, discussed next, we use a similar approach but rely on the special structure of the relevant functionalities to gain better efficiency.

4.2 Applications to Threshold Cryptography

As mentioned above, low-degree polynomials that take only random inputs, possibly along with other inputs that have been pre-shared, can be securely computed in only one round using our techniques. In this section, we show that this efficiently extends to functions defined over finite groups of prime order, involving exponents that are low degree polynomials. It will later become clear how such functions can be used to handle problems in threshold cryptography without interaction.

We will assume that we work in a fixed finite group G of prime order q, such as a subgroup of Z_p^*, where q divides $p-1$. Furthermore, we assume we have an ideal implementation of a function that chooses a vector $X = (x_1, ..., x_u) \in Z_q^u$, secret-shares these values according to Shamir's scheme with polynomials of degree $\leq t$, and outputs shares $x_{i,j}$, $j = 1..u$, to each server P_i. Finally, the function chooses and distributes seeds as required for the PRSS and PRZS protocols we described earlier. This corresponds to the key generation phase in a threshold cryptosystem.

We assume as usual an adversary that corrupts at most t servers. For simplicity, we assume first that the adversary is static and passive. Consider a randomized function $\Phi()$, which we will define using a fixed set of multivariate degree 2 polynomials $Q_1(X, R), ..., Q_w(X, R)$ (the following extends trivially to small degrees larger than 2, but degree 2 is all we will need in the following). To compute the function, all servers input the shares they received earlier, while an input client broadcasts elements $g_1, ..., g_w \in G$ to the servers. We need that the shares supplied by servers uniquely determine X. This is always the case if the adversary is passive, and is also the case for an active adversary, provided $t < n/3$. The function outputs to all players the element

$$\Phi(g_1, ..., g_w, \{x_{i,j}|\ i = 1..n, j = 1..u\}) = g_1^{Q_1(X,R)} \cdots g_w^{Q_w(X,R)}$$

where $R = (r_1, ..., r_v) \in Z_q^v$ consists of v uniformly random numbers. We let $(\lambda_1, ..., \lambda_n)$ be chosen as Lagrange interpolation coefficients such that $f(0) = \sum_{i=1}^n \lambda_i f(i)$ for any polynomial f with $deg(f) < n$.

To build a protocol for evaluating $\Phi()$, we use a technique similar to what we used before, but we now put everything "in the exponent": assuming players have shares of the x_i's and r_i's, they can compute the "same" expression as in the definition of Φ, using their respective shares in place of the x_i's and r_i's. When each local result is broadcast, one can find the answer using interpolation "in the exponent". And even though the shares of the r_i's are not predistributed, we can create them non-interactively using PRSS. Also as previously, we need to randomize the degree $2t$ polynomial that is (implicitly) revealed, using the PRZS technique.

Protocol. Compute $\Phi()$

1. Each server P_i computes a share $r_{j,i}$ of a pseudorandom value r_j, for $j = 1..v$, as well as a share t_i in a degree-$(2t)$ sharing of 0, as described in protocols PRSS and PRZS. Let $R_i = (r_{1,i}, ..., r_{v,i})$ and $X_i = (x_{1,i}, ..., x_{u,i})$. He sends to the output client(s) the group element

$$G_i = g_1^{Q_1(X_i, R_i)} \cdots g_w^{Q_w(X_i, R_i)} g_1^{t_i}$$

2. The output client(s) compute the output as $\prod_{i=1}^n G_i^{\lambda_i}$.

For the above protocol, one can prove the following:

Theorem 4. *Assuming a passive, static adversary, $t < n/2$, and that $\psi.(\cdot)$ (used in the pseudo-random secret-sharing) is a pseudo-random function, the above protocol computes $\Phi()$ securely.*

Proof sketch. Rewriting the definition of $\Phi()$ by expressing all group elements as powers of some fixed generator g of G, it is clear that the output value is of form $g^{Q(X,R)}$ for some multivariate degree 2 polynomial. Hence, from an information theoretic point of view, we are in fact computing $Q(X, R)$ using exactly the protocol we saw earlier. Therefore, essentially the same proof as for the earlier protocol applies here. □

Note that since we need to assume that the number of servers is small in order to use PRSS in the first place, the results from [10] imply that this same protocol is also adaptively secure.

For the case of an active, static adversary, we can use the same protocol, provided that $t < n/4$ and that we make a modification in the final step. In the earlier protocol for computing low-degree polynomials, we could use standard error correction, but this will not work here. We are faced with elements

$G_1, ..., G_n$ and all we know is that all but t of them are of form $G_i = g^{f(i)}$ for some polynomial f of degree at most $2t$. Since we cannot expect to compute discrete logs, direct error correction does not apply.

Since the number of servers is small, one option is to find the correct subset by exhaustive search. But this may not be entirely satisfactory. We do have to assume that servers have enough memory and computing power to handle such a problem (in order to carry out the PRSS protocol). But in a real application, we may not want to assume this about the clients.

We sketch a solution to this using results from [13]. We will assume two-level sharings, that is the x_j's have been shared and then the shares have themselves been shared using degree t polynomials. If $x_{j,i}$ is P_i's share, P_i also receives as part of his share the polynomial used for sharing $x_{j,i}$. We will use the same type of two-level sharing for the random r_j's and for the degree $2t$ sharing of 0. This can all be done non-interactively by our results on general share conversion, because such a two-level sharing is a linear scheme.

Hence, when P_i claims that the value G_i he contributes really satisfies $G_i = g_1^{Q_1(X_i, R_i)} \cdots g_w^{Q_w(X_i, R_i)} g_1^{t_i}$, we can assume that the exponents are shared among the servers using degree-$(2t)$ polynomials and P_i knows the polynomials that have been used. Therefore the value can non-interactively be shown to be correct using a straightforward generalization of the techniques from [13].

Theorem 5. *Assuming an active, static adversary, $t < n/4$, and that $\psi(\cdot)$ (used in the pseudo-random secret-sharing) is a pseudo-random function, the above protocol, modified as described for active adversaries, computes $\Phi()$ securely (and non-interactively).*

We expect that this general technique will be useful in many contexts. Below we give a few examples for applications to some concrete threshold cryptosystems.

Threshold Cramer-Shoup. Canetti and Goldwasser [11] proposed a threshold version of the Cramer-Shoup cryptosystem, the first really efficient public-key system that could proved secure under chosen ciphertext attacks, without assuming random oracles.

This scheme works in a group G of order q as we did above. The private key is (x_1, x_2, y_1, y_2, z), all chosen at random in Z_q. The public key consists of a number of elements in G, namely $g_1, g_2, c = g_1^{x_1} g_2^{x_2}, d = g_1^{y_1} g_2^{y_2}, h = g_1^z$.

A ciphertext is a 4-tuple of elements (u_1, u_2, e, v). To decrypt, one computes a value α from the ciphertext using a public hash function. Then we set $v' = u_1^{x_1+y_1\alpha} u_2^{x_2+y_2\alpha}$. We choose $r \in Z_q$ at random and compute $b = u_1^z (v'v^{-1})^r$. Finally, the output (which will be the decrypted message if the ciphertext was valid) is eb^{-1}.

The randomization introduced by r is a modification of the original scheme suggested in [11] to maintain CCA security even if one defines the decryption algorithm to output b always - instead of an error message in case the ciphertext was invalid. Clearly, if one can securely compute b assuming that x_1, x_2, y_1, y_2, z

have been pre-shared, a secure threshold version of the scheme follows. In [11] this was done non-interactively, essentially assuming that a number of random values had been pre-shared, so they could play the role of r. This resource runs out quickly since r-values cannot be reused, so this is not a satisfactory solution. Indeed Canetti and Goldwasser asked whether it was possible to create shares of r pseudorandomly without interaction. This is exactly possible using our techniques. Indeed by rewriting, the value we need to compute is

$$b = u_1^{z+x_1 r + \alpha y_1 r} u_2^{x_2 r + \alpha y_2 r} v^{-r}$$

It should be clear that this expression is a special case of the (class of) function(s) $\Phi()$. Hence a protocol for computing b securely follows immediately from the protocols for computing Φ, both in the passive and active adversary case. We also obtain the same bounds on t as in [11] [6].

Threshold Signatures. It is known how to obtain efficient non-interactive threshold signatures in the random oracle model based on RSA, see for instance [29]. If we drop the random oracle assumption, things seem to be much more difficult. We do know efficient secure signature schemes that need no random oracles [15, 17], but it is not at all clear how one could design a non-interactive threshold version of those schemes.

However, we can make use of the fact that Boneh and Boyen in [8] suggested a fully secure ID based encryption scheme without random oracles. A more efficient scheme was suggested by Waters in [30].

Briefly, an ID based encryption scheme has a public key pk, and a master secret key sk. Each user has an identity ID, and using the master key one can derive a secret key sk_{ID} for this user. Knowing only the ID, it is possible to encrypt a message such that only the user who knows sk_{ID} can decrypt.

Our interest in this comes from the fact that any such scheme implies a signature scheme, with public key pk and private key sk. To sign a string, one thinks of it as an identity and uses sk to extract sk_{ID} which now plays the role of a signature. The security properties of the ID based scheme imply security of the signature scheme in a natural way.

The scheme of [30] works in a prime order group G equipped with a bilinear mapping (which we do not have to consider explicitly here). Keys are generated as follows: fix a generator g of G, choose a random $\alpha \in Z_q$ and set $g_1 = g^\alpha$. Also pick random elements $g_2, u', u_1, ..., u_l$ where l is the length of identities. Then the public key is $g, g_1, g_2, u', u_1, ..., u_l$ and the master secret key is g_2^α. The secret key corresponding to identity v where the i'th bit of v is v_i is constructed by choosing r at random and setting sk_{ID} to be the pair

$$sk_{ID} = (g_2^\alpha (u' \prod_{i \in V} u_i)^r, \ g^r)$$

---

[6] In the active case, it was claimed in [11] that their solutions work for $t < n/3$ non-interactively and for $t < n/2$ with interaction, but this is not correct. The authors of [11] have confirmed that the correct bounds are $t < n/4$ and $t < n/3$, respectively.

where V is the set of indices such that $v_i = 1$. Now, since $u' \prod_{i \in V} u_i$ is an element all players can compute by themselves, also this expression is a special case of our function Φ. It follows that the protocol for computing Φ can be used to compute, non-interactively and securely, a signature in the scheme derived from [7].

This is the first non-interactive threshold signature scheme that can be shown secure without random oracles. We note that concurrently and independently from our work, Boneh and Boyen have recently found a different technique for distributing this signature scheme. This method is tailored to their scheme and does not use share conversion based techniques. It scales better w.r.t. the number of players than our method, but is less general. For instance, our technique also applies directly to distribute non-interactively the recent signature scheme by Camenisch and Lysyanskaya [12]. This leads to a distributed signature scheme with a much smaller public key than starting from Waters' scheme, and it also implies a distributed implementation of the authorities issuing credentials in their anonymous credential scheme.

References

1. B. Applebaum, Y. Ishai, and E. Kushilevitz. Computationally private randomizing polynomials and their applications. Manuscript, 2004.
2. D. Beaver, S. Micali, and P. Rogaway. The round complexity of secure protocols (extended abstract). In *Proc. of 22nd STOC*, pages 503–513, 1990.
3. D. Beaver and A. Wool. Quorum based secure multi-party computation. In *Proc. of EUROCRYPT '98, LNCS 1403, Springer Verlag*, pages 375–390, 1998.
4. A. Beimel. *Secure schemes for secret sharing and key distribution*. PhD thesis, Technion, 1996.
5. A. Beimel, Y. Ishai, E. Kushilevitz, and J. F. Raymond. Breaking the $O(n^{1/(2k-1)})$ Barrier for Information-Theoretic Private Information Retrieval. In *Proceedings of the 43rd IEEE Conference on the Foundations of Computer Science (FOCS '02)*, pages 261–270, 2002.
6. J. Benaloh and J. Leichter. Generalized secret sharing and monotone functions. In *Proc. of CRYPTO '88, LNCS 403, Springer Verlag*, pages 27–35, 1990.
7. D. Boneh and X. Boyen. Efficient Selective Identity-based Encryption. In *Proc. of Eurocrypt '04*.
8. D. Boneh and X. Boyen. Secure Identity-Based Encryption Without Random Oracles. In *Proc. of Crypto '04*.
9. R. Canetti. Security and composition of multiparty cryptographic protocols. In *J. of Cryptology*, 13(1), 2000.
10. R. Canetti, I. Damgård, S. Dziembowski, Y. Ishai, and T. Malkin. On Adaptive vs. Non-adaptive Security of Multiparty Protocols. In *J. of Cryptology*, 17(3), 2004. Preliminary version in Eurocrypt '01.
11. R. Canetti and S. Goldwasser. An efficient threshold public-key cryptosystem secure against adaptive chosen ciphertext attacks. In *Proc. of Eurocrypt '99*.
12. J. Camenisch and A. Lysyanskaya. Signature Schemes and Anonymous Credentials from Bilinear Maps. In *Proc. of Crypto 2004*.
13. R. Cramer and I. Damgård. Secret-Key Zero-Knowledge and Non-interactive Verifiable Exponentiation. In *Proc. TCC '04*.

14. R. Cramer, I. Damgård, and J. Nielsen. Multiparty computation from threshold homomorphic encryption. In *Proc. of EUROCRYPT '01*, LNCS 2045, pp. 280-299, 2001.
15. R. Cramer and V. Shoup. Signature Schemes Based on the Strong RSA Assumption. In *Proc. ACM Conference on Computer and Communications Security*, 1999.
16. M. Fitzi, S. Wolf and J. Wullschleger. Pseudo-signatures, broadcast, and multiparty computation from correlated randomness. In *Proc. Crypto '04*.
17. R. Gennaro, S. Halevi and T. Rabin. Secure Hash-and-Sign Signatures Without Random Oracles. In *Proc. of Eurocrypt '99*.
18. R. Gennaro, Y. Ishai, E. Kushilevitz and T. Rabin. The Round Complexity of Verifiable Secret Sharing and Secure Multicast. In *Proceedings of the 33rd ACM Symp. on Theory of Computing (STOC '01)*, pages 580-589, 2001.
19. R. Gennaro, Y. Ishai, E. Kushilevitz and T. Rabin. On 2-round secure multiparty computation. In *Proc. Crypto '02*.
20. N. Gilboa and Y. Ishai. Compressing cryptographic resources. In *Proc. of CRYPTO '99*.
21. Y. Ishai and E. Kushilevitz. Randomizing polynomials: A new representation with applications to round-efficient secure computation. In *Proc. 41st FOCS*, pp. 294–304, 2000.
22. Y. Ishai and E. Kushilevitz. Perfect constant-round secure computation via perfect randomizing polynomials. In *Proc. 29th ICALP*, pp. 244–256, 2002.
23. M. Ito, A. Saito, and T. Nishizeki. Secret sharing schemes realizing general access structures. In *Proc. IEEE Global Telecommunication Conf., Globecom 87*, pages 99–102, 1987.
24. M. Karchmer and A. Wigderson. On span programs. In *Proc. of 8th IEEE Structure in Complexity Theory*, pages 102–111, 1993.
25. U. Maurer. Secure multi-party computation made simple. In *Proc. of SCN '02*.
26. S. Micali and R. Sidney. A simple method for generating and sharing pseudorandom functions with applications to clipper-like key escrow systems. In *Proc. of CRYPTO '95*, LNCS 963, Springer Verlag, pages 185–196, 1995.
27. M. Naor, B. Pinkas, and O. Reingold. Distributed pseudo-random functions and KDCs. In *Proc. of EUROCRYPT '99*, LNCS 1592, Springer Verlag, pages 327–346, 1999.
28. A. Shamir. How to share a secret. *Commun. ACM*, 22(6):612–613, June 1979.
29. V. Shoup. Practical Threshold Signatures. In *Proc. of Eurocrypt '00*.
30. B. R. Waters. Efficient Identity-Based Encryption Without Random Oracles. Eprint report 2004/180.
31. A. C. Yao. How to generate and exchange secrets. In *Proc. 27th FOCS*, pp. 162–167, 1986.

Toward Privacy in Public Databases*

Shuchi Chawla[1], Cynthia Dwork[2], Frank McSherry[2],
Adam Smith[3,**], and Hoeteck Wee[4]

[1] Carnegie Mellon University
shuchi@cs.cmu.edu
[2] Microsoft Research SVC
{dwork, mcsherry}@microsoft.com
[3] Weizmann Institute of Science
adam.smith@weizmann.ac.il
[4] University of California, Berkeley
hoeteck@cs.berkeley.edu

In Memoriam
Larry Joseph Stockmeyer
1948–2004

Abstract. We initiate a theoretical study of the *census problem*. Informally, in a census individual respondents give private information to a trusted party (the census bureau), who publishes a *sanitized* version of the data. There are two fundamentally conflicting requirements: *privacy* for the respondents and *utility* of the sanitized data. Unlike in the study of secure function evaluation, in which privacy is preserved to the extent possible given a specific functionality goal, in the census problem privacy is paramount; intuitively, things that cannot be learned "safely" should not be learned at all.

An important contribution of this work is a definition of privacy (and privacy compromise) for statistical databases, together with a method for describing and comparing the privacy offered by specific sanitization techniques. We obtain several privacy results using two different sanitization techniques, and then show how to combine them via cross training. We also obtain two utility results involving clustering.

1 Introduction

We initiate a theoretical study of the *census problem*. Informally, in a census individual respondents give private information to a trusted party (the *census bureau*), who publishes an altered or *sanitized* version of the data. There are two fundamentally conflicting requirements: *privacy* for the respondents and *utility*

---

* A full version of this paper may be found on the World Wide Web at http://research.microsoft.com/research/sv/DatabasePrivacy/.
** This research was done while A.S. was a student at MIT, partially supported by a Microsoft fellowship and by US ARO Grant DAAD19-00-1-0177.

of the sanitized data. While both require formal definition, their essential tension is clear: perfect privacy can be achieved by publishing nothing at all – but this has no utility; perfect utility can be obtained by publishing the data exactly as received from the respondents, but this offers no privacy. Very roughly, the sanitization should permit the data analyst to identify strong stereotypes, while preserving the privacy of individuals.

This is not a new problem. Disclosure control has been studied by researchers in statistics, algorithms and, more recently, data mining. However, we feel that many of these efforts lack a sound framework for stating and proving guarantees on the privacy of entries in the database. The literature is too extensive to survey here; we highlight only a few representative approaches in Section 1.2.

1.1 Summary of Our Contributions and Organization of the Paper

Definitions of Privacy and Sanitization. We give rigorous definitions of privacy and sanitization (Sections 2 and 3, respectively). These definitions, and the framework they provide for comparing sanitization techniques, are a principal contribution of this work.

For concreteness, we consider an abstract version of the database privacy problem, in which each entry in the database—think of an individual, or a particular transaction—is an unlabeled point in high-dimensional real space $\mathbb{R}^d$. Two entries (points) that are close in $\mathbb{R}^d$ (say, in Euclidean distance) are considered more similar than two entries that are far.

Our first step was to search the legal and philosophical literature to find a good English language definition of privacy relevant to statistical databases. The phrase "protection from being brought to the attention of others" in the writings of Gavison [19] resonated with us. As Gavison points out, not only is such protection inherent in our understanding of privacy, but when this is compromised, that is, when we have been brought to the attention of others, it invites further violation of privacy, as now our every move is examined and analyzed. This compelling concept – protection from being brought to the attention of others – articulates the common intuition that our privacy is protected to the extent that we blend in with the crowd; moreover, we can convert it into a precise mathematical statement: intuitively, we will require that, from the adversary's point of view, every point be indistinguishable from at least $t-1$ other points, where t is a threshold chosen according to social considerations. Sweeney's seminal work on k-anonymity is similarly motivated [31]. In general, we think of t as much, much smaller than n, the number of points in the database.

In analogy to semantic security [21], we will say that a sanitization technique is secure if the adversary's probability of breaching privacy does not change significantly when it sees the sanitized database, even in the presence of auxiliary information (analogous to the history variable in the definition of semantic security). As noted below, auxiliary information is (provably) extremely difficult to cope with [14].

Histograms Preserve Privacy. A *histogram* for a database is a partition of $\mathbb{R}^d$ along with the exact counts of the number of database points present in each region. Histograms are prevalent in official statistics, and so they are a natural technique to consider for sanitization. We analyze a *recursive histogram sanitization*, in which the space is partitioned recursively into smaller regions (called cells) until no region contains $2t$ or more real database points. Exact counts of the numbers of points in each region are released. The intuition is that we reveal more detailed information in regions where points are more densely clustered. We prove a strong result on the privacy provided when the data are drawn uniformly from the d-dimensional unit cube, in the absence of auxiliary information. Generalizations are discussed in Remark 4 and Section 7.

Density-Based Perturbation Provides Utility. Section 5 describes a simple *input perturbation* technique, in which noise from a spherically symmetric distribution (such as a Gaussian) is added to database points. The magnitude of the noise added to a real database point is a function of the distance to the point's t-th nearest neighbor. The intuition for this sanitization is two-fold. On one hand, we are "blending a point in with its crowd," so privacy should be preserved. On the other hand, points in dense regions should be perturbed much less than points in sparse regions, and so the sanitization should allow one to recover a lot of useful information about the database, especially information about clusters and local density.

We formalize the intuition about utility via two results. First, we show a worst-case result: an algorithm that approximates the optimal clustering of the sanitized database to within a constant factor gives an algorithm that approximates the optimal clustering of the real database to within a constant factor. Second, we show a distributional result: if the data come from a mixture of Gaussians, then this mixture can be learned from the perturbed data. Our algorithmic and analysis techniques necessarily vary from previous work on learning mixtures of Gaussians, as the noise that we add to each point depends on the sampled data itself.

The intuition about privacy—namely, that this style of perturbation blends a point in with its crowd—is significantly harder to turn into a proof. We explain why privacy for this type of sanitization is tricky to reason about, and describe some simplistic settings in which partial results can be proven.

Privacy of Perturbed Data via Cross-Training. In Section 6 we describe a variant for which we can again prove privacy when the distribution is uniform over the d-dimensional unit cube. The idea is to use cross-training to get the desirable properties of histograms and spherical perturbations. The real database points are randomly partitioned into two sets, A and B. First, a recursive histogram sanitization is computed for set B. As stated above, this information can be released safely. Second, for each point in set A we add random spherical noise whose magnitude is a function of the histogram for B (it is based on the diameter of the B-histogram cell into which the point falls). We release the histogram for set B and the perturbed points from set A. Since the only information about the first set used for the perturbation is information provably safe to reveal, the privacy of the points in the first set is not compromised. An additional argument is used to prove privacy of the points in the second set. The intuition for the utility of the spherical perturbations is the same as before: points which lie in dense regions will lie in small histogram cells, and so will be perturbed little, thus preserving clusters approximately.

Open Questions. Our work suggests a rich set of fascinating questions, several of which we have begun exploring together with other researchers. We mention some of these in Section 7.

1.2 Related Work

We briefly highlight some techniques from the literature. Many additional references appear in the full paper (see the title page of this paper for the URL).

Suppression, Aggregation, and Perturbation of Contingency Tables. Much of the statistics literature is concerned with identifying and protecting sensitive entries in contingency tables (see, e.g., [12, 22]). For example, in the 2-dimensional case, for discrete data, these are frequency counts, or histograms, indicating how many entries in the database match each possible combination of the values of two particular attributes (e.g. number of cars and number of children). It is common to regard an entry as sensitive if it corresponds to a pair of attribute values that occurs at most a fixed number of times (typically 1 or 2) in the database. One reason for regarding low-count cells as sensitive is to prevent linkage with other databases: if a given pair of attribute values uniquely identifies an individual, then these fields can be used as a key in other databases to retrieve further information about the individual.

Input Perturbation. A second broad approach in the literature is to perturb the data (say via swapping attributes or adding random noise) before releasing the entire database, or some subset thereof (such raw, unaggregated entries are typically called *microdata*). Various techniques have been considered in the statistics [34, 35, 28] and data mining [4, 1, 16] communities. In some cases, privacy is measured by how (un)successfully existing software re-identifies individuals in the database from the sanitized data and a small amount of auxiliary information. In other cases, the notion of privacy fails to take into account precisely this kind of auxiliary information. The work of Efvimievski et al. [16] is a significant, encouraging exception, and is discussed below.

Imputation. A frequent suggestion is for the census bureau to learn $\mathcal{D}$ and then publish artificial data obtained by sampling from the learned distribution (see, e.g., [29]). We see two difficulties with this approach: (1) we want our sanitized database to reflect (possibly statistically insignificant) "facts on the ground". For example, if a municipality is deciding where to run a bus line, and a certain geographic region of the city has a higher density of elderly residents, it may make sense to run the bus line through this region. Note that such "blips" in populations can occur even when the underlying distribution is uniform; (2) any model necessarily eliminates information; we feel it is not reasonable to assume that the sanitizer can predict (privacy-respecting) statistical tests that may be invented in the future.

k-Anonymity, Input Aggregation, and Generalization. Similarly to input perturbation, one can also suppress or aggregate fields from individual records to reduce the amount of identifying information in the database. A database is said to be k-anonymized if every modified entry in the sanitized database is the same as at least k others [32]. The intuition is that privacy is protected in this way by guaranteeing that each released record will relate to at least k individuals. This requirement on the sanitization does not directly relate to what can and cannot be learned by the adversary. For example, the definition may permit information to be leaked by the choice of which records to aggregate (there are many aggregations that will make a database k-anonymous), or from the fact that certain combination of attribute values does not exist in the database. Information may also be gleaned based on the underlying distribution on data (for example, if the suppressed attribute is sex and the number of identical records with sex suppressed is two).

Interactive Solutions. In *query monitoring,* queries to an online database are audited to ensure that, even in the context of previous queries, the responses do not reveal sensitive information. This is sometimes computationally intractable [25], and may even fail to protect privacy, for example, in the setting in which the adversary knows even one real database record [11].

A related approach is *output perturbation,* in which a query control mechanism receives queries, computes exact answers, and then outputs a perturbed answer as the response to the query [9, 5]. This approach can sometimes be insecure, intuitively, because noise added in response to multiple queries can cancel out (see [2, 11]). The limitation can be shown to be inherent: if the number of queries to the database is large (even polynomial in the number of entries (rows) in the database), the amount of noise added to answers must be large [11]. By restricting the total number of queries allowed, one can in fact circumvent this and get strong privacy guarantees while adding much less noise [11, 15]. This approach is not available in our context: in an interactive solution, the query interceptor adds fresh noise to the response to each query; in our context, a noisy version of the database is constructed and published once and for all. Although this seems to make the problem more difficult, there are obvious advantages: the sanitization can be done off-line; the real data can be deleted or locked in a vault, and so may be less vulnerable to bribery of the database administrator.

A Recent Definitional Approach. The definitions of privacy in [11, 15, 16] (written concurrently with this work – see [13]) are consonant with our point of view, in that they provide a precise, meaningful, provable guarantee. All three follow the same paradigm: for every record in the database, the adversary's confidence in the values of the given record should not significantly increase as a result of interacting with or exposure to the database. The assumption is that the adversary can name individual records in the database; this captures the setting in which the database is multi-attribute, and the adversary has somehow, out of band, figured out enough about some individual to construct a query that effectively names the individual. Even in such a setting, it should be impossible to learn the value of even a single additional binary data field or the value of any predicate of the data tuple.

The work of Evfimievsky et al. [16] is in our model, ie, it describes a sanitization method (in this case, for transactions). Specifically, it is an input-perturbation technique, in which items are randomly deleted from and added to transactions. As we understand their work, both their specific technique and their definitions only consider applying the same, fixed, perturbation to each point in the database, independent of the other points in the database. Neither our definitions nor our techniques make this assumption. This both enhances utility and complicates privacy arguments.

Cryptographic Approaches. Much work in cryptography has focused on topics closely related to database privacy, such as private information retrieval and secure function evaluation (see, e.g., [18] and [20]). These problems are somewhat orthogonal to the one considered here. In secure function evaluation, privacy is preserved only to the extent possible given a specific functionality goal; but which functions are "safe" in the context of statistical databases? The literature is silent on this question.

2   A Formal Definition of Privacy

2.1   What do *We* Mean by "Privacy"?

As mentioned above, our notion of privacy breach is inspired by Gavison's writing on protection from being brought to the attention of others. This phrase articulates the common intuition that our privacy is protected to the extent that we blend in with the crowd. To convert this intuition into a precise mathematical statement we must abstract the concept of a database, formulate an adversary (by specifying the information to which it has access and its functionality), and define what it means for the adversary to succeed.

2.2   Translation into Mathematics

Under our abstraction of the database privacy problem, the real database (RDB) consists of n unlabeled points in high dimensional space $\mathbb{R}^d$, each drawn independently from an underlying distribution $\mathcal{D}$. Intuitively, one is one's collection of attributes. The census bureau publishes a sanitized Database (SDB), containing some n' points, possibly in a different space. This is a very general paradigm; it covers the case in which the SDB contains only summary information.

To specify security of a cryptographic primitive we must specify the power of the adversary and what it means to break the system. Since the goal of our adversary is to "single out" a record in the database, we call the adversary an *isolator*. The isolator takes two inputs – the sanitized database and auxiliary information (the auxiliary information is analogous to the history variable in the definition of semantic security). The isolator outputs a single point $q \in \mathbb{R}^d$. This completes the description of the functionality of the adversary. Note that the definition admits adversaries of unbounded computational power. Our results require no complexity-theoretic assumptions[1].

We next define the conditions under which the adversary is considered to have succeeded in isolating. The definition is parameterized with two values: a *privacy threshold* t, intuitively, the size of the "crowd" with which one is supposed to blend in, and an *isolation parameter* c, whose use will be clear in a moment. Roughly, c helps to formalize "blending in".

For a given isolating adversary $\mathcal{I}$, sanitized database SDB, and auxiliary input z, let $q = \mathcal{I}(\text{SDB}, z)$ ($\mathcal{I}$ may be a randomized algorithm). Let δ be the distance from q to the nearest real database point, and let x be an RDB point at distance δ from q. Let $B(p, r)$ denote a ball of radius r around point p. If the d-dimensional ball of radius $c\delta$ and centered at q contains at least t real database points, that is, if $|RDB \cap B(q, c\delta)| \geq t$, then the adversary *fails* to isolate x. Otherwise, the adversary succeeds.

We will give a slightly more general definition shortly. First we give some intuition for the definition. The adversary's goal is to single out someone (*i.e.*, some RDB point) from the crowd, formalized by producing a point that is much closer to some $x \in$ RDB than to $t-1$ other points in the RDB. The most likely victim x is the RDB point closest to q. So q "looks something like" x. On the other hand, if $B(q, c\delta)$ contains at least t RDB points, then q also looks almost as similar to lots of (*i.e.*, $t-1$) other RDB points, so x hasn't really been singled out.

Note that the definition of isolation is a relative one; the distance requirement for success varies according to the local density of real database points. This makes sense: if we name an intersection in New York City, there are perhaps a few hundred people living at or very near the intersection, so our point is "close to" many points (people) in the RDB. In contrast, if we name an intersection in Palo Alto, there are perhaps 10 people living near the intersection[2]. More generally, we have the following definition:

Definition 1. $((c,t)$-**isolation**$)$ *Let y be any RDB point, and let $\delta_y = \|q - y\|$. We say that q (c,t)-isolates y if $B(q, c\delta_y)$ contains fewer than t points in the RDB, that is, $|B(q, c\delta_y) \cap \text{RDB}| < t$.*

We frequently omit explicit mention of t, and speak of c-isolation. It is an easy consequence of the definitions that if $q = \mathcal{I}(\text{SDB}, z)$ fails to c-isolate the nearest RDB point to q, then it fails to c-isolate even one RDB point.

For any point p (not necessarily in the RDB), we let τ_p be the minimum radius so that $B(p, \tau_p)$ contains t RDB points. We call this the *t-radius of p*.

---

[1] We do not object to using complexity-theoretic assumptions. We simply have not yet had a need to employ them.
[2] This analogy was suggested by Helen Nissenbaum.

3   Definition of Sanitization

Suppose one of us publishes all our information on the web — that is, we publish our RDB point x where the adversary can find it — so that the point is part of the adversary's auxiliary information. Clearly, the adversary can isolate x by setting $q = x$ (in which case $\delta_x = 0$ and $B(q, c\delta_x)$ contains only x — we assume no two points are identical). It seems unfair to blame the sanitizing procedure for this isolation; indeed, there is an adversary simulator that, without access to the sanitized database, can also isolate x, since x is part of the auxiliary information. We are therefore concerned with how much seeing the sanitized database helps the adversary to succeed at isolating even one RDB point. Intuitively, we do not want that seeing the SDB should help "too much". Our notion of "too much" is fairly relaxed. Letting ε denote the probability that isolation may occur, we tend to think in terms of, say, $\varepsilon = 1/1000$. This says that about one out of every 1,000 sanitized databases created may be vulnerable to an isolation event[3]. The parameter ε can be a function of d and n. Note, however, that ε cannot depend only on n – otherwise privacy could be improved simply by the introduction of additional points.

More formally, a *database sanitizer*, or simply *sanitizer* for short, is a randomized algorithm that takes as input a real database of some number n of points in $\mathbb{R}^d$, and outputs a sanitized database of some number n' of points, in a possibly different space $\mathbb{R}^{d'}$.

A sanitizer is *perfect* if for every distribution $\mathcal{D}$ over $\mathbb{R}^{n \times d}$ from which the real database, for all isolating adversaries $\mathcal{I}$, and points are drawn, there exists an adversary simulator $\mathcal{I}'$ such that with high probability over choice of RDB, for all auxiliary information strings z, the probability that $\mathcal{I}(\text{SDB}, z)$ succeeds minus the probability that $\mathcal{I}'(z)$ succeeds is small. The probabilities are over the coin tosses of the sanitization and isolation algorithms. We allow the sanitizer to depend on the parameters c, t, and also allow $\mathcal{I}, \mathcal{I}'$ to have access to $\mathcal{D}$.

More precisely, let ε be a parameter (for example, $\varepsilon = 2^{-d/2}$). We require that for all $\mathcal{I}$ there exists an $\mathcal{I}'$ such that, if we first pick a real database RDB $\in_R \mathcal{D}$, then with overwhelming probability over RDB, for all z,

$$\forall S \subseteq \text{RDB} \, |Pr[\exists x \in S : I(SDB, z) \text{ isolates } x] - Pr[\exists x \in S : I'(z) \text{ isolates } x]| < \varepsilon$$

where the probabilities are over the choices made by $\mathcal{I}, \mathcal{I}'$, and the sanitization algorithm.

Remark 1. The use of the sets S in the definition is for the following reason. Suppose the adversary knows, as part of its auxiliary information, some point $y \in$ RDB. For every $x \neq y \in$ RDB, we want x's chances of being isolated to remain more or less unchanged when the adversary is given access to the sanitized database. Thus, if we were to write the more natural

---

[3] The adversary has no oracle to tell it when it has succeeded in isolating an RDB point.

$|\Pr[\exists x \; \mathcal{I}(\text{SDB}, z) \text{ isolates } x] - \Pr[\exists x \; \mathcal{I}'(z) \text{ isolates } x]| < \varepsilon$, then we might have $z = y \in \text{RDB}$, $\mathcal{I}'(z) = y$, and $\mathcal{I}(\text{SDB}, z)$ could (somehow) isolate a different point $x \neq y \in \text{RDB}$ with probability one. This is clearly unsatisfactory.

This is excessively ambitious, and in fact a nontrivial perfect sanitizer does not exist [14]. However, by specifying the ideal we can begin to articulate the value of specific sanitization techniques. For a given technique, we can ask what can be proved about the types of distributions and auxiliary information for which it can ensure privacy, and we can compare different techniques according to these properties.

4 Histograms

Consider some partition of our space $\mathbb{R}^d$ into disjoint cells. The histogram for a dataset RDB is a list describing how many points from the dataset fall into each cell. A sanitization procedure is *histogram-based* if it first computes the histogram for the dataset and bases the output only on that information.

For example, in one dimension the cells would typically be sub-intervals of the line, and the histogram would describe how many numbers from the dataset fall into each of the intervals. In higher dimensions, the possibilities are more varied. The simplest partition divides space into cubes of some fixed side-length (say at most 1). That is, each cell is a cube $[a_1, a_1 + 1] \times \cdots \times [a_d, a_d + 1]$ for integers $a_1, ..., a_d$.

Our principal result for histograms is that, if the original data set RDB consists of n points drawn independently and uniformly from some large cube $[-1, 1]^d$, then the following sanitization procedure preserves privacy:

> **Recursive Histogram Sanitization:** Divide the cube into 2^d equal-sized subcubes in the natural way, *i.e.*, by bisecting each side at the midpoint. Then, as long as there exists a subcube with at least $2t$ points from RDB, further subdivide it into 2^d equal-sized cubes. Continue until all subcubes have fewer than $2t$ points, and release the exact number of points in each subcube.

Theorem 1. *Suppose that* RDB *consists of n points drawn i.i.d. and uniformly from the cube $[-1, 1]^d$. There exists a constant c_{secure} (given in Lemma 2) such that the probability that an adversary, given a recursive histogram sanitization as described above, can c_{secure}-isolate an* RDB *point is at most $2^{-\Omega(d)}$.*

The proof is quite robust (in particular, the error bound does not depend on n). Using the techniques developed in this section, several variants on the partitioning described above can be shown to preserve privacy, even under less rigid assumptions about the underlying distribution. Thus, our goals are to prove privacy of histogram sanitizations, to illustrate techniques that are useful for proving privacy, and to establish a result which we will need when we deal with cross-training-based sanitizers later on.

The technical heart of the proof of Theorem 1 is following proposition:

Proposition 2. Suppose that the adversary knows only that the dataset consists of n points drawn i.i.d. from the cube $[-1,1]^d$, and that we release the exact histogram (cell counts) for the natural partition of this cube into 2^d subcubes of side-length 1. Then the probability that the adversary succeeds at c-isolating a point for $c > 121$ is at most $2^{-\Omega(d)}$, as long as $t = 2^{o(d)}$.

The constant 121 in the proposition can in fact be improved significantly, to approximately 30, with minor changes to the proofs in this section.

This result is strong—it essentially states that for any point q which the adversary might produce *after* seeing the histogram, the distance to q's nearest neighbor is at most a constant less than the distance between q and its $2^{o(d)}$-th nearest neighbor. When $n = 2^{o(d)}$, the result is perhaps less surprising: the distance between q and its nearest neighbor is $\Omega(\sqrt{d})$ with high probability, and $2\sqrt{d}$ is an upper bound on the distance from q to its farthest neighbor (assuming q is in the large cube $[-1,1]^d$). For very large values of n (say $2^{\Omega(d)}$), the proof becomes much more involved.

Remark 2. We would like to understand the probability that the adversary isolates a point after seeing the sanitization, given reasonable assumptions about the adversary's a priori view of the database. Currently we assume that the underlying distribution is uniform on a d-dimensional hypercube. The following example shows that such a "smoothness" condition is necessary to obtain a bound on the adversary's probability of success, when a histogram of the data is released.

Consider the following distribution. In each of the 2^d subcubes of the hypercube, there is an infinite sequence of points $p_1, p_2, \ldots$. The probability density at point p_i is $\frac{1}{2^d}\frac{1}{2^i}$. That is, each subcube has equal mass, and within a subcube, mass is distributed over the infinite sequence of points in an exponentially decreasing manner. Now, if the adversary knows the number of points in a subcube, say m, then she produces the point $q = p_{\log m}$ in that subcube. With a constant probability, there are at least one, but no more than t, points at q, and the adversary succeeds. On the other hand, without knowledge of the number of points in each subcube (as given by the histogram), the adversary simulator I' has an exponentially low probability of succeeding.

The next subsections sketch the proof of Proposition 2. The full version of the paper contains the details of the proof, as well as extensions to cover finer subdivisions of the cube and the recursive sanitization described above.

4.1 Simplifying the Adversary

We distinguish two related definitions of isolation. The adversary is always given the sanitized database SDB as input (the adversary may also receive side information about the real database—typically, in our setting, the distribution from which the points in the real database are drawn).

- A **ball adversary** produces a pair (q, r) where $q \in \mathbb{R}^n$ is a point in space and $r \in \mathbb{R}^+$ is a non-negative real number. The adversary succeeds if $B(q, r)$,

the ball of radius r centered at q, contains at least one point in RDB, but $B(q, cr)$ contains fewer than t points in RDB (equivalently, $r < \tau_q/c$).
- A **point adversary** produces only a point $q \in \mathbb{R}^n$. The adversary succeeds at c-isolation if there is a point in D within distance τ_q/c of q, i.e. if there exists some r for which the corresponding ball adversary would have won.

We first prove Proposition 2 for ball adversaries since their behavior is easier to understand. At the end of the proof, we show that point adversaries do essentially no better than ball adversaries in this context, and so the restriction to ball adversaries is made without loss of generality.

4.2 Proof Sketch for Proposition 2

We sketch here the proof of Proposition 2. Recall that the points in the real database RDB are drawn uniformly from the large cube $[-1, 1]^d$, and the sanitization consists of the number of points from RDB contained in each of the cubes obtained by dividing $[-1, 1]^d$ once along each dimension. That is, a cell C is a d-dimensional hypercube of side-length and volume 1, which has one vertex at $(0, 0, ..., 0)$ and the opposite vertex in the set $\{-1, +1\}^d$. The total number of points in the database is n, and we denote the number of points appearing in a cell C by n_C. The sanitization is simply the list of all 2^d values n_C.

Define a function $f : \mathbb{R}^n \to \mathbb{R}^+$ which captures the adversary's view of the data.

$$f(x) = \frac{n_C}{n} \cdot \frac{1}{\text{Vol}(C)} \quad \text{for } x \in C. \tag{1}$$

The function f is a probability density function. The adversary does not see the data as being drawn i.i.d. according to f, but the function is useful nonetheless for bounding the adversary's probability of success.

Lemma 1. *If a ball adversary succeeds at c-isolation with probability ε, then there exists a pair (q, r) such that $\Pr_f[B(q, r)] \geq \varepsilon/n$ and $\Pr_f[B(q, cr)] \leq (2t + 8\log(1/\varepsilon))/n$.* [4]

The intuition for the proof of Lemma 1 is simple: it is sufficient to show that if one considers only the number of points landing in a particular region, there is almost no difference between the adversary's view of RDB and a set of n points sampled i.i.d. from f.

Roughly, Lemma 1 means that for a ball adversary to succeed, a necessary (but not sufficient) condition is that:

$$\frac{\Pr_f[B(q, r)]}{\Pr_f[B(q, cr)]} \geq \frac{\varepsilon/n}{(2t + 8\log(1/\varepsilon))/n} = \varepsilon/(2t + 8\log(1/\varepsilon)). \tag{2}$$

---

[4] We are assuming that n is at least $2t + 8\log(1/\varepsilon)$. The assumption is not necessary, but simplifies the proofs. In fact, when n is small one can use completely different proofs from the ones described here which are much simpler.

This means that it is sufficient to bound the ratio on the left-hand side above by some negligible quantity to prove privacy against ball adversaries (in fact, the upper bound need only hold as long as $\Pr_f[B(q,r)]$ is itself not too large). The better the bound on the ratio in Eqn. (2), the better the result on privacy. To get the parameters described in the statement of Proposition 2, it will be sufficient to prove a bound of $2^{-\Omega(d)}$ for the ratio: we can then think of ε as $2^{-\gamma d}$, for a constant $\gamma < 1$, and think of t as being $2^{-o(d)}$.

The upper bound, and the proof of Proposition 2, rely on the following lemma:

Lemma 2. *There is a constant $1/60 < \beta < 1$ such that, for any point q, radius $r > 0$ and cell C for which $B(q,r) \cap C \neq 0$, we have:*

1. *If $r \leq \beta\sqrt{d}$, then*

$$\frac{\mathsf{Vol}(B(q,r) \cap C)}{\mathsf{Vol}(B(q,3r) \cap C)} \leq 2^{-\Omega(d)}. \tag{3}$$

2. *If $r > \beta\sqrt{d}$, then for all cells C' neighboring C:*

$$C' \subseteq B(q, c_{\text{secure}} r) \tag{4}$$

where $c_{\text{secure}} \leq (2/\beta)+1$. A neighboring cell is a cube of the same side-length which shares at least one vertex with C.

Remark 3. The value of the constant β can be made much larger, at least $1/15$. Obtaining this bound requires more careful versions of the proofs below.

A detailed proof is in the full version. Roughly, to prove Part 1, we need to estimate the probability that a point chosen from a ball lies is a cube. To this end, we approximate sampling from a ball by sampling from an appropriately chosen spherical Gaussian. This allows us to analyze behavior one coordinate at a time. Our (rather involved) analysis only holds for radii $r \leq \beta\sqrt{d}$. It is possible that a different analysis would yield a better bound on β and hence on β.

Part 2 is much simpler; it follows from the fact that the diameter of the cube $[-1,1]^d$ is $2\sqrt{d}$.

We can now prove Proposition 2, which states that releasing the histogram preserves privacy, with the adversary's success probability bounded by $2^{-\Omega(d)}$. We first give a proof for ball adversaries, and then observe that (almost) the same proof works for point adversaries too.

Proof (of Proposition 2). Ball adversaries: Assume there is a ball adversary who chooses the best possible pair (q,r) based on SDB.

First, suppose that $r \leq \beta\sqrt{d}$ (the constant is from Lemma 2). In that case, we will actually show that 3-isolation (not 121-isolation!) is possible only with very small probability. Our proof relies on Part 1 of the lemma. We can write the mass of $B(q,r)$ under f as a weighted sum of the volume of its intersections with all the possible cubes of C:

$$\Pr_f[B(q,r)] = \sum_C \frac{n_C}{n} \cdot \mathsf{Vol}(B(q,r) \cap C)$$

We can bound each of these intersections as an exponentially small fraction of the mass of $B(q, 3r)$:

$$\Pr_f[B(q,r)] \leq \sum_C \frac{n_C}{n} \cdot 2^{-\Omega(d)} \cdot \mathsf{Vol}(B(q,3r) \cap C) = 2^{-\Omega(d)} \cdot \Pr_f[B(q,3r)]$$

Now the mass of $B(q, 3r)$ is at most 1, which means that the ratio in Eqn. (2) is at most $2^{-\Omega(d)}$, and so $\varepsilon/(2t + 8\log(1/\varepsilon)) \leq 2^{-\Omega(d)}$. This is satisfied by $\varepsilon = 2^{-\Omega(d)}$ (for essentially the same constant in the Ω-notation), and so in this case, 3-isolating the point q is possible only with probability $2^{-\Omega(d)}$.

Now consider the case where $r > \beta\sqrt{d}$. If $B(q, r)$ doesn't intersect any cells C, then we are done since the ball captures a point with probability 0. If there is a cell C which intersects $B(q, r)$, then, by Part 2 of Lemma 2, $B(q, c_{\mathsf{secure}}r)$ (for $c_{\mathsf{secure}} \leq (2/\beta) + 1$) contains all the cells C' which are neighbors to C, in particular all of $[-1, 1]^d$. (Recall that our points are initially uniform in $[-1, 1]^d$, and we consider a subdivision which splits the cube into 2^d axis-parallel subcubes of equal size). The adversary succeeds with probability zero at c_{secure}-isolation, since all n points will be within distance $c_{\mathsf{secure}}r$.

Point adversaries: Suppose the adversary outputs a particular point q, and let r be the smallest radius such that $\Pr_f[B(q, r)] = \varepsilon$. By the previous discussion, $B(q, c_{\mathsf{secure}}r)$ contains mass at least $(2t + 8\log(1/\varepsilon))/n$. Thus, with probability at least $1 - 2\varepsilon$, there is no point inside $B(q, r)$ and there are t points inside $B(q, c_{\mathsf{secure}}r)$ (by the proof of Lemma 1). The ratio between the distances to the t-th nearest point and to the nearest point to q is then at most $c_{\mathsf{secure}}r/r = c_{\mathsf{secure}}$. The point adversary succeeds at c_{secure}-isolating a point with probability at most 2ε.

Because $\beta > 1/60$, the constant c_{secure} is at most 121, and so the adversary fails to 121-isolate any point in the database. □

Remark 4. The proof technique of this section is very powerful and extends in a number of natural ways. For example, it holds even if the adversary knows an arbitrary number of the points in the real database, or (with a worse isolation constant), if the adversary knows a constant fraction of the attributes of a database point. The analysis holds if the underlying distribution is a mixture of (sufficiently separated) hypercubes.

Recent work also indicates that histogram sanitization, at least to a limited depth of recursion, can be constructed for "round" distributions such as the sphere or the ball [6]. Together, these techniques yield privacy for sufficiently separated mixtures of round and square distributions.

5 "Round" Perturbation Sanitizations

Perturbation via additive noise is a common technique in the disclosure control literature. In this section, we present a variant on this technique in which the magnitude of the noise added to a point depends on the local density of the database near the point. We consider three perturbation sanitizers that are very similar when the dimension d is large. In these sanitizers, $d' = d$ and $n' = n$ (that is, the sanitized database consists of points in the same space as the real database, and the numbers of points in the real and sanitized databases are identical). As before, let $B(p,r)$ denote the ball of radius r around p, let $S(p,r)$ denote the corresponding sphere, or the surface of $B(p,r)$. Let $\mathcal{N}(\mu, \sigma^2)$ denote a d-dimensional Gaussian with mean μ and variance σ^2 in every dimension. For $x \in \mathbb{R}^d$, the t-radius τ_x is the minimum radius such that $B(x, \tau_x)$ contains t RDB points (x need not be an RDB point.)

1. **The Ball Sanitizer:** For $x \in RDB$, $\text{BSan}(x, RDB) \in_R B(x, \tau_x)$.
2. **The Sphere Sanitizer:** For $x \in RDB$, $\text{SSan}(x, RDB) \in_R S(x, \tau_x)$.
3. **The Gaussian Sanitizer:** For $x \in RDB$, $\text{GSan}(x, RDB) \in_R \mathcal{N}(x, \tau_x^2/d)$.

We will refer to these as *round*, or *spherical* sanitizers, because of the shape of the noise distribution. The intuition for these sanitizations is three-fold: we are blending a point in with a crowd of size t, so privacy should be preserved; points in dense regions should be perturbed much less than points in sparse regions, and so the sanitization should allow one to recover a lot of useful information about the database, especially information about clusters and local density; we are added noise with mean zero, so data means should be preserved.

Round sanitizations have been studied before, typically with independent, identically distributed noise added to each point in the database. This approach implicitly assumes that the density of the data is more or less uniform in space for the entire data set. Even with data drawn i.i.d. from a uniform distribution on a fixed region, this need not be the case. Indeed, Roque [28] showed that (in low dimensions) re-identification software defeats this i.i.d. spherical noise, though the standard packages fail if the noise is not spherical (say, drawn from from a mixture of Gaussians). Kargupta et al. [24] argue that independent additive perturbation may have limited application to preserving privacy insofar as certain informative features of the data set (e.g.: the principal components) are largely unaffected by such perturbations. Their argument assumes (critically) that the sanitizer applies a fixed distribution to each element, and ultimately describes how to reconstruct the covariance matrix of the *attributes*. In this work, we apply data-driven distributions to the elements, and prove privacy guarantees for the *individuals*. Moreover, we conjecture it is possible to exploit what Kargupta et al. perceive as a weakness (reconstructing the covariance matrix, which we can do), while provably maintaining privacy (which we conjecture)[5]. Finally, the data-dependent noise distribution provides more potential functionality than a fixed noise distribution [4, 1, 16], at the cost of a more difficult analysis.

---

[5] Specifically, given recent results in constructing histograms for round distributions [6], we conjecture it will be possible to obtain cross-training results for mixtures of Gaussians, analogous to our cross-training results for the hypercube described in Section 6 below.

5.1 Results for Round Sanitizers

We have obtained several results on the privacy and utility of round sanitizations. Our most powerful result is concerned with learning a mixture of Gaussians from the sanitized data. This result is of independent interest, and is described in the Section 5.2. We first summarize our results.

Utility. The task of extracting information from a database whose points have been spherically perturbed is essentially one of learning from noisy data. Standard techniques do *not* apply here, since the noise distribution actually depends on the data. Nonetheless, we prove two results using the intuition that round perturbations preserve expectations (on average) and that our particular strategy is suited to clustering.

1. When the data are drawn uniformly from a mixture of Gaussians $\mathcal{D}$, there is an efficient algorithm that learns $\mathcal{D}$ from the Gaussian sanitization. Learning mixtures of Gaussians has already been heavily investigated [3, 8, 33], however existing analyses do not apply in our setting. The algorithm and its analysis are sketched in Section 5.2.
2. For any distribution, suppose we are given an algorithm to find k clusters, each of cardinality at least t, minimizing the maximum diameter of a cluster, and assume the data are sanitized with either BSan or SSan. Then running the algorithm on the sanitized data does a good job of clustering the original data. More precisely, any algorithm that approximates the optimal clustering of the sanitized database to within a constant factor gives an algorithm that approximates the optimal clustering of the real database to within a constant factor, and the maximum diameter of a cluster exceeds the maximum diameter of an optimal k-clustering on the RDB by at most a factor of 3.

Privacy. The intuition for privacy is significantly harder to turn into a complete proof than is the one for utility. We analyze two special cases, and give a lower bound showing that high-dimensionality is necessary for the privacy of this type of sanitization. The proofs of the results below appear in the full version of the paper.

1. The database consists of only two points, x and y, which are sanitized with respect to each other, and the underlying distribution is the unit sphere in d dimensions. That is, $t = 2$ and each of x and y is perturbed using SSan with perturbation radius $\|x - y\|$. The adversary is given $\|x - y\|$, and the sanitizations x' and y'. We show that the probability of 4-isolation is exponentially small in d, with overwhelming probability over the choice of x and y. The proof is by symmetry: we construct many pairwise-distant "decoy" pairs x', y' which are equiprobable in the adversary's view.
2. The real database consists of n sanitized points, drawn from the d-dimensional unit sphere. The adversary is given all but one point in the clear, together with a sanitization of the final point using SSan. The adversary's goal is to 4-isolate the last point. Intuitively, privacy holds because the

hidden point can lie in any direction from its sanitization, while any point q produced by the adversary can only isolate points lying in an exponentially small fraction of these directions. The result is proved for $t = 2$.

3. Sanitization cannot be made arbitrarily safe: for any distribution, if sanitization is done using BSan, then there is a polynomial time adversary $\mathcal{I}$ requiring no auxiliary information, such that the probability that the adversary succeeds is $\Omega(\exp(-d)/\log(n/t))$.

Remark 5. The second result above highlights the delicacy of proving privacy for this type of sanitization. Contrary to intuition, it is *not* the case that seeing the sanitizations of the remaining $n-1$ points, rather than their exact values, gives less information. The reason is that the sanitization of y, implicitly contains information about the t-neighborhood of y. This sort of dependency is notoriously hard to deal with in cryptography, e.g. in the *selective decommitment problem*. We have not yet proved or disproved the viability of the above-mentioned sanitizations; instead we circumvent the difficulty via cross-training.

5.2 Learning Mixtures of Gaussians

In this section we look at an algorithm for mining sanitized data. We address the well-studied problem of learning a mixture of Gaussians, with the twist that the samples have been sanitized using one of the round sanitizers discussed above. The distribution that results from the sanitization is no longer a mixture of Gaussians (samples are not even independent!) and traditional algorithms for learning mixtures of Gaussians do not apply. Nonetheless, we will see that the core properties that make Gaussian mixtures learnable remain intact, and prove that the mixtures can be read from an optimal low rank approximation.

We assume there are k mean vectors μ_i, each with an associated mixing weight w_i. Let $w_{\min}$ denote the minimum of the mixing weights. Each point in the data set is independently produced by selecting a μ_i with probability proportional to the w_i, and applying independent, normally distributed noise to each coordinate. We assume that the Gaussians are spherical, in that every Gaussian has an associated variance that is used for each of the coordinates. Let σ_1^2 denote the maximum such variance. We assume that the sanitization process amounts to applying an additive perturbation established by choosing a point uniformly at random from the unit sphere, which is then scaled in length by a random variable at most $2\sigma_2\sqrt{d}$, where σ_2 may *depend on the sampled data and t*. Notice that this is sufficiently general to capture all the perturbation based sanitizations described above – SSan, BSan, and GSan – the latter two using a random variable for the scaling factor.

For the purposes of analysis, we assume that we have access to two data sets $\overline{A}$ and $\overline{B}$ that have been independently sanitized. Each is assumed to result from the same underlying set of means, and to be independent of the other. We use $\overline{A}_u$ to denote the sanitized vector associated with u, $\widehat{A}_u$ the unsanitized vector associated with u, and A_u the original mean vector associated with u. We form the matrices $\overline{A}$, $\widehat{A}$, and A by collecting these columns for each point in the data set. Let w_u denote the mixing weight associated with the Gaussian with mean A_u. The matrices $\overline{B}, \widehat{B}$, and B and their associated columns are analogous, though they represent an independently sanitized disjoint data set. While this setting is not difficult for the sanitization process to accommodate, the motivation is simply for the clarity of analysis and it is unclear whether disjoint data sets are necessary in practice.

The main linear algebraic tool that we use is a matrix's *optimal rank k projection*. For every matrix M, this is a projection matrix P_M, such that for all rank k matrices D, we have $\|M - P_M M\|_2 \leq \|M - D\|_2$. Computing the optimal projection is not difficult; in most cases it is an $O(dn \log(dn))$ operation. We also make use of the *single linkage* clustering algorithm [30]. For our purposes, given a collection of points, single linkage repeatedly inserts an edge between the closest pair of non-adjacent points until the resulting graph has k connected components. Our actual algorithm can be stated succinctly:

Cluster($\overline{A}, \overline{B}, k$)
1. Compute $P_{\overline{A}}$ and $P_{\overline{B}}$, and form $C = [P_{\overline{B}}\overline{A} | P_{\overline{A}}\overline{B}]$.
2. Apply single linkage to C, forming k clusters.

Cluster takes a pair of data sets, and uses the structure define by each data set to filter the noise from the points in the other. If the mean vectors μ_i have sufficient separation, all inter-cluster distances in C will exceed all intra-cluster distances in C, and single linkage will associate exactly those points drawn from the same mean.

Theorem 3. *Assume that $d < n/2$. If for each pair of means μ_i, μ_j,*

$$\|\mu_i - \mu_j\| \geq 4(\sigma_1 + \sigma_2)(16/w_{\min}^{1/2} + \sqrt{k \log(kn/\delta)})$$

then with probability $1 - 2(e^{-nw_{\min}/8} + 2^{-\log^6 n/2} + \delta)$, Cluster partitions the columns of $\overline{A}$ and $\overline{B}$ according to the underlying Gaussian mixture.

Proof. The proof is conducted by bounding $\|\mu_u - C_u\|$ for each u. Assume, without loss of generality, that $\mu_u = A_u$ and $C_u = P_{\overline{B}}\overline{A}_u$. Notice that

$$\|\mu_u - C_u\| = \|A_u - P_{\overline{B}}\overline{A}_u\| \leq \|A_u - P_{\overline{B}}A_u\| + \|P_{\overline{B}}A_u - P_{\overline{B}}\overline{A}_u\|$$

In Lemmas 3 and 4 below, we bound these two terms, so that their sum is at most $1/4$ the assumed separation of the mean vectors. With such separation, all inter-cluster distances are at least $1/2$ the mean separation, and all intra-cluster distances are at most $1/2$ the mean separation.

Although the above result requires a uniform bound on the pairwise separation between the means of the clusters, by using a more sophisticated clustering algorithm than Single-Linkage on the low-dimensional projection of the data, we can improve the results such that the requisite separation between a pair of means depends only on the variances of the corresponding Gaussians.

Lemma 3 (Systematic Error). *Assume $d < n/2$. With probability at least $1 - (e^{-nw_{\min}/8} + 2^{-\log^6 n/2})$, for all u*

$$\|(I - P_{\overline{B}})A_u\| \leq 16(\sigma_1 + \sigma_2)/w_u^{1/2}$$

Proof. (Sketch) In previous work using cross training techniques, [26] shows that

$$\|(I - P_{\overline{B}})A_u\| \leq 4\|B - \overline{B}\|_2/nw_u$$

To continue, [26] uses a result of Furedi and Komlos [17], which places a high probability bound on the norm of zero mean random matrices with independent entries of bounded variance. While $B - \overline{B}$ does not have independent entries – the sanitization process depends on the result of the random events producing $\widehat{B}$ – each of the matrices $B - \widehat{B}$ and $\widehat{B} - \overline{B}$ do. As such, we get that with probability at least $1 - 2^{-\log^6 n/2}$

$$\|B - \overline{B}\|_2 \leq \|B - \widehat{B}\|_2 + \|\widehat{B} - \overline{B}\|_2 \leq 4(\sigma_1 + \sigma_2)\sqrt{n}$$

Lemma 4 (Random Error). *For each u,*

$$Pr[\|P_{\overline{B}}(A_u - \overline{A}_u)\| > c(\sigma_1 + \sigma_2)\sqrt{k}] \leq 2ke^{-c^2}$$

6 Combining Histograms and Perturbations: Cross-Training

We are drawn to spherical sanitizations because of their apparent utility (see the discussion in Section 5.1). However, as noted in Remark 5, we have some concerns regarding the privacy offered: it is not the privacy of the perturbed points that concerns us, but the privacy of the points in the t-neighborhood of the perturbed points (since the sanitization radius itself leaks information about these points). In this section, we combine a histogram-based sanitization with a spherical sanitization to obtain a provably private spherical sanitization for $n = 2^{o(d)}$ points, (again in the absence of auxiliary information).

We randomly divide the dataset into two sets — A and B[6]. We construct a recursive histogram on B (as in Section 4). We then sanitize points in A using only their position in the histogram on B. We release the sanitizations of points in A, along with the exact count, for every cell in the histogram, of points in A and B lying in that cell. We also assume that the adversary knows for every sanitized point $v' \in SDB$, the cell in the histogram that its pre-image $v \in A$ lies in (this only helps the adversary).

For a point $v \in A$, let C be the cell containing v in the recursive histogram constructed for B. Let $P(C)$ be the parent cell of C. $P(C)$ has twice the side-length of C, and contains at least t points. Consider the following sanitization procedure:

---

[6] In fact, our proof only requires that A contain at most $2^{o(d)}$ points.

Cross-Training Round Sanitization: Let ρ_v be the side-length of the cube C. Select a point N_v at random from a spherical Gaussian which has variance ρ_v^2 in each coordinate. Output $v' = v + N_v$.

As shown above, this procedure protects the privacy of points in B since the information released about these points depends only on the recursive histogram of the set B. In this section we prove that it also protects the privacy of points in A, under the assumption that from the adversary's point of view, the *a priori* distribution of points in the database is uniform when restricted to the cell C.

Consider a particular point $v \in A$. Suppose the side-length of the cell C containing v is ρ_v. Lemma 5 below shows that with probability $1 - 2^{-\Omega(d)}$ over the choice of RDB and the coins of the sanitization algorithm, the following occurs: for any point q which the adversary might produce, the distance between q and v will be $\Omega(\rho_v\sqrt{d})$. Since the t-radius of v is $O(\rho_v\sqrt{d})$, this implies that adversary c-isolates v with probability at most $2^{-\Omega(d)}$ (for some constant c).

The result is quite useful. If A contains $2^{o(d)}$ points, then a union bound shows that with probability at least $1 - 2^{-(\Omega(d)-o(d))}$, the sanitization is "good": that is, the adversary can succeed at isolating some point with probability at most $2^{-\Omega(d)}$.

Below, we state the main lemma of this section; the proof, omitted for lack of space, is in the full version.

Lemma 5. *Suppose that v is uniformly distributed in the cube C, and q is the output of the adversary on input $v' = v + N_v$. There is a constant $\alpha < 9$ such that with probability $1 - 2^{-\Omega(d)}$, the ball $B(q, (\alpha+1)\|v-q\|)$ contains the entire parent cell $P(C)$. The probability is over choice of the real database RDB and the coins of the sanitization algorithm.*

7 Future Work

Isolation in few dimensions. Many have raised the case in which the adversary, studying the sanitized data, chooses a small set of attributes and outputs values that uniquely identify a point in the RDB (no other point in the RDB agrees well with the given point on this particular set of attributes). This may not be a privacy breach as we have defined it, since the adversary may have very bad luck at guessing the remaining attribute values, and therefore the point q that the adversary produces may not be particularly close to any point in the RDB. However, as M. Sudan has pointed out, the adversary may know the difference between the attributes on which it is guessing and the ones it has learned from the sanitized data.

We are uncertain exactly what it means for the adversary to "know" this difference. Our notion of privacy breach essentially says we don't care about such things: after all, releasing a histogram cell count of 1 says there is a unique individual in a certain subcube, but we prove that the advesary cannot isolate this individual. However, the question is provocative.

The attack corresponds imprecisely to identification of a short key for a population unique (see the discussion in Section 1.2). Alternatively, the adversary may know a key to a population unique and the worry is that the sanitization may permit the learning of additional attribute values. On the one hand, we note that our definition of a perfect sanitization precludes either of these possibilities: roughly speaking, if it were possible to learn the key to a population unique then there is a choice for the auxiliary information that would permit the remaining attribute values to be learned, which would constitute an isolation. On the other hand, we have already noted that perfect sanitizers cannot exist [14], and our privacy results have been proved, for the most part, without permitting the adversary auxiliary information.

With this in mind, one may extend the definition of isolation to allow the adversary to approximate a real point in only a few attributes. Note however, that as the number of attributes estimated by the adversary decreases, the notion of isolation must become more and more stringent. This corresponds to an increase in the parameter c in our definition of c-isolation.

This suggests the following extended definition. The adversary, upon receiving the SDB, outputs a k-dimensional axis-parallel hyperplane H ($k \leq d$), and a point q in this hyperplane. Let $\Pi_H(y)$ denote the projection of an RDB point y onto the hyperplane H. Let y be the RDB point which is closest to q under the projection Π_H. For a given function $\phi(k)$, we say that q (ϕ, c, t)-**isolates** y in H iff $\Pi_H(y)$ is ($\phi(k)c, t$)-isolated by q in the projected space H. Recursive histogram sanitizations are safe with respect to ($\phi, O(1), 2^{o(d)}$)-isolation for $\phi(k) = 2^{d/k}$.

We believe that understanding these issues is the most important conceptual challenge arising from this work.

Histogram Sanitizations of Round Distributions and of Mixtures. An immediate focus of future work will be to investigate histogram sanitizations in the context of the "round" (spherical, ball, and Gaussian) distributions. (Chawla *et al.* [6] prove privacy of a first-level histogram for balls and spheres, in which the distribution is partitioned into $exp(d)$ regions, but as of this writing the results only extend to a constant number of recursive steps). Together with a cross-training result for round distributions, such a result would nicely complement our learning algorithm for mixtures of Gaussians.

The results extend immediately to the case in which the underlying distribution is a mixture of sufficiently separated "nice" distributions such as hypercubes, balls, and spheres.

Utility. Another pressing direction is to further explore utility, in particular, a method for assessing the validity of results obtained by applying a given statistical test to sanitized data. The statistics literature on imputed data (e.g., [27] should be helpful in this endeavor.

Changes over Time. An important aspect of any sanitization technique is to consider its application in an online setting, where the database changes over time. We feel that sanitizations of points should not be recomputed independently as the database changes, because an adversary collecting information over time may be able to gather enough to filter out the noise. However, in situations such as when one of the t-nearest neighbors of a point dies, one may be forced to recompute the sanitization. We believe that in such a case the new sanitization should be conditioned on the previous one appropriately, so as to prevent leakage of extra information. A related open area is to extend the definitions and techniques to multiple databases.

Real-Life Data. Then, there are the more obvious questions: how to cope with discrete data, or even non-numerical data. In general, to draw a connection to real life data, we will need to scale different attributes appropriately, so that the data are well-rounded. This requires some formal treatment.

Impossibility Results. M. Naor has suggested studying impossibility results, for example, searching for utilities that cannot be obtained while maintaining privacy. Initial investigations, already mentioned in the paper, have been fruitful. This is a subject of work in progress [14].

Acknowledgements

We have benefited enormously from numerous conversations with many people, and are particularly grateful to Dimitris Achlioptas, Gagan Aggarwal, Jin-Yi Cai, Petros Drineas, John Dunagan, Amos Fiat, Michael Freedman, Russel Impagliazzo, Michael Isard, Anna Karlin, Moni Naor, Helen Nissenbaum, Kobbi Nissim, Anna Redz, Werner Steutzle, Madhu Sudan, and Luca Trevisan.

Not knowing of his terminal illness, but feeling his interest in research diminish, Larry Stockmeyer withdrew from this paper in April, 2004. We will always think of him as our co-author, and we dedicate this work to his memory.

References

1. D. Agrawal and C. Aggarwal, On the Design and Quantification of Privacy Preserving Data Mining Algorithms, *Proceedings of the 20th Symposium on Principles of Database Systems*, 2001.
2. N. R. Adam and J. C. Wortmann, Security-Control Methods for Statistical Databases: A Comparative Study, *ACM Computing Surveys* 21(4): 515-556 (1989).
3. S. Arora and R. Kannan. Learning mixtures of arbitrary Gaussians. *ACM STOC*, 2001.
4. R. Agrawal and R. Srikant, Privacy-preserving data mining, *Proc. of the ACM SIGMOD Conference on Management of Data*, pp. 439–450, 2000.
5. Beck, L., A security machanism for statistical database, *ACM Transactions on Database Systems (TODS)*, 5(3), p.316-3338, 1980.
6. Chawla S., Dwork, C., McSherry, F., Talwar, K., On the Utility of Privacy-Preserving Histograms, *in preparation*, November, 2004.
7. Cox, L. H., New Results in Disclosure Avoidance for Tabulations, *International Statistical Institute Proceedings of the 46th Session*, Tokyo, 1987, pp. 83-84.
8. S. Dasgupta, Learning mixtures of Gaussians, *IEEE FOCS*, 1999.
9. Denning, D., Secure statistical databases with random sample queries, *ACM Transactions on Database Systems (TODS)*, 5(3), p.291-315, 1980.

10. P. Diaconis and B. Sturmfels, Algebraic Algorithms for Sampling from Conditional Distributions, *Annals of Statistics* 26(1), pp. 363–397, 1998
11. I. Dinur and K. Nissim, Revealing information while preserving privacy, *Proceedings of the Symposium on Principles of Database Systems*, pp. 202-210, 2003.
12. A. Dobra and S.E. Fienberg, and M. Trottini, Assessing the risk of disclosure of confidential categorical data, *Bayesian Statistics 7*, pp. 125–14, Oxford University Press, 2000.
13. C. Dwork, A Cryptography-Flavored Approach to Privacy in Public Databases, lecture at Aladdin Workshop on Privacy in DATA, March, 2003; http://www.aladdin.cs.cmu.edu/workshops/privacy/slides/pdf/dwork.pdf)
14. C. Dwork, M. Naor, et al., Impossibility Results for Privacy-Preserving Data Sanitization, *in preparation*, 2004.
15. C. Dwork and K. Nissim, Privacy-Preserving Datamining on Vertically Partitioned Databases, *Proc. CRYPTO 2004*.
16. A. V. Evfimievski, J. Gehrke and R. Srikant, Limiting privacy breaches in privacy preserving data mining, *Proceedings of the Symposium on Principles of Database Systems*, pp. 211-222, 2003.
17. Füredi, Zoltán and Komlós, János, The eigenvalues of random symmetric matrices, *Combinatorica*, 1:3, 1981, pages 233–241.
18. W. Gasarch, A Survey on Private Information Retrieval. *BEATCS Computational Complexity Column*, 82, pp. 72-107, Feb 2004.
19. R. Gavison, Privacy and the Limits of the Law, in Deborah G. Johnson and Helen Nissenbaum, editors, *Computers, Ethics, and Social Values*, pp. 332–351. Prentice Hall, 1995.
20. O. Goldreich, *The Foundations of Cryptography - Volume 2*. Cambridge University Press, 2004.
21. S. Goldwasser and S. Micali, Probabilistic Encryption, *JCSS* 28(2), pp. 270–299, 1984.
22. D. Gusfield, A Graph Theoretic Approach to Statistical Data Security, *SIAM Journal on Computing* 17(3), pp. 552–571, 1988
23. P. Indyk and R. Motwani. Approximate Nearest Neighbor: Towards Removing the Curse of Dimensionality. *Proceedings of the 30th Annual ACM Symposium on Theory of Computing*, 1998.
24. H. Kargupta, S. Datta, Q. Wang, K. Sivakumar. On the Privacy Preserving Properties of Random Data Perturbation Techniques. *Proceedings of the Third ICDM IEEE International Conference on Data Mining*, 2003.
25. J. M. Kleinberg, C. H. Papadimitriou, and P. Raghavan, Auditing Boolean Attributes, *J. Comput. Syst. Sci.* 66(1), pp. 244–253, 2003
26. F. McSherry, Spectral Partitioning of Random Graphs, *Proc. 42nd FOCS*, pp. 529 – 537, 2001
27. T.E. Raghunathank, J.P. Reiter, and D.B. Rubin, Multiple Imputation for Statistical Disclosure Limitation, *J. Official Statistics* 19(1), 2003, pp. 1–16.
28. G. Roque. Application and Analysis of the Mixture-of-Normals Approach to Masking Census Public-use Microdata. Manuscript, 2003.
29. D. B. Rubin, Discussion: Statistical Disclosure Limitation, *Journal of Official Statistics* 9(2), 1993, pp. 461–468.
30. Sibson, R, SLINK: an optimally efficient algorithm for the single-link cluster method, In *the Computer Journal* Vol. 16, No. 1, 1973, pages 30–34.
31. L. Sweeney, k-anonymity: a model for protecting privacy. *International Journal on Uncertainty, Fuzziness and Knowledge-based Systems*, 10 (5), 2002; 557-570.

32. L. Sweeney, Achieving k-anonymity privacy protection using generalization and suppression. *International Journal on Uncertainty, Fuzziness and Knowledge-based Systems, 10* (5), 2002; 571-588.
33. S. Vempala and G. Wang, A spectral algorithm for learning mixtures of distributions, *IEEE FOCS*, 2002.
34. W. E. Winkler. Masking and Re-identification Methods for Public-Use Microdata: Overview and Research Problems. In *Proc. Privacy in Statistical Databases* 2004, Springer LNCS 3050.
35. W. E. Winkler. Re-identification Methods for Masked Microdata. In *Proc. Privacy in Statistical Databases* 2004, Springer LNCS 3050.

The Universal Composable Security of Quantum Key Distribution

Michael Ben-Or[1,4,6], Michał Horodecki[2,6], Debbie W. Leung[3,4,6], Dominic Mayers[3,4], and Jonathan Oppenheim[1,5,6]

[1] Institute of Computer Science,
The Hebrew University, Jerusalem, Israel
[2] Institute of Theoretical Physics and Astrophysics,
University of Gdańsk, Poland
[3] Institute of Quantum Information,
California Institute of Technology, Pasadena, USA
[4] Mathematical Science Research Institute, Berkeley, USA
[5] DAMTP, University of Cambridge, Cambridge, UK
[6] Isaac Newton Institute, University of Cambridge, Cambridge, UK
benor@cs.huji.ac.il, fizmh@univ.gda.pl,
{wcleung, dmayers}@cs.caltech.edu, J.Oppenheim@damtp.cam.ac.uk

Abstract. The existing unconditional security definitions of quantum key distribution (QKD) do not apply to joint attacks over QKD and the subsequent use of the resulting key. In this paper, we close this potential security gap by using a universal composability theorem for the quantum setting. We first derive a composable security definition for QKD. We then prove that the usual security definition of QKD still implies the composable security definition. Thus, a key produced in any QKD protocol that is unconditionally secure in the usual definition can indeed be safely used, a property of QKD that is hitherto unproven. We propose two other useful sufficient conditions for composability. As a simple application of our result, we show that keys generated by repeated runs of QKD degrade slowly.

1 Introduction

Quantum cryptography differs strikingly from its classical counterpart. On one hand, quantum effects are useful in the construction of many cryptographic schemes. On the other hand, dishonest parties can also employ more powerful quantum strategies when attacking cryptographic schemes.

The Security of Quantum Key Distribution. One of the most important quantum cryptographic applications is quantum key distribution (QKD) [1, 2, 3]. The goal of key distribution (KD) is to allow two *remote* parties, Alice and Bob, to share a *secret* bit string. Classically, KD cannot be unconditionally secure (i.e. secure against all possible classical attacks) (see Sect. 2). Furthermore, the security of existing KD schemes is based on assumptions in computation complexity or limitations of the memory space of the adversary, Eve. In contrast, QKD is based on an intrinsic property of quantum mechanics, "extracting information about an unknown quantum state inevitably disturbs it," [4] which allows eavesdropping activities to be detected in principle. Indeed, QKD can be *unconditionally secure*, i.e., against Eve whose capability is only limited by quantum mechanics [5, 6, 7, 8, 9, 10, 11]. Furthermore, QKD remains secure even if the quantum states are sent through a noisy quantum channel, as long as the observed error rates are below certain threshold values.

In what sense is QKD secure? We will describe the assumptions and security definitions more formally in Sect. 2. In QKD, Alice and Bob are assumed to start with a small initial key K_i (for authentication purposes). They have access to uncorrelated randomness that is not controlled by Eve. They may exchange quantum and classical messages in both directions via channels that are completely under the control of Eve, and may perform local quantum operations and measurements. Based on their measurement outcomes, Alice and Bob either abort QKD or generate their respective keys K_A, K_B. Correspondingly, we say that the QKD test is failed or passed, and the events can be described as $M=0$ or $M>0$, where M is the length of the key generated. Eve also obtains quantum and classical data (her "view" or "transcript") from which she extracts classical data K_E via a measurement. What happens during a specific run of QKD depends on Eve's strategy as well as the particular outcomes of the coins and quantum measurements of all the parties. However, the security of QKD can still be captured by requiring that (1) the conditional mutual information $I(K_E : K_A, K_B | M)$ is negligible and (2) for all eavesdropping strategies with nonnegligible $\Pr(M>0)$, K_A, K_B are near-uniform and $\Pr(k_A \neq k_B)$ is negligible. Throughout the paper, we use capitalized letters $K_A, K_B, K_E,$ and M to denote the random variables, and uncapitalized letters to denote specific outcomes.

The Security Problem of Using QKD. Proofs of security of QKD (in the sense described above) address all attacks on the QKD scheme allowed by quantum mechanics. The problem is that QKD is *not* the only occasion for attack — further attack may occur when Alice and Bob use the keys generated. In particular, Eve may never have made a measurement during QKD to obtain any K_E. Eve's transcript is a quantum state. She could have delayed measurements until after more attack during the application, a strategy with power that has no classical counterpart. In other words, security statements in QKD that revolve around bounding $I(K_E : K_A, K_B | M)$ is *not applicable* if the key is to be used!

The limitations of mutual-information-based security statements were known as a folklore for some time (for example, see Sec. 4.2 in [11]). One of the earliest known security problems in QKD is the "key degradation problem" [12]: QKD requires a key for authentication, which in turns may come from previous runs of QKD. Since each run of QKD is slightly imperfect, repeated runs of QKD produce less and less secure keys. A conclusive analysis on the degradation has been elusive, since joint attacks over all runs of QKD have to be considered.

As it turns out, joint attacks on QKD and the subsequent use of the generated key have to be considered in many other occasions. For example, suppose Alice and Bob perform QKD to obtain a key, and then use the key to encrypt quantum states [13, 14]. Eve eavesdrops during both QKD and encryption and performs a collective measurement on the two eavesdropped states. It is well-known that such a collective measurement may yield more *accessible information* than the sum of information obtained in two separate measurements [15].

Our current study is further motivated by the results in [16, 17], which show that there are ensembles of quantum states that provide little accessible information on their own, but can provide *much more* information when a little more *classical* data is available. The extra information can be arbitrarily large compared to both the initial information and the amount of extra classical data. Such strange property reveals a new, unexpected, inadequacy of mutual-information-based statements. In particular, in the context of QKD, the usefulness of bounding the initial accessible information of Eve becomes very questionable, if Eve delays her measurement until further data is available during the application of the key — the security of the key is questionable even in *classical* applications!

The goal of the current paper is to study the security of using a key generated by QKD, i.e., the composability of QKD.

The Universal Composability Approach. Composability is an active area of research that is concerned with the security of composing cryptographic primitives in a possibly complex manner. The simplest example is the security of using a cryptographic primitive as a subroutine in another application. We will follow the *universal composability* approach. For a specific task (functionality), a primitive that realizes the task is said to be universal composable if any application using the primitive is about as secure as one using the ideal functionality. A security definition that ensures *universal* composability was recently proposed by Canetti [18]. A simpler model in the quantum setting and a corresponding universal composable security definition were reported by by some of us [19, 20]. Universal composable security definitions are useful because they are in terms of the ideal functionality only, without reference to the potential application. The security of a complex protocol can then be analyzed in terms of the security of each individual component in a systematic and error-proof manner. In the quantum setting, universal composability provides the only existing systematic technique for analyzing security in the presence of subtleties including entanglement and collective attacks. In this paper, universal composability provides the precise framework for proving the security of using the keys generated from QKD, a problem that appears intractable at first sight.

An alternative approach to composability in the classical setting was obtained in [21], with a generalization to the quantum setting studied in [22, 23].

Main Results. We have pointed out a serious potential security problem in using the keys generated from QKD. We will address the problem in the rest of the paper. We derive a new security definition for QKD that is universal composable. The essence is that QKD and certain ideal KD should be indistinguishable from the point of view of potential adversaries. Then, we prove that the original mutual-information-based security definition implies the new composable definition. Other simple sufficient conditions for the composable security of QKD will be discussed. One of these conditions, high singlet-fidelity, has always been an intermediate step in the widely-used "entanglement-based" security proofs of QKD. We show that high singlet-fidelity is much more closely related to composable security than the usual security definition, and we obtain much better security bounds for known QKD schemes. We thus prove the security of using a key generated by QKD in various ways, and provide simple criteria for future schemes. As a corollary, we analyze the extent of key-degradation in repeated use of QKD [12].

Our work also has non-cryptographic applications in the study of correlations in quantum systems. The various security conditions are tied to correlation measures in quantum systems. Each derivation for the composable security for QKD is based on relating a pair of correlation measures.

Related Work. Since the current result was initially presented [24, 25], various related results were reported. The composable security of generic classes of QKD schemes were proved in [26, 27], following a different approach of showing the composable security of certain privacy amplification procedures against quantum adversaries [26]. These related works share the concerns raised in this paper, with results complementary to ours.

Organization of the Paper. We end this section by introducing some basic elements in the quantum setting. We review QKD in Sect. 2, stating our definitions and assumptions more formally. In Sect. 3, we review the quantum universal composability theorem. We will restrict ourselves to the simpler case concerning unconditional security. We start describing our main results in Sect. 4, which contains a derivation of a simple criteria for the universal composable security for QKD. In Sect. 5, we prove that the usual security definition for QKD implies the universal composable security. In addition, we demonstrate two other sufficient conditions for composable security. One is based on bounding the Holevo information of Eve on the key. The other is based on bounding the singlet-fidelity in security proofs using entanglement-purification. The latter implies much better security of existing QKD protocols than is generically implied by the usual security definition. We conclude with lessons learnt from the current results. Frequently used notations and functions are listed in the appendix.

Basic Elements of Quantum Mechanics. A quantum system or register is associated with a Hilbert space $\mathbb{H}$. We only consider finite dimensional Hilbert spaces. Let $\mathcal{B}(\mathbb{H})$ and $\mathbb{U}(\mathbb{H})$ denote, respectively, the set of bounded operators and the unitary group acting on $\mathbb{H}$. We loosely refer to the system as $\mathbb{H}$ also. A composite quantum system is associated with the tensor product of the Hilbert spaces associated with the constituent systems.

The state of $\mathbb{H}$ is specified by a positive semidefinite *density matrix* $\rho \in \mathcal{B}(\mathbb{H})$ of unit trace. A density matrix is a convex combination of rank-1 projectors (commonly called *pure states*) and represents a probabilistic mixture of pure states. Pure states can be represented as vectors in $\mathbb{H}$, up to a physically unobservable phase. $|\psi\rangle$ and $|\psi\rangle\langle\psi|$ denote the vector and rank-1 projector respectively.

A measurement $\mathcal{M}$ on $\mathbb{H}$ is specified by a POVM — a set of positive semidefinite operators $\{O_k\}$ such that $\sum_k O_k = I$. If the state is initially ρ, the measurement $\mathcal{M}$ yields the outcome k with probability $\text{Tr}(O_k\rho)$ and changes the state to $\sqrt{O_k}\rho\sqrt{O_k}/\text{Tr}(O_k\rho)$. $\mathcal{M}$ is said to be along a basis $\{|k\rangle\}$ if $\{O_k\} = \{|k\rangle\langle k|\}$. Measuring an unknown state generally disturbs it.

The most general evolution of the state is given by a trace-preserving completely-positive (TCP) linear map $\mathcal{E}$ acting on $\mathcal{B}(\mathbb{H})$. Any such $\mathcal{E}$ can be implemented by preparing a pure state in some ancillary system $\mathbb{H}'$, applying a joint unitary operator $U \in \mathbb{U}(\mathbb{H} \otimes \mathbb{H}')$, and discarding $\mathbb{H}'$ (i.e., a partial trace over $\mathbb{H}'$).

We mention two distance measures for quantum states. First, the trace distance $\|\rho_1 - \rho_2\|_1$ between two density matrices is twice the maximum probability of distinguishing between the two states. Second, the fidelity is $F(\rho_1, \rho_2) = \max_{|\psi_1\rangle, |\psi_2\rangle} |\langle\psi_1|\psi_2\rangle|^2$, where $\rho_{1,2} \in \mathcal{B}(\mathbb{H})$, $|\psi_{1,2}\rangle \in \mathbb{H} \otimes \mathbb{H}'$ are "purifications" of $\rho_{1,2}$ (i.e., $\text{Tr}_{\mathbb{H}'}|\psi_{1,2}\rangle\langle\psi_{1,2}| = \rho_{1,2}$), and $\langle \cdot | \cdot \rangle$ is the inner product in $\mathbb{H}$.

We refer our readers to the excellent textbook by Nielsen and Chuang [28] for a more comprehensive review of the quantum model of information processing.

2 Quantum Key Distribution

The goal of key distribution (KD) is to allow two *remote* parties, Alice and Bob, to share a *secret* bitstring such that no third party, Eve, will have much information about the bitstring. KD is impossible unless Alice and Bob can identify one another and detect alterations of their communication. In other words, the task of *message authentication* is necessary for KD. There are unconditionally secure methods for authenticating a classical message with a much shorter key [29]. Thus, KD uses authentication as a subroutine, and achieves key expansion (producing a key using a much shorter initial key).

Classically, unconditionally secure KD between two remote parties is impossible. Classical physics permits an eavesdropper to have exact duplicates of all communications in any KD procedure without being detected. In contrast, while quantum key distribution (QKD) cannot prevent eavesdropping, it can detect eavesdropping. This allows Alice and Bob to avoid generating compromised keys with high probability. The usefulness of QKD is to avoid Alice and Bob being fooled into having a false sense of security. It is worth emphasizing what QKD does not offer. First, QKD does not promise to always produce a key, since Eve can cause QKD to be aborted with high probability by intense eavesdropping. Second, there is a vanishing but non-zero chance that Eve is undetected, so that one cannot make simple security statements conditioned on not aborting QKD.

How and Why QKD Works, Through an Example. Various QKD schemes have been proposed and we only name a few here: BB84 [1], E91 [2], B92 [3], and the six-state scheme [30, 31]. We illustrate the general features and principles behind QKD by describing the class of "prepare-&-measure schemes." Recall that Alice and Bob are given secure local coin tosses. Step 1: Alice first generates a random bitstring, encodes it in some quantum state ρ_A, and sends ρ_A to Bob through an insecure quantum channel controlled by Eve. During this time, Eve can manipulate the message (system $\mathbb{A}$) in any way allowed by quantum mechanics. Eventually, she will have to give some quantum message ρ_B to Bob for QKD to proceed. Mathematically, Eve's most general operation can be described as attaching a private system $\mathbb{E}$ in the state $|0\rangle\langle 0|_E$, applying a joint unitary operation U to produce a joint state $\rho = U(\rho_A \otimes |0\rangle\langle 0|_E) U^\dagger$, and passing system $\mathbb{A}$ to Bob (relabeled as system $\mathbb{B}$). Thus, Bob and Eve share the joint state ρ, and $\rho_B := \text{Tr}_\mathbb{E}\,\rho$, $\rho_E := \text{Tr}_\mathbb{B}\,\rho$ are their respective reduced density matrices. Meanwhile, Bob measures ρ_B (according to his coin tosses). Step 2: Bob acknowledges to Alice receipt of the quantum message. Step 3: Only *after* Alice hears from Bob will further classical discussion be conducted over a public but authenticated channel. Step 4: At the end, based on their measurement outcomes and discussions, Alice and Bob either abort QKD ($m = 0$), or generate keys K_A and K_B ($m > 0$), and they announce m. Eve will have access to all the classical communication between Alice and Bob, besides the state ρ_E. She can measure ρ_E at any time to obtain a classical string K_E, though it is to her advantage to wait until after she receive the classical communication. See Fig. 1 for a schematic diagram for the class of prepare-&-measure QKD schemes.

The principle behind QKD is that, in quantum mechanics, one can only reversibly extract information from an unknown quantum state if the state is drawn from an orthogonal set [4]. Thus in the prepare-&-measure scheme de-

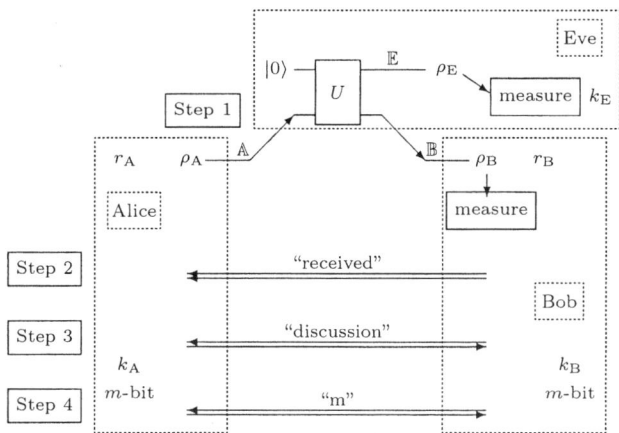

Fig. 1. Schematic diagram for the class of prepare-&-measure QKD schemes. The classical messages, represented by double lines, are available to Eve. Eve can make her measurement any time after step 1. Dashed boxes represent private laboratory spaces. Outcomes of Alice and Bob's local coins are represented by r_A, r_B

scribed above, if Alice encodes her message using a random basis chosen from several nonorthogonal possibilities, and Eve is to obtain any information on the outcomes of K_A, K_B, then $\rho_B \neq \rho_A$. To detect the disparity, Bob measures some of the received qubits (the "test-qubits" chosen randomly to avoid Eve tailoring her attack) and discusses with Alice to check if his measurement outcomes are consistent with what Alice has sent. This intuition can be turned into a provably secure procedure. Alice and Bob estimate various error rates on the test-qubits. If the observed error rates are above certain thresholds, Alice and Bob abort QKD. Otherwise, error reconciliation and privacy amplification are applied to the untested qubits to extract bitstrings k_A and k_B for Alice and Bob respectively. It is unlikely that the observed error rates are low while untested qubits have higher error rates. QKD remains secure whether the observed noise is due to natural channel noise or due to eavesdropping.

General Features of Any QKD Scheme. There are other QKD schemes besides prepare-&-measure schemes, for example, the entanglement-based QKD schemes (see [2, 7, 32]). Unless otherwise stated, our discussion applies to all QKD schemes. The basic ingredients are still secure local coins, completely insecure quantum communication, and authenticated public classical communication between Alice and Bob. In the most general QKD scheme, the ingredients may be used in any possible way. Alice and Bob still obtain some bitstrings as the output keys, k_A and k_B, of certain length m. Eve's view is still given by some quantum and classical data, denoted collectively by ρ_{E,k_A,k_B}, with explicit dependence on k_A, k_B. (Her view is a draw from an ensemble.)

We emphasize a limitation in QKD. Eve can be "lucky." For example, she may have attacked only the untested qubits, or may have attacked every qubit without causing inconsistency in Alice and Bob's measurements. Thus, it is unlikely, but still possible, for Eve to have a lot of information on the generated key without being detected. QKD does not promise that "*conditioned* on passing the test, the keys K_A, K_B will be so-and-so." With the above limitation in mind, there are several approaches to a proper security statement. The approach that is most commonly used in existing security proofs is to bound the probability that Alice and Bob generate bitstrings that are not equal, uniform, or private. We will use a more compact statement in the following.

Let n be a security parameter in QKD (for example, the number of qubits transmitted from Alice to Bob). Fix an arbitrary eavesdropping strategy. The attack induces a distribution $\Pr(M{=}m)$ on the key length M. The average value of M is typically a small fraction of n. The outcome m in a particular run of QKD depends on the outcome of the coins and measurements by Alice and Bob. We can assume that m is made *public* at the end of QKD. Recall $m > 0$ if the QKD test is passed and $m = 0$ if QKD is aborted.

Let $p_{\text{qkd}}^{(m)}$ denote the distribution of K_A, K_B generated in QKD conditioned on $|K_A| = |K_B| = m$, i.e.,

$$p_{\text{qkd}}^{(m)}(k_A, k_B) = \Pr(K_A = k_A, K_B = k_B | M = m) . \tag{1}$$

Let $p_{\text{ideal}(m)}$ be the following distribution over two m-bit strings,

$$\begin{cases} p_{\text{ideal}}^{(m)}(l,l) = 2^{-m} \\ p_{\text{ideal}}^{(m)}(l,l') = 0 \quad \text{if } l \neq l'. \end{cases} \quad (2)$$

Let $\mathcal{V}$ denote the set of exponentially decaying functions of n. With these notations, a simple statement for the security condition can be made.

Usual Security Definition for QKD. A QKD scheme is said to be secure if the following properties hold for all eavesdropping strategies.

- *Equality-and-uniformity:* $\exists \mu_1 \in \mathcal{V}$ s.t.

$$\sum_{m=0} \Pr(m) \, \| p_{\text{ideal}}^{(m)} - p_{\text{qkd}}^{(m)} \|_1 \leq \mu_1 \quad (3)$$

- *Privacy:* $\exists \mu_2 \in \mathcal{V}$ s.t.

$$\sum_{m=0} \Pr(m) \times I(K_E : K_A, K_B \,|\, M = m) \leq \mu_2 \quad (4)$$

where I denotes the mutual information [33] between K_E and K_A, K_B conditioned on $M = m$. Using the equality condition, we only need to focus on $k_A =: k$ in (4). In particular,

- *Privacy:* $\exists \mu_2' \in \mathcal{V}$ s.t.

$$\sum_{m=0} \Pr(m) \times I(K_E : K \,|\, M = m) \leq \mu_2' \quad (5)$$

The above security conditions revolve around bounding expressions that can be interpreted as deviations from the desired properties, averaged over m. The product in each summand is bounded, precisely capturing the security requirement that an undesired event occurs with low probability. Note that the $m=0$ terms do not contribute, as $\| p_{\text{ideal}}^{(m)} - p_{\text{qkd}}^{(m)} \|_1 = 0$ and $I(K_E : K_A, K_B | M = 0) = 0$.

3 Quantum Universal Composability Theorem

Cryptographic protocols often consist of a number of simpler components. A single primitive is rarely used alone. A strong security definition for the primitive should thus reflect the security of using it within a larger application. This allows the security of a complex protocol to be based only on the security of the components and how they are put together, but not in terms of the details of the implementation.

A useful approach is to consider the *universal composability* of cryptographic primitives [18, 19, 20]. The first ingredient is to ensure the security of a *basic composition*. We need a security definition stated for a single execution of the primitive that still guarantees security of composition with other systems. This definition involves a description of some ideal functionality of the primitive (i.e. the ideal task the primitive should achieve). More concretely, we want a security definition such that, if σ is a secure realization of an ideal subroutine σ_I, and a protocol $\mathcal{P}$ using σ_I, written as $\mathcal{P}+\sigma_I$, is a secure realization of $\mathcal{P}_I$ (the ideal functionality of $\mathcal{P}$), $\mathcal{P}+\sigma$ is also a secure realization of $\mathcal{P}_I$. Throughout the paper, we denote the associated ideal functionality of a protocol by adding a subscript I, and we denote a protocol $\mathcal{P}$ calling a subprotocol σ as $\mathcal{P}+\sigma$ (this last expression stretches the meaning of $\mathcal{P}$ a little bit to refer to the module of $\mathcal{P}$ calling σ). The second ingredient is a universal composability theorem stating how a complex protocol can be built out of secure components. It is simply a recipe on how to securely perform basic composition recursively.

The Simplifications in Analyzing the Composable Security of QKD. Our goal is to analyze the unconditional security of QKD using known results in quantum universal composability [19, 20]. The setting for QKD is simpler than that considered in [19, 20] in two important aspects. First, we are only concerned with unconditional security. Second, in QKD, Alice and Bob are known to be honest, and Eve is known to be adversarial, and no party is corrupted unpredictedly. The formal corruption rules are not used in our derivation of a composable security definition for QKD. We will describe a simplified model that is sufficient for our derivation of a universal composable security definition for QKD. This definition is applicable in the general setting considered in [19, 20] – so long as an appropriate model is used for analyzing the rest of the application when applying Theorems 1 and 2.

The Simplified Model. We first describe the model for quantum protocols and other concepts involved in the quantum composable security definition. We base our discussion on the (acyclic) quantum circuit model (see, for example, [34, 35]), with an important extension [20] (see also the endnotes [36]). Throughout the paper, we only consider circuits in the extended model.

1. *Structure of a protocol.* A (cryptographic) protocol $\mathcal{P}$ can be viewed as a quantum circuit in the extended model [20, 36], consisting of inputs, outputs, a set of registers, and some partially ordered operations. A protocol may consist of a number of subprotocols and parties. Each subprotocol consists of smaller units called "unit-roles," within each the operations are considered "local." For example, the operations and registers of each party in each subprotocol form a unit-role. Communications between unit-roles within a subprotocol represent *internal communications*; those between unit-roles in different subprotocols represent input/output of data to the subprotocols. A channel is modeled by an ordered pair of operations by the sender and receiver on a shared register. The channel available for the communication determines its security features.

2. *The game: security in terms of indistinguishability from the ideal functionality.* Let $\mathcal{P}_I$ denote the ideal functionality of $\mathcal{P}$. Intuitively, $\mathcal{P}$ is secure (in a sense defined by $\mathcal{P}_I$) if $\mathcal{P}$ and $\mathcal{P}_I$ behave similarly under any adversarial attack. "Similarity" between $\mathcal{P}$ and $\mathcal{P}_I$ is modeled by a game between *an environment $\mathcal{E}$*

and *a simulator* $\mathcal{S}$. These are sets of registers and operations to be defined, and they are sometimes personified in our discussion. In general, $\mathcal{P}$ and $\mathcal{P}_I$ have very different internal structures and are very distinguishable, and the simulator $\mathcal{S}$ is added to $\mathcal{P}_I$ to make an extended ideal protocol $\mathcal{P}_I+\mathcal{S}$ that is less distinguishable from $\mathcal{P}$. $\mathcal{E}$ consists of the adversaries that act against $\mathcal{P}$ and an application protocol that calls $\mathcal{P}$ as a subprotocol. At the beginning of the game, $\mathcal{P}$ or $\mathcal{P}_I+\mathcal{S}$ are picked at random. $\mathcal{E}$ will call and act against the chosen protocol, and will output a bit Γ at the end of the game. The similarity between $\mathcal{P}$ and $\mathcal{P}_I+\mathcal{S}$ (or the lack of it) is captured in the statistical difference in the output bit Γ.

3. Valid $\mathcal{E}$. The application and adversarial strategy of $\mathcal{E}$ are first chosen (the same whether it is interacting with $\mathcal{P}$ or $\mathcal{P}_I+\mathcal{S}$). $\mathcal{E}$ has to obey quantum mechanics, but is otherwise unlimited in computation power. If $\mathcal{P}$ is chosen in the game, $\mathcal{E}$ can (I) control the input/output of $\mathcal{P}$, (II) attack insecure internal communication as allowed by the channel type, (III) direct the adversarial parties to interact with the honest parties in $\mathcal{P}$. $\mathcal{E}+\mathcal{P}$ has to be an acyclic circuit in the extended model [20, 36].

4. Valid $\mathcal{P}_I$ and $\mathcal{S}$. If $\mathcal{P}_I+\mathcal{S}$ is chosen in the game, $\mathcal{E}$ (I) controls the input/output of $\mathcal{P}_I$ as before. However, the interaction given by (II) and (III) above will now occur between $\mathcal{E}$ and $\mathcal{S}$ instead. ($\mathcal{S}$ is impersonating or simulating $\mathcal{P}$.) The strategy of $\mathcal{S}$ can depend on the strategy of $\mathcal{E}$. $\mathcal{P}_I$ should have the same input/output structure as $\mathcal{P}$, but is otherwise arbitrary. (Of course, the security definition is only useful if $\mathcal{P}_I$ carries the security features we want to prove for $\mathcal{P}$.) In particular, $\mathcal{P}_I$ may be defined with internal channels and adversaries different from those of $\mathcal{P}$. $\mathcal{S}$ can (II') attack insecure internal communication of $\mathcal{P}_I$ and (III') direct the adversarial parties to interact with the honest parties in $\mathcal{P}_I$. Thus, $\mathcal{P}_I$ exchanges information with $\mathcal{S}$, and this can modified the security features of $\mathcal{P}_I$. To $\mathcal{E}$, $\mathcal{S}$ acts like part of $\mathcal{P}_I$, "padding" it to look like $\mathcal{P}$, while to $\mathcal{P}_I$, $\mathcal{S}$ acts like part of $\mathcal{E}$. It is amusing to think of $\mathcal{S}$ as making a "man-in-the-middle" attack between $\mathcal{E}$ and $\mathcal{P}_I$. Finally, $\mathcal{E}+\mathcal{P}_I+\mathcal{S}$ has to be an acyclic circuit in the extended circuit model [20, 36]. See Fig. 2 for a summary of the game and the rules.

With a slight abuse of language, the symbols $\mathcal{P}$ and $\mathcal{P}_I+\mathcal{S}$ are also used to denote the respective events of their being chosen at the beginning of the game. We can now state the universal composable security definition.

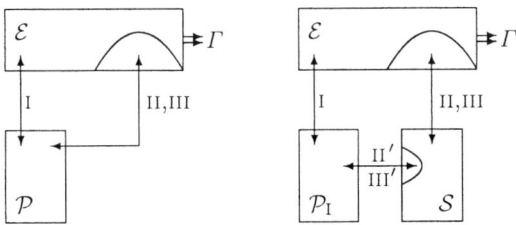

Fig. 2. The game defining the composable security definition. The curved region in $\mathcal{E}$ represents the adversaries against $\mathcal{P}$, and the curved region in $\mathcal{S}$ represents the adversaries against $\mathcal{P}_I$. We label the types of interactions as described in the text

Definition 1: $\mathcal{P}$ is said to ϵ-securely realize $\mathcal{P}_\mathrm{I}$ (shorthand $\mathcal{P}$ ϵ-s.r. $\mathcal{P}_\mathrm{I}$) if

$$\forall \mathcal{E} \ \exists \mathcal{S} \ \text{s.t.} \ \left| \Pr(\Gamma{=}0|\mathcal{P}) - \Pr(\Gamma{=}0|\mathcal{P}_\mathrm{I}{+}\mathcal{S}) \right| \leq \epsilon . \qquad (6)$$

We call ϵ in (6) the *distinguishability-advantage* between $\mathcal{P}$ and $\mathcal{P}_\mathrm{I}$. This security definition (in the model described) is useful because security of basic composition follows "by definition" [19, 20]. We have the following simple version of a universal composability theorem.

Theorem 1. *Suppose a protocol $\mathcal{P}$ calls a subroutine σ. If σ ϵ_σ-s.r. σ_I and $\mathcal{P}{+}\sigma_\mathrm{I}$ $\epsilon_\mathcal{P}$-s.r. $\mathcal{P}_\mathrm{I}$, then $\mathcal{P}{+}\sigma$ ϵ-s.r. $\mathcal{P}_\mathrm{I}$ for $\epsilon \leq \epsilon_\mathcal{P} + \epsilon_\sigma$.*

Theorem 1 can be generalized to any arbitrary protocol with a proper *modular structure*. An example of an improper modular structure is one with a security deadlock, in which the securities of two components are interdependent.

Proper modular structures can be characterized as follows. Let $\mathcal{P}{+}\sigma_1{+}\sigma_2{+}\cdots$ be any arbitrary protocol using a number of subprotocols. This can be represented by a 1-level tree, with $\mathcal{P}$ being the parent and $\{\sigma_i\}$ the children. For each σ_i that uses other subprotocols, replace the corresponding node by an appropriate 1-level subtree. This is done recursively, until the highest-level subprotocols (the leaves) call no other subprotocols. These are the primitives. It was proved in [20] that more general modular structures, represented by an acyclic directed graph, can be transformed to a tree. The following composability theorem relates the security of a protocol $\mathcal{P}$ to the security of all the components in the tree.

Theorem 2. *Let $\mathcal{P}$ be a protocol and $T_\mathcal{P}$ its associated tree. Let $\mathcal{M}$ be a subprotocol corresponding to any node in $T_\mathcal{P}$ with subprotocols $\{\mathcal{N}_i\}_{i=1,\cdots,l}$. Then, if $\mathcal{M}{+}\mathcal{N}_\mathrm{II}{+}\cdots{+}\mathcal{N}_\mathrm{lI}$ $\epsilon_\mathcal{M}$-s.r. $\mathcal{M}_\mathrm{I}$, we have $\mathcal{P}$ ϵ-s.r. $\mathcal{P}_\mathrm{I}$ for $\epsilon \leq \sum_\mathcal{M} \epsilon_\mathcal{M}$.*

Theorem 2 is obtained by recursive use of Theorem 1 and the triangle inequality. The idea is to recursively replace each subprotocol by its ideal functionality, from the highest to the lowest level toward the root. The distinguishability-advantage between $\mathcal{P}$ and $\mathcal{P}_\mathrm{I}$ is upper bounded by the sum of all the individual distinguishability-advantages between pairs of protocols before and after each replacement. See Fig. 4 for an example of $T_\mathcal{P}$ that describes repeated QKD.

In the next section, we analyze QKD in the composability framework. This is part of our main result and it also illustrates the composability framework.

4 Universal Composable Security Definition of QKD

We first describe a general QKD scheme in the composability framework. Then, we tailor an ideal functionality for KD that resembles QKD. Finally, the universal composable security definition of QKD is restated as a distinguishability criteria.

4.1 QKD in the Game Defining Security

Our discussion relies on the existence of authentication schemes that are universal composable in the quantum setting. Furthermore, the authentication scheme should use a key much shorter than the message to be authenticated (so that QKD indeed expands a key). For example, the scheme in [29] satisfies such conditions (composability is proved in [37]). Let α denote any such authentication scheme and let α_I denote ideal authentication. Let $\kappa+\alpha$ denote QKD using authentication scheme α and let κ_I denote an ideal KD protocol to be defined. By Theorem 1, we can focus on the security of $\kappa+\alpha_I$, i.e., QKD using perfectly authenticated classical channels. The initial key requirement is embedded in the subroutine α_I. In this case, QKD has no input and outputs some bitstrings k_A, k_B of certain length m to Alice and Bob, with $m = 0$ if and only if QKD is aborted. (We can assume that m is a publicly announced output of QKD.) Eve's view (including both quantum and classical data) is given by the state ρ_{E,k_A,k_B}.

We now turn to the game defining the composable security definition of QKD. Eve is an adversary that is part of the environment $\mathcal{E}$. Following the discussion in Sect. 3, $\mathcal{E}$ will fix an arbitrary strategy. Since there is no input to QKD, the optimal application in $\mathcal{E}$ is simply to receive the output keys from $\kappa+\alpha_I$ or κ_I. $\mathcal{E}$ will also consist of the action of Eve and other circuits that compute Γ. A schematic diagram is given in Fig. 3.

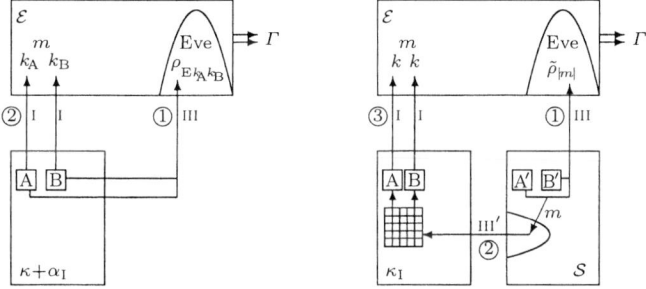

Fig. 3. The game defining the composable security definition of QKD, with our choice of ideal KD and simulator. An ordering of the interactions is given in circles. We also label the types of interactions (see rules 3 and 4 in Sect. 3) explicitly. Upon an input m, the checkered box generates a perfect key of length m to Alice and Bob

If $\mathcal{E}$ is interacting with $\kappa+\alpha_I$, $\mathcal{E}$ will: (I) receive the output bitstrings k_A, k_B, and $m = |k_A| = |k_B|$, and (III) obtain ρ_{E,k_A,k_B} which depends on Eve's strategy and k_A, k_B. Altogether, $\mathcal{E}$ will be in possession of the state

$$\rho_{\text{qkd}} = \sum_{k_A, k_B} \Pr(k_A, k_B) |k_A, k_B\rangle\langle k_A, k_B| \otimes \rho_{E,k_A,k_B} \quad (7)$$

in which ρ_{E,k_A,k_B} and k_A, k_B can be correlated. We have omitted an explicit register for m, because the information is redundant given k_A, k_B.

4.2 Ideal KD and the Simulator

We now define the ideal functionality for QKD. In general, when formulating an ideal functionality, one need not be concerned with how the functionality is realized. What is important is to impose the essential security features while mimicking the analyzed protocol from the point of view of $\mathcal{E}$.

Our ideal KD functionality κ_I has to model both the possibility to generate a perfect key, and the possibility for Eve to cause QKD to be aborted. Besides Alice and Bob, κ_I has a box that accepts a value m from an adversary "Devil" and outputs a perfect m-bit key K to Alice and Bob ($m = 0$ means abort). When κ_I is run, Devil sends m to the box, which sends K to Alice and Bob. This formulation of κ_I satisfies the security conditions (3) and (5) perfectly ($\mu_1, \mu_2 = 0$). See Fig. 3 for a schematic diagram.

Consider the following simulator $\mathcal{S}$. $\mathcal{S}$ runs a "fake QKD" with fake Alice' and Bob'. They interact with Eve (in $\mathcal{E}$) and run verification procedure as in QKD. A value m is announced for the fake QKD, but the fake output keys are unused and kept secret in $\mathcal{S}$. The Devil in $\mathcal{S}$ then sends m to the box in κ_I, which generates a perfect m-bit key string k to Alice and Bob in κ_I, who forward their outputs to $\mathcal{E}$. Let

$$\tilde{\rho}_m = \sum_{k_A, k_B : |k_A| = |k_B| = m} \Pr(k_A, k_B | M{=}m)\, \rho_{E, k_A, k_B} \;. \tag{8}$$

Then, at the end of the game, $\mathcal{E}$ will be in possession of the state

$$\rho_{\text{ideal}} = \sum_k \Pr(M{=}|k|)\, 2^{-|k|}\, |k,k\rangle\langle k,k| \otimes \tilde{\rho}_{|k|} \;. \tag{9}$$

How $\kappa_I + \mathcal{S}$ interacts with $\mathcal{E}$ is summarized in Fig. 3.

4.3 Universal Composable Security Definition and Simple Privacy Condition

Recall that at the beginning of the game, one of κ and $\kappa_I + \mathcal{S}$ is chosen at random to interact with $\mathcal{E}$. The distinguishability-advantage is upper bounded by the trace distance of the two possible final states of $\mathcal{E}$ right before Γ is computed,

$$\left| \Pr(\Gamma{=}0 \,|\, \kappa) - \Pr(\Gamma{=}0 \,|\, \kappa_I{+}\mathcal{S}) \right| \leq \tfrac{1}{2} \left\| \rho_{\text{qkd}} - \rho_{\text{ideal}} \right\|_1 \leq \tag{10}$$

$$\leq \tfrac{1}{2} \left\| \rho_{\text{qkd}} - \rho_{\text{qi1}} \right\|_1 + \tfrac{1}{2} \left\| \rho_{\text{qi1}} - \rho_{\text{qi2}} \right\|_1 + \tfrac{1}{2} \left\| \rho_{\text{qi2}} - \rho_{\text{ideal}} \right\|_1 \;, \tag{11}$$

where ρ_{qi1} and ρ_{qi2} are hybrid, intermediate, states between ρ_{qkd} and ρ_{ideal} defined as

$$\rho_{\text{qi1}} = \sum_k \Pr(M{=}|k|)\, 2^{-|k|} |k,k\rangle\langle k,k| \otimes \rho_{E,k,k} \;, \tag{12}$$

$$\rho_{\text{qi2}} = \sum_k \Pr(M{=}|k|)\, 2^{-|k|} |k,k\rangle\langle k,k| \otimes \tilde{\rho}_{|k|} \;, \tag{13}$$

with $\bar{\rho}_m = \frac{1}{2^m} \sum_{k:|k|=m} \rho_{E,k,k}$. The sum of the first and the last terms in (11) can be bounded by μ_1 in the equality-and-uniformity condition ((3) in Sect. 2) as follows. Using (7) and (12),

$$\left\| \rho_{\text{qkd}} - \rho_{\text{qi1}} \right\|_1 = \left\| \sum_{k_A \neq k_B} \Pr(k_A, k_B) |k_A, k_B\rangle\langle k_A, k_B| \otimes \rho_{E,k_A,k_B} \right. \\ \left. + \sum_k \left[\Pr(k,k) - \Pr(|k|) 2^{-|k|} \right] |k,k\rangle\langle k,k| \otimes \rho_{E,k,k} \right\|_1 \leq \mu_1 .$$

Using (9) and (13),

$$\left\| \rho_{\text{qi2}} - \rho_{\text{ideal}} \right\|_1 \leq \sum_m \Pr(M=m) \left\| \bar{\rho}_m - \tilde{\rho}_m \right\|_1 \leq \mu_1$$

where we have used

$$\bar{\rho}_m = \sum_{k_A, k_B} p_{\text{ideal}}^{(m)}(k_A, k_B) \rho_{E,k_A,k_B} \; , \quad \tilde{\rho}_m = \sum_{k_A, k_B} p_{\text{qkd}}^{(m)}(k_A, k_B) \rho_{E,k_A,k_B} \; , \quad (14)$$

and the equality-and-uniformity condition (3) for the last inequality. The remaining term in the composable security condition (11) is given by

$$\frac{1}{2} \left\| \rho_{\text{qi1}} - \rho_{\text{qi2}} \right\|_1 = \frac{1}{2} \left\| \sum_k \Pr(M=|k|) \, 2^{-|k|} \, |k,k\rangle\langle k,k| \otimes \left[\bar{\rho}_{|k|} - \rho_{E,k,k} \right] \right\|_1$$

$$\leq \frac{1}{2} \sum_k \Pr(M=|k|) \, 2^{-|k|} \left\| \bar{\rho}_{|k|} - \rho_{E,k,k} \right\|_1 , \quad (15)$$

which can be interpreted as a new privacy condition.

We have thus compartmentalized the quantity governing the composable security definition for QKD, (10) or (11), into two parts: a term governed by the equality-and-uniformity condition (3) and a new term (15) related to privacy, a bound of which will be called a "composable privacy condition" for QKD. Once (15) is bounded by some μ_2^*, QKD using ideal authentication $\kappa + \alpha_I$ ϵ_κ-securely realizes the ideal KD κ_I, if $\mu_1 + \mu_2^* \leq \epsilon_\kappa$. Following Theorems 1 and 2, one can use the key "as if it were perfect." Proving such a bound on (15) is relatively straightforward, as compared to a direct proof of the security of using a slightly imperfect key from QKD (without the composability theorem).

In the following section, we prove several bounds for (15). First, we show that for any QKD scheme satisfying the usual privacy condition (5), (15) can be bounded as well, albeit with a potentially large but manageable degradation. Second, we prove a tighter bound on (15) assuming a privacy condition in terms of Eve's Holevo information on the key. Finally, we propose a new, tight, sufficient condition for bounding (10) (the full composable security condition) based on the singlet-fidelity considered in most existing security proofs for QKD. This bypasses (5) and incorporates all of equality, uniformity, and privacy. As an application, we obtain sharp upper bounds for (10) for existing QKD schemes.

5 Universal Composability of QKD

We state some composable security results of QKD; proofs can be found in [38].

Usual Privacy Condition Implies Composable Privacy Condition.
Given the usual privacy condition (5), $\sum_{m=0} \Pr(m) \times I(K_E : K \mid M = m) \leq \mu_2$, the following bound for (15) holds, ensuring composable privacy:

$$\left\| \rho_{\text{qi1}} - \rho_{\text{qi2}} \right\|_1 \leq 2^{\max(m)/2+1} \sqrt{\mu_2} \, . \tag{16}$$

Typically, $\max(m)$ is a small fraction of n, the security parameter such as the number of qubits communicated. Since $\mu_2 \in \mathcal{V}$, the set of exponentially decaying functions of n, bounding the key rate m/n ensures the above is in $\mathcal{V}$ also.

Small Holevo Information Implies Composable Privacy. Suppose, instead of the usual privacy condition (5) in terms of the accessible information, we have

- *Privacy:* $\exists \mu_2' \in \mathcal{V}$ s.t.

$$\sum_m \Pr(M=m) \times \chi(\mathcal{F}_m) \leq \mu_2' \tag{17}$$

where χ is the Holevo information [39], and $\mathcal{F}_m$ is the ensemble $\{2^{-m}, \rho_{E,k,k}\}_{|k|=m}$. Equation (17) is more stringent than (5) since the Holevo information is an upper bound for the accessible information. In fact, (17) implies

$$\left\| \rho_{\text{qi1}} - \rho_{\text{qi2}} \right\|_1 \leq \sqrt{2 (\ln 2) \, \mu_2'} \tag{18}$$

which does not have an overhead exponential in the length of the key generated.

A New Sufficient Condition for Composable Security. We can easily analyze the composable security of any QKD scheme that has a security proof based on entanglement purification protocol. All existing QKD schemes have such security proofs. The final keys K_A, K_B are outcomes of Alice and Bob's measurements on a shared state ρ_{AB}^m for some m, and ρ_{AB}^m is supposed to be $\Phi^{\otimes m}$ in the absence of eavesdropping. Here, m is again the key length and $\Phi = \frac{1}{2}(|00\rangle + |11\rangle)(\langle 00| + \langle 11|)$. The usual privacy condition (5) is often obtained by showing the following.

- *High fidelity:* $\exists \mu_2'' \in \mathcal{V}$ s.t.

$$\sum_m \Pr(m) \left[1 - F(\rho_{AB}^m, \Phi^{\otimes m}) \right] \leq \mu_2'' \tag{19}$$

(See Sect. 1 for the definition of F.) The above turns out implying a sharp bound on (15):

$$\frac{1}{2} \left\| \rho_{\text{qkd}} - \rho_{\text{ideal}} \right\|_1 \leq \sqrt{\mu_2''} \, . \tag{20}$$

Equation (19) is thus a good new sufficient condition for *composable security*, being part of the standard QKD proof and implying a tight bound on (10) simultaneously. It also implies *both* equality-and-uniformity and privacy (unlike a bound on Holevo information or mutual information which only implies the composable privacy condition).

6 Discussions and Applications

We have motivated this work with a discussion of the potential gap between the desired security of using a key generated by QKD and the security promised by the privacy condition (5) used in many previous studies of "unconditional security" of QKD. Then, we apply the universal composability theorem to obtain a new security condition that will guarantee the security of using a key generated from QKD. We propose a new composable privacy condition based on bounding (15), and we propose useful sufficient conditions such as bounds on (17) or (19). Most interesting of all, we show that a bound on the singlet-fidelity (19) directly implies the composable security condition (a bound on (10)). These are our main contributions (in the context of cryptography).

We also provide a proof that the existing privacy condition (5) does imply composable security (a bound on (15)) though the bound degrades exponentially in the key size. Despite the existence of such connections, we emphasize that future security proofs should bound (10), (15), (17), or (19) directly. We also provide a sharp bound on (15) based on Holevo's information (17) or singlet-fidelity (19). We show that most existing security proofs for QKD imply sharp bounds on (10), when bypassing the usual privacy condition (5). Outside the context of cryptography, these connections between various privacy conditions can be useful for the study of correlations in quantum systems.

The pathologies of the accessible information exhibited recently [16, 17] suggest a conjecture that, when going from (5) to (15), the degradation of the security parameter exponential in the key size is necessary.

As a final application, we analyze the security of repeating QKD t times, without assuming the availability of an authenticated classical channel. (Note that t is a fixed parameter that does not grow with the problem size.) Each run of QKD κ calls a composable authentication scheme α as a subroutine, and each run of α requires a composably secure key, which is provided by the previous round of κ (as a subroutine to α). Call the t rounds of QKD our protocol $\mathcal{P}$. The associated tree for $\mathcal{P}$, and the ideal realization $\mathcal{P}_I$ are given in the far left and right of Fig. 4.

If $\kappa + \alpha_I$ ϵ_κ-s.r. κ_I (as in (10)) and if $\alpha + \kappa_I$ ϵ_α-s.r. α_I, $\mathcal{P}$ $t(\epsilon_\kappa + \epsilon_\alpha)$-s.r. $\mathcal{P}_I$. In other words, each extra around of QKD degrades the overall distinguishability-advantage by an additive constant $(\epsilon_\kappa + \epsilon_\alpha)$. The same result can be obtained by using Theorem 2, or conversely, this simple exercise illustrates the idea behind Theorem 2.

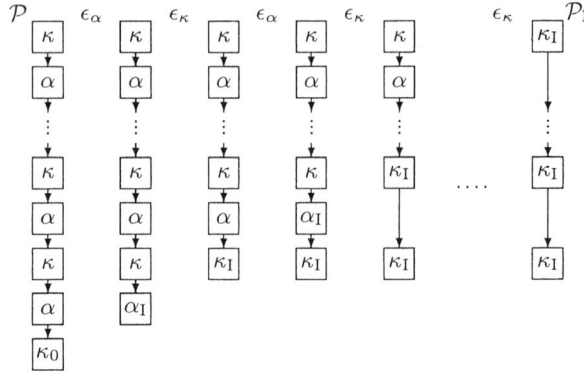

Fig. 4. Associated tree for t rounds of κ in the left. κ_0 represents some initially shared key. The arrows point from parents to children. Each tree to the right is obtained by replacing one node by its ideal functionality. The distinguishability-advantage of each pair of consecutive schemes is marked between their trees near the roots. Authentication is omitted in the ideal functionality $\mathcal{P}_\mathrm{I}$

Acknowledgement. We thank Charles Bennett, Daniel Gottesman, Aram Harrow, and John Smolin for interesting discussions on the security concerns of using a key obtained from QKD. We also thank Dominique Unruh and Jörn Müller-Quade for interesting discussions on their alternative framework of composability.

Part of this work was completed while MH and JO were visiting the MSRI program on quantum information, Berkeley, 2002. MB acknowledges the support of the Israel Science Foundation and a research grant from the Israeli Ministry of Defense. MH is supported by EU grants RESQ (IST-2001-37559) and QUPRODIS (IST-2001-38877). DL acknowledges the support from the Tolman Foundation and the Croucher Foundation. DL and DM acknowledge support from the US NSF under grant no. EIA-0086038. JO is supported by an EU grant PROSECCO (IST-2001-39227) and a grant from the Cambridge-MIT Institute.

A Notations

We gather notations frequently used in the paper, roughly in order of first appearance:
- KD: key distribution
- QKD: quantum key distribution
- Alice and Bob: two honest parties trying to establish a common key
- Eve: an active adversary
- A, B, E: subscripts labelling objects related to Alice, Bob, and Eve
  $\mathbb{A}, \mathbb{B}, \mathbb{E}$: labels of their respective quantum systems
- **Capitalized letters denote random variables and the corresponding uncapitalized letters denote particular outcomes**

- K_A, k_A, K_B, k_B: output keys for Alice and Bob
- K, k: $k := k_A$ when $k_A = k_B$
- M, m: length of key generated by QKD, with $M = 0$ iff QKD is aborted
- K_E, k_E: classical data possibly extracted by Eve at the end of QKD by measuring her quantum state
- $\Pr(\cdot)$: probability of the event "$\cdot$"
- log: logarithm in base 2
- For random variables X, Y, Z:
  $H(X) := -\sum_x \Pr(x) \log \Pr(x)$ is the entropy of X
  $I(X{:}Y) := H(X)+H(Y)-H(XY)$ is the mutual information between X, Y
  $I(X{:}Y|Z{=}z)$ is the mutual information between X, Y conditioned on $Z{=}z$
  $I(X{:}Y|Z) := \sum_z \Pr(z) I(X{:}Y|Z{=}z)$ is the conditional mutual information
- ρ: generic symbol for a density matrix
- $|\cdot\rangle$: a vector in a Hilbert space, with label "$\cdot$"
  $|\cdot\rangle\langle\cdot|$: the projector onto the subspace spanned by $|\cdot\rangle$, also known as "outer-product" of $|\cdot\rangle$ and $\langle\cdot|$
- $\text{Tr}(\cdot)$: the trace
- $\text{Tr}_{\mathbb{H}_1}(\cdot)$: the partial trace over the system $\mathbb{H}_1$. Let ρ_{12} be the density matrix for a joint state on $\mathbb{H}_1$ and $\mathbb{H}_2$. $\text{Tr}_{\mathbb{H}_1}(\rho_{12})$ is the state after $\mathbb{H}_1$ is discarded.
- $\|\cdot\|_1$: the trace distance, which can be taken as the sum of the singular values
- F: the fidelity. For two states ρ_1, ρ_2 in H, $F(\rho_1, \rho_2) = \max_{|\psi_1\rangle,|\psi_2\rangle} |\langle\psi_1|\psi_2\rangle|^2$ where $|\psi_{1,2}\rangle \in \mathbb{H} \otimes \mathbb{H}'$ are "purifications" of $\rho_{1,2}$ (i.e., $\text{Tr}_{\mathbb{H}'}|\psi_{1,2}\rangle\langle\psi_{1,2}| = \rho_{1,2}$), and $\langle\cdot|\cdot\rangle$ is the inner product. Here, we can take $\dim(\mathbb{H}') = \dim(\mathbb{H})$.
- ρ_{E,k_A,k_B}: Eve's view (both quantum and classical data) when the key outputs to Alice and Bob are k_A, k_B.
- n: security parameter such as the number of qubits communicated in QKD
- $p_{\text{qkd}}^{(m)}$: $p_{\text{qkd}}^{(m)}(k_A, k_B) = \Pr(K_A{=}k_A, K_B{=}k_B | M{=}m)$, i.e., the distribution of K_A, K_B generated in QKD conditioned on $|K_A| = |K_B| = m$
- $p_{\text{ideal}}^{(m)}$: the distribution over two m-bit strings l, l' defined as $p_{\text{ideal}}^{(m)}(l, l') = 0$ if $l \neq l'$, $p_{\text{ideal}}^{(m)}(l, l) = 2^{-m}$.
- $\mathcal{V}$: the set of exponentially decaying functions of n
- $\sigma, \mathcal{P}, \sigma_I, \mathcal{P}_I$: σ and $\mathcal{P}$ are generic labels for protocols, with σ possibly used as a subroutine. The symbol of a protocol with a subscript I denotes the ideal functionality of the protocol. $\mathcal{P}{+}\sigma$: a protocol $\mathcal{P}$ calling a subroutine σ.
- $\mathcal{E}, \mathcal{S}$: the environment and the simulator. These are sets of registers and operations and they are sometimes personified in our discussion.
- Γ: output bit of $\mathcal{E}$
- ϵ-s.r. : $\mathcal{P}$ ϵ-s.r. $\mathcal{P}_I$ is a shorthand for $\mathcal{P}$ ϵ-securely realizes $\mathcal{P}_I$ (see mathematical definition in (6)). ϵ is called the *distinguishability-advantage* between $\mathcal{P}$ and $\mathcal{P}_I$.
- $T_\mathcal{P}$: the associated tree for a protocol $\mathcal{P}$
- α, α_I: universal composable authentication with negligible key requirement and its ideal functionality
- $\kappa{+}\alpha$, $\kappa{+}\alpha_I$, κ_I: QKD using authentication α, QKD using ideal authentication α_I, and ideal KD defined in Sect. 4.2

- Devil: an adversary that determines the key length m generated by κ_{I}
- ρ_{qkd}: state possessed by $\mathcal{E}$ after interacting with $\kappa + \alpha_{\mathrm{I}}$, see (7)
- ρ_{ideal}: state possessed by $\mathcal{E}$ after interacting with κ_{I}, see (9)
- $\rho_{\mathrm{qi1}}, \rho_{\mathrm{qi2}}$: hybrid, intermediate, states between ρ_{qkd} and ρ_{ideal}, see (12), (13)
- $\tilde{\rho}_m$: Eve's state when $M = m$, averaged over K_{A}, K_{B}. See (8)
- $\bar{\rho}_m$: uniform average of $\rho_{\mathrm{E},k,k}$ for $|k| = m$. See the line right after (13)
- **Ensemble $\{q_x, \varrho_x\}_x$: a distribution $\{q_x\}_x$ of quantum states ϱ_x**
- I_{acc}: accessible information of an ensemble $\{q_x, \varrho_x\}_x$, i.e., the maximum mutual information between X and outcome Y obtained from measuring a specimen ϱ_x
- $\mathcal{F}_m$: the ensemble $\{2^{-m}, \rho_{\mathrm{E},k,k}\}_{|k|=m}$
- $\chi(\{q_x, \varrho_x\})$: Holevo information of the ensemble $\{q_x, \varrho_x\}$, given by $S(\sum_x q_x \varrho_x) - \sum_x q_x S(\varrho_x)$ where $S(\cdot) = \mathrm{Tr}(\cdot \log(\cdot))$ is the von Neumann entropy
- ρ_{AB}^m: state on which measurements by Alice and Bob output K_{A}, K_{B} in QKD-security-proofs based on entanglement purification
- Φ: a perfect EPR pair $\frac{1}{2}(|00\rangle + |11\rangle)(\langle 00| + \langle 11|)$
- Singlet fidelity: $F(\rho_{\mathrm{AB}}^m, \Phi^{\otimes m})$. Note that "singlet" usually refers to a state that is only unitarily equivalent to Φ, but we borrow the term in this paper.

References

1. C. Bennett and G. Brassard. Quantum cryptography: Public key distribution and coin tossing. In *Proceedings of IEEE International Conference on Computers, Systems and Signal Processing*, pages 175–179, New York, 1984. IEEE. Bangalore, India, December 1984.
2. A. Ekert. Quantum cryptography based on Bell's theorem. *Phys. Rev. Lett.*, 67(6):661–663, 1991.
3. C. Bennett. Quantum cryptography using any two nonorthogonal states. *Phys. Rev. Lett.*, 68(21):3121–3124, 1992.
4. C. Bennett, G. Brassard, R. Jozsa, D. Mayers, A. Peres, B. Schumacher, and W. Wootters. Reduction of quantum entropy by reversible extraction of classical information. *Journal of Modern Optics*, 41(12):2307–2314, 1994.
5. D. Mayers. Quantum key distribution and string oblivious transfer in noisy channels. In *Advances in Cryptography–Proceedings of Crypto'96*, pages 343–357, New York, 1996. Springer-Verlag.
6. D. Mayers. Unconditional security in quantum cryptography. *J. Assoc. Comp. Mach*, 48:351, 2001. quant-ph/9802025.
7. H.-K. Lo and H. F. Chau. Unconditional security of quantum key distribution over arbitrarily long distances. *Science*, 283:2050–2056, 1999. quant-ph/9803006.
8. E. Biham, M. Boyer, P. Boykin, T. Mor, and V. Roychowdhury. A proof of the security of quantum key distribution. In *Proceedings of the 32nd Annual ACM Symposium on Theory of Computing (STOC)*, pages 715–724, New York, 2000. ACM. quant-ph/9912053.
9. P. Shor and J. Preskill. Simple proof of security of the bb84 quantum key distribution protocol. *Phys. Rev. Lett.*, 85:441–444, 2000. quant-ph/0003004.
10. K. Tamaki, M. Koashi, and N. Imoto. Unconditionally secure key distribution based on two nonorthogonal states. *Phys. Rev. Lett.*, 90:167904, 2003. quant-ph/0212162.

11. G. Gottesman and H.-K. Lo. Proof of security of quantum key distribution with two-way classical communications. *IEEE Transactions on Information Theory*, 49(2):457–475, 2003. quant-ph/0105121.
12. C. Bennett and J. Smolin first suggested the key degradation problem to one of us, and A. Harrow has obtained partial results.
13. A. Ambainis, M. Mosca, A. Tapp, and R. de Wolf. Private quantum channels. In *IEEE Symposium on Foundations of Computer Science (FOCS)*, pages 547–553, 2000. quant-ph/0003101.
14. P. Boykin and V. Roychowdhury. Optimal encryption of quantum bits. quant-ph/0003059.
15. A. Peres and W. Wootters. Optimal detection of quantum information. *Phys. Rev. Lett.*, 66:1119–1122, 1991.
16. D. DiVincenzo, M. Horodecki, D. Leung, J. Smolin, and B. Terhal. Locking classical correlation in quantum states. *Phys. Rev. Lett.*, 92:067902, 2004. quant-ph/0303088.
17. P. Hayden, D. Leung, P. Shor, and A. Winter. Randomizing quantum state: constructions and applications. quant-ph/0307104.
18. R. Canetti. Universal composable security: A new paradigm for cryptographic protocols. In *Proceedings of the 42nd IEEE Symposium on Foundations of Computer Science (FOCS)*, pages 136–145. IEEE, 2001.
19. M. Ben-Or and D. Mayers. Composability theorem. Part I of presentation by D. Mayers, QIP 2003, MSRI, Berkeley. See http://www.msri.org/publications/ln/msri/2002/qip/mayers/1/index.html .
20. M. Ben-Or and D. Mayers. Composing quantum and classical protocols. quant-ph/0409062.
21. M. Backes, B. Pfitzmann, and M. Waidner. A general composition theorem for secure reactive systems. In *First Theory of Cryptography Conference (TCC)*, pages 336–354, 2004.
22. D. Unruh. Relating formal security for classical and quantum protocols. Presentation at the Special week on Quantum crytography, Isaac Newton Institute for Mathematical Sciecnes, September 2004. Available at http://www.unruh.de/DniQ/publications.
23. D. Unruh. Simulation security for quantum protocols. quant-ph/0409125.
24. M. Ben-Or, M. Horodecki, D. Leung, D. Mayers, and J. Oppenheim. Composability of QKD. Part II of presentation by D. Mayers, QIP 2003, MSRI, Berkeley. See http://www.msri.org/publi-cations/ln/msri/2002/qip/mayers/1/index.html .
25. M. Ben-Or, M. Horodecki, D. Leung, D. Mayers, and J. Oppenheim. Composability of quantum proocols - applications to quantum key distribution and quantum authentication. Part II of presentation by D. Leung, QIP 2004, IQC, University of Waterloo. See http://www.iqc.ca/conferences/qip/presentations/leung-.pdf.
26. R. Renner and Konig. Universally composable privacy amplification against quantum adversaries. quant-ph/0403133.
27. M. Christandl, R. Renner, and A. Ekert. A generic security proof for quantum key distribution. quant-ph/0402131.
28. M. Nielsen and I. Chuang. *Quantum computation and quantum information*. Cambridge University Press, Cambridge, U.K., 2000.
29. M. Wegman and J. Carter. New hash functions and their use in authentication and set equality. *Journal of Computer and System Sciences*, 22:265–279, 1981.
30. D. Bruss. Optimal eavesdropping in quantum cryptography with six states. *Phys. Rev. Lett.*, 81:3018–3021, 1998.

31. H. Bechmann-Pasquinucci and N. Gisin. Incoherent and coherent eavesdropping in the 6-state protocol of quantum cryptography. quant-ph/9807041.
32. D. Deutsch, A. Ekert, R. Jozsa, C. Macchiavello, S. Popescu, and A. Sanpera. Quantum privacy amplification and the security of quantum cryptography over noisy channels. *Phys. Rev. Lett.*, 77:2818, 1996. quant-ph/9604039.
33. T. Cover and J. Thomas. *Elements of Information Theory*. John Wiley and Sons, New York, 1991.
34. A. Yao. Quantum circuit complexity. *Proc. of the 34th Ann. IEEE Symp. on Foundations of Computer Science*, pages 352–361, 1993.
35. D. Aharonov, A. Kitaev, and N. Nisan. Quantum circuits with mixed states. quant-ph/9806029.
36. An acyclic circuit is a partially ordered set of gates. However, associating the circuit with constraints on the timing of the adversarial attack is a delicate issue. Suppose the circuit contains conditional gates controlled by random public classical registers. The gates on the target may or may not be applied depending on the values of the control registers. When the gates are not applied, the associated time-constraints of the adversarial attack disappear. In the extension to the usual acyclic circuit model, we consider all possible values of the control registers and the resulting sets of *nontrivial* partially ordered operations, and the corresponding constraints on the adversarial attack.
37. P. Hayden, D. Leung, and D. Mayers. On the composability of quantum message authentication and key recycling.
38. M. Ben-Or, M. Horodecki, D. Leung, D. Mayers, and J. Oppenheim. The universal composable security of quantum key distribution. quant-ph/0409078.
39. A. Holevo. Information-theoretical aspects of quantum measurement. *Problemy Peredachi Informatsii*, 9(2):31–42, 1973. [A. S. Kholevo, *Problems of Information Transmission*, vol. 9, pp. 110-118 (1973)].

Universally Composable Privacy Amplification Against Quantum Adversaries

Renato Renner and Robert König

Swiss Federal Institute of Technology (ETH), Zürich, Switzerland
{renner, rkoenig}@inf.ethz.ch

Abstract. Privacy amplification is the art of shrinking a partially secret string Z to a highly secret key S. We show that, even if an adversary holds quantum information about the initial string Z, the key S obtained by two-universal hashing is secure, according to a universally composable security definition. Additionally, we give an asymptotically optimal lower bound on the length of the extractable key S in terms of the adversary's (quantum) knowledge about Z. Our result has applications in quantum cryptography. In particular, it implies that many of the known quantum key distribution protocols are universally composable.

1 Introduction

1.1 Privacy Amplification

Consider two parties having access to a common string Z about which an adversary might have some partial information. *Privacy amplification*, introduced by Bennett, Brassard, and Robert [10], is the art of transforming this partially secure string Z into a highly secret key S by public discussion. A good technique is to compute S as the output of a publicly chosen two-universal hash function[1] F applied to Z. Indeed, it has been shown [10, 21, 9] that, if the adversary holds purely classical information W about Z, this method yields a secure key S and, additionally, is asymptotically optimal with respect to the length of S. For instance, if both the initial string Z and the adversary's knowledge W consist of many independent and identically distributed parts, the number of extractable key bits roughly equals the conditional Shannon entropy $H(Z|W)$.

The analysis of privacy amplification can be extended to a situation where the adversary holds quantum instead of only classical information about Z. This generalizes the classical setting in a non-trivial way. In particular, the adversary might store her quantum information until she learns the hash function F (which is publicly chosen) and then perform a measurement depending on F. This might allow her to obtain more information about the function output (i.e., the resulting key S) than if she had measured her state at the beginning (independently of F).

---

[1] See Section 2.1 for a definition of two-universal functions.

1.2 Universal Composability

Cryptographic primitives (such as a secret key or an authentic communication channel) are often used as components within a larger system (e.g., a system for secure message transmission usually makes use of a secret key for encryption). It is thus natural to require that the security of these components is not compromised when they are used in any (arbitrarily complex) scheme. This requirement is captured by the notion of universal composability. Roughly speaking, a cryptographic primitive is said to provide *universally composable security* if it is secure in *any* arbitrary context. For instance, the universally composable security of a secret key S guarantees that any bit of S remains secret even if some other part of S is given to an adversary.

In the past few years, composable security has attracted a lot of interest and led to important new definitions and proofs (see, e.g., the framework of Canetti [11] or Pfitzmann and Waidner [27]). Recently, Ben-Or and Mayers [5, 6] and Unruh [30] have generalized the notion of universal composability to the quantum world. Universally composable security definitions are usually based on the idea of characterizing the security of a cryptographic scheme by its distance to an ideal system which (by definition) is perfectly secure. For instance, a secret key S is said to be secure if it is close to an independent and almost uniformly distributed string U. This implies that any cryptosystem which is proven secure when using a perfect key U remains secure when U is replaced by the (real) key S.

Unfortunately, most of the existing security definitions in quantum cryptography do not provide universal composability. For instance, the security of the key S generated by a quantum key distribution (QKD) scheme is usually defined by the requirement that the mutual information between S and the classical outcome W obtained from an arbitrary measurement of the adversary's quantum system be small (for a formal definition, see, e.g., [26] or [18]). This, however, does not necessarily imply composability. Indeed, an adversary might wait with the measurement of her quantum state until she learns some of the bits of S, which possibly allows her to obtain information about the remaining bits (cf. Section 3).

1.3 Contributions

We address the problem of privacy amplification in a setting where an adversary holds quantum information. We show that, by two-universal hashing, one can obtain a key S which is secure according to a universally composable security definition. This means that, in any context, S is virtually as secure as a perfect key, i.e., a uniformly distributed string U which is completely independent of the adversary's knowledge. This has implications in quantum cryptography. In particular, since the security of many of the known QKD protocols such as BB84 [8] or B92 [7] can be proven based on the security of privacy amplification (cf. [13] and [22], or [23]), it follows immediately from our results that these protocols provide universal composability (cf. Section 4.5).

Our main technical result (Section 4) is an easily computable lower bound on the length of the extractable key S in terms of (smooth) Rényi entropy (see Section 2.4 for a definition of smooth Rényi entropy). The bound is asymptotically tight if the initial information Z as well as the adversary's (quantum) knowledge consist of n independent pieces, for n approaching infinity (Section 4.4).

1.4 Related Work

The problem of privacy amplification against quantum adversaries has first been studied for the case where the adversary can only store a certain limited number of qubits. Based on a result on communication complexity [1], Ben-Or [2] argued that it is possible to extract at least one secret bit from a uniformly distributed string Z, if Z is sufficiently longer than the size of the adversary's storage device. In [22], it is shown that two-universal hashing allows for the extraction of a secure key S whose length roughly equals the difference between the entropy of the original string Z and the number of qubits stored by the adversary. The security definition used in [22] does, however, not provide universal composability. Simultaneously, Devetak and Winter [15] gave a full analysis of privacy amplification for the special case where the initial string Z as well as the adversary's information consist of many independent pieces. Interestingly, their result can be reproduced from our general bound (Section 4.4).

Ben-Or, Horodecki, Leung, Mayers, and Oppenheim [3, 4] were the first to address the problem of universal composability in the context of QKD. Our security definition (cf. Definition 3 in Section 3) is essentially equivalent to the definitions proposed in [3, 4], which are based on the framework developed in [6]. More precisely, if S is ε-secure according to our definition, it satisfies the security definition of [3] for some parameter ε' depending on ε. It is thus an immediate consequence of the results in [3] that our security definition provides universal composability with respect to the framework of [6].

2 Preliminaries

2.1 Random Functions and Two-Universal Functions

A *random function* F from $\mathcal{X}$ to $\mathcal{Y}$ is a random variable taking values from the set of functions with domain $\mathcal{X}$ and range $\mathcal{Y}$. F is called a *two-universal* (random) function if $\Pr_{f \leftarrow P_F}[f(x) = f(x')] \leq \frac{1}{|\mathcal{Y}|}$, for any distinct $x, x' \in \mathcal{X}$.[2] In particular, F is two-universal if, for any distinct $x, x' \in \mathcal{X}$, the random variables $F(x)$ and $F(x')$ are independent and uniformly distributed. For instance, the random function distributed uniformly over the set of all functions from $\mathcal{X}$ to $\mathcal{Y}$ is two-universal. Examples of two-universal functions requiring less randomness can, e.g., be found in [12] and [31].

---

[2] In the literature, two-universality is usually defined for families $\mathcal{F}$ of functions: A family $\mathcal{F}$ is called *two-universal* if the random function F with uniform distribution over $\mathcal{F}$ is two-universal.

2.2 Density Operators and Random States

Let $\mathcal{H}$ be a Hilbert space. We denote by $\mathcal{P}(\mathcal{H})$ the set of non-negative (hermitian) operators ρ on $\mathcal{H}$ with $\text{tr}(\rho) \leq 1$, and call its elements *density operators*. We say that $\rho \in \mathcal{P}(\mathcal{H})$ is *normalized* if $\text{tr}(\rho) = 1$. A normalized density operator $\rho \in \mathcal{P}(\mathcal{H})$ is called *pure* if it has rank 1, i.e., $\rho = P_{|\phi\rangle}$ for some vector $|\phi\rangle \in \mathcal{H}$ (where $P_{|\phi\rangle}$ denotes the projector along $|\phi\rangle$).

We will be concerned with settings involving both classical and quantum information. More precisely, we will consider a situation where the state $\rho_x \in \mathcal{P}(\mathcal{H})$ of a quantum system depends on the value x of a classical random variable X with range $\mathcal{X}$. Note that ρ_X is then itself a random variable with range $\mathcal{P}(\mathcal{H})$. In the following, we call such a random variable with range $\mathcal{P}(\mathcal{H})$ a *random state* on $\mathcal{H}$, and denote it by a bold symbol $\boldsymbol{\rho}$. We say that the random state $\boldsymbol{\rho}$ is *normalized* if $\text{tr}(\boldsymbol{\rho}) \equiv 1$.

It is often convenient to represent classical information as a state of a quantum system. Let $\mathcal{X}$ be a set and let $\mathcal{H}$ be a Hilbert space with orthonormal basis $\{|x\rangle\}_{x \in \mathcal{X}}$. The *state representation* of $x \in \mathcal{X}$, denoted $\{x\}$, is defined as the projector along $|x\rangle$, i.e., $\{x\} := P_{|x\rangle}$. In particular, for a random variable X on $\mathcal{X}$, $\{X\}$ is a random state on $\mathcal{H}$.

Consider a quantum system described by a random state $\boldsymbol{\rho}$ on $\mathcal{H}$, i.e., if the random variable $\boldsymbol{\rho}$ takes the value ρ, then the system is in state ρ. For an observer which is ignorant of the value of the random variable $\boldsymbol{\rho}$, the system is described by the density operator $[\boldsymbol{\rho}]$ defined as the expectation value of $\boldsymbol{\rho}$,

$$[\boldsymbol{\rho}] := \mathop{\text{E}}_{\rho \leftarrow P_{\boldsymbol{\rho}}} [\rho] = \sum_{\rho \in \mathcal{P}(\mathcal{H})} P_{\boldsymbol{\rho}}(\rho) \rho ,$$

where $P_{\boldsymbol{\rho}}$ is the probability distribution of $\boldsymbol{\rho}$. More generally, for any event $\mathcal{E}$, we define

$$[\boldsymbol{\rho}|\mathcal{E}] := \mathop{\text{E}}_{\rho \leftarrow P_{\boldsymbol{\rho}|\mathcal{E}}} [\rho] .$$

Let X be a random variable and let $\boldsymbol{\rho}$ be a random state. The random state $\{X\} \otimes \boldsymbol{\rho}$ then describes a system consisting of both a state representation of X and a quantum subsystem which is in state $\rho_x := [\boldsymbol{\rho}|X = x]$ whenever X takes the value x. The density operator $[\{X\} \otimes \boldsymbol{\rho}]$ of the overall system is thus given by

$$[\{X\} \otimes \boldsymbol{\rho}] = \mathop{\text{E}}_{x \leftarrow P_X} [P_{|x\rangle} \otimes \rho_x] = \sum_{x \in \mathcal{X}} P_X(x) P_{|x\rangle} \otimes \rho_x . \tag{1}$$

In particular, $[\{X\} \otimes \boldsymbol{\rho}] = [\{X\}] \otimes [\boldsymbol{\rho}]$ if and only if X is independent of $\boldsymbol{\rho}$.

2.3 Distance Measures and Non-uniformity

The *variational distance* between two probability distributions P and Q over the same range $\mathcal{X}$ is defined by

$$\delta(P, Q) := \frac{1}{2} \sum_{x \in \mathcal{X}} |P(x) - Q(x)| .$$

The variational distance between P and Q can be interpreted as the probability that two random experiments described by P and Q, respectively, are different. This is formalized by the following lemma.

Lemma 1. *Let P and Q be two probability distributions. Then there exists a joint probability distribution $P_{XX'}$ such that $P_X = P$, $P_{X'} = Q$, and*

$$\Pr_{(x,x') \leftarrow P_{XX'}}[x \neq x'] = \delta(P,Q) \ .$$

The *trace distance* between two density operators ρ and σ on the same Hilbert space $\mathcal{H}$ is defined as

$$\delta(\rho, \sigma) := \frac{1}{2} \mathrm{tr}(|\rho - \sigma|) \ .$$

The trace distance is a metric on the set of density operators $\mathcal{P}(\mathcal{H})$. We say that ρ is ε-*close* to σ if $\delta(\rho, \sigma) \leq \varepsilon$, and denote by $\mathcal{B}^\varepsilon(\rho)$ the set of density operators which are ε-close to ρ, i.e., $\mathcal{B}^\varepsilon(\rho) = \{\sigma \in \mathcal{P}(\mathcal{H}) : \delta(\rho,\sigma) \leq \varepsilon\}$.

The trace distance is subadditive with respect to the tensor product, i.e., for any $\rho, \sigma \in \mathcal{P}(\mathcal{H})$ and $\rho', \sigma' \in \mathcal{P}(\mathcal{H}')$,

$$\delta(\rho \otimes \rho', \sigma \otimes \sigma') \leq \delta(\rho,\sigma) + \delta(\rho',\sigma') \ , \tag{2}$$

with equality if $\rho' = \sigma'$ is normalized,

$$\delta(\rho \otimes \rho', \sigma \otimes \rho') = \delta(\rho,\sigma) \ . \tag{3}$$

Moreover, $\delta(\cdot, \cdot)$ cannot increase when the same quantum operation $\mathcal{E}$ is applied to both arguments, i.e.,

$$\delta(\mathcal{E}(\rho), \mathcal{E}(\sigma)) \leq \delta(\rho,\sigma) \ . \tag{4}$$

The variational distance can be seen as a (classical) special case of the trace distance. Let X and Y be random variables. Then the variational distance between the probability distributions of X and Y equals the trace distance between the state representations $[\{X\}]$ and $[\{Y\}]$, i.e.,

$$\delta(P_X, P_Y) = \delta([\{X\}], [\{Y\}]) \ .$$

In particular, it follows directly from (4) that the trace distance between two normalized density operators ρ and σ is an upper bound for the variational distance between the probability distributions P_X and P_Y of the outcomes when applying the same measurement to ρ and σ, respectively, i.e.,

$$\delta(P_X, P_Y) \leq \delta(\rho,\sigma) \ . \tag{5}$$

The trace distance between two density operators involving a state representation of the same classical random variable X can be written as the expectation of the trace distance between the density operators conditioned on X.

Lemma 2. *Let X be a random variable and let ρ and σ be random states. Then*

$$\delta(\lbrack\{X\}\otimes\rho\rbrack,\lbrack\{X\}\otimes\sigma\rbrack) = \operatorname*{E}_{x \leftarrow P_X}\lbrack\delta(\rho_x, \sigma_x)\rbrack$$

where $\rho_x := \lbrack\rho|X=x\rbrack$ and $\sigma_x := \lbrack\sigma|X=x\rbrack$.

Proof. Using (1) and the orthogonality of the vectors $|x\rangle$, we can write

$$\delta(\lbrack\{X\}\otimes\rho\rbrack,\lbrack\{X\}\otimes\sigma\rbrack) = \frac{1}{2}\operatorname{tr}\left(\left| \operatorname*{E}_{x \leftarrow P_X}\lbrack P_{|x\rangle} \otimes (\rho_x - \sigma_x)\rbrack\right|\right)$$
$$= \frac{1}{2}\operatorname{tr}\left(\operatorname*{E}_{x \leftarrow P_X}\lbrack|P_{|x\rangle} \otimes (\rho_x - \sigma_x)|\rbrack\right) .$$

The assertion then follows from the linearity of the trace and the fact that $\operatorname{tr}(|P_{|x\rangle} \otimes (\rho_x - \sigma_x)|) = \operatorname{tr}(|\rho_x - \sigma_x|)$. □

In Section 3, we will see that a natural measure for characterizing the secrecy of a key is its trace distance to a uniform distribution. This motivates the following definition.

Definition 1. *Let X be a random variable with range $\mathcal{X}$ and let ρ be a random state. The* non-uniformity *of X given ρ is defined by*

$$d(X|\rho) := \delta(\lbrack\{X\}\otimes\rho\rbrack, \lbrack\{U\}\rbrack\otimes\lbrack\rho\rbrack)$$

where U is a random variable uniformly distributed on $\mathcal{X}$.

Note that $d(X|\rho) = 0$ if and only if X is uniformly distributed and independent of ρ.

2.4 (Smooth) Rényi Entropy

Let $\rho \in \mathcal{P}(\mathcal{H})$ be a density operator and let $\alpha \in [0,\infty]$. The *Rényi entropy of order α of ρ* is defined by[3]

$$S_\alpha(\rho) := \frac{1}{1-\alpha}\log(\operatorname{tr}(\rho^\alpha))$$

with the convention $S_\alpha(\rho) := \lim_{\beta \to \alpha} S_\beta(\rho)$ for $\alpha \in \{0, 1, \infty\}$.[4] In particular, for $\alpha = 0$, $S_0(\rho) = \log(\operatorname{rank}(\rho))$ and, for $\alpha = \infty$, $S_\infty(\rho) = -\log(\lambda_{\max}(\rho))$ where $\lambda_{\max}(\rho)$ denotes the maximum eigenvalue of ρ. Note that, for a classical random variable X, the Rényi entropy $S_\alpha(\lbrack\{X\}\rbrack)$ of the state representation of X corresponds to the (classical) Rényi entropy $H_\alpha(X)$ of X [29].

The notion of ε-*smooth* Rényi entropy H_α^ε has been introduced in [28] for the classical case, and can be seen as a generalization of (conventional) Rényi

---

[3] All logarithms in this paper are binary.
[4] Note that, for this definition, the density operator ρ must not necessarily be normalized.

entropy H_α (see Appendix C for a definition). Smooth Rényi entropy is useful for characterizing basic properties of random variables such as the amount of extractable randomness or the minimum encoding length. Moreover, it has natural properties similar to Shannon entropy.

Definition 2 below generalizes classical smooth Rényi entropy H_α^ε to density operators. This quantum version of smooth Rényi entropy will be useful to state our main results.

Definition 2. *Let $\rho \in \mathcal{P}(\mathcal{H})$ and let $\varepsilon \geq 0$. The ε-smooth Rényi entropy of order α of ρ is defined by[5]*

$$S_\alpha^\varepsilon(\rho) := \frac{1}{1-\alpha} \log \left(\inf_{\sigma \in \mathcal{B}^{\varepsilon/2}(\rho)} \left(\text{tr}(\sigma^\alpha) \right) \right) ,$$

for $\alpha \in (0,1) \cup (1,\infty)$, and $S_\alpha^\varepsilon(\rho) := \lim_{\beta \to \alpha} S_\beta^\varepsilon(\rho)$, for $\alpha \in \{0,\infty\}$.

The classical definition of smooth Rényi entropy can be seen as a special case of Definition 2. In particular, the smooth Rényi entropy $H_\alpha^\varepsilon(X)$ of a classical random variable X is equal to the smooth Rényi entropy $S_\alpha^\varepsilon([\{X\}])$ of the state representation of X. On the other hand, the smooth Rényi entropy of a density operator ρ can be expressed in terms of the classical smooth Rényi entropy of its eigenvalues. Formally,

$$S_\alpha^\varepsilon(\rho) = H_\alpha^\varepsilon(P) , \qquad (6)$$

where P is the (not necessarily normalized) probability distribution defined by the eigenvalues $\lambda_1, \ldots, \lambda_d$ of ρ, i.e., $P(i) = \lambda_i$, for $i \in \{1, \ldots, d\}$.

It is important to note that equation (6) provides an efficient method for *computing* the smooth Rényi entropy $S_\alpha^\varepsilon(\rho)$ of a given density operator ρ. In particular, since the smooth Rényi entropy $H_\alpha^\varepsilon(P)$ of a classical probability distribution P can be calculated in a simple way (see Appendix C), it is also easy to compute $S_\alpha^\varepsilon(\rho)$ if the eigenvalues of ρ are known.

The following lemma is a direct generalization of the corresponding statement for classical smooth Rényi entropy (see Lemma 15 in Appendix C) saying that the smooth Rényi entropy $H_\alpha^\varepsilon(Z^n)$ of a random variable Z^n consisting of many independent and identically distributed pieces asymptotically equals its Shannon entropy $H(Z^n)$.

Lemma 3. *Let ρ be a normalized density operator. Then, for any $\alpha \in [0,\infty]$,*

$$\lim_{\varepsilon \to 0} \lim_{n \to \infty} \frac{1}{n} S_\alpha^\varepsilon(\rho^{\otimes n}) = S(\rho) ,$$

where $S(\rho)$ denotes the von Neumann entropy of ρ.

---

[5] Recall that $\mathcal{B}^{\varepsilon/2}(\rho)$ denotes the set of non-negative operators $\sigma \in \mathcal{P}(\mathcal{H})$ such that $\delta(\sigma,\rho) \leq \frac{\varepsilon}{2}$, i.e., $\text{tr}(|\sigma - \rho|) \leq \varepsilon$.

3 Secret Keys and Composability

A very intuitive way of defining the security of a *real* cryptographic protocol is to compare it with an *ideal* functionality. The ideal functionality of a secret key S is simply an independent and uniformly distributed random variable U (in particular, U is fully independent of the adversary's information). This motivates the following definition.

Definition 3. *Let S be a random variable, let ρ be a random state, and let $\varepsilon \geq 0$. S is said to be ε-secure with respect to ρ if $d(S|\rho) \leq \varepsilon$.*

Consider a situation where S is used as a secret key and where the adversary's information is given by a random state ρ. If S is ε-secure with respect to ρ then it is guaranteed that this situation is ε-close—with respect to the trace distance—to an ideal setting where S is replaced by a perfect key U which is uniformly distributed and independent of ρ. Since the trace distance does not increase when appending an additional quantum system (cf. (2) or (3)) or when applying any arbitrary quantum operation (cf. (4)), this also holds for any further evolution of the system. In particular, it follows from (5) and Lemma 1 that the real and the ideal setting can be considered to be identical with probability at least $1 - \varepsilon$.

Note that our security definition can be seen as a natural generalization of classical security definitions based on the variational distance (which is the classical analogue of the trace distance). Indeed, if the adversary's knowledge is purely classical, Definition 3 is equivalent to a security definition as it is, e.g., used in [17].

The security of a key S according to Definition 3 implies that S is also secure according to many of the widely used security definitions in quantum cryptography. One of the most popular security requirements for a key S with respect to an adversary holding information ρ is that S be almost independent of the classical outcome W resulting from any arbitrary measurement of ρ.[6] Obviously, if a key S is ε-secure with respect to ρ (according to our definition), the probability distribution P_{SW} is ε-close (with respect to the variational distance) to a product distribution. Note, however, that the converse is not true: Even if S and W are almost independent for any measurement of ρ, the quantum state ρ might still strongly depend on S.

Indeed, security definitions which are formulated in terms of the adversary's measurement results W do not necessarily provide universal composability: If it is only known that a key S is almost independent of the classical outcome W obtained from measuring the quantum state ρ—for any measurement strategy chosen independently of S—, one cannot necessarily use S in any arbitrary cryptosystem, e.g., as a one-time pad. Consider for instance a cryptographic application where S consists of two parts S_1 and S_2, and where S is used in such a way that an adversary learns S_1. Hence, the adversary can let the measurement of her quantum system depend on the specific value of S_1. This might provide

---

[6] See, e.g., [26], and the references therein.

4 Main Result

4.1 Theorem and Proof

Consider a situation where an adversary holds quantum information ρ about a classical string Z. Additionally, let S be a key of length s computed by applying a (publicly chosen) two-universal function F to Z, that is, $S := F(Z)$. Theorem 1 below states that, if the length s is chosen to be sufficiently smaller than $\bar{s} := S_2([\{Z\} \otimes \rho]) - S_0([\rho])$, then the key S is ε-secure with respect to ρ (for ε decreasing exponentially fast in the difference $\bar{s} - s$). In other words, a two-universal function F can be used to turn a partially secure string Z into a highly secure key S of length roughly $\bar{s}$. In Section 4.3, we will discuss this application in more detail.

Theorem 1. *Let Z be a random variable with range $\mathcal{Z}$, let ρ be a random state, and let F be a two-universal function from $\mathcal{Z}$ to $\mathcal{S} = \{0,1\}^s$ which is independent of Z and ρ. Then*

$$d(F(Z)|\{F\} \otimes \rho) \leq \frac{1}{2} 2^{-\frac{1}{2}(S_2([\{Z\} \otimes \rho]) - S_0([\rho]) - s)} .$$

Let us state some technical lemmas to be used for the proof of Theorem 1.

Lemma 4. *Let Z be a random variable with range $\mathcal{Z}$, let ρ be a random state, and let F be a random function on $\mathcal{Z}$ which is independent of Z and ρ. Then*

$$d(F(Z)|\{F\} \otimes \rho) = \mathop{\mathrm{E}}_{f \leftarrow P_F}[d(f(Z)|\rho)] .$$

Proof. Let U be a random variable uniformly distributed on the range of F and independent of F and ρ. Then

$$d(F(Z)|\{F\} \otimes \rho) = \delta([\{F(Z)\} \otimes \rho \otimes \{F\}], [\{U\} \otimes \rho \otimes \{F\}]) .$$

Now, applying Lemma 2 to the random states $\{F(Z)\} \otimes \rho$ and $\{U\} \otimes \rho$ gives the desired result since

$$[\{F(Z)\} \otimes \rho | F = f] = [\{f(Z)\} \otimes \rho]$$
$$[\{U\} \otimes \rho | F = f] = [\{U\}] \otimes [\rho] ,$$

which holds because F is independent of Z, ρ, and U. □

---

[7] The effect of side information on the maximum classical correlation that can be obtained by measurements has been studied in different contexts [16, 19]. A simple example which demonstrates that classical information is indeed helpful for choosing a "good" measurement is as follows: Let S_1 and S_2 be random bits and let ρ be the state of a two-dimensional quantum system obtained by encoding the bit S_2 using either the rectilinear basis (if $S_1 = 0$) or the diagonal basis (if $S_1 = 1$). Clearly, if S_1 is known, S_2 can easily be determined by applying the appropriate measurement to ρ. On the other hand, the probability of correctly guessing S_2 from the outcome of any measurement chosen independently of S_1 is bounded away from 1.

The following lemmas can most easily be formalized in terms of the square of the Hilbert-Schmidt distance. For two density operators ρ and σ, let

$$\Delta(\rho, \sigma) := \operatorname{tr}((\rho - \sigma)^2) \ .$$

Moreover, for a random variable X and a random state ρ, we define

$$D(X|\rho) := \Delta([\{X\} \otimes \rho], [\{U\}] \otimes [\rho])$$

where U is a random variable uniformly distributed on $\mathcal{X}$.

Lemma 5. *Let ρ and σ be two density operators on $\mathcal{H}$. Then*

$$\delta(\rho, \sigma) \leq \frac{1}{2}\sqrt{\operatorname{rank}(\rho - \sigma) \cdot \Delta(\rho, \sigma)} \ .$$

Proof. The assertion follows directly from Lemma 11 (cf. Appendix A) and the definition of the distance measures $\delta(\cdot, \cdot)$ and $\Delta(\cdot, \cdot)$. □

Lemma 6. *Let X be a random variable with range $\mathcal{X}$ and let ρ be a random state. Then*

$$d(X|\rho) \leq \frac{1}{2} 2^{\frac{H_0(X) + S_0([\rho])}{2}} \sqrt{D(X|\rho)} \ .$$

Proof. Note that the rank of $[\{X\} \otimes \rho] - [\{U\}] \otimes [\rho]$ is bounded by $2^{H_0(X) + S_0([\rho])}$. The assertion thus follows as an immediate consequence of the definitions and Lemma 5. □

Lemma 7. *Let X be a random variable with range $\mathcal{X}$ and let ρ be a random state. Then*

$$D(X|\rho) = \operatorname{tr}\left(\left(\sum_{x \in \mathcal{X}} P_X(x)^2 \rho_x^2\right) - \frac{1}{|\mathcal{X}|}[\rho]^2\right)$$

where $\rho_x := [\rho|X = x]$, for any $x \in \mathcal{X}$.

Proof. From (1) and the fact that $\operatorname{tr}(P_{|x\rangle} P_{|x'\rangle}) = \delta_{x,x'}$ (where $\delta_{x,x'}$ is the Kronecker delta which equals 1 if $x = x'$ and 0 otherwise), we find

$$D(X|\rho) = \operatorname{tr}\left(\left(\sum_{x \in \mathcal{X}} P_X(x) P_{|x\rangle} \otimes \rho_x - \frac{1}{|\mathcal{X}|} \sum_{x \in \mathcal{X}} P_{|x\rangle} \otimes [\rho]\right)^2\right)$$

$$= \operatorname{tr}\left(\sum_{x \in \mathcal{X}} \left(P_X(x) \rho_x - \frac{1}{|\mathcal{X}|}[\rho]\right)^2\right)$$

$$= \operatorname{tr}\left(\sum_{x \in \mathcal{X}} P_X(x)^2 \rho_x^2 - \frac{2}{|\mathcal{X}|}[\rho] \sum_{x \in \mathcal{X}} P_X(x) \rho_x + \frac{1}{|\mathcal{X}|}[\rho]^2\right) \ .$$

Inserting the identity

$$[\rho] = \sum_{x \in \mathcal{X}} P_X(x) \rho_x$$

concludes the proof. □

Lemma 8. *Let Z be a random variable, let ρ be a random state, and let F be a two-universal function on $\mathcal{Z}$ chosen independently of Z and ρ. Then*

$$\mathop{\mathrm{E}}_{f \leftarrow P_F}[D(f(Z)|\rho)] \leq 2^{-S_2([\{Z\} \otimes \rho])} .$$

Proof. Let us define $\rho_z := [\rho|Z=z]$ for every $z \in \mathcal{Z}$ and let $\mathcal{S}$ be the range of F. With Lemma 7, we obtain

$$\mathop{\mathrm{E}}_{f \leftarrow P_F}[D(f(Z)|\rho)] = \mathrm{tr}\left(\mathop{\mathrm{E}}_{f \leftarrow P_F}\left[\sum_{s \in \mathcal{S}} P_{f(Z)}(s)^2 [\rho|f(Z)=s]^2\right]\right) - \frac{1}{|\mathcal{S}|}\mathrm{tr}([\rho]^2) ,\tag{7}$$

where we have used the linearity of the trace. Note that

$$P_{f(Z)}(s) \cdot [\rho|f(Z)=s] = \sum_{z \in f^{-1}(\{s\})} P_Z(z) \rho_z .$$

Using this identity and rearranging the summation order, we get

$$\sum_{s \in \mathcal{S}} P_{f(Z)}(s)^2 [\rho|f(Z)=s]^2 = \sum_{z,z' \in \mathcal{Z}} P_Z(z) P_Z(z') \rho_z \rho_{z'} \delta_{f(z),f(z')} .$$

Taking the expectation value over the random choice of F then gives

$$\mathop{\mathrm{E}}_{f \leftarrow P_F}\left[\sum_{s \in \mathcal{S}} P_{f(Z)}(s)^2 [\rho|f(Z)=s]^2\right] = \sum_{z,z' \in \mathcal{Z}} P_Z(z) P_Z(z') \rho_z \rho_{z'} \mathop{\mathrm{Pr}}_{f \leftarrow P_F}[f(z)=f(z')] .$$

Similarly, we obtain

$$[\rho]^2 = \sum_{z,z' \in \mathcal{Z}} P_Z(z) P_Z(z') \rho_z \rho_{z'} .$$

Inserting this into (7), we get

$$\mathop{\mathrm{E}}_{f \leftarrow P_F}[D(f(Z)|\rho)] = \sum_{z,z' \in \mathcal{Z}} P_Z(z) P_Z(z') \left(\mathop{\mathrm{Pr}}_{f \leftarrow P_F}[f(z)=f(z')] - \frac{1}{|\mathcal{S}|}\right) \mathrm{tr}(\rho_z \rho_{z'}) .$$

As we assumed that F is two-universal, all summands with $z \neq z'$ are not larger than zero and we are left with

$$\mathop{\mathrm{E}}_{f \leftarrow P_F}[D(f(Z)|\rho)] \leq \sum_{z \in \mathcal{Z}} P_Z(z)^2 \mathrm{tr}(\rho_z^2) = \mathrm{tr}([\{Z\} \otimes \rho]^2)$$

from which the assertion follows by the definition of the Rényi entropy S_2. □

Proof (Theorem 1). Using Lemma 4 and Lemma 6, we get

$$d(F(Z)|\{F\} \otimes \rho) = \mathop{\mathrm{E}}_{f \leftarrow P_F}[d(f(Z)|\rho)]$$

$$\leq \frac{1}{2} 2^{\frac{s+S_0([\rho])}{2}} \mathop{\mathrm{E}}_{f \leftarrow P_F}[\sqrt{D(f(Z)|\rho)}]$$

$$\leq \frac{1}{2} 2^{\frac{s+S_0([\rho])}{2}} \sqrt{\mathop{\mathrm{E}}_{f \leftarrow P_F}[D(f(Z)|\rho)]} ,$$

where the last inequality follows from Jensen's inequality and the convexity of the square root. Applying Lemma 8 concludes the proof. □

4.2 A Bound in Terms of Smooth Rényi Entropy

The goal of this section is to reformulate Theorem 1 in terms of smooth Rényi entropy (cf. Corollary 1 below). Since, e.g., $S_0([\rho])$ is generally larger than $S_0^\varepsilon([\rho])$, this gives a better bound on the length of the extractable key. Indeed, for the situation where Z and ρ are obtained from many repetitions of the same random experiment, the bound in terms of smooth Rényi entropy is asymptotically optimal (cf. Section 4.4), which is not true if conventional Rényi entropy is used instead.

The following derivation is based on the idea that, for any normalized density operator ρ with smooth Rényi entropy $S_\alpha^\varepsilon(\rho)$, there exists a (not necessarily normalized) density operator ρ' which is ε-close to ρ such that the (conventional) Rényi entropy of ρ', $S_\alpha(\rho')$, is equal to $S_\alpha^\varepsilon(\rho)$.[8]

Lemma 9. *Let X be a random variable and let ρ be a normalized random state. Then, for any $\varepsilon \geq 0$, there exists a random variable X' and a random state ρ' with $\delta([\{X'\} \otimes \rho'], [\{X\} \otimes \rho]) \leq 2\sqrt{\varepsilon}$ such that, for any $\alpha > 1$,*

$$S_\alpha([\{X'\} \otimes \rho']) - S_0([\rho']) \geq S_\alpha^\varepsilon([\{X\} \otimes \rho]) - S_0^\varepsilon([\rho]) \ .$$

Proof. Let P be the projector onto the minimum subspace which corresponds to eigenvalues of $[\rho]$ with total weight (at least) $1 - \varepsilon$, i.e.,

$$\mathrm{tr}(P[\rho]P^\dagger) \geq 1 - \varepsilon \ . \tag{8}$$

It is easy to verify that $\log(\mathrm{rank}(P)) = S_0^\varepsilon([\rho])$. Similarly, there exists a random variable X' and a random state σ with $\mathrm{tr}([\sigma]) \leq \mathrm{tr}([\rho]) = 1$ such that

$$S_\alpha([\{X'\} \otimes \sigma]) = S_\alpha^\varepsilon([\{X\} \otimes \rho])$$

and

$$\delta([\{X'\} \otimes \sigma], [\{X\} \otimes \rho]) \leq \frac{\varepsilon}{2} \ . \tag{9}$$

Let ρ' be the random state defined by $\rho' := P\sigma P^\dagger$. Then,

$$S_0([\rho']) \leq \log(\mathrm{rank}(P)) = S_0^\varepsilon([\rho])$$

and by Lemma 14 (see Appendix B), since $[\{X'\} \otimes \rho']$ is the projection of $[\{X'\} \otimes \sigma]$ (with respect to the projection operation $(\mathrm{id} \otimes P)$),

$$S_\alpha([\{X'\} \otimes \rho']) \geq S_\alpha([\{X'\} \otimes \sigma]) = S_\alpha^\varepsilon([\{X\} \otimes \rho]) \ .$$

It thus remains to be shown that

$$\delta([\{X'\} \otimes \rho'], [\{X\} \otimes \rho]) \leq 2\sqrt{\varepsilon} \ . \tag{10}$$

---

[8] Note that $S_\alpha(\rho')$ is also defined for density operators ρ' with $\mathrm{tr}(\rho') < 1$.

Since the trace distance cannot increase when applying the projection P (cf. (4)), we obtain from (9)
$$\operatorname{tr}(|P[\sigma]P^\dagger - P[\rho]P^\dagger|) = 2\delta(P[\sigma]P^\dagger, P[\rho]P^\dagger) \leq 2\delta([\sigma],[\rho]) \leq \varepsilon .$$
Hence, with (8),
$$\operatorname{tr}([\rho']) = \operatorname{tr}([P\sigma P^\dagger]) \geq \operatorname{tr}([P\rho P^\dagger]) - \operatorname{tr}(|P[\rho]P^\dagger - P[\sigma]P^\dagger|) \geq 1 - 2\varepsilon$$
and thus, from Lemma 12 (cf. Appendix A),
$$\delta([\{X'\}\otimes\rho'],[\{X'\}\otimes\sigma]) \leq \sqrt{\operatorname{tr}([\sigma])(\operatorname{tr}([\sigma])-\operatorname{tr}([\rho']))} \leq \sqrt{1-\operatorname{tr}([\rho'])} \leq \sqrt{2\varepsilon} .$$

Using once again (9) and applying the triangle inequality for the trace distance implies (10) and thus concludes the proof. □

Using Lemma 9, the following corollary of Theorem 1 follows directly from the triangle inequality for the trace distance.

Corollary 1. *Let Z be a random variable with range $\mathcal{Z}$, let ρ be a normalized random state, let F be a two-universal function from $\mathcal{Z}$ to $\mathcal{S} = \{0,1\}^s$ which is independent of Z and ρ, and let $\varepsilon \geq 0$. Then*
$$d(F(Z)|\{F\}\otimes\rho) \leq \frac{1}{2} 2^{-\frac{1}{2}(S_2^\varepsilon([\{Z\}\otimes\rho])-S_0^\varepsilon([\rho])-s)} + 4\sqrt{\varepsilon} .$$

Note that the smooth Rényi entropies occurring in the bound of Corollary 1 can easily be computed from the eigenvalues of the density operators $[\{Z\}\otimes\rho] = \sum_z P_Z(z) P_{|z\rangle}\otimes\rho_z$ and $[\rho] = \sum_z P_Z(z)\rho_z$, where $\rho_z = [\rho|Z=z]$ (cf. Section 2.4).

4.3 Privacy Amplification Against Quantum Adversaries

We now apply the results of the previous section to show that privacy amplification by two-universal hashing is secure (with respect to the universally composable security definition of Section 3) against an adversary holding quantum information. Consider two distant parties which are connected by an authentic, but otherwise fully insecure classical communication channel. Additionally, they have access to a common random string Z about which an adversary has some partial information represented by the state ρ of a quantum system. The two legitimate parties can apply the following simple *privacy amplification protocol* to obtain a secure key S of length s. Let F be a two-universal random function from the range of Z to $\{0,1\}^s$. First, one of the parties randomly chooses an instance of F and announces his choice to the other party using the public communication channel. Then, both parties compute $S = F(Z)$.

Note that, during the execution of this protocol, the adversary might learn F. The final key S must thus be secure with respect to both $\{F\}$ and ρ. It is an immediate consequence of Corollary 1 that, for any $\varepsilon \geq 0$, the key S generated by the described privacy amplification protocol is ε-secure with respect to $\rho \otimes \{F\}$ if its length s is not larger than
$$s_\varepsilon = S_2^{\bar\varepsilon}([\{Z\}\otimes\rho]) - S_0^{\bar\varepsilon}([\rho]) - 2\log(1/\varepsilon) , \qquad (11)$$
where $\bar\varepsilon = (\varepsilon/8)^2$.

4.4 Asymptotic Optimality

We now show that the bound (11) is asymptotically optimal, i.e., that the right hand side of (11) is (in an asymptotic sense) also an upper bound for the number of key bits that can be extracted by any protocol. Consider a setting where both the initial information $Z^{(n)}$ as well as the adversary's state $\rho^{(n)}$ consist of n independent pieces, for $n \in \mathbb{N}$. Formally, let $Z^{(n)} = (Z_1, \ldots, Z_n)$ and $\rho^{(n)} = \rho_1 \otimes \cdots \otimes \rho_n$ where the pairs (Z_i, ρ_i) are independent and identically distributed. Let $s(n)$ be the length of the key $S^{(n)}$ that can be extracted from $Z^{(n)}$ by an optimal privacy amplification protocol. Using Lemma 3, we conclude from (11) that

$$s(n) \geq H(Z^{(n)}|\rho^{(n)}) + o(n) \qquad (12)$$

where, for any Z and ρ, $H(Z|\rho)$ is defined in terms the von Neumann entropy $S(\cdot)$ by

$$H(Z|\rho) := S([\{Z\} \otimes \rho]) - S([\rho]) \; .$$

To derive an upper bound for $s(n)$, consider an arbitrary privacy amplification protocol for generating a key $S^{(n)}$ from $Z^{(n)}$. Let $C^{(n)}$ be the whole communication exchanged over the public channel during the execution of the protocol, and let $f_{C^{(n)}}$ be the function depending on $C^{(n)}$ which describes how the final key $S^{(n)}$ is computed from $Z^{(n)}$, that is, $S^{(n)} = f_{C^{(n)}}(Z^{(n)})$.

It is a direct consequence of Definition 3 that the von Neumann entropy of an ε-secure key $S^{(n)}$ virtually cannot be smaller than its length $s(n)$, i.e.,

$$s(n) \leq H(f_{C^{(n)}}(Z^{(n)})|\rho^{(n)} \otimes \{C^{(n)}\}) + o(n) \; . \qquad (13)$$

Using some well-known properties of the von Neumann entropy, it is easy to see that the quantity $H(Z|\rho)$ can only decrease when applying any function f to its first argument or when introducing an additional random variable in the second argument. We thus have

$$H(f_{C^{(n)}}(Z^{(n)})|\rho^{(n)} \otimes \{C^{(n)}\}) \leq H(Z^{(n)}|\rho^{(n)} \otimes \{C^{(n)}\}) \leq H(Z^{(n)}|\rho^{(n)}) \; . \qquad (14)$$

Hence, combining (12), (13), and (14), we obtain an expression for the maximum number $s(n)$ of extractable key bits,

$$s(n) = H(Z^{(n)}|\rho^{(n)}) + o(n) \; .$$

In particular, the maximum rate $R := \lim_{n \to \infty} \frac{s(n)}{n}$ at which secret key bits can be generated—from independent realizations of Z about which the adversary has information given by ρ—is

$$R = S([\{Z\} \otimes \rho]) - S([\rho]) = H(Z|\rho) \; . \qquad (15)$$

This exactly corresponds to the expression for the secret key rate obtained by Devetak and Winter [15].

In the purely classical case, i.e., if the adversary's information is given by a classical random variable W, expression (15) reduces to

$$R = H(ZW) - H(W) = H(Z|W) \;,$$

which is a well known result of Csiszár and Körner [14] (see also [24]).[9]

4.5 Applications to QKD

Theorem 1 has interesting implications for quantum key distribution (QKD). Recently, a generic protocol for QKD has been presented and proven secure against general attacks [13] (see also [23]). Moreover, it has been shown that many of the known protocols, such as BB84 or B92, are special instances of this generic protocol, i.e., their security directly follows from the security of the generic QKD protocol. Since the result in [13] is based on the security of privacy amplification, the strong type of security implied by Theorem 1 immediately carries over to this generic QKD protocol. In particular, the secret keys generated by the BB84 and the B92 protocol satisfy Definition 3 and thus provide universal composability.

Acknowledgment

The authors would like to thank Ueli Maurer for many inspiring discussions, and Dominic Mayers as well as anonymous referees for very useful comments. This project was partially supported by the Swiss National Science Foundation, project No. 200020-103847/1.

A Some Useful Identities

Lemma 10 (Schur's inequality). *Let A be a linear operator on a d-dimensional Hilbert space $\mathcal{H}$ and let $\lambda_1, \ldots, \lambda_d$ be its eigenvalues. Then*

$$\sum_{i=1}^{d} |\lambda_i|^2 \leq \operatorname{tr}(AA^\dagger) \;,$$

with equality if and only if A is normal (i.e., $AA^\dagger = A^\dagger A$).

Proof. See, e.g., [20].

Lemma 11. *Let A be a normal operator with rank r. Then*

$$\operatorname{tr}|A| \leq \sqrt{r}\sqrt{\operatorname{tr}(AA^\dagger)} \;.$$

---

[9] In the setting of [14], the two parties are connected by a channel which leaks partial information to an adversary. As shown in [24], the result of [14] also applies if the two parties are connected by a completely public channel, but start with some common information Z about which an adversary has partial knowledge W.

Proof. Let $\lambda_1, \ldots, \lambda_r$ be the r nonzero eigenvalues of A. Since the square root is concave, we can apply Jensen's inequality leading to

$$\operatorname{tr}|A| = \sum_{i=1}^{r} |\lambda_i| = \sum_{i=1}^{r} \sqrt{|\lambda_i|^2} \leq \sqrt{r} \sqrt{\sum_{i=1}^{r} |\lambda_i|^2} \ .$$

The assertion then follows from Schur's inequality (Lemma 10). □

Lemma 12. *Let $\rho \in \mathcal{P}(\mathcal{H})$ and let P be a projection on $\mathcal{H}$, i.e., $P \circ P = P$. Then, for $\rho' := P\rho P^\dagger$,*

$$\delta(\rho, \rho') \leq \sqrt{\operatorname{tr}(\rho)(\operatorname{tr}(\rho) - \operatorname{tr}(\rho'))} \ .$$

Proof. We first show that the assertion holds for normalized pure states $\rho = P_{|\phi\rangle}$. Since P is a projection, there exist $a, b \in \mathbb{R}$ with $a^2 + b^2 = 1$ and two orthogonal vectors $|\alpha\rangle, |\beta\rangle$ with $P|\alpha\rangle = |\alpha\rangle$ and $P|\beta\rangle = 0$ such that $|\phi\rangle = a|\alpha\rangle + b|\beta\rangle$. In particular, $\rho' = a^2 P_{|\alpha\rangle}$. It then follows by a straightforward calculation that

$$\delta(\rho, \rho') = \delta(P_{a|\alpha\rangle + b|\beta\rangle}, a^2 P_{|\alpha\rangle}) \leq b = \sqrt{1 - \operatorname{tr}(\rho')} \ .$$

To prove the assertion for general density operators $\rho \in \mathcal{P}(\mathcal{H})$, let

$$\rho = \sum_{i \in \mathcal{I}} p_i \rho_i$$

where, for any $i \in \mathcal{I}$, $p_i \geq 0$ and ρ_i is a normalized pure state. In particular, $\sum_{i \in \mathcal{I}} p_i = \operatorname{tr}(\rho)$. By linearity, we have

$$\rho' = \sum_{i \in \mathcal{I}} p_i \rho_i' \ ,$$

where $\rho_i' := P \rho_i P^\dagger$. Hence, using the convexity of the trace distance,

$$\delta(\rho, \rho') \leq \sum_{i \in \mathcal{I}} p_i \delta(\rho_i, \rho_i') \leq \sum_{i \in \mathcal{I}} p_i \sqrt{1 - \operatorname{tr}(\rho_i')} \ .$$

The assertion then follows from Jensen's inequality. □

B Rényi Entropy and Quantum Operations

The following lemma states that the Rényi entropy of a density operator ρ can only increase when applying a quantum operation $\mathcal{E}$ on ρ.

Lemma 13. *Let $\mathcal{E} : \rho \mapsto \sum_i E_i \rho E_i^\dagger$ be a doubly stochastic quantum operation on $\mathcal{H}$, i.e., E_i are linear operators on $\mathcal{H}$ satisfying $\sum_i E_i^\dagger E_i = \operatorname{id}$ and $\sum_i E_i E_i^\dagger = \operatorname{id}$. Then, for any $\rho \in \mathcal{P}(\mathcal{H})$ and $\alpha \in [0, \infty]$,*

$$S_\alpha(\mathcal{E}(\rho)) \geq S_\alpha(\rho) \ .$$

Proof. See, e.g., [25] (Theorem 5.1 together with Theorem 4.2, applied to the function S_α).

Lemma 13 can be used to show that, for $\alpha > 1$, the Rényi entropy of a density operator ρ can only increase when applying a projector P to ρ.

Lemma 14. *Let $\rho \in \mathcal{P}(\mathcal{H})$ and let P be a projection on $\mathcal{H}$, i.e., $P \circ P = P$. Then, for $\alpha > 1$,*
$$S_\alpha(P\rho P^\dagger) \geq S_\alpha(\rho) .$$

Proof. Consider the quantum operation $\mathcal{E}$ defined by
$$\mathcal{E}: \quad \rho \longmapsto P\rho P^\dagger + (\mathrm{id} - P)\rho(\mathrm{id} - P)^\dagger .$$

It is easy to verify that $\mathcal{E}$ is doubly stochastic. Hence, from Lemma 13,
$$S_\alpha(\rho' + \rho'') \geq S_\alpha(\rho) ,$$

where $\rho' := P\rho P^\dagger$ and $\rho'' := (\mathrm{id} - P)\rho(\mathrm{id} - P)^\dagger$. The assertion then follows from the fact that, because ρ' and ρ'' are orthogonal,
$$\mathrm{tr}\big((\rho' + \rho'')^\alpha\big) \geq \mathrm{tr}\big((\rho')^\alpha\big) ,$$

and the definition of S_α.

C Smooth Rényi Entropy of Classical Distributions

Smooth Rényi entropy has been introduced in [28] as a generalization of Rényi entropy. For any set $\mathcal{Z}$, let $\bar{\mathcal{P}}(\mathcal{Z})$ be the set of non-negative functions P on $\mathcal{Z}$ such that $\sum_{z \in \mathcal{Z}} P(z) \leq 1$, i.e., $\bar{\mathcal{P}}(\mathcal{Z})$ contains all (not necessarily normalized) probability distributions on $\mathcal{Z}$. For any $P \in \bar{\mathcal{P}}(\mathcal{Z})$, let $\mathcal{B}^\varepsilon(P)$ be the set of functions $Q \in \bar{\mathcal{P}}(\mathcal{Z})$ such that $\delta(P,Q) := \frac{1}{2}\sum_z |P(z) - Q(z)| \leq \varepsilon$.

Definition 4. *Let $P \in \bar{\mathcal{P}}(\mathcal{Z})$ and let $\varepsilon \geq 0$. The ε-smooth Rényi entropy $H_\alpha^\varepsilon(P)$ of order α of P is defined by*
$$H_\alpha^\varepsilon(P) := \frac{1}{1-\alpha} \log\left(\inf_{Q \in \mathcal{B}^{\varepsilon/2}(P)} \left(\sum_{z \in \mathcal{Z}} Q(z)^\alpha \right) \right) ,$$

for $\alpha \in (0,1) \cup (1,\infty)$, and $H_\alpha^\varepsilon(P) := \lim_{\beta \to \alpha} H_\beta^\varepsilon(P)$, for $\alpha \in \{0, \infty\}$.

For a random variable Z with probability distribution P_Z, we also write $H_\alpha^\varepsilon(Z)$ instead of $H_\alpha^\varepsilon(P_Z)$.

It turns out that, for $\alpha < 1$, the logarithm on the right hand side of this definition takes its minimum for the function $Q \in \mathcal{B}^{\varepsilon/2}(P)$ which is obtained from P by setting the smallest probabilities to zero. Similarly, for $\alpha > 1$, the minimum is taken for the function Q obtained by cutting the largest probabilities of P. The smooth Rényi entropy $H_\alpha^\varepsilon(P)$ can thus easily be computed from the probabilities $P(z)$, for $z \in \mathcal{Z}$.

Smooth Rényi entropy has many natural properties which are similar to the properties of Shannon entropy. In particular, the smooth Rényi entropy of many independent and uniformly distributed random variables is close to the Shannon entropy.

Lemma 15. *Let* $Z_1, \ldots, Z_n$ *be independent random variables distributed according to* P_Z. *Then, for any* $\alpha \neq 1$,

$$\lim_{\varepsilon \to 0} \lim_{n \to \infty} \frac{1}{n} H_\alpha^\varepsilon(Z_1 \cdots Z_n) = H(Z) .$$

For a discussion of further properties and applications of smooth Rényi entropy, see [28].

References

1. A. Ambainis, L. J. Schulman, A. Ta-Shma, U. Vazirani, and A. Wigderson. The quantum communication complexity of sampling. In *Proceedings of the 39th Annual Symposium on Foundations of Computer Science*, pages 342–351, 1998.
2. M. Ben-Or. Security of BB84 QKD Protocol. Slides available at http://www.msri.org/publications/ln/msri/2002/quantumintro/ben-or/2/, 2002.
3. M. Ben-Or, M. Horodecki, D. Leung, D. Mayers, and J. Oppenheim. Composability of QKD. Slides available at http://www.msri.org/publications/ln/msri/2002/qip/mayers/1/ (Part II), 2002.
4. M. Ben-Or, M. Horodecki, D. Leung, D. Mayers, and J. Oppenheim. The universal composable security of quantum key distribution. In *Proceedings of TCC 2005*, 2005.
5. M. Ben-Or and D. Mayers. Quantum universal composability. Slides available at http://www.msri.org/publications/ln/msri/2002/quantumcrypto/mayers/1/banner/01.html, 2002.
6. M. Ben-Or and D. Mayers. General security definition and composability for quantum & classical protocols. Available at http://arxiv.org/abs/quant-ph/0409062, 2004.
7. C. H. Bennett. Quantum cryptography using any two nonorthogonal states. *Physical Review Letters*, 68(21):3121–3124, 1992.
8. C. H. Bennett and G. Brassard. Quantum cryptography: Public-key distribution and coin tossing. In *Proceedings of IEEE International Conference on Computers, Systems and Signal Processing*, pages 175–179, 1984.
9. C. H. Bennett, G. Brassard, C. Crépeau, and U. Maurer. Generalized privacy amplification. *IEEE Transaction on Information Theory*, 41(6):1915–1923, 1995.
10. C. H. Bennett, G. Brassard, and J.-M. Robert. Privacy amplification by public discussion. *SIAM Journal on Computing*, 17(2):210–229, 1988.
11. R. Canetti. Universally composable security: A new paradigm for cryptographic protocols. In *Proceedings of the 42nd IEEE Symposium on Foundations of Computer Science*, pages 136–145, 2001.
12. J. L. Carter and M. N. Wegman. Universal classes of hash functions. *Journal of Computer and System Sciences*, 18:143–154, 1979.

13. M. Christandl, R. Renner, and A. Ekert. A generic security proof for quantum key distribution. Available at http://arxiv.org/abs/quant-ph/0402131, February 2004.
14. I. Csiszár and J. Körner. Broadcast channels with confidential messages. *IEEE Transactions on Information Theory*, 24:339–348, 1978.
15. I. Devetak and A. Winter. Distillation of secret key and entanglement from quantum states. Available at http://arxiv.org/abs/quant-ph/0306078, June 2003.
16. D. DiVincenzo, M. Horodecki, D. Leung, J. Smolin, and B. Terhal. Locking classical correlation in quantum states. *Physical Review Letters*, 92, 067902, 2004.
17. S. Dziembowski and U. Maurer. Optimal randomizer efficiency in the bounded-storage model. *Journal of Cryptology*, 17(1):5–26, 2004. Conference version appeared in Proc. of STOC '02.
18. D. Gottesman and H.-K. Lo. Proof of security of quantum key distribution with two-way classical communications. *IEEE Transactions on Information Theory*, 49(2):457–475, 2003.
19. P. Hayden, D. Leung, P. W. Shor, and A. Winter. Randomizing quantum states: Constructions and applications *Communications in Mathematical Physics*, 250(2):371–391, 2004.
20. R. A. Horn and C. R. Johnson. *Matrix analysis*. Cambridge University Press, 1985.
21. R. Impagliazzo, L. A. Levin, and M. Luby. Pseudo-random generation from one-way functions (extended abstract). In *Proceedings of the Twenty-First Annual ACM Symposium on Theory of Computing*, pages 12–24, 1989.
22. R. König, U. Maurer, and R. Renner. On the power of quantum memory. Available at http://arxiv.org/abs/quant-ph/0305154, May 2003.
23. B. Kraus, N. Cisin, and R. Renner. Lower and upper bounds on the secret key rate for QKD protocols using one-way classical communication. Available at http://arxiv.org/abs/quant-ph/0410215, 2004.
24. U. M. Maurer. Secret key agreement by public discussion from common information. *IEEE Transactions on Information Theory*, 39(3):733–742, 1993.
25. M. A. Nielsen. Majorization and its applications to quantum information theory. Available at http://www.qinfo.org/talks/1999/06-maj/maj.pdf, June 1999.
26. M. A. Nielsen and I. L. Chuang. *Quantum computation and quantum information*. Cambridge University Press, 2000.
27. B. Pfitzmann and M. Waidner. Composition and integrity preservation of secure reactive systems. In *7th ACM Conference on Computer and Communications Security*, pages 245–254. ACM Press, 2000.
28. R. Renner and S. Wolf. Smooth Rényi entropy and applications. In *Proceedings of the 2004 IEEE International Symposium on Information Theory*, page 233, 2004.
29. A. Rényi. On measures of entropy and information. In *Proceedings of the 4th Berkeley Symp. on Math. Statistics and Prob.*, volume 1, pages 547–561. Univ. of Calif. Press, 1961.
30. D. Unruh. Simulatable security for quantum protocols. Available at http://arxiv.org/abs/quant-ph/0409125, 2004.
31. M. N. Wegman and J. L. Carter. New hash functions and their use in authentication and set equality. *Journal of Computer and System Sciences*, 22:265–279, 1981.

A Universally Composable Secure Channel Based on the KEM-DEM Framework

Waka Nagao[1], Yoshifumi Manabe[1,2], and Tatsuaki Okamoto[1,2]

[1] Graduate School of Informatics, Kyoto University,
Yoshida-honmachi, Kyoto, 606-8501 Japan
[2] NTT Labs, Nippon Telegraph and Telephone Corporation,
1-1 Hikari-no-oka Yokosuka, 239-0847 Japan

Abstract. For ISO standards on public-key encryption, Shoup introduced the framework of KEM (Key Encapsulation Mechanism), and DEM (Data Encapsulation Mechanism), for formalizing and realizing *one-directional* hybrid encryption; KEM is a formalization of asymmetric encryption specified for key distribution, and DEM is a formalization of symmetric encryption. This paper investigates a more general hybrid protocol, *secure channel*, using KEM and DEM, such that KEM is used for distribution of a session key and DEM, along with the session key, is used for multiple *bi-directional* encrypted transactions in a session. This paper shows that KEM semantically secure against adaptively chosen ciphertext attacks (IND-CCA2) and DEM semantically secure against adaptively chosen plaintext/ciphertext attacks (IND-P2-C2) along with secure signatures and ideal certification authority are sufficient to realize a *universally composable* (UC) secure channel. To obtain the main result, this paper also shows several equivalence results: UC KEM, IND-CCA2 KEM and NM-CCA2 (non-malleable against CCA2) KEM are equivalent, and UC DEM, IND-P2-C2 DEM and NM-P2-C2 DEM are equivalent.

1 Introduction

1.1 Background

Key Encapsulation Mechanism (KEM) is a key distribution mechanism in public-key cryptosystems, that was proposed by Shoup for ISO standards on public-key encryption [11].

The difference between KEM and public-key encryption (PKE) is as follows: PKE's encryption procedure, on input plaintext M and receiver R's public-key PK_R, outputs ciphertext C, while KEM's encryption procedure, on input receiver R's public-key PK_R, outputs ciphertext C and key K, where C is sent to R, and K is kept secret inside the sender, and employed in the subsequent process of data encryption. PKE's decryption procedure, on input C and secret-key SK_R, outputs plaintext M, while KEM's decryption procedure, on input C and secret-key SK_R, outputs key K. Although KEM is a mechanism for key distribution and the applications of KEM are not specified, the most typical application is hybrid encryption, where a key shared via a KEM is employed for symmetric-key encryption. Shoup also formulated the symmetric-key encryption as the Data Encapsulation Mechanism (DEM)[11].

Shoup defined the security, "indistinguishable (semantically secure) against adaptively chosen-ciphertext attacks," for KEM and DEM, respectively, (we call them IND-CCA2-KEM and IND-CCA2-DEM, respectively), and showed that hybrid encryption (HPKE) implemented by combining KEM with IND-CCA2-KEM and DEM with IND-CCA2-DEM is a PKE with IND-CCA2-PKE [8, 11].[1]

Since the KEM-DEM hybrid encryption specified by Shoup is *one-directional* (or equivalent to public-key encryption in functionality), it is applicable for secure email and single direction transactions. However, in many secure protocols (e.g., SSL, IPSec, SSH), asymmetric and symmetric encryption schemes are employed in a different manner as a *secure channel* such that an asymmetric encryption scheme is used for distribution of a session key while a symmetric encryption scheme with the session key is used for many bi-directional encrypted transactions in a session.

The KEM-DEM framework can be modified for such a hybrid usage, secure channel; KEM can be used for key distribution of a session key and DEM with the session key is used for secure communications in a session. Since the KEM-DEM framework will be standardized in a near future, it is a promising way to employ the above-mentioned modified KEM-DEM framework to realize a secure channel. However, no research has been done on the security requirements of KEM and DEM such that a secure channel based on the modified KEM-DEM framework can guarantee a sufficient level of security, although KEM with IND-CCA2-KEM and DEM with IND-CCA2-DEM have been shown to be sufficient for an IND-CCA2-PKE single-directional KEM-DEM-hybrid scheme [8, 11]. That is, we have the following problems:

- What are the security requirements of KEM and DEM to construct a secure channel?
- How to define the satisfactory level of security of a secure channel? (since it cannot be characterized by just public-key encryption, but should require more complicated security definition.)

1.2 Our Results

This paper answers the above-mentioned problems:

- This paper shows that KEM with IND-CCA2-KEM and DEM with IND-P2-C2-DEM along with secure signatures and ideal certification authority are sufficient to realize a universally composable secure channel.
- We follow the definition of a *universally composable* secure channel by Canetti and Krawczyk [6]. There are two major merits in using the universal composability paradigm. Firstly, the paradigm provides a clear and unified (or standard) approach to defining the security of any cryptographic functionality including a *secure channel*. Second, our concrete construction of a secure channel based on the KEM-DEM

---

[1] Originally, the notion of IND-CCA2 was defined for PKE. The way to provide analogous definitions and to use the same name, "indistinguishable (semantically secure) against adaptively chosen-ciphertext attacks", for KEM and DEM follows that of [8]. In this paper, however, we explicitly distinguish them by the terms, IND-CCA2-PKE, IND-CCA2-KEM, and IND-CCA2-DEM.

framework guarantees not only stand-alone security but also universal composable security. Since a secure protocol like SSL, IPSec and SSH is often employed as an element of a large-scale security system, the universal composability of a secure protocol is especially important.

In order to obtain the above-mentioned main result, we firstly show that UC KEM, IND-CCA2 KEM and NM-CCA2 KEM are equivalent, and that UC DEM, IND-P2-C2 DEM and NM-P2-C2 DEM are equivalent. We then present that UC KEM and UC KEM as well as UC signatures and ideal certification authority are sufficient for realizing a UC secure channel.

Although in this paper we consider only protocols for a single session, the same result for the multi-session case is obtained automatically via the UC with joint state (JUC) [7].

1.3 Related Works

Canetti and Krawczyk [6] showed a UC secure channel protocol consisting of an authenticated Diffie-Hellman key exchange scheme, message authentication code, and pseudorandom generator. Accordingly, their results are specific to their construction, which uses an authenticated Diffie-Hellman key exchange scheme, message authentication code and pseudorandom generator. Our result is based on the general notions of KEM, DEM and signatures, but not on any specific scheme.

The equivalence of UC PKE and IND-CCA2 PKE has been suggested by Canetti [3], and the equivalence of NM-CCA2 PKE and IND-CCA2 PKE has been shown by Bellare et.al. [1, 2]. The relationship among several security notions of symmetric encryptions has been investigated by Katz and Yung [10]. However, no results have been reported on the equivalence among UC KEM, IND-CCA2 KEM and NM-CCA2 KEM, and that among UC DEM, IND-CCA2 DEM and NM-CCA2 DEM.

2 The KEM-DEM Framework

We describe probabilistic algorithms and experiments with standard notations and conventions. For probabilistic algorithm A, $A(x_1, x_2, \cdots ; r)$ is the result of running A that takes as inputs $x_1, x_2, \cdots$ and coins r. We let $y \leftarrow A(x_1, x_2, \cdots)$ denote the experiment of picking r at random and letting y equal the output of $A(x_1, x_2, \cdots ; r)$. If S is a finite set, then $x \leftarrow S$ denotes the experiment of assigning to x an element uniformly chosen from S. If α is neither an algorithm nor a set, then $x \leftarrow \alpha$ indicates that we assign α to x. We say that y can be output by $A(x_1, x_2, \cdots)$ if there is some r such that $A(x_1, x_2, \cdots ; r) = y$.

2.1 Key Encapsulation Mechanism

Formally, a key encapsulation mechanism KEM is given by the triple of algorithms $KEM.KeyGen()$, $KEM.Encrypt(pk, options)$ and $KEM.Decrypt(sk, C_0)$, where:

1. $KEM.KeyGen()$, the key generation algorithm, is a polynomial time and probabilistic algorithm that takes a security parameter $k \in N$ (provided in unary) and returns a pair (pk, sk) of matching public and secret keys.
2. $KEM.Encrypt(pk, options)$, the encryption algorithm, is a polynomial time and probabilistic algorithm that takes as input a public key pk, along with an optional $options$ argument, and outputs a key/ciphertext pair (K, C_0). The role of $options$ is analogous to that in public-key encryption.
3. $KEM.Decrypt(sk, C_0)$, the decryption algorithm, is a polynomial time and deterministic algorithm that takes as input secret key sk and ciphertext C_0, and outputs key K or special symbol $\bot$ ($\bot$ implies that the ciphertext was invalid).

We require that for all (pk, sk) output by $KEM.KeyGen(1^k)$, and for all C_0 output by $KEM.Encrypt(pk, options)$, $KEM.Decrypt(sk, C_0) = K$ ($|K|$ is denoted $KEM.OutputKeyLen$ — the length of the key output by $KEM.Encrypt$ and $KEM.Decrypt$). A function $\epsilon : N \to R$ is negligible if for every constant $c \geq 0$ there exists an integer k_c such that $\epsilon(k) \leq k^{-c}$ for all $z \geq k_c$. We write vectors in boldface, as in $\boldsymbol{x}$. We also denote the number of components in $\boldsymbol{x}$ by $|\boldsymbol{x}|$, and the i-th component by $\boldsymbol{x}[i]$, so that $\boldsymbol{x} = (\boldsymbol{x}[1], \cdots, \boldsymbol{x}[|\boldsymbol{x}|])$. Additionally, we denote a component of a vector as $x \in \boldsymbol{x}$ or $x \notin \boldsymbol{x}$, which mean, respectively, mean that x is in or is not in the set $\{\boldsymbol{x}[i] : 1 \leq i \leq |\boldsymbol{x}|\}$. Such notions provide convenient descriptions. For example, we can simply write $\boldsymbol{x} \leftarrow KEM.Decrypt(\boldsymbol{y})$ as the shorthand form of $1 \leq i \leq |\boldsymbol{y}|$ do $\boldsymbol{x}[i] \leftarrow KEM.Decrypt(\boldsymbol{y}[i])$. We will consider relations of amity t where t is polynomial in the security parameter k. Rather than writing $R(x_1, \cdots, x_t)$ we write $R(x, \boldsymbol{x})$, meaning the first argument is special and the rest are bunched into vector $\boldsymbol{x}$ with $|\boldsymbol{x}| = t - 1$.

Attack Types of KEM. We state following three attack types of KEM. First, we state CPA (Chosen Plaintext Attack). CPA is an attack type that an adversary is allowed to access to only encryption oracle but not decryption oracle. Secondly, we state CCA1 (Chosen Ciphertext Attack). CCA1 is an attack type that an adversary is allowed to access to both encryption and decryption oracle. However the adversary cannot access to decryption oracle after getting target ciphertext. Thirdly, we state CCA2 (Adaptive Chosen Ciphertext Attack). CCA2 is an attack type that an adversary is allowed to access to both encryption and decryption oracle even if after the adversary gets target ciphertext.

Indistinguishability of KEM. We use IND-ATK-KEM to describe the security notion of indistinguishability for KEM against ATK $\in$ {CPA, CCA1, CCA2}[11]. We redescribe the security notion of IND-CCA2-KEM by considering following attack scenario. First, the key generation algorithm is run to generate the public and private key for the protocol. The adversary can get the public key, but not the private key. Secondly, the adversary generates some queries of plaintexts/ciphertexts and sends the queries to encryption/decryption oracle. Each oracle encrypts/decrypts the queries and returns the results of ciphertexts/plaintexts to the adversary. If the algorithm fails, this information is informed to the adversary, and the attack continues. Thirdly, encryption oracle does the following:

1. Runs the encryption algorithm, generating pair (K^*, C_0^*).
2. Generates a random string $\widetilde{K}$ of length $KEM.OutputKeyLen$.
3. Chooses $b \in \{0,1\}$ at random.
4. If $b = 0$, outputs (K^*, C_0^*), otherwise outputs $(\widetilde{K}, C_0^*)$.

Fourth, the adversary generates plaintexts/ciphertexts to get information from each oracle on the condition of the ciphertext $C_0 \neq C_0^*$. Finally, the adversary outputs $\hat{b} \in \{0,1\}$.

Let $\Pi_{\text{KEM}} = (KEM.KeyGen, KEM.Encrypt, KEM.Decrypt)$ be an encryption protocol and let A be an adversary. The advantage of Π_{KEM} for adversary A, $Adv_{A,\Pi_{\text{KEM}}}^{\text{IND-ATK}}$ is defined as follows:

$$Adv_{A,\Pi_{\text{KEM}}}^{\text{IND-ATK}}(k) = |\Pr[\hat{b} = b] - \tfrac{1}{2}|.$$

Π_{KEM} is secure in the sense of IND-ATK if $Adv_{A,\Pi_{\text{KEM}}}^{\text{IND-ATK}}(k)$ is negligible for any PPT adversary A.

Non-malleability of KEM. We state formal definition of non-malleability for KEM in Fig.1 following [1], which we call NM-KEM. We also use NM-ATK-KEM to describe the security notion of non-malleability for KEM against ATK $\in$ {CPA, CCA1, CCA2}. Let $A = (A_1, A_2)$ be an adversary. (We state two more definitions in the full paper version.)

$$Adv_{A,\Pi_{\text{KEM}}}^{\text{NM-ATK}}(k) \equiv \Pr[Expt_{A,\Pi_{\text{KEM}}}^{\text{NM-ATK}}(k) = 1] - \Pr[\widetilde{Expt}_{A,\Pi_{\text{KEM}}}^{\text{NM-ATK}}(k) = 1]$$

where

$Expt_{A,\Pi_{\text{KEM}}}^{\text{NM-ATK}}(k)$

$(pk, sk) \leftarrow KEM.KeyGen(1^k)$
$(\mathcal{K}, s) \leftarrow A_1^{O_1}(pk)$
$(K^*, C_0^*) \leftarrow KEM.Encrypt(pk) \wedge K^* \in \mathcal{K}$
$(R, \boldsymbol{C_0}) \leftarrow A_2^{O_2}(s, C_0^*)$
$\boldsymbol{K} \leftarrow KEM.Decrypt(sk, \boldsymbol{C_0})$
return 1 iff $(C_0^* \notin \boldsymbol{C_0}) \wedge R(K^*, \boldsymbol{K})$

$\widetilde{Expt}_{A,\Pi_{\text{KEM}}}^{\text{NM-ATK}}(k)$

$(pk, sk) \leftarrow KEM.KeyGen(1^k)$
$(\mathcal{K}, s) \leftarrow A_1^{O_1}(pk)$
$K^* \leftarrow \mathcal{K}$
$(\widetilde{K}, \widetilde{C_0}) \leftarrow KEM.Encrypt(pk) \wedge \widetilde{K} \in \mathcal{K}$
$(R, \widetilde{\boldsymbol{C_0}}) \leftarrow A_2^{O_2}(s, \widetilde{C_0})$
$\widetilde{\boldsymbol{K}} \leftarrow KEM.Decrypt(sk, \widetilde{\boldsymbol{C_0}})$
return 1 iff $(\widetilde{C_0} \notin \widetilde{\boldsymbol{C_0}}) \wedge R(K^*, \widetilde{\boldsymbol{K}})$

and
If ATK = CPA then $O_1 = \varepsilon$ and $O_2 = \varepsilon$.
If ATK = CCA1 then $O_1 = KEM.Decrypt(sk, \cdot)$ and $O_2 = \varepsilon$.
If ATK = CCA2 then $O_1 = KEM.Decrypt(sk, \cdot)$ and $O_2 = KEM.Decrypt(sk, \cdot)$.

Fig. 1. NM-KEM Definition

Π_{KEM} is secure in the sense of NM-ATK-KEM, where ATK $\in$ {CPA, CCA1, CCA2}, if for every polynomial $p(k)$, A runs in $p(k)$, outputs a valid key space $\mathcal{K}$ in $p(k)$, and outputs relation R computable in $p(k)$, and $Adv_{A,\Pi_{\text{KEM}}}^{\text{NM-ATK}}(k)$ is negligible. We insist that the adversary is unsuccessful if some ciphertext $\boldsymbol{C_0}[i]$ does not have a valid decryption (that is, $\bot \in \boldsymbol{K}$).

Equivalence Results. We can obtain the equivalence of all three formal definitions and a following Theorem 1 between IND-CCA2-KEM and NM-CCA2-KEM. (See more details and proofs in the full paper version.)

Theorem 1. *(IND-CCA2-KEM ⇔ NM-CCA2-KEM)*
If encryption scheme Π_{KEM} is secure in the sense of IND-CCA2-KEM, then Π_{KEM} is secure in the sense of NM-CCA2-KEM.

2.2 Data Encapsulation Mechanism

Formally, a data encapsulation mechanism DEM is given by a pair of algorithms $DEM.Encrypt(K, M)$ and $DEM.Decrypt(K, C)$, where:

1. The encryption algorithm $DEM.Encrypt(K, M)$ takes as input a secret key K, and a plaintext M. It outputs a ciphertext C. Here, K, M and C are byte strings, and M may have arbitrary length, and K's length is $DEM.KeyLen$.
2. The decryption algorithm $DEM.Decrypt(K, C)$ takes as input secret key K and ciphertext C. It outputs plaintext M.

DEM must satisfy the soundness, $DEM.Decrypt(K, DEM.Encrypt(K, M)) = M$.

Attack Types of DEM. We state following six attack types of DEM. In the first, we consider the first three attack types, these are for access to encryption oracle. First, we state P0, that is an attack type with no access to encryption oracle by adversary. Secondly, we state P1 (Chosen Plaintext Attack). P1 is an attack type with access to encryption oracle. However the adversary cannot access to encryption oracle after getting target ciphertext. Thirdly, we state P2 (Adaptive Chosen Plaintext Attack). In this type, an adversary can access to encryption oracle even if after the adversary gets target ciphertext. Moreover, we consider the last three attack types, these are for access to decryption oracle. First, we state C0, that is an attack type with no access to decryption oracle by adversary. Secondly, we state C1 (Chosen Ciphertext Attack). C1 is an attack type with access to decryption oracle. However the adversary cannot access to decryption oracle after getting target ciphertext. Thirdly, we state C2 (Adaptive Chosen Ciphertext Attack). In this type, an adversary can access to decryption oracle even if after the adversary gets target ciphertext.

Indistinguishability of DEM. We state formal definition of indistinguishability for DEM in Fig.2 following [10], which we call IND-DEM. We also use IND-PX-CY-DEM to describe the security notion of indistinguishability for DEM against ATK ∈ {CPA, CCA1, CCA2}.

Let $\Pi_{\text{DEM}} = (DEM.Encrypt, DEM.Decrypt)$ be an encryption scheme over message space M and let $A = (A_1, A_2)$ be an adversary. We insist that $A_1(1^k)$ outputs $\{x_0, x_1\} \in M$ with $|x_0| = |x_1|$, where k is security parameter. Furthermore, when $Y = 2$, we insist that A_2 does not ask for the decryption of challenge ciphertext y.

Π_{DEM} is secure in the sense of IND-PX-CY for $\{X, Y\} \in \{0, 1, 2\}$ if $Adv_{A,\Pi_{\text{DEM}}}^{\text{IND-PX-CY}}(\cdot)$ is negligible for any PPT adversary A.

$$Adv_{A,\Pi_{\text{DEM}}}^{\text{IND-PX-CY}}(k) \equiv 2 \cdot \Pr[Expt_{A,\Pi_{\text{DEM}}}^{\text{IND-PX-CY}}(k)] - 1$$

where $Expt_{A,\Pi_{\text{DEM}}}^{\text{IND-PX-CY}}(k)$

$$K \leftarrow \{0,1\}^k; (x_0, x_1, s) \leftarrow A_1^{O_1,O_1'}(1^k); b \leftarrow \{0,1\}; y \leftarrow DEM.Encrypt(K, x_b);$$

$$g \leftarrow A_2^{O_2,O_2'}(1^k, s, y); \text{return 1 iff } g = b$$

and
If X = 0 then $O_1(\cdot) = \varepsilon$ and $O_2(\cdot) = \varepsilon$.
If X = 1 then $O_1(\cdot) = DEM.Encrypt(K, \cdot)$ and $O_2(\cdot) = \varepsilon$.
If X = 2 then $O_1(\cdot) = DEM.Encrypt(K, \cdot)$ and $O_2(\cdot) = DEM.Encrypt(K, \cdot)$.
If Y = 0 then $O_1'(\cdot) = \varepsilon$ and $O_2'(\cdot) = \varepsilon$.
If Y = 1 then $O_1'(\cdot) = DEM.Decrypt(K, \cdot)$ and $O_2'(\cdot) = \varepsilon$.
If Y = 2 then $O_1'(\cdot) = DEM.Decrypt(K, \cdot)$ and $O_2'(\cdot) = DEM.Decrypt(K, \cdot)$.

Fig. 2. IND-DEM Definition

Non-malleability of DEM. We state formal definition of non-malleability for DEM in Fig.3 following Bellare[2] and Katz[10], which we call NM-DEM. We also use NM-

$$Adv_{A,\Pi_{\text{DEM}}}^{\text{NM-PX-CY}}(k) \equiv \Pr[Expt_{A,\Pi_{\text{DEM}}}^{\text{NM-PX-CY}}(k) = 1] - \Pr[\widetilde{Expt}_{A,\Pi_{\text{DEM}}}^{\text{NM-PX-CY}}(k) = 1]$$

where

$Expt_{A,\Pi_{\text{DEM}}}^{\text{NM-PX-CY}}(k)$	$\widetilde{Expt}_{A,\Pi_{\text{DEM}}}^{\text{NM-PX-CY}}(k)$
$K \leftarrow \{0,1\}^k$	$K \leftarrow \{0,1\}^k$
$(M, s) \leftarrow A_1^{O_1,O_1'}(1^k)$	$(M, s) \leftarrow A_1^{O_1,O_1'}$
$x \leftarrow M$	$(x, \tilde{x}) \leftarrow M$
$y \leftarrow DEM.Encrypt(K, x)$	$\tilde{y} \leftarrow DEM.Encrypt(K, \tilde{x})$
$(R, \boldsymbol{y}) \leftarrow A_2^{O_2,O_2'}(s, y)$	$(R, \widetilde{\boldsymbol{y}}) \leftarrow A_2^{O_2,O_2'}(s, \widetilde{y})$
$\boldsymbol{x} \leftarrow DEM.Decrypt(K, \boldsymbol{y})$	$\widetilde{\boldsymbol{x}} \leftarrow DEM.Decrypt(K, \widetilde{\boldsymbol{y}})$
return 1 iff $(y \notin \boldsymbol{y}) \wedge R(x, \boldsymbol{x})$	return 1 iff $(\widetilde{y} \notin \widetilde{\boldsymbol{y}}) \wedge R(x, \widetilde{\boldsymbol{x}})$

and
If X = 0 then $O_1(\cdot) = \varepsilon$ and $O_2(\cdot) = \varepsilon$.
If X = 1 then $O_1(\cdot) = DEM.Encrypt(K, \cdot)$ and $O_2(\cdot) = \varepsilon$.
If X = 2 then $O_1(\cdot) = DEM.Encrypt(K, \cdot)$ and $O_2(\cdot) = DEM.Encrypt(K, \cdot)$.
If Y = 0 then $O_1'(\cdot) = \varepsilon$ and $O_2'(\cdot) = \varepsilon$.
If Y = 1 then $O_1'(\cdot) = DEM.Decrypt(K, \cdot)$ and $O_2'(\cdot) = \varepsilon$.
If Y = 2 then $O_1'(\cdot) = DEM.Decrypt(K, \cdot)$ and $O_2'(\cdot) = DEM.Decrypt(K, \cdot)$.

Fig. 3. NM-DEM Definition

PX-CY-DEM to describe the security notion of non-malleability for DEM for $\{X, Y\} \in \{0, 1, 2\}$.

In Fig.3, M is a distribution over messages and R is some relation and k is security parameter. We require that $|x| = |x'|$ for all x, x' in the support of M. We also require that the vector of ciphertexts y output by A_2 should be non-empty. Furthermore, when $Y = 2$, we insist that A_2 does not ask for the decryption of y.

Π_{DEM} is secure in the sense of NM-PX-CY for $\{X, Y\} \in \{0, 1, 2\}$ if $Adv_{A, \Pi_{\text{DEM}}}^{\text{NM-PX-CY}}(k)$ is negligible for any PPT adversary A.

We obtain that the two above security notions of DEM yield the following Theorem 2. (Proof is in the full paper version.)

Theorem 2. *(NM-P2-C2-DEM $\Leftrightarrow$ IND-P2-C2-DEM)*
Encryption scheme Π_{DEM} is secure in the sense of NM-P2-C2 if and only if Π_{DEM} is secure in the sense of IND-P2-C2.

3 Universally Composable KEM Is Equivalent to IND-CCA2 KEM

3.1 The Key Encryption Mechanithm Functionality $\mathcal{F}_{\text{KEM}}$

We define key encapsulation mechanism (KEM) functionality $\mathcal{F}_{\text{KEM}}$ in Fig.4. $\mathcal{F}_{\text{KEM}}$ is a functionality of KEM-key-generation, KEM-encryption and KEM-decryption. Here note that there is no functionality of data transmission between parties in $\mathcal{F}_{\text{KEM}}$.

3.2 UC KEM Is Equivalent to IND-CCA2 KEM

Let KEM= $(KEM.KeyGen, KEM.Encrypt, KEM.Decrypt)$ be a key encapsulation mechanism. Consider the following transformation from KEM to protocol π_{KEM} that is constructed for realizing $\mathcal{F}_{\text{KEM}}$:

1. Upon input (KEM.KeyGen, sid) within some party P_j, P_j obtains the public key pk and secret key sk by running the algorithm $KEM.KeyGen()$, then outputs (KEM Key, sid, pk).
2. Upon input (KEM.Encrypt, sid, pk') within some party P_i, P_i obtains pair (K^*, C_0^*) of a key and a ciphertext by running the algorithm $KEM.Encrypt(pk')$ and outputs (Encrypted Shared Key, sid, pk', K^*, C_0^*). (Note that it does not necessarily hold that $pk'= pk$).
3. Upon input (KEM.Decrypt, sid, C_0^*) within P_j, P_j obtains $K^* = KEM.Decrypt$ (sk, C_0^*) and output (Shared Key, sid, K^*).

Theorem 3. *π_{KEM} securely realizes $\mathcal{F}_{\text{KEM}}$ with respect to non-adaptive adversaries if and only if KEM is indistinguishable against adaptive chosen ciphertext attacks (IND-CCA2 KEM).*

Proof. ("only if" part) Because NM-CCA2-KEM equals to IND-CCA2-KEM by Theorem 1, we prove that if π_{KEM} is not NM-CCA2-KEM secure, then π_{KEM} does not

---

Functionality $\mathcal{F}_{\mathrm{KEM}}$

$\mathcal{F}_{\mathrm{KEM}}$ proceed as follows, running with parties $P_1, \ldots, P_n$ and an adversary S.

KEM.KeyGen
In the first activation, expect to receive (KEM.KeyGen, sid) from some party P_j. Then,

1. Send (KEM.KeyGen, sid) to S.
2. Upon receiving (KEM Key, sid, pk) from S, send (KEM Key, sid, pk) to P_j.
3. If this is the first activation then record the pair (P_j, pk), otherwise pk is discarded.

KEM.Encrypt
Upon receiving (KEM.Encrypt, sid, pk') from some party P_i, proceed as follows:

- Check the memory, if $pk' = pk$, and if P_j is not corrupted, then proceeds as follows:
  1. Send (KEM.Encrypt, sid, pk') to S.
  2. Receive (Encrypted Shared Key, sid, pk', C_0) from S.
  3. If C_0 is stored in memory then halt.
  4. Choose Shared Key $K \xleftarrow{R} \{0,1\}^*$ randomly.
  5. Send (Encrypted Shared Key, sid, pk', K, C_0) to P_i.
  6. Store the pair (K, C_0) in memory.
- Otherwise (includes $pk' \neq pk$ or pk is not yet recorded, or P_j is corrupted),
  1. Send (KEM.Encrypt with Key, sid, pk') to S.
  2. Receive (Encrypted Shared Key, sid, pk', K, C_0) from S.
  3. Send (Encrypted Shared Key, sid, pk', K, C_0) to P_i.

KEM.Decrypt
Upon receiving (KEM.Decrypt, sid, C_0') from P_j (and P_j only), hand (KEM.Decrypt, sid, C_0') to S. Upon receiving (Shared Key, sid, K') from S, proceed as follows:

1. If a pair (K, C_0') exists in memory, send (Shared Key, sid, K) to P_j.
2. Otherwise, send (Shared Key, sid, K') to P_j.

Fig. 4. The Key Encapsulation Mechanism Functionality

securely realize $\mathcal{F}_{\mathrm{KEM}}$. More details, we prove that we can construct an environment Z and a real life adversary A such that for any ideal process adversary (simulator) S, Z can tell whether it is interacting with A and π_{KEM} or with S in the ideal process for $\mathcal{F}_{\mathrm{KEM}}$ by using the adversary G that breaks NM-CCA2-KEM.

Z proceeds as follows:

1. Activates key receiver P_j with (KEM.KeyGen, sid), and obtains pk.
2. Activates P_i with (KEM.Encrypt, sid, pk), and obtains (K^*, C_0^*).
3. Activates G with pk and C_0^*, obtains $(R, \boldsymbol{C_0})$, where R is some relation.
4. Activates P_j with (KEM.Decrypt, sid, $\boldsymbol{C_0}[i]$) for each i, and obtains $\boldsymbol{K'}[i]$.
5. Return 1 iff $R(K^*, \boldsymbol{K'})$.

When Z interacts with A and π_{KEM}, Z obtains corresponding pair (K^*, C_0^*) in Step 2. In this case, Z returns 1 in Step 5. On the other hand, Z interacts with S in the ideal process for $\mathcal{F}_{\text{KEM}}$, Z obtains non-corresponding pair $(K^\star, C_0^*)$ in Step 2, where $K^\star \xleftarrow{R} \{0,1\}^*$ by $\mathcal{F}_{\text{KEM}}$ and C_0^* is generated by S. For C_0^*, G successfully obtains $(R, \boldsymbol{C_0})$. However Z cannot output 1 in Step 5 because there is no relation $R(K^\star, \boldsymbol{K'})$.

("if" part) We show that if π_{KEM} does not securely realize $\mathcal{F}_{\text{KEM}}$, then π_{KEM} is not IND-CCA2-KEM. More details, we assume that for any simulator S there is an adversary and an environment Z that can distinguish with non-negligible probability whether it interacts with S in the ideal process for $\mathcal{F}_{\text{KEM}}$ or with parties running π_{KEM} and the adversary A in the real-life world. Then we prove that π_{KEM} is not IND-CCA2-secure by using the distinguishable environment Z.

We will show that Z can distinguish only when receiver P_j is not corrupted. We discuss all the cases as follows.

(Case 1: Receiver P_j is corrupted.) In this case, we can make simulator S such that the environment Z cannot distinguish the real life world from the ideal process world. Once A corrupts P_j, simulator S corrupts dummy party $\widetilde{P_j}$. However receiver P_i is not corrupted, that is, P_i is honest. Simulator S proceeds as follows:

1. When S receives (KEM.KeyGen, sid), it obtains (pk, sk) by running KEM.KeyGen(), and returns pk to $\mathcal{F}_{\text{KEM}}$.
2. When S receives (KEM.Encrypt with Key, sid, pk), then S generates a corresponding pair (K, C_0) and returns C_0 to $\mathcal{F}_{\text{KEM}}$.
3. When S receives (KEM.Decrypt, sid, C_0), S generates key K and returns K to $\mathcal{F}_{\text{KEM}}$.

In this case Z cannot distinguish the real world from the ideal world because S can reconstruct by using the simulated copy of A. Note that, A can do stopping the protocol π_{KEM}. Even if this situation happens, Z cannot distinguish the real world from the ideal world, because S can also stop the protocol.

(Case 2: P_j is not corrupted.) We look at the generated key and ciphertext by P_i in each world.

- In the real life world, π_{KEM} runs among the honest parties, P_i generates corresponding pair (K^*, C_0^*) by running the algorithm $KEM.Encrypt(pk)$.
- In the ideal process world, when $\widetilde{P_i}$ sends (KEM.Encrypt, sid, pk) to $\mathcal{F}_{KEM}$, $\mathcal{F}_{KEM}$ obtains C_0 from S, and $\mathcal{F}_{KEM}$ chooses shared key $K \xleftarrow{R} \{0,1\}^*$ at random. Then sends (Encrypted Shared Key, sid, pk, K, C_0) to P_i.

It is easily seen that C_0 is not concerned to the key K (because $\mathcal{F}_{KEM}$ randomly generates the key K). In the real world, Z obtains the corresponding pair (K^*, C_0^*). However, in the ideal world, Z obtains the non-corresponding pair (K, C_0). Consequently, we can construct environment Z that can distinguish the real world from the ideal world.

Recall the formal settings, there are three types of messages between Z and A. That is, Z sends A a message either to corrupt parties, or to report on messages sending, or to deliver some message. In this protocol, no party corruption occurs during execution since we consider non-adaptive adversaries. Furthermore, parties don't send messages each other. Therefore, there are no request to report on or deliver messages. So, the way that S affects the output of Z is only the communication via $\mathcal{F}_{\text{KEM}}$. As a result, S proceeds as follows:

1. When S receives a message (KEM.KeyGen, sid) from $\mathcal{F}_{\text{KEM}}$, it runs the key generation algorithms KEM.KeyGen(), obtains the public key pk and the secret key sk, and returns pk to $\mathcal{F}_{\text{KEM}}$.
2. When S receives a message (KEM.Encrypt, sid, pk) from $\mathcal{F}_{\text{KEM}}$, then it generates C_0 from the output of the algorithm $KEM.Encrypt(pk)$, and returns C_0 to $\mathcal{F}_{\text{KEM}}$.
3. When S receives a message (KEM.Encrypt with Key, sid, pk) from $\mathcal{F}_{\text{KEM}}$, then it generates key $(K, C_0) = KEM.Encrypt(pk)$, and returns (K, C_0) to $\mathcal{F}_{\text{KEM}}$.
4. When S receives a message (KEM.Decrypt, sid, C_0) from $\mathcal{F}_{\text{KEM}}$, it obtains $K = KEM.Decrypt(sk, C_0)$ and returns K to $\mathcal{F}_{\text{KEM}}$.

We assume that there is an environment Z that can distinguish the interaction in the real life world from that in the ideal process world. We prove that we can construct an adversary F that breaks IND-CCA2-KEM by using the distinguishable environment Z. Precisely, for some value of the security parameter z for Z, we assume that there is an environment Z such that $IDEAL_{F,S,Z}(z) - REAL_{\pi_{\text{KEM}},A,Z}(z) > \sigma$, then we show that F correctly guesses the bit b with probability $\frac{1}{2} + \frac{\sigma}{2l}$ in the CCA2 game, where l is the total number of times invoking encryption oracle.

F is given a public key pk, and is allowed to query to decryption oracle and encryption oracle. First, F chooses a number $h \xleftarrow{R} \{1, \ldots, l\}$ at random. Secondly, F simulates Z on the following simulated interaction with a system running π_{KEM}. Let K_i and C_{0i} denote the i-th key and ciphertext that Z asks to encrypt in this simulation, respectively.

1. When Z activates some party P_j with (KEM.KeyGen, sid), F lets P_j output the value pk from F's input.
2. For the first $h - 1$ times that Z asks some party P_i to generate shared key K_i, F lets P_i return (K_i, C_{0i}) by using algorithm $(K_i, C_{0i}) = $ KEM.Encrypt(pk).
3. The h-th time that Z asks to generate key K_h, F queries its encryption oracle with pk, then obtains corresponding pair $X = (K_h, C_{0h})$ or non-corresponding pair $X = (K'_h, C_{0h})$ from encryption oracle. Accordingly, F hands X to Z as the test pair.
4. For the remaining $l - h$ times that Z asks P_i to generate shared key K_i, F lets P_i return (K_i, C_{0i}), where $K_i \xleftarrow{R} \{0,1\}^*$ randomly and C_0 from the output of algorithm $KEM.Encrypt(pk)$.
5. Whenever Z activates decryptor P_j with (KEM.Decrypt, sid, C_0), where $C_0 = C_{0i}$ for some i, F lets P_i return the corresponding key K_i for any i. If C_0 is different from all the C_{0i}'s, then F queries C_0 to its decryption oracle, obtains value v, and lets P_j return v to Z.
6. When Z halts, F outputs whatever Z outputs and halts.

We apply a standard hybrid argument for analyzing the success probability of F. Let the random variable D_i denote the output of Z from an interaction that is identical to an interaction with S in the ideal process, except that the first i pairs are computed with correctly generation, and the last pair are computed with non-corresponding generation. We can see that D_0 is identical to the output of Z in the ideal process world, and D_l is identical to the output of Z in the real life world. (This follows from the fact that the mechanism KEM guarantees that KEM.Decrypt(sk, C_0) = K, where $C_0 = $

KEM.Encrypt(pk), this is called "soundness".) Furthermore, in the simulation of F, if the value C_{0h} that F obtains from its encryption oracle is an encryption of K_h then the output of the simulated Z has the distribution of D_{h-1}. If C_{0h} does not correspond to the encryption of the key then the output of the simulated Z has the distribution of D_h. As discussed above, we can construct attacker F by using the distinguishable environment Z. We can conclude that if π_{KEM} does not securely realize $\mathcal{F}_{\text{KEM}}$, then π_{KEM} is not IND-CCA2-KEM. □

4 Universally Composable DEM Is Equivalent to IND-P2-C2 DEM

4.1 The KEM-DEM Functionality $\mathcal{F}_{\text{KEM-DEM}}$

We define KEM-DEM functionality $\mathcal{F}_{\text{KEM-DEM}}$ in Fig.5 and Fig.6. $\mathcal{F}_{\text{KEM-DEM}}$ is a functionality of hybrid usage of KEM and DEM, KEM-key-generation, KEM-encryption, KEM-decryption, DEM-encryption and DEM-decryption. Information obtained in KEM-encryption and KEM-decryption is transfered to DEM-encryption and DEM-decryption inside $\mathcal{F}_{\text{KEM-DEM}}$. Here note that there is no functionality of data transmission between parties in $\mathcal{F}_{\text{KEM-DEM}}$.

4.2 UC DEM Is Equivalent to IND-P2-C2 DEM

First, we define a protocol $\pi_{\text{KEM-DEM}}$ in Fig.7 that is constructed on an algorithm DEM = $(DEM.Encrypt, DEM.Decrypt)$ in the $\mathcal{F}_{\text{KEM}}$-hybrid model. We say that the underlying DEM is UC secure if and only if $\pi_{\text{KEM-DEM}}$ securely realizes $\mathcal{F}_{\text{KEM-DEM}}$ in the $\mathcal{F}_{\text{KEM}}$-hybrid model.

Therefore, the following theorem implies that UC DEM is equivalent to IND-P2-C2 DEM.

Theorem 4. *Protocol $\pi_{\text{KEM-DEM}}$ securely realizes $\mathcal{F}_{\text{KEM-DEM}}$ with respect to non-adaptive adversaries in the $\mathcal{F}_{\text{KEM}}$-hybrid model if and only if DEM is indistinguishable against adaptive chosen plaintext/ciphertext attacks(IND-P2-C2 DEM).*

Proof. (sketch) ("only if" part) Because NM-P2-C2-DEM equals to IND-P2-C2-DEM by Theorem 2, we prove that if π_{DEM} is not NM-P2-C2-DEM secure, then $\pi_{\text{KEM-DEM}}$ does not securely realize $\mathcal{F}_{\text{KEM-DEM}}$ in the $\mathcal{F}_{\text{KEM}}$ - hybrid model. More details, we prove that we can construct an environment Z and a real life adversary A such that for any ideal process adversary (simulator) S, Z can tell whether it is interacting with A and $\pi_{\text{KEM-DEM}}$ or with S in the ideal process for $\mathcal{F}_{\text{KEM-DEM}}$ by using the adversary which breaks NM-P2-C2-DEM. Note that A corrupts no party and Z sends no messages to A. We assume that there exists a successful attacker G for π_{DEM} in the sense of NM-P2-C2-DEM. Environment Z proceeds as usual, except that Z runs a copy of G.

Z proceeds as above, except that Z runs a simulated copy of G. For more details:

1. Activates key receiver P_j with (KEM.KeyGen, sid), then obtains pk.
2. Activates key encrypter P_i with (KEM.Encrypt, sid, pk), then obtains $C_0{}^*$.
3. Activates P_j with (KEM.Decrypt, sid, C_0).

> **Functionality** $\mathcal{F}_{\text{KEM-DEM}}$
>
> $\mathcal{F}_{\text{KEM-DEM}}$ proceeds as follows, running with parties $P_1, \ldots, P_n$ and an adversary S.
>
> **KEM.KeyGen**
> In the first activation, expect to receive (KEM.KeyGen, sid) from some party P_j. Then,
>
> 1. Send (KEM.KeyGen, sid) to S.
> 2. Upon receiving (KEM Key, sid, pk) from S, send (KEM Key, sid, pk) to P_j.
>
> **KEM.Encrypt**
> Upon receiving (KEM.Encrypt, sid, pk') from some party P_i, proceed as follows:
>
> – If an entry (P_i, C, **active**) is not in memory for any C,
>   1. Send (KEM.Encrypt, sid, pk') to S, and receive (Encrypted Shared Key, sid, pk', C_0) from S.
>   2. Send (Encrypted Shared Key, sid, pk', C_0) to P_i, and store the pair (pk', C_0) and (P_i, C_0, **active**) in memory.
> – Otherwise, do nothing.
>
> **KEM.Decrypt**
> Upon receiving (KEM.Decrypt, sid, C_0') from P_j (and P_j only), hand (KEM.Decrypt, sid, C_0') to S. Upon receiving ok from S, proceed as follows:
>
> – If an entry (P_j, C, **active**) is not in memory for any C, send ok to P_j and store the pair (P_j, C_0', **active**) in memory.
> – Otherwise, do nothing.
>
> **DEM.Encrypt**
> Upon receiving (DEM.Encrypt, sid, m) from party P_e ($e \in \{i, j\}$ only), proceed as follows:
>
> – If (P_e, C_0, **active**) is stored in memory.
>   • If both P_e are uncorrupted, then proceeds as follows:
>   1. Send (DEM.Encrypt, sid, $|m|$) to S, where $|m|$ denotes the length of m and receive (DEM.Ciphertext, sid, c') from S.
>   2. Send (DEM.Ciphertext, sid, c') to P_e, and store the entry (m, c', C_0) in memory.
>   • Otherwise, proceeds as follows:
>   1. Send(DEM.Encrypt, sid, m) to S, and receive (DEM.Ciphertext, sid, c') from S.
>   2. Send (DEM.Ciphertext, sid, c') to P_e, and store the entry (m, c', C_0) in memory.
> – Otherwise, do nothing.

Fig. 5. The KEM-DEM Functionality

4. Activates message encrypter P_i with (DEM.Encrypt, sid, m), then obtains c.
5. Activates G on c, obtains (R, $\boldsymbol{c}$), where R is some relation.
6. Activates P_j with (DEM.Decrypt, sid, $\boldsymbol{c}[i]$) for each i, and obtains $\boldsymbol{m'}[i]$.
7. Return 1 iff $R(m, \boldsymbol{m'})$.

Functionality $\mathcal{F}_{\text{KEM-DEM}}$ (continued)

DEM.Decrypt

Upon receiving (DEM.Decrypt, sid, c') from P_e ($e \in \{i,j\}$ only), hand (DEM.Decrypt, sid, c') to S. Upon receiving (DEM.Plaintext, sid, ϕ) from S, proceed as follows:

- If an entry (P_e, C, **active**) exists in memory for some C:
  1. If the entry (m, c', C) is stored in the memory, then send (DEM.Plaintext, sid, m) to P_j.
  2. Else, if P_i and P_j is not corrupted, and if (m, c', C) doesn't recorded in the memory, then store the entry ($\perp$, c', C) and send (DEM.Plaintext, sid, $\perp$) to P_e.
  3. Else, if an entry ($\perp$, c', C) is recorded, then send (DEM.Plaintext, sid, $\perp$) to P_e.
  4. Otherwise, send (DEM.Plaintext, sid, ϕ) to P_e, and record the entry (ϕ, c', C) in memory.
- Otherwise, do nothing.

Fig. 6. The KEM-DEM Functionality

When Z interacts with A and $\pi_{\text{KEM-DEM}}$, Z obtains ciphertext c in Step 4. In this case, Z return 1 in Step 7. Therefore when Z interacts with A and $\pi_{\text{KEM-DEM}}$, Z outputs 1 with non-negligible probability. On the other hand, Z interacts with S in the ideal process for $\mathcal{F}_{\text{KEM}}$, Z also obtains ciphertext c in Step 4. For ciphertext c, G successfully obtains (R, c). However Z cannot output 1 in Step 7 because there is no relation $R(m, m')$.

("if" part) We prove that if $\pi_{\text{KEM-DEM}}$ does not securely realize $\mathcal{F}_{\text{KEM-DEM}}$, then π_{DEM} is not IND-P2-C2-DEM. More details, we assume that there is an adversary A such that for any simulator S, there is an environment Z can tell with non-negligible probability whether it is interacting with $\mathcal{F}_{\text{KEM-DEM}}$ and S in the ideal process world or with parties running $\pi_{\text{KEM-DEM}}$ and the adversary A in the real life world. Then, we prove that there is adversary F breaks IND-P2-C2-DEM by using distinguishable Z. Note that there are three cases of party corruption since we take account of non-adaptive adversaries.

Recall the formal settings, there are three types of messages between Z and A. That is, Z sends A a message either to corrupt parties, or to report on messages sending, or to deliver some message. In this protocol, no party corruption occurs during execution since we consider non-adaptive adversaries. Furthermore, parties don't send messages each other. Therefore, there are no request to report on or deliver messages. In fact, there is no communication between Z and A at all. So, the way that S affects the output of Z is only the communication via $\mathcal{F}_{\text{KEM-DEM}}$.

We will show that Z can distinguish is only when both sender P_i and receiver P_j are not corrupted. We discuss all the cases for the following simulator S as follows:

1. When S receives (KEM.KeyGen, sid), S obtains (pk, sk) by running KEM.KeyGen(), and returns (KEM Key, sid, pk) to $\mathcal{F}_{\text{KEM-DEM}}$.
2. When S receives (KEM.Encrypt, sid, pk), S generates a corresponding pair (K, C_0), and returns (Encrypted Shared Key, sid, pk, C_0) to $\mathcal{F}_{\text{KEM-DEM}}$.

Protocol $\pi_{\text{KEM-DEM}}$

Key Encapsulation Mechanithm KEM

KEM.KeyGen

1. Upon input (KEM.KeyGen, sid), P_j sends (KEM.KeyGen, sid_1) to $\mathcal{F}_{\text{KEM}}$.
2. Upon receiving (KEM Key, sid_1, pk) from $\mathcal{F}_{\text{KEM}}$, P_j outputs pk.

KEM.Encrypt
Upon input (KEM.Encrypt, sid, pk) within party P_i,

– If boolean variable **active** is not set,
  1. P_i sends (KEM.Encrypt, sid_1, pk) to $\mathcal{F}_{\text{KEM}}$.
  2. Upon receiving (Encrypted Shared key, sid_1, pk, K, C_0) from $\mathcal{F}_{\text{KEM}}$, then P_i outputs C_0 and stores the key K in memory and sets a boolean variable **active** in memory.
– Otherwise, do nothing.

KEM.Decrypt
Upon input (KEM.Decrypt, sid, C_0) within P_j,

– If boolean variable **active** is not set,
  1. P_j sends (KEM.Decrypt, sid_1, C_0) to $\mathcal{F}_{\text{KEM}}$.
  2. Upon receiving (Shared Key sid_1, K), P_j stores K in memory and outputs ok and sets a boolean variable **active** in memory.
– Otherwise, do nothing.

Data Encapsulation Mechanithm DEM

DEM.Encrypt
Upon input (DEM.Encrypt, sid, m) from P_e ($e \in \{i, j\}$), proceeds as follows:

– If the boolean variable is **active** in P_e's memory, P_e obtains ciphertext c = DEM.Encrypt(K, m) and outputs (DEM Ciphertext, sid, c).
– Otherwise do nothing.

DEM.Decrypt
Upon input (DEM.Decrypt, sid, c) from P_e ($e \in \{i, j\}$), proceeds as follows:

– If the boolean variable is **active** in P_e's memory, P_e obtains m = DEM.Decrypt (K, c) and outputs (DEM Plaintext, sid, m).
– Otherwise do nothing.

Fig. 7. The KEM-DEM Protocol

3. When S receives (KEM.Decrypt, sid, C_0), S obtains key K by KEM.Decrypt(sk, C_0), and returns ok to $\mathcal{F}_{\text{KEM-DEM}}$.
4. When S receives (DEM.Encrypt, sid, $|m|$), S generates c' by output of DEM.Encrypt(K, $0^{|m|}$), and returns (DEM.Ciphertext, sid, c') to $\mathcal{F}_{\text{KEM-DEM}}$.

5. When S receives (DEM.Encrypt, sid, m), S generates c' by the output of DEM.Encrypt(K, m) and returns (DEM.Ciphertext, sid, c') to $\mathcal{F}_{\text{KEM-DEM}}$.
6. When S receives (DEM.Decrypt, sid, c'), S generates ϕ by DEM.Decrypt(K, c'), and sends (DEM.Plaintext, sid, ϕ).

(Case 1: Sender P_i is corrupted.) In this case, once A corrupts P_i, simulator S corrupts dummy party $\widetilde{P}_i$. However receiver P_j is not corrupted, that is, P_j is honest. Environment Z cannot distinguish the real life world from the ideal process world for the above simulator S because S can reconstruct by using the simulated copy of A. Note that, A can do stopping the protocol $\pi_{\text{KEM-DEM}}$. Even if this situation is happened, Z cannot distinguish the real world from the ideal world, because S can also stop the protocol.

(Case 2: Receiver P_j is corrupted.) In this case, once A corrupts P_j, simulator S corrupts dummy party $\widetilde{P}_j$. However sender P_i is not corrupted, that is, P_i is honest. Environment Z cannot distinguish the real life world from the ideal process world by the above simulator S because simulator S can reconstruct by using the simulated copy of A.

(Case 3: No party is corrupted.) In this case, sender P_i and receiver P_j are not corrupted i.e., they are honest parties. We look at the generated key and ciphertext by P_i in each world.

- In the real life world, $\pi_{\text{KEM-DEM}}$ runs among the honest parties, P_i generates c by running the algorithm $DEM.Encrypt(K, m)$. Note that c is corresponding to m.
- In the ideal process world, $\mathcal{F}_{\text{KEM-DEM}}$ send (DEM.Encrypt, sid, $|m|$) to S. P_i obtains c' from S via $\mathcal{F}_{\text{KEM-DEM}}$. Note that c is non-corresponding to m because S sees only the length of m.

By applying a hybrid argument similar to the one in the proof of Theorem 3, we can obtain adversary F that attacks IND-P2-C2-DEM by using the environment Z that can distinguish the real world from the ideal world. □

5 A Universally Composable Secure Channel Based on the KEM-DEM Framework

To realize secure channel functionality, $\mathcal{F}_{\text{SC}}$, defined in [4], we define a secure channel protocol π_{SC} in Fig.8 in the ($\mathcal{F}_{\text{KEM-DEM}}$, $\mathcal{F}_{\text{SIG}}$, $\mathcal{F}_{\text{CA}}$)-hybrid model, where $\mathcal{F}_{\text{SIG}}$ is a signature functionality [4], and $\mathcal{F}_{\text{CA}}$ is certification authority functionality [4]. (Due to the page limitation, we omit the description of $\mathcal{F}_{\text{SIG}}$ and $\mathcal{F}_{\text{CA}}$. See [4] for the definitions.)

Combining with the previous theorems, the following theorem implies that IND-CCA2 KEM, IND-P2-C2 DEM, secure signatures and ideal CA are sufficient to securely realize $\mathcal{F}_{\text{SC}}$.

Theorem 5. *Protocol π_{SC} securely realizes $\mathcal{F}_{\text{SC}}$ in the ($\mathcal{F}_{\text{KEM-DEM}}$, $\mathcal{F}_{\text{SIG}}$, $\mathcal{F}_{\text{CA}}$)-hybrid model.*

Protocol π_{SC}

Session Set-up

1. Upon input (Establish-session, sid, P_j, initiator), P_i sends (KEM.KeyGen, sid_1) to $\mathcal{F}_{\text{KEM-DEM}}$, and stores (sid, P_j).
2. Upon receiving (KEM Key, sid_1, PK_i) from $\mathcal{F}_{\text{KEM-DEM}}$, P_i sends (Register, P_i, PK_i) to $\mathcal{F}_{\text{CA}}$.
3. Upon input (Establish-session, sid, P_i, responder), P_j sends (Retrieve, P_i) to $\mathcal{F}_{\text{CA}}$.
4. Upon receiving (Retrieve, P_i, PK_i) from $\mathcal{F}_{\text{CA}}$, P_j sends (KEM.Encrypt, sid_1, PK_i) to $\mathcal{F}_{\text{KEM-DEM}}$, and receives (Encrypted Shared key, sid_1, PK_i, C_0) from $\mathcal{F}_{\text{KEM-DEM}}$.
5. P_j sends (KeyGen, (P_j, sid')) to $\mathcal{F}_{\text{SIG}}$, receives (Verification Key, $(P_j, sid'), PK_j$).
6. P_j sends (Register, P_j, PK_j) to $\mathcal{F}_{\text{CA}}$, then sends (Sign, P_j, C_0) to $\mathcal{F}_{\text{SIG}}$, receives (Signature, $(P_j, sid'), C_0, \sigma$) from $\mathcal{F}_{\text{SIG}}$.
7. P_j sends (sid, C_0, σ, P_j) to P_i, and set a boolean variable **active**.
8. Upon receiving (sid, C_0, σ, P_j), P_i checks whether (sid, P_j) is stored. If it is not stored, discard the message. Otherwise, P_i sends (Retrieve, P_j) to $\mathcal{F}_{\text{CA}}$ and receives (Retrieve, P_j, PK_j), then sends (Verify, $(P_j, sid'), C_0, \sigma, PK_j$) to $\mathcal{F}_{\text{SIG}}$ and receives (Verified, $(P_j, sid'), C_0, f$). If f is 1 then P_i goes to next step. Else finish the protocol.
9. P_i sends (KEM.Decrypt, sid_1, C_0) to $\mathcal{F}_{\text{KEM-DEM}}$. If ok is returned from $\mathcal{F}_{\text{KEM-DEM}}$, set a boolean variable **active**.

Data Exchange

1. Upon input (Send, sid, m), to P_e, if P_e is **active** (i.e., $e \in \{i, j\}$), P_e sends the message (DEM.Encrypt, sid_1, m) to $\mathcal{F}_{\text{KEM-DEM}}$.
2. Upon receiving (DEM.Ciphertext, c) from $\mathcal{F}_{\text{KEM-DEM}}$, P_e sends c to $P_{\bar{e}}$.
3. Upon receiving c, if $P_{\bar{e}}$ is **active** (i.e., $\bar{e} \in \{i, j\}$), $P_{\bar{e}}$ sends (DEM.Decrypt, sid_1, c) to $\mathcal{F}_{\text{KEM-DEM}}$.
4. $P_{\bar{e}}$ receives (DEM.Plaintext, m) from $\mathcal{F}_{\text{KEM-DEM}}$ and outputs m.

Session Ending

1. Upon input (Expire-session, sid), P_e sends (Expire-session, sid) to $P_{\bar{e}}$ and erases the session state (including all keys and local values) and terminates this protocol.
2. Upon receiving (Expire-session, sid), $P_{\bar{e}}$ erases the session state (including all keys and local values) and terminates this protocol.

Fig. 8. The Secure Channel Protocol π_{SC}

Proof. (sketch) Let A be an adversary that interacts with parties running π_{SC} in the ($\mathcal{F}_{\text{KEM-DEM}}, \mathcal{F}_{\text{SIG}}, \mathcal{F}_{\text{CA}}$)-hybrid model, and S be an ideal process adversary (simulator) that interacts with the ideal process for $\mathcal{F}_{SC}$. We construct S such that any environment Z cannot tell whether it is interacting with A in π_{SC} or with S in the ideal process for $\mathcal{F}_{SC}$. S invokes a simulated copy of A, and proceeds as follows:

1. Inputs from Z are forwarded to A and outputs from A are forwarded to Z.
2. **(Simulating the interaction of A in the session set-up)** Upon receiving a message (sid, P_i, P_j) from $\mathcal{F}_{SC}$ (which means that P_i and P_j have set-up a session),

simulates for A the process of exchanging shared key between P_i and P_j. That is, play functionalities, $\mathcal{F}_{\text{CA}}$, $\mathcal{F}_{\text{KEM-DEM}}$, $\mathcal{F}_{\text{SIG}}$, for A as follows: send to A (in the name of $\mathcal{F}_{\text{KEM-DEM}}$) the message (KEM.KeyGen, sid_1, PK_i), obtain the response (KEM Key, sid_1, PK_i) from A; send to A (in the name of $\mathcal{F}_{\text{CA}}$) the message (Registered, P_i, PK_i), obtain the response ok from A; send to A (in the name of $\mathcal{F}_{\text{CA}}$) the message (Retrieve, P_i, P_j), obtain the response ok from A; send to A (in the name of $\mathcal{F}_{\text{KEM-DEM}}$) the message (KEM.Encrypt, sid_1, PK_i), obtain the response (Encrypted Shared key, sid_1, PK_i, C_0) from A; send to A (in the name of $\mathcal{F}_{\text{SIG}}$) the message (KeyGen, (P_j, sid')), obtain the response (Verification Key, $(P_j, sid'), PK_j$) from A; send to A (in the name of $\mathcal{F}_{\text{CA}}$) the message (Registered, P_j, PK_j), obtain the response ok from A; send to A (in the name of $\mathcal{F}_{\text{SIG}}$) the message (Sign, $(P_j, sid'), C_0$), obtain the response (Signature, $(P_j, sid'), C_0, \sigma$) from A; send to A (in the name of $\mathcal{F}_{\text{CA}}$) the message (Retrieve, P_j, P_i), obtain the response ok from A; send to A (in the name of $\mathcal{F}_{\text{SIG}}$) the message (Verify, (P_j, sid'), C_0, σ, PK_j), obtain the response (Verified, $(P_j, sid'), C_0, \phi$) from A; send to A (in the name of $\mathcal{F}_{\text{KEM-DEM}}$) the message (KEM.Decrypt, sid_1, C_0, PK_i), obtain the response ok from A.
3. (**Simulating the interaction of A in the data exchange**) Upon receiving a message (sid, P_e, u) ($e \in \{i, j\}$) from $\mathcal{F}_{\text{SC}}$ (which means that P_e sent a message of length u to $P_{\bar{e}}$), simulates for A the process of exchanging shared key between P_i and P_j. That is, play functionality $\mathcal{F}_{\text{KEM-DEM}}$ for A as follows: send to A (in the name of $\mathcal{F}_{\text{KEM-DEM}}$) the message (DEM.Encrypt, $sid_1, |m|$), obtain the response (DEM.Ciphertext, c) from A; send to A (in the name of $\mathcal{F}_{\text{KEM-DEM}}$) the message (DEM.Decrypt, sid_1, c), obtain the response (DEM.Plaintext, ψ) from A.
4. (**Simulating the interaction of a corrupted party**) Simulating the interaction of a corrupted party can be done by simulating the functionalities and transmissions in the natural way. So, we omit the precise description here.
5. (**Simulating party corruption**) When A corrupts a party, S corrupts that party in the ideal process, and forwards the obtained information to A. This poses no problem since none of the parties maintains any secret information.

It is straightforward to verify that the simulation is perfect. That is, for any environment Z and A, it holds that the view of Z interacting with S and $\mathcal{F}_{\text{SC}}$ is distributed identically to the view of Z interacting with A and parties running protocol π_{SC} in the ($\mathcal{F}_{\text{KEM-DEM}}$, $\mathcal{F}_{\text{SIG}}$, $\mathcal{F}_{\text{CA}}$)-hybrid model. □

6 Conclusion

The KEM-DEM framework is a promising formulation for hybrid encryption based on symmetric and asymmetric encryption, and will be standardized in ISO in the near future. This paper studied the possibility of constructing a *UC secure channel* using the KEM-DEM framework. We presented that IND-CCA2 KEM and IND-P2-C2 DEM along with secure signatures and ideal certification authority are sufficient to realize a UC secure channel. This paper also shows several equivalence results: UC KEM, IND-CCA2 KEM and NM-CCA2 KEM are equivalent, and UC DEM, IND-P2-C2 DEM and NM-P2-C2 DEM are equivalent.

References

1. M.Bellare, A.Desai, D.Pointcheval, and P.Rogaway, "Relations Among Notions of Security for Public-Key Encryption Schemes, Crypto'98 LNCS 1462.
2. M.Bellare and A.Sahai, "Non-Malleable Encryption: Equivalence between Two Notions, and an Indistinguishability-Based Characterisation, Crypto'99 LNCS 1666.
3. R. Canetti, "Universally Composable Security: A New paradigm for Cryptographic Protocols, 42nd FOCS, 2001. Full version available at http://eprint.iacr.org/2000/067.
4. R. Canetti, "Universally Composable Signature, Certification, and Authentication, August, 2004. http://eprint.iacr.org/2003/239/.
5. R. Canetti and H. Krawczyk, "Analysis of Key-Exchange Protocols and Their Use for Building Secure Channels, Eurocrypt 01, 2001. Full version at http://eprint.iacr.org/2001.
6. R. Canetti and H. Krawczyk, "Universally Composable Notions of Key Exchange and Secure Channels, Eurocrypt 02, LNCS, Springer, 2002. http://eprint.iacr.org/2002.
7. R. Canetti and T. Rabin, "Universal Composition with Joint State," Proceedings of Crypto 03, LNCS, Springer, 2003. available at http://eprint.iacr.org/2002.
8. R.Cramer and V.Shoup, "Design and analysis of practical public-key encryption schemes secure against adaptive chosen ciphertext attack, http://shoup.net/papers/, 2001 Dec.
9. D.Dolev, C.Dwork, and M.Naor, "Non-Malleable Cryptography, 23rd STOC, 1991. Also Technical Report CS95-27, Weizmann Institute of Science, 1995.
10. J. Katz and M.Yung, "Characterization of Security Notions for Probabilistic Private-Key Encryption," to appear. Full version available at http://www.cs.umd.edu/~jkatz/.
11. V.Shoup, "A Proposal for an ISO Standard for Public Key Encryption (version 2.1), ISO/IEC JTC1/SC27, N2563, http://shoup.net/papers/, 2001 Dec.

Sufficient Conditions for Collision-Resistant Hashing

Yuval Ishai[1,*], Eyal Kushilevitz[1,**], and Rafail Ostrovsky[2,***]

[1] Computer Science Department, Technion, Haifa 32000, Israel
{yuvali, eyalk}@cs.technion.ac.il
[2] Computer Science Department, UCLA
rafail@cs.ucla.edu

Abstract. We present several new constructions of *collision-resistant hash-functions* (CRHFs) from general assumptions. We start with a simple construction of CRHF from any *homomorphic encryption*. Then, we strengthen this result by presenting constructions of CRHF from two other primitives that are implied by homomorphic-encryption: one-round *private information retrieval* (PIR) protocols and *homomorphic one-way commitments*.

Keywords: Collision-resistant hash functions, homomorphic encryption, private information-retrieval.

1 Introduction

Collision resistant hash-functions (CRHFs) are an important cryptographic primitive. Their applications range from classic ones such as the "hash-and-sign" paradigm for signatures, via efficient (zero-knowledge) arguments [14, 17, 2], to more recent applications such as ones relying on the non-black-box techniques of [1].

In light of the importance of the CRHF primitive, it is natural to study its relations with other primitives and try to construct it from the most general assumptions possible. It is known that CRHFs can be constructed from claw-free pairs of permutations [5] (which in turn can be based on the intractability of discrete logarithms or factoring) and under lattice-based assumptions [10]. On the other hand, Simon [20] rules out a black-box construction of CRHF from one-way permutations; thus, there is not much hope to base CRHF on very general assumptions involving one-wayness alone.

In practice, when people are in need for CRHFs in various cryptographic protocols, they often use constructions such as SHA1, MD5 and others. However,

---
[*] Partially supported by Israel Science Foundation grant 36/03.
[**] Partially supported by BSF grant 2002-354 and by Israel Science Foundation grant 36/03.
[***] Partially supported by BSF grant 2002-354 and by a gift from Teradata, Intel equipment grant, OKAWA research award and NSF Cybertrust grant.

recent weaknesses found in some of these constructions (such as MD5) [22] only provide further evidence for the value of theoretically sound constructions.

Our Results. In this paper, we present several new constructions of CRHFs from general assumptions. We start by describing a simple construction of CRHF from any *homomorphic encryption* (HE) scheme. A homomorphic encryption is a semantically secure encryption in which the plaintexts are taken from a group, and given encryptions of two group elements it is possible to efficiently compute an encryption of their sum. For instance, the Goldwasser-Micali scheme [11] is homomorphic over the group Z_2. We note that this notion does not impose any algebraic structure on the space of ciphertexts, but only on the space of plaintexts.

We then weaken the above assumption in two incomparable ways. First, we show how to construct CRHF from any (single-server, sublinear-communication) one-round PIR protocol [15]. Since PIR is implied by homomorphic encryption [15, 21, 16], this is indeed a weaker assumption. This result strengthens the result of [3], that constructs unconditionally hiding commitment (UHC) from PIR, as it is known that CRHF imply UHC [6, 12].

Second, we obtain a construction of CRHFs from *homomorphic one-way commitments* (HOWC). Such a commitment does not provide semantic security for the committed value x but only "one-way" security, guaranteeing that x is hard to find. For instance, a simple deterministic HOWC is defined by $C(x) = g^x$, where g is a generator of a group in which finding discrete logarithms is hard.

The relation between the different primitives discussed above is summarized in Figure 1.

One way to think of our results is the following. It is known how to build CRHFs from all major (specific) assumptions used in public-key cryptography. These (specific) assumptions also have an algebraic structure that usually implies homomorphic properties. The results of this work suggest that this is not a coincidence, establishing a rather general link between "homomorphic" prop-

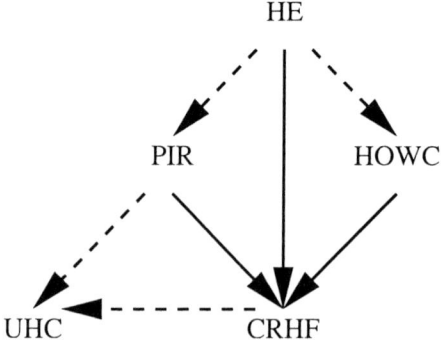

Fig. 1. Solid arrows stand for implications shown in this paper. Dashed arrows stand for implications that were shown in other papers (or that follow directly from the definition).

erties and collision resistance. First results in this direction were given in [18]; see below.

Related Work. As mentioned, Damgård [5] shows how to construct CRHFs based on any claw-free pair of permutations (and based on specific assumptions such as the hardness of factoring or discrete-log). Russell [19] shows that this is essentially the best one can do, as the existence of CRHFs is equivalent to the existence of a related primitive that he terms "claw-free pair of pseudo-permutations"; this characterization is not satisfactory in the sense that this primitive is not a well-studied one and its relations with other primitives are not known. Hsiao and Reyzin [13] consider two variants of the definition of CRHF (that differ in whether or not the security of the CRHF depends on the secrecy of the random coins used by the key-generation algorithm) and show some relations between the two variants. Simon [20] shows, by demonstrating an appropriate separation oracle, that one-way permutations are unlikely to imply CRHFs (see [7] for some stronger versions of this result). In contrast with [20], Ogata and Kurosawa [18] show that a stronger version of one-way permutations, i.e. *homomorphic one-way permutations*, can be used to construct claw-free permutations and hence also CRHFs. While this result gives an indication for the usefulness of homomorphic properties for constructing CRHFs, their construction heavily relies on the function being a *permutation*; our results, on the other hand, do not impose such structural constraints on the underlying primitives. To illustrate the significance of the extra generality, consider the question of basing CRHF on lattice-related intractability assumptions. Combining our results with the lattice-based PIR scheme from [16], we can obtain CRHFs whose security is based on a standard lattice-related assumption (providing an alternative to [10]). In contrast, there are no known constructions of one-way *permutations* (let alone homomorphic ones) from such assumptions.

Organization. In Section 2, we provide some necessary definitions (in particular that of CRHF). The first construction of CRHF, presented in Section 3, is based on the existence of homomorphic encryption. In Sections 4 and 5, we strengthen this result by describing constructions that are based on (computational) PIR and on homomorphic one-way commitment (respectively).

2 Preliminaries

We start with a formal definition of collision-resistant hash-functions (CRHFs). In fact, the definition applies to a *family* of functions,[1] and uses the terminology of secret-coin CRHFs from [13].

Definition 1. *Let $\ell, \ell' : \mathbb{N} \to \mathbb{N}$ be such that $\ell(n) > \ell'(n)$ and let $I \subseteq \{0,1\}^*$. A collection of functions $\{H_s\}_{s \in I}$ is called* (secret-coin) collision-resistant hash family *(with index-set I) if the following holds:*

---

[1] Speaking of a *single* collision-resistant is meaningless if one allows the adversary to be non-uniform.

1. There exists a probabilistic polynomial-time key-generation algorithm, GEN, that on input 1^k outputs an index $s \in I$ (of a function H_s). The function H_s maps strings of length $\ell(k)$ to strings of length $\ell'(k)$.
2. There exists a probabilistic polynomial-time evaluation algorithm that on input $s \in I, x \in \{0,1\}^{\ell(k)}$ computes $H_s(x)$.
3. Collisions are hard to find. Formally, a pair x, x' is called a collision for a function H_s if $x \neq x'$ but $H_s(x) = H_s(x')$. The collision-resistance requirement states that every probabilistic polynomial-time algorithm B, that is given input $s = \text{GEN}(1^k)$, succeeds in finding a collision for the function H_s with a negligible probability (where the probability is taken over the coin tosses of both GEN and B).

Remark 1. Various variants of the above definition are possible. For example, one can consider hash-functions that can be applied to strings of arbitrary length (and not just to strings of the specified length $\ell(k)$); such functions can be obtained from the more restricted functions, defined above, by using standard techniques such as block-chaining or hash-trees (where the restricted function is applied repeatedly); cf. [9–Sec. 6.2.3].

Example 1. Let p be a prime and q be a "large" divisor of $p-1$. Let $h_1, h_2 \in Z_p^*$ be two elements of order q. Let $H_{p,h_1,h_2}(x) = h_1^{x_L} \cdot h_2^{x_R} \bmod p$, where $x = (x_L, x_R) \in Z_q \times Z_q$, and consider the family of all these functions. Each such function maps strings of length $2 \log q$ to strings of length $\log p$. An algorithm that can find a collision for such a function (i.e., x, x' such that $H_{p,h_1,h_2}(x) = H_{p,h_1,h_2}(x')$) can be used to compute the discrete-log $\text{DLOG}_{h_1}(h_2) \bmod p$.

3 CRHF from Homomorphic Encryption

In this section, we present the simplest of our constructions. This construction is based on a stronger assumption than what we use in subsequent sections; namely, the existence of *homomorphic encryption* schemes. In fact, we never use the standard requirement from encryption schemes that decryption can also be performed in polynomial-time (but just the fact that decryption is possible). Therefore, we actually work with the weaker assumption that *homomorphic commitment* exists. Informally speaking, a homomorphic commitment scheme is a (semantically-secure, perfectly binding, non-interactive) commitment scheme C (cf., [8–Sec. 4.1.1]) that has the additional property that from commitments $C(x), C(x')$ it is possible to compute efficiently a commitment to $x + x'$, where $+$ is the operation of some group G.

Below, we formally define the notion of "homomorphic commitment scheme". We stress that this definition is *not* necessarily the most general definition that is possible here; instead, it is aimed at the simplicity of the presentation. Later in the paper, these results are strengthened in various ways.

Definition 2. *A (semantically secure) homomorphic commitment scheme consists of a (group-size) function $L(k) : \mathbb{N} \to \mathbb{N}$ and a triplet of algorithms* (GEN, COMMIT, ADD) *as follows.*

1. GEN is a probabilistic polynomial-time key-generation algorithm; on input 1^k it outputs a public-key PK.
2. The commitment algorithm COMMIT is a probabilistic polynomial-time algorithm that takes input 1^k, the public-key PK and a string x which is an element of the group $Z_{L(k)}$ (where $L(k)$ is a prime); it outputs a string COMMIT$_{PK}(x)$ of some length $p(k)$. On one hand, this string hides the value x; i.e., given COMMIT$_{PK}(x)$, the value x is semantically secure. (Note that the notation COMMIT$_{PK}(x)$ hides the fact that the algorithm COMMIT is probabilistic. When we wish to emphasize this fact, we sometimes use the notation COMMIT$_{PK}(x, \cdot)$. In other cases, we may wish to obtain a deterministic value by fixing some randomness r to the algorithm COMMIT; in such a case we use the notation COMMIT$_{PK}(x, r)$.) On the other hand, the commitment is perfectly binding; i.e., given PK, the commitment string COMMIT$_{PK}(x)$ uniquely determines x.[2]
3. The composition algorithm ADD is a probabilistic polynomial-time algorithm that takes input 1^k, the public-key PK and two commitments COMMIT$_{PK}(x)$, COMMIT$_{PK}(x')$ and computes a commitment to $x + x'$; i.e., COMMIT$_{PK}(x + x', r)$, where $+$ refers to addition operation in the group $Z_{L(k)}$ and r is any possible randomness for the commitment algorithm, COMMIT. Finally, we require that commitments can be re-randomized.[3] That is, there is a probabilistic polynomial-time algorithm RERAND that, given any commitment COMMIT$_{PK}(x, r)$, outputs a re-randomized commitment distributed according to COMMIT$_{PK}(x, \cdot)$ (of the same string x).

Example 2. A simple example is the quadratic-residuosity based probabilistic encryption of [11]; in this case the group that is used is Z_2. For an additional example, consider the ElGamal commitment: Let p be a prime, and let g be a generator of a subgroup $G \subseteq Z_p^*$ of prime order q in which the discrete-log problem is "hard". Let PK $= (p, q, g, g^a)$, for some a. The commitment is defined by COMMIT$_{PK}(x, b) = (g^b, g^x \cdot g^{ab})$. Note that by taking the product of two commitments, i.e. COMMIT$_{PK}(x, b) \odot$ COMMIT$_{PK}(x', b')$, we get

$$(g^b, g^x \cdot g^{ab}) \odot (g^{b'}, g^{x'} \cdot g^{ab'}) = (g^{b+b'}, g^{x+x'} \cdot g^{a(b+b')}) = \text{COMMIT}_{PK}(x+x', b+b').$$

Also note that if we work directly with ElGamal encryption (i.e., with x instead of g^x) then this allows decryption, but the product gives a value that corresponds to $x \cdot x'$ rather than to $x + x'$.

Remark 2. Observe that the definition guarantees also that, for any integer $c \geq 0$, a commitment to cx (i.e., a value COMMIT$_{PK}(cx, r)$ for some r) can be

---
[2] i.e., for all $(x, r), (x', r')$ such that $x \neq x'$, we have COMMIT$_{PK}(x, r) \neq$ COMMIT$_{PK}(x', r')$; this is the analogue of the (perfect) correctness property in the terminology of *encryption*.
[3] This requirement, which is standard for most applications of homomorphic encryption, is actually not used in this section but will be needed in Section 5.

efficiently computed (using repeated doubling) from $\text{COMMIT}_{\text{PK}}(x)$ by applying the algorithm ADD $O(\log_2 c)$ times.

Construction: Given an arbitrary homomorphic commitment scheme, i.e. a triplet (GEN, COMMIT, ADD), we construct a CRHF family as follows. The key-generation algorithm of the hash-family GEN′, on input 1^k, works by first applying GEN(1^k) to obtain a public-key PK and then choosing at random an $n_1 \times n_2$ matrix M (where n_1, n_2 are specified below) whose elements are in $Z_{L(k)}$. The index for the hash-function that GEN′ outputs is $s = (\text{PK}, \text{COMMIT}_{\text{PK}}(M))$, where $\text{COMMIT}_{\text{PK}}(M)$ consists of commitments to each of the $n_1 \cdot n_2$ elements of the matrix M. The function H_s, on input $x = (x_1, \ldots, x_{n_2})$ (where each x_i is $l(k)$-bit string and $l(k) = \lfloor \log_2 L(k) \rfloor$), is defined as follows:

$$H_{\text{PK}, \text{COMMIT}_{\text{PK}}(M)}(x) \stackrel{\text{def}}{=} \text{COMMIT}_{\text{PK}}(M \cdot x, r),$$

where the commitment to $M \cdot x$ can be efficiently computed from s and x using ADD and Remark 2 above. (Here r is the randomness implicitly defined by this computation.) It remains to prove that collisions are hard to find. Assume towards a contradiction, that there exists an algorithm B that, given s, finds (with high probability) a pair x, x' that forms a collision for H_s; i.e., $H_s(x) = H_s(x')$ or alternatively $\text{COMMIT}_{\text{PK}}(M \cdot x, r) = \text{COMMIT}_{\text{PK}}(M \cdot x', r')$. It follows, by the perfect binding property, that $M \cdot x = M \cdot x'$ or that $M(x - x') = 0$. This contradicts the semantic security of COMMIT, as we found a vector y in the kernel of the committed matrix M. According to the semantic security, this should have been possible only with probability which is very close to the a-priori probability; if we choose $n_1 = \lceil k/l(k) \rceil$ then this a-priori probability is at most $1/2^k$. [4] Finally, the parameter n_2 is chosen such that the output of H_s (whose length is $n_1 \cdot p(k)$) is shorter than its input (whose length is $n_2 \cdot \log_2 L(k)$).

We summarize the above discussion with the following theorem:

Theorem 1. *If there exists a homomorphic commitment scheme then there exists a family of CRHFs.*

To conclude this section, we would like to offer (in an informal manner) a slightly more general view of the above construction. Assume that we are given homomorphic commitment scheme, as above, and in addition a *linear MAC*. We construct a family of CRHFs as follows. The index of each function s consists of a public-key for the commitment, PK, and an "encryption" (by applying $\text{COMMIT}_{\text{PK}}$) of a MAC-key SK. The function $H_s(x)$ is defined by

$$H_s(x) \stackrel{\text{def}}{=} \text{COMMIT}_{\text{PK}}(\text{MAC}(x), r).$$

As before, computing the commitment to $\text{MAC}(x)$ can be done (without knowing the secret-keys) based on the linearity of the MAC and the homomorphic properties of the commitment. Now, assume that an adversary can

---

[4] Note that if the group-size is sufficiently large then n_1 might be as small as 1.

efficiently come up with a collision x, x' such that $\text{COMMIT}_{PK}(\text{MAC}(x), r) = \text{COMMIT}_{PK}(\text{MAC}(x'), r')$ (for some r, r'). Again, by the perfect binding, it follows that $\text{MAC}(x) = \text{MAC}(x')$ which contradicts the security of the MAC. Hence, if the MAC is secure then the only other possibility is that the adversary, by examining s, could obtain information about the MAC secret-key; this, in turn, contradicts the security of COMMIT (which is used to "encrypt" this key).

4 CRHF from PIR

In this section, we show a construction of CRHFs based on (computationally) private information retrieval (PIR) schemes. (In fact, our construction can also use PIR schemes where the user's reconstruction is unbounded.) Since PIR is implied by homomorphic encryption [15, 21, 16] (and unbounded PIR by homomorphic commitment), this result is stronger than the result presented in Section 3. The nature of the construction presented in this section is combinatorial, as opposed to the algebraic nature of the constructions presented in Section 3 and Section 5.

Definition 3. *A (computational, 1-round) PIR scheme is a protocol for two parties: a user, $\mathcal{U}$, and a server, $\mathcal{S}$. The server holds a database $x \in \{0,1\}^n$ and the user holds an input $i \in [n]$. The goal of a PIR scheme is for the user to learn the value of the bit x_i while keeping the value of i hidden from the server. The protocol uses only one round of interaction: $\mathcal{U}$ sends to the server a query, $q = \text{QUERY}(1^n, i, \rho)$, where ρ is the user's random input, and it gets in return an answer $a = \text{ANS}(x, q)$. The user then applies a reconstruction algorithm REC to compute $x_i = \text{REC}(a, i, \rho)$. The 3 algorithms $(\text{QUERY}(\cdot), \text{ANS}(\cdot), \text{REC}(\cdot))$ that define the PIR scheme are polynomial-time algorithms that should satisfy the following two requirements:*

1. *(Correctness) The user always retrieve x_i correctly (where the probability is over the choice of the user's random input ρ).*
2. *(Privacy) For every two indices $i, j \in [n]$ the corresponding distributions of queries, $\text{QUERY}(1^n, i, \cdot)$ and $\text{QUERY}(1^n, j, \cdot)$, are indistinguishable. (Alternatively, it will be useful to talk about semantic security of the query rather than about indistinguishability; namely, no adversary can gain a significant advantage in guessing a predicate of i given $q = \text{QUERY}(1^n, i, \cdot)$.)*

The main complexity measure for PIR schemes is their communication complexity. Specifically, we denote by $\alpha(n)$ the (worst-case) query length (over all $x \in \{0,1\}^n$ and all possible choices of ρ) and by $\beta(n)$ the (worst-case) answer length.

Our construction will use, in addition to the PIR scheme, the standard (non-cryptographic) primitive of error correcting code. Specifically, we will use any error correcting code $\text{ECC}(\cdot)$ that expands $x \in \{0,1\}^k$ to $y \in \{0,1\}^n$, where $n = c \cdot k$ (for a constant c) and that can correct up-to $\lambda \cdot n$ errors (for a constant λ).[5]

Construction: Given a PIR scheme (QUERY, ANS, REC) and an error correcting code ECC, we construct a CRHF family as follows. The key-generation algorithm of the hash-family, GEN, on input 1^k, works by choosing $t = \omega(\log k)$ queries $q_1, \ldots, q_t$ (this is done by choosing t random strings $\rho_1, \ldots, \rho_t$ and t random indices $i_1, \ldots, i_t \in_R [n]$, where as above $n = c \cdot k$, and computing $q_j = \text{QUERY}(1^n, i_j, \rho_j)$). The hash-index is $s = (q_1, \ldots, q_t)$. The function H_s, on input $x \in \{0, 1\}^k$, is defined as

$$H_s(x) = (\text{ANS}(y, q_1), \ldots, \text{ANS}(y, q_t)),$$

where $y = \text{ECC}(x)$. Clearly, H_s is computable in polynomial time. It maps strings of length k to strings of length $t \cdot \beta(n)$ (a possible choice of parameters is $t = \text{polylog}(k)$ and $\beta(n) = n^\epsilon = (ck)^\epsilon$; in such a case H_s indeed shrinks its input). Next, we argue that the resulting family is indeed collision-resistant. Suppose that an adversary can find a collision for H_s; i.e., it can find different strings x, x' such that $H_s(x) = H_s(x')$ or, equivalently, such that $(\text{ANS}(y, q_1), \ldots, \text{ANS}(y, q_t)) = (\text{ANS}(y', q_1), \ldots, \text{ANS}(y', q_t))$, where $y = \text{ECC}(x)$ and $y' = \text{ECC}(x')$. This implies (by the correctness of the PIR) that $y_{i_j} = y'_{i_j}$, for $1 \leq j \leq t$. However, since y, y' are distinct codewords of the error-correcting code then the distance between y, y' is at least $2\lambda n$; since each i_j is random, the probability that for a certain i_j we have $y_{i_j} = y'_{i_j}$ is constant (specifically, 2λ) and the probability that $y_{i_j} = y'_{i_j}$ for all j is (by the choice of t) negligible. By the semantic security of QUERY, finding such y, y' given q should be possible only with a negligible probability. This gives the desired contradiction.

Thus, we have:

Theorem 2. *If there exists a 1-round (single-server) PIR scheme with communication complexity $O(n^c)$ for some $c < 1$ then there exists a family of CRHFs.*

Remark 3. Fischlin [7] shows the impossibility of a black-box transformation from one-way trapdoor permutations to (one-round, computational) PIR. Our transformation from PIR to CRHF, together with the results of [20], yields a completely different way to obtain the same result.

5 CRHF from Homomorphic One-Way Commitment

The construction of CRHF from homomorphic encryption (or even from homomorphic commitment), presented in Section 3, seem to rely heavily on the semantic-security of the underlying commitment. In this section, we show that

---

[5] It suffices for us that the encoding algorithm ECC($\cdot$) will work in polynomial time. It is not needed for us that the error correction will be efficient; we will only rely on the "large" distance between codewords.

this is not really essential. Namely, we consider a primitive that we term *homomorphic one-way commitment*. In this case, the security of the committed value COMMIT(x) does not guarantee that no information about x is leaked but only that it is hard (for a randomly chosen x) to "invert" the commitment and find x. Note however that it does not suffice to require that COMMIT$_{\text{PK}}(x,r)$ is a one-way function, as we not only require that finding a pre-image (x,r) is hard but that even finding x alone is hard.

Definition 4. *A* one-way homomorphic commitment *is defined as homomorphic commitment (Definition 2), except for the security requirement:*

- *(One-Wayness) Every probabilistic polynomial-time algorithm I that is given* COMMIT$_{\text{PK}}(x,r)$ *has a negligible probability of finding x, where the probability is over a random choice of $x \in Z_{L(k)}$, the choice of r by* COMMIT, *and the internal random choices of I. (Note that, by the binding property, for every value* COMMIT$_{\text{PK}}(x,r)$, *there is a unique pre-image x.)*

Remark 4. The one-wayness requirement in particular implies that the size of the group from which x is taken, i.e. $L(k)$, needs to be "large". This is in contrast with the definition of "standard" homomorphic commitment where the group might be as small as Z_2. On the other hand, any (standard) homomorphic commitment where $L(k) = k^{\omega(1)}$ is immediately also a one-way homomorphic commitment. We can turn any standard homomorphic commitment (over an arbitrarily small group) into a one-way homomorphic commitment by concatenating "sufficiently many" copies of the original scheme (where $\omega(\log k)$ copies are always enough). Such a concatenation yields a group which is a product group, and in particular is not a cyclic group. It is possible to extend our results to such groups as well.

Construction: Given an arbitrary one-way homomorphic commitment scheme (GEN, COMMIT, ADD), we construct a CRHF family as follows. The key-generation algorithm of the hash-family GEN', on input 1^k, works by first applying GEN(1^k) to obtain a public-key PK. Then, it chooses m random elements $x_1, \ldots, x_m \in_R Z_{L(k)}$ (where m, as before, is chosen so that the output length is shorter than the input length) and m random strings $r_1, \ldots, r_m$ to be used by the commitment algorithm. It finally computes m values $y_i = $ COMMIT$_{\text{PK}}(x_i, r_i)$. The index of the hash function that GEN' outputs is $s = (\text{PK}, y_1, \ldots, y_m)$. The function H_s, on input $\boldsymbol{a} = (a_1, \ldots, a_m)$ (where each a_i is an $l(k)$-bit integer and, as before, $l(k) = \lfloor \log_2 L(k) \rfloor$), is defined as follows:

$$H_{\text{PK}, y_1, \ldots, y_m}(\boldsymbol{a}) \stackrel{\text{def}}{=} \text{COMMIT}_{\text{PK}}(\sum_{i=1}^{m} a_i x_i, r),$$

where, as in the construction of Section 3, we observe that by using algorithm ADD (and Remark 2) this commitment can be efficiently computed (without knowledge of $x_1, \ldots, x_m$) from s and $\boldsymbol{a}$ (and r is the corresponding randomness). It remains to prove that collisions are hard to find. Assume towards a contradiction, that there exists an algorithm B that, given s, finds (with high probability) a pair $\boldsymbol{a}, \boldsymbol{a}'$ that forms a collision for H_s; i.e., $H_s(\boldsymbol{a}) = H_s(\boldsymbol{a}')$. This means that $\text{COMMIT}_{\text{PK}}(\sum_{i=1}^{m} a_i x_i, r) = \text{COMMIT}_{\text{PK}}(\sum_{i=1}^{m} a'_i x_i, r')$ which, by the perfect binding of the commitment, implies that $\sum_{i=1}^{m} a_i x_i = \sum_{i=1}^{m} a'_i x_i$. Therefore, the vector $\boldsymbol{d} = \boldsymbol{a} - \boldsymbol{a}'$ (which is easily computable from the collision) is such that $\boldsymbol{d} \cdot \boldsymbol{x} = 0$. We want to use the procedure that finds such vectors $\boldsymbol{d}$ in order to construct an inverter I for the commitment (i.e., an algorithm that finds $\boldsymbol{x}$ from $\text{COMMIT}(\boldsymbol{x})$, for a uniformly random $\boldsymbol{x}$) in contradiction to the one-wayness. While each such $\boldsymbol{d}$ gives some information about $\boldsymbol{x}$, applying the procedure repeatedly should be done with some care to avoid getting vectors $\boldsymbol{d}$ which are linearly dependent and are therefore useless. Next, we describe an inverter that uses the above ideas in a more careful way.

The inverter: The algorithm I (the inverter) that we construct gets as input a public-key PK and a vector $\boldsymbol{z}$ of m commitments (where PK, $\boldsymbol{z}$ and the randomness are all chosen at random with the appropriate distributions) and it finds, with non-negligible probability, the vector $\boldsymbol{x}$ of m committed values.[6] The inverter I repeats the following at most M times (where $M = O(m \cdot q(k))$ and $q(k)$ is the polynomial such that B succeeds with probability at least $1/q(k)$) or until I collects m linearly independent equations about $\boldsymbol{x}$. In the jth iteration, I picks at random an $m \times m$ matrix C_j and a length m vector $\boldsymbol{b}_j$. The elements of both C_j and b_j are taken from $Z_{L(k)}$ (and recall that we assume here that $L(k)$ is a prime number). Denote $\boldsymbol{x}_j = C_j \cdot \boldsymbol{x} + \boldsymbol{b}_j$ (of course the inverter does not compute this value as $\boldsymbol{x}$ is not available to it; we use $\boldsymbol{x}_j$ to simplify notation). The inverter computes $\boldsymbol{w}_j = \text{COMMIT}_{\text{PK}}(\boldsymbol{x}_j) = \text{COMMIT}_{\text{PK}}(C_j \cdot \boldsymbol{x} + \boldsymbol{b}_j)$ (that can be computed from $\boldsymbol{z}$ using the algorithm ADD) and finally it re-randomizes this vector of commitments (using the algorithm RERAND); i.e., it computes $\boldsymbol{y}_j = \text{RERAND}_{\text{PK}}(\boldsymbol{w}_j)$. The inverter provides $s_j = (\text{PK}, \boldsymbol{y}_j)$ to algorithm B and in return it gets a collision $\boldsymbol{a}_j, \boldsymbol{a}'_j$ for the hash function H_{s_j} (note that it is easy to check whether this pair is indeed a collision, simply by applying the function H_{s_j}; we can therefore assume that B itself either outputs a collision or the value "fail"). If B returns "fail" the current iteration is terminated and I proceeds to the next iteration; otherwise, I sets $\boldsymbol{d}_j = \boldsymbol{a}_j - \boldsymbol{a}'_j$ (and, as above, the vector $\boldsymbol{d}_j$ satisfies $\boldsymbol{d}_j \cdot \boldsymbol{x}_j = 0$). The iteration ends by computing the vector $\boldsymbol{u}_j = \boldsymbol{d}_j \cdot C_j$ and the scalar $\lambda_j = \boldsymbol{d}_j \cdot \boldsymbol{b}_j$. Note that $\boldsymbol{u}_j \cdot \boldsymbol{x} + \lambda_j = \boldsymbol{d}_j \cdot C_j \cdot \boldsymbol{x} + \boldsymbol{d}_j \cdot \boldsymbol{b}_j = \boldsymbol{d}_j(C_j \cdot \boldsymbol{x} + \boldsymbol{b}_j) = \boldsymbol{d}_j \cdot \boldsymbol{x}_j = 0$. Hence, if all goes well, the inverter ends the jth iteration with a new linear constraint about $\boldsymbol{x}$. Moreover, we will argue that after M iterations I is likely to have m linearly independent constraints; this allows solving the system of equations and to find $\boldsymbol{x}$; i.e., to invert the commitment.

It remains to prove that I succeeds in inverting the commitment $\boldsymbol{z}$ with non-negligible probability (assuming that B succeeds in finding collisions with non-negligible probability). For this, we make several simple observations. First,

---

[6] Note that this is slightly stronger than what we need, since we "invert" m commitments at once; however we get this feature "for free".

note that I invokes B several times, all with the same PK (which is part of all the indices s_j). We call a public-key PK *good* if B, when given a randomly chosen index of a hash function that includes this public-key (i.e., $s = (\text{PK}, \boldsymbol{y})$ for a randomly chosen $\boldsymbol{y}$) succeeds with non-negligible probability. Since B finds a collision with non-negligible probability over a randomly chosen s, it follows that a non-negligible fraction of the public-keys are good. Therefore, the probability that PK that is given to the inverter is good is non-negligible. From now on, we will assume that this is indeed the case. Fix any $\boldsymbol{z}$ (and hence also $\boldsymbol{x}$). Next, we note that, for every j, the vector $\boldsymbol{x}_j$ (whose choice is determined by the random choice of $C_j, \boldsymbol{b}_j$) is totally random and, moreover, these vectors are all independent. Hence, when B is given s_j (that includes PK and $\boldsymbol{y}_j$ – a rerandomized commitment to $\boldsymbol{x}_j$), by the assumption that PK is good, B succeeds in finding a collision with non-negligible probability. Finally, we argue that the vector $\boldsymbol{u}_j$ obtained from this collision is random; if this is true then indeed $O(m)$ successful iterations suffice (with high probability) and therefore total of M iterations are enough. This is so because $\boldsymbol{x}_j$ (and hence also the input for B; i.e., $\boldsymbol{y}_j$) is independent of C_j (because no matter what $\boldsymbol{x}, C_j$ are, the choice of $\boldsymbol{b}_j$ yields a uniformly random $\boldsymbol{x}_j$). Therefore $\boldsymbol{d}_j$ (the output of B) is also independent of the matrix C_j; hence, computing $\boldsymbol{u}_j = \boldsymbol{d}_j \cdot C_j$ (where $\boldsymbol{d}_j \neq \boldsymbol{0}$) yields a random vector.

To summarize, we state the following theorem:

Theorem 3. *If there exists a homomorphic one-way commitment scheme then there exists a family of CRHFs.*

Acknowledgements. We thank the anonymous referees for helpful comments and pointers.

References

1. B. Barak. How to Go Beyond the Black-Box Simulation Barrier. *Proc. of 42nd FOCS*, pp. 106–115, 2001.
2. B. Barak, and O. Goldreich. Universal Arguments and their Applications. *Proc. of 17th Conference on Computational Complexity*, pp. 194-203, 2002.
3. A. Beimel, Y. Ishai, E. Kushilevitz, and T. Malkin. One-Way Functions Are Essential for Single-Server Private Information Retrieval. *Proc. of 31st STOC*, pp. 89–98, 1999.
4. C. Cachin, S. Micali, and M. Stadler. Computationally private information retrieval with polylogarithmic communication. *Proc. of IACR EUROCRYPT*, LNCS 1592, pp. 402–414, 1999.
5. I. Damgård: Collision Free Hash Functions and Public Key Signature Schemes. In *Proc. of EUROCRYPT*, pages 203-216, 1987.
6. I. Damgard, T. P. Pedersen and B. Pfitzmann. On the existence of statistically hiding bit commitment schemes and fail-stop signatures. In *Proc. of IACR Crypto*, LNCS 773, pp. 250–265, 1993.
7. M. Fischlin. On the Impossibility of Constructing Non-interactive Statistically-Secret Protocols from Any Trapdoor One-Way Function. *Proc. of CT-RSA*, pp. 79-95, 2002.

8. O. Goldreich. *Foundations of Cryptography. Volume I: Basic Tools.* Cambridge University Press, 2001.
9. O. Goldreich. *Foundations of Cryptography. Volume II: Basic Applications.* Cambridge University Press, 2004.
10. O. Goldreich, S. Goldwasser, and S. Halevi. Collision-Free Hashing from Lattice Problems. ECCC TR-42, 1996.
11. S. Goldwasser, and S. Micali. Probabilistic Encryption. *Journal of Computer and systems sciences* 28, 270-299, 1984.
12. S. Halevi, and S. Micali, Practical and Provably-Secure Commitment Schemes from Collision-Free Hashing. In *Proc. of IACR Crypto*, LNCS 1109, pp. 201-215, 1996.
13. C.Y. Hsiao and L. Reyzin. Finding Collisions on a Public Road, or Do Secure Hash Functions Need Secret Coins? In *Proc. of IACR Crypto*, 2004.
14. J. Kilian. A Note on Efficient Zero-Knowledge Proofs and Arguments. *Proc. of 24th STOC*, pp. 723–732, 1992.
15. E. Kushilevitz and R. Ostrovsky. Replication is Not Needed: Single Database, Computationally-Private Information Retrieval. In *Proc. of 38th FOCS*, pages 364–373, 1997.
16. E. Mann. Private access to distributed information. Master's thesis, Technion – Israel Institute of Technology, Haifa, 1998.
17. S. Micali. CS Proofs. *SIAM J. Computing*, Vol. 30(4), pp. 1253-1298, 2000. (Early version appeared in FOCS 1994.)
18. W. Ogata, and K. Kurosawa. On Claw Free Families. IEICE Trans., Vol.E77-A(1), pp. 72-80, 1994. (Early version appeared in AsiaCrypt'91.)
19. A. Russell. Necessary and Sufficient Conditions for Collision-Free Hashing. *J. Cryptology*, Vol. 8(2), pages 87-100, 1995. (Early version in CRYPTO92).
20. D. Simon. Finding Collisions on a One-Way Street: Can Secure Hash Functions Be Based on General Assumptions? In *Proc. of EUROCRYPT*, pages 334-345, 1998.
21. J. P. Stern. A new and efficient all-or-nothing disclosure of secrets protocol. In *Advances in Cryptology – ASIACRYPT '98*, volume 1514 of *Lecture Notes in Computer Science*, pages 357–371. Springer, 1998.
22. X. Wang, D. Feng, X. Lai, and H. Yu. Collisions for Hash Functions MD4, MD5, HAVAL-128 and RIPEMD. Cryptology ePrint Archive TR-199, 2004.

The Relationship Between Password-Authenticated Key Exchange and Other Cryptographic Primitives

Minh-Huyen Nguyen[*]

Harvard University, Cambridge, MA
mnguyen@eecs.harvard.edu

Abstract. We consider the problem of password-authenticated key exchange (PAK) also known as session-key generation using passwords: constructing session-key generation protocols that are secure against active adversaries (person-in-the-middle) and only require the legitimate parties to share a low-entropy password (e.g. coming from a dictionary of size poly(n)).

We study the relationship between PAK and other cryptographic primitives. The main result of this paper is that password-authenticated key exchange and public-key encryption are *incomparable* under black-box reductions. In addition, we strengthen previous results by Halevi and Krawczyk [14] and Boyarsky [5] and show how to build key agreement and semi-honest oblivious transfer from any PAK protocol that is secure for the Goldreich-Lindell (GL) definition [11].

We highlight the difference between two existing definitions of PAK, namely the indistinguishability-based definition of Bellare, Pointcheval and Rogaway (BPR) [1] and the simulation-based definition of Goldreich and Lindell [11] by showing that there exists a PAK protocol that is secure for the BPR definition and only assumes the existence of one-way functions in the case of exponential-sized dictionaries. Hence, unlike the GL definition, the BPR definition does not imply semi-honest oblivious transfer for exponental-sized dictionaries under black-box reductions.

1 Introduction

The problem of *password-authenticated key exchange* (PAK), also known as *session-key generation using passwords*, is to enable private communication between two legitimate parties over an insecure channel in the setting where the legitimate parties have only a small amount of shared information, i.e. a low-entropy key such as an ATM pin or a human-chosen password. In addition to its practical implications, the problem of session-key generation using passwords is quite natural as it focuses on the minimal amount of information that two

---

[*] Supported by NSF grants CCR-0205423, CNS-0430336, and ONR grant N00014-04-1-0478.

parties must share in order to perform non-trivial cryptography. A recent series of works [1, 6, 11, 17, 8, 20, 7] has focused on our theoretical understanding of this PAK problem by proposing several definitions of security as well as secure protocols. Bellare, Pointcheval and Rogaway [1] proposed a definition based on the indistinguishability of the session key. Following the simulation paradigm for secure multi-party computation, Boyko, MacKenzie and Patel [6] and Goldreich and Lindell [11] gave their own simulation-based definitions. However, it is not clear how these existing definitions of security for PAK relate to one another.

The first protocols for the password-authenticated key exchange problem were proposed in the security literature, based on informal definitions and heuristic arguments (e.g. [4, 24]). The first rigorous proofs of security were given in the random oracle model by [1, 6]. Only recently were rigorous solutions without random oracles given, in independent works by Goldreich and Lindell [11] (under the assumption that trapdoor permutations exist) and Katz, Ostrovsky, and Yung [17] (under number-theoretic assumptions). Subsequently, the protocol of [11] was simplified in [20] and the protocol of [17] was generalized in [8, 7].

What is the minimal assumption needed to solve PAK? How does this problem relate to other basic cryptographic primitives such as key agreement and oblivious transfer? These are natural questions to ask when considering the problem of password-authenticated key exchange. The goal of this paper is to study the relationship between PAK and other cryptographic primitives as well as try to explain how the existing definitions of security for PAK relate to one another. Next, we informally describe the problem of password-authenticated key exchange.

Password-Authenticated Key Exchange. The problem of session-key generation using passwords suggested by Bellovin and Merritt [3] considers the situation where Alice and Bob share a password, i.e. an element chosen uniformly at random from a small dictionary $\mathcal{D} \subseteq \{0,1\}^n$. This dictionary can be very small, e.g. $|\mathcal{D}| = \text{poly}(n)$, and in particular it may be feasible for an adversary to exhaustively search it. The aim is to construct a protocol enabling Alice and Bob to generate a "random" session key $K \in \{0,1\}^n$ which they can subsequently use for standard private-key cryptography. We consider an active adversary that completely controls the communication channel between Alice and Bob and in particular can attempt to impersonate either party through a person-in-the-middle attack.

The goal of a PAK protocol is that, even after the adversary mounts such an attack, Alice and Bob will generate a session key that is indistinguishable from uniform even given the adversary's view. However, our ability to achieve this goal is limited by two unpreventable attacks. First, the adversary can block all communication, so it can prevent one or both of the parties from completing the protocol and obtaining a session key. Second, the adversary can choose a password $\tilde{w}$ uniformly at random from $\mathcal{D}$ and attempt to impersonate one of the parties. With probability $1/|\mathcal{D}|$, the guess equals the real password (i.e., $\tilde{w} = w$), and the adversary will succeed in the impersonation and therefore learn the session key. Thus, we revise the goal to effectively limit the adversary to these two attacks.

Our Results. Our goal in this paper is to understand the relationship between session-key generation using passwords and other well-known cryptographic primitives. Doing so will help us characterize the complexity of PAK and place this problem within our current view of cryptography. In this work we study the relationship of PAK to *public-key encryption* (PKE), *oblivious transfer* (OT) and *key agreement* (KA). We provide positive results, e.g. exhibit a reduction of KA to PAK, as well as negative results, e.g. prove that PAK does not imply PKE under black-box reductions.

Following the oracle separation paradigm of [15], we first separate PAK and PKE by constructing an oracle Γ relative to which PAK exists but PKE does not.

Theorem 1. *There is no "black-box" construction of PKE from PAK for the Goldreich-Lindell (GL) definition [11].*

Loosely speaking, a black-box construction of the primitive Q from the primitive P is a construction of Q out of P which does not use the code of the implementation of P[1]. We note that similarly to most separation results, Theorem 1 and our other separation results only apply to uniform adversaries. We actually prove Theorem 1 using a definition of PAK which is stronger than the GL definition in order to strengthen the result. This separation result can also be seen in a positive way since it provides a direction for proving implications. In order to prove that PAK implies PKE, one must use non-black-box techniques, for example by using the code of an adversary for the PKE protocol.

We then exhibit a reduction of *semi-honest* OT[2] to PAK for the GL definition.

Theorem 2. *The existence of a PAK-protocol that is secure for the GL definition implies semi-honest OT (via a black-box reduction). Moreover, this reduction does not depend on the size of the dictionary $\mathcal{D}$ and holds even for dictionaries of exponential size such as $\mathcal{D} = \{0,1\}^n$.*

The proof of Theorem 2 actually uses only the weaker definition of [20] where the security holds for a *specific* dictionary $\mathcal{D}$ and the probability of breaking is bounded by $\frac{1}{\omega(\log n)}$ instead of $O\left(\frac{1}{|\mathcal{D}|}\right)$, which strengthens the result.

Combining Theorem 2 and the result of Gertner et al. [9] that there is no black-box construction of semi-honest OT from PKE, we obtain the following corollary:

Corollary 1. *There is no black-box construction of GL-secure PAK from PKE.*

Putting Theorem 1 and Corollary 1 together, we obtain that PAK and PKE are *incomparable* under black-box reductions. This is similar to the result of [9] that

---

[1] We refer the reader to Section 2.3 and [22] for a more formal definition of black-box reductions. In the taxonomy of [22], we are considering *semi* black-box reductions.
[2] In the honest (but curious) or semi-honest model, the parties Alice and Bob are guaranteed to follow the protocol but might use their views of the interaction in order to compute some additional information.

OT and PKE are incomparable under black-box reductions, and thus provides an additional motivation to try and establish the equivalence of PAK and OT, as conjectured in [5]. Indeed, the protocol proposed by Goldreich and Lindell is actually based on the existence of oblivious transfer and one-way permutations. Theorem 2 shows that if one can bypass the use of one-way permutations (for example by using one-way functions instead of one-way permutations) and build a secure PAK protocol from oblivious transfer only, then PAK and OT are equivalent[3].

The question of the relationship between PAK to KA is particularly interesting as the PAK and KA problems are very similar in essence: both problems consider honest parties A and B who wish to generate a common random session key K. In the case of PAK, the honest parties have to withstand an active adversary and share a low-entropy password whereas in the case of KA, the honest parties have to withstand a passive adversary and share no prior information. Combining Theorem 2 and the previous result by Gertner et al. [9] that semi-honest OT is strictly stronger than KA under black-box reductions, we obtain the following corollary:

Corollary 2. *The existence of a PAK protocol that is secure for the GL definition implies KA (via a black-box reduction). Moreover, this reduction does not depend on the size of the dictionary $\mathcal{D}$ and holds even for dictionaries of exponential size such as $\mathcal{D} = \{0,1\}^n$.*

Combining Corollary 1 and the previous result by Gertner et al. [9] that PKE implies KA (via a black-box reduction), we obtain the following corollary:

Corollary 3. *There is no black-box construction of GL-secure PAK from KA.*

Again, Corollary 3 can be seen in a positive way: to prove that KA implies PAK, one must use non-black-box techniques.

Theorem 2 also enables us to understand the relationship between existing definitions of security and in particular to highlight a difference between the simulation-based definition of [11] and the indistinguishability-based definition of [1]. Indeed, we have the following result:

Theorem 3. *If one-way functions exist, there exists a PAK protocol that is secure for the Bellare-Pointcheval-Rogaway (BPR) definition [1] for the dictionary $\mathcal{D} = \{0,1\}^n$.*

Hence, unlike the GL definition, the BPR definition does not imply honest OT in the case of exponential-sized dictionaries under black-box reductions. However we conjecture that any PAK protocol that is secure for the BPR definition for *polynomial-sized dictionaries* implies semi-honest OT.

---

[3] This equivalence would be non-black-box as the known construction of OT from honest OT is non-black-box since it uses the zero-knowledge proofs of [12] (see [9]).

Related Work. Although the relationship between PAK and other cryptographic primitives has not been explicitly studied before, some results are known for the related problem of *password-based authentication*, where the legitimate parties only want to be convinced that they are talking to one another (but not generate a common session key). Assuming the existence of one-way functions, it is known that one can transform a PAK protocol into a protocol for password-based authentication using two additional messages [2, 1, 16, 11, 17].

Halevi and Krawczyk [14] showed that a secure protocol for password-based authentication[4] can be used to implement KA. We see Corollary 2 as a strengthening of their result, since our result holds even for dictionaries of exponential size whereas their reduction only holds for polynomial-sized dictionaries.

Boyarsky [5] states without proof that password-based authentication[5] implies OT, which is similar to Theorem 2. However, [5] does not provide a formal definition of PAK for which this implication holds, and indeed, our results show that the relationship between PAK and OT *is* sensitive to the choice of definition. Moreover, our black-box construction of semi-honest OT from a secure PAK protocol holds even if we relax the security of the PAK protocol in two respects. First, it holds even if the PAK protocol is secure only for a fixed dictionary of exponential size, e.g. $\mathcal{D} = \{0,1\}^n$. Second, we only require that the probability of breaking the PAK protocol be bounded by $\frac{1}{\omega(\log n)}$ (on security parameter 1^n) instead of $O\left(\frac{1}{|\mathcal{D}|}\right)$.

2 Preliminaries

We denote by n the security parameter, by U_n the uniform distribution over strings of length n, by $\mathrm{neg}(n)$ a negligible function and write $x \stackrel{R}{\leftarrow} S$ when x is chosen uniformly from the set S. We use the abbreviation "PPT" for probabilistic polynomial-time algorithms.

Since we will prove our results for uniform adversaries, our definitions are for the uniform model of computation. An ensemble $X = \{X_n\}_{n \in \mathbb{N}}$ is *(polynomial-time) samplable* if there exists a PPT algorithm M such that for every n, the random variables $M(1^n)$ and X_n are identically distributed.

Let S be a set of strings. For a function $\gamma : \mathbb{N} \to [0,1]$, we say that the probability ensembles $\{X_w\}_{w \in S}$ and $\{Y_w\}_{w \in S}$ are $(1-\gamma)$-*indistinguishable* (denoted by $\{X_w\} \stackrel{\gamma}{\equiv} \{Y_w\}$) if for every PPT algorithm D, for all sufficiently large n, for every $w \in \{0,1\}^n \cap S$,

$$|\Pr[D(X_w, w) = 1] - \Pr[D(Y_w, w) = 1]| < \gamma(n) + \mathrm{neg}(n)$$

---

[4] Their result is for password-based one-way authentication where one clients tries to authenticate itself to a server.
[5] This result is for password-based mutual authentication where two honest parties try to authenticate each other.

In the proofs, we will slightly abuse notation when talking about a distribution's index w by writing "for every $w \in S$" and omitting the index w as an input to the distinguisher D. We say that $\{X_w\}$ and $\{Y_w\}$ are *computationally indistinguishable*, which we denote by $X_w \stackrel{c}{\equiv} Y_w$, if they are 1-indistinguishable.

2.1 Cryptographic Primitives

Two-Party Protocols. The following is an informal presentation of two-party computation which will suffice for our purposes. Recall that we are interested in protocols for *semi-honest* oblivious transfer and key-agreement for which we are guaranteed that the two parties follow the protocol. We refer the reader to [10] for more details.

A two-party protocol problem is defined by specifying a (possibly probabilistic) functionality $f : \{0,1\}^* \times \{0,1\}^* \to \{0,1\}^* \times \{0,1\}^*, (x,y) \to (f_1(x,y), f_2(x,y))$ which maps pairs of inputs to pairs of outputs. A two-party protocol is a pair of probabilistic polynomial-time algorithms (A, B) which represent the strategies of the two parties, i.e. functions that map a party's input, private randomness and the sequence of messages received so far to the next message to be sent. The view of a party consists of its input, its random-tape and the sequence of messages received. We measure the amount of interaction in a protocol by its number of rounds, where a round consists of a single message sent from one party to another. Whenever we consider a protocol for securely computing a functionality f, we assume that the protocol correctly computes f when both parties follow the protocol, i.e. the joint output distribution of the protocol played by parties following the protocol on input pair (x,y) equals the distribution of $f(x,y)$.

Semi-Honest Oblivious Transfer. In the semi-honest model, the two parties A and B are guaranteed to follow the protocol but might use their views of the interaction in order to learn some additional information. As noted in [9], in the semi-honest model, one can transform an OT protocol for bits into an OT protocol for strings without increasing the number of rounds. We will therefore focus on the version of OT where s_0 and s_1 are bits rather than strings. 1-out-of-2 oblivious transfer (OT) is the following two-party functionality:

- Inputs: A has the security parameter 1^n and two secret bits s_0 and s_1. B has the security parameter 1^n and a selection bit c.
- Outputs: A outputs nothing, B outputs s_c.

A protocol (A, B) for semi-honest OT is *secure* if there exists a pair of PPT $(\tilde{A}, \tilde{B})$ such that:

- *Receiver's privacy:* for every s_0, s_1, c, $\tilde{A}(1^n, s_0, s_1)$ is computationally indistinguishable from A's view of the interaction $(A(1^n, s_0, s_1), B(1^n, c))$
- *Sender's privacy:* for every s_0, s_1, c, $\tilde{B}(1^n, b, s_b)$ is computationally indistinguishable from B's view of the interaction $(A(1^n, s_0, s_1), B(1^n, c))$

Key Agreement. Key agreement (KA) is the following two-party functionality:
- Inputs: A and B have the security parameter 1^n.
- Outputs: A and B output the same string K of length n

A KA protocol is *secure* if we have $(T, K) \stackrel{c}{\equiv} (T, U_n)$ where T is the transcript of the interaction $(A(1^n), B(1^n))$ and K is the common output of A and B in the interaction $(A(1^n), B(1^n))$. In other words, the session key K will be computationally indistinguishable from a truly random string given the view of a passive adversary.

2.2 Password-Authenticated Key Exchange

Password-authenticated key exchange (PAK) or session-key generation using passwords is similar to key agreement in that two honest parties A and B want to generate a session key K of length n that is indistinguishable from uniform even given the adversary's view. However, PAK differs from KA in two important respects. First, A and B have as input a shared password w which is chosen at random from a dictionary $\mathcal{D} \subseteq \{0,1\}^n$. Second, the adversary is not passive but completely controls the communication channel between A and B.

The Goldreich-Lindell Definition and Its Variants. The definition of PAK in [11] follows the standard paradigm for secure computation: define an ideal functionality (using a trusted third party) and require that every adversary attacking the real protocol can be simulated by an ideal adversary attacking the ideal functionality. In the real protocol, an active adversary can prevent one or both of the parties from completing the protocol. Thus, in the ideal model, we will allow C_{ideal} to specify an input bit dec_B, which determines whether B obtains a session key or not[6]. We can therefore cast PAK as a three-party functionality which is described in the ideal model as follows.

Ideal Model.

- Inputs: A and B receive a security parameter 1^n and a joint password $w \stackrel{R}{\leftarrow} \mathcal{D}$.
- A and B both send w to the trusted party. C_{ideal} sends a decision bit dec_B to the trusted party to indicate whether B's execution is successful or not.
- Outputs: The trusted party chooses $K \stackrel{R}{\leftarrow} \{0,1\}^n$ and sends it to A. If $\text{dec}_B = 1$, then the trusted party sends K to B; otherwise it sends $\perp$ to B.

The ideal distribution of inputs and outputs is defined by:

$$\text{IDEAL}_{C_{\text{ideal}}}(\mathcal{D}) = (w, \text{output}(A), \text{output}(B), \text{output}(C_{\text{ideal}}))$$

Real Model. Let A, B be the honest parties and let C be any PPT real adversary. In an initialization stage, A and B receive $w \stackrel{R}{\leftarrow} \mathcal{D}$. The real protocol is executed by A and B communicating via C. We will augment C's view of the protocol with B's decision bit, denoted by dec_B, where $\text{dec}_B = \texttt{reject}$ if $\text{output}(B) = \perp$,

---

[6] We will adopt the convention that A always completes the protocol and accepts.

and $\text{dec}_B = \texttt{accept}$ otherwise (indeed in typical applications, the decision of B will be learned by the real adversary C: if B obtains a session key, then it will use it afterwards; otherwise, B will stop communication or try to re-initiate an execution of the protocol). C's augmented view is denoted by $\text{view}(C^{A(w),B(w)})$.

The real distribution of inputs and outputs is defined by:

$$\text{REAL}_C(\mathcal{D}) = (w, \text{output}(A), \text{output}(B), \text{view}(C^{A(w),B(w)}))$$

One might want to say that a PAK protocol is secure if the above ideal and real distributions are computationally indistinguishable. Unfortunately as mentioned above, an active adversary can guess the password and successfully impersonate one of the parties with probability $\frac{1}{|\mathcal{D}|}$. This implies that the real and ideal distributions are always distinguishable with probability at least $\frac{1}{|\mathcal{D}|}$ so we will only require that the distributions be distinguishable with probability at most $O\left(\frac{1}{|\mathcal{D}|}\right)$. In the case of a passive adversary, we require that the real and ideal distributions be computationally indistinguishable (for all subsequent definitions, this requirement will be implicit):

Definition 1. *[11] A protocol for password-based session-key generation is secure if for every samplable dictionary $\mathcal{D} \subseteq \{0,1\}^n$, for every real adversary C, there exists an ideal adversary C_{ideal} such that the ideal and real distributions are $\left(1 - O\left(\frac{1}{|\mathcal{D}|}\right)\right)$-indistinguishable[7].*

Although standard definitions of security for PAK protocols require that the security hold for *every* dictionary, we will consider two variants of the standard GL definition (Definition 1) where we change the security to hold for a *specific* dictionary $\mathcal{D}$ instead of every dictionary. Moreover, we will only require that the distributions be distinguishable with probability at most γ, where γ is a function of the dictionary size $|\mathcal{D}|$ and the security parameter n, and not necessarily $O\left(\frac{1}{|\mathcal{D}|}\right)$.

Although these variants of Definition 1 are weaker, a PAK protocol which is secure for a specific dictionary is still interesting since it corresponds to the setting where the honest parties are restricted to choose their passwords from a specific dictionary, such as in the case of ATM pin numbers[8]. Moreover, as noted in [20], such a PAK protocol can be converted into one for arbitrary dictionaries in the common reference string model (using the common reference string as the seed of a randomness extractor [21]).

---

[7] As pointed out by Rackoff, this basic definition is actually not completely satisfactory and needs to be augmented to take into account any use of the key K by one party while the other party has not completed the protocol. Our results will hold for the augmented definition as well but we will not handle the augmented definition explicitly.

[8] By restricting our attention to a specific dictionary, it may be possible to obtain a more efficient protocol, such as the [20] simplification of [11].

Definition 2. *[20] Let $\mathcal{D} \subseteq \{0,1\}^n$ be a samplable dictionary. A protocol for password-based session-key generation is $(1-\gamma)$-GL-secure for the dictionary $\mathcal{D}$ (where γ is a function of the dictionary size $|\mathcal{D}|$ and n) if for every real adversary C, there exists an ideal adversary C_{ideal} such that the ideal and real distributions are $(1-\gamma)$-indistinguishable.*

Our goal is to make γ as small as possible. Ideally, we would like $\gamma = O\left(\frac{1}{|\mathcal{D}|}\right)$. Note that Definition 2 guarantees that the password w is $(1-\gamma)$ indistinguishable from a random password $\tilde{w} \xleftarrow{R} \mathcal{D}$ since C_{ideal} learns nothing about the password w which is explicitly in the ideal distribution.

Security with Respect to Password Guesses. A stronger definition of security can be obtained by allowing the ideal adversary some number of password guesses but requiring that the ideal and real distributions be computationally indistinguishable. We will therefore modify the ideal model by adding the following steps after A and B receive their inputs:

- C_{ideal} sends its (possibly adaptive) guesses for the password $w_1, \cdots, w_\alpha$ to the trusted party. The trusted party answers whether the guesses are correct or not.
- If the adversary C_{ideal} guesses the password correctly, C_{ideal} can force the outputs of A and B to be whatever it wants.

The modified ideal distribution for α password guesses is defined by:

$$\text{IDEAL}_{C_{\text{ideal}}}^{\text{Guess}}(\mathcal{D}) = (w, \text{output}(A), \text{output}(B), \text{output}(C_{\text{ideal}}))$$

Definition 3 (Security with respect to α password guesses). *Let $\mathcal{D} \subseteq \{0,1\}^n$ be a samplable dictionary. A protocol for password-based session-key generation is secure with respect to α password guesses for the dictionary $\mathcal{D}$ if for every real adversary C, there exists an ideal adversary C_{ideal} making at most α password guesses such that the ideal and real distributions are computationally indistinguishable.*

Note that the ideal model in the definition of security with respect to password guesses can be simulated by the ideal model in Definition 2 with probability $\left(1 - \frac{\alpha}{|\mathcal{D}|}\right)$. Hence we obtain that the definition of security with respect to password guesses is stronger than Definition 2:

Proposition 1. *Security with respect to α password guesses implies GL-security with $\gamma = \frac{\alpha}{|\mathcal{D}|}$.*

In Section 3, we will show that even the *stronger* definition of security with respect to password guesses does not imply PKE under black-box reductions. In Section 4, we will show that the *weaker* GL definition (Definition 2) implies semi-honest OT.

Other Definitions. Bellare, Pointcheval and Rogaway [1] introduced a definition based on the *indistinguishability* of the session key. In this model, there are not just two honest parties as in the previous definitions but rather a set of honest parties (called principals) that are either a *client* or a *server*. Each client has some password $w \xleftarrow{R} \mathcal{D}$ and each server has the passwords of the clients.

The interaction of the adversary with the principals is modeled using oracle queries. Each principal is modeled by a collection of oracles that represent all possible actions, such as passive eavesdropping (the adversary sees the transcript of a protocol execution between a client and a server), corruption of a party (the adversary obtains the client's password), loss of session keys (the adversary learns the session key generated by a protocol execution) and person-in-the-middle attack (the adversary sends messages of its choosing to a principal). The adversary is allowed to make these oracle queries to any principal and there might be several instances of the same principal U that model concurrent executions.

The adversary chooses a *test* concerning the instance i of an uncorrupted principal U: a bit b is chosen uniformly from $\{0,1\}$. If $b=0$, then the adversary is given the session key output by the instance i of the principal U. If $b=1$, the adversary is given a truly random key. A PAK protocol is secure for the dictionary $\mathcal{D}$ according to the BPR definition if after mounting at most q person-in-the-middle attacks, the adversary has advantage at most $O\left(\frac{q}{|\mathcal{D}|}\right) + \text{neg}(n)$ in distinguishing the true session key from a random key in this test.

Boyko, MacKenzie and Patel [6] proposed a *simulation-based* definition which allows the ideal adversary to make a constant number of password guesses to the trusted party. The BMP definition is similar to the definition of security with respect to password guesses (in fact, the definition of security with respect to password guesses was inspired by the BMP definition) but their model differs from ours in two important respects. First, there are not just two honest parties executing the protocol but rather a set of honest users. Each user may have several instances that model concurrent executions of the protocol. Second, the ideal and real distributions in this model *do not include the passwords*. Loosely speaking, a PAK protocol is secure according to the BMP definition if for every real adversary, there exists an ideal adversary such that the ideal and real distributions are computationally indistinguishable.

Both the BPR and BMP definitions present some advantages over the GL definition because they handle concurrent executions easily. However, unlike the GL definition, these definitions do not explicitly guarantee that the password w remain pseudorandom after an execution. For example, the first bit of the password w could be revealed during an execution. This distinction is important as we will show that unlike the GL definition, the BPR definition does not imply semi-honest OT for exponential-sized dictionaries. Indeed, in Section 5 we exhibit a PAK protocol that is secure according to the BPR definition for the dictionary $\mathcal{D} = \{0,1\}^n$ but only assumes the existence of one-way functions. In particular, the password w does not remain pseudorandom after an execution of this protocol.

2.3 Black-Box Reductions

We give an informal presentation of black-box reductions that will suffice for our purposes. For more details, we refer the reader to [22]. The function (or algorithm) $f : \{0,1\}^* \to \{0,1\}^*$ is an implementation of a primitive P if it satisfies the structural requirements of the primitive (for example, in the case of one-way permutations, we require that f be a length-preserving permutation). We do not require that the implementation f satisfy some security requirements.

A black-box reduction of Q to P is the construction of two PPT oracle machines G and S such that:

- If f is an implementation of P (not necessarily efficient), then G^f is an implementation of Q.
- For every adversary A (not necessarily efficient) that breaks the implementation G^f, $S^{A,f}$ breaks the implementation f.

A black-box reduction relativizes, hence to show that there are no black-box reductions of Q to P, it suffices to construct an oracle relative to which P exists but Q does not.

3 There Is no Black-Box Construction of PKE from PAK

3.1 Overview of the Result

Theorem 4. *There exists an oracle Γ relative to which PAK exists but PKE does not.*

The oracle Γ we will use is composed of the following parts:

- f_1, f_2 and f_3 are three uniformly distributed length-tripling injective functions.
- The function R is defined to satisfy $R(w, s, \alpha) = K$ whenever $\alpha = f_3(w, K, r, f_2(w, s, f_1(w, K, r)))$ for some $|K| = |r| = |s| = |w|$, $\bot$ otherwise (R is well-defined since the f_i's are injective).
- a **PSPACE**-complete oracle

We now describe a PAK protocol using Γ:

Protocol 1. 1. Inputs: A and B have a security parameter 1^n and a joint password $w \in \mathcal{D} \subseteq \{0,1\}^n$, where $\mathcal{D}$ is samplable.
2. A chooses two n-bit strings $K_A, r_A \xleftarrow{R} \{0,1\}^n$ and sends $\alpha_1 \stackrel{\text{def}}{=} f_1(w, K_A, r_A)$. B receives β_1.
3. B chooses $r_B \xleftarrow{R} \{0,1\}^n$ and sends $\beta_2 \stackrel{\text{def}}{=} f_2(w, r_B, \beta_1)$. A receives α_2.
4. A sends $\alpha_3 \stackrel{\text{def}}{=} f_3(w, K_A, r_A, \alpha_2)$. B receives β_3.
5. Outputs: A outputs K_A. B outputs $R(w, r_B, \beta_3)$.

Note that in this protocol A always accepts and B accepts iff $R(w, r_B, \beta_3) \neq \bot$.

We prove Theorem 4 via the following two lemmas. The first lemma establishes that relative to Γ, PAK exists.

Lemma 1. *Protocol 1 is secure with respect to 2 password guesses for the dictionary $\mathcal{D}$, i.e. for every real adversary C, there exists an ideal adversary C_{ideal} with 2 password guesses such that $\text{REAL}_C(\mathcal{D}) \stackrel{c}{\equiv} \text{IDEAL}_{C_{\text{ideal}}}^{\text{Guess}}(\mathcal{D})$, where the probabilities are also taken over the random choice of Γ.*

The proof of Lemma 1 is quite involved and can be found in the full version of the paper [19]. We try to give the main idea of the proof in the section below.

It is known that PKE and 2-round KA are equivalent [9]. Thus, to prove Theorem 4, it suffices to prove that relative to Γ, there is no secure 2-round KA protocol.

Lemma 2. *For every 2-round KA protocol (A, B), for every polynomial p, there exists a passive adversary E such that the probability over Γ and the random tapes of A, B and E that E outputs the session key is at least $1 - \frac{1}{n^2 p(n)}$.*

In other words, with overwhelming probability, any 2-round KA protocol is not secure since there exists a passive adversary E that is able to distinguish the session key from a truly random string. The proof of Lemma 2 is very similar to that of [15, 23] and can be found in the full version of the paper [19]. Using Lemmas 1 and 2, we show that with probability 1 over the random choice of Γ, Protocol 1 is secure with respect to 2 password guesses and there exists no secure 2-round KA protocol. This establishes Theorem 4.

3.2 Relative to Γ, PAK Exists

We give some intuition on how to prove that Protocol 1 is secure with respect to 2 password guesses. For every real adversary C, we need to exhibit an ideal adversary with 2 password guesses which simulates C's view. We will follow the paradigm of [6] and show how to transform some of the real adversary's queries to the oracle Γ into password guesses for the ideal adversary.

The ideal adversary C_{ideal} will run the real adversary C and simulate the honest parties A and B. Using the queries made by C to the oracle Γ and the messages sent by C, C_{ideal} will determine if password guesses need to be made and if so, forward these password guesses to the trusted party. The output of C_{ideal} will be C's view of this simulated execution and we show that C's view of this execution simulated by C_{ideal} produces a view which is computationally indistinguishable from C's view of a real execution with A and B.

- As long as no password guess has been successful, C_{ideal} will simulate the honest parties by sending random strings of appropriate length. Intuitively, in this case, the messages sent by the honest parties A and B in a real execution are computationally indistinguishable from random strings with respect to the real adversary's view.
- If a password guess has been successful, C_{ideal} will have the password w and intuitively C will simulate the honest parties A and B perfectly.

We now show how to transform some of the real adversary's queries to Γ into password guesses for the ideal adversary. When C makes a query to the oracle

Γ, C_{ideal} makes this query to Γ and records the query/answer pair. Recall that an active adversary C can mount a person-in-the-middle attack that effectively gives two concurrent executions of the PAK protocol, one between A and C and one between C and B. We denote by α_i the ith message in the (A,C) interaction and by β_i the ith message in the (C,B) interaction. We define *password guesses in a real interaction* of C with A and B as follows.

- Password guess in the (C,B) interaction: C impersonates A on input $w' \in \mathcal{D}$ by sending to B the messages $\beta_1 = f_1(w', K_C, r_C)$ for some pair $(K_C, r_C) \in \{0,1\}^n \times \{0,1\}^n$ and $\beta_3 = f_3(w', K_C, r_C, \beta_2)$. C's guess w' is correct if $dec_B = 1$.
- Password guess in the (A,C) interaction: C impersonates B on input $w'' \in \mathcal{D}$ by sending to A the message $\alpha_2 = f_2(w'', r_C, \alpha_1)$ for some string $r_C \in \{0,1\}^n$. C's guess w'' is correct if $R(w'', r_C, \alpha_3) \neq \bot$

We can turn these cases in the real model into *password guesses in the ideal model*:

- Password guess in the simulated (C,B) interaction:
  If C sends $\beta_1 = f_1(w', K_C, r_C)$ for some previous query (w', K'_C, r_C) made by C to f_1, C_{ideal} sends its guess w' to the trusted party.
- Password guess in the simulated (A,C) interaction:
  If C sends $\alpha_2 = f_2(w'', r_C, \alpha_1)$ for some previous query (w'', r_C, α_1) made by C to f_2, C_{ideal} sends its guess w'' to the trusted party.

4 GL-Security for PAK Implies Semi-honest OT

Theorem 5. *Let $\mathcal{D} = \{\mathcal{D}_n\}_{n \in \mathbb{N}}$ be a samplable ensemble such that $\mathcal{D}_n \subseteq \{0,1\}^{\text{poly}(n)}$. The existence of a $(1-\gamma)$-GL-secure PAK protocol for the dictionary $\mathcal{D}_n$ on security parameter 1^n such that $\gamma \leq \frac{1}{5f(n)}\left(1 - \frac{1}{t(n)}\right)$ for some function $f(n) = \omega(\log n)$ and some polynomial t implies semi-honest OT (via a black-box reduction).*

Note that we do not require many dictionaries for a single security parameter 1^n, but rather a single fixed dictionary $\mathcal{D}_n$ for a given security parameter 1^n.

In order to prove Theorem 5, we consider a protocol for "*Weak OR*" (WOR). A WOR protocol (A,B) computes the functionality of the standard OR but its security is weak. More formally, a protocol (A,B) for weak OR is $(1-\eta)$-secure if

- B's privacy: A's view of $(A(1^n, 1), B(1^n, 0))$ is $(1-\eta)$-indistinguishable from $(A(1^n, 1), B(1^n, 1))$.
- A's privacy: B's view of $(A(1^n, 0), B(1^n, 1))$ is $(1-\eta)$ indistinguishable from $(A(1^n, 1), B(1^n, 1))$.

As suggested by Boyarsky [5], we establish Theorem 5 in two steps:

1. We first prove that a PAK protocol that is $(1-\gamma)$-GL-secure can be used to build a WOR protocol that is $(1-5\gamma)$-secure.

2. We then show that a WOR protocol that is $(1-\eta)$-secure for $\eta \leq \frac{1}{f(n)}$ $\left(1 - \frac{1}{t(n)}\right)$ can be used to build a secure protocol for semi-honest OT. Our proof that WOR implies semi-honest OT is similar to the proof of Kilian [18] that OR implies OT but the two results are incomparable since we restrict our focus to the semi-honest setting but are given a weaker OR primitive.

4.1 GL-Security for PAK Implies Weak OR

Given a $(1 - \gamma)$-GL-secure PAK protocol (A_P, B_P) for the samplable dictionary $\mathcal{D} \subseteq \{0,1\}^n$, we build the following WOR protocol.

Protocol 2. 1. Inputs: A has a bit a, B has a bit b.
2. A chooses $w, w' \xleftarrow{R} \mathcal{D}$ and sends w to B. B chooses $w'' \xleftarrow{R} \mathcal{D}$. (This is where we use the assumption that $\mathcal{D}$ is samplable.)
3. A and B run the PAK protocol (A_P, B_P) on inputs w_A and w_B respectively, where w_A and w_B are defined as follows:
   - If $a = 0$, A_P sets its password w_A to be w. Otherwise $w_A = w'$.
   - If $b = 0$, B_P sets its password w_B to be w. Otherwise $w_B = w''$.
   At the end of the PAK protocol, B sends its decision bit dec_B to A.
4. Outputs: If $a = 1$ and $\text{dec}_B = 1$, then A sends a message to B to set the output to be 1. Similarly, if $b = 1$ and $\text{dec}_B = 1$, then B sends a message to A to set the output to be 1. Otherwise, the common output of the execution is set to be $(1 - \text{dec}_B)$.

Analysis Sketch. Note that Protocol 2 computes the OR functionality correctly. If $a = b = 0$, then the passwords w_A and w_B are both equal to w and by definition of the PAK protocol, B will accept and the common output of A and B will be $1 - \text{dec}_B = 0$. If $a = 1$ or $b = 1$, we know that either B rejects (and the common output will be $1 - \text{dec}_B = 1$) or one of the parties will send an additional message and set the output to be 1.

If $a = 1$, then $\text{OR}(a, b) = 1$ regardless of the value of b so A should not learn B's input. Indeed, we will show that A's view of the interaction $(A(1), B(0))$ is $(1 - 5\gamma)$-indistinguishable from A's view of the interaction $(A(1), B(1))$ (the reasoning for B is similar).

We first consider A's view of the interaction $(A(1), B(b))$ when the possible additional message sent by B in Step 4 is not included. This possibly truncated view of A of the interaction $(A(1), B(b))$ is $(w, A_P(w')^{B_P(w_B)})$ where the second component refers to A_P's view of the PAK protocol and w_B is either w or w''. Because w' is independent of w_B, we can think of $A_P(w')$ as a real adversary C for the PAK protocol that interacts with the honest party $B_P(w_B)$. Since the PAK protocol is $(1 - \gamma)$-GL-secure, we can show that even if the adversary C is given w, C cannot distinguish the case $w_B = w$ from the case $w_B = w''$ with probability greater than 2γ. This is because in the ideal model, an ideal adversary learns nothing about the password w_B.

If B sends an additional message after the execution of the PAK protocol, then we know that B's input is $b = 1$, which makes A's views of $(A(1), B(0))$

and $(A(1), B(1))$ distinguishable. Recall that B sends an additional message iff B_P accepts in an execution of the PAK protocol where A_P has input $w' \xleftarrow{R} \mathcal{D}$ and B_P has input $w'' \xleftarrow{R} \mathcal{D}$. Because w' is independent of w'', we can think of $A_P(w')$ as a real adversary C for the PAK protocol that interacts with the honest party $B_P(w'')$. Since the PAK protocol is $(1-\gamma)$-GL-secure, we can show that an adversary C makes B accept (and B sends an additional message after the execution of the PAK protocol) with probability at most 3γ.

4.2 Weak OR Implies Semi-honest OT

Lemma 3. *The existence of a WOR protocol that is $(1-\eta)$-secure for $\eta \leq \frac{1}{f(n)}\left(1 - \frac{1}{t(n)}\right)$ for some function $f(n) = \omega(\log n)$ and some polynomial t (where 1^n is the security parameter) implies honest OT (via a black-box reduction).*

In order to prove Lemma 3, we introduce a two-party functionality called "Weak OT" (WOT). A protocol (A, B) for weak OT is similar to a protocol for OT except that

- B does not choose which secret bit it will obtain. That is, B has no input except for the security parameter 1^n and when interacting with $A(1^n, s_0, s_1)$, B's output is (c, s_c) for a random bit c.
- For every s_0, s_1 and a random bit c, $\tilde{B}(1^n, c, s_c)$ is $(1-\epsilon)$ indistinguishable from B's view of the interaction $(A(1^n, s_0, s_1), B(1^n))$, where $\epsilon \leq 1 - \frac{1}{t(n)}$ for some polynomial t. In other words, the sender's privacy only holds with probability $(1-\epsilon)$.

Kilian [18] showed how to build a protocol for OT from a secure protocol for OR in two steps:

1. Using a secure protocol for OR, we first build a protocol for weak OT
2. Using a protocol for weak OT, we then build a protocol for OT

To prove Lemma 3, we adapt these two steps to our weak OR primitive. We strengthen the first step of [18] and show how to build a protocol for weak OT given a weak OR protocol that is $(1-\eta)$-secure. More precisely, we first show how to use a weak OR protocol to build a protocol for a functionality called "very weak OT" and then we show how use a protocol for very weak OT to implement weak OT. For the second step, we can use Kilian's result:

Lemma 4. *[18] The existence of a protocol for weak OT implies (honest) OT[9].*

---

[9] This lemma uses the reduction of OT to weak OT given in [18], Section 2.4. The analysis is slightly different from the original analysis: we use the uniform version of Yao's XOR lemma to guarantee the sender's privacy and a hybrid argument to guarantee the receiver's privacy.

Weak OR Implies Very Weak OT. A protocol (A, B) for *very weak OT* is similar to a protocol for OT except that both the sender's privacy and the receiver's privacy hold with low probability. More formally, a protocol for very weak OT is $(1 - \eta)$-secure if the following conditions hold:

- *Receiver's privacy:* If $s_0 = s_1 = 1$, A's view of $(A(1^n, s_0, s_1), B(1^n, 0))$ is $(1 - 2\eta)$-indistinguishable from A's view of $(A(1^n, s_0, s_1), B(1^n, 1))$.
- *Sender's privacy:* For every s_0, s_1, c, B's view of $(A(1^n, s_c, s_{\bar{c}}), B(1^n, c))$ is $(1 - \eta)$-indistinguishable from B's view of $(A(1^n, s_c, \overline{s_{\bar{c}}}), B(1^n, c))$

Given a WOR protocol that is $(1-\eta)$-secure, we build the following protocol for very weak OT.

Protocol 3. 1. Inputs: A has the security parameter 1^n and two secret bits s_0 and s_1. B has the security parameter 1^n and a selection bit c.
2. A sets $a_0 = s_0$ and $a_1 = s_1$. B sets $b_c = 0$ and $b_{\bar{c}} = 1$.
3. A and B run the WOR protocol to obtain $c_j = OR(a_j, b_j)$ for $j \in \{0, 1\}$.[10]
4. Outputs: B computes the secret bit $s_c = OR(a_c, b_c)$, A outputs nothing.

Analysis Sketch. Note that Protocol 3 computes the OT functionality correctly: B obtains the secret bit a_c because $a_c = OR(a_c, b_c) = OR(a_c, 0)$.

Since $b_{\bar{c}} = 1$, the security of the WOR protocol implies that B has advantage at most η in distinguishing the case $a_{\bar{c}} = 0$ from the case $a_{\bar{c}} = 1$.

If $s_0 = s_1 = 1$, i.e. $a_c = a_{\bar{c}} = 1$, the security of the WOR protocol implies that A has advantage at most 2η in distinguishing the case $c = 0$ from the case $c = 1$ (by the security of the WOR protocol). Note that in Protocol 3 if $a_0 = 0$ or $a_1 = 0$, then A learns B's selection bit c.

Very Weak OT Implies Weak OT. Given a $(1 - \eta)$-secure protocol for very weak OT, where $\eta \leq \frac{1}{f(n)} \left(1 - \frac{1}{t(n)}\right)$, $f(n) = \omega(\log n)$ and t is a polynomial, we build a protocol for weak OT. In order to amplify the receiver's privacy, we will repeat the protocol for very weak OT $f(n)$ times and apply a secret sharing scheme to B's selection bit.

Protocol 4. 1. Inputs: A has the security parameter 1^n and two secret bits s_0 and s_1. B has 1^n.
2. For $1 \leq i \leq f(n)$, A uniformly chooses $a_0^i, a_1^i \in \{0, 1\}$ and B uniformly chooses $c^i \in \{0, 1\}$.
3. For $1 \leq i \leq f(n)$, A and B execute the protocol for very weak OT on (a_0^i, a_1^i, c^i).
4. A uniformly chooses $z^1, \cdots, z^{f(n)} \in \{0, 1\}$ and sends to B the following values

---

[10] The executions of the WOR protocol can be done in parallel. Indeed, the executions of the WOR protocol are independent because the parties are assumed to be honest. Hence the privacy condition still holds with probability $1 - \eta$ for each execution of the WOR protocol (otherwise an adversary could violate the privacy condition for a single execution of the protocol by simulating an independent execution on its own).

- for $1 \leq i \leq f(n)$, $q_0^i = z^i \oplus a_0^i$
- for $1 \leq i \leq f(n)$, $q_1^i = z^i \oplus a_1^i \oplus (s_0 \oplus s_1)$
- $Q = s_0 \oplus (\oplus_{i=1}^{f(n)} z^i)$

5. B computes for every $i \in [f(n)]$, $v^i = q_{c^i}^i \oplus a_{c^i}^i$, $c = \oplus_{i=1}^{f(n)} c^i$ and outputs $s_c = Q \oplus \left(\oplus_{i=1}^{f(n)} v^i\right)$.

Analysis Sketch. By the correctness of the protocol for very weak OT, we know that for every i, B learns the value of $a_{c^i}^i$. Thus we can show that B computes the secret s_c correctly in Protocol 4.

Intuitively, in order to know c, A needs to know the values of all the c^i's. By the security of the protocol for very weak OT, we know that for every i, the probability that A distinguishes the case $c^i = 1$ from the case $c^i = 0$ is at most $3/4 + 2\eta$. Using the uniform version of Yao's XOR Lemma [13], we can show that the probability that A distinguishes the case $c = 0$ from the case $c = 1$ is negligible.

Intuitively, in order to know $s_{\bar{c}}$, B needs to know the value of one of the $a_{c^i}^i$. Using a hybrid argument, we can show that the probability that B distinguishes the case $s_{\bar{c}} = 0$ from the case $s_{\bar{c}} = 1$ is at most $\eta \cdot f(n) \leq 1 - \frac{1}{t(n)}$ for some polynomial t.

5 On the Different Definitions of PAK

We highlight the difference between the indistinguishability-based definition of [1] with the simulation-based definition of [11] by showing that unlike the GL definition, the BPR definition does not imply semi-honest OT in the case of exponential-sized dictionaries. Bellare, Pointcheval and Rogaway started with the model and definition of [2] for authenticated key exchange and modified them appropriately to take into account passwords instead of high-entropy keys. In particular, the definition of security of [1] for password-authenticated key exchange for the dictionary $\mathcal{D} = \{0,1\}^n$ (when we do not guarantee forward secrecy) is *exactly* the original definition of [2] for plain (=non-password-based) authenticated key exchange.

Consider the following protocol that was proposed in [2]. The legitimate parties share a password $w \xleftarrow{R} \mathcal{D} = \{0,1\}^n$ that we can see as two $n/2$-bit strings (w_1, w_2). The first part is taken as the key to a pseudorandom function family $\mathcal{F} = \{f_{w_1} : \{0,1\}^{2n} \to \{0,1\}^n\}_{w_1 \in \{0,1\}^{n/2}}$. The second part is taken as the key to a pseudorandom permutation family $\mathcal{G} = \{g_{w_2} : \{0,1\}^n \to \{0,1\}^n\}_{w_2 \in \{0,1\}^{n/2}}$.

Protocol 5. 1. Inputs: A and B have a security parameter 1^n and a joint password $w = (w_1, w_2) \in \mathcal{D} = \{0,1\}^n$.

2. A chooses $r_A \xleftarrow{R} \{0,1\}^n$ and sends $\alpha_1 = r_A$ to B. B receives β_1.

3. B chooses $r_B \xleftarrow{R} \{0,1\}^n$ and sends $\beta_2 \stackrel{\text{def}}{=} (r_A, r_B, f_{w_1}(r_A, r_B))$ to A. A receives α_2.

4. If $\alpha_2 \neq (r_A, r_B, f_{w_1}(r_A, r_B))$ (which A can check using its password), then A chooses $K_A \xleftarrow{R} \{0,1\}^n$. Otherwise, A sends $\alpha_3 \stackrel{\text{def}}{=} (r_B, f_{w_1}(r_B))$ to B. B receives β_3.
5. Outputs: If α_2 was of the form $(r_A, r_B, f_{w_1}(r_A, r_B))$, A outputs $g_{w_2}(r_B)$; otherwise, it outputs K_A. If $\beta_3 \neq (r_B, f_{w_1}(r_B))$, B rejects. Otherwise, B outputs $g_{w_2}(r_B)$.

Lemma 5. *[2] If one-way functions exist, Protocol 5 is a secure authenticated key exchange protocol for the definition of [2]. In other words, Protocol 5 is a secure PAK protocol for the dictionary $\mathcal{D} = \{0,1\}^n$ for the BPR definition.*

Hence, unlike the GL definition, the BPR definition does not imply semi-honest OT for dictionaries of exponential size under black-box reductions ([15,9]). Intuitively, the difference is that the BPR definition does not guarantee that the password w remain pseudorandom after an execution of a secure PAK protocol. Indeed, we can see that the password w will not remain pseudorandom even with respect to a passive adversary's view of an execution of Protocol 5 since the adversary learns the pair $(r_B, f_{w_1}(r_B))$. The pseudorandomness property required by the GL definition makes a secure PAK protocol a strong enough primitive to imply semi-honest OT, even in the case of dictionaries of exponential size (which corresponds to plain authenticated key exchange). The guarantee that the password w remain pseudorandom after an execution of a PAK protocol is indeed important if one intends to also use the password in a protocol other than the PAK protocol.

This does not necessarily mean a PAK protocol that is secure for the BPR definition is not a "strong" primitive. Indeed, we conjecture that one can implement semi-honest OT using a PAK protocol that is secure for the BPR definition for *all dictionaries* (including poly-sized dictionaries which is the case of most interest). Another open question is the relationship between the simulation-based definition of Boyko, MacKenzie and Patel [6] and the BPR and GL definitions.

As noted in [7], *"settling on a "good" definition of security for password-based authentication has been difficult and remains a challenging problem"*. However, a study of the relationship between each definition of security for PAK and other cryptographic primitives provides a better understanding of the tradeoffs and advantages offered by one definition of security over another.

Acknowledgments. We thank Salil Vadhan for suggesting this problem and for many helpful discussions and detailed comments. Many thanks to Yehuda Lindell for his help in reconstructing [5]. We are grateful to Alex Healy and Omer Reingold for helpful conversations on this subject and to the anonymous reviewers for their insightful comments.

References

1. Bellare, M., Pointcheval, D., Rogaway, P.: Authenticated Key Exchange Secure against Dictionary Attacks. Lecture Notes in Computer Science **1807** (2000) 139–155

2. Bellare, M., Rogaway, P.: Entity Authentication and Key Distribution. Lecture Notes in Computer Science **773** (1994) 232–249
3. Bellovin, S., Merritt, M.: Encrypted Key Exchange: Password-Based Protocols Secure Against Dictionary Attacks. ACM/IEEE Symposium on Research in Security and Privacy (1992) 72–84
4. Bellovin, S., Merritt, M.: Augmented Encrypted Key Exchange: A Password-Based Protocol Secure against Dictionary Attacks and Password File Compromise. ACM Conference on Computer and Communications Security (1993) 244–250
5. Boyarsky, M.: Public-Key Cryptography and Password Protocols: The Multi-User Case. ACM Conference on Computer and Communications Security (1999) 63–72
6. Boyko, V., MacKenzie, P., Patel, S.: Provably Secure Password-Authenticated Key Exchange Using Diffie-Hellman. Lecture Notes in Computer Science **1807** (2000) 156–171
7. Canetti, R., Halevi, S., Katz, J., Lindell, Y., MacKenzie, P.: Universally Composable Password-Based Key Exchange. Unpublished manuscript (2004)
8. Gennaro, R., Lindell, Y.: A Framework for Password-Based Authenticated Key Exchange. Lecture Notes in Computer Science **2656** (2003) 524–543
9. Gertner, Y., Kannan, S., Malkin, T., Reingold, O., Viswanathan, M.: The Relationship between Public-Key Encryption and Oblivious Transfer. IEEE Symposium on the Foundations of Computer Science (2001) 325–335
10. Goldreich, O.: Foundations of Cryptography, Volume 2. Cambridge University Press (2004)
11. Goldreich, O., Lindell, Y.: Session-Key Generation Using Human Passwords Only. Lecture Notes in Computer Science **2139** (2001) 408–432
12. Goldreich, O., Micali, S., Wigderson, A.: Proofs that yield nothing but their validity or all languages in NP have zero-knowledge proofs. Journal of the ACM **38:3** (1991) 691–729
13. Goldreich, O., Nisan, N., Wigderson, A.: On Yao's XOR Lemma. Electronic Colloquium on Computational Complexity (1995) TR95-050
14. Halevi, S., Krawczyk, H.: Public-Key Cryptography and Password Protocols. ACM Conference on Computer and Communications Security (1998) 122–131
15. Impagliazzo, R., Rudich, S.: Limits on the Provable Consequences of One-way Permutations. ACM Symposium on Theory of Computing (1989) 44–61
16. Katz, J.: Efficient Cryptographic Protocols Preventing 'Man-in-the-Middle' Attacks. Ph.D. Thesis. Columbia University (2002)
17. Katz, J., Ostrovsky, R., Yung, M.: Efficient Password-Authenticated Key Exchange Using Human-Memorable Passwords. Lecture Notes in Computer Science **2045** (2001) 475–494
18. Kilian, J.: A General Completeness Theorem for Two-Party Games. ACM Symposium on Theory of Computing (1991) 553–560
19. Full version of this paper at http://www.people.fas.harvard.edu/~mnguyen
20. Nguyen, M.-H., Vadhan, S.: Simpler Session-Key Generation from Short Random Passwords. Lecture Notes in Computer Science **2951** (2004) 428–445
21. Nisan, N., Zuckerman, D.: Randomness is Linear in Space. Journal of Computer and System Sciences **52:1** (1996) 43–52
22. Reingold, O., Trevisan, L., Vadhan, S.: Notions of Reducibility between Cryptographic Primitives. Lecture Notes in Computer Science **2951** (2004) 1–20
23. Rudich, S.: The Use of Interaction in Public Cryptosystems. Lecture Notes in Computer Science **576** (1992) 242–251
24. Steiner, M., Tsudik, G., Waidner, M.: Refinement and Extension of Encrypted Key Exchange. Operating Systems Review **29:3** (1995) 22–30

On the Relationships Between Notions of Simulation-Based Security*

Anupam Datta[1], Ralf Küsters[2], John C. Mitchell[1], and Ajith Ramanathan[1]

[1] Computer Science Department, Stanford University,
Stanford CA 94305-9045, USA
{danupam, jcm, ajith}@cs.stanford.edu
[2] Institut für Informatik,
Christian-Albrechts-Universität zu Kiel, 24098 Kiel, Germany
kuesters@ti.informatik.uni-kiel.de

Abstract. Several compositional forms of simulation-based security have been proposed in the literature, including universal composability, black-box simulatability, and variants thereof. These relations between a protocol and an ideal functionality are similar enough that they can be ordered from strongest to weakest according to the logical form of their definitions. However, determining whether two relations are in fact identical depends on some subtle features that have not been brought out in previous studies. We identify the position of a "master process" in the distributed system, and some limitations on transparent message forwarding within computational complexity bounds, as two main factors. Using a general computational framework, we clarify the relationships between the simulation-based security conditions.

1 Introduction

Several current projects use ideal functionality and indistinguishability to state and prove compositional security properties of protocols and related mechanisms. The main projects include work by Canetti and collaborators on an approach called *universal composabiliity* [8, 10, 11, 12, 13] and work by Backes, Pfitzmann, and Waidner on a related approach that also uses *black-box simulatability* [22, 7, 4, 5]. Other projects have used the notion of equivalence in process calculus [16, 18, 19], a well-established formal model of concurrent systems. While some process-calculus-based security studies [2, 3, 1] abstract away probability

---

* This work was partially supported by the DoD University Research Initiative (URI) program administered by the Office of Naval Research under Grant N00014-01-1-0795, by OSD/ONR CIP/SW URI "Trustworthy Infrastructure, Mechanisms, and Experimentation for Diffuse Computing" through ONR Grant N00014-04-1-0725, by NSF CCR-0121403, Computational Logic Tools for Research and Education, and by NSF CyberTrust Grant 0430594, Collaborative research: High-fidelity methods for security protocols. Part of this work was carried out while the second author was at Stanford University supported by the "Deutsche Forschungsgemeinschaft (DFG)".

and computational complexity, at least one project [20, 17, 21, 23] has developed a probabilistic polynomial-time process calculus for security purposes. The common theme in each of these approaches is that the security of a real protocol is expressed by comparison with an ideal functionality or ideal protocol. However, there are two main differences between the various approaches: the precise relation between protocol and functionality that is required, and the computational modeling of the entities (protocol, adversary, simulator, and environment). All of the computational models use probabilistic polynomial-time processes, but the ways that processes are combined to model a distributed system vary. We identify two main ways that these computational models vary: one involving the way the next entity to execute is chosen, and the other involving the capacity and computational cost of communication. We then show exactly when the main security notions differ or coincide.

In [8], Canetti introduced universal composability (UC), based on probabilistic polynomial-time interacting Turing machines (PITMs). The UC relation involves a real protocol and ideal functionality to be compared, a real and ideal adversary, and an environment. The real protocol realizes the ideal functionality if, for every attack by a real adversary on the real protocol, there exists an attack by an ideal adversary on the ideal functionality, such that the observable behavior of the real protocol under attack is the same as the observable behavior of the ideal functionality under attack. Each set of observations is performed by the same environment. In other words, the system consisting of the environment, the real adversary, and the real protocol must be indistinguishable from the system consisting of the environment, the ideal adversary, and the ideal functionality. The scheduling of a system of processes (or ITMs) is *sequential* in that only one process is active at a time, completing its computation before another is activated. The default process to be activated, if none is designated by process communication, is the environment. In the present work, we use the term *master process* for the default process in a system that runs when no other process has been activated by explicit communication.

In [22], Pfitzmann and Waidner use a variant of UC and a notion of blackbox simulatability (BB) based on probabilistic polynomial-time IO automata (PIOA). In the BB relation between a protocol and ideal functionality, the UC ideal adversary is replaced by the combination of the real adversary and a simulator that must be chosen independently of the real adversary. Communication and scheduling in the PIOA computational model are sequential as in the PITM model. While the environment is the master process in the PITM studies, the adversary is chosen to be the master process in the Pfitzmann-Waidner version of UC. In the Pfitzmann-Waidner version of BB the master process is the adversary or the simulator [22]. In a later version of the PIOA model (see, e.g., [4]), the environment is also allowed to serve as the master process, subject to the restriction that in any given system it is not possible to designate both the adversary/simulator and the environment as the master process. In proofs in cryptography, another variant of BB is often considered in which the simulator

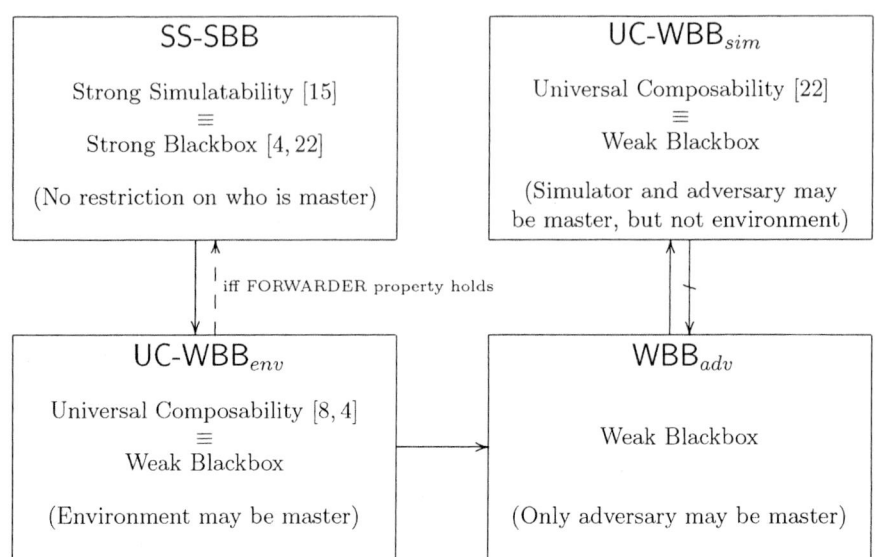

Fig. 1. Equivalences and implications between the security notions in SPPC

may depend on the real adversary or its complexity. We call this variant Weak BB (WBB) and the previous one Strong BB (SBB).

In [17, 21, 23, 24], Mitchell et al. have used a form of process equivalence, where an environment directly interacts with the real and ideal protocol. The computational model in this work is a probabilistic polynomial-time processes calculus (PPC) that allows concurrent (non-sequential) execution of independent processes. The process equivalence relation gives rise to a relation between protocols and ideal functionalities by allowing a simulator to interact with the ideal functionality, resulting in a relation that we call strong simulatability, SS [15]. The difference between SS and SBB is that in SBB, the environment and the adversary are separated while the SS environment also serves as the adversary.

Contribution of the Paper. In this paper, we clarify the relationships between UC, SBB, WBB, SS under different placements of the master process and an additional issue involving the ability to define a "forwarding" process that forwards communication from one process to another. While it seems intuitively reasonable that such a forwarder can be placed between two processes without changing the overall behavior of the system, this may violate complexity bounds if a polynomial-time forwarder must be chosen before the sending or receiving process. If the time bound of the sender, for example, exceeds the time bound of the forwarder, then some sent messages may be lost because the time bound of the forwarder has been exhausted. This is relevant to our study because some equivalence proofs require the existence of forwarders that cannot be exhausted.

Our main results are summarized in Figure 1. Each of the four boxes in this figure stands for a class of equivalent security notions. Specifically, if a real and ideal protocol are related by one notion in this class, then they are also related by all other notions in this class. A solid arrow from one class to another indicates that relations in the first class imply relations in the second class. The implication indicated by the dashed arrow is contingent on whether the aforementioned forwarding property holds for the processes in question.

The proofs of equivalence and implication between security notions are axiomatic, using a relatively small set of clearly stated equivalence principles involving processes and distributed systems. This approach gives us results that carry over to a variety of computational models. Our axiomatic system is proved sound for a specific computational model, a sequential probabilistic polynomial-time process calculus (SPPC), developed for the purpose of this study. SPPC is a sequential model, allowing only one process to run at a time. When one process completes, it sends an output indicating which process will run next. This calculus is close to PIOA and PITM in expressiveness and spirit, while (1) providing a syntax for writing equations between systems of communicating machines and (2) being flexible enough to capture different variants of security notions, including all variants of SS, SBB, WBB, and UC discussed in this paper. Our results about these security notions formulated over SPPC are:

1. Equivalences between security notions.
   (a) The different forms of Strong Simulatability and Strong Blackbox obtained by varying the entity that is the master process are all equivalent. This equivalence class, denoted SS-SBB, is depicted in the top-left box in Figure 1 and includes placements of the master process as considered for Strong Blackbox in [4, 22]
   (b) All variants of Universal Composability and Weak Blackbox in which the environment may be the master process are equivalent. This equivalence class, denoted UC-WBB$_{env}$, is depicted in the bottom-left box in Figure 1 and includes placements of the master process as considered for Universal Composability in [8, 4].
   (c) All variants of Universal Composability and Weak Blackbox in which the simulator and the adversary may be the master process, but not the environment are equivalent. This equivalence class, denoted UC-WBB$_{sim}$, is depicted in the top-right box in Figure 1 and includes placements of the master process as considered for Universal Composability in [22].
   (d) All variants of Weak Blackbox where the adversary may be the master process, but neither the environment nor the simulator may play this role are equivalent. This equivalence class, denoted WBB$_{adv}$, is depicted in the bottom-right box in Figure 1.
2. Implications between the classes.
   (a) SS-SBB implies UC-WBB$_{env}$. In particular, Strong Blackbox with placements of the master process as considered in [4, 22] implies Universal Composability with placements of the master process as considered in [8, 4].
   (b) UC-WBB$_{env}$ implies WBB$_{adv}$.

(c) WBB_{adv} implies UC-WBB_{sim}. In particular, Strong Blackbox with placements of the master process as considered in [4, 22] and Universal Composability with placements of the master process as considered in [8, 4] implies Universal Composability with placements of the master process as considered in [22].

3. Separations between the classes.
    (a) The security notions in UC-WBB_{env} are strictly weaker than those in SS-SBB in any computational model where the forwarding property (expressed precisely by the FORWARDER axiom) fails. Since this property fails in the PITM model [8] and the buffered PIOA model [4], it follows that UC-WBB_{env} does not imply SS-SBB in these models. This contradicts a theorem claimed in [4]. However, the forwarding property holds in SPPC and the buffer-free PIOA model for most protocols of interest. In these cases, UC-WBB_{env} implies SS-SBB.
    (b) The security notions in UC-WBB_{sim} are strictly weaker than the notions in WBB_{adv}, and hence, the notions in UC-WBB_{env} and SS-SBB. In particular, the Universal Composability relation with placements of the master process as considered in [22] does neither imply the Strong Blackbox relations with placements of the master process as considered in [4, 22] nor Universal Composability relations with placements of the master process as considered in [8, 4].

These results all show that the relationship between universal composability and black-box simulatability is more subtle than previously described. One consequence is that when proving compositional security properties by a black-box reduction, care must be taken to make sure that the computational model gives appropriate power to the environment. In particular, the composability theorem of Canetti [8] does not imply that blackbox simulatability is a composable security notion, over any computational model in which the forwarding property (expressed by the FORWARDER axiom) is not satisfied.

Outline of the Paper. Section 2 defines the sequential polynomial-time process calculus SPPC, with security relations defined precisely in Section 3. The main results are given in Section 4, with consequences for PIOA and PITM models developed in Section 5. In Section 6, we briefly consider a less prominent security notion, called *reactive simulatability* in [5] and *security with respect to specialized simulators* in [9], and relate it to the other notions.

Full definitions, proofs, and further explanations are provided in a technical report [14]. This technical report also proves a composition theorem for SPPC that is similar to the composition theorem for ITMs established by Canetti [8].

2 Sequential Probabilistic Process Calculus

In this section, we introduce Sequential Probabilistic Process Calculus (SPPC) as a language-based computational model for studying security notions (see [14] for a detailed technical presentation and further explanation). We start by discussing

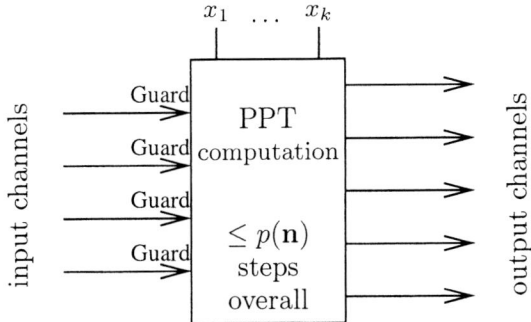

Fig. 2. Probabilistic polynomial-time machines in SPPC

how individual probabilistic polynomial-time machines are modelled in SPPC and then explain how to build and execute systems of interacting machines. Our exposition parallels that of related models [8, 22, 5].

Single Probabilistic Polynomial-Time Machines. In SPPC, single machines are of the form as depicted in Figure 2. For the time being, let us ignore the "guards" and the variables $x_1, \ldots, x_k$. Conceptually, a single machine is a black-box with internal state that receives inputs, performs polynomially-bounded computation and then produces outputs. Inputs are received on input channels and outputs are written on output channels. More precisely, single machines are restricted to receiving one input and producing at most one output at a time. While this at first might appear to be a restriction, it is not really a problem since any machine that sends multiple messages can be converted to a machine that stores internally (possibly using internal buffers) the messages it wants to send, and then sends the messages one at a time on request. In fact, this style of communication corresponds exactly to the manner in which communication is defined in other *sequential* models, notably the PIOA and PITM models [8, 22]. Also, just as in these models, the overall runtime of a machine is bounded by a polynomial in the security parameter and does not depend on the number or length of inputs sent to the machine.

The channels of a single machine in SPPC correspond to ports in the PIOA model and to tapes in the PITM model. However, while messages on channels (and ports) are removed when read, this is not the case for tapes. Nevertheless, tapes can be modelled by adding machines, one for each input channel, which simulate the tapes in the obvious way. The "main machine" will then receive its input from the "tape machines". In the PIOA model, buffer machines serve a similar purpose. Note that while in SPPC and the PIOA model, the number of input and output channels/ports is not restricted, in Canetti's PITM model only one pair of input/output and input/output communication tapes is considered.

In SPPC, machines can preprocess their input using *guards* (see Figure 2) which are deterministic polynomial-time machines that are placed on input channels. Given an input on the channel, a guard may accept or reject the input. If rejected, the process does no computation. If accepted, the process receives the output of the guard. This may be different from the input, e.g., a guard can eliminate unnecessary information or transform data. The computation performed by the guard may depend on the current internal state of the process. Its runtime is polynomially-bounded in the security parameter per invocation and is not factored into the overall runtime of the process using the guard. In particular, a guard can be invoked an unbounded number of times. Since guards allow a process to discard messages without incurring a computation cost, attempts to "exhaust" a process by sending many useless messages to the process can be defeated. Additionally, using guards we can simulate an unbounded number of "virtual" channel names by prefixing each message with a session id and/or party name and then stipulating that the guards accept only those messages with the right header information. Such an ability is required for systems with a polynomial number of machines, e.g., multiparty protocols, or with multiple instances of the same protocol. While mechanisms analogous to guards are absent in other models, notably [22, 8], a newer version of PIOA [6] has a length function that, when set to zero, prevents messages from being received by the machine. This corresponds to a guard which rejects all inputs and so can be used to help avoid exhaustion attacks. However, it does not help in the creation of a mechanism analogous to virtual channels.

As mentioned above, guards can be invoked an unbounded number of times without being exhausted and in every invocation their runtime is bounded by a polynomial in the security parameter—the runtime could even depend on the length of the input. Hence, the runtime of a single machine including the guards is polynomially bounded in the security parameter *and* the number of invocations. However, the overall runtime of a single machine excluding the guards is polynomially bounded in the security parameter alone, and hence, such a machine can produce at most polynomially many output messages overall in the security parameter. Now, since guards can only be triggered by messages sent by single machines, it follows that in a system of polynomially many machines guards are only invoked a polynomial number of times in the security parameter. As shown in [14], from this we can conclude that such systems can be simulated by a probabilistic polynomial time Turing machine.

In SPPC, a machine may have auxiliary input, just like auxiliary input can be given to the interacting Turing machines in Canetti's model. This input is written on specific tapes before a (system of) machines is run. If such auxiliary input is used, it results in a non-uniform computational model. The tapes are represented by $x_1, \ldots, x_k$ (see Figure 2). Just like in Canetti's model, we only allow the environment machine to use auxiliary input. However, whether the environment machine is uniform or not does not affect the results presented in this paper.

Formally, in SPPC a single machine is defined by a *process expression* $\mathcal{P}$. Such an expression corresponds to a description of an interacting Turing machine in the PITM model or an I/O automaton in the PIOA model. A process expression is always parameterized by the security parameter **n** and possibly so-called free variables $x_1, \ldots, x_k$, which represent the tapes for the auxiliary input mentioned above. Therefore, we sometimes write $\mathcal{P}(x_1, \ldots, x_k)$ instead of $\mathcal{P}$. A process expression with value i chosen for the security parameter and values $\vec{a}$ (the auxiliary inputs) substituted for its free variables $\vec{x}$ yields a *process* $\mathcal{P}(\vec{a})^{n \leftarrow i}$. A process corresponds to an interacting Turing machine where the security parameter is written on the security parameter tape and the auxiliary input is written on the input tape. Hence, a process can perform computations as soon as it receives input on the input channels. As an expositional convenience, we will use the terms 'process expression' and 'process' interchangeably. A process expression is called *open* if it has free variables, and *closed* otherwise. Hence, open process expressions correspond to non-uniform machines and closed expressions to uniform ones.

Systems of Interacting Machines. In SPPC, a system of interacting machines is simply a multiset of single machines where an output channel of one machine connects directly to an identically-named input channel of another machine. The manner in which these machines are wired together is uniquely determined by the channel names since we stipulate that no two machines have the same input and output channel names respectively. After a machine M_1 has sent a message on an output channel, the machine waits to receive input on an input channel. The message sent on the output channel is immediately received by the machine M_2 that has an identically-named input channel. If the guards on the input channel of this machine accepts the message, then M_2 may perform some computation and produce one output message. While M_2 now waits for new input on its input channels, the output message (if any) is processed by the next receiving machine, and so on. If there is no receiving machine, or the guard of the receiving machine rejects the message, or no output message is produced, computation would halt since no machine is triggered. To avoid this, in a system of machines, one machine is always declared to be a master machine, also called *master process*, and this machine is triggered if no other machine is.

In SPPC, given process expressions $\mathcal{P}_1, \ldots, \mathcal{P}_n$, each representing a single machine, the combined system of machines is denoted by the process expression $\mathcal{P}_1 \mid \cdots \mid \mathcal{P}_n$. Instead of interpreting $\mathcal{P}_1 \mid \cdots \mid \mathcal{P}_n$ as a system of n single machines, one can consider this system as a single machine (consisting of n submachines). This corresponds to the transformation, in the PIOA model, of a system of fixed, finite number of machines into a single machine. However, in SPPC we can apply such transformations to systems containing a polynomial number of machines as well.

With the bounded replication operator $!_{q(\mathbf{n})}\, \mathcal{P}$, where $q(\mathbf{n})$ is some polynomial in the security parameter and $\mathcal{P}$ is a process expression (representing a single machine or a system of machines), systems containing a polynomial number of machines can be described. The process expression $!_{q(\mathbf{n})}\, \mathcal{P}$ stands for a $q(\mathbf{n})$-fold parallel composition $\mathcal{P} \mid \cdots \mid \mathcal{P}$. Note that in such a system, different copies of $\mathcal{P}$ have the same input and output channels. However, as discussed earlier, guards allow us to send messages to (virtual) channels of particular copies of a protocol. Bounded replication can be combined with parallel composition to build bigger systems such as $!_{q_1(\mathbf{n})} (\mathcal{P}_1 \mid \mathcal{P}_2 \mid !_{q_3(\mathbf{n})} \mathcal{P}_3)$.

We note that the details of the communication model, such as a specific activation order of entities, the communication primitives available (such as insecure, authenticated, or secure channels), and specific forms of buffering, are not explicitly modelled in SPPC. The driving philosophy behind the design of SPPC is to move such details into the specification of the protocol rather than explicitly encoding them into the model. This makes SPPC simple and flexible, thereby allowing easy formulation of a variety of security notions.

As described earlier, since our execution model is sequential, computation may not proceed if currently executing machine produces no output, or a receiving machine rejects an input. In order to ensure that computation proceeds even in this case, we identify a master process by using a special input channel start. In case no output is produced by a machine, a fixed value is written on start thereby triggering the master process. The master process is also the first machine to be activated when execution starts.

Additionally, in studying security notions, it will be useful to define the output of a system. We do so by writing a bit, the output, onto an output channel named decision. The machine containing this channel is called the *decision process*. Given a process expression $\mathcal{R}(\vec{x})$ with free variables $\vec{x}$, we denote by $\text{Prob}[\mathcal{R}(\vec{a})^{\mathbf{n} \leftarrow i} \leadsto 1]$ the probability that $\mathcal{R}$ with security parameter i and substitution of values $\vec{a}$ for its variables $\vec{x}$ outputs a 1 on decision. Recall that $\mathcal{R}(\vec{a})^{\mathbf{n} \leftarrow i}$ denotes the process obtained from the process expression $\mathcal{R}$ by replacing the security parameter $\mathbf{n}$ by a value i and replacing the variables $\vec{x}$ by values $\vec{a}$. Two process expressions $\mathcal{P}(\vec{x})$ and $\mathcal{Q}(\vec{x})$ are called *equivalent* or *indistinguishable*, written $\mathcal{P}(\vec{x}) \equiv \mathcal{Q}(\vec{x})$, iff for every polynomial $p(\mathbf{n})$ there exists i_0 such that $|\text{Prob}[\mathcal{P}(\vec{a})^{\mathbf{n} \leftarrow i} \leadsto 1] - \text{Prob}[\mathcal{Q}(\vec{a})^{\mathbf{n} \leftarrow i} \leadsto 1]| \leq 1/p(i)$ for every $i \geq i_0$ and every tuple $\vec{a}$ of bit strings.

We call machines which are neither master nor decision processes *regular*. A machine which is both master and decision is called a *master decision process*. In what follows, by **R**, **M**, **D**, and **MD** we denote the set of all closed regular processes, closed master processes, open or closed decision processes, and open or closed master decision processes, respectively.

3 The Security Notions and Their Variants

In this section, we formulate the security notions and their variants. In order to do so, we must first define which SPPC expressions constitute well-formed systems of interacting machines. We will do so by specifying how machines are connected together.

We start by defining the communication interfaces of individual processes. A process uses directional external channels—*input* and *output external channels*—to communicate with other machines. These channels are partitioned into two types: *network channels* and *IO channels*. The channels start and decision are respectively defined to be input and output IO channels. Input channels connect to identically-named output channels of the same type.

If $\mathcal{P}$ represents a protocol, then an adversary $\mathcal{A}$ connects to the network channels of $\mathcal{P}$ while an environment $\mathcal{E}$ connects to the IO channels of $\mathcal{P}$ and $\mathcal{A}$. Formally, two processes, $\mathcal{P}$ and $\mathcal{Q}$, are *compatible* if they have the same set of external channels. We say that two processes are *IO-compatible* if they have the same set of IO channels and disjoint sets of network channels. A process expression $\mathcal{Q}$ is *connectible* for $\mathcal{P}$ if each common external channel of $\mathcal{P}$ and $\mathcal{Q}$ has the same type in both and complementary directions. A process expression $\mathcal{A}$ is *adversarially connectible* for $\mathcal{P}$ if $\mathcal{A}$ is connectible for $\mathcal{P}$ and the set of external channels of $\mathcal{A}$ is disjoint from the set of IO channels of $\mathcal{P}$. Thus an adversary can only connect on the network channels of a protocol. Similarly, $\mathcal{E}$ is *environmentally connectible* for $\mathcal{P}$ if it can only connect on the IO channels of $\mathcal{P}$.

We can now define what it means for a process to be an adversary, environment, or simulator. We do so in a parametric fashion so that we can succinctly represent the variants of a security notion. Given a set of processes $\mathbf{C}$ we define: a) $Env_{\mathbf{C}}(\mathcal{P})$ to be the set of all processes in $\mathbf{C}$ that are environmentally connectible for $\mathcal{P}$, b) $Adv_{\mathbf{C}}(\mathcal{P})$ to be the set of all processes in $\mathbf{C}$ that are adversarially connectible for $\mathcal{P}$, c) $Sim_{\mathbf{C}}(\mathcal{P}, \mathcal{F})$ to be the set of all processes $\mathcal{S}$ in $\mathbf{C}$ that are adversarially connectible for $\mathcal{F}$ and such that $\mathcal{S} \upharpoonright \mathcal{F}$ is compatible with $\mathcal{P}$, d) $Con_{\mathbf{C}}(\mathcal{P})$ to be the set of all processes in $\mathbf{C}$ that are connectible for $\mathcal{P}$.

Definition 1. *Let* **A** *(real adversaries),* **I** *(ideal adversaries),* **E** *(environments), and* **S** *(simulators) be sets of process expressions, and* $\mathcal{P}$ *(the real protocol) and* $\mathcal{F}$ *(the ideal functionality/protocol) be IO-compatible process expressions.*

Strong Simulatability (SS): $\mathrm{SS}_{(\mathbf{S},\mathbf{E})}(\mathcal{P}, \mathcal{F})$ *iff* $\exists \mathcal{S} \in Sim_{\mathbf{S}}(\mathcal{P}, \mathcal{F}) \forall \mathcal{E} \in Con_{\mathbf{E}}(\mathcal{P}) : \mathcal{E} \upharpoonright \mathcal{P} \equiv \mathcal{E} \upharpoonright \mathcal{S} \upharpoonright \mathcal{F}$, *i.e., there exists a simulator such that no environment can distinguish whether it is interacting with the real protocol or the ideal functionality-simulator combination.*

Strong Blackbox Simulatability (SBB): $\mathrm{SBB}_{(\mathbf{A},\mathbf{S},\mathbf{E})}(\mathcal{P}, \mathcal{F})$ *iff* $\exists \mathcal{S} \in Sim_{\mathbf{S}}(\mathcal{P}, \mathcal{F})$ $\forall \mathcal{A} \in Adv_{\mathbf{A}}(\mathcal{P}) \forall \mathcal{E} \in Env_{\mathbf{E}}(\mathcal{A} \upharpoonright \mathcal{P}) : \mathcal{E} \upharpoonright \mathcal{A} \upharpoonright \mathcal{P} \equiv \mathcal{E} \upharpoonright \mathcal{A} \upharpoonright \mathcal{S} \upharpoonright \mathcal{F}$, *i.e., there exists a simulator such that for all adversaries, no environment can distinguish whether it is interacting with the real protocol-adversary combination or the ideal functionality-simulator-adversary combination.*

Weak Blackbox Simulatability (WBB): $\mathrm{WBB}_{(\mathbf{A},\mathbf{S},\mathbf{E})}(\mathcal{P}, \mathcal{F})$ *iff* $\forall \mathcal{A} \in Adv_{\mathbf{A}}(\mathcal{P}) \exists \mathcal{S} \in Sim_{\mathbf{S}}(\mathcal{P}, \mathcal{F}) \forall \mathcal{E} \in Env_{\mathbf{E}}(\mathcal{A} \upharpoonright \mathcal{P}) : \mathcal{E} \upharpoonright \mathcal{A} \upharpoonright \mathcal{P} \equiv \mathcal{E} \upharpoonright \mathcal{A} \upharpoonright \mathcal{S} \upharpoonright \mathcal{F}$, *i.e., for each adversary there exists a simulator such that no environment can distinguish whether it is interacting with the real protocol-adversary combination or the ideal functionality-simulator-adversary combination.*

Universal Composability (UC): $\mathrm{UC}_{(\mathbf{A},\mathbf{I},\mathbf{E})}(\mathcal{P}, \mathcal{F})$ *iff* $\forall \mathcal{A} \in Adv_{\mathbf{A}}(\mathcal{P})$ $\exists \mathcal{I} \in Sim_{\mathbf{I}}(\mathcal{A} \upharpoonright \mathcal{P}, \mathcal{F})$ $\forall \mathcal{E} \in Env_{\mathbf{E}}(\mathcal{A}|\mathcal{P}) : \mathcal{E} \upharpoonright \mathcal{A} \upharpoonright \mathcal{P} \equiv \mathcal{E} \upharpoonright \mathcal{I} \upharpoonright \mathcal{F}$, *i.e., for each real adversary there exists an ideal adversary such that no environment can distinguish whether it is interacting with the real protocol-real adversary combination or the ideal functionality-ideal adversary combination.*

4 Relationships Between the Security Notions

In this section, we examine the relationships between the security notions introduced in the previous section. The instances of each security notion are obtained by assigning roles (decision, master, regular) to the various entities (environment, real and ideal adversary, simulator, real and ideal protocol). The environment is always the decision process, and the real and ideal protocols are always regular processes. So, the variants of each security notion differ only wrt the entity that assumes the role of the master process. Formally, the variants are obtained by defining the sets $\mathbf{A}$, $\mathbf{I}$, $\mathbf{S}$, and $\mathbf{E}$ to be one of the sets $\mathbf{R}$, $\mathbf{M}$, $\mathbf{D}$, and $\mathbf{MD}$. For example, the security notion considered in [8] is $\text{UC}_{(\mathbf{R},\mathbf{R},\mathbf{MD})}(\mathcal{P}, \mathcal{F})$. Here, $\mathcal{P}$ and $\mathcal{F}$ are UC with the environment as the master decision process and the real and ideal adversary as regular processes. Another variant of UC considered in [4] is $\text{UC}_{(\mathbf{M},\mathbf{M},\mathbf{MD})}(\mathcal{P}, \mathcal{F})$. Here, the environment is the master decision process and the real and ideal adversaries are the master processes. Notice that although both the environment $\mathcal{E}$ and the real/ideal adversary $\mathcal{A}/\mathcal{I}$ may play the role of the master process, in any specific setting, according to our definitions, exactly one of $\mathcal{E}$ or $\mathcal{A}/\mathcal{I}$ will actually be the master process. If the real adversary is the master process, then the ideal adversary must be master as well; furthermore, the environment cannot be master (as it would not be environmentally valid if it also had the start channel). Conversely, if the adversary is not the master (i.e., does not have the start channel), then the environment may be the master. Note that combinations without a master process or a decision process do not make sense since no computation can take place in their absence or no decision can be generated, i.e., no process can write on the channel decision. Henceforth, we omit such combinations. We will consider combinations where we *require* a certain entity to play the role of the master by saying that this entity is a process expression in $\mathbf{M} \setminus \mathbf{R}$. In variants of these notions where the simulator $\mathcal{S}$ plays the role of the master process, we allow the simulator to hand over control to $\mathcal{A}$ or $\mathcal{E}$ via a channel start', which replaces start in $\mathcal{A}$ and $\mathcal{E}$. Additionally, we also consider a variant of WBB where the simulator $\mathcal{S}$ only depends on the complexity of $\mathcal{A}$ rather than on $\mathcal{A}$ in its entirety. We can easily show that these two variants are equivalent [14], and so we will not distinguish between them in what follows.

In the following theorems, the security notions are considered to be binary relations over the set of regular processes $\mathbf{R}$.

Theorem 1. *All variants of Strong Simulatability and Strong Blackbox obtained by varying the entity that is the master process are equivalent, i.e., the following identities hold:*

$$\begin{aligned}
\text{SS}_{(\mathbf{R},\mathbf{MD})} &= \text{SS}_{(\mathbf{M},\mathbf{MD})} = \text{SBB}_{(\mathbf{R},\mathbf{R},\mathbf{MD})} = \text{SBB}_{(\mathbf{M},\mathbf{R},\mathbf{MD})} = \\
\text{SBB}_{(\mathbf{M},\mathbf{R},\mathbf{D})} &= \text{SBB}_{(\mathbf{M}\setminus\mathbf{R},\mathbf{R},\mathbf{D})} = \text{SBB}_{(\mathbf{M}\setminus\mathbf{R},\mathbf{R},\mathbf{MD})} = \text{SBB}_{(\mathbf{M},\mathbf{M},\mathbf{MD})} = \\
\text{SBB}_{(\mathbf{R},\mathbf{M},\mathbf{MD})} &= \text{SBB}_{(\mathbf{M},\mathbf{M},\mathbf{D})}.
\end{aligned}$$

We call this class of security notions SS-SBB. It includes placements of the master process as considered for Strong Blackbox in [4] and [22]. In [4], the environment, the adversary, and the simulator may play the role of the master process, and hence, this corresponds to the notion $\text{SBB}_{(\text{M},\text{M},\text{MD})}$. In [22], only the adversary and the simulator may be the master process, but not the environment, and hence, this corresponds to the notion $\text{SBB}_{(\text{M},\text{M},\text{D})}$.

Recall that the difference between SS and SBB is that, in the latter notion, the environment and the adversary are separate entities, while in the former they are combined into one. Since the adversary and the environment can communicate freely, it is perhaps expected that the two notions should be equivalent. Theorem 1 bears out this intuition, and shows that the equivalences among the notions are independent of which entity plays the role of the master process. Consequently, there appears to be no technical benefit from treating the adversary and environment as two separate entities as in the SBB setting. We point out that in order to prove Theorem 1, it is important that situations in which the simulator is a master process do not differentiate between SS and SBB. This is true because such situations yield degenerate relations; for instance, $\text{SS}_{(\text{M}\backslash\text{R},\text{MD})}$ is an empty relation. The reason is that the environment can exhaust a master simulator since whenever execution defaults to the simulator, it triggers the environment immediately. Hence the environment can repeatedly "ping" the simulator until its time-bound is exhausted. In a model in which the runtime of the simulator may depend on how often it is invoked by the environment, such exhaustion attacks would not be viable. In such a model, the notions SS and SBB may differ depending on whether or not the simulator plays the role of the master. Since such extensions are not studied in published work, we defer a detailed study to future work.

Theorem 2. *All variants of Universal Composability and Weak Blackbox in which the environment may be the master process are equivalent, i.e., the following identities hold:*

$$\text{UC}_{(\text{R},\text{R},\text{MD})} = \text{UC}_{(\text{M},\text{M},\text{MD})} = \text{WBB}_{(\text{R},\text{R},\text{MD})} = \text{WBB}_{(\text{M},\text{R},\text{MD})} = \text{WBB}_{(\text{M},\text{M},\text{MD})} = \text{WBB}_{(\text{R},\text{M},\text{MD})}.$$

We call this class of security notions UC-WBB_{env}. It includes placements of the master process as considered for Universal Composability in [8] and [4]. While in [8] only the environment may play the role of the master process, corresponding to the notion $\text{UC}_{(\text{R},\text{R},\text{MD})}$, in [4] the adversary may play this role as well, corresponding to $\text{UC}_{(\text{M},\text{M},\text{MD})}$.

The fact that WBB implies UC follows simply by combining the simulator and real adversary to produce an ideal adversary. To go in the reverse direction, we consider what happens when instantiating the real adversary with a process that simply forwards messages between the protocol and the environment. The corresponding ideal adversary then serves as the simulator in the definition of WBB. We note that it is important that the runtime of the simulator be allowed to depend on the complexity of the real adversary. Note that as in the class SS-SBB, equivalence among the security notions in UC-WBB_{env} holds independently of whether or not the simulator may be the master process.

Theorem 3. *All variants of Universal Composability and Weak Blackbox in which the simulator and the adversary may be the master process and the environment is not the master process are equivalent, i.e., the following identities hold:*

$$UC_{(M,M,D)} = UC_{(M\backslash R, M\backslash R, D)} = UC_{(M\backslash R, M\backslash R, MD)} = WBB_{(M,M,D)} = WBB_{(M\backslash R, M, D)}.$$

We call this class of security notions UC-WBB$_{sim}$. It includes the placements of the master process as considered for Universal Composability in [22], which corresponds to the notion $UC_{(M,M,D)}$.

Equivalence among the notions in UC-WBB$_{sim}$ is established similarly to the class UC-WBB$_{env}$. Note that UC-WBB$_{sim}$ does not contain a version of WBB where the simulator is restricted to be regular. As we will see, restricting the simulator in this way, yields a strictly stronger notion.

Theorem 4. *All variants of Weak Blackbox where the adversary is the master process and neither the environment nor the simulator is the master process are equivalent, i.e., the following identities hold:* $WBB_{(M,R,D)} = WBB_{(M\backslash R, R, D)}$.

We call this class of security notions WBB$_{adv}$. We now study the relationships between the four classes. We say that a class of equivalent security notions $\mathcal{C}$ implies another class $\mathcal{C}'$ of equivalent security notions ($\mathcal{C} \Rightarrow \mathcal{C}'$), if a notion in $\mathcal{C}$ (and hence, every notion in $\mathcal{C}$) implies a notion in $\mathcal{C}'$ (and hence, every notion in $\mathcal{C}'$).

Theorem 5. *SS-SBB $\Rightarrow$ UC-WBB$_{env}$ $\Rightarrow$ WBB$_{adv}$ $\Rightarrow$ UC-WBB$_{sim}$, but the class UC-WBB$_{sim}$ does not imply the other classes, i.e., UC-WBB$_{sim}$ $\not\Rightarrow$ WBB$_{adv}$, and hence, UC-WBB$_{sim}$ $\not\Rightarrow$ UC-WBB$_{env}$ and UC-WBB$_{sim}$ $\not\Rightarrow$ SS-SBB*

In particular, we have that the Strong Blackbox relation with the placements of the master process as considered in [22,4] implies the Universal Composability relation with the placements of the master process as considered in [8,22,4]. Also, the Universal Composability relation with the placement of the master process as considered in [22] is strictly weaker than the Universal Composability relation with the placements of the master process as considered in [8] and [4].

The argument from SS-SBB to UC-WBB$_{env}$ relies on the order of quantification over the entities. The fact that UC-WBB$_{env}$ implies WBB$_{adv}$ relies on the observation that making the environment the master intuitively gives the environment more discriminatory power. The final implication follows from the fact that the set of simulators considered in WBB$_{adv}$ is a subset of the set of simulators considered in UC-WBB$_{sim}$.

To show that UC-WBB$_{sim}$ $\not\Rightarrow$ WBB$_{adv}$ we provide a concrete example. Consider a protocol $\mathcal{P}$ that receives a bit on an IO channel and forwards it on a network channel. The ideal functionality $\mathcal{F}$ does the same but only forwards the bit if it is 1. We can show that $UC_{(M,M,D)}(\mathcal{P}, \mathcal{F})$ but not $WBB_{(M,R,D)}(\mathcal{P}, \mathcal{F})$. It is open whether WBB$_{adv}$ implies UC-WBB$_{env}$.

The following theorem identifies a necessary and sufficient condition—the forwarder axiom (defined below)—for the equivalence of UC-WBB$_{env}$ and SS-SBB.

As a consequence, we can see that the strongest variant of UC (where the environment is the master process) implies SBB just when the forwarder axiom holds.

Theorem 6. *Let* **C** *be a class of regular processes closed under channel-renaming. Then, restricting the relations to* **C**, *we obtain that*

$$(\textit{UC-WBB}_{env} \implies \textit{SS-SBB}) \textit{ iff } (\textit{FORWARDER holds for all processes in } \mathbf{C}).$$

In particular, the Universal Composability relations with the placements of the master process as considered in [8] and [4] are strictly weaker than the Strong Blackbox relations with the placements of the master process as considered in [4] and [22] in any computational model in which the forwarding property given by the FORWARDER axiom does not hold. The axiom FORWARDER(**C**) for a class of regular processes **C** is stated as follows:

FORWARDER(**C**). Given any process $\mathcal{P} \in \mathbf{C}$ with network channels *net*, there exists a process $\mathcal{D}$ with network channels $net \cup (net' = \{c' | c \in net\})$ such that for all $\mathcal{E}$ whose only shared channels with $\mathcal{P}$ are external channels of $\mathcal{P}$:

$$\mathcal{E} \upharpoonright \mathcal{P} \equiv \mathcal{E} \upharpoonright \mathcal{D} \upharpoonright [net'/net]\mathcal{P}$$

where $[net'/net]\mathcal{P}$ denotes the process obtained from $\mathcal{P}$ by replacing the channels in *net* by those in *net'*.

Intuitively, this axiom allows us to invisibly plug a communication medium $\mathcal{D}$ between two entities connected over network channels.

While the forwarder property appears believable, it turns out that for arbitrary protocols, the forwarder property does not hold. The problem lies in the fact that the forwarder is chosen independently of the environment $\mathcal{E}$. As a result its runtime is fixed a priori and the environment can exhaust the forwarder by sending it many useless messages. Then, the presence of the forwarder can be easily detected by the environment. All is not lost. For a class of protocols including those commonly studied in the literature (c.f., [22, 8, 5]) the forwarder property holds in SPPC. We shall refer to these protocols as *standard* protocols. The specific way in which the exhaustion problem is avoided for standard protocols involves using guards to reject the spurious messages. We cannot do this for every protocol because it is important that the forwarder knows the communication structure of the protocol (see [14] for details). The following corollary to Theorem 6 is now immediate.

Corollary 1. *SS-SBB* $\Leftrightarrow$ *UC-WBB*$_{env}$ *for the class of standard protocols.*

In particular, the Strong Blackbox relation with the placement of the master process as considered in [22, 4] and the Universal Composability relation with the placement of the master process as considered in [8, 4] are equivalent for standard protocols in SPPC (see also Section 5.1).

Although in this extended abstract, we have only given intuitions behind some of the proofs, we emphasize that the actual proofs are carried out using

$\mathcal{P} \uparrow \mathcal{Q} \equiv \mathcal{Q} \uparrow \mathcal{P}$ \hfill COM
$\mathcal{P} \uparrow (\mathcal{Q} \uparrow \mathcal{R}) \equiv (\mathcal{P} \uparrow \mathcal{Q}) \uparrow \mathcal{R}$ \hfill ASC
$\mathcal{P} \equiv \mathcal{Q}, \mathcal{Q} \equiv \mathcal{R} \implies \mathcal{P} \equiv \mathcal{R}$ \hfill TRN
$\mathcal{P} \equiv \mathcal{Q} \implies \mathcal{Q} \equiv \mathcal{P}$ \hfill SYM
$\mathcal{P} \equiv [d/c]\mathcal{P}$ where $c, d \notin \{\texttt{start}, \texttt{decision}\}, d \notin Channels(\mathcal{P})$ \hfill RENAME

Fig. 3. A representative fragment of SPPC's reasoning system

an equational reasoning system for SPPC. A small representative fragment of the axiom system is given below. These axioms capture simple structural properties like commutativity, associativity, transitivity, and symmetry of process equivalence. It also allows structural operations such as channel-renaming.

These axioms can also serve as an abstract specification of a "reasonable" computational model for simulation-based security.

5 Implications for Other Models

We now study the relationships of the security notions for the PIOA model [22, 5, 4] and the PITM model [8, 9]. The simplicity of SPPC's axiom system enables us to carry over our results to these other computational models. Furthermore, the counter-examples used to demonstrate that certain notions are strictly stronger than others are quite simple and easily translate into the related models.

5.1 The PIOA Model

Most of the axioms used to prove the relationships among the security notions also hold in the different versions of the PIOA model. Therefore, the relationships given in the previous section mostly carry over to PIOA. In particular, we obtain that all the security notions in SS-SBB, UC-WBB$_{sim}$, and WBB$_{adv}$ are equivalent, respectively, and their relationships are as depicted in Figure 1.

However, an axiom used to prove that the security notions in UC-WBB$_{env}$ are equivalent does not hold in PIOA. This axiom captures a property similar to the forwarder property discussed in Section 4. This axiom essentially states that there exists a forwarder $\mathcal{D}$ that is allowed to depend on the complexity of the protocol $\mathcal{P}$ and the adversary $\mathcal{A}$ such that $\mathcal{E} \uparrow \mathcal{A} \uparrow \mathcal{P} \equiv \mathcal{E} \uparrow \mathcal{A}' \uparrow \mathcal{D} \uparrow \mathcal{P}$ (where $\mathcal{A}'$ is $\mathcal{A}$ with some renamed network channels). The axiom fails in PIOA because machines always have to communicate through buffers and buffers are triggered by machines other than the one writing into the buffer. In fact, we show that UC does *not* imply WBB in PIOA when the environment is master. This failure of equivalence seems counterintuitive. The problem vanishes if the PIOA model is modified so that machines always trigger their own buffers. In effect, this is equivalent to not having buffers at all, which is why we call this fragment of the PIOA model the *buffer-free PIOA model (BFPIOA)*. This fragment is essentially as expressive as PIOA and this fragment can be embedded into SPPC (see [14]). In particular, *all* axioms (except the forwarder property of the previous section) are satisfied in BFPIOA and the examples used to prove separation results can also be expressed in BFPIOA. As mentioned in Section 2, starting from the work [6] PIOA (and thus, BFPIOA) has a restricted form of guards. Similar to SPPC, this mechanism suffices to satisfy the forwarder property for standard protocols, but just as in SPPC, there are protocols expressible in BFPIOA which do not satisfy this property. In summary, we obtain for BFPIOA exactly the same relationships as for SPPC (see Figure 1).

In [22], the security notions $\text{UC}_{(M,M,D)}(\mathcal{P},\mathcal{F})$ and $\text{SBB}_{(M,M,D)}(\mathcal{P},\mathcal{F})$ were introduced for the PIOA model, while in [4] the notions $\text{UC}_{(M,M,MD)}(\mathcal{P},\mathcal{F})$ and $\text{SBB}_{(M,M,MD)}(\mathcal{P},\mathcal{F})$ were considered. Our results clarify the relationships between these security notions: while the two variants of SBB are equivalent (they both belong to the class SS-SBB), these notions are different from the two variants of UC. Also, the two variants of UC are not equivalent. Our results contradict the claim in [4] that $\text{SBB}_{(M,M,MD)}(\mathcal{P},\mathcal{F})$ and $\text{UC}_{(M,M,MD)}(\mathcal{P},\mathcal{F})$ are equivalent.

5.2 The PITM model

The PITM model [8] is tailored towards defining UC where the environment is a master process and the adversaries are regular processes i.e., $\text{UC}_{(R,R,MD)}(\mathcal{P},\mathcal{F})$. Depending on which entities are involved, different computational models are defined: the real model (involving the environment, the real adversary, and the real protocol), the ideal model (involving the environment, the ideal adversary, and the ideal functionality together with dummy parties), and the hybrid model which is a combination of the previous two models.

Therefore, it is not immediately clear how the security notions SS, SBB, and WBB, which involve a simulator, would be defined in PITM. Different variants are possible, and as we have seen, differences in the definitions may affect the relationships between the security notions. It is out of the scope of this paper, to extend PITM in order to define SS, SBB, and WBB. However, some general points can be made. The version of PITM in [8] does not have a mechanism, like the guards of SPPC, that will enable the forwarder property to be satisfied. Without this property, UC is a strictly weaker notion than SBB. In ongoing work [9], Canetti allows PITMs to depend on the number of invocations as well as the length of messages on the IO tapes. This mechanism could enable PITM to satisfy the forwarder property whence UC would imply SBB. However, this is speculative since the details of the model are still being developed.

We finally note that in [8], Canetti introduces a special case of UC where the adversary merely forwards messages between the environment and the parties. Canetti proves UC and this notion equivalent. This notion can easily be formulated in SPPC and proved equivalent to UC along the lines of the proof which shows that $\text{UC}_{(R,R,MD)}(\mathcal{P},\mathcal{F})$ implies $\text{WBB}_{(M,R,MD)}(\mathcal{P},\mathcal{F})$.

6 Reactive Simulatability and Extensions of SPPC

In this section, we consider another security notion, called *reactive simulatability* in [5] and *security with respect to specialized simulators* in [9]. This notion has not drawn as much attention as the others studied in the present work because, to our best knowledge, a general composition theorem along the lines of [22, 4, 8, 14], has not been proved for reactive simulatability (see, however, [9]). Therefore, in the previous sections, we have concentrated on the other security notions and only very briefly cover reactive simulatability here. In our terminology, reactive simulatability is defined as follows:

Reactive Simulatability: $\text{RS}_{(\mathbf{A},\mathbf{I},\mathbf{E})}(\mathcal{P},\mathcal{F})$ iff $\forall \mathcal{A} \in Adv_\mathbf{A}(\mathcal{P}) \forall \mathcal{E} \in Env_\mathbf{E}(\mathcal{A} \upharpoonright \mathcal{P})$
$\exists \mathcal{I} \in Adv_\mathbf{I}(\mathcal{F}) : \mathcal{E} \upharpoonright \mathcal{A} \upharpoonright \mathcal{P} \equiv \mathcal{E} \upharpoonright \mathcal{I} \upharpoonright \mathcal{F}$.

The only difference between reactive simulatability and universal composability (UC) is that in the former the ideal adversary is allowed to depend on the environment. It has been pointed out by Canetti [9] that reactive simulatability is equivalent to UC if the runtime of the environment may depend on the length of the message on its input tape.[3] In such a model, the notion of indistinguishability has to be slightly modified. The idea of the proof of equivalence is that one can define a *universal environment* which interprets part of its input as an encoding of another environment. The ideal adversary corresponding to this environment, in effect, works for all environments. It is straightforward to extend SPPC in a way that the runtime of *open* processes (recall that the environment is modeled as an open process), may depend on the length of the messages substituted for the free variables. Thus, the same proof also works in SPPC. We note that the argument goes through regardless of whether the environment may or may not play the role of the master process. In the former case, reactive simulatability is equivalent to the notions in the class UC-WBB$_{env}$, and in the latter case, it is equivalent to the notions in UC-WBB$_{sim}$. This result also carries over to an appropriate extension of BFPIOA.

7 Conclusion

We have carried out a thorough study of the relationships among various notions of simulation-based security, identifying two properties of the computational model that determine equivalence between these notions. Our main results are that all variants of SS (strong simulatability) and SBB (strong black box simulatability) are equivalent, regardless of the selection of the master process, and they imply UC (universal composability) and WBB (weak black box simulatability). Conditions UC and WBB are equivalent as long as the role (master process or not) of the environment is the same in both. However, the variant of UC in which the environment may be a master process (as in [8, 4]) is strictly stronger than the variants in which the environment must not assume this role (as in [22]). In addition, the weaker forms of WBB do not imply SS/SBB. Finally,

---

[3] In his new model, Canetti allows every interacting Turing machine to depend on the number of invocations on input tapes and the length of the messages on input tapes. However, to prove the equivalence, it suffices to require this only for the environment.

we prove a necessary and sufficient condition for UC/WBB to be equivalent to SS/SBB, based on the ability to define forwarders. These results all show that the relationship between universal composability and black-box simulatability is more subtle than previously described. In particular, the composability theorem of Canetti [8] does not necessarily imply that blackbox simulatability is a composable security notion over any computational model in which the forwarding property is not satisfied. Another technical observation is that making the environment the master process typically yields a stronger security notion. Hence, we recommend that in subsequent developments of the various models, the environment is always assigned the role of the master process.

Since our proofs are carried out axiomatically using the equational reasoning system developed for SPPC, we are able to apply the same arguments to suitably modified versions of the alternative computational models. We emphasize that the our suggested modifications to the other systems are motivated by the failure, in those systems, of simple equational principles. In particular, it seems reasonable to adopt a buffer-free variant of PIOA.

While our study concentrates on models where the runtime of processes is bounded by a polynomial in the security parameter, it would be interesting to consider those models where the runtime may depend on the number of invocations and the length of inputs (e.g., [9]). We believe that most of our results carry over also to these models as they seem to satisfy the axioms that we use in our proofs. However, the issue remains open since the details of these models have not yet been fixed.

Acknowledgments. We thank Michael Backes, Ran Canetti, Birgit Pfitzmann, Andre Scedrov and Vitaly Shmatikov for helpful discussions.

References

1. Martín Abadi and Cédric Fournet. Mobile values, new names, and secure communication. In *POPL 2001*, pages 104–115, 2001.
2. Martín Abadi and Andrew D. Gordon. A bisimulation method for cryptographic protocol. In *Proc. ESOP 98*, Lecture notes in Computer Science. Springer, 1998.
3. Martín Abadi and Andrew D. Gordon. A calculus for cryptographic protocols: the spi calculus. *Information and Computation*, 143:1–70, 1999. Expanded version available as SRC Research Report 149 (January 1998).
4. M. Backes, B. Pfitzmann, and M. Waidner. A General Composition Theorem for Secure Reactive Systems. In *TCC 2004*, volume 2951 of *LNCS*, pages 336–354. Springer, 2004.
5. M. Backes, B. Pfitzmann, and M. Waidner. Secure asynchronous reactive systems. Technical Report 082, Eprint, 2004.
6. Michael Backes, Birgit Pfitzmann, Michael Steiner, and Michael Waidner. Polynomial fairness and liveness. In *CSFW-15 2002*, pages 160–174, 2002.
7. Michael Backes, Birgit Pfitzmann, and Michael Waidner. Reactively secure signature schemes. In *Proceedings of 6th Information Security Conference*, volume 2851 of *LNCS*, pages 84–95. Springer, 2003.

8. Ran Canetti. Universally composable security: A new paradigm for cryptographic protocols. In *FOCS 2001*. IEEE, 2001. Full version available at http://eprint.iacr.org/2000/067/.
9. Ran Canetti. Personal communication, 2004.
10. Ran Canetti and Marc Fischlin. Universally composable commitments. In *Proc. CRYPTO 2001*, volume 2139 of *LNCS*, pages 19–40, 2001. Springer.
11. Ran Canetti and Hugo Krawczyk. Universally composable notions of key exchange and secure channels. In *EUROCRYPT 2002*, volume 2332 of *LNCS*, pages 337–351. Springer, 2002.
12. Ran Canetti, Eyal Kushilevitz, and Yehuda Lindell. On the limitations of universally composable two-party computation without set-up assumptions. In *EUROCRYPT 2003*, volume 2656 of *LNCS*, pages 68–86. Springer, 2003.
13. Ran Canetti, Yehuda Lindell, Rafail Ostrovsky, and Amit Sahai. Universally composable two-party and multi-party secure computation. In *STOC 2002*, pages 494–503, 2002.
14. Anupam Datta, Ralf Küsters, John C. Mitchell, and Ajith Ramanathan. Sequential probabilisitic process calculus and simulation-based security. 2004. An extended version is available as a technical report at http://www.ti.informatik.uni-kiel.de/~kuesters/publications_html/DattaKuestersMitchellRamanathan-TR-SPPC-2004.ps.gz.
15. Anupam Datta, Ralf Küsters, John C. Mitchell, Ajith Ramanathan, and Vitaly Shmatikov. Unifying equivalence-based definitions of protocol security. In *ACM SIGPLAN and IFIP WG 1.7, 4th Workshop on Issues in the Theory of Security*, 2004.
16. C.A.R. Hoare. *Communicating Sequential Processes*. Prentice-Hall, 1985.
17. Patrick D. Lincoln, John C. Mitchell, Mark Mitchell, and Andre Scedrov. Probabilistic polynomial-time equivalence and security protocols. In *Formal Methods World Congress, vol. I*, number 1708 in LNCS, pages 776–793, 1999. Springer.
18. Robin Milner. *A Calculus of Communicating Systems*. Springer, 1980.
19. Robin Milner. *Communication and Concurrency*. International Series in Computer Science. Prentice Hall, 1989.
20. John C. Mitchell, Mark Mitchell, and Andre Scedrov. A linguistic characterization of bounded oracle computation and probabilistic polynomial time. In *FOCS 1998*, pages 725–733, 1998. IEEE.
21. John C. Mitchell, Ajith Ramanathan, Andre Scedrov, and Vanessa Teague. A probabilistic polynomial-time calculus for the analysis of cryptographic protocols (preliminary report). In *17th Annual Conference on the Mathematical Foundations of Programming Semantics, 2001*, volume 45. ENTCS, 2001.
22. B. Pfitzmann and M. Waidner. A Model for Asynchronous Reactive Systems and its Application to Secure Message Transmission. In *IEEE Symposium on Security and Privacy*, pages 184–200. IEEE Computer Society Press, 2001.
23. Ajith Ramanathan, John C. Mitchell, Andre Scedrov, and Vanessa Teague. Probabilistic bisimulation and equivalence for security analysis of network protocols. Unpublished, see http://www-cs-students.stanford.edu/~ajith/, 2004.
24. Ajith Ramanathan, John C. Mitchell, Andre Scedrov, and Vanessa Teague. Probabilistic bisimulation and equivalence for security analysis of network protocols. In *FOSSACS 2004*, 2004. Summarizes results in [23].

A New Cramer-Shoup Like Methodology for Group Based Provably Secure Encryption Schemes

María Isabel González Vasco[1], Consuelo Martínez[2], Rainer Steinwandt[3], and Jorge L. Villar[4]

[1] Área de Matemática Aplicada, Universidad Rey Juan Carlos,
c/Tulipán, s/n, 28933 Madrid, Spain
migonzalez@escet.urjc.es
[2] Departamento de Matemáticas, Universidad de Oviedo,
c/Calvo Sotelo, s/n, 33007 Oviedo, Spain
chelo@pinon.ccu.uninovi.es
[3] IAKS, Arbeitsgruppe Systemsicherheit Prof. Beth, Fakultät für Informatik,
Universität Karlsruhe,
76128 Karlsruhe, Germany
steinwan@ira.uka.de
[4] Departamento de Matemática Aplicada IV,
Universitat Politècnica de Catalunya, Campus Nord,
c/Jordi Girona, 1–3, 08034 Barcelona, Spain
jvillar@mat.upc.es

Abstract. A theoretical framework for the design of—in the sense of IND-CCA—provably secure public key cryptosystems taking non-abelian groups as a base is given. Our construction is inspired by Cramer and Shoup's general framework for developing secure encryption schemes from certain language membership problems; thus all our proofs are in the standard model, without any idealization assumptions. The skeleton we present is conceived as a guiding tool towards the construction of secure concrete schemes from finite non-abelian groups (although it is possible to use it also in conjunction with finite abelian groups).

1 Introduction

In the last few years, the outrageous development of cryptanalytic techniques has encouraged the search for theoretical models allowing for mathematical proofs of security. Ideally, a security model should take into account all possible attacks, including those performed on the physical device where the scheme is implemented (such as timing attacks, differential power analysis or attacks relying on the induction of faults) or those that could be carried out with non-standard computing resources like a quantum computer.

One step behind such an ideal model, the nowadays standard notion of security for public key encryption schemes (IND-CCA) abstracts the implementation-dependent characteristics and models the attacker in terms of probabilistic polynomial-time (ppt) algorithms. Building on ideas of Naor and Yung [19], IND-CCA security was introduced by Rackoff and Simon [20], who also presented a scheme secure in this sense. A scheme with similar properties was afterwards designed by Dolev, Dwork, and Naor [9]. Note that, equivalently, instead of IND-CCA sometimes the term IND-CCA2 or the notion of *semantic security against adaptive chosen ciphertext attacks* is used. As a standard reference for further details on formal security notions like IND-CCA1, NM-CPA, etc. we mention the paper of Bellare et al. [3].

Unfortunately, developing practical cryptosystems which can be proven to be IND-CCA secure is a highly non-trivial task, and therefore, idealized models of computation have been introduced in order to obtain simpler proofs yet reasonable security guarantees [10, 4]. The first 'realistic' (that is, practical) proposal without idealization hypothesis was that of Cramer and Shoup [7, 8], which uses the Decision Diffie-Hellman assumption as a base. The same authors gave later a very general construction which in particular led to the design of IND-CCA group theoretic schemes constructed from certain group based primitives called *group systems*. Essentially, such primitives are derived from hard subgroup membership problems of suitable abelian groups.

On the other hand, group theory has lately attracted a lot of attention as a potential source of cryptographic primitives. Having in mind the existing quantum algorithms for factoring integers and computing discrete logarithms, it is indeed worthwhile to explore different areas of mathematics in search of hard problems. Several proposals to use hard problems in non-abelian groups for public key encryption have been made, some based on word or factorization problems [22, 11, 18] and others on variants of the conjugacy problem in braid groups [2, 1, 15, 16]. Unfortunately, almost all of these have been proven insecure in some sense [13, 5, 14, 6]. In [12] common properties of some of these schemes have been exploited to identify a security flaw according to one of the standard security notions (malleability). A sound design framework could be very helpful to prevent this kind of flaws when developing new schemes based on non-abelian groups.

With this purpose in mind, below a theoretical framework for constructing IND-CCA secure public key schemes using finite not necessarily abelian groups is described. Our design is inspired by that of Cramer and Shoup [8, 7], but it is not a generalization of it. It is our aim to provide precise guidelines for developing group-based schemes with a sound theoretical basis, and we hope that the design presented here leads to practical and secure constructions as soon as reasonable hardness assumptions for certain group-based problems are identified.

2 Main Tools of Cramer and Shoup's Construction

The main building blocks of the public key cryptosystem introduced by Cramer and Shoup are so-called *projective hash families, subset membership problems* and *universal hash proof systems*. We include an informal summary of these notions and refer to [7, 8] for the corresponding definitions.

2.1 Projective Hash Families

Let X, Π be finite non-empty sets, and K some finite index set. Consider a family $H = \{H_k : X \longrightarrow \Pi\}_{k \in K}$ of mappings from X into Π, and let $\alpha : K \longrightarrow S$ be a map from K into some finite non-empty set S.

With this notation, for a given subset $L \subset X$, we refer to the tuple $\mathbf{H} = (H, K, X, L, \Pi, S, \alpha)$, as *projective hash family* (PHF) for (X, L) if for all $k \in K$ the restriction of H_k to L is determined by $\alpha(k)$, i.e., for all $x \in L$ and $k_1, k_2 \in K$ the equality $\alpha(k_1) = \alpha(k_2)$ implies $H_{k_1}(x) = H_{k_2}(x)$.

Next, we consider three concepts to limit the amount of information about the behavior of a map H_k on $X \setminus L$, given by $\alpha(k)$:

- We say that $\mathbf{H}$ is ε-*universal* if for any $x \in X \setminus L$ and for a uniformly at random chosen $k \in K$, the probability of correctly guessing $H_k(x)$ from x and $\alpha(k)$ is at most ε. In other words, $\alpha(k)$ determines $H_k \mid_L$ completely, but gives (almost) no information about $H_k \mid_{X \setminus L}$.
- We say $\mathbf{H}$ is ε-*universal*$_2$ if even knowing (besides $H_k \mid_L$) the value of H_k in some $x^* \in X \setminus L$, for any $x \in X \setminus (L \cup \{x^*\})$ the value of $H_k(x)$ can be guessed correctly with probability at most ε.
- Finally, we say that $\mathbf{H}$ is ε-*smooth* if the probability distributions of $(x, s, H_k(x))$ and (x, s, π), where k, x and π are chosen uniformly at random in K, $X \setminus L$ and Π, respectively, and $s = \alpha(k)$, are ε-close.

2.2 Subset Membership Problems

Many cryptosystems base their semantic security on a decisional assumption such as the Decision Diffie-Hellman (DDH) assumption or the Quadratic Residuosity (QR) assumption. Most of these assumptions can be formulated in terms of indistinguishability of two probability distributions. Namely, the uniform distribution on a set X and the uniform distribution on a subset $L \subset X$. For instance, if G is a cyclic group of prime order p and g_1 and g_2 are two randomly selected generators of G, the DDH assumption on G is formalized by setting $X = G \times G$ and $L = \langle (g_1, g_2) \rangle$.

Since computational assumptions are in nature complexity theoretical statements, a complexity parameter $l \in \mathbb{N}_0$, (as the binary length of p in DDH) must be taken into account. Also, for each value of l, there are some possible instances of the same problem. The (random) choice of a particular instance for complexity parameter l is modelled by a samplable probability distribution I_l on the set of instance descriptions. In addition to a set X along with a subset $L \subset X$, an instance description Λ specifies a binary relation $\mathcal{R} \subseteq L \times W$, where W is a so-called *witness test* whose elements provide 'proofs of belonging' to the elements in L, that is, given $x \in L$, there is always a $w \in W$ that can be used to prove that x belongs to L.

Now, a *subset membership problem* $\mathcal{M}$ specifies a collection of distributions $(I_l)_{l \in \mathbb{N}_0}$ on the set of instance descriptions along with several sampling and verifying algorithms:

- a ppt algorithm called the *instance generator* that on input 1^l, outputs a description $\Lambda = \Lambda[X, L, W, \mathcal{R}]$ as just described;
- a ppt algorithm which, upon input of 1^l and a certain instance $\Lambda = \Lambda[X, L, W, \mathcal{R}]$, outputs a random $x \in L$ and a witness $w \in W$ for x (*subset sampling algorithm*);
- a deterministic polynomial time algorithm that takes as input 1^l, an instance $\Lambda = \Lambda[X, L, W, \mathcal{R}]$ and a binary string ζ, and checks whether ζ is a valid encoding of an element $x \in X$.

Moreover, $\mathcal{M}$ is *hard* if the probability distributions (Λ, x) and (Λ, x'), where $\Lambda = \Lambda[X, L, W, \mathcal{R}]$ is the output of the instance generator and x, x' are uniformly distributed on L and $X \setminus L$ respectively, are polynomially indistinguishable.

2.3 Universal Hash Proof Systems

A *hash proof system* (HPS) $\mathcal{P}$ is a rule which for a subset membership problem $\mathcal{M}$ associates to each instance $\Lambda = \Lambda[X, L, W, \mathcal{R}]$ of $\mathcal{M}$ a projective hash family $(H, K, X, L, \Pi, S, \alpha)$ for (X, L). In addition, $\mathcal{P}$ provides the following sampling and verifying algorithms which are polynomial in the complexity parameter l:

- a probabilistic algorithm that on input 1^l and Λ (with non-zero probability according to the corresponding distribution I_l) outputs $k \in K$ chosen uniformly at random;
- a deterministic algorithm that on input 1^l, Λ and k as above, outputs $\alpha(k) \in S$;
- a deterministic *private evaluation algorithm* that on input l, Λ, k as above, and $x \in X$ outputs $H_k(x) \in \Pi$;
- a deterministic *public evaluation algorithm* that on input 1^l, Λ as above, $s \in \alpha(K)$ and $x \in L$ together with a witness $w \in W$ for x, outputs $H_k(x) \in \Pi$ (where $\alpha(k) = s$);
- a deterministic algorithm that on input 1^l, Λ as above and a bitstring ζ determines if ζ is a valid encoding of an element of Π.

A hash proof system $\mathcal{P}$ is referred to as ε-*universal*, if the PHFs it associates to the instances of a subset membership problem $\mathcal{M}$ are 'almost' ε-*universal*. Namely, consider $\varepsilon : \mathbb{N}_0 \longrightarrow \mathbb{R}_{>0}$, a function of the complexity parameter l. Then we call $\mathcal{P}$ ε-*universal* (resp. *universal*$_2$, *smooth*) if there exists a negligible function $\delta(l)$ such that for all $l \in \mathbb{N}_0$ and all instances Λ of $\mathcal{M}$, the PHF **H** associated to Λ by $\mathcal{P}$ is $\delta(l)$-close to an $\varepsilon(l)$-universal (resp. universal$_2$, smooth) PHF. Moreover, if this is the case, and $\varepsilon(l)$ is a negligible function, then we say that $\mathcal{P}$ is *strongly* universal (resp. universal$_2$, smooth). Finally, it is convenient to provide an extended notion of hash proof systems obtained by simply replacing the sets X and L by $X \times E$ and $L \times E$ for a a suitable finite set E. Also, in these *extended hash proof systems* a value $e \in E$ is passed as an additional input to both the private and the public evaluation algorithm.

It is worth noticing that if a HPS is strongly universal and the underlying subset membership problem is hard, then the problem of evaluating $H_k(x)$ for random $k \in K$ and arbitrary $x \in X$ given only x and $\alpha(k)$ is also hard. Thus, the role of the witness in the public evaluation algorithm becomes clear: without w there is no way to efficiently compute $H_k(x)$.

2.4 Cramer and Shoup's IND-CCA Secure Public Key Encryption Scheme

Roughly speaking, in the scheme proposed by Cramer and Shoup [7, 8] a message $m \in \Pi$ is encrypted by using $H_k(x)$ as a one time pad; while the value of k is kept secret, x and $\alpha(k)$ are made public. More precisely, given a strongly smooth HPS for a hard subset membership problem, the secret key of the encryption scheme is $k \in K$, and the public key consists of $s = \alpha(k)$ along with the instance description. The message space is Π. To encrypt a message $m \in \Pi$, first a random pair $(x, w) \in L \times W$ is generated, so that w is a witness for x. Next, by means of the public evaluation algorithm, the value $H_k(x)$ is computed; the ciphertext is the pair $(x, m \cdot H_k(x))$, where $\cdot$ is a suitable group operation. Implicitly, it is assumed that Π is a group where elements can be efficiently inverted and multiplied.

Clearly, the holder of k can retrieve $H_k(x)$ by using the private evaluation algorithm, and therewith the message. On the other hand, since the subset membership problem is hard, there is no way for a polynomially bounded adversary to distinguish between a well-formed ciphertext and a fake ciphertext obtained by choosing $x \in X \setminus L$ instead of $x \in L$. However, due to the smoothness of the HPS, since k is unknown, $H_k(x)$ is close to be uniformly distributed on Π, so the message is nearly perfectly hidden. Therefore, no information about the plaintext can be obtained in polynomial time by a passive adversary.

IND-CCA security is achieved by appending to the ciphertext a 'proof of integrity' obtained from a strong universal$_2$ extended HPS. The set E in the definition of this extended HPS is just the message space Π. More formally: Let $\mathcal{M}$ be a hard subset membership problem and $\mathcal{P}, \hat{\mathcal{P}}$ be two HPSs for $\mathcal{M}$, strongly smooth and strongly universal$_2$ extended respectively. An instance of these objects is described by an instance $\Lambda[X, L, W, \mathcal{R}]$ of $\mathcal{M}$ and two instances $\mathbf{H} = (H, K, X, L, \Pi, S, \alpha)$ and $\hat{\mathbf{H}} = (\hat{H}, \hat{K}, X \times \Pi, L \times \Pi, \hat{\Pi}, \hat{S}, \hat{\alpha})$ of $\mathcal{P}$ and $\hat{\mathcal{P}}$, respectively. Note that the instances of $\mathcal{P}$ and $\hat{\mathcal{P}}$ must share the sets X, L and W and the sampling algorithm. Once the above parameters are fixed, the algorithms of the encryption scheme can be described as follows:

Key Generation Choose $k \in K$ and $\hat{k} \in \hat{K}$ uniformly at random, compute $s = \alpha(k) \in S$, $\hat{s} = \hat{\alpha}(\hat{k}) \in \hat{S}$ and output two pairs $(s, \hat{s})$—the public key—and $(k, \hat{k})$—the private key.

Encryption To encrypt a plaintext $m \in \Pi$, first generate $x \in L$ and a corresponding witness $w \in W$ by means of the subset sampling algorithm provided by $\mathcal{M}$. Then compute

- $\pi = H_k(x)$ (from x, s and w, by using the public evaluation algorithm provided by $\mathcal{P}$)
- $e = m \cdot \pi \in \Pi$ and $\hat{\pi} = \hat{H}_{\hat{k}}(x, e)$ (from $\hat{s}, x, e$ and w, by using the public evaluation algorithm provided by $\hat{\mathcal{P}}$).

The output ciphertext is the tuple $(x, e, \hat{\pi})$.

Decryption Algorithm To decrypt the received ciphertext $(x, e, \hat{\pi})$,
- compute $\hat{\pi}' = \hat{H}_{\hat{k}}(x, e) \in \hat{\Pi}$ (by means of the private evaluation algorithm of $\hat{\mathcal{P}}$),
- check whether $\hat{\pi} = \hat{\pi}'$ and, if not, output *reject* and halt. Otherwise, compute $\pi = H_k(x) \in \Pi$ (by means of the private evaluation algorithm of $\mathcal{P}$) as well as the plaintext $m = e \cdot \pi^{-1} \in \Pi$.

This algorithm is also supposed to recognize and reject bitstrings that do not correspond to properly formed ciphertexts, i.e., bitstrings that do not encode an element of $X \times \Pi \times \hat{\Pi}$.

3 Main Tools of a Non-abelian Construction Based on Group Automorphisms

In [8], Cramer and Shoup give a group-theoretic construction for deriving universal projective hash families from so-called *group systems*. Their construction is based on the use of finite abelian groups, and they prove that, if the group system has certain properties, then the corresponding PHF is ε-universal$_2$. We establish the same result for a different group-based primitive, which we call *automorphism group system*.

3.1 Automorphism Group Systems

Let X be a (not necessarily abelian) group. Multiplicative notation will be used for all groups, thus the unit element will be denoted by 1. Let H be a finite subgroup of $\mathrm{Aut}(X)$, S some finite group and $\chi: H \longrightarrow S$ a group homomorphism. Note that for any $\phi \in H$, $\chi(\phi)$ gives some (limited) information about ϕ.

Definition 1. *Let X, H, S and χ be defined as above. Then the tuple (X, H, χ, S) is called an* automorphism group system.

For any $\phi \in H$, let $[\phi] = \chi^{-1}(\chi(\phi))$ denote the class of ϕ in $H/\ker\chi$. Obviously, $|[\phi]| = |\ker\chi|$, and for any $x \in X$ and $\phi \in H$ we have

$$[\phi](x) = \{\psi(x) \mid \psi \in [\phi]\} = \phi((\ker\chi)(x)).$$

Denoting the orbit of x under the action of $\ker\chi$ by $[x]$, we have $|[\phi](x)| = |\phi([x])| = |[x]|$, as ϕ is a bijection. Clearly, $x \in [x]$ and hence $|[x]| \geq 1$; denote by L the set $\{x \in X \mid |[x]| = 1\}$, that is $\{x \in X \mid [x] = \{x\}\}$. Then it is trivial to check that L is a subgroup of X. Note also that, if $x, y \in X$ are in the same class modulo L, i.e., if $xL = yL$, then $|[x]| = |[y]|$.

Observe that the restriction of ϕ to L only depends on $\chi(\phi)$ and that $\ker\chi \subseteq \mathrm{Stab}(L)$ although they are not necessarily equal.

As the systems above will be useful for us if χ gives little information about the action of H on $X \setminus L$, we will be particularly interested in those systems for which the $(\ker \chi)$-orbits of elements in $X \setminus L$ are large.

Definition 2. *Let $p > 1$ be a positive integer. The automorphism group system (X, H, χ, S) is p-diverse if $|[x]| \geq p$ for all $x \in X \setminus L$.*

Lemma 1. *Let (X, H, χ, S) be an automorphism group system, and let p be the smallest prime dividing $|\ker \chi|$. Then (X, H, χ, S) is p-diverse.*

Proof. Note that $\ker \chi$ acts on X, and thus $|[x]|$ divides $|\ker \chi|$, so if $x \in X \setminus L$ (i.e., if $|[x]| \neq 1$) then $|[x]|$ is at least p. □

To get a better intuition of the notion of automorphism group system, we conclude this section with a simple (abelian) example in a setting analogue to [7–Section 7.4.2 Example 2]:

Example 1. Denote by X some cyclic group of composite order $a = b \cdot b'$ with $b < b'$ being different prime numbers, and let L be the (unique) subgroup of X of order b. Then X is isomorphic to $\mathbb{Z}/b\mathbb{Z} \times \mathbb{Z}/b'\mathbb{Z}$, and the automorphism group $H := \mathrm{Aut}(X)$ can be identified with $\mathbb{Z}/(b-1)\mathbb{Z} \times \mathbb{Z}/(b'-1)\mathbb{Z}$.

Thus, using this identification, define χ as the corresponding natural projection
$$\chi : \begin{array}{l} H \longrightarrow S := \mathbb{Z}/(b-1)\mathbb{Z} \\ (h_1, h_2) \longmapsto h_1 \end{array}.$$

Thus, the kernel of χ is isomorphic to $\mathbb{Z}/(b'-1)\mathbb{Z}$, and obviously each element of L is stabilized by $\ker \chi$. Moreover, one easily checks that any element having only a single image under $\ker \chi$ is already contained in L. In other words (X, H, χ, S) is an automorphism group system in the sense of Definition 2, and $L = \{x \in X \mid |[x]| = 1\}$. It is also easy to check that this automorphism group system is $(b'-1)$-diverse.

Remark 1. Note that Example 1 can easily be generalized to the case $X = A \times B$ for some not necessarily abelian finite groups A and B, $H = \mathrm{Aut}(A) \times \mathrm{Aut}(B)$, $S = \mathrm{Aut}(A)$ and χ the corresponding projection. Actually, in Example 1 we have $\gcd(|A|, |B|) = 1$, and therefore $H = \mathrm{Aut}(A) \times \mathrm{Aut}(B) = \mathrm{Aut}(X)$.

3.2 Automorphism Group Projective Hash Families

As it was the case for abelian group systems [7,8], a projective hash family can be built from an automorphism group system by providing some additional elements:

Let us consider an automorphism group system (X, H, χ, S), and denote by $\hbar : K \to H$ a bijection from a suitable index set K (which will later serve as the private key space). Noting that $\chi(\hbar(k))$ determines the action of $\hbar(k)$ on L completely, it is easy to see that the tuple $(H, K, X, L, X, S, \chi \circ \hbar)$ is a projective hash family.

Definition 3. *Any PHF constructed from an automorphism group system as described above is called* automorphism group projective hash family *(APHF)*.

An automorphism group projective hash family is made explicit by the tuple $(X, H, K, S, \chi, \hbar)$.

It is our aim to prove that, if the automorphism group projective hash family has certain nice properties, the resulting APHF will be ε-universal for some $\varepsilon > 0$. We start by demonstrating that for any $x \in X$, choosing $k \in K$ uniformly at random (that is, choosing uniformly at random a homomorphism in H), given $\chi(\hbar(k))$, there are exactly $|[x]|$ equally probable candidates for $(\hbar(k))(x)$.

Lemma 2. *Let (X, H, χ, S) be an automorphism group system and let $x \in X$. If $\phi \in H$ is chosen uniformly at random, once $s = \chi(\phi)$ is given then ϕ is uniformly distributed on the coset $\chi^{-1}(s)$ and $\phi(x)$ is uniformly distributed on the set $\{\psi(x) \mid \psi \in \chi^{-1}(s)\}$, that is, on a set of cardinality equal to $|[x]|$.*

Proof. Clearly, as ϕ is chosen uniformly at random, once we fix $s = \chi(\phi)$, the resulting distribution is uniform on $\chi^{-1}(s)$. Moreover, for any $x \in X$, $\phi(x)$ is uniformly distributed on

$$\{\psi(x) \mid \psi \in \chi^{-1}(s)\}$$

provided that the sets

$$S_y = \{\psi \in \chi^{-1}(s) \mid \psi(x) = y\}$$

for all $y \in \{\psi(x), \psi \in \chi^{-1}(s)\}$ are of the same size. But this is straightforward to see, as all S_y are left cosets modulo $\ker \chi \cap \mathrm{Stab}(\{x\})$. □

Proposition 1. *Let $\mathbf{H} = (X, H, K, S, \chi, \hbar)$ be an automorphism group projective hash family.*

If the underlying automorphism group system (X, H, χ, S) is p-diverse then $\mathbf{H}$ is $1/p$-universal.

Proof. From Lemma 2, for any $x \in X \setminus L$, the probability of guessing the right value of $(\hbar(k))(x)$ for a random choice of $k \in K$ given $\chi(\hbar(k))$ is $1/|[x]|$, that is at most $1/p$. □

In [8] a generic method to obtain a smooth projective hash family from any universal projective hash family, taking advantage of the Leftover Hash Lemma, is described. Nevertheless, in some special cases, the smoothness can be guaranteed directly.

Proposition 2. *Let $\mathbf{H} = (X, H, K, S, \chi, \hbar)$ be an automorphism group projective hash family. If the whole set $X \setminus L$ is a single orbit under the action of $\ker \chi$ then $\mathbf{H}$ is $|L|/|X|$-smooth.*

Proof. Let $x \in X \setminus L$. From Lemma 2, $(\hbar(k))(x)$ is uniformly distributed on a set of size $|[x]| = |X \setminus L|$. Then, the statistical distance between $(\hbar(k))(x)$ and the uniform distribution on X is

$$\frac{1}{2} \sum_{x \in X \setminus L} \left| \frac{1}{|X| - |L|} - \frac{1}{|X|} \right| + \frac{1}{2} \sum_{x \in L} \frac{1}{|X|} = \frac{|L|}{|X|},$$

thus the probability distribution of $(\hbar(k))(x)$ is $|L|/|X|$-close to the uniform distribution on X □

3.3 Universal$_2$ Extended Projective Hash Families

In [8], the authors outline a generic transformation from any ε-universal projective hash family to an ε-universal$_2$ extended projective hash family. But in the case of automorphism group projective hash families there is a more efficient way to achieve this goal.

Let $\mathbf{H} = (X, H, K, S, \chi, \hbar)$ be an automorphism group projective hash family such that the underlying automorphism group system (X, H, χ, S) is p-diverse. Let q be the smallest prime factor of $|H|$. Further on, denote by n a positive integer and by E a finite set. Let us define a new extended projective hash family $\hat{\mathbf{H}}$ by means of $n+1$ independent copies of $\mathbf{H}$ and a "gluing" function $g_\gamma^H : H^{n+1} \to H$ defined by:

$$g_\gamma^H(\phi_0, \ldots, \phi_n) := \phi_0 \circ \phi_1^{\gamma_1} \circ \cdots \circ \phi_n^{\gamma_n}$$

where $\gamma = (\gamma_1, \ldots, \gamma_n) \in \mathbb{Z}^n$ and $\phi_i^{\gamma_i}(x) := \underbrace{\phi_i \circ \cdots \circ \phi_i}_{\gamma_i}(x)$.

Similarly, we define $g_\gamma^S : S^{n+1} \to S$ by

$$g_\gamma^S(s_0, \ldots, s_n) := \chi(g_\gamma^H(\phi_0, \ldots, \phi_n)) = s_0 s_1^{\gamma_1} \cdots s_n^{\gamma_n},$$

where $\phi_j \in \chi^{-1}(s_j)$ for all $j = 0, \ldots, n$.

Now, $\hat{K} = K^{n+1}$, $\hat{S} = S^{n+1}$ and the natural extensions $\hat{\chi}$ of χ and $\hat{\hbar}$ of $\hbar$ are used. The set X is extended to $\hat{X} = X \times E$. Further on, given $\hat{k}$, we define $\Phi_{\hat{k}} : X \times E \longrightarrow X$ by

$$\Phi_{\hat{k}}(x, e) := g_{\Gamma(x,e)}^H(\hat{\hbar}(\hat{k}))(x),$$

where $\Gamma : (x, e) \mapsto (\Gamma_1(x, e), \ldots, \Gamma_n(x, e))$ is an injective map from $X \times E$ into $\{0, \ldots, q-1\}^n$. Let us denote by $\hat{H}$ the set $\{\Phi_{\hat{k}} \mid \hat{k} \in \hat{K}\}$.

The soundness of our construction will rely on the commutativity of the following diagram:

$$\begin{array}{ccc} H^{n+1} & \xrightarrow{g_\gamma^H} & H \\ \hat{\chi} \downarrow & & \downarrow \chi \\ \hat{S} & \xrightarrow{g_\gamma^S} & S \end{array}$$

It can be shown that

$$\hat{\mathbf{H}} = (\hat{H}, \hat{K}, X \times E, L \times E, X, \hat{S}, \hat{\chi} \circ \hat{\hbar})$$

is a $1/p$-universal$_2$ projective hash family. Recall that this actually means that for any $x \in X \setminus L$ and $e \in E$ if $\hat{k} \in \hat{K}$ is chosen uniformly at random and $\hat{\chi}(\hat{\hbar}(\hat{k}))$,

$\Phi_{\hat{k}}(x^*, e^*)$ are known (for some $x^* \in X \setminus (L \cup \{x\})$ and $e^* \in E$), the probability of guessing $\Phi_{\hat{k}}(x, e)$ correctly is smaller than $1/p$.

We start by obtaining an analogue of Lemma 2.

Lemma 3. *Let $\hat{H}$ be as above, $x \in X$ and $e \in E$. Then, if $\hat{\phi} \in H^{n+1}$ is chosen uniformly at random, once $\hat{s} = \hat{\chi}(\hat{\phi})$ is fixed, then $\phi = g^H_{\Gamma(x,e)}(\hat{\phi})$ is uniformly distributed on the coset $\chi^{-1}(s)$, where $s = g^S_{\Gamma(x,e)}(\hat{s})$. Moreover, $\phi(x)$ is uniformly distributed on the set $\{\psi(x) \mid \psi \in \chi^{-1}(s)\}$, that is, on a set of cardinality equal to $\|[x]\|$.*

Proof. It is clear that in the conditional probability space, ϕ is uniformly distributed on the set $g^H_{\Gamma(x,e)}(\hat{\chi}^{-1}(\hat{s}))$. Let us show that this set is just the coset $\chi^{-1}(s)$. It is clear that $g^H_{\Gamma(x,e)}(\hat{\chi}^{-1}(\hat{s})) \subseteq \chi^{-1}(s)$ since $\chi(g^H_{\Gamma(x,e)}(\hat{\chi}^{-1}(\hat{s}))) = g^S_{\Gamma(x,e)}(\hat{\chi}^{-1}(\hat{\chi}(\hat{s}))) = s$. Conversely, $g^H_{\Gamma(x,e)}(\hat{\chi}^{-1}(\hat{s}))$ contains a whole coset modulo $\ker \chi$. To see this, pick an element $\psi \in g^H_{\Gamma(x,e)}(\hat{\chi}^{-1}(\hat{s}))$. Then, there exists $\hat{\psi} = (\psi_0, \psi_1, \ldots, \psi_n) \in \hat{\chi}^{-1}(\hat{s})$ such that $\psi = g^H_{\Gamma(x,e)}(\hat{\psi})$. For each $\eta \in \ker \chi$, $\eta \circ \psi = g^H_{\Gamma(x,e)}(\eta \circ \psi_0, \psi_1, \ldots, \psi_n)$ that is also in $\hat{\chi}^{-1}(\hat{s})$. From this point, the proof proceeds exactly as in Lemma 2. □

Proposition 3. *If (X, H, χ, S) is p-diverse then $\hat{H}$ is a $1/p$-universal projective hash family.*

Proof. From Lemma 3, for any $x \in X \setminus L$ and $e \in E$, the probability of guessing the right value of $\Phi_{\hat{k}}(x, e) = g^H_{\Gamma(x,e)}(\hat{h}(\hat{k}))(x)$ for a random choice of $\hat{k} \in \hat{K}$ given $\hat{\chi}(\hat{h}(\hat{k}))$ is $1/\|[x]\|$, that is at most $1/p$. □

The next proposition shows that $\hat{H}$ is also universal$_2$ (see Appendix A for a proof):

Proposition 4. *If (X, H, χ, S) is p-diverse then $\hat{H}$ is a $1/p$-universal$_2$ projective hash family.*

Equipped with these results, we can now mimic Cramer and Shoup's (abelian) construction. Given a hard subset membership problem $\mathcal{M}$ and suitable automorphism group systems, we can construct, analogously as it is done in [7,8], two HPSs for $\mathcal{M}$, $\mathcal{P}$ and $\hat{\mathcal{P}}$, strongly smooth and strongly universal$_2$ extended respectively. Then, with the same arguments as in the security proof of the general Cramer and Shoup construction, we obtain:

Proposition 5. *Let $\mathcal{M}$ be a hard subset membership problem, $\mathcal{P}$ and $\hat{\mathcal{P}}$ strongly smooth resp. strongly universal$_2$ extended HPSs for $\mathcal{M}$ constructed from automorphism group systems.*

Then the public key encryption scheme described in Section 2.4 is secure in the sense of IND-CCA.

4 Deriving Examples of Provably Secure Public Key Encryption Schemes

As pointed out, e.g., by Shpilrain in [21], some investigation should still be devoted to the construction of group theoretical schemes with satisfactory security guarantees. At the moment we cannot provide a practical new provably secure public key scheme based on non-abelian groups and the above framework. In the following, we restrict to outlining a possible methodology for designing a cryptosystem fitting our framework. One plausible approach to deriving examples is as follows:

Find a Suitable Decisional Problem. Take, e.g., the decisional Diffie-Hellman problem in a cyclic group $G = \langle g \rangle$ of prime order q.
Represent it as a Subset Membership Problem. For instance: $X = G \times G$ and $L = \langle (g, g^c) \rangle$, for some secret $c \in \{1, \ldots, q-1\}$. Thus, L can be seen as a line in $GF(q)^2$ generated by the vector $(1, c)$.
Study a Related Automorphism Group Which Would Fix the Subset Elements. Take, for the case above, the subgroup of $GL(2, q)$ that fixes L. That is, the group formed by the matrices that fix the vector $(1, c)$. This will act as the kernel of the homomorphism χ. As this subgroup is not necessarily normal, we take as H its normalizer in $GL(2, q)$, which has order $q(q-1)^2$.
Construct χ Accordingly.

Of course, all these steps have to be done in such a way that the final construction is computationally feasible, so that the required sampling and evaluation algorithms for the encryption scheme can be provided.

The above automorphism group system can be used directly to derive a projective hash family which would however be neither universal nor smooth, but some slight modifications allow to achieve these two properties. Nevertheless, we do not encourage the construction of a hash proof system from it due to the lack of efficiency of some of the required algorithms.

Also, the example above is in some sense 'close' to the abelian case (which, in the end, inspires this construction). However, based on the above methodology one can also think of similar constructions that are genuinely non-abelian. To this aim, we recall the definition of a logarithmic signature, first given by [17]:

Definition 4. *Let L be a finite group. Next, denote by $\xi = [\xi_1, \ldots, \xi_s]$ a sequence of length $s \in \mathbb{N}_0$ such that each ξ_i ($1 \leq i \leq s$) is itself a sequence $\xi_i = [\xi_{i0}, \ldots, \xi_{ir_i-1}]$ with $\xi_{ij} \in L$ ($0 \leq j < r_i$) and $r_i \in \mathbb{N}_0$. Then we call ξ a logarithmic signature for L if each $g \in L$ is represented uniquely as a product*

$$g = \xi_{1j_1} \cdots \xi_{sj_s} \tag{1}$$

with $\xi_{ij_i} \in \xi_i$ ($1 \leq i \leq s$).

Example 2. Suppose we have at hand a hard subset membership problem $\mathcal{M}$ which for each input $l \in \mathbb{N}_0$ selects an instance constructed as follows: Let X be a non-abelian group, $H \leq \text{Aut}(X)$ and $\xi = [\xi_1, \ldots, \xi_s]$ a logarithmic signature for a subgroup L of X, H-invariant (i.e., $\phi(L) = L \; \forall \phi \in H$). Suppose that factoring elements according to ξ is a hard computational problem.

Moreover, let $W := A_{r_1} \times \cdots \times A_{r_s}$ where $|\xi_i| = r_i$ and A_r stands for the set $\{0, \ldots, r-1\}$. Define the bijection

$$\beta: \quad W \quad \longrightarrow \quad L$$
$$(w_1, \ldots, w_s) \longmapsto \xi_{1w_1} \cdots \xi_{sw_s}.$$

The sampling algorithm just chooses a random $w \in W$ and computes $x = \beta(w)$. Now let us describe an automorphism group system for X and L: Assume H is such that that given $\phi \in H$ the images $\phi(\xi_{ij})$, $j = 0, \ldots, r_i - 1$, $i = 1, \ldots, s$ give no information about the action of ϕ on $X \setminus L$. Suppose also that $\phi(\xi)$ induces a polynomial time factorization of $\phi(L)$ for all $\phi \in H^1$. Let $\hbar$ be an efficiently computable bijection defined between some index set K and H.

Moreover, take $S := H|_L$ and $\chi : H \longrightarrow S$ the natural projection, i.e. $\chi(\phi) := \phi|_L$. Note that the image $\chi(\phi)(x)$ can be efficiently computed for any $x \in L$ given a witness $(w_1, \ldots, w_s)$ for x and the images $\phi(\xi_{iw_i})$ for $i = 1, \ldots, s$. Thus, in practice, $\chi(\phi)$ may be specified by $\phi(\xi)$. Clearly, (X, H, χ, S) is an automorphism group system (see Section 3.1).

Now, from a good enough automorphism group system (i.e., p-diverse for some large prime p), two PHFs, $\mathbf{H}$ and $\hat{\mathbf{H}}$, can be constructed as in Sections 3.2 and 3.3. Then, if there exist efficient algorithms for sampling, public and private evaluation, the resulting encryption scheme will be secure in the sense of IND-CCA.

As a final remark on this example, let us suppose the group H is a subgroup of $\text{Inn}(X)$, that is, for each $\phi \in H$ there exists a certain $a \in X$ so that $\phi(x) = axa^{-1}$. For the scheme to be secure, a special kind of *simultaneous conjugacy problem* must be hard to solve in L. Also, it must be possible to produce *hard* logarithmic signatures of L which could be used as parts of the public keys.

Examples of schemes already proposed relying on similar assumptions are the MST_2 scheme [18] and the key exchange proposed by Anshel et al. in [2]. However, even if the underlying mathematical problems used as a base could be considered hard, such constructions would not be provably secure in the sense of IND-CCA.

5 Conclusions

We have given a theoretical framework which, if sound hardness assumptions are identified, may lead to the construction of IND-CCA public key encryption schemes based on non-abelian groups. The main tool we introduced are *automorphism group systems* for deriving projective hash families from non-abelian

---
[1] This last condition could be avoided using the generic transformation from [8].

groups. The idea used here parallels Cramer and Shoup's abelian construction based on *group systems*. As in their framework, we give criteria for choosing suitable automorphism group systems in order to obtain useful (i.e. universal) projective hash families. In principle, our model may also help in developing new examples of IND-CCA secure schemes based on abelian groups; it is however especially interesting as a design guide for developing new tools in non-abelian cryptography. Up until now, cryptosystems based on non-abelian groups often turned out to have security flaws which are independent of the soundness of the underlying mathematical assumptions; it is our aim that this design supplies a useful tool to overcome such problems. Unfortunately, so far we cannot offer a practical example of a new public key encryption scheme derived from non-abelian groups in our framework. Having in mind the goal of identifying new mathematical primitives offering provably secure encryption schemes, however, we think it is certainly worthwhile to explore the existence of automorphism group systems and hard subset membership problems based on non-abelian groups fitting our framework.

References

1. I. Anshel, M. Anshel, B. Fisher, and D. Goldfeld. New Key Agreement Protocols in Braid Group Cryptography. In *CT-RSA 2001*, volume 2020 of *Lecture Notes in Computer Science*, pages 13–27, Berlin, Heidelberg, 2001. Springer.
2. I. Anshel, M. Anshel, and D. Goldfeld. An algebraic method for public-key cryptography. *Mathematical Research Letters*, 6:1–5, 1999.
3. M. Bellare, A. Desai, D. Pointcheval, and P. Rogaway. Relations Among Notions of Security for Public-Key Encryption Schemes. In *Advances in Cryptology, Proceedings of CRYPTO '98*, volume 1462 of *Lecture Notes in Computer Science*, pages 26–45. Springer, 1998.
4. M. Bellare and P. Rogaway. Random oracles are practical: A paradigm for designing efficient protocols. In *ACM Conference on Computer and Communications Security*, pages 62–73, 1993.
5. J.-M. Bohli, M.I. González Vasco, C. Martínez, and R. Steinwandt. Weak Keys in MST_1. *Designs, Codes and Cryptography*, to appear.
6. J.H. Cheon and B. Jun. Diffie-Hellman Conjugacy Problem on Braids. Cryptology ePrint Archive: Report 2003/019, 2003. Electronically available at http://eprint.iacr.org/2003/019/.
7. R. Cramer and V. Shoup. Universal Hash Proofs and a Paradigm for Adaptive Chosen Ciphertext Secure Public-Key Encryption. Cryptology ePrint Archive: Report 2001/085, 2001. Electronically available at http://eprint.iacr.org/2001/085/.
8. R. Cramer and V. Shoup. Universal Hash Proofs and a Paradigm for Adaptive Chosen Ciphertext Secure Public-Key Encryption. In Lars Knudsen, editor, *Advances in Cryptology — EUROCRYPT 2002*, volume 2332 of *Lecture Notes in Computer Science*, pages 45–64. Springer, 2002.
9. D. Dolev, C. Dwork, and M. Naor. Non-malleable cryptography. *SIAM Journal on Computing*, 30:391–437, 2000.
10. A. Fiat and A. Shamir. How to prove yourself: practical solutions to identification and signature problems. In *Advances in cryptology—CRYPTO '86*, volume 263 of *Lecture Notes in Computer Science*, pages 186–194. Springer, 1987.

11. M. Garzon and Y. Zalcstein. The Complexity of Grigorchuk groups with application to cryptography. *Theoretical Computer Science*, 88:83–98, 1991.
12. M.I. González Vasco, C. Martínez, and R. Steinwandt. Towards a Uniform Description of Several Group Based Cryptographic Primitives. *Designs, Codes and Cryptography*, 33:215–226, 2004.
13. M.I. González Vasco and R. Steinwandt. Reaction Attacks on Public Key Cryptosystems Based on the Word Problem. *Applicable Algebra in Engineering, Communication and Computing*, 14:335–340, 2004.
14. D. Hofheinz and R. Steinwandt. A Practical Attack on Some Braid Group Based Cryptographic Primitives. In *Public Key Cryptography, 6th International Workshop on Practice and Theory in Public Key Cryptosystems, PKC 2003 Proceedings*, volume 2567 of *Lecture Notes in Computer Science*, pages 187–198. Springer, 2003.
15. K.H. Ko, S.J. Lee, J.H. Cheon, J.W. Han, J. Kang, and C. Park. New Public-Key Cryptosystem using Braid Groups. In *Advances in Cryptology. Proceedings of CRYPTO 2000*, volume 576 of *Lecture Notes in Computer Science*, pages 166–183. Springer, 2000.
16. H.K. Lee, H.S. Lee, and Y.R. Lee. An Authenticated Group Key Agreement Protocol on Braid Groups. *Cryptology ePrint Archive: Report 2003/018*, 2003. Electronically available at http://eprint.iacr.org/2003/018/.
17. S.S. Magliveras and N.D. Memon. Algebraic properties of cryptosystem PGM. *Journal of Cryptology*, 5:167–183, 1992.
18. S.S. Magliveras, D.R. Stinson, and T. Trung. New approaches to designing public key cryptosystems using one-way functions and trap-doors in finite groups. *Journal of Cryptology*, 15:285–297, 2002.
19. M. Naor and M. Yung. Public-key Cryptosystems Provably Secure against Chosen Ciphertext Attacks. In *Proceedings of the twenty-second annual ACM symposium on Theory of computing*, pages 427–437. ACM Press, 1990.
20. C. Rackoff and D. Simon. Non-Interactive Zero-Knowledge Proof of Knowledge and Chosen Ciphertext Attack. In *Advances in Cryptology — CRYPTO'91*, volume 576 of *Lecture Notes in Computer Science*, pages 433–444. Springer, 1992.
21. V. Shpilrain. Assessing security of some group based cryptosystems. *Cryptology ePrint Archive: Report 2003/123*, 2003. Electronically available at http://eprint.iacr.org/2003/123/.
22. N.R. Wagner and M.R. Magyarik. A Public Key Cryptosystem Based on the Word Problem. In *Advances in Cryptology: Proceedings of CRYPTO 84*, volume 196 of *Lecture Notes in Computer Science*, pages 19–36. Springer, 1985.

A Proof of Proposition 4

Proof. Let us suppose as above that $\hat{k} \in \hat{K}$ is selected uniformly at random and $\hat{s} = (s_0, \ldots, s_n) = \hat{\chi}(\hat{h}(\hat{k}))$ is given. Then $\hat{\phi} = \hat{h}(\hat{k}) = (\phi_0, \ldots, \phi_n)$ is also uniformly distributed on $\hat{\chi}^{-1}(\hat{s})$.

In order to guarantee that $\hat{\mathbf{H}}$ is $1/p$-universal$_2$, it suffices to show the independence of the two random variables $\phi = g^H_{\Gamma(x,e)}(\hat{\phi})$ and $\phi^* = g^H_{\Gamma(x^*,e^*)}(\hat{\phi})$, for any $e, e^* \in E$, $x \in X \setminus L$ and $x^* \in X \setminus \{x\}$.

From Lemma 3, ϕ and ϕ^* are uniformly distributed on $\chi^{-1}(s)$ and $\chi^{-1}(s^*)$, respectively, where $s := g^S_{\Gamma(x,e)}(\hat{s})$ and $s^* := g^S_{\Gamma(x^*,e^*)}(\hat{s})$. Now let i be the smallest integer such that $\Gamma_i(x,e) \neq \Gamma_i(x^*, e^*)$, that surely exists since Γ is injective. Now, for any fixed values $\phi_j \in \chi^{-1}(s_j)$ for $j = 1, \ldots, i-1, i+1, \ldots, n$ let us consider the map

$$\triangle_i : \chi^{-1}(s_0) \times \chi^{-1}(s_i) \longrightarrow \chi^{-1}(s) \times \chi^{-1}(s^*)$$
$$(\phi_0, \phi_i) \longrightarrow (\phi, \phi^*),$$

where, as above, $\phi = g^H_{\Gamma(x,e)}(\hat{\phi})$ and $\phi^* = g^H_{\Gamma(x^*,e^*)}(\hat{\phi})$. By defining

$$\psi_L = \phi_1^{\Gamma_1(x,e)} \circ \cdots \circ \phi_{i-1}^{\Gamma_{i-1}(x,e)} = \phi_1^{\Gamma_1(x^*,e^*)} \circ \cdots \circ \phi_{i-1}^{\Gamma_{i-1}(x^*,e^*)},$$
$$\psi_R = \phi_{i+1}^{\Gamma_{i+1}(x,e)} \circ \cdots \circ \phi_n^{\Gamma_n(x,e)} \quad \text{and}$$
$$\psi_R^* = \phi_{i+1}^{\Gamma_{i+1}(x^*,e^*)} \circ \cdots \circ \phi_n^{\Gamma_n(x^*,e^*)}$$

we can write

$$\triangle_i(\phi_0, \phi_i) = (\phi_0 \circ \psi_L \circ \phi_i^{\Gamma_i(x,e)} \circ \psi_R, \phi_0 \circ \psi_L \circ \phi_i^{\Gamma_i(x^*,e^*)} \circ \psi_R^*).$$

The map $\triangle_i$ is injective. Indeed, consider two pairs (ϕ_0, ϕ_i) and $(\bar{\phi}_0, \bar{\phi}_i)$ in $\chi^{-1}(s_0) \times \chi^{-1}(s_i)$ such that $\triangle_i(\phi_0, \phi_i) = \triangle_i(\bar{\phi}_0, \bar{\phi}_i)$. Then, $\phi_0 \circ \psi_L \circ \phi_i^{\Gamma_i(x,e)} = \bar{\phi}_0 \circ \psi_L \circ \bar{\phi}_i^{\Gamma_i(x,e)}$ and $\phi_0 \circ \psi_L \circ \phi_i^{\Gamma_i(x^*,e^*)} = \bar{\phi}_0 \circ \psi_L \circ \bar{\phi}_i^{\Gamma_i(x^*,e^*)}$. Combining these two equalities, we obtain

$$\phi_i^{\Gamma_i(x^*,e^*) - \Gamma_i(x,e)} = \bar{\phi}_i^{\Gamma_i(x^*,e^*) - \Gamma_i(x,e)},$$

that leads to $\phi_i = \bar{\phi}_i$ and then to $\phi_0 = \bar{\phi}_0$.[2] Thus, $\triangle_i$ is injective.

Then, as $\chi^{-1}(s_0) \times \chi^{-1}(s_i)$ and $\chi^{-1}(s) \times \chi^{-1}(s^*)$ have the same (finite) cardinality, $\triangle_i$ is a bijection. So, if (ϕ_0, ϕ_i) is chosen uniformly at random in $\chi^{-1}(s_0) \times \chi^{-1}(s_i)$ then (ϕ, ϕ^*) is uniformly distributed on $\chi^{-1}(s) \times \chi^{-1}(s^*)$, for any choice of ϕ_j, $j = 1, \ldots, i-1, i+1, \ldots, n$. Then, the same occurs when the whole tuple $\hat{\phi}$ is chosen uniformly at random in $\hat{\chi}^{-1}(\hat{s})$. Consequently, ϕ and ϕ^* are independent uniformly distributed random variables. In particular, this independence implies that the knowledge of $\Phi_{\hat{k}}(x^*, e^*) = \phi^*(x^*)$ does not affect the probability distribution of $\Phi_{\hat{k}}(x, e) = \phi(x)$. Thus, by Lemma 3, $\Phi_{\hat{k}}(x, e)$ is uniformly distributed on a set of size $|[x]|$. Then, $\hat{\mathbf{H}}$ is $1/p$-universal$_2$. □

---

[2] Note that, as $|\Gamma_i(x^*, e^*) - \Gamma_i(x, e)| < q$, we have $\gcd(\Gamma_i(x^*, e^*) - \Gamma_i(x, e), |H|) = 1$. So there are $a, b \in \{0, \ldots, |H|-1\}$ such that $a(\Gamma_i(x^*, e^*) - \Gamma_i(x, e)) = 1 + b|H|$, and, consequently, $\phi_i^{a(\Gamma_i(x^*,e^*) - \Gamma_i(x,e))} = \phi_i^{1+b|H|} = \phi_i$.

Further Simplifications in Proactive RSA Signatures

Stanisław Jarecki and Nitesh Saxena

School of Information and Computer Science,
UC Irvine, Irvine, CA 92697, USA
{stasio, nitesh}@ics.uci.edu

Abstract. We present a new robust proactive (and threshold) RSA signature scheme secure with the optimal threshold of $t < n/2$ corruptions. The new scheme offers a simpler alternative to the best previously known (static) proactive RSA scheme given by Tal Rabin [36], itself a simplification over the previous schemes given by Frankel et al. [18, 17]. The new scheme is conceptually simple because all the sharing and proactive re-sharing of the RSA secret key is done modulo a *prime*, while the reconstruction of the RSA signature employs an observation that the secret can be recovered from such sharing using a simple equation over the *integers*. This equation was first observed and utilized by Luo and Lu in a design of a simple and efficient proactive RSA scheme [31] which was not proven secure and which, alas, turned out to be completely *insecure* [29] due to the fact that the aforementioned equation leaks some partial information about the shared secret. Interestingly, this partial information leakage can be proven harmless once the polynomial sharing used by [31] is replaced by top-level *additive* sharing with second-level polynomial sharing for back-up.

Apart of conceptual simplicity and of new techniques of independent interests, efficiency-wise the new scheme gives a factor of 2 improvement in speed and share size in the general case, and almost a factor of 4 improvement for the common RSA public exponents 3, 17, or 65537, over the scheme of [36] *as analyzed in* [36]. However, we also present an improved security analysis and a generalization of the [36] scheme, which shows that this scheme remains secure for smaller share sizes, leading to the same factor of 2 or 4 improvements for that scheme as well.

1 Introduction

The idea of distributing a cryptosystem so that to secure it against corruption of some threshold, e.g. a minority, of participating players is known as *threshold cryptography*. It was introduced in the works of Desmedt [13], Boyd [4], Croft and Harris [9], and Desmedt and Frankel [14], which built on the *polynomial secret-sharing* technique of Shamir [39]. A threshold signature scheme [14] is an example of this idea. It allows a group of n players to share the private signature key in such a way that the signature key remains secret, and the signature scheme remains secure, as long as no more than t of the players are corrupt.

Simultaneously, as long as at least $n-t$ players are honest, these players can efficiently produce correct signatures on any message even if the other t players act in an arbitrarily malicious way.

Proactive signature schemes [28, 27] are threshold signature schemes which offer an improved resistance against player corruptions. Time is divided into *update rounds*, and the proactive signature scheme offers the same combination of security and robustness even in the presence of so-called *mobile* faults [34], where a potentially new group of up to t players becomes corrupted in each update round. Technically, this is done by the players randomly re-sharing the shared private key at the beginning of each update round. A proactive signature scheme offers stronger security guarantee then a threshold scheme, especially in an application which might come under repeated attacks, like a certification authority or a timestamping service. Moreover, a proactive scheme offers more secure management of a system whose size and make-up need to change throughout its lifetime. Efficiency of the distributed signature protocol involved in a proactive signature scheme is very important in some applications, like in a timestamping service, or in the decentralized control of peer-to-peer groups, ad-hoc groups, or sensor networks [30, 37]. An efficient proactive scheme for RSA signatures is especially important because RSA signatures are widely used in practice, and because verification of RSA signatures is several orders of magnitude faster than verification of other signatures.

Prior Work on Threshold and Proactive RSA. While the work of Herzberg et al. [28, 27] and Gennaro et al. [25, 26] quickly yielded efficient secure proactive signature schemes for discrete-log based schemes like Schnorr [38] or DSS [33] signatures, the work on secure proactive RSA schemes progressed more slowly, and the initial threshold RSA scheme of Desmedt and Frankel [14] was robust only against crashes and not malicious faults, and had only heuristic security. The difficulty in adopting Shamir's polynomial secret-sharing technique to threshold RSA was caused by the fact that the RSA private key d is an element of a group $\mathbb{Z}_{\phi(N)}$, where $\phi(N)$ needs to remain hidden from all players because it allows immediate computation of the private key d from the RSA public exponent e. This difficulty was overcome by the schemes of Frankel et al. [16, 12] which provided a proof of security but used secret shares which were elements of a polynomial extension field of $\mathbb{Z}_n$, which increased the cost of the signature operation by a factor of at least t. These schemes were then extended to provide robustness against malicious faults by [19, 24]. Subsequently, Victor Shoup [40] presented a threshold RSA signature scheme which was robust and provably secure with optimal adversarial threshold $t < n/2$, and which did away with the extension field representation of the shares, thus making the cost of the signature operation for each participating player comparable to the standard RSA signature generation.

Proactive RSA scheme is a harder problem because it requires the players to re-share the private key d in each update round even if no single player is allowed to know the secret modulus $\phi(N)$. The first proactive RSA scheme of Frankel et al. [18] solved this problem using additive secret sharing over integers in conjunction with combinatorial techniques which divide the group of n players into two levels of families and sub-families. However, the resulting proactive protocol did not achieve optimal adversarial threshold $t < n/2$ and did not scale well with the group size n. These shortcomings were later overcome by the same authors [17], who showed that the RSA private key d can be shared over integers using polynomials with specially chosen large integer coefficients that simultaneously allowed interpolation without knowing $\phi(N)$ and unpredictability of the value d given any t polynomial shares. In this solution, even though the underlying secret sharing was polynomial, the players need to create a one-time additive sharing for every group of players participating in threshold signature generation. A simpler and more efficient proactive RSA scheme was then given by Tal Rabin [36]. Her solution also used sharing of the private key over integers, and employed shares of size about twice the length of the private key. The new idea was that the secret d was shared additively among the players, every share was backed-up by a secondary level of polynomial secret sharing, and the proactive update consisted of shuffling and re-sharing of the additive shares.

Limitations and Open Problems in Proactive RSA. The proactive RSA schemes of [18, 17, 36] leave at least two important problems unaddressed. While the new proactive RSA scheme we present in this paper does not solve these problems either, the techniques we present might help solve these problems in the future. The first problem is that of handling *adaptive* rather than *static* adversaries. The static adversary model assumes that the adversary decides which player to corrupt obliviously to the execution of the protocol, while the adaptive model allows the adversary to decide which player to corrupt based on his view of the protocol execution. This difference in the adversarial model is not known to be crucial for the security of the above protocols in practice. However, the above protocols are not known to be adaptively secure, while the known adaptively secure RSA schemes [20, 7, 21, 22] are significantly less efficient.

The second problem is that of requiring some form of additive rather than polynomial secret-sharing. The additive sharing implies that the shares of all temporarily unavailable players need to be reconstructed by the active players that participate in the signature generation protocol. This hurts both the efficiency and the resilience of a scheme in applications where one player might be temporarily unavailable to another without an actual corruption by the adversary. Since the threshold (but not proactive) RSA signature schemes discussed above do not resort to additive sharing, this is a disadvantage of the currently known proactive RSA schemes.

This somewhat unsatisfactory state of the known proactive RSA schemes led a group of network security researchers to design a proactive RSA scheme [30] which attempted to solve these problems by using polynomial secret sharing. The technique they employed was very simple. It relied on an observation that the secret sharing of the private key d modulo any modulus which has only large prime factors enables efficient reconstruction of d over *integers*. Consequently, by the homomorphic properties of exponentiation, it also enables reconstruction of the RSA signature $m^d \mod N$. This scheme, however, did not come with a security proof, and indeed upon a closer examination [29], the proposed interpolation over the integers leaks a few most significant bits of the shared private key d, which together with the adversarial ability to manipulate the choice of shares in the proactive update protocol allows the threshold attacker to stage a binary search for d. Nevertheless, the above technique, which was utilized by the insecure scheme of [30], can be corrected, resulting in the *provably secure* proactive RSA scheme we present here.

Our Contribution: Further Simplification and Efficiency Improvements in Proactive RSA. Based on the corrected use of a technique discovered by Lu and Luo [31], we present a new robust and provably secure optimal-threshold proactive RSA scheme. Our scheme is known to be secure only in the static model, and it employs top-level additive sharing similarly as the Rabin's scheme [36], but it is interesting for the following reasons: (1) It is simpler than the previous schemes; (2) It offers factor of 2 improvement in share size and signature protocol efficiency for general RSA public keys, and factor of 4 improvement for the common case of public exponents like 3, 17, or 65537, over the most efficient previously known proactive RSA scheme [36] *as originally analyzed* by [36]; (3) The new scheme led us to a tighter security analysis of the [36] scheme, which resulted in similar, up to a logarithmic factor, efficiency improvements for the [36] scheme; (4) The new scheme offers an interesting case of a technique, invented by Lu and Luo [31], which makes a distributed protocol run faster but leaks some partial information about the shared secret. This partial information leakage led to an efficient key-recovery attack [29] on the original scheme of [31]. Yet, with some fixes, this partial information leakage can be provably neutralized and the same technique results in a provably secure scheme presented here; (5) Finally, our scheme offers new techniques which could aid in overcoming the two problems that still haunt proactive RSA solutions, namely achieving efficient adaptive security and the removal of additive sharing.

Paper Organization. Section 2 describes our adversarial model; section 3 presents the new scheme; section 4 contains the security proof; and section 5 shows an efficiency improvement for the proactive RSA scheme of [36].

2 Our Computation Model and the Adversarial Model

We work in the standard model of threshold cryptography and distributed algorithms known as synchronous, secure links, reliable broadcast, trusted dealer, static, and proactive adversary model. This is the same model as employed for example in [28, 27, 18, 17, 36] discussed in the introduction, with the exception that the first two did not need a trusted dealer (but did not handle RSA).

This model involves n players $M_1, ..., M_n$ equipped with synchronized clocks and an ability to erase information. The players are connected by weakly synchronous communication network offering secure point-to-point channels and a reliable broadcast. The time is apriori divided into evenly spaced update rounds, say of length of one day. We assume the presence of the so-called "mobile" adversary, modeled by a probabilistic polynomial time algorithm, who can *statically*, i.e., at the beginning of the life time of the scheme, schedule up to $t < n/2$ arbitrarily malicious faults among these n players, independently for every update round. We also assume a trusted dealer who initializes the distributed scheme by picking an RSA key and securely sharing the private key among the players. Since the adversary attacks a proactive *signature* scheme, the adversary can also stage a chosen-message attack [CMA], i.e. it can ask any of the n players to run a signature protocol on any message it chooses. The adversary's goal is to either (1) forge a signature on a message he did not request a signature on, exactly as in the CMA attack against a standard (non-threshold) signature scheme, or (2) to prevent the efficient generation of signatures on messages which at least $t+1$ uncorrupted players want to sign.

3 The New Proactive RSA Signature Scheme

3.1 Overview of the Proposed Scheme

The sharing of the private RSA key d is done additively modulo a *prime* q s.t. $q \geq r2^{|N|+\tau}$, where r is the maximal number of rounds in the lifetime of the system, $|N|$ is the bit length of the RSA modulus N, and τ is a security parameter, e.g. $\tau = 80$. Namely each player M_i holds a share d_i which is a random number in $\mathbb{Z}_q$ s.t. $d_1 + ... + d_n = d \bmod q$. Each of these top-level additive shares is also polynomially shared for backup reconstruction of d_i in case M_i is corrupted, using the information-theoretically secret verifiable secret sharing (VSS) of Pedersen [35], similarly as in the proactive RSA scheme of Rabin [36]. In order to handle the common case of a small public RSA exponent e more efficiently, the most significant $l = \frac{|N|}{2}$ bits of the private key d can be publicly revealed as d_{pub}, and only the remaining portion of the private key d, namely $d - 2^{|N|-l}d_{pub}$, is shared as above modulo q, for any $q \geq r2^{|N|-l+\tau}$.

The proactive update is very easy in this setting, adopting the original proactive secret sharing of [28] to additive sharing. Such method was used before e.g. in [7]. To re-randomize the the sharing, each M_i picks random partial shares d_{ij} in $\mathbb{Z}_q$ s.t. $d_i = d_{i1} + ... + d_{in} \bmod q$, and sends d_{ij} to M_j. Each M_j computes then his new share as $d'_j = d_{1j} + ... + d_{nj} \bmod q$, and shares it polynomially for backup again. All this can be easily verified using Pedersen's VSS, and the new shares sum to the same secret d modulo q.

For the threshold signature protocol, we use the observation of [31] that if $\sum_{j=1}^{n} d_j = d \pmod{q}$ and $0 \leq d_j \leq q-1$ for all j's, then

$$d = \sum_{j=1}^{n} d_j - \alpha q \quad \text{(over the integers)} \qquad (1)$$

for some integer $\alpha \in \{0, ..., n-1\}$. Consequently, if $\forall_j$, $s_j = m^{d_j} \bmod N$ then

$$m^d = (\prod_{j=1}^{n} s_j) m^{-\alpha q} \pmod{N}$$

Therefore the signature $m^d \bmod N$ can be reconstructed if players submit their partial signatures as $s_j = m^{d_j} \pmod{N}$, and the correct value of α is publicly reconstructed by cycling over the possible n choices of α, which adds at most $2n$ modular exponentiations to the cost of the signature generation protocol. (Note that in most applications $n < 100$.) In the (rare) case of a malicious fault causing a failure in this procedure, each player has to prove in zero-knowledge that it used a correct value d_i in its partial signature, i.e. the value committed in the Pedersen VSS that shares this d_i. Efficient zero-knowledge proofs to handle such statement were given by Camenisch and Michels [5], and Boudot [2], and while not blazing fast, they have constant number of exponentiations, and they are practical. This procedure is more expensive than the robustness procedure in [36], but we believe that this efficiency difference does not matter since active corruptions of this type should be unlikely, as active faults are rare in general and the adversary would not gain much by using his corrupted player in this way.

In the attack [29] on a similar scheme involving *polynomial* rather than additive top-level sharing, the adversary uses the fact that the above procedure reveals whether d is greater or smaller than some value in the $[0, q]$ interval which the adversary can easily compute from his shares. Since the adversary can perfectly control his shares in the proactive update protocol for this (top-level) polynomial secret sharing scheme, the adversary can use this partial information leakage to stage a binary search for the shared secret d.

However, the scheme we present fixes the above problem. Assume that the adversary corrupts players $M_1, ..., M_t$. Giving the adversary the extra knowledge of shares $d_{t+1}, ..., d_{n-1}$, the only information about the secret key revealed by value α is, by equation (1), whether or not the secret d is smaller or larger than $R = (D \bmod q)$ where $D = d_1 + ... + d_{n-1}$. Since the adversary does not have enough control over the shares created by our "additive" proactive update protocol, shares $d_{t+1}, ..., d_{n-1}$ are random in $\mathbb{Z}_q$, and hence so is value R. Therefore, if q is significantly larger than the maximal value of d, then the α value almost never reveals anything about d, because d is almost always smaller than R. For this reason, if $q \geq r2^{|N|+\tau}$ then the modified scheme keeps d indistinguishable from a value uniform in $\mathbb{Z}_n$, with the statistical difference of $2^{-\tau}$. The additional factor r in the bound on q appears because of the linear increase in the statistical difference with every update round. This captures the security proof of our scheme in a nutshell.

We now give the detailed description of our scheme.

3.2 Setup Procedure

We require a trusted dealer to securely set up the system. The dealer generates RSA private/public key pair, i.e. an RSA modulus N, public exponent e, and private key $d = e^{-1} \bmod \phi(N)$. Optionally, $l \leq \frac{|N|}{2}$ most significant bits of d can be publicly revealed as d_{pub} (otherwise $d_{pub} = 0$ and $l = 0$). The dealer also chooses an instance of Pedersen commitment [35], i.e. primes p and q s.t. $q|(p-1)$, and two random elements g, h of order q in $\mathbb{Z}_p^*$, for $|q| = \log_2 r + |N| - l + \tau + 1$, where τ is a security parameter ($\tau \geq 80$) and r is the number of rounds the system is expected to run. The dealer then runs the sharing protocol of Figure 1.

Input: private key $d \in \mathbb{Z}_{\phi(n)}$, public value d_{pub} corresponding to l MSBs of d, public RSA modulus N, Pedersen commitment instance (p, q, g, h).

1. Select shares $d_j \in \mathbb{Z}_q$ uniformly at random for $j = 1, \ldots, n-1$ and set $d_n = d - 2^{|N|-l} d_{pub} - \sum_{j=1}^{n-1} d_j \pmod{q}$.
2. Share each d_j using Pedersen's VSS protocol [35]. Namely, select random polynomials $f_j(z) = d_j + f_{j1} z + \cdots + f_{jt} z^t$ and $f'_j(z) = d'_j + f'_{j1} z + \cdots + f'_{jt} z^t$ over $\mathbb{Z}_q$ of degree t s.t. $f_j(0) = d_j$. Compute and publish the witnesses $w_{j0} = g^{d_j} h^{d'_j} \pmod{p}$ and $w_{jk} = g^{f_{jk}} h^{f'_{jk}} \pmod{p}$ for $k = 1, \ldots, t$.
3. Compute the secret shares ss_{ij} and ss'_{ij} as $ss_{ij} = f_j(i) \pmod{q}$ and $ss'_{ij} = f'_j(i)$, deliver d_i, d'_i, ss_{ij} and ss'_{ij} ($\forall j$) to each M_i over a secure channel.

Fig. 1. Trusted Dealer's Protocol: Sharing of the Private Key d

3.3 Threshold Signature Protocol

The goal of the threshold RSA signature protocol is to generate in a distributed manner an RSA signature $s = m^d \pmod{N}$ under the secret-shared key d, where $m \in \mathbb{Z}_n^*$ is some hashed/padded function of the signed message, e.g. $m = H(M)$ for the Full Domain Hash RSA [1]. Our protocol consists of two parts. First each player M_j creates its *partial signature* on the intended message $s_j = m^{d_j} \bmod N$, and sends it to the signature recipient. The recipient then locally reconstructs the RSA signature from these partial signatures using the n-bounded reconstruction algorithm of [31]. The threshold signature generation and reconstruction protocol is summarized in Figure 2, and we explain the details of the reconstruction algorithm below.

Input: (hashed) message $m \in \mathbb{Z}_n^*$, outputs of the Setup procedure

1. Player M_i broadcasts its partial signature $s_i = m^{d_i} \pmod{N}$.
2. If M_i fails to provide its partial signature, all players reconstruct d_i and compute $s_i = m^{d_i} \pmod{N}$.
3. Reconstruct RSA signature using the n-bounded offsetting algorithm (see below).
4. If signature reconstruction fails, trace the faulty signer(s) by executing the protocol $ZKPK(d_i : w_{i0} = g^{d_i} h^{d'_i} \pmod{p} \wedge s_i = m^{d_i} \pmod{N}) \wedge d_i \in [0, q-1])$ with each M_i (see Appendix).
5. If M_i fails this proof, any set of $t+1$ players reconstruct d_i and compute and broadcast $s_i = m^{d_i} \pmod{N}$.

Fig. 2. Signature Generation and Reconstruction

Signature Reconstruction with n-Bounded Offsetting. On receiving n partial signatures s_j from the n players, the signature recipient reconstructs the RSA signature s using the *n-bounded-offsetting* algorithm [30] which works as follows. Since $\sum_{j=1}^{n} d_j = d - 2^{|N|-l} d_{pub}$ (mod q) and $0 \le d_j \le q-1$ for all j's, therefore

$$d = 2^{|N|-l} d_{pub} + \sum_{j=1}^{n} d_j - \alpha q \quad \text{(over the integers)} \qquad (2)$$

for some integer $\alpha \in \{0, \ldots, n-1\}$, which implies that

$$s = m^d = m^{2^{|N|-l} d_{pub}} (\prod_{j=1}^{n} s_j) m^{-\alpha q} \pmod{N}$$

Since there can be at most n possible values of α, the signature recipient can recover $s = m^d$ (mod N) by trying each of the n possible values $Y_\alpha = Y(m^{-q})^\alpha$ (mod N) for $Y = m^{2^{|N|-l} d_{pub}} (\prod_{j=1}^{n} s_j)$ and $\alpha = 0, \ldots, n-1$, and returning $s = Y_\alpha$ if $(Y_\alpha)^e = m$ (mod N). The decisive factor in the cost of this procedure is the cost of the full exponentiation m^q mod N, where q can be e.g. 613-bit long for $N = 1024$, $e = 3$, $l = |N|/2$, $\tau = 80$, and $r \le 2^{20}$.

As discussed in the overview subsection above, this procedure reveals value α which contains some partial information on the shared secret d. Namely, granting to the adversary some extra knowledge and assuming he knows shares $d_1, \ldots, d_{n-1}$, the α value reveals whether $d \in \mathbb{Z}_{\phi(n)}$ lies in the interval $[0, R[$ or in $[R, N]$, where $R = (D \bmod q)$ and $D = d_1 + \ldots + d_{n-1}$, if $l = 0$. More generally, α reveals if d is smaller or larger than $R + 2^{|N|-l} d_{pub}$.

Robustness Mechanisms. In case some player M_u does not issue a partial signature, share d_u of M_u needs to be reconstructed to recover partial signature $s_u = m^{d_u}$ (mod N). In reconstruct d_u, every player M_i broadcasts its shares ss_{iu}, ss'_{iu} of d_u. The validity of these shares can be ascertained by checking

$$g^{ss_{iu}} h^{ss'_{iu}} = \prod_{k=0}^{t} (w_{uk})^{i^k} \pmod{p}.$$

Share d_u can then be recovered using the interpolation

$$d_u = \sum_{j \in G} ss_{ju} l_j(u) \pmod{q}$$

where G is a subgroup of $t+1$ players who broadcast valid shares and $l_j(u) = \prod_{j \in G, j \ne i} \frac{(u-j)}{i-j}$ mod q is the Lagrange interpolation polynomial computed at u.

If all the partial signatures are present but the above n-bounded signature reconstruction algorithm fails, then at least one out of n players did not issue a correct partial signature. The signature recipient must then trace the faulty players(s) by verifying the correctness of each partial signature. Once a player is detected as faulty, the share(s) of the faulty player(s) can be reconstructed as above. To prove correctness of its partial signature, each M_i proves in zero-knowledge that there is a pair of integers (d_i, d'_i) s.t.

$$w_{i0} = g^{d_i} h^{d'_i} \bmod p \quad , \quad s_i = m^{d_i} \bmod N \quad , \quad 0 \le d_i < q$$

It is crucial that the range of d_i is checked because otherwise player M_i can submit its partial signature as $m^{d'_i} \bmod N$ where $d'_i = d_i + kq$ for some k. An efficient zero-knowledge proof system for the proof of equality of discrete logarithms (and representations) in two different groups was given in [3,6], and the efficient proof that a committed number lies in a given range appeared in [2]. The resulting ZKPK proof system is in the appendix. It is non-interactive in the random oracle model and involves a (small) constant amount of exponentiations.

3.4 Proactive Update Protocol

At the beginning of every update round, the players perform the share update protocol of Figure 3 to re-randomize the sharing of d.

Input: Outputs of the Setup procedure or the previous Update protocol.
Let $r \ge 1$ be the round number. Denote current values $d_i^{(r-1)}, d'^{(r-1)}_i, w_{ij}^{(r-1)}$, etc.

1. Each player M_i selects (sub)shares d_{ij} and $d'_{ij} \in \mathbb{Z}_q$, uniformly at random for $j = 1, \ldots, n-1$, and sets $d_{in} = d_i^{(r-1)} - \sum_{k=1}^{n-1} d_{ik} \pmod{q}$ and $d'_{in} = d'^{(r-1)}_i - \sum_{k=1}^{n-1} d'_{ik} \pmod{q}$. M_i broadcasts witness values $w_{ij}^{(r)} = g^{d_{ij}} h^{d'_{ij}} \pmod{p}$, and hands (d_{ij}, d'_{ij}) to M_j ($\forall j$) over a secure channel.
2. M_j verifies the validity of the received shares using witness values as $w_{ij}^{(r)} = g^{d_{ij}} h^{d'_{ij}} \pmod{p}$, and ascertains whether the sub-shares in fact sum up to the previous share of M_i by checking that $\prod_{j=1}^{n} w_{ij}^{(r)} = w_{i0}^{(r-1)} \pmod{p}$.
3. M_j computes its new additive shares as $d_j^{(r)} = \sum_{i=1}^{n} d_{ij} \pmod{q}$ and $d'^{(r)}_j = \sum_{i=1}^{n} d'_{ij} \pmod{q}$. (Note that $\sum_{j=1}^{n} d_j^{(r)} = d - 2^{|N|-l} d_{pub} \pmod{q}$.)
4. M_j shares its new additive shares $d_j^{(r)}, d'^{(r)}_j$ using Pedersen's VSS, as in the setup phase described in Section 3.2. In order to check if M_j is indeed sharing its new additive share, every player checks that the witness value in this VSS instance corresponding to the shares $d_j^{(r)}, d'^{(r)}_j$ equals to $\prod_{i=1}^{n} w_{ij}^{(r)} \pmod{p}$.

Fig. 3. Proactive Share Update

4 Security Analysis of the New Proactive RSA Scheme

Theorem 1 (Security). *If there is a t-threshold proactive adversary for $t < n/2$, which in time T succeeds with probability β in a chosen-message attack against our new proactive (full domain hash) RSA signature scheme running for up to r rounds, for any $l \le |N|$ and prime $q \ge r2^{|N|-l+\tau}$, then there is a CMA attack against the standard (full domain hash) RSA signature scheme,* which succeeds in time $T + poly(n, |N|)$ with probability $\beta - 2^{-\tau}$ given the l most significant bits of the secret key d as an additional public input.

Proof. We show that if the adversary succeeds in staging the CMA attack on our (Full Domain Hash) proactive RSA signature scheme in time T with probability β, then there is also an efficient CMA attack against the standard (non-threshold) FDH-RSA signature which given the l most significant bits of d succeeds in time comparable to T by an amount polynomial in $|N|$ and n, with probability no worse than $\beta - 2^{-\tau}$. We show it by exhibiting a very simple simulator, which the adversary against the standard FDH-RSA scheme can run to interact with the proactive adversary which (T, β)-succeeds in attacking the proactive scheme. We will argue that the statistical difference between the view presented by this simulator on input of the public RSA parameters, l MSBs of d, and (message,signature) pairs acquired by the CMA attacker from the CMA signature oracle, and the adversarial view of the run of the real protocol on these parameters, for any value of the private key d with these l most significant bits, is at most $2^{-\tau}$, which will complete the proof.

The simulator $\mathcal{SIM}$ is described in Figure 4. The simulation procedure is very simple. The simulator picks a random value $\hat{d}$ in $\mathbb{Z}_n$ with the given l most-significant bits, and runs the secret-sharing protocol in the setup stage using this $\hat{d}$. Similarly in every update, the simulator just runs the actual protocol, but on the simulated values which we denote $\hat{d}_i$, $\hat{d}_{ij}$, etc. The only deviation from the protocol is that in the simulation of the threshold signature protocol, assuming w.l.o.g. that the M_n is an uncorrupted player, the simulator runs the actual protocol for all uncorrupted players except of M_n, i.e. it outputs $\hat{s}_j = m^{\hat{d}_j}$ for each uncorrupted M_j, $j \neq n$. The simulator then determines the $\hat{\alpha}$ value, which is an approximation to the actual value α the adversary would see in the protocol, by computing $D = \sum_{j=1}^{n-1} \hat{d}_j$, and taking $\hat{\alpha} = \lfloor D/q \rfloor + 1$. In this way we have $D = (\hat{\alpha} - 1)q + R$ where $R = (D \mod q)$. Finally, the simulator computes the missing partial signature $\hat{s}_n$ corresponding to the player M_n as $\hat{s}_n = s * m^{\hat{\alpha}q} / (m^{2^{|N|-l}d_{pub}} \prod_{j=1}^{n-1} \hat{s}_j) \pmod{N}$. In this way, partial signatures $\hat{s}_j$ add up to a valid RSA signature $\hat{s}$, and value $\hat{\alpha}$ the adversary sees in the *simulation* of the signature reconstruction algorithm is equal to the above α with an overwhelmingly high probability.

For ease of the argument, assume that the adversary corrupts players $M_1, ..., M_t$ throughout the lifetime of the scheme. We will argue that the adversarial views of the protocol and the simulation are indistinguishable with the statistical difference no more than $2^{-\tau}$, even if the adversary additionally sees shares $d_{t+1}, ..., d_{n-1}$ *and* the shared secret key d.

Setup Procedure: Since d_i and d'_i in the protocol and $\hat{d}_i$ and $\hat{d}'_i$ in the simulation are all picked uniformly from $\mathbb{Z}_q$ for $i = 1, \ldots, n - 1$, the two ensembles $(d, \{d_i, d'_i\}_{i=1,\ldots,n-1})$ and $(d, \{\hat{d}_i, \hat{d}'_i\}_{i=1,\ldots,n-1})$ have identical distributions.

By the information theoretic secrecy of Pedersen VSS, the second-layer shares and the associated verification values visible to the adversary are also distributed identically in the protocol and in the simulation.

Input: Pedersen commitment instance (p, q, g, h), RSA public parameters (N, e), optional values $l > 0$ and $d_{pub} < 2^l$ (otherwise set $l = d_{pub} = 0$).
Additionally, for every simulation of the threshold signature protocol, the simulator gets pair (m, s) s.t. $s = m^d \mod N$.

Setup Procedure
Pick random $\hat{d} \in \mathbb{Z}_n$ and proceed as in the Setup of the actual protocol:

1. Select random shares $\hat{d}_j, \hat{d}'_j \in \mathbb{Z}_q$, for $j = 1, \ldots, n-1$, and set $\hat{d}_n = \hat{d} - 2^{|N|-l}d_{pub} - \sum_{i=1}^{n-1} \hat{d}_j \pmod{q}$, as in step 1 of the Setup procedure.
2. Share each $\hat{d}_j$ and $\hat{d}'_j$ using the Pedersen's VSS: Choose random polynomials $\hat{f}_j(z) = \hat{d}_j + \hat{f}_{j1}z + \cdots + \hat{f}_{jt}z^t$ and $\hat{f}'_j(z) = \hat{d}'_j + \hat{f}'_{j1}z + \cdots + \hat{f}'_{jt}z^t$ over $\mathbb{Z}_q$ of degree t; compute and publish the witnesses $\hat{w}_{j0} = g^{\hat{d}_j}h^{\hat{d}'_j} \pmod{p}$ and $\hat{w}_{jk} = g^{\hat{f}_{jk}}h^{\hat{f}'_{jk}} \pmod{p}$ for $k = 1, \ldots, t$.
3. Compute the secret shares $\hat{s}s_{ij}$ and $\hat{s}s'_{ij}$ as $\hat{s}s_{ij} = \hat{f}_j(i) \pmod{q}$ and $\hat{s}s'_{ij} = \hat{f}'_j(i)$ and distribute $\hat{d}_i, \hat{d}'_i, \hat{s}s_{ij}$ and $\hat{s}s'_{ij}$ ($\forall j$) to each M_i over a secure channel.

Threshold Signature Protocol (on additional input (m, s)):

1. Generate partial signatures $\hat{s}_i$ for $i = 1, \ldots, n-1$ as $\hat{s}_i = m^{\hat{d}_i} \pmod{N}$. Compute $D = \hat{d}_1 + \ldots + \hat{d}_{n-1}$, and $\hat{\alpha} = \lfloor D/q \rfloor + 1$. Compute $\hat{s}_n = s * m^{\hat{\alpha}q}/(m^{2^{|N|-l}d_{pub}}\prod_{j=1}^{n-1} \hat{s}_j) \pmod{N}$.
2. Output values $\hat{s}_i$ on behalf of the uncorrupted players M_i.
3. If needed, execute the ZKPK proof for $M_i \neq M_n$, and simulate it for M_n.

Proactive Update
Proceed in exactly the same manner as the Proactive Update protocol:

1. At the beginning of round r, for all uncorrupted players M_i, select (sub)shares $\hat{d}_{ij}$ and $\hat{d}'_{ij}$ uniformly in $\mathbb{Z}_q$ for $j = 1, \ldots, n-1$, and set $\hat{d}_{in} = \hat{d}_i^{(r-1)} - \sum_{k=1}^{n-1} \hat{d}_{ik} \pmod{q}$ and $\hat{d}'_{in} = \hat{d}_i'^{(r-1)} - \sum_{k=1}^{n-1} \hat{d}'_{ik} \pmod{q}$. Broadcast witness values $\hat{w}_{ij}^{(r)} = g^{\hat{d}_{ij}}h^{\hat{d}'_{ij}} \pmod{p}$, and hand $(\hat{d}_{ij}, \hat{d}'_{ij})$ to M_j ($\forall j$) over a secure channel.
2. Compute M_j's new secret shares $\hat{d}_j^{(r)} = \sum_{i=1}^n \hat{d}_{ij} \pmod{q}$ and $\hat{d}_j'^{(r)} = \sum_{i=1}^n \hat{d}'_{ij} \pmod{q}$, as in the Proactive Update protocol.
3. Re-share the new additive share $\hat{d}_j^{(r)}, \hat{d}_j'^{(r)}$ using Pedersen's VSS, as in the Proactive Update protocol.

Fig. 4. Simulator Construction ($\mathcal{SIM}$)

Threshold Signature Protocol: Since d_i and $\hat{d}_i$, for $i = 1, \ldots, n-1$, have the identical distributions, therefore distributions of the corresponding partial signatures s_i and $\hat{s}_i$, are also identical. However, values s_n and $\hat{s}_n$ are the same only in the event that value α in the protocol and value $\hat{\alpha}$ in the simulation are the same. Recall that $\hat{\alpha}$ in the simulation is computed as $\hat{\alpha} = \lfloor D/q \rfloor + 1$ where $D = \sum_{j=1}^{n-1} \hat{d}_j$. Note that $D = (\hat{\alpha} - 1)q + R$ where $R = (D \mod q)$. By equation (2), value α computed by the protocol would satisfy equation

$$d = 2^{|N|-l}d_{pub} + D + d_n - \alpha q = 2^{|N|-l}d_{pub} + R + d_n + (\hat{\alpha} - \alpha - 1)q$$

because $d_1, ..., d_{n-1}$ are distributed identically to $\hat{d}_1, ..., \hat{d}_{n-1}$.

Since d_n and R are elements in $\mathbb{Z}_q$ for $q \geq 2^{|N|-l+\tau+\log r}$, and since $d \in [2^{|N|-l}d_{pub}, 2^{|N|-l}d_{pub} + 2^{|N|-l}]$, the above equation implies that there are only two possible cases: $\alpha = \hat{\alpha} - 1$ and $\alpha = \hat{\alpha}$. The first case happens if $d \geq 2^{|N|-l}d_{pub} + R$ and the second if $d < 2^{|N|-l}d_{pub} + R$. However, the probability that $d < 2^{|N|-l}d_{pub} + R$, and hence that $\alpha = \hat{\alpha}$, is at least $1 - 2^{-(\tau+\log r)}$ because the probability of the other case is at most the probability that R is less than $2^{|N|-l}$, which, given that R is a uniformly distributed element in $[0, q]$, is at most $2^{-(\tau+\log r)}$.

Note that value α stays the same in all instances of the threshold signature protocol in any given update round. Since the same holds for the $\hat{\alpha}$ value in the simulation, the probability that the adversary's view of all these protocol instances is different from the view of all the simulation instances remains at most $2^{-(\tau+\log r)}$. In other words, the statistical difference between the adversary's view of the real execution and the simulation in any update round, is at most $(1/r)2^{-\tau}$.

Proactive Update Protocol: Since values $\{d_i\}_{i=1..n-1}$ and $\{\hat{d}_i\}_{i=1..n-1}$ are distributed identically, the only difference in the execution and the simulation of the update protocol can come from sharing of the d_n value in the protocol and $\hat{d}_n$ in the simulation. However, since this sharing is a "additive" equivalent of Pedersen VSS, and the second-layer sharing of the shares of the d_n or $\hat{d}_n$ value is done with Pedersen VSS too, the whole protocol hides the shared value d_n perfectly, and hence the adversarial view in the simulation of the update protocol is identical to the adversarial view of the actual protocol.

Since the statistical difference between the protocol and the simulation is zero in the setup stage and in any proactive update stage, and at most $(1/r)2^{-\tau}$ in any single update round, given r rounds the overall difference between adversarial view of the protocol execution and its simulation is at most $2^{-\tau}$, which completes our argument.

Theorem 2 (Robustness). *Under the Discrete Logarithm and Strong RSA assumptions, our proactive signature scheme is robust against a t-threshold proactive adversary for $t < n/2$.*

Proof. Note that the only way robustness can be broken is if some malicious player M_i cheats either in the proactive update protocol, by re-sharing a value different than its proper current share d_i committed in Pedersen commitment $w_i = g^{d_i}h^{d_i} \mod p$, or M_i cheats in the signature protocol, by proving correct the wrong partial signature $s_i \neq m^{d_i} \mod N$. Since the first type of cheating is infeasible under the discrete logarithm assumption and the second type is infeasible under the strong RSA assumption, the claim follows.

4.1 Security Implications

Taking $l = 0$, Theorem 1 implies that the new proactive signature scheme is as secure as the standard RSA:

Corollary 1. *Under the RSA assumption in the Random Oracle Model, our scheme is a secure t-threshold proactive signature, for $l = 0$ and $q \geq r2^{|N|+80}$.*

On the other hand, note that the RSA adversary can always correctly guess the most significant half of the bits of d with probability $1/(e-1)$.[1] Together with theorem 1, this implies the following corollary:

Corollary 2. *Under the RSA assumption (in the Random Oracle Model), the time T_{PRSA} to break the new proactive signature scheme for $e = 2^i+1$, $l = |N|/2$ and $q \geq r2^{|N|/2+80}$, is at least $T_{PRSA} \geq 2^{-i} T_{RSA}$, where T_{RSA} is the time required to break the CMA security of the standard (FDH) RSA signature scheme for modulus of length $|N|$.*

For the most popular value of $e = 3$, this implies that if the 1024-bit modulus RSA has a 2^{80} security then our proactive RSA scheme running on the same modulus for $l = 512$ and $q \geq r2^{512+80}$ would have at least 2^{79} security. For $e = 17$ the provable security would be 2^{76}. Of course, our scheme could be executed with slightly larger N to compensate for the 2^i factor in security degradation, but with key shares sizes still limited by $q < r2^{|N|/2+80}$. The efficiency of the resulting schemes resulting from Corollary 2 should be compared with the straightforward settings implied by Corollary 1, where same 2^{80} security is given by 1024 bit N but with larger bound of $r2^{|N|+80}$ on the share size q.

However, since there are no known attacks against RSA which speed up the factorization of N when half of the most significant bits of d are revealed for small values of e, it can be plausibly hypothesized that for small e's, the proposed proactive RSA scheme remains as secure as standard RSA for the same modulus size even with half of the most significant bits of d are revealed.

Finally we remark that the security analysis of our scheme given in Theorem 1 grants the adversary the knowledge of $n-1$ shares instead of just t shares he can see in the protocol, which suggests that our security analysis can be improved and that our scheme is possibly secure using smaller share sizes than our analysis recommends.

5 Improved Security Analysis of Rabin's Proactive RSA

Overview of Proactive RSA Scheme of [36]. During the setup, a trusted dealer generates the RSA public (N, e) and private $(d, \hat{p}, \hat{q})$ key pairs. The signature

---

[1] Note that $ed = 1 \pmod{\phi(N)}$ implies that $d = 1/e(1 + k\phi(N))$ for some integer $k = 1, ..., e-1$. Therefore, since $N - \phi(N) < \sqrt{N}$, it follows that $0 \leq \hat{d}_k - d < \sqrt{N}$ for $\hat{d}_k = \lfloor 1/e(1+kN) \rfloor$ for one of the $e-1$ choices of k. Thus any adversary facing the RSA cryptosystem can with probability $1/(e-1)$ guess the $|N|/2$ most significant bits of d by picking the right k and computing $\hat{d}_k$ as above.

key d is shared additively among the players. Each M_i gets a share d_i, chosen uniformly in $[-R, R]$ where $R = nN^2$, and the dealer publishes public value d_{public} such that

$$d_{public} = d - \sum_{i=1}^{n} d_i \quad (\text{over } \mathbb{Z}) \qquad (3)$$

This can be easily extended, so that like our new scheme, l most significant bits of d are publicly revealed and added to the d_{pub} value, and only the remaining $(|N|-l)$-bit value $d - 2^{|N|-l}d_{pub}$ is shared as above. The witness value $w_i = g^{d_i}$ (mod N) corresponding to each d_i is published, where g is an element of high order in $\mathbb{Z}_n^*$. Each share d_i is then itself shared using the Feldman VSS [15] over $\mathbb{Z}_n$. To sign a message m, each player M_i, generates a partial signature $s_i = m^{d_i}$ (mod N). Since the signature key d is shared over integers, the RSA signature can be easily reconstructed by simply multiplying n partial signatures, i.e.,

$$s = m^{d_{public}} \prod_{i=1}^{n} s_i \quad (\text{mod } N)$$

The detection of faults during the signing process can be performed using the protocols of [24, 19]. The secret share of the faulty player is then reconstructed by pooling in the shares of any $t+1$ players using a special variant of polynomial interpolation (refer to [36] for details). In the share update protocol each M_i additively re-shares its secret share d_i with (sub)shares $d_{ij} \in [-R/n, R/n]$ and

$$d_{i,public} = d_i - \sum_{j=1}^{n} d_{ij} \quad (\text{over } \mathbb{Z})$$

is made a public value. The new secret share for $d_i^{(r)}$ of M_i is then computed as $d_i^{(r)} = \sum_{j=1}^{n} d_{ji}$, and M_i shares it using Feldman VSS over $\mathbb{Z}_n$.

Improved Security Analysis and Improved Performance. First, we note that the simulator for the setup phase presented in [36] has a small error. That simulator for the key distribution protocol picks random shares $\hat{d}_i \in [-R, R]$, for $i = 1, \ldots, n-1$, and it picks $\hat{d}_{public}$ *uniformly* at random in $[-nR, nR+N]$. However, values generated in this way are not statistically indistinguishable from the values in the protocol, because if the d_i values are chosen uniformly in $[-R, R]$, then by equation (3), value d_{public} has a normal probability distribution, which is immediately *distinguishable* from the uniform distribution of $\hat{d}_{public}$.

The corrected simulation of the key distribution (and the subsequent update protocols) works exactly in the same manner as the actual protocol. The simulator should choose some secret value $\hat{d} \in [0, N-1]$ at random, and share this new value in exactly the same manner as in the protocol. After r update rounds, the overall statistical difference between the view of the adversary interacting with the protocol and the view of the adversary interacting with the (new) simulator is at most rN/R. This difference is negligible if $R = rN2^\tau$, where $\tau \geq 80$, instead of the $R = nN^2$ value recommended in [36].

This shows that secret shares can be picked from range $[-rN2^\tau, rN2^\tau]$, instead of range $[-nN^2, nN^2]$ of the original scheme, which means an almost factor of 2 improvement in the share size. Since the computational cost of this scheme is driven by cost of the exponentiation $s_i = m^{d_i} \mod N$ done by each player, factor of 2 improvement in the size of d_i speeds the signature generation by the same factor.

References

1. M. Bellare and P. Rogaway. Random oracles are practical: A paradigm for designing efficient protocols. In *ACM Conference on Computer and Communications Security*, pages 62–73, 1993.
2. F. Boudot. Efficient proofs that a committed number lies in an interval. In *EUROCRYPT'00*, volume 1807 of *LNCS*, pages 431–444, 2000.
3. F. Boudot and J. Traor. Efficient Publicly Verifiable Secret Sharing Schemes with Fast or Delayed Recovery. In *Second International Conference on Information and Communication Security (ICICS)*, pages 87–102, November 1999.
4. C. Boyd. Digital multisignatures. In *Cryptography and Coding*, pages 241–246. Claredon Press, May 1989.
5. J. Camenisch and M. Michels. Proving in zero-knowledge that a number is the product of two safe primes. In *EUROCRYPT'99*, volume 1592 of *LNCS*, pages 107–122, 1999.
6. J. Camenisch and M. Michels. Separability and efficiency for generic group signature schemes. In *CRYPTO'99*, volume 1666 of *LNCS*, pages 106–121, 1999.
7. R. Canetti, R. Gennaro, S. Jarecki, H. Krawczyk, and T. Rabin. Adaptive security for threshold cryptosystems. In *CRYPTO'99*, volume 1666 of *LNCS*, pages 98–115, 1999.
8. A. Chan, Y. Frankel, and Y. Tsiounis. Easy come - easy go divisible cash. In *EUROCRYPT'98*, volume 1403 of *LNCS*, pages 561–575, 1998.
9. R. Croft and S. Harris. Public-key cryptography and re-usable shared secrets. In *Cryptography and Coding*, pages 189–201. Claredon Press, May 1989.
10. I. Damgård. Efficient Concurrent Zero-Knowledge in the Auxiliary String Model. In *EUROCRYPT'00*, volume 1807 of *LNCS*, pages 418–430, 2000.
11. I. Damgård and E. Fujisaki. A statistically-hiding integer commitment scheme based on groups with hidden order. In *ASIACRYPT'02*, volume 2501 of *LNCS*, pages 125–142. Springer, 2002.
12. A. De Santis, Y. Desmedt, Y. Frankel, and M. Yung. How to share a function securely. In *Proc. 26th ACM Symp. on Theory of Computing*, pages 522–533, Montreal, Canada, 1994.
13. Y. Desmedt. Society and Group Oriented Cryptosystems. In *CRYPTO '87*, number 293 in LNCS, pages 120–127, 1987.
14. Y. Desmedt and Y. Frankel. Threshold cryptosystems. In *CRYPTO '89*, number 435 in LNCS, pages 307–315, 1990.
15. P. Feldman. A practical scheme for non-interactive verifiable secret sharing. In *28th Symposium on Foundations of Computer Science (FOCS)*, pages 427–437, 1987.
16. Y. Frankel and Y. Desmedt. Parallel reliable threshold multisignature. Technical Report TR-92-04-02, Dept. of EE and CS, U. of Winsconsin, April 1992.

17. Y. Frankel, P. Gemmell, P. D. MacKenzie, and M. Yung. Optimal-resilience proactive public-key cryptosystems. In *38th Symposium on Foundations of Computer Science (FOCS)*, pages 384–393, 1997.
18. Y. Frankel, P. Gemmell, P. D. MacKenzie, and M. Yung. Proactive RSA. In *Crypto'97*, volume 1294 of *LNCS*, pages 440–454, 1997.
19. Y. Frankel, P. Gemmell, and M. Yung. Witness-based cryptographic program checking and robust function sharing. In *Proc. 28th ACM Symp. on Theory of Computing*, pages 499–508, Philadelphia, 1996.
20. Y. Frankel, P. MacKenzie, and M. Yung. Adaptively-secure distributed threshold public key systems. In *Proceedings of ESA 99*, 1999.
21. Y. Frankel, P. MacKenzie, and M. Yung. Adaptively-secure optimal-resilience proactive RSA. In *ASIACRYPT'99*, volume 1716 of *LNCS*, 1999.
22. Y. Frankel, P. D. MacKenzie, and M. Yung. Adaptive security for the additive-sharing based proactive rsa. In *Public Key Cryptography 2001*, volume 1992 of *LNCS*, pages 240–263, 2001.
23. E. Fujisaki and T. Okamoto. Statistical Zero Knowledge Protocols to Prove Modular Polynomial Relations. In *CRYPTO '97*, volume 1294 of *LNCS*, pages 16–30, 1997.
24. R. Gennaro, S. Jarecki, H. Krawczyk, and T. Rabin. Robust and Efficient Sharing of RSA Functions. In *CRYPTO '96*, volume 1109 of *LNCS*, pages 157–172, 1996.
25. R. Gennaro, S. Jarecki, H. Krawczyk, and T. Rabin. Robust Threshold DSS Signatures. In *EUROCRYPT '96*, number 1070 in LNCS, pages 354–371, 1996.
26. R. Gennaro, S. Jarecki, H. Krawczyk, and T. Rabin. Secure distributed key generation for discrete log based cryptosystems. In *EUROCRYPT'99*, volume 1592 of *LNCS*, pages 295–310, 1999.
27. A. Herzberg, M. Jakobsson, S. Jarecki, H. Krawczyk, and M. Yung. Proactive public key and signature systems. In *ACM Conference on Computers and Communication Security*, pages 100–110, 1997.
28. A. Herzberg, S. Jarecki, H. Krawczyk, and M. Yung. Proactive secret sharing, or how to cope with perpetual leakage. In *CRYPTO '95*, volume 963 of *LNCS*, pages 339–352, 1995.
29. S. Jarecki, N. Saxena, and J. H. Yi. An Attack on the Proactive RSA Signature Scheme in the URSA Ad Hoc Network Access Control Protocol. In *ACM Workshop on Security of Ad Hoc and Sensor Networks (SASN)*, pages 1–9, October 2004.
30. J. Kong, P. Zerfos, H. Luo, S. Lu, and L. Zhang. Providing Robust and Ubiquitous Security Support for MANET. In *IEEE 9th International Conference on Network Protocols (ICNP)*, pages 251–260, 2001.
31. H. Luo and S. Lu. Ubiquitous and Robust Authentication Services for Ad Hoc Wireless Networks. Technical Report TR-200030, Dept. of Computer Science, UCLA, 2000. Available online at http://citeseer.ist.psu.edu/luo00ubiquitous.html.
32. D. Micciancio and E. Petrank. Simulatable Commitments and Efficient Concurrent Zero-Knowledge. In *EUROCRYPT'03*, volume 2656 of *LNCS*, pages 140–159, 2003.
33. NIST. Digital signature standard (DSS). Technical Report 169. National Institute for Standards and Technology, August 30, 1991.
34. R. Ostrovsky and M. Yung. How to withstand mobile virus attacks. In *10th ACM Symp. on the Princ. of Distr. Comp.*, pages 51–61, 1991.
35. T. Pedersen. Non-interactive and information-theoretic secure verifiable secret sharing. In *Crypto 91*, volume 576 of *LNCS*, pages 129–140, 1991.
36. T. Rabin. A Simplified Approach to Threshold and Proactive RSA. In *CRYPTO '98*, volume 1462 of *LNCS*, pages 89 – 104, 1998.

37. N. Saxena, G. Tsudik, and J. H. Yi. Admission Control in Peer-to-Peer: Design and Performance Evaluation. In *ACM Workshop on Security of Ad Hoc and Sensor Networks (SASN)*, pages 104–114, October 2003.
38. C. P. Schnorr. Efficient signature generation by smart cards. *Journal of Cryptology*, 4(3):161–174, 1991.
39. A. Shamir. How to share a secret. *Commun. ACM*, 22(11):612–613, Nov. 1979.
40. V. Shoup. Practical Threshold Signatures. In *EUROCRYPT'00*, volume 1807 of *LNCS*, pages 207–220, 2000.

A Zero Knowledge Proof of Partial Signature Correctness

For the purpose of proving the correctness of partial signatures in the proposed proactive RSA scheme, we apply the zero knowledge proofs for the equality of committed numbers in two different groups and for the range of a committed number. All these proofs are honest verifier zero-knowledge and can be converted either into standard zero-knowledge proof either at the expense of 1-2 extra rounds using techniques of [10, 11, 32], or into a non-interactive proof in the random oracle model using the Fiat-Shamir heuristic. We adopt the notation of [5] for representing zero-knowledge proof of knowledge protocols. For example, $\boxed{ZKPK\{x : R(x)\}}$ represents a ZKPK protocol for proving possession of a secret x which satisfies statement $R(x)$. In the protocols to follow, u (≥ 80) and v (≥ 40) are security parameters.

Protocol for proving the correctness of a partial signature:

$$\boxed{ZKPK\{d_i, d_i' : w_{i0} = g^{d_i} h^{d_i'} \ (\text{mod } p) \ \wedge \ s_i = m^{d_i} \ (\text{mod } N) \ \wedge \ d_i \in [0, q-1]\}}$$

The signer (or prover) M_i proves to the verifier the possession of its correct secret share d_i by using the following zero-knowledge proof system. The verifier can either be one of the players or an outsider who has inputs $w_{i0}, g, h, p, s_i, m, N, q$. All the protocols run in parallel, and failure of these protocols at any stage implies the failure of the whole proof.

1. The verifier follows the setup procedure of the Damgard-Fujisaki-Okamoto commitment scheme [23, 11], e.g. it picks a safe RSA modulus n and two elements G, H in $\mathbb{Z}_n^*$ whose orders are greater than 2. (We refer to [11] for the details of this commitment scheme.) If N is a safe RSA modulus then set $n = N$, $G = (G')^2 \bmod N$, $H = (H')^2 \bmod N$ for random $G', H' \in \mathbb{Z}_n^*$.
2. The prover computes the commitment $C = G^{d_i} H^R \ (\text{mod } n)$, where R is picked randomly from $[0, 2^v(q-1)]$ and uses **Protocol (1)** (see below), by substituting $(x, x_1', x_2', g_1, h_1, g_2, h_2, n_1, n_2, w_1, w_2, b, b')$ with $(d_i, R, d_i', G, H, g, h, n, p, C, w_{i0}, q-1, 2^v(q-1))$, respectively, to execute:
   $ZKPK\{d_i, R, d_i' : C = G^{d_i} H^R \ (\text{mod } n) \ \wedge \ w_{i0} = g^{d_i} h^{d_i'} \ (\text{mod } p)\}$.
3. The prover then uses **Protocol (1)** (see below), by substituting $(x, x_1', x_2', g_1, h_1, g_2, h_2, n_1, n_2 w_1, w_2, b, b')$ with $(d_i, R, 0, G, H, m, m, n, N, C, s_i, q-1, 2^v(q-1))$, respectively, to execute:
   $ZKPK\{d_i, R : C = G^{d_i} H^R \ (\text{mod } n) \ \wedge \ s_i = m^{d_i} \ (\text{mod } N)\}$.
4. The prover uses **Protocol (2)** (see below), by substituting (x, x', b) with $(d_i, R, q-1)$, respectively, to execute:
   $ZKPK\{d_i, R : C = G^{d_i} H^R \ (\text{mod } n) \ \wedge \ d_i \in [0, q-1]\}$

Protocol (1). $\boxed{ZKPK\{x, x_1', x_2' : w_1 = g_1^x h_1^{x_1'} \pmod{n_1} \wedge w_2 = g_2^x h_2^{x_2'} \pmod{n_2}\}}$

Assumption: $x, x_2' \in [0, b]$ and $x_1' \in [0, b']$.

This protocol is from [5], [2], and is perfectly complete, honest verifier statistical zero-knowledge and sound under the strong RSA assumption [23] with the soundness error 2^{-u+1}, given than (g_1, h_1, n_1) is an instance of the Damgard-Fujisaki-Okamoto commitment scheme [23, 11].

1. The prover picks random $r \in [1, \ldots, 2^{u+v}b - 1]$, $\eta_1 \in [1, \ldots, 2^{u+v}b' - 1]$, $\eta_2 \in [1, \ldots, 2^{u+v}b - 1]$ and computes $W_1 = g_1^r h_1^{\eta_1} \pmod{n_1}$ and $W_2 = g_2^r h^{\eta_2} \pmod{n_2}$. It then sends W_1 and W_2 to the verifier V.
2. The verifier selects a random $c \in [0, \ldots, 2^u - 1]$ and sends it back to the prover.
3. The prover responds with $s = r + cx$ (in $\mathbb{Z}$), $s_1 = \eta_1 + cx_1'$ (in $\mathbb{Z}$) and $s_2 = \eta_2 + cx_2'$ (in $\mathbb{Z}$)
4. The verifier verifies as $g_1^s h_1^{s_1} = W_1 w_1{}^c \pmod{n_1}$ and $g_2^s h_2^{s_2} = W_2 w_2{}^c \pmod{n_2}$.

Protocol (2). $\boxed{ZKPK\{x, x' : C = G^x H^{x'} \pmod{n} \wedge x \in [0, b]\}}$

Assumption: $x \in [0, b]$ and $x' \in [0, 2^v b]$.

This protocol (from [2]) is an exact range proof, honest verifier statistical zero-knowledge, complete with a probability greater than $1 - 2^{-v}$, and sound under the strong RSA assumption given that (G, H, n) is an instance of the Damgard-Fujisaki-Okamoto commitment scheme, similarly as in protocol (1).

1. The prover sets $T = 2(u + v + 1) + |b|, X = 2^T x, X' = 2^T x', \beta = 2^{u+v+1}\sqrt{b}$ and $C_T = G^X H^{X'} \pmod{n}$.
2. The prover uses **Protocol (3)** (see below), by substituting (x, x', com, B, γ) with $(X, X', C_T, 2^T b, 2^{T/2}\beta)$, respectively, to execute the following (note that $X \in [0, 2^T b]$):
$$ZKPK\{X, X' : C_T = G^X H^{X'} \pmod{n} \wedge X \in [-2^{T/2}\beta, 2^T b + 2^{T/2}\beta]\}$$
Proving that $X \in [-2^{T/2}\beta, 2^T b + 2^{T/2}\beta]$ implies that $x \in [0, b]$, since $2^{T/2}\beta < 2^T$.

Protocol (3). $\boxed{ZKPK\{x, x' : com = G^x H^{x'} \pmod{n} \wedge x \in [-\gamma, B + \gamma]\}}$
Here $\gamma = 2^{u+v+1}\sqrt{B}$.

Assumption: $x \in [0, B]$ and $x' \in [0, 2^v B]$.

This proof was proposed in [2] and is honest verifier statistical zero-knowledge, complete with a probability greater than $1 - 2^{-v}$, and sound under the strong RSA assumption just like protocol (2).

1. The prover executes $ZKPK\{x, x' : com = G^x H^{x'} \pmod{n}\}$
2. The prover sets $x_1 = \lfloor\sqrt{x}\rfloor, x_2 = x - x_1^2, \hat{x}_1 = \lfloor\sqrt{B-x}\rfloor, \hat{x}_2 = B - x - \hat{x}_1^2$, and chooses randomly $r_1, r_2, \hat{r}_1, \hat{r}_2$ in $[0, 2^v B]$, such that $r_1 + r_2 = x'$ and $\hat{r}_1 + \hat{r}_2 = -x'$.
3. The prover computes new commitments $e_1 = G^{x_1^2} H^{r_1} \pmod{n}, \hat{e}_1 = G^{\hat{x}_1^2} H^{\hat{r}_1} \pmod{n}, e_2 = G^{x_2} H^{r_2} \pmod{n}, \hat{e}_2 = G^{\hat{x}_2} H^{\hat{r}_2} \pmod{n}$, and sends e_1 and $\hat{e}_1$ to the verifier.
4. The verifier computes $e_2 = com/e_1 \pmod{n}$ and $\hat{e}_2 = G^B/(com * \hat{e}_1) \pmod{n}$.

5. The prover uses **Protocol (4)** (see below), by substituting (x, x', com_{sq}) with (x_1, r_1, e_1) and then with $(\hat{x}_1, \hat{r}_1, \hat{e}_1)$, to execute the following:
   $ZKPK\{x_1 : e_1 = G^{x_1^2} H^{r_1} \pmod{n}\}$
   $ZKPK\{\hat{x}_1 : \hat{e}_1 = G^{\hat{x}_1^2} H^{\hat{r}_1} \pmod{n}\}$
   This proves that e_1 and $\hat{e}_1$ hide a square.
6. The prover uses **Protocol (5)** (see below), by substituting (x, x', com_2, B_1) with $(x_2, r_2, e_2, 2\sqrt{B})$, respectively and then with $(\hat{x}_2, \hat{r}_2, \hat{e}_2, 2\sqrt{B})$, respectively, to execute the following (note that x_2 and $\hat{x}_2 \in [0, 2\sqrt{B}]$):
   $ZKPK\{x_2 : e_2 = G^{x_2} H^{r_2} \pmod{n} \wedge x_2 \in [-\gamma, \gamma]\}$
   $ZKPK\{\hat{x}_2 : \hat{e}_2 = G^{\hat{x}_2} H^{\hat{r}_2} \pmod{n} \wedge \hat{x}_2 \in [-\gamma, \gamma]\}$
   This proves that e_2 and $\hat{e}_2$ hide numbers belonging to $[-\gamma, \gamma]$.

Steps 2, 5 and 6 above, imply that $x \in [-\gamma, B + \gamma]$.

Protocol (4). $\boxed{ZKPK\{x, x' : com_{sq} = G^{x^2} H^{x'} \pmod{n}\}}$

This protocol first appeared in [23], generalized (and corrected) in [11] and proves that a committed number is a square. The protocol is honest verifier statistical zero-knowledge, perfectly complete, and sound under the strong RSA assumption just like protocol (2).

Protocol (5). $\boxed{ZKPK\{x, x' : com_2 = G^x H^{x'} \pmod{n} \wedge x \in [-2^{u+v} B_1, 2^{u+v} B_1]\}}$

Assumption: $x \in [0, B_1]$, and $x' \in [0, 2^v B_1]$.

This proof was proposed in [8], allows a prover to prove the possession of a discrete logarithm x lying in the range $[-2^{u+v} B_1, 2^{u+v} B_1]$ given x which belongs to a smaller interval $[0, B_1]$. Using the commitment scheme of [23, 11], this proof is honest verifier statistical zero-knowledge, complete with a probability greater than $1 - 2^{-v}$, and sound under the strong RSA assumption with soundness error 2^{-u+1}.

Proof of Plaintext Knowledge for the Ajtai-Dwork Cryptosystem

Shafi Goldwasser[1,2] and Dmitriy Kharchenko[2]

[1] CSAIL, Massachusetts Institute of Technology,
Cambridge, MA 02139, USA
[2] Department of Computer Science and Applied Mathematics,
Weizmann Institute of Science, Rehovot 76100, Israel

Abstract. Ajtai and Dwork proposed a public-key encryption scheme in 1996 which they proved secure under the assumption that the unique shortest vector problem is hard in the worst case. This cryptosystem and its extension by Regev are the only one known for which security can be proved under a worst case assumption, and as such present a particularly interesting case to study.

In this paper, we show statistical zero-knowledge protocols for statements of the form "plaintext m corresponds to ciphertext c" and "ciphertext c and c' decrypt to the same value" for the Ajtai-Dwork cryptosystem. We then show a interactive zero-knowledge proof of plaintext knowledge (PPK) for the Ajtai-Dwork cryptosystem, based directly on the security of the cryptosystem rather than resorting to general interactive zero-knowledge constructions. The witness for these proofs is the randomness used in the encryption.

Keywords: Lattices, Verifiable Encryption, Ajtai-Dwork Cryptosystem, Worst Case Complexity Assumption, Proof of Plaintext Knowledge.

1 Introduction

There is much to celebrate in the progress made by cryptography on many fronts: rigorous definitions of security of natural cryptographic tasks, constructions of schemes achieving security based on general assumptions, new and seemingly contradictory possibilities such as zero-knowledge proofs and secure multi-party computations.

Still, during all this time, the implementations of this progress or rather the assumptions that underly all implementations, remain almost exclusively the intractability of factoring integers and of computing discrete logarithms which go back to the original papers of [9, 25] (often even stronger versions of these assumptions are utilized to gain better efficiency, such as higher quadratic residuosity, DDH, Strong-RSA). There are a couple of exceptions: computational problems over Elliptic Curves and computational problems over Integer Lattices. Whereas the computational problems over Elliptic curves do not seem to be inherintely harder than the analogous problems over finite fields, the use of computational problems over lattices seem to present a new frontier. Due to the pioneering work of Ajtai[], these problems certainly show the greatest promise from a *theoretical treatment* point of view.

In particular, in 1996, Ajtai and Dwork proposed [1] a public-key cryptosystem which is secure under the assumption that the unique shortest vector problem in integer lattices is hard in the worst case. The Ajtai-Dwork cryptosystem (and its extension by Regev [24]) are the only known public-key cryptosystems with the property that breaking a random instance of it is as hard as solving the worst-case instance of problem on which the system security is based. As such it present a particularly interesting and unique system to study from a complexity theoretic point of view.

Much study has been dedicated to the number theory based encryption systems (e.g. Cramer-Shoup, Paillier, RSA), showing how to incorporate them efficiently into larger protocols (e.g. designated confirmer signatures, e-cash protocols), extending their basic functionality (e.g. threshold decryption, verifiable encryption, group encryption, key-escrow versions), and extending them to achieve stronger security definitions (e.g. chosen cipher-text security, interactive encryption with efficient proofs of plaintext knowledge).

In contrast, the work on AD cryptosystems has been restricted to attempting cryptanalysis of the original scheme([23], showing chosen cipher text attacks [19], and proofs tightening the worst case versus average security reductions [24]. To date, there has been no protocol work involving the usage of AD encryption.

We can only speculate why this study is missing. Possibly, since the mathematics underlying the AD systems seemingly does not lend itself to simple treatment as in the case of the number theoretic schemes. Possibly, because AD is viewed largely of interest as a theoretical case study rather than one envisioned useful within other application. Or, perhaps, because it is a secondary order concern which naturally will follows the basic study of security. In any case, as by enlarge all existing number theoretic cryptosystems stand and fall together whereas the security of AD seems unrelated and could hold even if the former does not, we feel it is time to begin such treatment. Certainly, we will only be able develop intuition about the usability of this system, by attempting to do so. We initiate this study in this paper.

We begin with investigating very simple questions, which seem fundamental to many applications of public-key encryption schemes.

- First, we show how AD can be augmented to be a verifiable encryption scheme, by providing statistical zero knowledge proofs for basic statements about the plaintext of AD ciphertexts, such as 'ciphertexts c and c' decrypt to the same plaintext' and 'ciphertext c decrypts to plaintext m'. The witness for these proofs is the randomness used in the encryption.
- Second, we show a zero-knowledge interactive proof of plaintext knowledge for AD ciphertexts. Again the witness for this proof is the randomness used in the encryption. The construction is simple and direct, and does not utilize general ZK interactive proof constructions or general tools such as the existence of one-way functions. Rather it exploits the statistical zero knowl-

edge protocols constructed above to prove statements which arise within the interactive proof of plaintext knowledge. The computational zero knowledge property is proved assuming the security of the AD cryptosystem itself. The existence of a zero-knowledge interactive proof of plaintext knowledge, establishes in turn an interactive encryption variant of AD cryptosystem which is CCA1 secure ([13, 16]) costing reasonable overhead beyond the complexity of AD encryption itself. In contrast Hall, Goldberg, Schneier [19] showed that the secret key of the AD cryptosystem can be recovered using a CCA1 attack.

Previously, computational zero knowledge protocols for all the statements we prove were only known by utilizing general ZK interactive proofs for NP [17].

Throughout our work, instead of using the original Ajtai-Dwork construction which has non-zero decryption error probability, we use the decryption-error-free variant of Goldreich, Goldwasser and Halevi [15]. The semantic security of the modified cryptosystem holds under under the same assumption as the original cryptosystem. We refer to it as the AD cryptosystem throughout.

We make technical use of two prior works. The work of Micciancio and Vadhan[21] which shows a statistical zero-knowledge protocol with efficient provers for approximate versions of the SVP and CVP problems where the witness is a short vector in a lattice (or a point close to the target in the CVP case). And the work of Nguyen and Stern [23] which show how to use a CVP oracle to cryptanalyze the AD cryptosystem. Although Nguyen and Stern's work was aimed at cryptanalysis and showed that AD cryptosystem is no harder to break than the CVP problem, we use it as a positive result, using it as a tool to generate 'good instances' of an AD public key and ciphertexts for our verifiable encryption protocols for which our protocols will work. This continues the traditional pattern of research on lattices in cryptography, where progress on lattice research is used on one hand to cryptanalyze existing schemes and on the other hand to provide security proofs for lattice based cryptographic schemes.

We proceed to elaborate on related work and concepts, and our results in some detail.

1.1 Related Results and Conepts

VERIFIABLE ENCRYPTION. *Verifiable encryption* was introduced by Stadler in [26] in the context of publicly verifiable secret sharing, and in more general form by Asokan, Shoup and Waidner in [2] for the purpose of fair exchange of digital signatures. In the verifiable encryption setting, there are three parties. A party who generates the secret/public key pair (PK, SK), an encryptor which we refer to as the prover who creates a ciphertext of some plaintext, and a verifier who on input a public-key and a ciphertext verifies some application-driven properties of the plaintext. Verifiable encryption is defined with respect to some binary relation R defined on plaintext messages. Informally, a *verifiable encryption* with respect to relation R is a zero-knowledge protocol which, on public inputs ciphertext c, δ, and PK allows a prover to convince a verifier that the ciphertext c is an encryption of a message m with public key PK such that $(m, \delta) \in R$ (as in [4]). The prover uses the randomness which was used to generate the ciphertext c as auxilary input.

By using zero knowledge interactive proofs for NP [17], it is clearly possible to turn all known encryption schemes into verifiable encryption schemes for any $R \in NP$. However, for specific relations R of interest we may be able to get much more efficient protocols, with stronger security properties (e.g. statistical vs. computational zero-knowledge). For example, in recent work of Camenisch and Shoup [5], they propose a modification of the Cramer-Shop cryptosystem [7] based on the Paillier's decision composite residuosity assumption, for which they show an efficient verifiable encryption scheme for the relation $R = \{(m, (\delta, \gamma)) | \gamma^m = \delta\}$. Namely, they demonstrate efficient statistical ZK proofs on input a public key, ciphertext c (of the modified encryption scheme), and γ, δ pair, that c is the encryption of an m for which $\gamma^m = \delta$.

PLAINTEXT PROOFS OF KNOWLEDGE Given an instance of a public-key encryption scheme with public key pk, a proof of plaintext knowledge(PPK) allows an encryptor (or prover) to prove knowledge of the plaintext m of some ciphertext $C \in E_{pk}(m)$ to a receiver. A proof of plaintext knowledge should guarantee that no additional knowledge about m is revealed to the receiver or an eavesdropper. Customarily, this requirement is captured by requiring the plaintext proof of knowledge to be a zero-knowledge proof.

For the Rabin, RSA, Goldwasser-Micali, Paillier, El-Gamal encryption schemes, well known 3-round zero-knowledge public-coin proofs of knowledge protocols (often referred to as Σ protocols) can be easily adapted to achieve efficient PPKs.

When both the sender and the receiver are on-line, interactive public-key encryption protocols may be used. Starting with an underlying semantically secure public-key encryption scheme which has a zero-knowledge proof of plaintext knowledge, the sender of the ciphertext c in addition engages in a proof of plaintext knowledge with the receiver. The result is a CCA1 secure public-key encryption scheme [13, 16]. Utilizing efficient PPKs for specific number theoretic based semantically secure public-key encryption schemes such as the Blum-Goldwasser, Paillier, and El Gamal scheme, thus yields efficient CCA1 secure interactive public-key encryption variants of these schemes. Better yet, Katz[20] shows how design efficient interactive *non-malleable* proofs of plaintext knowledge for the RSA, Rabin, Paillier, and El-Gamal encryption schemes. Using these, one obtains efficient CCA2 secure interactive public-key encryption variants of the underlying schemes.

Naturally, if one-way functions exist, PPKs can be achieved using completeness results [17] for interactive zero-knowledge proofs for NP, proofs of knowledge for NP[12], and non-malleable interactive zero knowledge PPK for NP[8]. However, these general constructions are prohibitively inefficient as they require as a preliminary step polynomial time reductions to instances of NP-complete problems.

For the Ajtai-Dwork cryptosystem, these general completeness constructions of PPK were the only one knows prior to our work.

Finally, we note that in contrast to the interactive case, known constructions of *non-interactive* zero-knowledge proofs (NIZK) [8] for NP languages (which are a central tool in constructing CCA2 secure non-interactive public-key encryption given semantically secure public-key encryption algorithms) require trapdoor permutations. The intractability assumption on which the security of the Ajtai-Dwork cryptosystem is based, however, is not known to imply the existence of trapdoor permutations. It remains a central open problem to find a non-interactive CCA2 secure public-key encryption algorithm (efficient or otherwise) based on the AD-cryptosystem assumption.

LATTICE TOOLS. Our work uses as tools the results of [21] and [23]. In [21] Micciancio and Vadhan provide a zero-knowledge proof system for the GapCVP$_\gamma$ problem for $\gamma = \Omega(\sqrt{\frac{n}{\log(n)}})$ where n is the dimension of the lattice. An instance of the GapCVP$_\gamma$ is a triple consisting of a lattice L, a vector x and a value t. An instance is a YES instance if the distance between the vector x and the lattice L is less than t. If the distance is greater than γt the instance is a NO instance. In the proof zero-knowledge system of Micciancio and Vadhan [21] a prover proves to a verifier that an instance of the GapCVP$_\gamma$ is a YES instance. If the instance is NO instance, the verifier rejects with high probability.

Nguyen and Stern showed in [23] how to use a CVP oracle to distinguish between ciphertexts of '0' and '1' of the Ajtai-Dwork cryptosystem (with decryption errors). For a random public key and a random ciphertext of the Ajtai-Dwork cryptosystem, Nguyen and Stern construct some lattice L and some vector x. They show that for ciphertexts of '0' the distance between the lattice L and the vector x is likely to be small, whereas for ciphertexts of '1' the distance is likely to be large.

1.2 Our Results in Detail

VERIFIABLE ENCRYPTION FOR THE AD CRYPTOSYSTEM. The first result of this paper is the design of statistical zero-knowledge protocol for proving that ciphertexts decrypt to given plaintexts for the AD public key cryptosystems. Namely, on public inputs ciphertext c, δ, and public-key PK a verifiable encryption scheme for the *equivalence relation* $R = \{(m, \delta)|m = \delta\}$.

The encryption method of Ajtai and Dwork is bit-by-bit. Thus, to prove statement of the form "c is the ciphertext corresponding to m" it suffices to construct two zero-knowledge protocols: one to prove that a ciphertext decrypts to '0' and the other is to prove that a ciphertext decrypts to '1'. We construct two separate but in principle similar protocols for these tasks.

Ciphertexts of the AD cryptosystem are vectors in some public key dependent domain. The decryption algorithm decrypts every vector of the domain to '0' or '1', but not all vectors can be obtained by encrypting '0' or '1'. We say that a ciphertext is *legal* if it can be legally obtained by running encryption algorithm. The protocol for proving that a ciphertext c decrypts to 'b' (for $b \in \{0, 1\}$ respectively) has the following properties of completeness and soundness: if c is a legal ciphertext of 'b', then the verifier always accepts; if the decryption of c is not 'b' (regardless whether c is a legal encryption of 'b' or not), then the verifier rejects with high probability. Thus, completeness holds only for c's which were obtained legally by applying the encryption algorithm, whereas soundness of the protocols holds for any input c from the prescribed domain.

We remark that the completeness of the protocols we present here requires some technical condition to hold for the public-key and the input ciphertext on which it is applied. Luckily, theorems proved in [23] show that with good probability, random public-keys produced by the AD key generation algorithm and random ciphertexts produced by the AD encryption algorithm obey these technical conditions. Moreover, it is easy to check if these conditions hold for a given public-key at key generation time, and for a given ciphertext at encryption time (using the randomness used by the algorithm to generate the ciphertext). Thus, we modify the AD key generation algorithm and encryption algorithm to ensure that all legally generated public-keys and ciphertext obey the desired conditions. We emphasize that the soundness of our protocols hold for all ciphertexts and public keys, regardless of whether they obey the said conditions.

The idea behind the protocol for proving that a ciphertext decrypts to '0' is as follows. We show a transformation of AD public-keys and ciphertexts to instances of the GapCVP$_\gamma$ problem, such that (1) a legal AD public key and legal AD ciphertext which decrypts to '0', transforms to a YES instance of the GapCVP$_\gamma$; and (2) any AD public key and any ciphertext which decrypts to '1' transforms to a NO instance of the GapCVP$_\gamma$. On common input, a public key and a ciphertext, the prover and verifier transform it to the appropriate instance of GapCVP$_\gamma$ and run the Micciancio and Vadhan [21] zero-knowledge protocol for proving that the constructed instance is a YES instance. The value of $\gamma = \Omega(\sqrt{\frac{n}{\log(n)}})$ where n is polynomially related to the value of the security parameter. The same approach is used to design the protocol proving that a ciphertext decrypts to '1'.

The second result of this paper is the design of a verifiable encryption scheme on inputs PK and ciphertext c for the *encrypted equivalence* relation $R_1 = \{(m, c')|c'$ is a legal AD encryption with public key PK of $m\}$. Again, as the AD cryptosystem is bit-by-bit, it will suffice to construct a statistical zero-knowledge protocol to prove that given two ciphertexts c and c', encryped with public key PK, decrypt to the same bit. The prover's auxilary inputs are the random bits used by the encryption algorithm to generate c and c'.

We take advantage of the observation that if c and c' are legal AD ciphertexts of the same bit under the same AD public-key PK, then with high probability $\bar{c} = (c + c')$ mod $P(w_1, \ldots, w_n))$ decrypts to '0' (where $P(w_1, \ldots, w_n)$ is the parallelepiped spanned by the w_i's specified in the public key PK, see section 2.2). Thus, the prover need only prove is that $\bar{c}$ decrypts to '0' , using the statistical zero-knowledge protocol above for proving that AD ciphertext decrypts to '0'. If c is a legal ciphertext which decrypts to the same bit as c' the prover will succeed, whereas for any c which does not decrypt to the same bit as c' the prover will fail with high probability. Due to lack of space in this extended abstract further treatment of this result is omitted.

ZK PROOFS OF PLAINTEXT KNOWLEDGE FOR AD CRYPTOSYSTEM. We provide a **direct** (without using general results about NP in Zero-knowledge) zero-knowledge interactive proof of knowledge of the plaintext(PPK) for the AD cryptosystem.

As AD cryptosystem is a bit-by-bit encryption scheme, it suffices to describe how to prove on input public key PK, and ciphertext c of a single-bit plaintext b that the prover 'knows' b.

We prove that if c and c' are legal encryptions of b and b' respectively under AD public key PK, then with high probability $c+c'$ mod $P(w_1 \ldots w_n)$ decrypts to $b \oplus b'$. The proof of plaintext knowledge for the AD cryptosystem follows naturally. On input (PK, c) where c is an encryption of b, the prover sends the verifier a random encryption c' of a random bit b'. The verifier then asks the prover to either prove that it knows the decryption of c' or to prove that it knows a decryption of $c + c'$ mod $P(w_1 \ldots w_n)$. The former can be done simply by revealing the randomness used to encrypt c' and the latter can be done by proving in statistical zero-knowledge that $c + c'$ decrypts to $b \oplus b'$ using the statistical zero knowledge protocols designed in the first part of this work.

We prove that the resulting protocol is computational zero-knowledge under the same worst case intractability ISVP assumption of the AD cryptosystem.

Assumption ISVP: (Infeasibility of Shortest Vector Problem): There is no polynomial time algorithm, which given an arbitrary basis for an n-dimensional lattice which has a "unique $poly(n)$-shortest" vector, finds the shortest non-zero vector in the lattice. By "unique $poly(n)$-shortest" vector we mean that any vector in the lattice of length at most "poly(n)" times bigger than the shortest vector, is parallel to the shortest vector.

Combining the zero-knowledge PPK protocol with the AD cryptosystem, where the sender/encryptor (along with sending the ciphertext) interactively proves to the receiver that he knows the plaintext, yields automatically an interactive encryption scheme which is CCA1 secure based on ISVP. Previousy, Hall, Goldberg, Schneier [19] show how to completely recover the secret key of AD cryptosystem under a CCA1 attack.[1]

We believe that addressing the smaller problem of zero-knowledge PPK for AD cryptosystem as we have done here, is a promising first step in the pursuit of an CCA2 secure lattice based public-key encryption scheme, possibly first in an interactive setting by extending our protocol to be non-malleable.

2 Preliminaries

2.1 Notations

We let $x \in_R S$ denote choosing x at random with uniform probability in set S.

Given a parallelepiped $P = P(w_1, \ldots, w_n)$ and a vector v, we *reduce v modulo P* by obtaining a vector $v' \in P$ so that $v' = v + \sum_i c_i w_i$, where the c_i are all integers. We denote it by $v' = v$ mod P.

---

[1] Their work explicitly addresses the [15] variant with eliminated decryption.

All distances in this paper, are the Euclidean distances in $\mathbb{R}^n$. Let $dist(v_1, v_2)$ denote the distance between vectors v_1 and v_2 in $\mathbb{R}^n$, and $dist(v, S)$ denote the distance between vector v and a set S in $\mathbb{R}^n$.

Let $v_1, \ldots, v_m$ be linearly independent vectors in $\mathbb{R}^n$. An m-dimensional lattice with the basis $\{v_1, \ldots, v_m\}$ is the set of all integer linear combinations of v_i's, $\{\sum_{i=1}^{m} a_i v_i : a_i \in \mathbb{Z}\}$.

For linearly independent vectors $w_1, \ldots, w_n$ in $\mathbb{R}^n$ the *parallelepiped* spanned by w_i's is the set

$$P(w_1, \ldots, w_n) = \left\{ \sum_{i=1}^{n} a_i w_i : a_i \in [0,1) \right\}.$$

The *width* of the parallelepiped $P(w_1, \ldots, w_n)$ is the maximum over i of distances between w_i and the subspace spanned by other w_i's.

For every $v \in \mathbb{R}^n$ there is only one $v' \in P(w_1, \ldots, w_n)$ such that $v - v' = \sum_{i=1}^{n} a_i w_i$ for some integers $a_1, \ldots, a_n$. We denote this by $v' = v \mod P(w_1, \ldots, w_n)$. Note, that we can consider n to be dimension of the lattice L. We can always consider a lattice to be enclosed in a subspace spanned by it's basis vectors.

For interactive protocols involving two parties A (the prover) and B (the verifier), we let the notation $(A(a), B(b))(x)$ be the random variable denoting whether B accepts or rejects common input x following an execution of the protocol where B has private private input b and A has private input a.

2.2 The Ajtai-Dwork Cryptosystem with Eliminated Decryption Errors

Let the security parameter be denoted by n.

In order to simplify the construction we present the scheme in terms of real numbers, but we always mean numbers with some fixed finite precision. We need to define several parameters which will be used throughout the paper. For a security parameter n let $m = n^3$, $\rho_n = 2^{n \log n}$. We denote by B_n the n-dimensional cube of side-length ρ_n. We also denote by S_n the n-dimensional ball of radius n^{-8}.

The errorless Ajati-Dwork cryptosystem [15] consists of three algorithms $(\mathcal{K}, \mathcal{E}, \mathcal{D})$, where $\mathcal{K}$ is a key generation algorithm, $\mathcal{E}$ is an encryption algorithm, and $\mathcal{D}$ is a decryption.

The encryption algorithm encrypts strings in a bit-by-bit fashion and thus in this paper we shall assume henceforth that all messages are single bits.

Key Generating algorithm $\mathcal{K}$ on input 1^n:

The private key SK = vector u chosen at random from the n-dimensional unit ball.

The public key $PK = \{w_1, \ldots, w_n, v_1, \ldots, v_m, k\}$, where $v_1, \ldots, v_m, w_1, \ldots, w_n$ are vectors in $\mathbb{R}^n$ generated as follows.

v's: For $i = 1 \ldots n$ (1) Pick vector a_i at random from the set $\{x \in B_n : \langle x, u \rangle \in \mathbb{Z}\}$; (2) For $j = 1, \ldots, n$ select δ_j at random in S_n; (3) Output $v_i = a + \sum_{j=1}^{n} \delta_j$.

w's: The vectors $w_1,\ldots,w_n$ are obtained according to the same procedure as vectors $v_1,\ldots,v_m$, subject to the additional constraint that the width of the parallelepiped $P(w_1,\ldots,w_n)$ is at least $n^{-2}\rho_n$. Remark: It is shown in [1] that the width of $P(w_1,\ldots,w_n)$ will be large enough with probability at least $1 - n^{-1/2}$.

k: Choose k at random from the set of $\{i : \langle a_i, u\rangle \text{ is an odd integer}\}$. We note that such an index exists with probability $1 - 2^{-\Omega(m)}$.

We let $(SK, PK) \in \mathcal{K}(1^n)$ denote picking a pair of keys according to generating algorithm $\mathcal{K}$ on input 1^n, and call such pair an *instance* of AD cryptosystem. In various definitions and theorems in this paper, given an instance (SK, PK) of the AD cryptosystem, we often refer directly to components of PK and SK as $u, v_1, \ldots, v_n$ etc.

At times our algorithms may take as input keys $K = \{w_1,\ldots,w_n, v_1,\ldots,v_m, k\}$ which may not have been generated by $\mathcal{K}$, in which case we refer to them as AD public-key's.

Encryption algorithm $\mathcal{E}$ on input public key PK and message bit b:
  Choose $r = r_1,\ldots,r_m$, $r_i \in_R \{0,1\}$.
  If $b = \text{`0'}$, set ciphertext $c = \sum_{i=1}^{m} r_i v_i \mod P(w_1,\ldots,w_n)$.
  If $b = \text{`1'}$, set ciphertext $c = (\frac{v_k}{2} + \sum_{i=1}^{m} r_i v_i) \mod P(w_1,\ldots,w_n)$.
Denote ciphertext c obtained by encrypting b under public key PK using randomness r, as $c = \mathcal{E}_{pk}(b; r)$.

Decryption algorithm $\mathcal{D}$ on input ciphertext c and secret key u:
  If $\text{dist}(\langle c, u\rangle, \mathbb{Z}) < \frac{1}{4}$, output `0', otherwise output `1'.
We let $\mathcal{D}_{SK}(c) = b$, denote the event that c decrypts to b, under secret key SK.

Note that the cryptosystem $(\mathcal{K}, \mathcal{E}, \mathcal{D})$ is errorless. Namely, a legal encryption of `0' will always be decrypted as `0' and analogously an encryption of `1' is always decrypted as `1'.

2.3 Generating Good Public-Keys and Ciphertexts

We note that completeness of the protocols we design in this paper, will only hold for public-keys and ciphertexts which obey certain 'good' technical conditions defined below.

By theorems proved by Nguyen and Stern in [23] (for the purposes of cryptanalysis of AD cryptosystem), it follows that such good public-keys and ciphertexts will come up with high probability in the natural course of running the generating algorithm $\mathcal{K}$ and encryption algorithm $\mathcal{E}$. Moreover, the parties who run $\mathcal{K}$ and $\mathcal{E}$ can check that the outputs are good, and if not repeat the process till a good output is computed.

We will thus modify the definition of algorithms $\mathcal{K}$ (for key generation) and $\mathcal{E}$ (for encryption) to to ensure they always output public-keys and ciphertexts which are good.

Formally,

Definition 21. Let $\varepsilon \in (0,1)$. We say that a public key $PK = \{w_1, \ldots, w_n, v_1, \ldots, v_m, k\}$ where $v_1, \ldots, v_m, w_1, \ldots, w_n$ are vectors in $\mathbb{R}^n$ of AD is ε-good if

$$E\left[\sum_{j=1}^{n}\left\langle \sum_{i=1}^{m}(b_i v_i), w_j^{\perp}\right\rangle^2\right] \leq \frac{n^4 \rho_n^2}{2\varepsilon}, \tag{2.1}$$

where $w_j^{\perp}$ is a unit vector orthogonal to the hyperplane spanned by other w_j's. Expectation is taken over independent uniform choices of $b_1, \ldots, b_m$ from $\{0,1\}$.

Claim 22. [23] For sufficiently large n, for any $\varepsilon \in (0,1)$, a public key PK of AD picked at random according to the key generating protocol of section (2.2) is ε-good with probability at least $1 - \varepsilon$.

Definition 23. Let $\varepsilon, \varepsilon_1 \in (0,1)$, and PK be an ε-good public key of AD. We say that a ciphertext c of '0' is $(\varepsilon, \varepsilon_1)$ – good if for a_i, b_i's such that $c = \sum_{i=1}^{m} b_i v_i + \sum_{i=1}^{n} a_i w_i$

$$\text{dist}\left(\begin{pmatrix} n^6 \sqrt{nc} \\ 0 \end{pmatrix}, B_{PK}(a_1, \ldots, a_n, b_1, \ldots, b_m)^t\right) \leq \sqrt{1 + \frac{1}{2\varepsilon\varepsilon_1}} n^4 \tag{2.2}$$

Claim 24. [23] For sufficiently large n, for any $\varepsilon, \varepsilon_1 \in (0,1)$ and an ε-good public key PK of AD the following holds: a random ciphertext c of '0' is $(\varepsilon, \varepsilon_1)$-good with probability at least $1 - \varepsilon_1$. Probability is taken over random bits used by the encryption algorithm $\mathcal{E}$ to encrypt c.

Definition 25. Let $\varepsilon, \varepsilon_1 \in (0,1)$ and PK be an ε-good public key of AD. We say that a ciphertext c of '1' is $(\varepsilon, \varepsilon_1)$ – good if and only if $(c - \frac{v_k}{2}) \mod P(w_1, \ldots, w_n)$ is a $(\varepsilon, \varepsilon_1)$-good ciphertext of '0'.

Since, a random ciphertext c of '1' $(c - \frac{v_k}{2}) \mod P(w_1, \ldots, w_n)$ is distributed as a random ciphertext of '0', we automatically get an analogous claim for random ciphertexts of '1'.

Claim 26. For sufficiently large n, for any $\varepsilon, \varepsilon_1 \in (0,1)$ and for an ε-good public key PK of AD the following holds: a random ciphertext c of '1' is $(\varepsilon, \varepsilon_1)$-good with probability at least $1 - \varepsilon_1$. Probability is taken over random bits used by the encryption algorithm to encrypt c.

2.4 Modified AD Key Generation and Encryption Algorithms

We modify $\mathcal{K}$ and $\mathcal{E}$ to enforce the output of $\mathcal{K}$ to be ε-good and the output of $\mathcal{E}$ to be $(\varepsilon, \varepsilon_1)$-good.

For the protocols of section 3 we need $\varepsilon, \varepsilon_1 \in (0,1)$ to satisfy

$$\sqrt{1 + \frac{1}{2\varepsilon\varepsilon_1}} \leq \left(\frac{1}{4} - \frac{2}{n^2}\right) \frac{n\sqrt{\log(n+n^3)}}{3\sqrt{2}}, \tag{2.3}$$

For the protocol of section 4 we need $\varepsilon, \varepsilon_1 \in (0,1)$ to satisfy

$$\sqrt{1 + \frac{1}{2\varepsilon\varepsilon_1}} \leq \left(\frac{1}{4} - \frac{2}{n^2}\right) \frac{n\sqrt{\log(n+n^3)}}{12\sqrt{2}}. \tag{2.4}$$

Modified Key Generating algorithm $\mathcal{K}'$ on input 1^n:
  Repeat
  Let $(SK, PK) \in_R \mathcal{K}(1^n)$
  Until $E\left[\sum_{j=1}^n \langle \sum_{i=1}^m (b_i v_i), w_j^\perp \rangle^2\right] \leq \frac{n^4 \rho_n^2}{2\varepsilon}$ (where $PK = \{w_1, \ldots, w_n, v_1, \ldots, v_m, k\}$)
  Output (SK, PK)
We let $(SK, PK) \in \mathcal{K}'(1^n)$ denote generating instance (SK, PK) according to key generation algorithm $\mathcal{K}'(1^n)$.

Modified Encryption algorithm $\mathcal{E}'$ on input public key PK and message bit b:
  Repeat
  Pick $r = r_1, \ldots, r_m$, $r_i \in_R \{0, 1\}$.
  Let $c = \sum_{i=1}^m r_i v_i \mod P(w_1, \ldots, w_n)$.
  Compute a_i's such that $c = \sum_{i=1}^m r_i v_i + \sum_{i=1}^n a_i w_i$.
  Until $\text{dist}\left(\begin{pmatrix} n^6 \sqrt{n}c \\ 0 \end{pmatrix}, B_{PK}(a_1, \ldots, a_n, b_1, \ldots, b_m)^t\right) \leq \sqrt{1 + \frac{1}{2\varepsilon\varepsilon_1}} n^4$
  Output $c + b\frac{v_k}{2} \mod P(w_1, \ldots, w_n)$.

We let $c \in \mathcal{E}'_{PK}(b)$ denote generating c by running algorithm $\mathcal{E}'$ on inputs PK and b, let $c \in \mathcal{E}'_{PK}(\cdot)$ denote c being in the domain of $\mathcal{E}'_{PK}$, and let $c = \mathcal{E}'_{PK}(b, r)$ denote generating c by running algorithm $\mathcal{E}'_{PK}$ on input b using randomness r.

2.5 Zero-Knowledge Proof System for Approximate Closest Vector Problem

The protocols presented in this paper, exploit heavily the recent zero-knowledge protocol with for promise closest vector problem presented by Micciancio and Vadhan in [21].

Definition 27. *For $\gamma > 1$ instances of the promise closest vector problem GapCVP_γ are tuples (L, t, x) where L is a lattice in $\mathbb{R}^n$ specified by its basis, $t > 0$, and vector x in $\mathbb{R}^n$.*

- *(L, t, x) is a YES instance of the GapCVP_γ if $\text{dist}(L, x) \leq t$*
- *(L, t, x) is a NO instance of the GapCVP_γ if $\text{dist}(L, x) > \gamma t$*

The promise is that an instance of the GapCVP_γ is restricted to be YES or NO instance, any other tuples are not instances of the GapCVP_γ.

In the protocol described by Micciancio and Vadhan [21] the prover proves to the verifier in zero-knowledge that a given instance of the GapCVP_γ is a YES instance.

The protocol is statistical zero-knowledge for $\gamma = \Omega(\sqrt{\frac{n}{\log(n)}})$, where n is the dimension of the vector space containing the lattice L. Moreover, for such a γ the prover runs in polynomial time.

3 Verifiable Encryption for AD Cryptosystem

The ultimate goal of this section is to present two zero-knowledge protocols which form verifiable encryption schema for the equivalence relation. The first protocol is for proving that a ciphertext of AD decrypts to '0', and the second is for proving that a ciphertext of AD decrypts to '1'. In both protocols a common input to the prover and the verifier is a pair (PK, c) – public key of AD and a ciphertext. In addition, the prover has access to an auxiliary input consisting of random bits used to encrypt the ciphertext.

We will show a mapping from a pair (PK, c) to an instance (L, t, x) of GapCVP$_\gamma$ such that for good public keys and ciphertexts of bit '0' the pair maps to a YES instance of GapCVP$_\gamma$, whereas for any ciphertext which descrypts to '1' the pair maps to a NO instance of GapCVP$_\gamma$. Then, to prove that c decrypts to '0', simply run the ZK protocol of [21] to prove that (L, t, x) is a YES instance of GapCVP$_\gamma$. The case of ciphertext which decrypts to '1', is similarly handled.

Throughout this section n denotes the security parameter, $m = n^3$, and $\gamma = \sqrt{\frac{n+m}{\log(n+m)}}$.

3.1 Mapping AD Ciphertexts to GapCVP Instances

We define a mapping from pairs (PK, c) consisting of a public key and a ciphertext of AD to instances of GapCVP$_\gamma$.

Definition 31. *Let $PK = \{w_1, \ldots, w_n, v_1, \ldots, v_m, k\}$ be a public key of AD. Let c be a vector from $P((w_1, \ldots, w_n))$. Define mapping $\mathcal{F}(PK, c) = (L_{PK}, t, x_c)$ where*

$$x_c = \begin{pmatrix} n^6\sqrt{n}c \\ 0 \end{pmatrix} \in \mathbb{R}^{n+2m}, \quad t = n^4\sqrt{1 + \frac{1}{2\varepsilon\varepsilon_1}} \quad (3.1)$$

And L_{PK} is an $(n+m)$-dimensional lattice in $\mathbb{R}^{2n+m}$ spanned by the columns of the following matrix B_{PK},

$$B_{PK} = \begin{pmatrix} n^6\sqrt{n}w_1 & \cdots & n^6\sqrt{n}w_n & n^6\sqrt{n}v_1 & \cdots & n^6\sqrt{n}v_m \\ 1 & 0 & & & 0 & \\ 0 & \ddots & & & & \\ \vdots & & 1 & & \vdots & \\ & & & \ddots & & \\ & & & n^2\sqrt{n} & & \\ & & & & \ddots & 0 \\ 0 & & & & 0 & n^2\sqrt{n} \end{pmatrix} \quad (3.2)$$

3.2 Connection Between AD Ciphertexts of '0' and the GapCVP$_\gamma$ Problem

We next state the theorem which forms a theoretical basis for the protocol for proving that a ciphertext decrypts to '0'. The theorem states that good public keys and ciphertexts of '0' map under $\mathcal{F}$ to a YES instance of GapCVP$_\gamma$, whereas any ciphertext which decrypts to '1', will map under $\mathcal{F}$ to a NO instance of GapCVP$_\gamma$.

Theorem 32. *For sufficiently large n,*

1. *For $(SK, PK) \in \mathcal{K}'(1^n)$ and $c \in \mathcal{E}'_{PK}(0)$, $\mathcal{F}(PK, c)$ is a YES instance of GapCVP$_\gamma$.*
2. *for any instance (SK, PK) of AD and $c \in P(w_1, \ldots, w_n)$ such that $\mathcal{D}_{SK}(c) = $ '1', $\mathcal{F}(PK, c)$ is a NO instance of GapCVP$_\gamma$.*

Proof. (1) The first statement directly follows from the definition of an $(\varepsilon, \varepsilon_1)$-good ciphertext of '0'.

(2) Let $c \in P(w_1, \ldots, w_n)$ be any vector which decrypts to '1'. Let $T = t\gamma$. From (2.3) it follows that

$$\frac{3T}{n^6\sqrt{n}} = 3\frac{\sqrt{1+\frac{1}{2\varepsilon\varepsilon_1}}\sqrt{n+n^3}}{n^2\sqrt{n}\sqrt{\log(n+n^3)}} < \frac{3\sqrt{1+\frac{1}{2\varepsilon\varepsilon_1}}\sqrt{2}}{n\sqrt{\log(n+n^3)}} \leq \frac{1}{4} - \frac{2}{n^2} < \frac{1}{4} < \mathrm{dist}\left(\langle c, u\rangle, \mathbb{Z}\right).$$

By theorem 33 (proved below) $\mathrm{dist}\left(\begin{pmatrix} n^6\sqrt{n}c \\ 0 \end{pmatrix}, L_{PK}\right) \leq T$ can not hold.

Thus $\left(L_{PK}, t, \begin{pmatrix} n^6\sqrt{n}c \\ 0 \end{pmatrix}\right)$ is a NO instance of the GapCVP$_\gamma$.

Theorem 33. *Let $T > 0$, PK be a public key of AD, and $c \in P(w_1, \ldots, w_n)$. For sufficiently large n,*

$$\text{If } \mathrm{dist}\left(\begin{pmatrix} n^6\sqrt{n}c \\ 0 \end{pmatrix}, L_{PK}\right) \leq T \text{ then } \mathrm{dist}(\langle u, c\rangle, \mathbb{Z}) \leq \frac{3T}{n^6\sqrt{n}} \quad (3.3)$$

Proof. Let $c \in P(w_1, \ldots, w_n)$ be such that $\mathrm{dist}\left(\begin{pmatrix} n^6\sqrt{n}c \\ 0 \end{pmatrix}, L_{PK}\right) \leq T$, hence there are integers $a_1, \ldots, a_n, b_1, \ldots, b_m$ such that
$$\left\| \begin{pmatrix} n^6\sqrt{n}c \\ 0 \end{pmatrix} - B_{PK}(a_1, \ldots, a_n, b_1, \ldots, b_m)^t \right\|^2 \leq T^2.$$

Observing the construction of the matrix B_{PK} (3.2) we get that for the vector $e = n^6\sqrt{n}c - n^6\sqrt{n}\left(\sum_{i=1}^n a_i w_i + \sum_{i=1}^m b_i v_i\right)$

$$\sum_{i=1}^n a_i^2 + \sum_{i=1}^m n^5 b_i^2 + \|e\|^2 \leq T^2. \quad (3.4)$$

$\|e\| \leq T$, thus $|\langle u, e\rangle| \leq T$. It follows that $\mathrm{dist}(\langle u, e\rangle, \mathbb{Z}) \leq T$.

Note that $c = \sum_{i=1}^{n} a_i w_i + \sum_{i=1}^{m} b_i v_i + \frac{e}{n^6 \sqrt{n}}$, hence

$$\text{dist}(\langle u, c \rangle, \mathbb{Z}) \leq \sum_{i=1}^{n} |a_i| \text{dist}(\langle u, w_i \rangle, \mathbb{Z}) + \sum_{i=1}^{m} |b_i| \text{dist}(\langle u, v_i \rangle, \mathbb{Z}) + \frac{T}{n^6 \sqrt{n}}. \quad (3.5)$$

Let us upper bound the first term of (3.5). According to the construction of AD for all $i = 1, \ldots, n$ $\text{dist}(\langle u, w_i \rangle, \mathbb{Z}) \leq \frac{1}{n^7}$. From (3.4) it follows that $\sum_{i=1}^{n} a_i^2 \leq T^2$. Thus, by the Cauchy-Schwartz inequality we have that $\sum_{i=1}^{n} |a_i| \text{dist}(\langle u, w_i \rangle, \mathbb{Z}) \leq \sqrt{\sum_{i=1}^{n} a_i^2} \times \sqrt{\sum_{i=1}^{n} \text{dist}(\langle u, w_i \rangle, \mathbb{Z})^2} \leq T\sqrt{n \times n^{-14}} = \frac{T}{n^6 \sqrt{n}}$.
Let us now upper bound the second term of (3.5). Similarly, for all $i = 1, \ldots, m$ $\text{dist}(\langle u, v_i \rangle, \mathbb{Z}) \leq \frac{1}{n^7}$. From (3.4) we have that $\sum_{i=1}^{m} b_i^2 \leq \frac{T^2}{n^5}$. Applying the Cauchy-Schwartz inequality we get that $\sum_{i=1}^{m} |b_i| \text{dist}(\langle u, v_i \rangle, \mathbb{Z}) \leq \sqrt{\sum_{i=1}^{m} b_i^2} \times \sqrt{\sum_{i=1}^{m} \text{dist}(\langle u, v_i \rangle, \mathbb{Z})^2} \leq \frac{T}{n^2 \sqrt{n}} \sqrt{n^3 \times n^{-14}} = \frac{T}{n^8} \leq \frac{T}{n^6 \sqrt{n}}$.
Combining all together we obtain that $\text{dist}(\langle c, u \rangle, \mathbb{Z}) \leq \frac{3T}{n^6 \sqrt{n}}$.

We are ready to present the protocol which form verifiable encryption schema for the equivalence relation when the claimed plaintext is '0'.

Protocol$_0$: proving that a ciphertext decrypts to '0'.
Let P_0 and V_0 denote the prover and the verifier. Let the common input to P_0 and V_0 be a pair (PK, c) where $PK = \{w_1, \ldots, w_n, v_1, \ldots, v_m, k\}$ is a public key of AD and c is a vector from $P(w_1, \ldots, w_n)$. The prover's auxiliary input is $b_1, \ldots, b_m \in \{0, 1\}$ such that $c = \sum_{i=1}^{m} b_i v_i \mod P(w_1, \ldots, w_n)$.

- **Prover** P_0 Calculates integers $a_1, \ldots, a_n$ such that $c = \sum_{i=1}^{m} b_i v_i + \sum_{i=1}^{n} a_i w_i$. Invokes the [21] prover (with auxiliary input $B_{PK}(a_1, \ldots, a_n, b_1, \ldots, b_m)^t$) to prove that input $\mathcal{F}(PK, c)$ is a YES instance of GapCVP$_\gamma$.
- **Verifier** V_0 Invoke the [21] verifier to verify that input $\mathcal{F}(PK, c)$ is a YES instance of GapCVP$_\gamma$.

Claim 34. *Protocol (P_0, V_0) satisfy the following completeness, soundness, and zero-knowledge properties:*

- **Completeness:** *If $(SK, PK) \in \mathcal{K}'(1^n)$ and $c \in \mathcal{E}'_{PK}(0)$, then $Prob((P_0, V_0)(PK, c) = accepts) = 1$.*
- **Soundness** *If (PK, SK) is an instance of AD and $c \in P(w_1, \ldots, w_n)$ such that $D_{SK}(c) = $ '1', then for all prover P_0', $Prob((P_0', V_0)(PK, c) = rejects) > \frac{1}{2}$.*
- **Zero-Knowledge :** *statistical zero-knowledge.*

Proof. The soundness condition relies on the part (2) of the theorem 32 and the soundness condition of the proof system from [21]. The completeness condition follows from the part (1) of the theorem 32 and completeness condition of the proof system from [21]. The lattice L_{PK} is an $(n + m)$-dimensional lattice, hence, the approximation factor $\gamma = \sqrt{\frac{n+m}{\log(n+m)}}$ is as required for statistical zero-knowledge property of the proof system from [21].

3.3 Connection Between AD '1' Ciphertexts and the GapCVP$_\gamma$ Problem

In this subsection we construct a zero-knowledge protocol for proving that a ciphertext of AD decrypts to '1'. We use the nice observation that for a random ciphertext of AD of '1' the distribution of vector $(c - \frac{v_k}{2})$ mod $P(w_1, \ldots, w_n)$ is the same as distribution of a random ciphertext of '0'. Thus, to prove that a ciphertext c decrypts to '1', we will prove that $(c - \frac{v_k}{2})$ mod $P(w_1, \ldots, w_n)$ decrypts to '0', by running *protocol*$_0$ on inputs PK and $(c - \frac{v_k}{2})$ mod $P(w_1, \ldots, w_n)$.

To prove soundness however, we must be careful, as we notice that for a c which decrypts to '0', $(c - \frac{v_k}{2})$ mod $P(w_1, \ldots, w_n)$ is *not* distributed as a random ciphertext of '1', however as shown by the the following theorem it is quite close to it.

Theorem 35. *For any (SK, PK) instance of AD, for any vector $c \in P(w_1, \ldots, w_n)$ such that $\mathcal{D}_{SK}(c) = $ '0', for sufficiently large n, the* dist $(\langle y, u \rangle, \mathbb{Z}) > \frac{1}{4} - \frac{2}{n^2}$ *for $y = (c - \frac{v_k}{2})$ mod $P(w_1, \ldots, w_n)$*

Proof. Let $c \in P(w_1, \ldots, w_n)$ decrypts to '0'.
There is a representation $(c - \frac{v_k}{2})$ mod $P(w_1, \ldots, w_n) = c - \frac{v_k}{2} + \sum_{i=1}^{n} a_i w_i$.

$$\text{dist}\left(\left\langle c - \frac{v_k}{2} + \sum_{i=1}^{n} a_i w_i, u \right\rangle, \mathbb{Z}\right) \geq \text{dist}\left(\left\langle \frac{v_k}{2}, u \right\rangle, \mathbb{Z}\right) -$$
$$\text{dist}(\langle c, u \rangle, \mathbb{Z}) - \text{dist}\left(\left\langle \sum_{i=1}^{n} a_i w_i, u \right\rangle, \mathbb{Z}\right) \quad (3.6)$$

Let us bound the terms of (3.6).

$$\text{dist}\left(\left\langle \sum_{i=1}^{n} a_i w_i, u \right\rangle, \mathbb{Z}\right) \leq \sum_{i=1}^{n} |a_i| \, \text{dist}(\langle w_i, u \rangle, \mathbb{Z}) \leq \frac{1}{n^7} \sum_{i=1}^{n} |a_i|. \quad (3.7)$$

Note, that $a_i = \lfloor \theta_i \rfloor$ for θ_i defined as $c - \frac{v_k}{2} = \sum_{i=1}^{n} \theta_i w_i$. Since the width of the parallelepiped $P(w_1, \ldots, w_n)$ is greater than $\frac{\rho_n}{n^2}$, (3.7) can be bounded by

$$\frac{1}{n^7}\sum_{i=1}^{n}|a_i| \leq \frac{1}{n^7}\sum_{i=1}^{n}|\theta_i| \leq \frac{1}{n^5\rho_n}\sum_{i=1}^{n}\left|\left\langle c - \frac{v_k}{2}, w_i^\perp \right\rangle\right| \leq \frac{1}{n^4\rho_n}\left\|c - \frac{v_k}{2}\right\| \leq \frac{1}{n^2}.$$

dist $(\langle c, u \rangle, \mathbb{Z}) \leq \frac{1}{4}$ and dist $(\langle \frac{v_k}{2}, u \rangle, \mathbb{Z}) \geq \frac{1}{2} - \frac{1}{n^7}$. Collecting all together we get that (3.6) is greater than $\frac{1}{2} - \frac{1}{n^7} - \frac{1}{4} - \frac{1}{n^2}$ which is greater than $\frac{1}{4} - \frac{2}{n^2}$ for sufficiently large n.

The following theorem forms the theoretical basis for the protocol for proving that a ciphertext decrypts to '1'

Theorem 36. *For sufficiently large n,*

- *If $(SK, PK) \in \mathcal{K}'(1^n)$ and $c \in \mathcal{E}'_{PK}(1)$, then $\mathcal{F}(PK, y)$ is a YES instance of the $GapCVP_\gamma$ for $y = (c - \frac{v_k}{2}) \mod P(w_1, \ldots, w_n)$.*
- *If (PK, SK) is an instance of AD cryptosystem and $c \in P(w_1, \ldots, w_n)$ such that $\mathcal{D}_{SK}(c) = $ '0', then $\mathcal{F}(PK, y)$ is a NO instance of the $GapCVP_\gamma$ for $y = (c - \frac{v_k}{2}) \mod P(w_1, \ldots, w_n)$.*

(1) The statement directly follows from the definition of an $(\varepsilon, \varepsilon_1) - good$ ciphertext of '1'.

(2) Let $c \in P(w_1, \ldots, w_n)$ be any vector which decrypts to '0'. Define $y = (c - \frac{v_k}{2}) \mod P(w_1, \ldots, w_n)$. From (2.3) it follows that

$$\frac{3t\gamma}{n^6\sqrt{n}} = 3\frac{\sqrt{1 + \frac{1}{2\varepsilon\varepsilon_1}}\sqrt{n+n^3}}{n^2\sqrt{n}\sqrt{\log(n+n^3)}} < \frac{3\sqrt{1+\frac{1}{2\varepsilon\varepsilon_1}}\sqrt{2}}{n\sqrt{\log(n+n^3)}} \leq \frac{1}{4} - \frac{2}{n^2} <$$

[By the theorem 35]$< \text{dist}(\langle y, u \rangle, \mathbb{Z})$.

Thus, by the theorem 33 $\text{dist}\left(\begin{pmatrix} n^6\sqrt{n}y \\ 0 \end{pmatrix}, L_{PK}\right) \leq t\gamma$ can not hold, and $\left(L_{PK}, t, \begin{pmatrix} n^6\sqrt{n}y \\ 0 \end{pmatrix}\right)$ is a NO instance of the $GapCVP_\gamma$.

We are ready to present the protocol for proving that a ciphertext decrypts to '1'.

Protocol$_1$: proving that a ciphertext decrypts to '1'.
Let P_1 and V_1 denote the prover and the verifier. Let the common input to P_1 and V_1 be a pair (PK, c) where $PK = \{w_1, \ldots, w_n, v_1, \ldots, v_m, k\}$ is a public key of AD and c is a vector from $P(w_1, \ldots, w_n)$. Let P_1 auxiliary input be $b_1, \ldots, b_m \in \{0, 1\}$ such that $c = (\frac{v_k}{2} + \sum_{i=1}^{m} b_i v_i) \mod P(w_1, \ldots, w_n)$.

- **Prover P_1:** Calculate $y = (c - \frac{v_k}{2}) \mod P(w_1, \ldots, w_n)$. Calculate integers $a_1, \ldots, a_n$ such that $y = \sum_{i=1}^{m} b_i v_i + \sum_{i=1}^{n} a_i w_i$. Invoke the [21] prover (with auxiliary input $B_{PK}(a_1, \ldots, a_n, b_1, \ldots, b_n)$) to prove that input $\mathcal{F}(PK, y)$ is a YES instance of $GapCVP_\gamma$.
- **Verifier V_1:** Calculate $y = (c - \frac{v_k}{2}) \mod P(w_1, \ldots, w_n)$. Invoke the [21] verifier to verify that $\mathcal{F}(PK, y)$ is a YES instance of $GapCVP_\gamma$.

It is evident that the soundness, completeness, and Zero-knowledge properties of $Protocol_1$ are similar to the soundness and Zero-Knowledge properties of $Protocol_0$.

4 Proof of AD Plaintext Knowledge

4.1 Definition of Proofs of Knowledge

We use the definition of a proof of knowledge from [18]

Definition 41. Let $Q(\cdot)$ be a polynomial, x the common input for the prover P and verifier V, and r a uniformly selected random tape of prver P. Run the protocol between P and V, $Q(|x|)$ times, each time runniing prover P on the same random tape r and the verifier V on a newly selected uniformly chosen random tape. Let (P, V, x, Q) denote the sequence of the verifier's views obtaind from the above execution. We call the distribution over such sequences a valid (P, V, x, Q) - distribution.

Definition 42. Let $\eta \in \{0,1\}$, an interactive protocol (P,V) with prover P and a verifier V is a proof of knowledge system with knowledge error η for a relation R if the following holds:

> *Completeness:* For every common input x for which there exists y such that $(x,y) \in R$ the verifier V always accepts interacting with the prover P.
> *Validity with error η:* There exists a polynomial time interacting oracle Turing machine Sample and a polynomial time algorithm Extract, a constant $c > 0$ and a polynomial $Q(\cdot)$ such that for every $x \in \{0,1\}^*$ such that $R(x) \neq \emptyset$ and for every prover P' the following holds:
> - Sample$^{P'}(x)$ outputs a valid (P', V, x, Q)- distribution of verifier's view.
> - Extract(Sample$^{P'}(x)) \in R(x) \cup \{\text{"fail"}\}$
> - $Pr[\text{Extract}(\text{Sample}^{P'}(x)) \in R(x)] \geq (p-\eta)^c$, where $p > \eta$ is a probability that V accepts while interacting with P' on common input x.

We call the pair (Sample, Extract) a knowledge extractor.

4.2 The Plaintext Knowledge Relation for AD Cryptosystem

Throughout the rest of section 4 we assume that n denotes the security parameter, and m, L_{PK}, and γ, are as defined in section 3 whereas $t = 4\sqrt{1 + \frac{1}{2\varepsilon\varepsilon_1}n^4}$. Define relation R_{AD} corresponding to knowing a plaintext of an AD ciphertext as follows.

Definition 43. Let $PK = \{w_1, \ldots, w_n, v_1, \ldots, v_m, k\}$ be a public key of AD, c and c' vectors from $P(w_1, \ldots, w_n)$, b' and $b'' \in \{0,1\}$, $r' \in \{0,1\}^m$, and p be a point from L_{PK}. We say that input (PK, c) and witness (c', b', r', b'', p) are in R_{AD} if:

- $c' = \mathcal{E}_{PK}(b'; r')$
- dist $\left(\begin{pmatrix} n^6\sqrt{n}((c'+c-b''\frac{v_k}{2}) \mod P(w_1,\ldots,w_n)) \\ 0 \end{pmatrix}, p\right) \leq \gamma t$ *(i.e. $(c+c')$ mod $P(w_1,\ldots,w_n)$) decrypts to b'')*

Intuitively, proving knowledge of a witness for (PK, c), implies knowledge of plaintext of c under PK. This is formally captured by the following theorem.

Theorem 44. Let (PK, SK) be an instance of the AD cryptosystem. If $((PK,c), w) \in R_{AD}$ for $w = (c', b', r', b'', p)$, then $b' \oplus b'' = D_{SK}(c)$.

Proof. Let $PK = \{w_1, \ldots, w_n, v_1, \ldots, v_m, k\}$.
Consider the case when $b'' = 0$. In this case

$$\text{dist}\left(\left(\begin{array}{c} n^6\sqrt{n}((c'+c) \bmod P(w_1,\ldots,w_n)) \\ 0 \end{array}\right), p\right) \leq \gamma t,$$

By theorem 33, $\text{dist}\left(\langle(c+c') \bmod P(w_1,\ldots,w_n), SK\rangle, \mathbb{Z}\right) \leq \frac{3T}{n^6\sqrt{n}} =$
$12\frac{\sqrt{1+\frac{1}{2\epsilon\epsilon_1}}\sqrt{n+n^3}}{n^2\sqrt{n}\sqrt{\log(n+n^3)}} \leq \frac{\sqrt{n+n^3}}{8n\sqrt{2}\sqrt{n}} \leq \frac{1}{8}$.

Suppose $b' = 0$. Since c' is a legal ciphertext, $\text{dist}(\langle c', SK\rangle, \mathbb{Z}) \leq \frac{1}{n}$ which implies that $\text{dist}(\langle c, SK\rangle, \mathbb{Z}) < \frac{1}{4}$ and $\mathcal{D}_{SK}(c) = $ '0'.

Suppose $b' = 1$. Since c' is a legal ciphertext, $\text{dist}(\langle c', SK\rangle, \mathbb{Z}) \geq \frac{1}{2} - \frac{1}{n}$ which implies that $\text{dist}(\langle c, SK\rangle, \mathbb{Z}) > \frac{1}{4}$ and $\mathcal{D}_{SK}(c) = $ '1'.

A similar case analysis follows when $b'' = 1$.

Note, that one can easily check whether a pair (PK, c) and a particular witness are in the relation R_{AD}. Since AD is semantically secure, for a public key PK of AD generated in random according to the key generating algorithm and a random ciphertext c of a uniformly chosen bit encrypted under the public key PK it is impossible to construct a witness for (PK, c) with non-negligible probability.

4.3 Protocol for Proof of Plaintext Knowledge for AD

Let us first provide a sketch of the protocol. For public key $PK = \{w_1, \ldots, w_n, v_1, \ldots, v_m, k\}$ and ciphertext c, we distill the following nice homomorphic properties of AD:

- If c is an encryption of the bit b, then $c + \frac{v_k}{2} \bmod P(w_1 \ldots w_n)$ is decrypted to $\bar{b}$
- If c, c' are encryptions of b, b' (respectively) then $c + c' \bmod P(w_1 \ldots w_n)$ is decrypted to $b \oplus b'$.

Using these properties, it is simple to design a proof of knowledge of bit b encrypted by ciphertext c: the prover sends a random encryption c' of a random bit b', and the verifier asks the prover to show either that it knows the decryption of c' or that it knows a decryption of $c + c'$. The former can be done simply revealing the randomness used to encrypt c' and the latter can be done by proving in zero-knowledge that $c + c'$ decrypts to $b \oplus b'$. This is achieved by utilizing a variant of the protocols of section 3.2 to show that $(c+c')$ decrypts to zero (in case of $b \oplus b' = 0$) or that $(c+c') + \frac{v_k}{2}$ decrypts to zero (when $b \oplus b' = $ '1').

Protocol PPK
Let P_{PPK} and V_{PPK} denote the prover and the verifier respectively. The common input to P_{PPK} and V_{PPK} is (PK, c) where $PK = \{w_1, \ldots, w_n, v_1, \ldots, v_m, k\}$ is an AD public-key and c is a vector from $P(w_1, \ldots, w_n)$. The prover's auxiliary input is plaintext b and randomness r such that $c = \mathcal{E}'_{PK}(b; r)$.

- **Step (P1):** P_{PPK} selects $b' \in_R \{0,1\}$, computes $c' \in_R \mathcal{E}'_{PK}(b')$ and sends c' to V_{PPK}.
- **Step (V1):** V_{PPK} sends a random challenge bit $\delta \in_R \{0,1\}$ to P_{PPK}.
- **Step (P2):**
  - If $\delta = 0$, P_{PPK} sends pair (b', r') where $c' = \mathcal{E}'_{PK}(b'; r')$ to V_{PPK}.
  - If $\delta = 1$, P_{PPK} computes $b'' = b \oplus b'$; sends b'' to verifier; lets $\bar{c} = (c+c') \mod P(w_1, \ldots, w_n))$ and runs the prover of $Protocol'_{b''}$ on input $(PK, \bar{c})$.
- **Step (V2):**
  - If $\delta = 0$, then (c', r') has been received in step (P2). V_{PPK} rejects if $c' \neq E(b'; r')$, else it accepts.
  - If $\delta = 1$, let b'' be bit received in step P2. V_{PPK} set $\bar{c} = (c + c') \mod P(w_1, \ldots, w_n))$; run the verifier of $Protocol'_{b''}$ on input $(PK, \bar{c})$.

The flow of message communication is presented in picture 1.

Protocol PPK (in steps P2,V2) makes calls to two zero-knowledge protocols $Protocol'_0$ and $Protocol'_1$ which enable the prover to prove that a given sum of two ciphertexts of AD decrypt to '0' (or '1' respectively). These protocols are identical in structure to the protocols of section 3.2 and 3.3, except for a slight difference in the YES instances of GapCVP$_\gamma$ constructed.

Define [2] mapping $\mathcal{G}(PK, c) = (L_{PK}, t, x_c)$ where $t = 4\sqrt{1 + \frac{1}{2\varepsilon \varepsilon_1}} n^4$ and x_c, L_{PK} are as in section 3.1

$Protocol'_0$ on input $(PK, \bar{c})$ is the statistical ZK protocol of [21] proving that input $\mathcal{G}(PK, \bar{c})$ is a YES instance of GapCVP$_\gamma$.

$Protocol'_1$ on input $(PK, \bar{c})$ is the statistical ZK protocol of [21] proving that input $\mathcal{G}(PK, (\bar{c} - \frac{v_k}{2}) \mod P(w_1, \ldots, w_n))$ is a YES instance of GapCVP$_\gamma$.

The following properties of these protocols are needed for larger protocol PPK. Note the similarity with theorem 32 and 36.

Claim 45. *For sufficiently large n,*

1. *If $(SK, PK) \in \mathcal{K}'(1^n)$, $c = (c_1 + c_2) \mod P(w_1, \ldots, w_n)$ such that $\mathcal{D}_{SK}(c) = $ '0' and $c_1, c_2 \in \mathcal{E}'_{PK}(\cdot)$, $\mathcal{G}(PK, c)$ is a YES instance of GapCVP$_\gamma$.*
2. *Let (SK, PK) be an instance of AD and $c \in P(w_1, \ldots, w_n)$. If $\text{dist}(\langle c, SK \rangle, \mathbb{Z}) > \frac{1}{8}$, then $\mathcal{G}(PK, c)$ is a NO instance of GapCVP$_\gamma$.*

Proof. We defer the proof to the end of the section.

Claim 46. *For sufficiently large n, the following holds:*

1. *For any $(SK, PK) \in \mathcal{K}'(1^n)$, for any $c = (c_1 + c_2) \mod P(w_1, \ldots, w_n)$ such that $\mathcal{D}_{SK}(c) = $ '1' and where $c_1, c_2 \in \mathcal{E}'_{PK}(\cdot)$, $\mathcal{G}(PK, y)$ is a YES instance of the GapCVP$_\gamma$ where $y = (c - \frac{v_k}{2}) \mod P(w_1, \ldots, w_n)$.*
2. *For any instance (SK, PK) of AD, and for any $c = P(w_1, \ldots, w_n)$ such that $\text{dist}(\langle c, SK \rangle, \mathbb{Z}) < \frac{3}{8}$, $\mathcal{G}(PK, y)$ is a NO instance of GapCVP$_\gamma$ for $y = (c - \frac{v_k}{2}) \mod P(w_1, \ldots, w_n)$.*

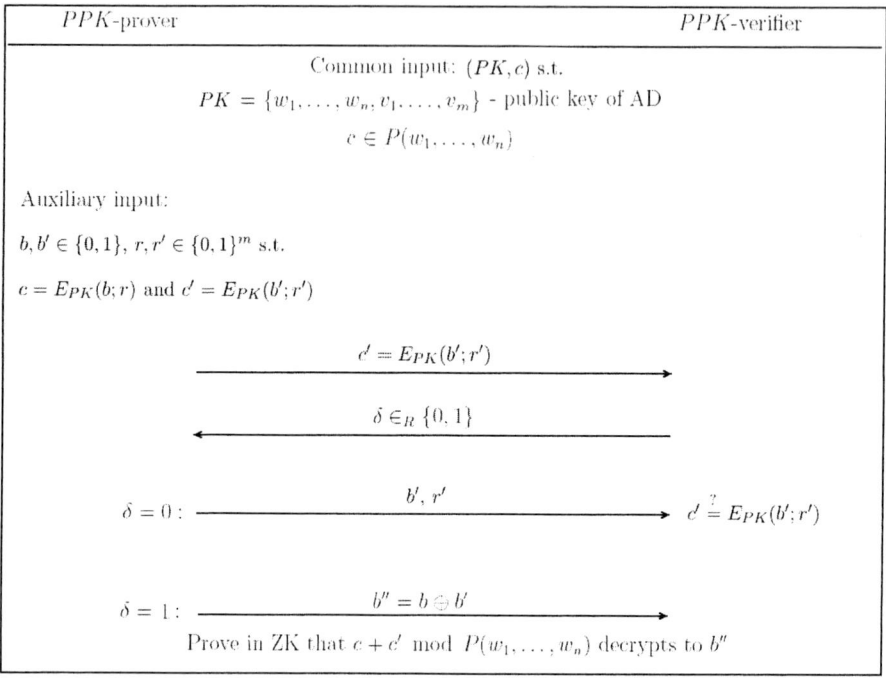

The proof is similar to the proof of theorem 45 and is omitted.

We are now ready to prove that protocol PPK forms a proof of knowledge system with error $\frac{3}{4}$ for binary relation R_{AD} which is zero-knowledge.

Theorem 47 (Completeness and Soundness of PPK). *Interactive protocol (P_{PPK}, V_{PPK}) is a proof of knowledge system with knowledge error $\frac{3}{4}$ for R_{AD}.*

Proof. First lets argue completeness. Namely, if PK is an ε-good ciphertext and c is an $\varepsilon, \varepsilon_1$-good ciphertext under PK then the V_{PPK} always accepts interacting with the P_{PPK}.

The completeness property becomes evident, due to the simple fact about ciphertexts of AD: for two legal ciphertexts c_1 and c_2 of AD with plaintexts b_1 and b_2 the vector $(c_1 + c_2)$ mod $P(w_1, \ldots, w_n)$ decrypts to $b_1 \oplus b_2$.

Second, lets argue validity with knowledge error $\frac{3}{4}$. We will present a knowledge extractor consisting of two algorithms `Sample` and `Extract` which satisfy the conditions of the definition of a proof of knowledge.

Let $PK = \{w_1, \ldots, w_n, v_1, \ldots, v_m, k\}$ be a public key of AD and $c \in P(w_1, \ldots, w_n)$. Let P' be an arbitrary prover making the V_{PPK} accept with probability $\frac{3}{4} + \sigma$, for $\sigma > 0$ on common input (PK, c).

The algorithm `Sample`: The algorithm `Sample` is an interactive Turing machine with oracle access to P'. The input of `Sample` is (PK, c). The algorithm

---

[2] The only difference between $\mathcal{G}$ and $\mathcal{F}$ of section 3 is in the value of t used

outputs three strings distributed as verifier's views at the end of the protocol between P' and V_{PPK} run on common input (PK, c) (i.e Sample outputs a valid $(P', V_{PPK}, (PK, c), 3)$-distribution of verifier's view). Sample chooses a random string r which will serve as a random tape for P'. Smaple outputs three verifiers views V_1, V_2, V_3 independently according to the following procedure: Set the random tape of P' to r. Generate a random bit δ which will be used for verifier's challenge. If $\delta = 1$ the prover and the verifier should be involved in one of the subprotocols ($Protocol'_0$ or $Protocol'_1$). Each subprotocol is a three-move interactive proof system with one-bit verifier's challenge. Generate a random bit δ_1 for the second verifier's challenge. Simulate the protocol between P' an the V_{PPK} on common input (PK, c) interacting with P' as a verifier and sending challenge bits δ and δ_1 (if needed). Output the verifiers view which consists of common input (PK, c), simulated transcript of the protocol and random bits δ and δ_1 (if needed).

The algorithm Extract: Input of the algorithm Extract consists of three verifier's views V_1, V_2, V_3 generated by Sample. Let transcripts of the protocol involved in the views be denoted as T_1, T_2 and T_3. If one of the transcript is not accepting, Extract outputs "fail" and halts. Since the probability that P' makes V_{PPK} accept is $\frac{3}{4} + \sigma$, Extarct continues with probability at least σ. The algorithm checks the following conditions:

Verifier's view V_1 involves $\delta = 0$.
Verifier's view V_2 involves $\delta = 1$ and $\delta_1 = 0$.
Verifier's view V_3 involves $\delta = 1$ and $\delta_1 = 1$.

If at least one of the conditions does not hold then Extract outputs "fail" and halts. If the algorithm continues, what happens with probability $\frac{1}{32}$, T_1, T_2 and T_3 has the following form:

$T_1 = (c', 0, b', r')$
$T_2 = (c', 1, b'', T'_1)$
$T_3 = (c', 1, b'', T'_2)$

Where T'_1 and T'_2 are transcripts of $Protocol'_0$ and $Protocol'_1$ respectively. Note, that the subprotocols are based on the proof system of Micciancio and Vadhan and actually are aimed to prove that

$$\left(L_{PK}, t, \begin{pmatrix} n^6\sqrt{n}((c' + c - b''\frac{v_k}{2}) \mod P(w_1, \ldots, w_n)) \\ 0 \end{pmatrix}\right)$$

is not a NO instance of the GapCVP$_\gamma$ problem for L_{PK}, γ and t as defined in this section. Assume T'_1 and T'_2 are accepting transcripts with the same prover's random tape and different verifier's challenges. Then, when $b'' = 0$ it is possible to obtain from T'_1 and T'_2 a point p in L_{PK} such that

$$\text{dist}\left(\begin{pmatrix} n^6\sqrt{n}((c' + c) \mod P(w_1, \ldots, w_n)) \\ 0 \end{pmatrix}, p\right) \leq \gamma t.$$

When $b'' = 1$ it is possible to obtain a point p such that

$$\text{dist}\left(\begin{pmatrix} n^6\sqrt{n}((c' + c - \frac{v_k}{2}) \mod P(w_1, \ldots, w_n)) \\ 0 \end{pmatrix}, p\right) \leq \gamma t.$$

Since T_1 is an accepting transcript, ciphertext c' and b',r' satisfy $c' = \mathcal{E}'_{PK}(b';r')$. Extract outputs the witness $(c',b',b'_1,\ldots,b'_m,b'',p)$. The algorithm succeeds with probability at least $\frac{1}{32}\sigma$.

We prove that the PPK protocol is computational zero-knowledge under the same intractability assumption of the AD cryptosystem.

Assumption ISVP:

(Infeasibility of Shortest Vector Problem): There is no polynomial time algorithm, which given an arbitrary basis for an n-dimensional lattice, having a "unique $poly(n)$-shortest" vector, finds the shortest non-zero vector in the lattice. By having a "unique $poly(n)$-shortest" vector we mean that any vector of length at most "poly(n)" times bigger than the shortest vector is parallel to the shortest vector.

Theorem 48 (Zero-Knowledge of PPK). *The protocol PPK is computational zero knowledge under the assumption ISVP.*

Proof. For every verifier V' we construct an expected polynomial time simulator S such that on input (PK,c) where PK an ε-good public key of AD and c is an $(\varepsilon,\varepsilon_1)$-good ciphertext encrypted under PK the output of the simulator is computationally indistinguishable from a transcript of the protocol between the P_{PPK} and the verifier V' on common input (PK,c).

The simulator S proceeds as follows:

Simulate prover's first step: Chose Δ uniformly from $\{0,1\}$. If $\Delta = 0$ uniformly select a random bit b' and generate a random $(\varepsilon,\varepsilon_1)$-good ciphertext c' of b' under PK using uniformly generated random string $r' \in \{0,1\}^m$. If $\Delta = 1$ uniformly select a bit b'' and generate a random $(\varepsilon,\varepsilon_1)$-good ciphertext $\bar{c}$ of b'', set $c' = (\bar{c}-c)$ mod $P(w_1,\ldots,w_n)$. Pass c' to the verifier V'.

Simulate verifiers's first step: Receive a challenge bit δ from V'.

Simulate prover's second step and output the transcript of the protocol:

If $\delta \neq \Delta$ go to the step "Simulate prover's first step".

Let us show that the simulator repeats the step "Simulate prover's first step" only an expected polynomial number of times. Let U be the uniform distribution in $P(w_1,\ldots,w_n)$. We assume that ISVP holds, hence according to the security property of AD if $\Delta = 0$ then c' is computationally indistinguishable from U; if $\Delta = 1$ then $c' = (\bar{c} - c)$ mod $P(w_1,\ldots,w_n)$ is also indistinguishable from U. c' generated for $\Delta = 0$ is computationally indistinguishable from c' generated for $\Delta = 1$. δ equal to Δ with probability less than $\frac{1}{2} + v(n)$ for some negligible function $v(n)$, otherwise verifier can distinguish between c' generated for $\Delta = 0$ and c' generated for $\Delta = 1$. Thus the expected number of repetitions of the step "Simulate prover's first step" is polynomial.

- If $\delta = 0$ send bits b' and r' to V' and receive a verifier's verdict v on acceptance or rejectance. Output the transcript (c',δ,b',r',v). Since

c' is indeed an $(\varepsilon, \varepsilon_1)$-good ciphertext of b' with random bits r', the simulator perfectly simulates a real transcript between P_{PPK}-prover and the verifier V'.

– Consider the case when $\delta = 1$. Note, that $\bar{c} = (c+c') \bmod P(w_1, \ldots, w_n)$, hence, according to zero-knowledge property of $Protocol'_0$ and $Protocol'_1$, there exist simulators S_1 and S_2 with the following properties: if $b'' = 0$ then an output of S_1 on input $(PK, \bar{c})$ is computationally indistinguishable from the real transcript of $Protocol'_0$ run between the P_{PPK} and the verifier V'. If $b'' = 1$ then output of S_2 on input $(PK, \bar{c})$ is indistinguishable from the real transcript of $Protocol'_1$. If $b'' = 0$ set $T = S_1(PK, \bar{c})$ otherwise set $T = S_2(PK, \bar{c})$. Output the transcript (c', δ, b'', T). Since the ISVP assumption holds, according to the security property of AD the distribution of $\bar{c}$ is computationally indistinguishable from U, hence the distribution of $c' = (\bar{c} - c) \bmod P(w_1, \ldots, w_n)$ is also indistinguishable from U which is indistinguishable from the distribution of c' generated by the P_{PPK}. Therefore, the generated transcript is computationally indistinguishable from a real transcript of the protocol between the P_{PPK} and V'.

missing proof of claim 46.

(1) The vector c can decrypts to '0' in two cases: when both c_1 and c_2 are ciphertexts of '0' and when both c_1 and c_2 are ciphertexts of '1'.

– Let c_1 and c_2 be $(\varepsilon, \varepsilon_1)$-good ciphertexts of '0'. For c_1 and c_2 equation (2.2) holds. Thus for $c = (c_1 + c_2) \bmod P(w_1, \ldots, w_2)$ by lemma 49 below

$$\text{dist}\left(\begin{pmatrix} n^6\sqrt{nc} \\ 0 \end{pmatrix}, L_{PK}\right) \leq 2\sqrt{1 + \frac{1}{2\varepsilon\varepsilon_1}} n^4 + \sqrt{n}$$

which is less then t for sufficiently large n. By the definition of a YES instance of the GapCVP_γ the statement of part (1) holds.

– Let c_1 and c_2 be $(\varepsilon, \varepsilon_1)$-good ciphertexts of '1'. By the definition of an $(\varepsilon, \varepsilon_1)$-good ciphertext of '1' the vectors $\overline{c_1} = (c_1 - \frac{v_k}{2}) \bmod P(w_1, \ldots, w_n)$ and $\overline{c_2} = (c_2 - \frac{v_k}{2}) \bmod$ are $(\varepsilon, \varepsilon_1)$-good ciphertexts of '0', thus for $\bar{c} = (\overline{c_1} + \overline{c_2}) \bmod P(w_1, \ldots, w_n)$ the following statement holds:

$$\text{dist}\left(\begin{pmatrix} n^6\sqrt{n\bar{c}} \\ 0 \end{pmatrix}, L_{PK}\right) \leq 2\sqrt{1 + \frac{1}{2\varepsilon\varepsilon_1}} n^4 + \sqrt{n}$$

The vector $c = (\bar{c} + v_k) \bmod P(w_1, \ldots, w_n)$ thus by lemma 410 below for sufficiently large n the following statement holds:

$$\text{dist}\left(\begin{pmatrix} n^6\sqrt{nc} \\ 0 \end{pmatrix}, L_{PK}\right) \leq 2\sqrt{1 + \frac{1}{2\varepsilon\varepsilon_1}} n^4 + \sqrt{n} + n^4 \quad (4.1)$$

Expression (4.1) is less than t for sufficiently large n.

(2) Let $c \in P(w_1, \ldots, w_n)$ be any vector which decrypts to '1'. Let $T = t\gamma$. From (2.4) it follows that $\frac{3T}{n^6\sqrt{n}} = 12\frac{\sqrt{1+\frac{1}{2\epsilon\epsilon_1}}\sqrt{n+n^3}}{n^2\sqrt{n}\sqrt{\log(n+n^3)}} \leq 12\sqrt{2}\frac{\sqrt{1+\frac{1}{2\epsilon\epsilon_1}}}{n\sqrt{\log(n+n^3)}} \leq \frac{1}{4} - \frac{2}{n^2} < \mathrm{dist}\left(\langle c, u\rangle, \mathbb{Z}\right)$.

Hence, by theorem 33 $\mathrm{dist}\left(\begin{pmatrix} n^6\sqrt{n}c \\ 0 \end{pmatrix}, L_{PK}\right) \leq T$ can not hold. Thus $\left(L_{PK}, t, \begin{pmatrix} n^6\sqrt{n}c \\ 0 \end{pmatrix}\right)$ is a NO instance of the GapCVP$_\gamma$.

The following lemmas complete the proof.

Lemma 49. *Let $PK = \{w_1, \ldots, w_n, v_1, \ldots, v_m, k\}$ be a public key of AD, p_1 and p_2 be points from L_{PK}. If for $c_1, c_2 \in P(w_1, \ldots, w_n)$*
$$\mathrm{dist}\left(\begin{pmatrix} n^6\sqrt{n}c_1 \\ 0 \end{pmatrix}, p_1\right) = D_1 \text{ and } \mathrm{dist}\left(\begin{pmatrix} n^6\sqrt{n}c_2 \\ 0 \end{pmatrix}, p_2\right) = D_2 \text{ then}$$
$$\mathrm{dist}\left(\begin{pmatrix} n^6\sqrt{n}((c_1+c_2) \mod P(w_1,\ldots,w_n)) \\ 0 \end{pmatrix}, L_{PK}\right) \leq D_1 + D_2 + \sqrt{n}.$$

Proof. We can represent $n^6\sqrt{n}((c_1+c_2) \mod P(w_1,\ldots,w_n)) = n^6\sqrt{n}(c_1+c_2+\sum_{i=1}^n a_i w_i)$. Since both vectors c_1 and c_2 belong to $P(w_1,\ldots,w_n)$ we can bound $|a_i| \leq 1$ for all i. Consider a vector $p_3 = B_{PK}(a_1,\ldots,a_n,0,\ldots,0)^t$ where B_{PK} is the matrix defined in (3.2).

$$\mathrm{dist}\left(\begin{pmatrix} n^6\sqrt{n}\sum_{i=1}^n a_i w_i \\ 0 \end{pmatrix}, p_3\right) = \sqrt{\sum_{i=1}^n a_i^2} \leq \sqrt{n}.$$

The lemma follows.

Lemma 410. *Let $PK = \{w_1,\ldots,w_n,v_1,\ldots,v_m,k\}$ be a public key of AD, and p be a point from L_{PK}. If for $c \in P(w_1,\ldots,w_n)$ $\mathrm{dist}\left(\begin{pmatrix} n^6\sqrt{n}c \\ 0 \end{pmatrix}, p\right) = D$ then for sufficiently large n*
$$\mathrm{dist}\left(\begin{pmatrix} n^6\sqrt{n}((c+v_k) \mod P(w_1,\ldots,w_n)) \\ 0 \end{pmatrix}, L_{PK}\right) \leq D + n^4.$$

Proof. We can represent $n^6\sqrt{n}((c+v_k) \mod P(w_1,\ldots,w_n)) = n^6\sqrt{n}(c+v_k+\sum_{i=1}^n a_i w_i)$. Consider a point p' from L_{PK} such that $p' = B_{PK}(a_1,\ldots,a_n,0,\ldots,0,1,0,\ldots,0)^t$ (with '1' at the $(n+k)$-th position). It is easy to see that

$$\mathrm{dist}\left(\begin{pmatrix} n^6\sqrt{n}(v_k + \sum_{i=1}^n a_i w_i) \\ 0 \end{pmatrix}, p'\right) \leq \sqrt{n^5 + \sum_{i=1}^n a_i^2}. \qquad (4.2)$$

Let us bound $\sum_{i=1}^n a_i^2$. Note, that $a_i = \lfloor \theta_i \rfloor$ for θ_i defined as $c + v_k = \sum_{i=1}^n \theta_i w_i$. Since the width of the parallelepiped $P(w_1,\ldots,w_n)$ is greater than $\frac{\rho_n}{n^2}$, we can bound $\sum_{i=1}^n a_i^2$ as follows:

$$\sum_{i=1}^n a_i^2 \leq \sum_{i=1}^n \theta_i^2 \leq \frac{n^4}{\rho_n^2} \sum_{i=1}^n \langle c+v_k, w_i^\perp \rangle^2 \leq \frac{n^5}{\rho_n^2} \|c+v_k\|^2 \leq \frac{n^5}{\rho_n^2}(\|c\|+\|v_k\|)^2 \leq 4n^7$$

Expression (4.2) is less than n^4 for sufficiently large n. The lemma follows.

5 Open Problems

There are a great deal of open problems. We highlight a few here.

VERIFIABLE DECRYPTION FOR AD CRYPTOSYSTEM. The AD cryptosystem is a probabilistic scheme for which in the process of decryption, the legal decryptor who knows the private key computes the plaintext without being able to recover the randomness used by the encryptor. This latter task, requires the ability to solve subset sum problem instances. A similar situation holds with respect to the El-Gamal and Cramer-Shoup cryptosystems [10, 6] in which a legal decryptor who knows the private key can decrypt, and yet cannot recover the randomness used by an encryptor, as that would require solving discrete log problem instances.

Such cryptosystems raise an interesting challenge: can a legal decryptor, who knows the private-key of the cryptosystem but does not know the randomness used in the computation of a given ciphertext, prove to a third party that a given ciphertext corresponds to a cleartext without revealing his private key?[3] A cryptosystem for which this can be done was named a *verifiable decryption* scheme by Camenisch and Shoup in [5]. The the challenge is to do this efficiently for the AD cryptosystem. In principle it is achievable based on the existence of one-way functions (which is implied in the context of encryption in any case) using general computational zero-knowledge proofs for NP statements [17].

NON MALLEABLE PROOFS OF PLAINTEXT KNOWLEDGE FOR THE AD CRYPTOSYSTEM. Katz[20] shows efficient non-malleable PPKs for the Blum-Goldwasser RSA and Rabin based encryption, Paillier and El-Gamal, and gets as an application CCA2 secure efficient interactive encryption schems. A promising open problem (although far from obvious) is to design an efficient **non-malleable** PPK for the AD cryptosysetm, and thus obtain a CCA2 secure efficient interactive encryption variant of the AD cryptosystem. One obstacle in tackling this problem is that Katz's protocol utilizes one-time signatures (which although exist in principle under ISVP) for which there are no efficient constructions under ISVP.

REGEV CRYTOSYSTEM. In this paper we addressed the AD cryptoststem. Design a PPK for the Regev cryptosystem, and address the above open problems for the Regev cryptosystem.

Acknowledgment. This work was supported in part by NSF Cybertrust 043045, a Minerva project grant 8495 and grant from Potters Wheel Foundation.

---

[3] In other words, is there a verifiable encryption scheme for the equivalence relation by a prover who does not know the randomness used to encrypt.

References

1. M. Ajtai and C. Dwork. A Public-Key Cryptosystem with Worst-Case/Average-Case Equivalence. *ECCC, TR96-065, Dec. 1996*.
2. N. Asokan, V. Shoup, M. Waidner. Optimistic Fair Exchange of Digital Signatures (Extended Abstract). *EUROCRYPT 1998: 591-606*
3. M. Bellare and M. Yung, Certifying Permutations: Noninteractive Zero-Knowledge Based on Any Trapdoor Permutation *J. Cryptology 9(3): 149-166 (1996)*
4. J. Camenisch and I. Damgard Verifiable Encryption, Group Encryption, and Their Applications to Separable Group Signatures and Signature Sharing Schemes. *ASIACRYPT 2000: 331-345*
5. J. Camenisch and V. Shoup. Practical Verifiable Encryption and Decryption of Discrete Logarithms. *CRYPTO 2003: 126-144*
6. R. Cramer, V. Shoup. A Practical Public Key Cryptosystem Provably Secure against Adaptive Chosen Ciphertext Attack. (1998)
7. R. Cramer, V. Shoup. Universal Hash Proofs and a paradigm for adaptive chosen ciphertext secure public key encryption. *Eurocrypt 2002*.
8. D. Dolev, C. Dwork, M. Naor: Nonmalleable Cryptography. SIAM J. Comput. 30(2): 391-437 (2000)
9. W. Diffie, M. E. Hellman. New Directions in Cryptography . IEEE Transactions on Information Theory 1976.
10. T. ElGamal. A Public Key cryptosystem and a signature scheme based on discrete logarithm. Proceedings of Crypto 84.
11. U. Feige, D. Lapidot, A. Shamir: Multiple NonInteractive Zero Knowledge Proofs Under General Assumptions. *SIAM J. Comput. 29(1): 1-28 (1999)*
12. U. Feige, A. Fiat, A. Shamir: Zero Knowledge Proofs of Identity. Journal of Cryptology1(2):77-94 (1988).
13. Zvi Galil, Stuart Haber, Moti Yung. Symmetric Public-Key Encryption. CRYPTO 1985: 128-137
14. O. Goldreich, S. Goldwasser. On the Limits of Nonapproximability of Lattice Problems. *JCSS 60(3): 540-563 (2000)*
15. O. Goldreich, S. Goldwasser, and S. Halevi. Eliminating decryption errors in the Ajtai-Dwork cryptosystem. In: *Advances of Cryptology, Proc of Crypto'97 Lecture Notes in Computer Science, 1997*.
16. O. Goldreich. Foundations of Cryptography: Basic Tools. Cambridge University Press, Cambridge, UK, 2001
17. O. Goldreich, S. Micali, and A. Wigderon. Proofs that Yield Nothing but their Validity or NP in Zero Knowledge. JACM 91.
18. S. Halevi and S. Micali. More on Proofs of Knowledge. *LCS Document Number: MIT-LCS-TM-578*
19. Chris Hall, Ian Goldberg, Bruce Schneier, Reaction Attacks Against Several Public-Key Cryptosystems. Proceedings of Information and Communication Security, ICICS'99
20. Jonathan Katz. Efficient and Non-Malleable Proofs of Plaintext Knowledge and Applications. Eurocrypt 2003 .
21. D. Micciancio and S. Vadhan. Statistical zero-knowledge proofs with efficient provers: lattice problems and more. *Advances in Cryptology - Crypto 2003. Santa Barbara, CA, USA, August 2003. LNCS 2729, Springer*.

22. M. Naor, M. Yung, Public-key cryptosystems provably secure against chosen ciphertext attacks, *Proceedings of the twenty-second annual ACM symposium on Theory of computing, p.427-437, May 13-17, 1990, Baltimore, Maryland, United States*
23. P. Nguyen and J. Stern. Cryptanalysis of the Ajtai-Dwork cryptosystem. *In Advances in Cryptology: Proceedings of Crypto '98, volume 1462 of Lecture Notes in Computer Science, pages 223-242. Springer-Verlag, 1998.*
24. O. Regev, *New Lattice Based Cryptographic Constructions, STOC 2003.*
25. R. L. Rivest, A. Shamir, L. Adleman Public key cryptography, CACM 21, 120-126, 1978.
26. M. Stadler. Publicly Verifiable Secret Sharing. *EUROCRYPT 1996: 190-199*

Entropic Security and the Encryption of High Entropy Messages[*]

Yevgeniy Dodis[1] and Adam Smith[2]

[1] New York University
dodis@cs.nyu.edu
[2] Weizmann Institute of Science
adam.smith@weizmann.ac.il

Abstract. We study *entropic security*, an information-theoretic notion of security introduced by Russell and Wang [24] in the context of encryption and by Canetti et al. [5,6] in the context of hash functions. Informally, a probabilitic map $Y = \mathcal{E}(X)$ (e.g., an encryption sheme or a hash function) is entropically secure if knowledge of Y does not help predicting any predicate of X, whenever X has high min-entropy from the adversary's point of view. On one hand, we strengthen the formulation of [5, 6, 24] and show that entropic security in fact implies that Y does not help predicting any *function* of X (as opposed to a predicate), bringing this notion closer to the conventioonal notion of semantic security [10]. On the other hand, we also show that entropic security is equivalent to *indistinguishability* on pairs of input distributions of sufficiently high entropy, which is in turn related to *randomness extraction* from non-uniform distributions [21].

We then use the equivalence above, and the connection to randomness extraction, to prove several new results on entropically-secure encryption. First, we give two general frameworks for constructing entropically secure encryption schemes: one based on expander graphs and the other on XOR-universal hash functions. These schemes generalize the schemes of Russell and Wang, yielding simpler constructions and proofs, as well as improved parameters. To encrypt an n-bit message of min-entropy t while allowing at most ϵ-advantage to the adversary, our best schemes use a shared secret key of length $k = n - t + 2\log\left(\frac{1}{\epsilon}\right)$. Second, we obtain lower bounds on the key length k for entropic security and indistinguishability. In particular, we show near tightness of our constructions: $k > n - t$. For a large class of schemes — including all the schemes we study — the bound can be strengthened to $k \geq n - t + \log\left(\frac{1}{\epsilon}\right) - O(1)$.

1 Introduction

If X and Y are random variables, the statement "Y leaks no information about X" is normally formalized by requiring that X and Y be almost statistically

---

[*] A more complete version of this paper may be found on IACR Cryptology ePrint Archive, report 2004/219, at http://eprint.iacr.org/2004/219/ [9].

independent. Equivalently, one can require that the Shannon mutual information $\mathbf{I}(X, Y)$ be very small. In this work we study situations where information leakage is unavoidable — that is, $\mathbf{I}(X, Y)$ is large — yet we still want a guarantee that no *useful* information about X is leaked by Y, even to a computationally unbounded adversary.

Consider an alternative notion of security, inspired by semantic security of encryptions [10]. We say Y *hides all functions of* X if for every function f, it is nearly as hard to predict $f(X)$ given Y as it is without Y, regardless of the adversary's computing power. If $Y = \mathcal{E}(X)$ for some probabilistic map $\mathcal{E}()$ (for example, an encryption scheme), then we say the map $\mathcal{E}$ is *entropically secure* if $\mathcal{E}(X)$ hides all functions of X whenever X has sufficiently high entropy.

A seemingly weaker variant of this definition has produced surprising results in at least two contexts so far: Canetti, Micciancio and Reingold [5, 6] constructed hash functions whose outputs leak no partial information about the input. Russell and Wang [24] showed how one can construct entropically-secure symmetric encryption schemes with keys much shorter than the length of the input, thus circumventing Shannon's famous lower bound on key length.

Our contributions can be divided into two areas.

- We elucidate the notion of entropic security. Our results apply to all entropically-secure primitives, including encryption schemes and hash functions. We provide two new variants on entropic security, one closer in spirit to semantic security of encryptions [10], and the other along the lines of indistinguishability of encryptions. The proofs that these various notions are equivalent give us new tools for handling entropic security and highlight a relationship with "randomness extraction" from non-uniform distributions.

- We use the connection to randomness extraction to obtain new constructions and lower bounds for encryption of high-entropy messages with a short key. First, we give two general frameworks for constructing entropically secure encryption schemes: one based on expander graphs and the other on XOR-universal hash functions. These schemes generalize the schemes of Russell and Wang, yielding simpler constructions and proofs, as well as improved parameters. Second, we obtain nearly tight lower bounds on the key length k for entropic security and indistinguishability.

1.1 Background

Although our general results apply to all entropically-secure primitives, we present entropic security (and our results) in the context of symmetric-key one-time encryption. Alice and Bob share a secret key K and Alice wants to securely send some message X to Bob over a public channel. X is assumed to come from some a-priori distribution on $\{0, 1\}^n$ (e.g., uniform), and the goal is to compute a ciphertext Y which: (a) allows Bob to extract X from Y using K; (b) reveals "no information" about X to the adversary Eve beyond what she already knew. Below, we write $Y \leftarrow \mathcal{E}(X, K)$ and $X = \mathcal{D}(Y, K)$.

Perfect and Computational Security. The first formalization of this problem came in a fundamental work of Shannon [25], who defined "no information" by requiring that X and Y be independent as random variables: using information theoretic notation, $\mathbf{I}(X;Y) = 0$, where $\mathbf{I}$ is the mutual information. He showed a lower bound on key length for his definition: encrypting messages of length n requires at least n bits of shared key (more formally, the Shannon entropy of the key must be at least that of message distribution: $\mathbf{H}_{sh}(K) \geq \mathbf{H}_{sh}(X)$). This bound is tight when the message is chosen uniformly from all strings of a fixed length n, since one can use a one-time pad. The bound was extended to the interactive setting by Maurer [19].

Goldwasser and Micali [10] relaxed the notion of perfect security to the *computational* setting: namely, any *efficient* Eve can extract only negligible "information" about X from Y. They had to properly redefine the notion of "information", since mutual information or conditional probabilities do not make much sense in a computationally-bounded world. They suggested two now classical definitions. Consider the following, equivalent version of Shannon's definition: for any two messages x_0 and x_1, the two corresponding distributions on ciphertexts should be identical, that is $\mathcal{E}(x_0) = \mathcal{E}(x_1)$ (as distributions). The first definition of Goldwasser and Micali, called *computational indistinguishability of encryptions*, generalizes this version of perfect security: they require that no efficient (polynomial-time adversary) can distinguish the encryptions of x_0 and x_1 with advantage more than ϵ over random guessing, where ϵ is some negligible quantity. Their second notion is called *semantic security*: for *any* distribution on messages X and any function $f()$, the adversary can predict $f(X)$ given $\mathcal{E}(X)$ with probability only negligibly better than she could without seeing $\mathcal{E}(X)$. The first definition is easier to work with, but the second definition seems to capture a stronger, more intuitive notion of security: for example, indistinguishability is the special case of semantic security when the message distribution X is restricted to uniform distributions over two points $\{x_0, x_1\}$. In fact, Goldwasser and Micali showed that the two definitions are equivalent. Thus, distributions with entropy 1 are in some sense the hardest to deal with for semantic security.

Statistical Security? A natural intermediate notion of security between perfect and computational security would be some kind of *statistical security*. Eve is computationally unbounded, as in the perfect setting, but can potentially recover some negligible amount of "information", as in the computational setting. At first glance, it seems like there is no gain in this notion, no matter how we interpret "information". For example, following Shannon's approach we could require that $\mathbf{I}(X;Y) \leq \epsilon$ instead of being 0. Unfortunately, Shannon's proof still implies that $\mathbf{H}_{sh}(K) \geq \mathbf{H}_{sh}(X) - \epsilon$. Similarly for indistinguishability: since the distribution $\mathcal{E}(x)$ should look almost the same for *any* fixed x, one can argue that $\mathbf{I}(Y;X) = \mathbf{H}_{sh}(\mathcal{E}(X)) - \mathsf{Exp}_x[\mathbf{H}_{sh}(\mathcal{E}(x))]$ still has to be negligible, and so the key must again have entropy almost $\mathbf{H}_{sh}(X)$.

In his original work Shannon envisioned applications where Eve has a lot of uncertainty about the message. To get a pessimistic bound that $\mathbf{H}_{sh}(K) \geq n$, one only has to take X to be uniformly distributed in $\{0,1\}^n$. In fact, in the perfect setting, security against the uniform distribution implies security against *any* distribution on messages. On the other hand, the notions of indistinguishability and semantic security primarily deal with entropy 1 distributions, and the straightforward extension of Shannon's bound to the statistical versions of these notions crucially uses this fact. Thus, it is natural to ask if we can meaningfully define (statistical) semantic security and/or indistinguishability for high entropy distributions (say, uniform), similar in spirit to the original work of Shannon. And if yes,

1. How do these notions relate to Shannon's (statistical) notion, $\mathbf{I}(X;Y) \leq \epsilon$? Most importantly, does the pessimistic bound on the key length still extend to these notions?

2. How do these notions relate to each other? Are semantic security and indistinguishability still equivalent when the message is guaranteed to have high entropy?

1.2 Entropic Security

Russell and Wang [24] introduced the idea of statistical security for encryption of high-entropy message spaces. They considered the first question above, though they focused on a weakened version of semantic security. Their definition, *entropic security of encryptions for predicates*, is natural: for any distribution X of min-entropy[1] at least t and any predicate $g : \{0,1\}^n \to \{0,1\}$, Eve can predict $g(X)$ using Y only negligibly better than she could without Y (here n is the message length and t is a parameter). Russell and Wang showed that Shannon's lower bound does *not* extend to this new notion: they presented two schemes beating Shannon's bound on key length, which we describe further below. Entropic security also arose earlier in work of Canetti [5] and Canetti, Micciancio and Reingold [6]. They constructed probabilistic hash functions whose output reveals no partial information about their input as long as it had sufficiently high entropy.

We discuss a stronger version of the definition of [5, 6, 24], which requires that the adversary gain no significant advantage at predicting any *function* of the input (as opposed to a predicate). One of our results is the equivalence of their notion of security to the one described here.

Definition 1 (Entropic Security). *The probabilistic map Y hides all functions of X with leakage ϵ if for every adversary $\mathcal{A}$, there exists some adversary $\mathcal{A}'$ such that for all functions f,*

$$\big| \Pr[\mathcal{A}(Y(X)) = f(X)] - \Pr[\mathcal{A}'() = f(X)] \big| \leq \epsilon.$$

The map $Y()$ is called (t, ϵ)-entropically secure if $Y()$ hides all functions of X, whenever the min-entropy of X is at least t.

---

[1] The *min-entropy* of a random variable A is a measure of the uncertainty of its outcome. It is the negative logarithm of the probability that one can predict A ahead of time: $\mathbf{H}_\infty(A) = -\log(\max_a \Pr(A = a))$.

One gets some insight about this definition by thinking of it as an information-theoretic reformulation of semantic security of encryptions [10], although restricted to high-entropy message spaces. Alternatively, it might be instructive to view this definition as saying that Y leaks no *a-priori* information about X. Here "a-priori" refers to the fact that the function f has to be specified *before* the pair (X,Y) is sampled. In other words, although f is arbitrary, it cannot depend on the outcome of Y. This should be contrasted with *a-posteriori* information, where first the pair (X,Y) is sampled, then the adversary is given the outcome y of Y, and can choose a function f_y which is supposedly easier to predict when given y. In this latter case it is not very hard to see that Y leaks almost no a-posteriori information about X if and only if X and Y are essentially independent, i.e. the quantity $\mathbf{I}(X;Y)$ is "low". Thus, the results of [24, 5, 6] could be interpreted by saying that leakage of no a-priori information — although for the moment restricted to predicates rather than general functions — can be achieved in situations where it is *impossible* to leak no a-posteriori information.

1.3 Contributions of This Paper

This paper carefully studies and eludicates the notion of entropic security, obtaining several new insights into this notions, as well as simplifying and improving previous results of [24, 5, 6].

A Strong Definition of Security. As we mentioned, the definition we propose (Definition 1) is seemingly stronger than previously studied formulations of entropic security [5, 6, 24], which only considered the adversary's ability to predict *predicates* instead of all possible functions of the secret input. This definition may not be quite satisfying from several points of view. First, it states only that no predicate of the input is leaked, and provides no explicit guarantees about other functions. In contrast, the original semantic security definition of Goldwasser and Micali held for all functions. Second, there is no guarantee that the new adversary $\mathcal{A}'()$ is polynomial time, even in the case where, say, $\mathcal{A}$ runs polynomial time and X is samplable in polynomial time. We show that (a) entropic security for predicates *does* imply that for arbitrary functions (see Lemma 2), and (b) in the definition of entropic security one can always set $\mathcal{A}'()$ to be $\mathcal{A}(U_n)$, where U_n is the uniform distribution on n bits.

The equivalence between predicates and functions is not trivial. Consider, for example, the special case where f is the identity function. One might hope that a good predictor for X must imply a good predictor for some bit X_i of X. However, this is false. As a counterexample, assume X is equal to U_n and Y is equal to X with probability $1/2$, and to the bit-wise complement of X otherwise. Clearly, Y reveals X with probability at least $1/2$ (which is much larger than 2^{-n}), although no physical bit of X can be predicted with probability better than its natural probability $1/2$. Of course, in this case one can predict any even parity of the bits of X with probability 1, but this shows that a more sophisticated approach is needed. As we show in Proposition 1, for this function we can choose a "Goldreich-Levin" predicate at random, that is we can use the predicate $g_r(x) = r \odot x$ where r is a random n-bit string and $\odot$ is the binary inner product $r \odot x = \sum_i r_i x_i \mod 2$. For general functions, a more complicated construction is required (Lemma 2). This general equivalence between predicting predicates and predicting arbitrary functions could be of independent interest, as it provides an information-theoretic converse to the Goldreich-Levin hardcore bit construction.

An Equivalence to Indistinguishability. We also define a new *indistinguishability* notion, analogous to indistinguishability of (computationally secure) encryptions [10]. Namely, we say that the map $Y()$ is t-indistinguishable, if for any distributions X_1 and X_2 of min-entropy at least t, the distribution $Y(X_1)$ is statistically close to $Y(X_2)$. A bit more formally, indistinguishability is stated in terms of the statistical difference $\mathbf{SD}(A, B)$ between a pair of random variables A, B. This is half the L_1 distance between the distributions, $\mathbf{SD}(A,B) = \frac{1}{2} \sum_z |\Pr[A=z] - \Pr[B=z]|$. It also has an operational meaning: given a sample from either A or B (at random), the optimal adversary's chance of correctly guessing which distribution the sample came from is exactly $\frac{1}{2} + \frac{1}{2}\mathbf{SD}(A,B)$. This distance measure satisfies the triangle inequality, and so all the distributions $Y(X)$ must actually be close to a single distribution $G = Y(U_n)$, where U_n is the uniform distribution. We arrive at the following:

Definition 2. *A randomized map $Y()$ is (t, ϵ)-indistinguishable if there is a random variable G such that for every distribution on messages X over $\{0,1\}^n$ with min-entropy at least t, we have*

$$\mathbf{SD}(Y(X), G) \leq \epsilon.$$

As we can see, the notion of entropic security seems to be well motivated, but hard to work with. On the other hand, indistinguishability seems to be a much easier definition to work with, but might be less intuitively meaningful. Our main result is that the definitions are in fact equivalent:

Theorem 1. *Let Y be a randomized map with inputs of length n. Then*

1. *(t, ϵ)-entropic security for predicates implies $(t-1, 4\epsilon)$-indistinguishability.*

2. *$(t-2, \epsilon)$-indistinguishability implies $(t, \epsilon/8)$-entropic security for **all functions** when $t \geq 2\log\left(\frac{1}{\epsilon}\right) + 1$.*

In particular, since entropic security with respect to predicates is trivially implied by entropic security for all functions, Theorem 1 states that all three notions of security discussed above are equivalent up to small changes in the parameters. Although this result is inspired by a similar looking result of Goldwasser and Micali [10] (for computational encryption), our proof is considerably different and does not seem to follow from the techniques in [10].

The equivalence not only reconciles two natural definitions, but has several nice implications. First, in Definition 1 we can always take $\mathcal{A}'()$ to be $\mathcal{A}(Y(U_n))$, where U_n is the uniform distribution on $\{0,1\}^n$. Thus, the "simulated" adversary is as efficient as the original.

Second, the equivalence provides a new application of *randomness extractors* [21] to cryptography. Recall that an extractor takes as input an arbitrary, high entropy random source and a tiny random seed, and outputs uniformly random bits. The output bits are guaranteed to be almost uniformly distributed as long as the min-entropy of the input is above some threshold t. In other words, an extractor Y is precisely a t-indistinguishable map — in the sense of Definition 2 — with G being the uniform distribution. Thus, Theorem 1 implies that an extractor for t-sources hides all a-priori information about sources of min-entropy at least $t+2$. From the constructive point of view, it also suggests that to design an appropriate entropically secure scheme for a given task, such as encryption, it is sufficient to design a "special purpose" randomness extractor. In the case of encryption the extractor should be invertible when given the seed, since the seed corresponds to the shared secret key.

Finally, and most importantly, our equivalence simplifies the design and analysis of entropically secure schemes, yielding improvements over known schemes, new lower bounds, simpler proofs, and a stronger security guarantee. We illustrate these points for the case of entropically secure encryption.

Encryption of High-Entropy Messages. As we mentioned, Russell and Wang [24] provided two constructions of entropically-secure encryption schemes which use short keys. Let ϵ denote the leakage — that is, the advantage which we allow the adversary. First, [24] give a deterministic scheme of the form $\mathcal{E}(X,K) = X \oplus p(K)$, which is secure only when X is uniformly distributed on $\{0,1\}^n$, where K has length only $k = 2\log n + 3\log\left(\frac{1}{\epsilon}\right) + O(1)$ and $p(K)$ is a random point in a δ-biased spaces [20] (where [24] used $\delta = \epsilon^{3/2}$). Thus, $p(K)$ could be viewed as a very sparse one-time pad which nevertheless hides any a-priori specified function $f(X)$. Second, for general min-entropy t, Russell and Wang gave a *randomized* scheme of the form $(\psi, \psi(X) + K) \leftarrow \mathcal{E}(X,K)$, where ψ is chosen at random from a family of 3-wise independent permutations (and the addition is defined over some appropriate space). The analysis in [24] shows that this second scheme needs key length $n - t + 3\log\left(\frac{1}{\epsilon}\right) + O(1)$. While less than n for nontrivial settings of t, this key length again becomes $\Omega(n)$ when $n - t = \Omega(n)$. [24] left it open whether such dependence on $n - t$ is necessary.
We obtain the following new results:

1. Lower bounds on the key length k for entropic security and indistinguishability. In particular, we show near tightness of Russell-Wang constructions: $k > n - t$. (In fact, for a large class of schemes $k \geq n - t + \log\left(\frac{1}{\epsilon}\right)$.)

2. Two general frameworks for designing entropically secure encryption schemes: one based on expander graphs and the other on XOR-universal hash functions. These schemes generalize the schemes of Russell and Wang, yielding simpler constructions and proofs, as well as improved parameters. Namely, both constructions can yield keys of size $k = n - t + 2\log\left(\frac{1}{\epsilon}\right)$.

Our Techniques. All our results critically use the equivalence between entropic security and indistinguishability.

On one hand, we use it to show that the general construction of Russell and Wang is nearly optimal: *any* entropically secure scheme must have $k > n - t$. In fact, for a special case of *public-coin* schemes, where the ciphertext contains the randomness used for encryption, we get an even stronger bound: $k \geq n - t + \log\left(\frac{1}{\epsilon}\right) - O(1)$. The latter result is proven by relating the notion of indistinguishability to that of *randomness extractors* [21]: namely, any indistinguishable public-coin scheme almost immediately yields a corresponding extractor. Using the optimal lower bounds on extractors [23], we get our stronger bound as well. In fact, if the ciphertext is statistically close to uniform (i.e., $G = U_n$ meaning that the encryption is actually a randomness extractor), we get a lower bound which *exactly* matches our upper bounds: $k \geq n - t + 2\log\left(\frac{1}{\epsilon}\right) - O(1)$. The schemes in [24] and this work are all public-coin and have random ciphertexts.

On the other hand, the indistinguishability view allows us to give a general framework for constructing entropically secure encryption schemes. Specifically, assume we have a d-regular expander G on 2^n vertices V with the property that for any subset T of 2^t vertices, picking a random vertex v of T and taking a random neighbor w, we obtain an almost uniform distribution on V. Then, we almost immediately get an encryption scheme with key length $k = \log d$ which is indistinguishable for message spaces of min-entropy t. Looking at this from another perspective, the above encryption scheme corresponds to a randomness extractor which takes a source X of length n and min-entropy t, invests $\log d$ extra random bits K, and extracts n almost random bits Y (with the additional property that the source X is recoverable from Y and K). From this description, it is clear that the key length of this paradigm must be at least $n - t$ (which we show is required in any entropically secure encryption scheme). However, using optimal expanders we can (essentially) *achieve* this bound, and in several ways. First, using Ramanujan expanders [17], we get the best known construction with key length $k = n - t + 2\log\left(\frac{1}{\epsilon}\right)$. Second, using δ-biased spaces [20] (for appropriate $\delta = \delta(\epsilon, n, t)$ explained later), we get a general construction with slightly larger but still nearly optimal key length $k = n - t + 2\log n + 2\log\left(\frac{1}{\epsilon}\right)$. This last result generalizes (and slightly improves) to any value of t the special case of the uniform message distribution ($n - t = 0$) obtained by Russell and Wang [24]. Our approach also gives clearer insight as to why small-biased spaces are actually useful for entropic security.

While the deterministic constructions above are nearly optimal and quite efficient, we also observe that one can get simpler constructions by allowing the encryption scheme to be *probabilistic*. In our approach, this corresponds to having a *family* of "average case" expanders $\{G_i\}$ with the property that for any set T of size at least 2^t, picking a random graph G_i, a random v in T and taking a random neigbor w of v in G_i, we get that w is nearly uniform, *even given the graph index i*. By using any family of pairwise independent hash functions h_i (resp. permutations ψ_i) and a new variant of the leftover hash lemma [15], we get a probabilistic scheme of the form $\langle i, X \oplus h_i(K) \rangle$ (resp. $\langle i, \psi_i(X) \oplus K \rangle$) with a nearly optimal key length $k = n - t + 2\log\left(\frac{1}{\epsilon}\right)$. As a concrete example of this approach, we get the following simple construction: $\mathcal{E}(X, K; i) = (i, X + i \cdot K)$, where the local randomness i is a random element in $GF(2^n)$, $K \in \{0,1\}^k$ is interpreted as belonging to $GF(2^k) \subseteq GF(2^n)$, and addition and multiplication are done in $GF(2^n)$.

Once again, the above result (with permutations ψ_i) improves and simplifies the intuition behind the second scheme of Russell and Wang [24]. Indeed, the latter work had to assume that the ψ_i's come from a family of 3-wise independent permutations — which are more compicated and less efficient than 2-wise independent permutations (or functions) — and presented a significantly more involved analysis of their scheme.

1.4 A Caveat: Composing Entropically-Secure Constructions

A desirable property of definitions of security of cryptographic primitives is *composability*: once some protocol or algorithm has been proven secure, you would like to be able to use it as a building block in other protocols with your eyes closed—without having to worry about effects that violate the intuitive notion of security, but which are not covered by the original definition.

Composability is difficult to guarantee, since it is not clear how to translate it into a mathemetical property. There are various formalizations of composability, most notably "Reactive Systems" [22], "Universal Composability" [7] and several frameworks based on logic algebras for automated reasoning (see [14] and the references therein). Finding protocols that are provably secure in these general frameworks is difficult, and sometimes provably impossible. A more common approach is to prove that a particular definition remains intact under a few straightforward types of composition, say by proving that it is still secure to encrypt the same message many times over.

The main weakness of entropic security, as defined above, is that it does not ensure composability, even in this straightforward sense. If $Y()$ and $Y'()$ are independent versions of the same entropically-secure mapping, then the map which outputs the pair $Y(X), Y'(X)$ may be insecure to the point of revealing X completely. In the case of encryption, this means that encrypting the same message twice may be problematic. (Given the first value $Y(X)$, the entropy of X may be too low for the security guarantee of $Y'()$ to hold).

For example, suppose that $Y(x)$ consists of the pair $\langle M, Mx \rangle$, where M is a random $\frac{3n}{4} \times n$ binary matrix M and $x \in \{0,1\}^n$. We will see later that $Y()$ is entropically secure whenever the entropy of X is close to n. However, the pair $Y(x), Y'(x)$ provides a set of $\frac{3n}{2}$ randomly chosen linear constraints on x. With high probability, these determine x completely, and so the pair $Y(), Y'()$ is insecure under any reasonable definition.

Given these issues, entropically-secure primitives must be used with care: one must ensure that the inputs truly have enough entropy for the security guarantee to hold. Requiring entropy is natural in many situations (e.g. when the input is a password), but the issue of composability nonetheless raises a number of interesting open questions for future research.

The generality and intuitive appeal of entropic security, as well as the variety of contexts in which it has arisen, make it an important concept to understand. We hope that the present work provides a major step in this direction.

2 Entropic Security and Indistinguishability

In this section we sketch the proof of Theorem 1, that is of the equivalence between entropic security for functions/predicates and indistinguishability.

First, some notation. Fix a distribution X on $\{0,1\}^n$. For a function $f : \{0,1\}^n \to \{0,1\}^*$, let $\mathsf{pred}_{f,X}$ be the maximum probability of any particular outcome, that is the maximum probability of predicting $f(X)$ without having any information about X: $\mathsf{pred}_{f,X} \stackrel{def}{=} \max_z \Pr[f(X) = z]$.
(When X is clear from the context, we may simply write pred_f.) We may rephrase entropic security as follows: for every function f and adversary $\mathcal{A}$, the probability of $\mathcal{A}$ predicting $f(X)$ given $Y(X)$ is at most $\mathsf{pred}_f + \epsilon$:

$$\Pr[\mathcal{A}(Y(X)) = f(X)] \leq \mathsf{pred}_{f,X} + \epsilon.$$

2.1 From Entropic Security to Indistinguishability

The first statement of Theorem 1 is the easier of the two to prove, and we give the intuition here: given two distributions X_0, X_1, we can define a predicate $g(x)$ which captures the question "is x more likely to have come from X_0 or X_1?" If X is a equal mixture of X_0 and X_1, then the adversary which makes the maximum likelihood guess at $g(X)$ given $Y(X)$ will have success probability $\frac{1}{2} + \frac{1}{2}\mathbf{SD}\left(Y(X_0), Y(X_1)\right)$. On the other hand, with no access to $Y(X)$, the adversary can succeed with probability at most $\mathsf{pred}_P = \frac{1}{2}$. Entropic security implies that the advantage over random guessing, and hence the statistical distance, must be small. The formal proof is more involved, and is given below.

Proof. It is sufficient to prove indistinguishability for all distributions which are uniform on some set of 2^{t-1} points. To see why, recall that any distribution of min-entropy at least $t - 1$ can be written as a convex combination of such *flat* distributions. If $X_0 = \sum \lambda_{0,i} X_{0,i}$ and $X_1 = \sum_j \lambda_{1,j} X_{1,j}$, where the $X_{0,i}$ and $X_{1,j}$ are all flat distributions, then the statistical distance $\mathbf{SD}\left(Y(X_0), Y(X_1)\right)$ is bounded above by $\sum_{i,j} \lambda_{0,i} \lambda_{1,j} \mathbf{SD}\left(Y(X_{0,i}), Y(X_{1,j})\right)$ (by the triangle inequality). If each of the pairs $Y(X_{0,i}), Y(X_{1,j})$ has distance at most ϵ, then the entire sum will be bounded by ϵ.

Now let X_0, X_1 be any two flat distributions over *disjoint* sets of 2^{t-1} points each (we will deal with non-disjoint sets below), and let X be an equal mixture of the two. That is, to sample from X, flip a fair coin B, and sample from X_B. Take g to be any predicate which is 0 for any sample from X_0 and 1 for any sample from X_1. A good predictor for g will be the adversary $\mathcal{A}$ who, given a string y as input, guesses as follows:

$$\mathcal{A}(y) = \begin{cases} 0 \text{ if } y \text{ is more likely under the distribution } Y(X_0) \text{ than under } Y(X_1) \\ 1 \text{ otherwise} \end{cases}$$

By the definition of statistical difference, this adversary guesses the predicate with probability exactly:

$$\Pr\left[\mathcal{A}(Y(X)) = B = g(X)\right] = \tfrac{1}{2} + \tfrac{1}{2}\mathbf{SD}\left(Y(X_0), Y(X_1)\right). \tag{1}$$

We can now apply the assumption that $Y()$ is (t, ϵ)-entropically secure to bound $\mathbf{SD}\left(Y(X_0), Y(X_1)\right)$. First, for any random variable G over $\{0, 1\}$ which is independent of X, the probability that $G = g(X)$ is exactly $\tfrac{1}{2}$. The distribution X has min-entropy t by construction, and so by entropic security the probability that $\mathcal{A}(y)$ can guess $g(X)$ is bounded:

$$\Pr[\mathcal{A}(Y(X)) = g(X)] \leq \max_G \{\Pr[G = g(X)]\} + \epsilon = \tfrac{1}{2} + \epsilon. \tag{2}$$

Combining the last two equations, the statistical difference $\mathbf{SD}\left(Y(X_0), Y(X_1)\right)$ is at most 2ϵ. This takes care of the case where X_0 and X_1 have disjoint supports.

To get the general indistinguishability condition, fix any $\tilde{X}_0$ as above (flat on 2^{t-1} points). For any other flat distribution $\tilde{X}_1$, there is some third flat distribution X' which is disjoint from both $\tilde{X}_0$ and $\tilde{X}_1$. By the previous reasoning, both $\mathbf{SD}\left(Y(\tilde{X}_0), Y(X')\right)$ and $\mathbf{SD}\left(Y(X'), Y(\tilde{X}_1)\right)$ are less than 2ϵ. By the triangle inequality $\mathbf{SD}\left(Y(X_0), Y(X_1)\right) \leq 4\epsilon$. A more careful proof avoids the triangle inequality and gives distance 2ϵ even when the supports of X_0, X_1 overlap. □

2.2 From Indistinguishability to Entropic Security

Proving that indistinguishability implies entropic security is considerably more delicate. We begin with an overview of the main ideas and notation.

The Case of Balanced Predicates. We say a function f is *balanced* (w.r.t. X) if it takes on all its possible values with equal probability, i.e. there are $\frac{1}{\text{pred}_f}$ possible values and each occurs with probability pred_f. The reductions we consider are much easier for balanced functions. In fact, we start with balanced *predicates*.

Namely, suppose that $g()$ is a balanced predicate for distribution X, that is $\Pr[g(X) = 0] = \Pr[g(X) = 1] = \tfrac{1}{2}$, and that that $\mathcal{A}$ is an adversary contradicting entropic security for min-entropy $t = \mathbf{H}_\infty(X)$, that is $\Pr[\mathcal{A}(Y(X)) = g(X)] = \tfrac{1}{2} + \epsilon$. For $b \in \{0, 1\}$, let X_b be the distribution of X conditioned on $g(X) = b$. The adversary's advantage over random guessing in distinguishing $Y(X_0)$ from $Y(X_1)$ is ϵ. However, that same advantage is also a lower bound for the statistical difference. We get:

$$\tfrac{1}{2} + \epsilon = \Pr[\mathcal{A}(Y(X)) = g(X)]$$
$$= \Pr[b \leftarrow \{0,1\} : \mathcal{A}(Y(X_b)) = b] \leq \tfrac{1}{2} + \tfrac{1}{2}\mathbf{SD}\left(Y(X_0), Y(X_1)\right),$$

and so the distance between $Y(X_0)$ and $Y(X_1)$ is at least $\epsilon/2$. To see that this contradicts indistinguishability, note that since $g(X)$ is balanced, we obtain X_0 and X_1 by conditioning on events of probability at least $\tfrac{1}{2}$. Probabilities are at most doubled, and so the min-entropies of both X_0 and X_1 are at most $\mathbf{H}_\infty(X) - 1$.

Balancing Predicates. If the predicate $g()$ is not balanced on X, then the previous strategy yields a poor reduction. For example, $\Pr[g(X) = 0]$ may be very small (potentially as small as ϵ). The probabilities in the distribution X_0 would then be a factor of $1/\epsilon$ bigger than their original values, leading to a loss of min-entropy of $\log(1/\epsilon)$. This argument therefore proves a weak version of Theorem 1: (t, ϵ) indistinguishability implies $(t + \log\left(\frac{1}{\epsilon}\right), 2\epsilon)$ entropic security for *predicates*.

This entropy loss is not necessary. We give a better reduction in Lemma 1 below. The idea is that to change the predicate $g()$ into a balanced predicate by flipping the value of the predicate on points on which the original adversary $\mathcal{A}$ performed poorly. By greedily choosing a set of points in $g^{-1}(0)$ of the right size, we show that there exists a balanced predicate $g'()$ on which the same adversary as before has advantage at least $\epsilon/2$, if the adversary had advantage ϵ for the original predicate.

Lemma 1. $(t - 2, 2\epsilon)$-*indistinguishability implies* (t, ϵ)-*entropic security for* **predicates** *for* $t \geq 2$.

Proof. Suppose that the scheme is not (t, ϵ)-entropically secure. That is, there is a message distribution X with min-entropy at least t, a predicate g and an adversary $\mathcal{A}$ such that

$$\Pr[\mathcal{A}(Y(X)) = g(X)] > \epsilon + \max_{i=0,1}\{\Pr[g(X) = i]\} \tag{3}$$

We wish to choose two distributions of min-entropy $t - 2$ and use the adversary to distinguish them, thus contradicting indistinguishability. It's tempting to choose the sets $g^{-1}(0)$ and $g^{-1}(1)$, since we know the adversary can predict g reasonably well. That attempt fails because one of the pre-images $g^{-1}(0), g^{-1}(1)$ might be quite small, leading to distributions of low min-entropy. Instead, we partition the support of X into sets of (almost) equal measure, making sure that the smaller of $g^{-1}(0)$ and $g^{-1}(1)$ is entirely contained in one partition.

Now let:

$$p = \Pr[h(X) = 1]$$
$$q_0 = \Pr[\mathcal{A}(Y(X)) = 1 | g(X) = 0]$$
$$q_1 = \Pr[\mathcal{A}(Y(X)) = 1 | g(X) = 1]$$

Suppose without loss of generality that $p \geq 1/2$, i.e. that $g(X) = 1$ is more likely than, or as likely as, $g(X) = 0$ (if $p < 1/2$, we can just reverse the roles of 0 and 1). The violation of entropic security (Eq. 3) can be re-written:

$$pq_1 + (1-p)(1-q_0) > p + \epsilon$$

In particular, $p - pq_1 > 0$ so we get:

$$(1-p)(q_1 - q_0) > \epsilon \tag{4}$$

Now we wish to choose two distributions A, B, each of min-entropy $t - 2$. For now, fix any set $\mathcal{S} \subseteq g^{-1}(1)$, where $g^{-1}(1) = \{x \in \{0,1\}^n | g(x) = 1\}$. We make the choice of $\mathcal{S}$ more specific below. Let $A_\mathcal{S}$ be the conditional distribution of X conditioned on $X \in \mathcal{S}$, and let $B_\mathcal{S}$ be distributed as X conditioned on $X \in \{0,1\}^n \setminus \mathcal{S}$. That is, $A_\mathcal{S}$ and $B_\mathcal{S}$ have disjoint supports and the support of $B_\mathcal{S}$ covers $g^{-1}(0)$ entirely.

The first property we will need from $\mathcal{S}$ is that it split the mass of X somewhat evenly. If the probability mass p' of $\mathcal{S}$ under X was exactly $1/2$, then the min-entropies of $A_\mathcal{S}$ and $B_\mathcal{S}$ would both be exactly $t - 1$. Depending on the distribution X, it may not be possible to have such an even split. Nonetheless, we can certainly get $\frac{1}{2} \leq p' < \frac{1}{2} + 2^{-t}$, simply by adding points one at a time to $\mathcal{S}$ until it gets just below $1/2$. The order in which we add the points is not important. For $t > 2$ (which is a hypothesis of this proof), we get $\frac{1}{2} \geq p' \geq \frac{3}{4}$. Hence, we can choose $\mathcal{S}$ so that the min-entropies of $A_\mathcal{S}$ and $B_\mathcal{S}$ are both at least $t - 2$.

We will also need that $\mathcal{S}$ have other properties. For every point x in the support of X, we define $q_x = \Pr[\mathcal{A}(Y(x)) = 1]$. The average over $x \leftarrow X$, restricted to $g^{-1}(1)$, of q_x is exactly q_1, that is

$$\mathsf{Exp}_{x \leftarrow X}[q_x] = q_1$$

If we now the choose the set $\mathcal{S}$ greedily, always adding points which maximize q_x, we are guaranteed that the average over X, conditioned on $X \in \mathcal{S}$, is at least q_1. That is, there exists a choice of $\mathcal{S}$ with mass $p' \in [\frac{1}{2}, \frac{3}{4}]$ such that

$$\Pr[\mathcal{A}(Y(A_\mathcal{S})) = 1] = \mathsf{Exp}_{x \leftarrow A_\mathcal{S}}[q_x] \geq q_1.$$

We can also now compute the probability that $\mathcal{A}(Y(B_\mathcal{S}))$ is 1:

$$\Pr[\mathcal{A}(Y(B_\mathcal{S})) = 1] = \frac{1-p}{1-p'} q_0 + \frac{p-p'}{1-p'} \Pr[\mathcal{A}(Y(X)) = 1 | X \notin \mathcal{S} \text{ and } g(X) = 0]$$

Now $\Pr[\mathcal{A}(Y(X)) = 1 | X \notin \mathcal{S} \text{ and } g(X) = 0]$ is at most q_1 (since by the greedy construction of $\mathcal{S}$, this is the average over elements in $g^{-1}(1)$ with the lowest values of q_x). Using $\mathcal{A}$ as a distinguisher for the distributions $Y(A_\mathcal{S})$ and $Y(B_\mathcal{S})$, we get:

$$\left| \Pr\left[\mathcal{A}(Y(A_\mathcal{S})) = 1 \right] - \Pr\left[\mathcal{A}(Y(B_\mathcal{S})) = 1 \right] \right| \geq q_1 - \frac{1-p}{1-p'} \cdot q_0 - \frac{p-p'}{1-p'} \cdot q_1$$

$$= \frac{1-p}{1-p'} \cdot (q_1 - q_0)$$

Since entropic security is violated (Eq. 4), we have $(1-p)(q_1 - q_0)/(1-p') > \epsilon/(1-p')$. By construction, we have $p' > \frac{1}{2}$ so the advantage of the predictor is at least 2ϵ, that is:

$$\mathsf{SD}\left(Y(A_\mathcal{S}), Y(B_\mathcal{S})\right) \geq \left| \Pr\left[\mathcal{A}(Y(A_\mathcal{S})) = 1\right] - \Pr\left[\mathcal{A}(Y(B_\mathcal{S})) = 1\right] \right| \geq 2\epsilon$$

Since A and B each have min-entropy at least $t-2$, this contradicts $(t-2, 2\epsilon)$-indistinguishability, completing the proof. □

From Predicates to Arbitary Functions. In order to complete the proof of Theorem 1, we need to show that entropic security for predicates implies entropic security for all functions. The reduction is captured by the following lemma, which states that for every function with a good predictor (i.e. a predictor with advantage at least ϵ), there exists a predicate for which nearly the same predictor does equally well. This is the main technical result of this section.

The reduction uses the predictor $\mathcal{A}(Y(X))$ as a black box, and so we will simply use the random variable $A = \mathcal{A}(Y(X))$.

Lemma 2 (Main Lemma). *Let X be any distribution on $\{0,1\}^n$ of min-entropy $t \geq \frac{3}{2}\log\left(\frac{1}{\epsilon}\right)$, and let A be any random variable (possibly correlated to X). Suppose there exists a function $f : \{0,1\}^n \to \{0,1\}^*$ such that $\Pr[A = f(X)] \geq \mathsf{pred}_f + \epsilon$. Then there exists a predicate $g : \{0,1\}^n \to \{0,1\}$ and an algorithm $B(\cdot)$ such that*

$$\Pr[B(A) = g(X)] \geq \mathsf{pred}_g + \epsilon/4.$$

Due to space limitations, the proof is only given in the full version [9]. We mention only that there are two main steps to proving the lemma:

- If A is a good predictor for an (arbitrary) function $f(\cdot)$, then there is a (almost) *balanced* function $f'(\cdot)$ and a good predictor A' of the form $g(A)$.
- If $f(\cdot)$ is a balanced (or almost balanced) function and A is a good predictor for $f(X)$, then there is a predicate $g(\cdot)$ of the form $g'(f(\cdot))$ such that $g'(A)$ is a good predictor for $g(X)$.

A More Efficient Reduction. Lemma 2 completes the proof of Theorem 1. However, it says nothing about the running time of $B(\cdot)$—in general, the reduction may yield a large circuit. Nonetheless, we may indeed obtain a polynomial-time reduction for certain functions f. If no value of f occurs with probability more than ϵ^2, then inner product with a random vector provides a good predicate. The idea behind the following proof has appeared in other contexts, e.g. [11].

Proposition 1. *Let X be any random variable distributed in $\{0,1\}^n$. Let $f : \{0,1\}^n \to \{0,1\}^N$ be a function such that $\mathsf{pred}_{f,X} \leq \epsilon^2/4$, and let A be a random variable with advantage ϵ at guessing $f(X)$. For $r \in \{0,1\}^N$, let $g_r(x) = r \odot f(x)$. If r is drawn uniformly from $\{0,1\}^N$, then*

$$\mathsf{Exp}_r\left[\Pr[r \odot A = g_r(X)] - \mathsf{pred}_{g_r}\right] \geq \epsilon/4.$$

In particular, there exists a value r and a $O(N)$-time circuit B which satisfy $\Pr[B(A) = g_r(X)] \geq \mathsf{pred}_{g_r} + \epsilon/4$.

Proof. We can calculate the expected advantage almost directly. Note that conditioned on the event $A = f(X)$, the predictor $r \odot A$ always agrees with $g_r(X)$. When $A \neq f(X)$, they agree with probability exactly $\frac{1}{2}$. Hence, we have

$$\mathsf{Exp}_r\left[\Pr[r \odot A = g_r(X)]\right] = \frac{1}{2} + \frac{1}{2}\Pr[A = f(X)] \geq \frac{1}{2}(1 + \mathsf{pred}_f + \epsilon)$$

We must still bound the expected value of pred_{g_r}. Let $r_z = (-1)^{z \odot r}$. For any particular, r, we can compute pred_{g_r} as $\frac{1}{2} + \frac{1}{2}|\sum_z p_z r_z|$. Using the fact $\text{Exp}\,[\|Z\|] \leq \sqrt{\text{Exp}\,[Z^2]}$ for any random variable Z, we get:

$$\text{Exp}_r\left[\text{pred}_{g_r}\right] = \frac{1}{2} + \frac{1}{2}\text{Exp}_r\left[\left|\sum_z p_z r_z\right|\right] \leq \frac{1}{2} + \frac{1}{2}\sqrt{\text{Exp}_r\left[\left(\sum_z p_z r_z\right)^2\right]}$$

By pairwise independence of the variables r_z, we have $\text{Exp}\,[r_z r_a]$ is 1 if $z = a$ and 0 otherwise.

$$\text{Exp}_r\left[\text{pred}_{g_r}\right] \leq \frac{1}{2} + \frac{1}{2}\sqrt{\sum_z p_z^2} \leq \frac{1}{2} + \frac{1}{2}\sqrt{\text{pred}_f}.$$

The last inequality holds since pred_f is the maximum of the values p_z, and the expression $\sum_z p_z^2$ is maximized when $p_z = \text{pred}_f$ for all z (note that this sum is the collision probability of $f(X)$). Combining the two calculations we have

$$\text{Exp}_r\left[\Pr[r \odot A = g_r(X)] - \text{pred}_{g_r}\right] \geq \frac{1}{2}\left(\text{pred}_f + \epsilon - \sqrt{\text{pred}_f}\right)$$

Using the hypothesis that $\text{pred}_f \leq \epsilon^2/4$, we see that the expected advantage is at least $\epsilon/4$. □

3 Encryption of High-Entropy Sources

In this section, we discuss the results on entropic security to the encryption of mesages which are guaranteed to come from a high-entropy distribution. Roughly: if the adversary has only a small chance of guessing the message ahead of time, then one can design information-theoretically secure encryption (in the sense of hiding all functions, Definition 1) using a much shorter key than is usually possible—making up for the small entropy of the key using the entropy inherent in the message.

3.1 Using Expander Graphs for Encryption

Formally, a symmetric encryption scheme is a pair of randomized maps $(\mathcal{E}, \mathcal{D})$. The encryption takes three inputs, an n-bit message x, a k-bit key κ and r random bits i, and produces a N-bit ciphertext $y = \mathcal{E}(x, \kappa; i)$. Note that the key and the random bits are expected to be uniform random bits, and when it is not necessary to denote the random bits or key explicitly we use either $\mathcal{E}(x, \kappa)$ or $\mathcal{E}(x)$. The decryption takes a key κ and ciphertext $y \in \{0,1\}^N$, and produces the plaintext $x' = \mathcal{D}(y, \kappa)$. The only condition we impose for $(\mathcal{E}, \mathcal{D})$ to be called an encryption scheme is completeness: for all keys κ, $\mathcal{D}(\mathcal{E}(x, \kappa), \kappa) = x$ with probability 1.

In this section, we discuss graph-based encryption schemes and show that graph expansion corresponds to entropically secure encryption schemes.

Graph-Based Encryption Schemes. Let $G = (V, E)$ be a d-regular graph, and let $N(v, j)$ denote the j-th neighbor of vertex v under some particular labeling of the edges. We'll say the labeling is *invertible* if there exists a map N' such that $N(v, j) = w$ implies $N'(w, j) = v$.

By Hall's theorem, every d-regular graph has an invertible labeling.[2] However, there is a large class of graphs for which this invertibility is much easier to see. The Cayley graph $G = (V, E)$ associated with a group $\mathcal{G}$ and a set of generators $\{g_1, ..., g_d\}$ consists of vertices labeled by elements of $\mathcal{G}$ which are connected when they differ by a generator: $E = \{(u, u \cdot g_i)\}_{u \in V, i \in [d]}$. When the set of generators contains all its inverses, the graph is undirected. For such a graph, the natural labeling is indeed invertible, since $N(v, j) = v \cdot j$ and $N'(w, j) = w \cdot j^{-1}$. All the graphs we discuss in this paper are in fact Cayley graphs, and hence invertibly labeled.

Now suppose the vertex set is $V = \{0, 1\}^n$ and the degree is $d = 2^k$, so that the neighbor function N takes inputs in $\{0, 1\}^n \times \{0, 1\}^k$. Consider the encryption scheme:

$$\mathcal{E}(x, \kappa) = N(x, \kappa). \tag{5}$$

Notice, $\mathcal{E}$ is a proper encryption scheme if and only if the labeling is invertible. In that case, $\mathcal{D}(y, \kappa) = N'(y, \kappa) = x$. For efficiency, we should be able to compute N and N' in polynomial time. We will show that this encryption scheme is secure when the graph G is a sufficiently good expander. The following definition is standard:

Definition 3. *A graph $G = (V, E)$ is a (t, ϵ)-extractor if, for every set S of 2^t vertices, taking a random step in the graph from a random vertex of S leads to a nearly uniform distribution on the whole graph. That is, let U_S be uniform on S, J be uniform on $\{1, ..., d\}$ and U_V be uniform on the entire vertex set V. Then for all sets S of size at least 2^t, we require that:*

$$\mathbf{SD}(\ N(U_S, J)\ ,\ U_V\ ) \leq \epsilon.$$

The usual way to obtain extractors as above is to use good expanders. This is captured by the following fact.

Fact 1 (Expander smoothing lemma [12]). *A graph G with second largest (normalized) eigenvalue $\lambda \leq \epsilon 2^{-(n-t)/2}$ is a (t, ϵ)-extractor.*

The equivalence between entropic security and indistinguishability (Theorem 1) gives us the following result:

---

[2] We thank Noga Alon for pointing out this fact. If $G = (V, E)$ is a d-regular undirected graph, consider the bipartite graph with $|V|$ vertices on each side and where each edge in E is replaced by the corresponding pair of edges in the bipartite graph. By Hall's theorem, there exist d disjoint matchings in the bipartite graph. These induce an invertible labeling on the original graph.

Proposition 2. *For a 2^k-regular, invertible graph G as above, the encryption scheme $(\mathcal{E}, \mathcal{D})$ given by N, N' is (t, ϵ)-entropically secure if G is a $(t-2, 2\epsilon)$-extractor (in particular, if G has second eigenvalue $\lambda \le \epsilon \cdot 2^{-(n-t-2)/2}$).*

Proof. By Theorem 1, it suffices to show that $(t-2, \epsilon)$-indistinguishability. And this immediately follows from the lemma above and the fact that any min-entropy $(t-2)$ distribution is a mixture of flat distributions. □

We apply this in two ways. First, using optimal expanders (Ramanujan graphs) we obtain the best known construction of entropically-secure encryption schemes (Corollary 1). Second, we give a simpler and much stronger analysis of the original scheme of Russell and Wang (Corollary 2).

Corollary 1. *There exists an efficient deterministic (t, ϵ)-entropically secure scheme with $k = n - t + 2\log\left(\frac{1}{\epsilon}\right) + 2$.*

Proof. We apply Proposition 2 to *Ramanujan graphs*. These graphs are optimal for this particular construction: they achieve optimal eigenvalue $\lambda = 2\sqrt{d-1}$ for degree d [17]. The bound on k now follows. □

The main drawback of Ramanujan graphs is that explicit constructions are not known for all sizes of graphs and degrees. However, large families exist (e.g. graphs with $q+1$ vertices and degree $p+1$, where p and q are primes congruent to 1 mod 4). Below we show why the construction from Russell and Wang [24] using small-biased spaces is actually a special case of Proposition 2.

Using Small-Biased Sets. A set S in $\{0, 1\}^n$ is δ-biased if for all nonzero $\alpha \in \{0, 1\}^n$, the binary inner product $\alpha \odot s$ is nearly balanced for s drawn uniformly in S:

$$\Pr_{s \leftarrow S}[\alpha \odot s = 0] \in \left[\frac{1-\delta}{2}, \frac{1+\delta}{2}\right] \quad \text{or, equivalently,} \quad \left|\mathsf{Exp}_{s \leftarrow S}\left[(-1)^{\alpha \odot s}\right]\right| \le \delta. \quad (6)$$

Alon et al. [1] gave explicit constructions of δ-biased sets in $\{0,1\}^n$ with size $O(n^2/\delta^2)$. Now suppose the δ-biased set is indexed $\{s_\kappa | \kappa \in \{0,1\}^k\}$. Consider the encryption scheme: $\mathcal{E}(x, \kappa) = x \oplus s_\kappa$. Russell and Wang introduced this scheme and showed that it is (n, ϵ)-entropically secure when $\delta = \epsilon^{3/2}$, yielding a key length of $k = 2\log n + 3\log\left(\frac{1}{\epsilon}\right)$. However, their analysis works only when the message is drawn uniformly from $\{0,1\}^n$.

We propose a different analysis: consider the Cayley graph for $\mathbb{Z}_2^n$ with generators S, where S is δ-biased. This graph has second eigenvalue $\lambda \le \delta$ [20, 2]. Hence, by Proposition 2 the scheme above is (t, ϵ)-entropically secure as long as $\delta \le \epsilon 2^{-(n-t-2)/2}$. This gives a version of the Vernam one-time pad for high-entropy message spaces, with key length $k = n - t + 2\log n + 2\log\left(\frac{1}{\epsilon}\right) + O(1)$. Unlike [24], this works for *all* settings of t, and also improves the parameters in [24] for $n = t$.

Corollary 2. *If $\{s_\kappa | \kappa \in \{0,1\}^k\}$ is a δ-biased set, then the encryption scheme $\mathcal{E}(x, \kappa) = x \oplus s_\kappa$ is (t, ϵ) indistinguishable when $\epsilon = \delta 2^{(n-t-2)/2}$. Using the costruction of [1], this yields a scheme with key length $k = n - t + 2\log\left(\frac{1}{\epsilon}\right) + 2\log(n) + O(1)$ (for any value of t).*

3.2 A Random Hashing Construction

This section presents a simpler construction of entropically secure encryption based on pairwise independent hashing. Our result generalizes the construction of Russell and Wang [24] for nonuniform sources, and introduces a new variant of the leftover-hash/privacy-amplification lemma [3, 15].

The idea behind the construction is that indistinguishability is the same as extraction from a weak source, except that the extractor must in some sense be invertible: given the key, one must be able to recover the message.

Let $\{h_i\}_{i \in I}$ be some family of functions $h_i : \{0,1\}^k \to \{0,1\}^n$, indexed over the set $I = \{0,1\}^r$. We consider encryption schemes of the form

$$\mathcal{E}(x, \kappa; i) = (\,i,\, x \oplus h_i(\kappa)\,) \quad \text{(for general functions } h_i\text{), or} \tag{7}$$
$$\mathcal{E}'(x, \kappa; i) = (\,i,\, h_i(x) \oplus \kappa\,) \quad \text{(when the functions } h_i \text{ are permutations)} \tag{8}$$

These schemes can be thought of as low-entropy, probabilistic one-time pads. Decryption is obviously possible, since the description of the function h_i is public. For the scheme to be (t, ϵ)-secure, we will see that it is enough to have $k = n - t + 2\log\left(\frac{1}{\epsilon}\right) + 2$, and for the function family to be pairwise independent. (This matches the result in Corollary 1.) In fact, a slightly weaker condition is sufficient: The following definition was introduced in the context of authentication [16]:

Definition 4 (XOR-universal function families). *A collection of functions $\{h_i\}_{i \in I}$ from n bits to n bits is XOR-universal if: $\forall a, x, y \in \{0,1\}^n, x \neq y$: $\Pr_{i \leftarrow I}[h_i(x) \oplus h_i(y) = a] \leq \frac{1}{2^n - 1}$.*

It is easy to construct XOR-universal families. Any (ordinary) pairwise independent hash family will do, or one can save some randomness by avoiding the "offset" part of constructions of the form $h(x) = ax + b$. Specifically, view $\{0,1\}^n$ as $\mathcal{F} = GF(2^n)$, and embed the key set $\{0,1\}^k$ as a subset of $\mathcal{F}$. For any $i \in \mathcal{F}$, let $h_i(\kappa) = i\kappa$, with multiplication in $\mathcal{F}$. This yields a family of linear maps $\{h_i\}$ with 2^n members. Now fix any $a \in \mathcal{F}$, and any $x, y \in \mathcal{F}$ with $x \neq y$. When i is chosen uniformly from $\{0,1\}^n$, we have $h_i(x) \oplus h_i(y) = i(x-y) = a$ with probability exactly 2^{-n}. If we restrict i to be nonzero, then we get a family of *permutations*, and we get $h_i(x) \oplus h_i(y) = a$ with probability at most $\frac{1}{2^n-1}$.

Proposition 3. *If the family $\{h_i\}$ is XOR-universal, then the encryption schemes*

$$\mathcal{E}(x, \kappa; i) = (i, x \oplus h_i(\kappa)) \quad \text{and} \quad \mathcal{E}'(x, \kappa; i) = (i, h_i(x) \oplus \kappa)$$

are (t, ϵ)-entropically secure, for $t = n - k + 2\log\left(\frac{1}{\epsilon}\right) + 2$. (However, $\mathcal{E}'$ is a proper encryption scheme only when $\{h_i\}$ is a family of permutations.)

This proposition proves, as a special case, the security of the Russell-Wang construction, with slightly better parameters (their argument gives a key length of $n - t + 3\log\left(\frac{1}{\epsilon}\right)$ since they used 3-wise independent permutations, which are also harder to construct). It also proves the security of the simple construction $\mathcal{E}(x, \kappa; i) = (i, x + i\kappa)$, with operations in $GF(2^n)$.

Proposition 3 follows from the following lemma of independent interest, which is closely related to the to the *leftover hash lemma* [13] (also called *privacy amplification*; see, e.g. [3,4]), and which conveniently handles both the $\mathcal{E}$ and the $\mathcal{E}'$ variants.

Lemma 3. *If A, B are independent random variables such that $\mathbf{H}_\infty(A) + \mathbf{H}_\infty(B) \geq n + 2\log\left(\frac{1}{\epsilon}\right) + 1$, and $\{h_i\}$ is a XOR-universal family, then*

$$\mathbf{SD}\left(\langle i, h_i(A) \oplus B \rangle, \langle i, U_n \rangle\right) \leq \epsilon,$$

where U_n and i are uniform on $\{0,1\}^n$ and $\mathcal{I}$.

Proof. Consider the collision probability of $(i, h_i(A) \oplus B)$. A collision only occurs if the same function h_i is chosen both times. Conditioned on that, one obtains a collision only if $h_i(A) \oplus h_i(A') = B \oplus B'$, for A', B' i.i.d. copies of A, B. We can use the XOR-universality to bound this last term:

$$\Pr[(i, h_i(A) \oplus B) = (i, h_i(A') \oplus B')]$$
$$= \Pr[i = i']\Big(\Pr[B = B'] \cdot \Pr[h_i(A) = h_i(A')]$$
$$+ \sum_{a \neq 0} \Pr[B \oplus B' = a] \cdot \Pr[h_i(A) \oplus h_i(A') = a]\Big) \quad (9)$$

Now let $t_a = \mathbf{H}_2(A)$, $t_b = \mathbf{H}_2(B)$. For $a \neq 0$, we have $\Pr[h_i(A) \oplus h_i(A') = a] \leq 1/(2^n - 1)$, by the conditions on $\{h_i\}$. On the other hand, by a union bound we have

$$\Pr[h_i(A) = h_i(A')] \leq \Pr[A = A'] + \frac{1}{2^n - 1} \leq 2^{-t_a} + \frac{1}{2^n - 1}$$

Hence, Eqn. 9 reduces to

$$\frac{1}{|\mathcal{I}|}\left(2^{-t_b}\left(2^{-t_a} + \frac{1}{2^n - 1}\right) + \frac{1}{2^n - 1}\left(\sum_{a \neq 0} \Pr[B \oplus B' = a]\right)\right)$$
$$\leq \frac{1}{|\mathcal{I}|2^n}\left(1 + 2^{n-t_a-t_b} + 2^{-t_b} + \frac{2}{2^n - 1}\right)$$

Now $2^{n-t_a-t_b} \leq \epsilon^2/2$ by assumption, and we also have $2^{-n} \leq 2^{-t_b} \leq \epsilon^2/2$, since $t_a, t_b \leq n$ and $t_a + t_b \geq n + 2\log\left(\frac{1}{\epsilon}\right)$ (similarly, $n \geq 2\log\left(\frac{1}{\epsilon}\right)$). Hence Eqn. 9 reduces to $(1 + 2\epsilon^2)/|\mathcal{I}|2^n$. Any distribution on a finite set S with collision probability $(1 + 2\epsilon^2)/|S|$ is at statistical distance at most ϵ from the uniform distribution [15]. Thus, $(i, h_i(A) \oplus B)$ is ϵ-far from uniform. □

Note that the lemma gives a special "extractor by XOR" which works for product distributions $A \times B$ with at least n bits of min-entropy between them.

3.3 Lower Bounds on the Key Length

Proposition 4. *Any encryption scheme which is (t, ϵ)-entropically secure for inputs of length n requires a key of length at least $n - t$.*

Proof. We can reduce our entropic scheme to Shannon-secure encryption of strings of length $n - t + 1$. Specifically, for every $w \in \{0,1\}^{n-t+1}$, let X_w be the uniform over strings with w as a prefix, that is the set $\{w\} \times \{0,1\}^{t-1}$. Since X_w has min-entropy $t - 1$, any pair of distributions $\mathcal{E}(X_w), \mathcal{E}(X_{w'})$ are indistinguishable, and so we can use $\mathcal{E}()$ to encrypt strings of length $n - t + 1$. When $\epsilon < 1/2$, we must have key length at least $(n - t + 1) - 1 = n - t$ by the usual Shannon-style bound (the loss of 1 comes from a relaxation of Shannon's bounds to statistical security). □

Bounds for Public-Coin Schemes via Extractors. In the constructions of Russell and Wang and that of Section 3.1 and Section 3.2, the randomness used by the encryption scheme (apart from the key) is sent *in the clear* as part of the ciphertext. That is, $\mathcal{E}(x, \kappa; i) = (i, \mathcal{E}'(x, \kappa; i))$. For these types of schemes, called *public-coin* schemes, the intuitive connection between entropic security and extraction from weak sources is pretty clear: encryption implies extraction. As a result, lower bounds on extractors [23] apply, and show that our construction is close to optimal.

Proposition 5. *Any public-coin, (t, ϵ)-entropically secure encryption has key length $k \geq n - t + \log\left(\frac{1}{\epsilon}\right) - O(1)$ (as long as $t > 2\log\left(\frac{1}{\epsilon}\right)$).*

To prove the result, we first reduce to the existence of extractors:

Lemma 4. *Assume $(\mathcal{E}, \mathcal{D})$ is a public-coin, (t, ϵ)-entropically secure encryption scheme with message length n, key length k and r bits of extra randomness. Then there exists an extractor with seed length $k + r$, input length n and output length $n + r - \log\left(\frac{1}{\epsilon}\right)$, such that for any input distribution of min-entropy $t + 1$, the output is within distance 3ϵ of the uniform distribution.*

Proof. We combine three observations. First, when U is uniform over all messages in $\{0,1\}^n$, the entropy of the distribution $\mathcal{E}(U)$ must be high. Specifically: $\mathbf{H}_\infty(\mathcal{E}(U)) = n + r$. To see this, notice that there is a function ($\mathcal{D}$) which can produce R, K, U from $K, \mathcal{E}(U, K; R)$. Since the triple (R, K, U) is uniform on $\{0,1\}^{r+k+n}$, it must be that $(K, \mathcal{E}(U, K))$ also has min-entropy $r + k + n$, i.e. that any pair (κ, c) appears with probability at most $2^{-(n-k-r)}$. Summing over all 2^k values of κ, we see that any ciphertext value c appears with probability at most $\sum_\kappa 2^{-n-r-k} = 2^{-n-r}$, as desired.

The second observation is that there is a deterministic function ϕ which maps ciphertexts into $\{0,1\}^{n+r-\log\left(\frac{1}{\epsilon}\right)}$ such that $\phi(\mathcal{E}(U))$ is within distance ϵ of the uniform distribution. In general, any *fixed* distribution of min-entropy t can be mapped into $\{0,1\}^{t-\log(1/\epsilon)}$ so that the result is almost uniform (Simply assign elements of the original distribution one by one to strings in $\{0,1\}^{t-\log(1/\epsilon)}$, so that at no time do two strings have difference of probability more than 2^{-t}. The total variation from uniform will be at most $2^{t-\log(1/\epsilon)} \cdot 2^{-t} = \epsilon$.). Note that ϕ need not be efficiently computable, even if both $\mathcal{E}$ and $\mathcal{D}$ are straightforward. This doesn't matter, since we are after a combinatorial contradiction.

Finally, by Theorem 1, for all distributions of min-entropy $t-1$, we have $\mathbf{SD}\left(\mathcal{E}(U),\mathcal{E}(X)\right) \leq 2\epsilon$, and so $\mathbf{SD}\left(\phi(\mathcal{E}(U)),\phi(\mathcal{E}(X))\right) \leq 2\epsilon$. By the triangle inequality, $\phi(\mathcal{E}(X))$ is within 3ϵ of the uniform distribution on $n+r-\log\left(\frac{1}{\epsilon}\right)$ bits, proving the lemma. □

We can now apply the lower bound of Radhakrishnan and Ta-Shma [23], who showed that any extractor for distributions of min-entropy t with error parameter δ and d extra random bits can extract at most $t+d-2\log(1/\delta)+O(1)$ nearly random bits. From Lemma 4, we get and extractor for min-entropy $t+1$, $\delta = 3\epsilon$, $k+r$ extra random bits, and output length $n+r-\log(1/\epsilon)$. Thus, $n+r-\log(1/\epsilon)$ is at most $t+1+k+r-2\log(1/\epsilon)+O(1)$, which immediately gives us Proposition 5.

Remark 1. We do not lose $\log(1/\epsilon)$ in the output length in Lemma 4 when the encryption scheme in indistinguishable from the uniform distribution (i.e., ciphertexts look truly random). For such public-coin schemes, we get $k \geq n-t+2\log\left(\frac{1}{\epsilon}\right)-O(1)$. Since all of our constructions are of this form, their parameters cannot be improved at all. In fact, we conjecture that $k \geq n-t+2\log\left(\frac{1}{\epsilon}\right)-O(1)$ for *all* entropically-secure schemes, public-coin or not.

Acknowledgements

We are grateful to many friends for helpful discussions on this work. We especially thank Noga Alon, Leonid Reyzin, Madhu Sudan, Salil Vadhan and Avi Wigderson for their insights.

References

1. Noga Alon, Oded Goldreich, Johan Håstad, René Peralta: Simple Constructions of Almost k-Wise Independent Random Variables. FOCS 1990: 544-553
2. Noga Alon and Yuval Roichman. Random Cayley graphs and expanders. *Random Structures & Algorithms* 5 (1994), 271–284.
3. C. Bennett, G. Brassard, and J. Robert. Privacy Amplification by Public Discussion. *SIAM J. on Computing*, 17(2), pp. 210–229, 1988.
4. C. Bennett, G. Brassard, C. Crépeau, and U. Maurer. Generalized Privacy Amplification. *IEEE Transactions on Information Theory*, 41(6), pp. 1915-1923, 1995.
5. R. Canetti. Towards realizing random oracles: Hash functions that hide all partial information. In *Crypto 1997*.
6. R. Canetti, D. Micciancio, O. Reingold. Perfectly One-Way Probabilistic Hash Functions. In *Proc. 30th ACM Symp. on Theory of Computing*, 1998, pp. 131–140.
7. Ran Canetti. Universally Composable Security: A New Paradigm for Cryptographic Protocols. *Proc. IEEE Symp. on Foundations of Computer Science*, 2001, pp. 136-145.
8. T. Cover, J. Thomas. *Elements of Information Theory*. Wiley series in telecommunication, 1991, 542 pp.

9. Y. Dodis, and A. Smith. Entropic Security and the Encryption of High Entropy Messages. Full version of this paper. Available at *IACR Cryptology ePrint Archive, report 2004/219*, at http://eprint.iacr.org/2004/219/.
10. S. Goldwasser and S. Micali. Probabilistic encryption. *JCSS*, **28**(2), pp. 270–299, April 1984.
11. Oded Goldreich, Salil Vadhan and Avi Wigderson. On Interactive Proofs with a Laconic Prover. *Computational Complexity*, 11(1-2): 1-53 (2002).
12. Oded Goldreich, Avi Wigderson: Tiny families of functions with random properties: A quality-size trade-off for hashing. Random Structures and Algorithms 11(4): 315-343 (1997)
13. J. Håstad, R. Impagliazzo, L. Levin, M. Luby. A Pseudorandom generator from any one-way function. In *Proc. 21st ACM Symp. on Theory of Computing*, 1989.
14. Jonathan Herzog. *Computational Soundness for Standard Assumptions of Formal Cryptography*. Ph.D. Thesis, Massachusetts Institute of Technology, May 2004.
15. R. Impagliazzo and D. Zuckerman. How to Recycle Random Bits. In *Proc. 30th IEEE Symp. on Foundations of Computer Science*, 1989.
16. H. Krawczyk. LFSR-Based Hashing and Authentication. In *Proc. CRYPTO '94*, p. 129–139, 1994.
17. A. Lubotzky, R. Phillips, P. Sarnak: Ramanujan graphs. Combinatorica 8(3): 261-277 (1988).
18. U. Maurer. Conditionally-Perfect Secrecy and a Provably-Secure Randomized Cipher. *J. Cryptology*, **5**(1), pp. 53–66, 1992.
19. U. Maurer. Secret Key Agreement by Public Discussion. *IEEE Trans. on Info. Theory*, 39(3):733–742, 1993.
20. J. Naor, M. Naor. Small-Bias Probability Spaces: Efficient Constructions and Applications. In *SIAM J. Comput.* 22(4): 838-856 (1993).
21. N. Nisan, D. Zuckerman. Randomness is Linear in Space. In *JCSS*, **52**(1), pp. 43–52, 1996.
22. B. Pfitzmann, M. Waidner. A Model for Asynchronous Reactive Systems and its Application to Secure Message Transmission. In *Proc. IEEE Symp. on Security and Privacy*, 2001, 184–200.
23. J. Radhakrishnan and A. Ta-Shma. Tight bounds for depth-two superconcentrators. In *Proc. 38th IEEE Symp. on Foundations of Computer Science*, 1997, pp. 585–594.
24. A. Russell and Wang. How to Fool an Unbounded Adversary with a Short Key. In *Advances in Cryptology — EUROCRYPT 2002*.
25. C. Shannon. Communication Theory of Secrecy systems. In *Bell Systems Technical J.*, 28:656–715, 1949. Note: The material in this paper appeared originally in a confidential report 'A Mathematical Theory of Cryptography', dated Sept. 1, 1945, which has now been declassified.

Error Correction in the Bounded Storage Model

Yan Zong Ding

College of Computing, Georgia Institute of Technology, 801 Atlantic Drive, Atlanta,
Georgia 30332-0280, USA
ding@cc.gatech.edu
Supported by NSF grant CCR-0205423

Abstract. We initiate a study of Maurer's *bounded storage model* (*JoC*, 1992) in presence of transmission errors and perhaps other types of errors that cause different parties to have *inconsistent views of the public random source*. Such errors seem inevitable in any implementation of the model. All previous schemes and protocols in the model assume a perfectly consistent view of the public source from all parties, and do not function correctly in presence of errors, while the private-key encryption scheme of Aumann, Ding and Rabin (*IEEE IT*, 2002) can be extended to tolerate only a $O(1/\log{(1/\varepsilon)})$ fraction of errors, where ε is an upper bound on the advantage of an adversary.

In this paper, we provide a general paradigm for constructing secure and error-resilient private-key cryptosystems in the bounded storage model that tolerate a *constant* fraction of errors, and attain the near optimal parameters achieved by Vadhan's construction (*JoC*, 2004) in the errorless case. In particular, we show that any *local fuzzy extractor* yields a secure and error-resilient cryptosystem in the model, in analogy to the result of Lu (*JoC*, 2004) that any local strong extractor yields a secure cryptosystem in the errorless case, and construct efficient local fuzzy extractors by extending Vadhan's sample-then-extract paradigm. The main ingredients of our constructions are *averaging samplers* (Bellare and Rompel, *FOCS '94*), *randomness extractors* (Nisan and Zuckerman, *JCSS*, 1996), *error correcting codes*, and *fuzzy extractors* (Dodis, Reyzin and Smith, *EUROCRYPT '04*).

1 Introduction

The *bounded storage model*, introduced by Maurer [Mau92], has seen increasing activities recently. In contrast to the standard complexity-based model for cryptography, this model imposes a bound on the storage space of an adversary rather than its running time. The model does not rely on complexity assumptions, and achieves information-theoretic security by employing a public source emitting random strings whose length exceeds the known space bound of the adversary. The security is guaranteed against a computationally unbounded adversary who stores almost all information about a public random string, while a legitimate user is only required to store a small number of public random bits. In a practical implementation, a good candidate for the public source is a system of high-speed satellites broadcasting random bits at a very high rate.

The bounded storage model has enjoyed success in private-key cryptography [Mau92, CM97b, AR99, ADR02, DR02, DM04b, Lu04, Vad04]. In particular, an important property known as *everlasting security* was observed in [ADR02, DR02], namely the private key can be reused exponentially many times under active attacks, and security is preserved even if after the execution of the protocol, the key is revealed to the adversary and the adversary becomes unbounded in both time and space. Subsequent works [DM04b, Lu04, Vad04] succeeded in constructing highly efficient (in terms of key length and storage requirement) cryptosystems in the model that attain everlasting security, culminating in the near optimal construction of Vadhan [Vad04]. Significant progress has also been made in oblivious transfer [CCM98, Din01, DHRS04] and key agreement [CM97b, DM04a] in the bounded storage model. More recently, it was shown that a primitive known as non-interactive timestamping, which is impossible in standard complexity-based cryptography, can be constructed in the bounded storage model [MST04].

All the above-mentioned works are based an ideal assumption that all the parties have a perfectly consistent view of the public random source. It seems, however, that in any implementation of the bounded storage model, transmission errors and perhaps other types of errors that cause different parties to have *inconsistent views of the public source*, are inevitable. The previous schemes and protocols do not function correctly in presence of such errors. Error-correcting the source might at the first glance appear as a natural solution, however this approach has several disadvantages, and in certain circumstances is infeasible, insufficient, or even impossible: (1) Error-correcting an entire string from the source is infeasible due to its huge size. (2) Encoding the source blockwise does not withstand worst-case adversarial errors that cause too many bits from a same block to be corrupted or erased. Worst-case adversarial errors may at first seem very unnatural. However, considering such errors is necessary, for instance in a setting where a system of several sources is employed, and the adversary compromises a fraction of the sources. (3) The practicality of the bounded storage model is based on the assumption that communications technology allows transmission of data at a rate higher than the storage rate of the adversary. Encoding the source by an error correcting code may significantly slow down the speed of transmission, thereby giving the adversary an advantage in storing information. (4) Error-correcting the source is impossible in implementations which use, for instance, existing natural sources of randomness that cannot be modified. Thus, the ability to cope with errors in the model itself, without an error-corrected source, is natural and fundamental for the bounded storage model.

It was noted by Rabin [Rab02] that the cryptosystem of [ADR02] (the ADR scheme for shorthand), which uses a long private key, can in fact be extended to tolerate a $O(1/\log(1/\varepsilon))$ fraction of errors, where ε is an upper bound on the advantage of an adversary. Throughout the paper, the error is measured by the maximum relative Hamming distance between the original public source and the source as perceived by a party. The ADR scheme extracts a one-time pad from the source where each bit of the one-time pad is the parity of $O(\log(1/\varepsilon))$ bits of the source at random positions. Thus, if the error in accessing the source is $O(1/\log(1/\varepsilon))$, then with high probability the fraction of corrupted bits in the one-time pad is a constant, and therefore correct decryption can be achieved by error-correcting the message using an asymptotically good error correcting code. It is also easy to see that $O(1/\log(1/\varepsilon))$ is an upper bound on the fraction of errors that can be tolerated by the extended ADR scheme. We note that by a slightly more careful analysis, it can be shown that a similar result also holds for the schemes of Lu [Lu04], which can be viewed as being obtained by derandomizing the ADR scheme.

In this paper, we provide a general paradigm for constructing secure and error-resilient private-key cryptosystems in the bounded storage model that tolerate a *constant* fraction of worst-case errors, and simultaneously attain the near optimal parameters achieved by Vadhan's construction [Vad04] in the errorless case. In particular, we show that any *local fuzzy extractor* yields a secure and error-resilient cryptosystem in the bounded storage model, in analogy to the results of Lu [Lu04] that any local strong extractor yields a secure cryptosystem in the errorless case, and construct efficient local fuzzy extractors by extending Vadhan's sample-then-extract paradigm [Vad04]. Further, for ensuring correct functionality in presence of errors, our cryptosystems only incur a communication overhead that can be made as small as any constant fraction. The main ingredients of our constructions are *averaging samplers* [BR94] and *randomness extractors* [NZ96], two powerful tools from the theory of pseudorandomness that are now standard in bounded-storage cryptography (c.f., [Lu04, Vad04, DHRS04]), as well as *error correcting codes*, and a new primitive known as *fuzzy extractors* recently introduced by Dodis, Reyzin and Smith [DRS04].

Averaging samplers, introduced by Bellare and Rompel [BR94], are procedures that approximate the average of a $[0,1]$-function by taking the average of the function evaluated at sampled points determined by a short random seed. Randomness extractors, introduced by Nisan and Zuckerman [NZ96], are functions that extract near perfect randomness from imperfect random sources using a short random seed. An extractor is *strong* if its output remains near uniform even if the seed is given. See the excellent surveys and tutorials of [NT99, Sha02, Vad02, Gol97] and references therein for constructions, connections, and applications of extractors and samplers.

Recently extractors and averaging samplers have proven fundamental in bounded-storage cryptography. Lu [Lu04] showed that *any* strong extractor yields a secure private-key cryptosystem in the bounded storage model, however due to the huge size of the source, the extractor is required to be *locally computable*, or simply *local*, namely the output of the extractor depends on only a few bits of the source. In [Vad04], Vadhan gave a general *sample-then-extract* paradigm for constructing local extractors from *any* averaging sampler and randomness extractor: first sample a small number of bits from the source using an averaging sampler, then apply an extractor to the sampled bits. By using strong extractors and samplers with near optimal parameters, the construction of [Vad04] yields near optimal local strong extractors.

Fuzzy extractors were introduced by Dodis, Reyzin and Smith [DRS04] recently, motivated by the problem of using biometrics for cryptography. The basic underlying ideas and techniques for constructing such objects have however already been used in the rich literature on information reconciliation and privacy amplification (c.f. [BBR88, BS93, BBCM95, CM97a]). The work of [DRS04] and this work can be seen as revisiting these ideas, using modern terminologies and techniques from the pseudorandomness literature. Informally speaking, a fuzzy extractor is a function which on input $x \xleftarrow{R} X$ where X is an imperfect random source, extracts a near uniform string Y together with a "fingerprint" P using a random seed K,[1] such that: (1) Y is near uniform even when given (K, P), and (2) there is a recovery algorithm that recovers Y from P, K, and any x' "sufficiently close" to x. Fuzzy extractors that allow recovery from a constant fraction of errors can be constructed using strong extractors and asymptotically good error correcting codes. ([DRS04]. See also Section 4.4 of this paper.)

1.1 An Overview of Our Constructions

We show that any fuzzy extractor yields a secure and error-resilient cryptosystem in the bounded storage model, and construct efficient *local* fuzzy extractors by extending Vadhan's sample-then-extract paradigm. Here the term *local* means that both extraction and recovery depend on a small number of bits from the input source, and further the positions of the bits read for both extraction and recovery are completely determined by the seed K and do not depend on the source X. Thus the positions of the bits read can be *preprocessed* using K by a *sampling algorithm*. Therefore we assume that both the extraction algorithm and the recovery algorithm proceed in two phases. In the first phase, both read bits from the source whose positions are determined by the seed. In the second phase, the actual extraction and recovery take place, on the bits read in the first phase along with other inputs. As the local extraction and recovery procedures do not access the entire source, we allow a small recovery error.

Construction of Local Fuzzy Extractors. A local fuzzy extractor LFE can be constructed from any given averaging sampler Samp and fuzzy extractor FE with recovery algorithm Rec, as follows. A seed for the resulting LFE is of form (K_1, K_2), where K_1 is a random seed for Samp, and K_2 is a random seed for FE. For local fuzzy extraction from X, one samples $W = X_{\text{Samp}(K_1)}$ from X, then computes and outputs $(Y, P) = \text{FE}(W, K_2)$. For local recovery of Y using P, (K_1, K_2), and a string X' that is sufficiently close to X in Hamming distance, one samples $W' = X'_{\text{Samp}(K_1)}$ from X', and recovers $Y = \text{Rec}(W', K_2, P)$. The security (or randomness) property of LFE follows from the fact that for almost

---
[1] Our definition of a fuzzy extractor differs slightly from the original definition in [DRS04] in that our fuzzy extractor explicitly uses a random seed, whereas that of [DRS04] does not make the seed explicit yet makes it part of the fingerprint.

all seeds K_1 of Samp, the sampler Samp essentially preserves the entropy rate of the source X (see [NZ96, Vad04]), and the security property of FE that output Y is near uniform even when K_2 and P are given. The local recovery property of LFE follows from the recovery property of FE, and the fact that for almost all seeds K_1 of Samp, the sampled substrings $X_{\text{Samp}(K_1)}$ and $X'_{\text{Samp}(K_1)}$ essentially preserve the relative Hamming distance between X and X', i.e. the fraction of positions at which X and X' differ. Details of our construction and analysis will be give in Section 4.3. In Sections 4.4, 4.5 and 4.6, we show that by proper choice of the underlying building blocks, our general paradigm yields a local fuzzy extractor that attains the near optimal seed length and sample complexity of Vadhan's strong local extractor, and produces a very short fingerprint needed for recovery from errors.

Private-Key Encryption from a Local Fuzzy Extractor. Given a local fuzzy extractor LFE together with a recovery algorithm REC that allows recovery from a constant fraction of errors, as well as a sampling procedure Samp (see the discussion at the beginning of Section 1.1), a *basic one-time* private-key encryption scheme in the bounded storage model that tolerates a constant fraction of errors can be constructed as follows: The sender Alice and the receiver Bob share a private-key K which is a random seed for LFE. While the public random string X is transmitted, Alice computes $(Y, P) = \text{LFE}(X^A, K)$, and Bob samples $W^B = X^B_{\text{Samp}(K)}$ from X^B required for the recovery of Y, where X^A and X^B are the views of X as perceived by Alice and Bob respectively. To encrypt a message M, Alice computes $C = M \oplus Y$, and sends (C, P) to Bob. Upon receiving (C, P), Bob decrypts by first recovering the one-time pad $Y = \text{REC}(W^B, K, P)$, then computing $M = C \oplus Y$.

Correct decryption (with high probability) of the resulting basic scheme follows directly from the recovery property of a local fuzzy extractor, and its security, in the case that the key K is used *just once* to encrypt one message, follows immediately from the security property of a local fuzzy extractor. However, an important question is *whether the key can be used many times* as in the errorless case, under the attack of an active space-bounded adversary who at each time step is also given the one-time pads and fingerprints from the past.[2] Recall that in the errorless case, the very general results of [Lu04] and [Vad04] show that *any* strong local extractor yields a cryptosystem in which the key is reusable and everlasting security is attained. In contrast, a moment's thought shows that one *cannot* hope to have such an analogous general result for an *arbitrary* local fuzzy extractor in the case of errors! Consider for instance the following (contrived) counter-example. Let LFE be a local fuzzy extractor constructed by the sample-then-extract paradigm described above, which takes as input a source X and a key $K = (K_S, K_E)$, and outputs $(Y, P) = \text{LFE}(X, K) \triangleq \text{FE}(X_{\text{Samp}(K_S)}, K_E)$, where Samp and FE are the given sampler and fuzzy extractor. Let REC be its

---

[2] The fingerprints are sent in the clear and are thus public to anyone, while the past one-time pads can be obtained by a chosen plaintext or chosen ciphertext attack.

recovery algorithm. Now let $\widehat{\text{LFE}}$ be obtained by modifying LFE as follows: on input (X, K), $\widehat{\text{LFE}}$ computes $(Y, P) = \text{LFE}(X, K)$, but outputs (Y, P') where $P' = P \circ K_S$ is the concatenation of P and K_S. Let $\widehat{\text{REC}}$ be the same as REC, except that $\widehat{\text{REC}}$ uses only $|P|$ bits of the fingerprint P'. It is not hard to see that the resulting $\widehat{\text{LFE}}$ is a local fuzzy extractor with recovery algorithm $\widehat{\text{REC}}$: As LFE is a local fuzzy extractor, by definition Y is near uniform even when given (K, P), and thus is also near uniform when given $(K, P \circ K_S)$.[3] The security property of $\widehat{\text{LFE}}$ follows. The recovery property, i.e. the correctness of $\widehat{\text{REC}}$ is obvious. However, if $\widehat{\text{LFE}}$ is employed in the above construction of a private-key encryption scheme, then from a fingerprint P' from a past time period the adversary gets K_S, the part of the key used for sampling. If the same key $K = (K_S, K_E)$ is reused, then from this point on, just as the sender and receiver the adversary need only store a small number of bits from the source as specified by K_S, and when he obtains K_E later he can simply decrypt just as the receiver. In general, the fingerprint P and the seed K are *dependent*. The definition of a local fuzzy extractor only guarantees that its first output Y is nearly uniform and independent of (K, P). The dependence between K and P renders a generic local fuzzy extractor non-reusable in this context, as the fingerprint P, sent in the clear, could give information about the seed K.

Note that the above counter-example only shows that a generic local fuzzy extractor does not yield a stateless cryptosystem with a reusable key, and does not answer the question whether the sample-then-extract paradigm, with a general averaging sampler and (non-local) fuzzy extractor, results in such a system. We believe that the answer to the latter question is also negative, for the following reason. First, it can be seen that if the sampled substring $W = X_{\text{Samp}(K_S)}$ were given, then an adversary who stores sufficient information about the source X and has the capability to introduce sufficient errors to X, could obtain substantial information about the seed K_S from W and his state. The fingerprint P is a function of W and thus gives partial information about W, which together with the adversary's state, may give adequate information about K_S.

However, we do note that a stateless encryption scheme under the sample-then-extract paradigm with a reusable key would result from a stronger type of fuzzy extractors, called entropically secure fuzzy extractors recently introduced by Dodis and Smith (see [Smi04]), which would result in a local fuzzy extractor where (K, Y) is essentially uniformly random even conditioned on the fingerprint P. Yet, the current constructions of entropically secure fuzzy extractors are not randomness-efficient enough to yield a desired value for key length.

Is there still any hope of using a generic local fuzzy extractor to construct a full-fledged error-resilient encryption scheme, where many messages can be encrypted? The answer is yes, if encryption and decryption are allowed to maintain a state. We circumvent the difficulty described above by *refreshing the key*, *instead of reusing it*, as follows. Let LFE be an arbitrary local fuzzy extractor,

---

[3] More generally, for any function f, Y is near uniform even when given $(K, f(P, K))$.

and let Alice and Bob share an initial key K_1. At each time t, we use the given local fuzzy extractor to extract a few more bits that will be used as the key for time $t+1$. That is, at time t, Alice computes $((Y_t^A, K_{t+1}^A), P_t) = \text{LFE}(X^A, K_t^A)$, where K_t^A is Alice's key for time t, Y_t^A is Alice's one-time pad for encrypting a (single) message at time t, and K_{t+1}^A is the new key Alice uses for time $t+1$. The fingerprint P_t is used by Bob to recover (Y_t^B, K_{t+1}^B), where Y_t^B and K_{t+1}^B are respectively Bob's one-time pad for decrypting a ciphertext at time t, and Bob's new key for time $t+1$. Ideally we would like to have $(Y_t^A, K_{t+1}^A) = (Y_t^B, K_{t+1}^B)$, although a small recovery error is inevitable. Intuitively, the resulting encryption scheme is secure as the new key K_{t+1}^A is a part of the first output of LFE, which by definition is near uniform given (K_t^A, P_t). Had there been no error from the source, security would have followed from known results [Vad04, Lu04]. The presence of error however does complicate the matter quite substantially, and a careful analysis of security and error-resilience is necessary.

Thus unlike the previous schemes in which the same key is reused, this scheme updates the key at each time step in a *forward-secure* manner (c.f. [And97]), and is therefore *stateful*. Such state-dependence may be viewed as a drawback in some cases. However, in communication settings where communication devices do maintain much state information (e.g. session IDs and counters), such a stateful encryption scheme is reasonable. On the other hand, it remains an interesting problem to construct a stateless error-resilient scheme matching the near optimal parameters achieved by the stateful construction. However, the general negative result described above suggests that resolving this issue may require resorting to and analyzing particular constructions of the building blocks, such as the underlying error correcting code. One promising approach is to derandomize the construction of entropically secure fuzzy extractors in [Smi04].

In Section 3 we carefully define the bounded storage model with errors, and give a definition of security and error-resilience. In Section 4.2 (Theorem 1), we will show that under the general forward-secure paradigm described above, *any* local fuzzy extractor yields a secure encryption scheme that achieves desired security and error correction properties simultaneously. More precisely, both the adversary's advantage and the probability of a single recovery error in the first T time periods, grow only linearly with T, essentially the best one can hope.

2   Preliminaries

We use the following standard notations in this paper. For a random variable X, the notation $x \xleftarrow{R} X$ denotes that x is chosen according to X. For a set S, $x \xleftarrow{R} S$ denotes that x is chosen uniformly from S. For an integer n, we denote by U_n a uniformly distributed random variable on the set $\{0,1\}^n$, and denote by $[n]$ the set $\{1,\ldots,n\}$. For a string $x \in \{0,1\}^n$ and a subset $S = \{i_1,\ldots,i_l\} \subseteq [n]$, $x_S \stackrel{\Delta}{=} x_{i_1} \ldots x_{i_l}$, where x_i is the i-th bit of x. We denote by $\text{Supp}(X)$ the support of a random variable X.

For two strings x and y of the same length, we use $\Delta(x,y)$ to denote their Hamming distance, i.e. the number of bit positions at which x and y differ.

We say that a function (e.g. an extractor, a sampler, or an error correcting code) is explicit if it can be computed by a polynomial-time algorithm.

In the remainder of this section, we give definitions of weak random sources and statistical distance.

Definition 1 ([CG88, Zuc96]). *For a random variable X on a finite set Ω, the* min-entropy *of X is defined by:* $H_\infty(X) = \min_{x \in \Omega} \log(1/\Pr[X = x])$. *We say that X is a k-source if $H_\infty(X) \geq k$. We say that a random variable X over $\{0,1\}^n$ has entropy rate α if X is an αn-source.*

Definition 2. *For random variables X and Y taking values in Ω, their* statistical distance *is defined as* $\mathrm{SD}(X,Y) \triangleq \max_{A \subseteq \Omega} |\Pr[X \in A] - \Pr[Y \in A]| = \frac{1}{2} \sum_{x \in \Omega} |\Pr[X=x] - \Pr[Y=x]|$. *We say X and Y are ε-close if $\mathrm{SD}(X,Y) \leq \varepsilon$.*

3 The Model and Definition of Security

In this section we take a closer look at the bounded storage model *with errors*, and define security in the model. In the presentation we use many terminologies and notations from [Vad04].

The Public Random Source. The bounded storage model (BSM) employs a public source of random strings, each of length exceeding the storage bound of the adversary. Throughout the paper, we use N to denote the length of a public random string. The public source is thus modeled as a sequence of random variables $X_1, X_2, \ldots, X_t, \ldots$, each distributed over $\{0,1\}^N$. We denote by βN the storage bound, where $\beta < 1$ is constant fraction, and call β the *storage rate* of the adversary.

The original work of Maurer [Mau92], as well as some early works (c.f., [AR99, ADR02, DR02]) assume that the public source is perfectly random, that is, each X_t is uniformly distributed and independent of the others. It was noted in [Lu04, Vad04] that each X_t need not be uniform, and it is sufficient (and necessary) that each X_t has entropy rate $\alpha > \beta$. Moreover, it was pointed out in [Vad04] that the X_t's need not be independent, and it is sufficient (and necessary) that the sequence of random variables $X_1, X_2, \ldots, X_t, \ldots$ form a *reverse block source*, which is the Chor-Goldreich [CG88] notion of a block source but backwards in time. Namely, in a reverse block source, each X_t has sufficient min-entropy conditioned on the future, whereas in a standard block source of [CG88] each X_t has sufficient min-entropy conditioned on the past. For the model with errors considered in this paper, we slightly strengthen the requirement on the public source by postulating that it be blockwise *both forward and backward*, i.e. it be both a standard block source and a reverse block source. The reason for imposing this forward blockwise structure in addition to its reverse counterpart is that the fingerprints $P_1, \ldots, P_{t-1}$ required for recovery from errors in the past time periods are exposed and depend on the $X_1, \ldots, X_{t-1}$. Therefore it is necessary that X_t has sufficient min-entropy conditioned on $P_1, \ldots, P_{t-1}$. This would certainly be satisfied if the source is (forward) blockwise, that is X_t has sufficient min-entropy conditioned on $X_1, \ldots, X_{t-1}$.

Definition 3. *Let $(X_t) = (X_1, X_2, \ldots)$ be a sequence of random variables each distributed over $\{0,1\}^n$. For each $t \in \mathbb{N}$, denote $X\backslash_t = (X_1, \ldots, X_{t-1}, X_{t+1}, X_{t+2}, \ldots)$. We say that (X_t) is a two-way block source of entropy rate α if for every $t \in \mathbb{N}$, and every $\boldsymbol{x} = (x_1, \ldots, x_{t-1}, x_{t+1}, x_{t+2}, \ldots) \in \mathrm{Supp}(X\backslash_t)$, the random variable $X_t|_{X\backslash_t = \boldsymbol{x}}$ is an αn-source.*

Intuitively, this means that X_t has αn bits of information that can not be predicted from the past and will be forgotten in the future. In the special case of $\alpha = 1$, $X_1, X_2, \ldots$ are uniform and independent.

BSM Randomness Extraction. An essential ingredient is a *bounded storage model randomness extraction scheme*,[4] or simply a *BSM extraction scheme*. In the errorless case, such an extraction scheme is a function of the form $\mathrm{EXT} : \{0,1\}^N \times \{0,1\}^d \to \{0,1\}^m$. In a private-key setting, such an extraction scheme is typically used as follows. A seed or a key $K \xleftarrow{R} \{0,1\}^d$ is chosen, and shared between two parties. At time t, the parties extract a common string $Y_t = \mathrm{EXT}(X_t, K)$, while the adversary A updates and stores his state $S_t = A(S_{t-1}, Y_1, \ldots, Y_{t-1}, X_t)$, where $|S_t| = \beta N$. The scheme EXT is secure if for every adversary A with storage rate β, Y_t is statistically close to uniform even when given the key K, all the previous $Y_1, \ldots, Y_{t-1}$, the adversary's state S_t, and the future public random strings $X_{t+1}, X_{t+2}, \ldots$.

In order to be used as a BSM primitive, the extraction scheme needs to be *locally computable*, that is $\mathrm{EXT}(X, K)$ depends only on a few bits of X whose positions are completely determined by K. As in the discussion on local fuzzy extractors in Section 1.1, here we also assume that a sampling procedure Samp precomputes positions $\mathrm{Samp}(K)$, the bits $W = X_{\mathrm{Samp}(K)}$ are read when X is transmitted, and the extraction algorithm EXT actually takes W and K as input, and computes $\mathrm{EXT}(W, K)$.

Incorporating Errors. We now incorporate errors into the model, and consider the case where two parties Alice and Bob have inconsistent views of the source as a result of errors. We consider *error-resilient BSM randomness extraction with forward security*, as motivated in Section 1.1 of the Introduction. Such an extraction scheme is a pair of algorithms (EXT, REC), where $\mathrm{EXT} : \{0,1\}^N \times \{0,1\}^d \to \{0,1\}^{m+d} \times \{0,1\}^\ell$ is a *local* extraction function, and $\mathrm{REC} : \{0,1\}^N \times \{0,1\}^d \times \{0,1\}^\ell \to \{0,1\}^{m+d}$ is a *local* recovery algorithm. The second output of EXT is a fingerprint that enables recovery of its first output, and the last d bits from the first output of EXT will be used as the key for the next time period.

---

[4] In [Vad04], such a scheme is called a BSM pseudorandom generator. We choose to call it a BSM randomness extraction scheme, because of the usual computational connotations of "pseudorandom generators".

Alice and Bob initially share a common random key $K_1 \xleftarrow{R} \{0,1\}^d$ for EXT. Let $K_1^A = K_1^B = K_1$.

We model errors by having an *unbounded* adversary who at time t for each $t \in \mathbb{N}$, on input $x_t \xleftarrow{R} X_t$, computes x_t^A and x_t^B such that $\Delta(x_t, x_t^A) \leq \delta N$ and $\Delta(x_t, x_t^B) \leq \delta N$, where $\delta < 1$ is a constant fraction, and sends x_t^A and x_t^B to Alice and Bob respectively. We call δ the *error rate*.

On input the corrupted string x_t^A and her key K_t^A for time t, Alice computes $((Y_t^A, K_{t+1}^A), P_t) = \mathrm{EXT}(x_t^A, K_t^A)$, where Y_t^A is Alice's extracted "one-time pad" for time t, K_{t+1}^A is Alice's key for time $t+1$, and P_t is a fingerprint needed by Bob to recover (Y_t^A, K_{t+1}^A). Meanwhile, on input x_t^B, Bob reads the substring w of x_t^B at positions in $\mathrm{Samp}(x_t^B, K_t^B)$ needed to recover (Y_t^A, K_{t+1}^A), where Samp is the sampling procedure. Upon receiving P_t from Alice, Bob recovers $(Y_t^B, K_{t+1}^B) = \mathrm{REC}(w, K_t^B, P_t)$. Ideally we would like to have $(Y_t^B, K_{t+1}^B) = (Y_t^A, K_{t+1}^A)$, although we allow a small recovery error which is inevitable.

As in [Vad04], we use $S_t \in \{0,1\}^{\beta N}$ to denote the state of the adversary at time t. For a sequence $Z_1, Z_2, \ldots$, we use the shorthand $Z_{[a,b]} = (Z_a, Z_{a+1}, \ldots, Z_b)$, and $Z_{[a,\infty)} = (Z_a, Z_{a+1}, \ldots)$. At time t, we allow the adversary access to the current corrupted strings x_t^A, x_t^B, all previous one-time pads $Y_{[1,t-1]}^A, Y_{[1,t-1]}^B$ and keys $K_{[1,t-1]}^A, K_{[1,t-1]}^B$ of *both* Alice and Bob, as well as $P_{[1,t-1]}$. With this information the adversary computes

$$S_t = \mathcal{A}(Y_{[1,t-1]}^A, Y_{[1,t-1]}^B, K_{[1,t-1]}^A, K_{[1,t-1]}^B, P_{[1,t-1]}, S_{t-1}, x_t^A, x_t^B),$$

with $|S_t| = \beta N$.

We now define the security and error correction properties of an error-resilient BSM randomness extraction scheme. In doing so, we use a *real-vs-ideal* paradigm as [Vad04] does.

The real experiment is a real execution of a protocol. For $T \in \mathbb{N}$, the output of our real experiment is $(Y_{[1,T]}^A, Y_{[1,T]}^B, K_{[1,T+1]}^A, K_{[1,T+1]}^B, P_{[1,T]}, S_T, X_{[T+1,\infty)})$, with each component defined above. The ideal experiment is a simulated execution of the protocol in an ideal setting that guarantees security.

In our ideal experiment, for each $t \in [T]$, we choose a uniform one-time pad $Y_t \xleftarrow{R} \{0,1\}^m$, and set $Y_t^A = Y_t^B = Y_t$. Similarly, for each $t \in [T+1]$, we choose a uniform key $K_t \xleftarrow{R} \{0,1\}^d$, and set $K_t^A = K_t^B = K_t$. Thus, in the output of the ideal experiment, each of $Y_{[1,T]}$ and $Y_{[1,T+1]}$ is *uniformly and independently* chosen, and further each Y_t and K_t are *replicated twice* to simulate Y_t^A, Y_t^B and K_t^A, K_t^B respectively, *as if there is no recovery error*. Hence proving security amounts to proving that the outputs of the real and ideal experiments are indistinguishable.

We now precisely define the real and ideal experiments. For both experiments, let $X_1, X_2, \ldots$ be the public random source, let $K_1 \xleftarrow{R} \{0,1\}^d$ be the initial shared key, let $K_1^A = K_1^B = K_1$, $S_0 = 0^{\beta N}$, and let $\mathcal{A}$ be the adversary's algorithm.

Real Experiment:

- For $t = 1, \ldots, T$: On $x_t \xleftarrow{R} X_t$:
  Let $(x_t^A, x_t^B) = \mathcal{A}(x_t, Y_{[1,t-1]}^A, Y_{[1,t-1]}^B, K_{[1,t-1]}^A, K_{[1,t-1]}^B, P_{[1,t-1]}, S_{t-1})$, where $\Delta(x_t^A, x_t) \leq \delta N$ and $\Delta(x_t^B, x_t) \leq \delta N$. In this step we allow $\mathcal{A}$ to be unbounded in both time and space.
  Let $((Y_t^A, K_{t+1}^A), P_t) = \text{EXT}(x_t^A, K_t^A)$, and $(Y_t^B, K_{t+1}^B) = \text{REC}(x_t^B, K_t^B, P_t)$.
  Let $S_t = \mathcal{A}(x_t^A, x_t^B, Y_{[1,t-1]}^A, Y_{[1,t-1]}^B, K_{[1,t-1]}^A, K_{[1,t-1]}^B, P_{[1,t-1]}, S_{t-1}) \in \{0,1\}^{\beta N}$.
- Output $Z_T^{\text{real}} = (Y_{[1,T]}^A, Y_{[1,T]}^B, K_{[1,T+1]}^A, K_{[1,T+1]}^B, P_{[1,T]}, S_T, X_{[T+1,\infty)})$.

Ideal Experiment:

- For $t = 1, \ldots, T$: On $x_t \xleftarrow{R} X_t$:
  Let $(x_t^A, x_t^B) = \mathcal{A}(x_t, Y_{[1,t-1]}, Y_{[1,t-1]}, K_{[1,t-1]}, K_{[1,t-1]}, P_{[1,t-1]}, S_{t-1})$, where $\Delta(x_t^A, x_t) \leq \delta N$ and $\Delta(x_t^B, x_t) \leq \delta N$.
  Let $((\tilde{Y}_t, \tilde{K}_{t+1}), P_t) = \text{EXT}(X_t^A, K_t)$.
  Choose uniformly and independently $Y_t \xleftarrow{R} \{0,1\}^m$ and $K_{t+1} \xleftarrow{R} \{0,1\}^d$.
  Let $S_t = \mathcal{A}(x_t^A, x_t^B, Y_{[1,t-1]}, Y_{[1,t-1]}, K_{[1,t-1]}, K_{[1,t-1]}, P_{[1,t-1]}, S_{t-1}) \in \{0,1\}^{\beta N}$.
- Output $Z_T^{\text{ideal}} = (Y_{[1,T]}, Y_{[1,T]}, K_{[1,T+1]}, K_{[1,T+1]}, P_{[1,T]}, S_T, X_{[T+1,\infty)})$.

Notation: From now on, we denote by X_t^A and X_t^B the induced sources at time t as perceived by Alice and Bob after errors are introduced to X_t.

Definition 4. *A BSM randomness extraction scheme* (EXT, REC) *is ε-secure for storage rate β, entropy rate α, and error rate δ if for every two-way block source* (X_t) *of entropy rate α, every adversary $\mathcal{A}$ with storage rate β, every means to introduce a δ-fraction of errors to the source* (X_t)*, and every* $T \in \mathbb{N}$*,* $\text{SD}(Z_T^{\text{real}}, Z_T^{\text{ideal}}) \leq T \cdot \varepsilon$*, where Z_T^{real} and Z_T^{ideal} are the outputs of the Real and Ideal Experiments respectively.*

We say that (EXT, REC) is *t-local* if for every key $K \in \{0,1\}^d$, both the extraction scheme $\text{EXT}(x, K)$ and its recovery algorithm $\text{REC}(x', K, P)$ depend on only t-bits of x and x' respectively, whose positions are completely determined by the key K.

We refer readers to the remarks after Definition 3.2 of [Vad04] for a discussion on the definition of everlasting security in the errorless model, which apply to the model with errors as well. Below are some more remarks that are important.

Remarks:

- A reader may notice that we have not explicitly defined the error correction property of a BSM randomness extraction scheme. However, by a careful inspection, it is not hard to see that *the security property as defined in Definition 4 implies error correction.* That is, if (EXT, REC) is ε-secure, then for every two-way block source (X_t) of entropy rate α, for

error rate δ, and every $T \in \mathbb{N}$, the probability of a single recovery error in the first T time periods in the Real Experiment, is at most $T\varepsilon$. More precisely, with probability at least $1 - T\varepsilon$ (over the source (X_t) and the initial common key $K_1 \xleftarrow{R} \{0,1\}^d$), we have that for each $t \in [T]$, $(Y_t^A, K_{t+1}^A) = (Y_t^B, K_{t+1}^B)$, where $((Y_t^A, K_{t+1}^A), P_t) = \text{EXT}(X_t^A, K_t^A)$, and $(Y_t^B, K_{t+1}^B) = \text{REC}(X_t^B, K_t^B, P_t)$. This is because in the output Z_T^{ideal} of the Ideal Experiment, each Y_t and K_t are *replicated* twice. Thus if the probability of a recovery error in the first T time periods in the Real Experiment is greater than $T\varepsilon$, then the distinguisher that simply compares the corresponding components of the two inputs, and outputs 1 if and only if they are the same, distinguishes between Z_T^{real} and Z_T^{ideal} with an advantage greater than $T\varepsilon$, contradicting the ε-security of (EXT, REC).
- From Definition 4, it is clear that the output Y_t of an error-resilient BSM extraction scheme can be used in place of a truly random string at time t for general cryptographic purposes. In particular, using each Y_t as a one-time pad for time t, such an extraction scheme yields an error-resilient BSM private-key encryption scheme *secure against chosen plaintext attacks and chosen ciphertext attacks* (c.f. [NY90]), with a small decryption error.

4 Local Fuzzy Extractors and BSM Extraction

In this section, we construct local fuzzy extractors and error-resilient BSM randomness extraction schemes.

4.1 Local Fuzzy Extractors

First we define fuzzy extractors, which were recently introduced by Dodis, Reyzin and Smith [DRS04]. We slightly modify the original definition in [DRS04] to suit our application.

Definition 5 ([DRS04] - modified). *A $(k, \varepsilon, \delta, \gamma)$-fuzzy extractor is a pair* $\text{FE} = (\text{EXT}, \text{REC})$ *of algorithms, where* $\text{EXT}: \{0,1\}^n \times \{0,1\}^d \to \{0,1\}^m \times \{0,1\}^\ell$ *is an extraction algorithm and* $\text{REC}: \{0,1\}^n \times \{0,1\}^d \times \{0,1\}^\ell \to \{0,1\}^m$ *is a recovery algorithm satisfying*

- **(Security)** *For every k-source X, (K, Y, P) is ε-close to (K, U_m, P), where $(Y, P) = \text{EXT}(X, K)$, $K \xleftarrow{R} \{0,1\}^d$ is a uniformly chosen seed independent of X, and U_m is independent of K and P.*
- **(Recovery)** *For every $x, x' \in \{0,1\}^n$ with $\Delta(x, x') \leq \delta n$, $\Pr[\text{REC}(x', K, P) = Y] \geq 1 - \gamma$, where $(Y, P) = \text{EXT}(x, K)$, and the probability is taken over $K \xleftarrow{R} \{0,1\}^d$.*

A fuzzy extractor $\text{FE} = (\text{EXT}, \text{REC})$ is *t-local* if for every seed $r \in \{0,1\}^d$, both the extraction algorithm $\text{EXT}(x, r)$ and the recovery algorithm $\text{REC}(x', r, p)$ depend on only t-bits of x and x' respectively, whose positions are completely determined by the seed r.

4.2 Error-Resilient BSM Extraction from Local Fuzzy Extractors

The following main theorem of this paper states that any t-local fuzzy extractor yields a t-local error-resilient BSM randomness extraction scheme.

Theorem 1. *For every $t \in \mathbb{N}$, if* LFE $=$ (EXT, REC) *is a t-local $(k, \varepsilon, 2\delta, \gamma)$-fuzzy extractor for $\gamma < 1/2$ and $k = (\alpha - \beta - \mathrm{H}(\delta))N - \log(1/\varepsilon)$, where $\mathrm{H}(\delta) \triangleq -\delta \log \delta - (1-\delta) \log (1-\delta)$ is the binary entropy function, and* EXT *is of the form* EXT $: \{0,1\}^N \times \{0,1\}^d \to \{0,1\}^{m+d} \times \{0,1\}^\ell$, *then* LFE *is a t-local $4(\varepsilon + \gamma)$-secure BSM randomness extraction scheme for storage rate β, entropy rate α, and error rate δ.*

Proof. (Sketch) Let LFE $=$ (EXT, REC) be a $(k, \varepsilon, 2\delta, \gamma)$-fuzzy extractor where $k = (\alpha - \beta - \mathrm{H}(\delta))N - \log(1/\varepsilon)$, $\gamma < 1/2$, and EXT is of the form EXT $: \{0,1\}^N \times \{0,1\}^d \to \{0,1\}^{m+d} \times \{0,1\}^\ell$. We prove the theorem by induction on T. The proof builds on the framework developed in [Vad04, Lu04].

As in [Vad04], we use superscripts to distinguish between random variables in the Real Experiment and the Ideal Experiment, e.g. K_t^{real} vs. K_t^{ideal}. We prove by induction on T that for every T, the random variable

$$Z_T^{\mathrm{real}} = (Y_{[1,T]}^A, Y_{[1,T]}^B, K_{[1,T+1]}^A, K_{[1,T+1]}^B, P_{[1,T]}^{\mathrm{real}}, S_T^{\mathrm{real}}, X_{[T+1,\infty)})$$

is $T \cdot 4(\varepsilon + \gamma)$-close to

$$Z_T^{\mathrm{ideal}} = (Y_{[1,T]}^{\mathrm{ideal}}, Y_{[1,T]}^{\mathrm{ideal}}, K_{[1,T+1]}^{\mathrm{ideal}}, K_{[1,T+1]}^{\mathrm{ideal}}, P_{[1,T]}^{\mathrm{ideal}}, S_T^{\mathrm{ideal}}, X_{[T+1,\infty)}),$$

where Z_T^{real} and Z_T^{ideal} are the output of the Real Experiment and the Ideal Experiment respectively.

Recall that for each t, $Y_t^{\mathrm{ideal}} \equiv U_m^{(t)}$ and $K_t^{\mathrm{ideal}} \equiv U_d^{(t)}$, where $U_m^{(t)}$ (resp. $U_d^{(t)}$) is an independent copy of U_d (resp. U_m). Note again that each Y_t^{ideal} and K_t^{ideal} are replicated twice in Z_T^{ideal}.

As the induction hypothesis, suppose that Z_{T-1}^{real} and Z_{T-1}^{ideal} are $(T-1) \cdot 4(\varepsilon+\gamma)$-close. It follows from the definition of the Real Experiment that Z_T^{real} is obtained from Z_{T-1}^{real} by applying the function f_T that:

- Computes
  $(X_T^A, X_T^B) = \mathcal{A}(X_T, Y_{[1,T-1]}^A, Y_{[1,T-1]}^B, K_{[1,T-1]}^A, K_{[1,T-1]}^B, P_{[1,T-1]}^{\mathrm{real}}, S_{T-1}^{\mathrm{real}})$,
  where $\Delta(X_T^A, X_T) \leq \delta N$ and $\Delta(X_T^B, X_T) \leq \delta N$.
- Computes $((Y_T^A, K_{T+1}^A), P_T^{\mathrm{real}}) = \mathrm{EXT}(X_T^A, K_T^A)$,
  and $(Y_T^B, K_{T+1}^B) = \mathrm{REC}(X_T^B, K_T^B, P_T^{\mathrm{real}})$.
- Updates state
  $S_T^{\mathrm{real}} = \mathcal{A}(X_T^A, X_T^B, Y_{[1,T-1]}^A, Y_{[1,T-1]}^B, K_{[1,T-1]}^A, K_{[1,T-1]}^B, P_{[1,T-1]}^{\mathrm{real}}, S_{T-1}^{\mathrm{real}}) \in \{0,1\}^{\beta N}$.
- Removes X_T.
- Outputs $Z_T^{\mathrm{real}} = (Y_{[1,T]}^A, Y_{[1,T]}^B, K_{[1,T+1]}^A, K_{[1,T+1]}^B, P_{[1,T]}^{\mathrm{real}}, S_T^{\mathrm{real}}, X_{[T+1,\infty)})$.

Applying the same function f_T to Z_{T-1}^{ideal}, we get the random variable $f_T(Z_{T-1}^{\mathrm{ideal}})$ as follows:

- Let $(X_T^A, X_T^B) = \mathcal{A}(X_T, Y_{[1,T-1]}^{\text{ideal}}, Y_{[1,T-1]}^{\text{ideal}}, K_{[1,T-1]}^{\text{ideal}}, K_{[1,T-1]}^{\text{ideal}}, P_{[1,T-1]}^{\text{ideal}}, S_{T-1}^{\text{ideal}})$,
  where $\Delta(X_T^A, X_T) \leq \delta N$ and $\Delta(X_T^B, X_T) \leq \delta N$.
- Let $((\tilde{Y}_T^A, \tilde{K}_{T+1}^A), P_T^{\text{ideal}}) = \text{EXT}(X_T^A, K_T^{\text{ideal}})$,
  and $(\tilde{Y}_T^B, \tilde{K}_{T+1}^B) = \text{REC}(X_T^B, K_T^{\text{ideal}}, P_T^{\text{ideal}})$.
- Update state $S_T^{\text{ideal}} =$
  $\mathcal{A}(X_T^A, X_T^B, Y_{[1,T-1]}^{\text{ideal}}, Y_{[1,T-1]}^{\text{ideal}}, K_{[1,T-1]}^{\text{ideal}}, K_{[1,T-1]}^{\text{ideal}}, P_{[1,T-1]}^{\text{ideal}}, S_{T-1}^{\text{ideal}}) \in \{0,1\}^{\beta N}$.
- Remove X_T.
- Output $f(Z_{T-1}^{\text{ideal}}) =$
  $(Y_{[1,T-1]}^{\text{ideal}}, \tilde{Y}_T^A, Y_{[1,T-1]}^{\text{ideal}}, \tilde{Y}_T^B, K_{[1,T]}^{\text{ideal}}, \tilde{K}_{T+1}^A, K_{[1,T]}^{\text{ideal}}, \tilde{K}_{T+1}^B, P_{[1,T]}^{\text{ideal}}, S_T^{\text{ideal}}, X_{[T+1,\infty)})$.

Therefore the *only places* where $f(Z_{T-1}^{\text{ideal}})$ and Z_T^{ideal} differ are Y_T^{ideal} vs. $\tilde{Y}_T^A$, Y_T^{ideal} vs. $\tilde{Y}_T^B$, K_{T+1}^{ideal} vs. $\tilde{K}_{T+1}^A$, and K_{T+1}^{ideal} vs. $\tilde{K}_{T+1}^B$.

Since $Z_T^{\text{real}} = f_T(Z_{T-1}^{\text{real}})$, and $\text{SD}(Z_{T-1}^{\text{real}}, Z_{T-1}^{\text{ideal}}) \leq (T-1) \cdot 4(\varepsilon + \gamma)$, by basic properties of statistical distance, we have

$$\begin{aligned}
\text{SD}(Z_T^{\text{real}}, Z_T^{\text{ideal}}) &= \text{SD}(f_T(Z_{T-1}^{\text{real}}), Z_T^{\text{ideal}}) \\
&\leq \text{SD}(f_T(Z_{T-1}^{\text{real}}), f_T(Z_{T-1}^{\text{ideal}})) + \text{SD}(f_T(Z_{T-1}^{\text{ideal}}), Z_T^{\text{ideal}}) \\
&\leq \text{SD}(Z_{T-1}^{\text{real}}, Z_{T-1}^{\text{ideal}}) + \text{SD}(f_T(Z_{T-1}^{\text{ideal}}), Z_T^{\text{ideal}}) \\
&\leq (T-1) \cdot 4(\varepsilon + \gamma) + \text{SD}(f_T(Z_{T-1}^{\text{ideal}}), Z_T^{\text{ideal}}).
\end{aligned}$$

Thus to prove that $\text{SD}(Z_T^{\text{real}}, Z_T^{\text{ideal}}) \leq T \cdot 4(\varepsilon + \gamma)$, it suffices to show that $\text{SD}(f_T(Z_{T-1}^{\text{ideal}}), Z_T^{\text{ideal}}) \leq 4(\varepsilon + \gamma)$.

Let

$$Z_T' \stackrel{\Delta}{=} f_T(Z_{T-1}^{\text{ideal}}) \setminus (\tilde{Y}_T^B, \tilde{K}_{T+1}^B),$$

that is, obtained from $f_T(Z_{T-1}^{\text{ideal}})$ by removing $\tilde{Y}_T^B$ and $\tilde{K}_{T+1}^B$. Let

$$Z_T'' \stackrel{\Delta}{=} (Y_{[1,T]}^{\text{ideal}}, Y_{[1,T-1]}^{\text{ideal}}, K_{[1,T+1]}^{\text{ideal}}, K_{[1,T]}^{\text{ideal}}, P_{[1,T]}^{\text{ideal}}, S_T^{\text{ideal}}, X_{[T+1,\infty]})$$

be obtained from Z_T^{ideal} by the same procedure, that is, by removing the second Y_T^{ideal} and the second K_{T+1}^{ideal} from Z_T^{ideal}. Thus, Z_T' and Z_T'' are respectively $f_T(Z_{T-1}^{\text{ideal}})$ and Z_T^{ideal} without simulating Bob's recovery of (Y_T^B, K_{T+1}^B), and the *only places* where Z_T' and Z_T'' differ are Y_T^{ideal} vs. $\tilde{Y}_T^A$, and K_{T+1}^{ideal} vs. $\tilde{K}_{T+1}^A$.

The next basic fact, which follows from simple counting, states that if a source X has "sufficient" entropy, and if a source X' is obtained from X by changing at most a δ fraction of bits in each $x \leftarrow X$, then as long as δ is not too large, X' still has sufficient entropy.

Proposition 1. *Let δ and α satisfy $0 \leq \delta < 1/2$ and $\text{H}(\delta) < \alpha \leq 1$, where $\text{H}(\cdot)$ is the binary entropy function. If X is an αN-source taking values in $\{0,1\}^N$, and source X' is obtained from X by changing at most δN bits of each $x \leftarrow X$, then X' is a $(\alpha - \text{H}(\delta))N$-source.*

By Proposition 1 and the two-way block structure of (X_t), we have

Corollary 1. *For each t, the random variable X_t^A, conditioned on all other $X_{t'}$ for $t' \neq t$, has entropy rate at least $\alpha - H(\delta)$.*

The following technical claims follow by manipulating statistical distance and weak random sources.

Claim 1 $\mathrm{SD}(Z_T', Z_T'') \leq 2\varepsilon$.

The proof of Claim 1 is similar to the reasoning in the proof of Lemma 3.3 of [Vad04]. Claim 1 follows from Corollary 1, the definition of the Ideal Experiment, the security property of a local fuzzy extractor, and basic properties of statistical distance and weak random sources.

Let $\mathcal{S}_T$ denote the event that $(\tilde{Y}_T^A, \tilde{K}_{T+1}^A) = (\tilde{Y}_T^B, \tilde{K}_{T+1}^B)$, i.e. the event of correct recovery at time T in the Ideal Experiment.

Claim 2 $\Pr[\mathcal{S}_T] \geq 1 - \gamma$.

Claim 2 follows from the definition of the Ideal Experiment, and the recovery property of a local fuzzy extractor. The next claim follows from Claims 1 and 2, as well as basic properties of statistical distance.

Claim 3 $\mathrm{SD}(f_T(Z_{T-1}^{\mathrm{ideal}})|_{\mathcal{S}_T}, Z_T^{\mathrm{ideal}}|_{\mathcal{S}_T}) < 4\varepsilon + 2\gamma$.

Therefore by Claims 2 and 3, and basic properties of statistical distance,

$$\mathrm{SD}(f_T(Z_{T-1}^{\mathrm{ideal}}), Z_T^{\mathrm{ideal}}) < 4\varepsilon + 2\gamma + \gamma = 4\varepsilon + 3\gamma < 4 \cdot (\varepsilon + \gamma),$$

and the theorem follows.

4.3 Construction of Local Fuzzy Extractors

In this section we construct a local fuzzy extractor from any given averaging sampler and fuzzy extractor.

Averaging Samplers. Averaging samplers are procedures that approximate the average of a $[0,1]$-function by taking the average of the function evaluated at sampled points determined by a random seed. We adopt the following variant of definition in [Vad04] that makes the dependence on μ explicit.

Definition 6 ([BR94, Vad04]). *A function* $\mathrm{Samp}: \{0,1\}^r \to [n]^t$ *is a* (μ, θ, γ)-*averaging sampler if for every function* $f: [n] \to [0,1]$ *with average value* $\bar{\mu} = \frac{1}{n} \cdot \sum_{i=1}^n f(i) \geq \mu$,

$$\Pr_{(i_1,\ldots,i_t) \leftarrow \mathrm{Samp}(U_r)}\left[\frac{1}{t} \cdot \sum_{j=1}^t f(i_j) < \bar{\mu} - \theta\right] \leq \gamma. \tag{1}$$

Samp has distinct samples if for every $x \in \{0,1\}^r$, $\mathrm{Samp}(x)$ produces t distinct samples.

The following result, analogous to Theorem 6.3 of [Vad04], states that combining an averaging sampler and a fuzzy extractor scheme yields a local fuzzy extractor.

Theorem 2. *Let $\alpha, \tau, \delta, \theta > 0$ be constants satisfying relations $\tau < \alpha/3$ and $\theta = \tau/\log(1/\tau) < 1 - \delta$. Let $\mathrm{Samp} : \{0,1\}^r \to [n]^t$ be a (μ, θ, γ)-averaging sampler with distinct samples with $\mu = \min\{(\alpha - 2\tau)/\log(1/\tau), 1 - \delta\}$, and let $\mathrm{FE} = (\mathrm{Ext}, \mathrm{Rec})$ be a $((\alpha - 3\tau)t, \varepsilon, \delta + \theta, \gamma')$-fuzzy extractor, where Ext is of the form $\mathrm{Ext} : \{0,1\}^t \times \{0,1\}^d \to \{0,1\}^m \times \{0,1\}^\ell$. Define $\mathrm{EXT} : \{0,1\}^n \times \{0,1\}^{r+d} \to \{0,1\}^m \times \{0,1\}^\ell$ as*

$$\mathrm{EXT}(x, (k_1, k_2)) \stackrel{\Delta}{=} \mathrm{Ext}(x_{\mathrm{Samp}(k_1)}, k_2),$$

and define $\mathrm{REC} : \{0,1\}^n \times \{0,1\}^{r+d} \times \{0,1\}^\ell \to \{0,1\}^m$ as

$$\mathrm{REC}(x', (k_1, k_2), p) \stackrel{\Delta}{=} \mathrm{Rec}(x'_{\mathrm{Samp}(k_1)}, k_2, p).$$

Then $(\mathrm{EXT}, \mathrm{REC})$ is a t-local $(\alpha n, \varepsilon + 2 \cdot (\gamma + 2^{-\Omega(\tau n)}), \delta, \gamma + \gamma')$-fuzzy extractor.

4.4 Construction of the Underlying Fuzzy Extractor

In this section, we describe a construction of (non-local) fuzzy extractors from any given strong extractor and linear error correcting code with an efficient syndrome decoding algorithm. The underlying ideas in the construction have already been used in information reconciliation and privacy amplification (c.f. [BBR88, BS93, BBCM95, CM97a]). This construction also appears in [DRS04].

Randomness Extractor. Randomness extractors are functions that extract near perfect randomness from imperfect random sources using a short random seed. An extractor is *strong* if its output remains near uniform even if the seed is given.

Definition 7 ([NZ96]). *A function $\mathrm{Ext} : \{0,1\}^n \times \{0,1\}^d \to \{0,1\}^m$ is a strong (k, ε)-extractor if for every k-source X, $(U_d, \mathrm{Ext}(X, U_d))$ is ε-close to (U_d, U_m).*

Syndrome Decoding. We quickly review syndrome decoding of a linear error correcting code. Background and details on error correcting codes can be found in standard texts (e.g. [vL99]). Let $C : \{0,1\}^k \to \{0,1\}^n$ be a linear code over $\mathbb{F}_2$ with minimum distance at least $2d + 1$. Let H be the $(n-k) \times n$ parity check matrix of C. For $x \in \{0,1\}^n$, the *syndrome* of x is defined as $\mathrm{Syn}_C(x) \stackrel{\Delta}{=} Hx$. It is clear that for any *codeword* $y \in C$ and any $e \in \{0,1\}^n$, $\mathrm{Syn}_C(y \oplus e) = \mathrm{Syn}_C(e)$, as $H(y \oplus e) = Hy \oplus He = He$. It is not hard to see that for any $e \in \{0,1\}^n$ with $\mathrm{wt}(e) \leq d$, for every $r \in \{0,1\}^n$ such that $\mathrm{Syn}_C(r) = \mathrm{Syn}_C(e)$, we have $\mathrm{wt}(r) > d \geq \mathrm{wt}(e)$. Hence for any $e \in \{0,1\}^n$ with $\mathrm{wt}(e) \leq d$, e is the *unique* (minimum-weight) vector whose syndrome is $\mathrm{Syn}_C(e)$ and whose weight is at most d. A *syndrome decoder* for C that decodes up to d errors is an algorithm

D that for every error pattern $e \in \{0,1\}^n$ with $\text{wt}(e) \le d$, on input $\text{Syn}_C(e)$, outputs $D(\text{Syn}_C(e)) = e$. It is well known that any decoder for a linear code can be converted to a syndrome decoder.

As an important application, syndrome decoding yields a communication efficient protocol for recovering a string x held by a remote party, using a string y that is sufficiently close to x in Hamming distance. Suppose Alice holds $x \in \{0,1\}^n$, Bob holds $y \in \{0,1\}^n$, and $\Delta(x,y) \le d$. Let $C : \{0,1\}^k \to \{0,1\}^n$ be a linear code over $\mathbb{F}_2$ with minimum distance at least $2d+1$, and an efficient syndrome decoding algorithm D that decodes up to d errors. In order for Bob to recover x,

1. Alice sends $\text{Syn}_C(x)$ to Bob.
2. Bob computes $s = \text{Syn}_C(x) \oplus \text{Syn}_C(y) = \text{Syn}_C(x \oplus y)$. Since $\Delta(x,y) \le d$, $\text{wt}(x \oplus y) \le d$.
3. Bob then decodes $x \oplus y = D(s)$, and recovers $x = x \oplus y \oplus y$.

Thus Alice sends only $|\text{Syn}_C(x)| = n - k$ bits, as opposed to n bits, to Bob. The correctness of the protocol follows from the correctness of the syndrome decoder D: For any $x, y \in \{0,1\}^n$ such that $\Delta(x,y) \le d$, $\text{wt}(x \oplus y) = \Delta(x,y) \le d$. Thus $D(\text{Syn}_C(x \oplus y)) = x \oplus y$, and correct recovery follows.

We use $\text{Rep}(D, p, y)$ to denote Bob's algorithm in Steps 2 and 3 above, i.e. on input s and y, $\text{Rep}(D, p, y)$ computes $s = p \oplus \text{Syn}_C(y)$, and outputs $D(s) \oplus y$.

Syndrome-Based Fuzzy Extractor. This communication efficient recovery protocol above suggests the following fuzzy extractor construction. We adopt an unconventional terminology and say that a code $C : \{0,1\}^{\rho n} \to \{0,1\}^n$ of rate ρ is a (n, ρ, δ)-code if it has minimum distance at least $2\delta n + 1$.

Lemma 1. *Let $C : \{0,1\}^{\rho n} \to \{0,1\}^n$ be a linear (n, ρ, δ)-code with an efficient syndrome decoder D that decodes up to δn errors. Let $\text{Ext} : \{0,1\}^n \times \{0,1\}^d \to \{0,1\}^m$ be a strong (k', ε)-extractor, where $k' = k - (1-\rho)n - \log(1/\varepsilon')$. Define $\text{EXT} : \{0,1\}^n \times \{0,1\}^d \to \{0,1\}^m \times \{0,1\}^{(1-\rho)n}$ as*

$$\text{EXT}(x, K) \triangleq (\text{Ext}(x, K), \text{Syn}_C(x)),$$

and define

$$\text{REC}(x', K, p) \triangleq \text{Ext}(\text{Rep}(D, p, x'), K),$$

where $\text{Rep}(.,.,.)$ is defined above. Then $\text{FE} = (\text{EXT}, \text{REC})$ is a $(k, \varepsilon + \varepsilon', \delta, 0)$-fuzzy extractor.

4.5 Choice of Ingredients

Averaging Sampler. We use the averaging sampler of Vadhan [Vad04] that is near optimal in both randomness and sample complexity for constant μ and θ.

Theorem 3 ([Vad04]). *For every $n \in \mathbb{N}$, $1 > \mu > \theta > 0$, $\gamma > 0$, there is an explicit (μ, θ, γ)-averaging sampler* Samp : $\{0,1\}^r \to [n]^t$ *that uses*

- *t distinct samples for any $t \in \left[O(\frac{1}{\theta^2} \cdot \log \frac{1}{\gamma}), n\right]$;*
- *$r = \log \frac{n}{t} + \log \frac{1}{\gamma} \cdot \text{poly}(\frac{1}{\theta})$ random bits.*

Strong Extractor. We use the near optimal extractor of Zuckerman [Zuc97] for constant entropy rate.

Theorem 4 ([Zuc97]). *For every constant $\alpha, \nu > 0$, for every n, and every $\varepsilon > \exp\left(-n/2^{O(\log^* n)}\right)$, there is an explicit strong $(\alpha n, \varepsilon)$-extractor* Ext : $\{0,1\}^n \times \{0,1\}^d \to \{0,1\}^m$ *with $d = O(\log n + \log \frac{1}{\varepsilon})$ and $m = (1-\nu) \cdot \alpha n$.*

Linear Code. We need an asymptotically good linear code with rate close to 1 and with an efficient syndrome decoder. Explicit constructions of such codes are well known. In particular, it has been shown in [CRVW02] that the expander codes of Sipser and Spielman [SS96], using a *lossless* expander of [CRVW02], achieve a constant rate ρ that is *arbitrarily* close to 1, and a constant $\delta < 1$.

Lemma 2 ([SS96, CRVW02]). *For every constant $\rho < 1$ and every $n \in \mathbb{N}$, there is an explicit linear $(n, \rho, \delta(\rho))$-code $C : \{0,1\}^{\rho n} \to \{0,1\}^n$, where $\delta = \delta(\rho)$ is a constant (depending on ρ). Further, C has a linear time syndrome decoder that decodes up to δn errors.*

4.6 Putting Pieces Together

In this section, we put all pieces together to yield our final local fuzzy extractor and BSM randomness extraction scheme. First as a corollary of Lemmas 1 and 2, and Theorem 4, we have our final (non-local) fuzzy extractor.

Lemma 3. *For every constant $1 \geq \alpha, \gamma, \nu > 0$, there is a constant $\delta > 0$ such that for every sufficiently large $n \in \mathbb{N}$, and every $\varepsilon > \exp\left(-n/2^{O(\log^* n)}\right)$, there is an explicit $(\alpha n, \varepsilon, \delta, 0)$-fuzzy extractor (EXT, REC), where EXT is of the form* EXT : $\{0,1\}^n \times \{0,1\}^d \to \{0,1\}^m \times \{0,1\}^\ell$, *with*

- *$d = O(\log n + \log(1/\varepsilon))$,*
- *$m = (1-\nu)\alpha n$, and*
- *$\ell \leq \gamma m$.*

Next, plugging into Theorem 2 the averaging sampler of Theorem 3 and the fuzzy extractor of Lemma 3, we have our final local fuzzy extractor.

Theorem 5. *For every constant $1 \geq \alpha, \gamma, \nu > 0$, there is a constant δ such that for every sufficiently large $N \in \mathbb{N}$, $\varepsilon > \exp\left(-m/2^{O(\log^* m)}\right)$, and $m \leq (1-\nu)\alpha N$, there is an explicit t-local $(\alpha N, \varepsilon, \delta, \varepsilon)$-fuzzy extractor $\text{FE} = (\text{EXT}, \text{REC})$, where EXT is of the form* EXT : $\{0,1\}^N \times \{0,1\}^d \to \{0,1\}^m \times \{0,1\}^\ell$, *with*

- seed length $d = \log N + O(\log m + \log(1/\varepsilon))$,
- sample size $t = (1+\nu)m/\alpha$, and
- fingerprint length $\ell \leq \gamma m$.

Theorem 5 is the "fuzzy" analogue of Theorem 8.5 of [Vad04]. The seed length and sample complexity (i.e. the value of t) of our local fuzzy extractor match those of Vadhan's (non-fuzzy) local extractor [Vad04], and thus are optimal up to constant factors.

Finally as a corollary of Theorem 1 and Theorem 5, we have

Theorem 6. *For every constant $\alpha > 0$, $\beta < \alpha$, $\gamma > 0$, and $\nu > 0$, there is a constant δ such that for every sufficiently large $N \in \mathbb{N}$, sufficiently large $m \leq (1-\nu)(\alpha - \beta - \mathrm{H}(\delta))N$, and $\varepsilon > \exp\left(-m/2^{O(\log^* m)}\right)$, there is an explicit ε-secure t-local BSM randomness extraction scheme* (EXT, REC) *for storage rate β, entropy rate α, and error rate δ, where* EXT *is of the form* EXT $: \{0,1\}^N \times \{0,1\}^d \to \{0,1\}^{m+d} \times \{0,1\}^\ell$, *with*

- key length $d = \log N + O(\log m + \log(1/\varepsilon))$,
- sample size $t = (1+\nu)m/\alpha'$, where $\alpha' = \alpha - \beta - \mathrm{H}(\delta)$, and
- fingerprint length $\ell \leq \gamma m$.

5 Conclusion

We initiate a study of the bounded storage with errors from the public random source that cause parties to have inconsistent view of the source. We provide a general paradigm for constructing error-resilient BSM cryptosystems based on averaging samplers and fuzzy extractors. By proper choice and construction of the underlying building blocks, our general paradigm yields BSM cryptosystems that tolerate a constant fraction of errors, attain near optimal key length and sample complexity (i.e. the number of bits read from the source), and incur a very small communication overhead. It is interesting to study whether the communication overhead can be further reduced.

The recovery property of our local fuzzy extractor can be further improved by taking advantage of the shared randomness between the extraction and the recovery algorithms. By the method of [Lan04], a local fuzzy extractor can be based on any explicit and *list decodable* (as opposed to uniquely decodable) asymptotically good linear code with rate arbitrarily close to 1, while the seed length increases by only $O(\log t + \log 1/\gamma)$ bits, where t is the number of bits read from the source, and γ is the recovery error.

Our general paradigm also yields efficient error-resilient message authentication codes (MAC) in the bounded storage model. By combining the BSM extraction scheme of Theorem 6 and an efficient information-theoretically secure MAC (c.f. that of Krawczyk [Kra95]), we obtain an efficient error-resilient BSM MAC that is secure against *chosen message attacks* [GMR89]. Our paradigm can also be used to construct efficient error-resilient protocols for other cryptographic primitives, such as oblivious transfer and key agreement in the bounded storage model. We leave details to the full version.

Our cryptosystems are stateful. That is, our cryptosystems do not reuse the key, but instead update the key in a forward-secure manner. It is an interesting open problem to construct a stateless error-resilient BSM cryptosystem with a reusable key that matches the near optimal parameters achieved by the stateful construction. One promising approach is to derandomize the construction of entropically secure fuzzy extractors in [Smi04].

Another interesting open problem is to construct efficient local fuzzy extractors for other natural metrics, such as editing distance, where the sample-then-extract paradigm fails.

Acknowledgment

I thank Michael Rabin and Salil Vadhan for numerous insightful discussions on the bounded storage model, in particular the one that brought up the issue of error correction. I thank Yevgeniy Dodis, Adam Smith, Salil Vadhan and anonymous TCC referees for very helpful comments on this paper.

References

[ADR02]   Yonatan Aumann, Yan Zong Ding, and Michael O. Rabin. Everlasting security in the bounded storage model. *IEEE Transactions on Information Theory*, 48(6):1668–1680, June 2002.

[And97]   Ross Anderson. Two remarks on public key cryptology. Invited Lecture. In *4th ACM Conference on Computer and Communications Security*, 1997.

[AR99]   Yonatan Aumann and Michael O. Rabin. Information theoretically secure communication in the limited storage space model. In *Advances in Cryptology - CRYPTO '99*, pages 65–79. Springer-Verlag, 1999.

[BBCM95]   C. Bennett, G. Brassard, C. Crépeau, and U. Maurer. Generalized privacy amplification. *IEEE Transactions on Information Theory*, 41(6):1915 – 1923, 1995.

[BBR88]   C. Bennett, G. Brassard, and J. Roberts. Privacy amplification by public discussion. *SIAM Journal on Computing*, 17(2):210–229, 1988.

[BR94]   Mihir Bellare and John Rompel. Randomness-efficient oblivious sampling. In *35th Annual IEEE Symposium on Foundations of Computer Science*, pages 276–287, November 1994.

[BS93]   Gilles Brssard and Louis Salvail. Secret-key reconciliation by public discussion. In *Advances in Cryptology - EUROCRYPT '93*, pages 410–423. Springer-Verlag, 1993.

[CCM98]   Christian Cachin, Claude Crépeau, and Julien Marcil. Oblivious transfer with a memory-bounded receiver. In *39th Annual IEEE Symposium on Foundations of Computer Science*, pages 493–502, November 1998.

[CG88]   Benny Chor and Oded Goldreich. Unbiased bits from sources of weak randomness and probabilistic communication complexity. *SIAM Journal on Computing*, 17(2):230–261, April 1988.

[CM97a] Christian Cachin and Ueli Maurer. Linking information reconciliation and privacy amplification. *Journal of Cryptology*, 10(2):97–110, 1997.

[CM97b] Christian Cachin and Ueli Maurer. Unconditional security against memory bounded adversaries. In *Advances in Cryptology - CRYPTO '97*, pages 292–306. Springer-Verlag, 1997.

[CRVW02] Michael R. Capalbo, Omer Reingold, Salil P. Vadhan, and Avi Wigderson. Randomness conductors and constant-degree lossless expanders. In *34th Annual ACM Symposium on the Theory of Computer Science*, pages 659–668, 2002.

[DHRS04] Yan Zong Ding, Danny Harnik, Alon Rosen, and Ronen Shaltiel. Constant-round oblivious transfer in the bounded storage model. In *1st Theory of Cryptography Conference – TCC '04*, pages 446–472, 2004.

[Din01] Yan Zong Ding. Oblivious transfer in the bounded storage model. In *Advances in Cryptology – CRYPTO '01*, pages 155–170. Springer-Verlag, August 2001.

[DM04a] Stefan Dziembowski and Ueli Maurer. On generating the initial key in the bounded storage model. In *Advances in Cryptology - EUROCRYPT '04*. Springer-Verlag, 2004.

[DM04b] Stefan Dziembowski and Ueli Maurer. Optimal randomizer efficiency in the bounded-storage model. *Journal of Cryptology*, 17(1):5–26, 2004.

[DR02] Yan Zong Ding and Michael O. Rabin. Hyper-encryption and everlasting security (extended abstract). In *19th Annual Symposium on Theoretical Aspects of Computer Science*, pages 1–26. Springer-Verlag, March 2002.

[DRS04] Yevgeniy Dodis, Leonid Reyzin, and Adam Smith. Fuzzy extractors and cryptography, or how to use your fingerprints. In *Advances in Cryptology - EUROCRYPT '04*. Springer-Verlag, 2004.

[GMR89] Shafi Goldwasser, Silvio Micali, and Charles Rackoff. The knowledge complexity of interactive proof systems. *SIAM Journal on Computing*, 18(1):186–208, February 1989.

[Gol97] Oded Goldreich. A sample of samplers: A computational perspective on sampling. Technical Report TR97-020, Electronic Colloquium on Computational Complexity, May 1997.

[Kra95] Hugo Krawczyk. New hash functions for message authentication. In *Advances in Cryptology - EUROCRYPT '95*, pages 301–310. Springer-Verlag, 1995.

[Lan04] Michael Langberg. Private codes or succinct random codes that are (almost) perfect. In *45th Annual Symposium on Foundations of Computer Science*, 2004.

[Lu04] Chi-Jen Lu. Encryption against space-bounded adversaries from on-line strong extractors. *Journal of Cryptology*, 17(1):27–42, 2004.

[Mau92] Ueli Maurer. Conditionally-perfect secrecy and a provably-secure randomized cipher. *Journal of Cryptology*, 5(1):53–66, 1992.

[MST04] Tal Moran, Ronen Shaltiel, and Amnon Ta-Shma. Non-interactive timestamping in the bounded storage model. In *Advances in Cryptology - CRYPTO '04*, pages 460–476. Springer-Verlag, 2004.

[NT99] Noam Nisan and Amnon Ta-Shma. Extracting randomness: A survey and new constructions. *Journal of Computer and System Sciences*, 58(1):148–173, 1999.

[NY90] Moni Naor and Moti Yung. Public-key cryptosystems provably secure against chosen ciphertext attacks. In *22nd Annual ACM Symposium on the Theory of Computer Science*, pages 427–437, 1990.

[NZ96]   Noam Nisan and David Zuckerman. Randomness is linear in space. *Journal of Computer and System Sciences*, 52(1):43–52, 1996.
[Rab02]   Michael O. Rabin. Personal communication, 2002.
[Sha02]   Ronen Shaltiel. Recent developments in explicit constructions of extractors. *Bulletin of the European Association for Theoretical Computer Science*, 77:67–95, 2002.
[Smi04]   Adam Smith. Maintaining secrecy when information leakage is unavoidable. Ph.D. Thesis, MIT, 2004.
[SS96]   Michael Sipser and Daniel A. Spielman. Expander codes. *IEEE Transactions on Information Theory*, 42(6):1710–1722, 1996.
[Vad02]   Salil P. Vadhan. Randomness extractors and their many guises. In *43rd Annual IEEE Symposium on Foundations of Computer Science*, pages 9–, November 2002. Presentation available at http://www.eecs.harvard.edu/~salil/extractor-focs.ppt.
[Vad04]   Salil P. Vadhan. Constructing locally computable extractors and cryptosystems in the bounded storage model. *Journal of Cryptology*, 17(1):43–77, 2004.
[vL99]   J.H. van Lint. *Introduction to Coding Theory*. Spring, 1999.
[Zuc96]   David Zuckerman. Simulating BPP using a general weak random source. *Algorithmica*, 16(4/5):367–391, 1996.
[Zuc97]   David Zuckerman. Randomness-optimal oblivious sampling. *Random Structures & Algorithms*, 11(4):345–367, 1997.

Characterizing Ideal Weighted Threshold Secret Sharing

Amos Beimel[1], Tamir Tassa[1,2], and Enav Weinreb[1]

[1] Dept. of Computer Science,
Ben-Gurion University, Beer Sheva, Israel
[2] Division of Computer Science,
The Open University, Ra'anana, Israel

Abstract. Weighted threshold secret sharing was introduced by Shamir in his seminal work on secret sharing. In such settings, there is a set of users where each user is assigned a positive weight. A dealer wishes to distribute a secret among those users so that a subset of users may reconstruct the secret if and only if the sum of weights of its users exceeds a certain threshold. A secret sharing scheme is ideal if the size of the domain of shares of each user is the same as the size of the domain of possible secrets (this is the smallest possible size for the domain of shares). The family of subsets authorized to reconstruct the secret in a secret sharing scheme is called an access structure. An access structure is ideal if there exists an ideal secret sharing scheme that realizes it.

It is known that some weighted threshold access structures are not ideal, while other nontrivial weighted threshold access structures do have an ideal scheme that realizes them. In this work we characterize all weighted threshold access structures that are ideal. We show that a weighted threshold access structure is ideal if and only if it is a hierarchical threshold access structure (as introduced by Simmons), or a tripartite access structure (these structures, that we introduce here, generalize the concept of bipartite access structures due to Padró and Sáez), or a composition of two ideal weighted threshold access structures that are defined on smaller sets of users. We further show that in all those cases the weighted threshold access structure may be realized by a linear ideal secret sharing scheme. The proof of our characterization relies heavily on the strong connection between ideal secret sharing schemes and matroids, as proved by Brickell and Davenport.

1 Introduction

A *threshold secret sharing scheme* enables a dealer to distribute a secret among a set of users, by giving each user a piece of information called a *share*, such that only large sets of users will be able to reconstruct the secret from the shares that they got, while smaller sets gain no information on the secret. Threshold secret sharing schemes were introduced and efficiently implemented, independently, by Blakley [6] and Shamir [26]. Efficient threshold secret sharing schemes were used in many cryptographic applications, e.g., Byzantine agreement [24], secure multiparty computations [4, 11], and threshold cryptography [13].

In this paper we deal with *weighted* threshold secret sharing schemes. In these schemes, considered already by Shamir [26], the users are not of the same status. That is, each user is assigned a positive weight and a set can reconstruct the secret if the sum of weights assigned to its users exceeds a certain threshold. As a motivation, consider sharing a secret among the shareholders of some company, each holding a different amount of shares. Such settings are closely related to the concept of *weighted threshold functions*, which play an important role in complexity theory and learning theory.

Ito, Saito, and Nishizeki [14] generalized the notion of secret sharing such that there is an arbitrary monotone collection of authorized sets, called the *access structure*. The requirements are that only sets in the access structure are allowed to reconstruct the secret, while sets that are not in the access structure should gain no information on the secret. A simple argument shows that in every secret sharing scheme, the domain of possible shares for each user is at least as large as the domain of possible secrets (see [17]). Shamir's threshold secret sharing scheme is *ideal* in the sense that the domain of shares of each user coincides with the domain of possible secrets. Ideal secret sharing schemes are the most space-efficient schemes. Some access structures do not have any ideal secret sharing schemes that realizes them [5]. Namely, some access structures demand share domains that are larger than the domain of secrets. Access structures that may be realized by an ideal secret sharing scheme are called ideal. Ideal secret sharing schemes and ideal access structures have been studied in, e.g., [1, 7, 8, 15, 18, 19, 21, 23, 25, 30, 33]. Ideal access structures are known to have certain combinatorial properties. In particular, there is a strong relation between ideal access structures and matroids [8].

While threshold access structures are ideal, weighted threshold access structures are not necessarily so. For example, the access structure on four users with weights 1, 1, 1, and 2, and a threshold of 3, has no ideal secret sharing scheme (see Example 1 for a proof). Namely, in any perfect secret sharing scheme that realizes this access structure, the share domain of at least one user is larger than the domain of secrets. On the other hand, there exist ideal weighted threshold access structures, other than the trivial threshold ones. For example, consider the access structure on nine users, where the weights are 16, 16, 17, 18, 19, 24 ,24 ,24, and 24 and the threshold is 92. Even though this access structure seems more complicated than the above access structure, it has an ideal secret sharing scheme (see the full version of this paper [2]). Another example of an ideal weighted threshold access structure is the one having weights 1, 1, 1, 1, 1, 3, 3, and 3 and threshold 6 (see Example 2).

We give a combinatorial characterization of ideal weighted threshold access structures. We show that a weighted threshold access structure is ideal if and only if it is a hierarchical threshold access structure (as introduced by Simmons [27]), or a tripartite access structure (these structures, that we introduce here, generalize the bipartite access structures of Padró and Sáez [23]), or a composition of two ideal weighted threshold access structures that are defined on smaller sets of users. We further show that in all those cases the weighted threshold access structure may be realized by a linear ideal secret sharing scheme. The present study generalizes the work of Morillo, Padró, Sáez, and Villar [21] who characterized the ideal weighted threshold access structures in which all the minimal authorized sets have at most two users. The proof of our characterization relies heavily on the strong connection between ideal secret sharing schemes and matroids, as presented in [8]. We utilize results regarding the structure of matroids to understand and characterize the structure of ideal weighted threshold access structures. An important tool in our analysis is composition of ideal access structures, previously studied in, e.g., [1, 5, 9, 12, 20, 32].

Efficiency of Secret Sharing Schemes. Secret sharing schemes for general access structures were defined by Ito, Saito, and Nishizeki in [14]. More efficient schemes were presented in, e.g., [5, 7, 16, 29]. We refer the reader to [28, 31] for surveys on secret sharing. However, for most access structures the known secret sharing schemes are highly inefficient, that is, the size of the shares is exponential in n, the number of users. It is not known if better schemes exist. For weighted threshold access structures the situation is better. In a recent work [3], secret sharing schemes were constructed for arbitrary weighted threshold access structures in which the shares are of size $O(n^{\log n})$. Furthermore, under reasonable computational assumptions, a secret sharing scheme with computational security was constructed in [3] for every weighted threshold access structure with a polynomial share size.

Organization. We begin in Section 2 by supplying the necessary definitions. Then, in Section 3, we state our characterization theorem and outline its proof. We proceed to describe in Section 4 the connection between matroids and ideal secret sharing, and then prove, in Section 5, several properties of matroids that are associated with weighted-threshold access structures. Thereafter, we discuss the connection between ideal weighted threshold access structures and two families of access structures: hierarchical threshold access structures in Section 6, and tripartite access structures in Section 7. Finally, in Section 8 we complete the proof of the characterization theorem by proving that if an ideal weighted threshold access structure is not hierarchical nor tripartite then it is a composition of two access structures on smaller sets of users. For lack of space, some proofs are omitted. All proofs may be found in the full version of the paper [2].

2 Definitions and Notations

Definition 1 (Access Structure). *Let $U = \{u_1, \ldots, u_n\}$ be a set of users. A collection $\Gamma \subseteq 2^U$ is* monotone *if $B \in \Gamma$ and $B \subseteq C$ imply that $C \in \Gamma$. An* access structure *is a monotone collection $\Gamma \subseteq 2^U$ of non-empty subsets of U. Sets in Γ are called* authorized, *and sets not in Γ are called* unauthorized. *A set B is called a* minterm *of Γ if $B \in \Gamma$, and for every $C \subsetneq B$, the set C is unauthorized.*

A user u is called self-sufficient *if* $\{u\} \in \Gamma$. A user is called redundant *if there is no minterm that contains it*. An access structure is called connected *if it has no redundant users*.

Definition 2 (Secret-Sharing Scheme). *Let S be a finite set of secrets, where $|S| \geq 2$. An n-user secret-sharing scheme Π with domain of secrets S is a randomized mapping from S to a set of n-tuples $\prod_{i=1}^{n} S_i$, where S_i is called the share-domain of u_i. A dealer shares a secret $s \in S$ among the n users of some set U according to Π by first sampling a vector of shares $\Pi(s) = (s_1, \ldots, s_n) \in \prod_{i=1}^{n} S_i$, and then privately communicating each share s_i to the user u_i. We say that Π realizes an access structure $\Gamma \subseteq 2^U$ if the following requirements hold:*

CORRECTNESS. *The secret s can be reconstructed by any authorized set of users. That is, for any set $B \in \Gamma$ (where $B = \{u_{i_1}, \ldots, u_{i_{|B|}}\}$), there exists a reconstruction function* $\text{RECON}_B : S_{i_1} \times \ldots \times S_{i_{|B|}} \to S$ *such that for every $s \in S$ and for every possible value of $\Pi_B(s)$, the restriction of $\Pi(s)$ to its B-entries,* $\text{RECON}_B(\Pi_B(s)) = s$.

PRIVACY. *Every unauthorized set can learn nothing about the secret (in the information theoretic sense) from their shares. Formally, for any set $C \notin \Gamma$, for every two secrets $a, b \in S$, and for every possible $|C|$-tuple of shares $\langle s_i \rangle_{u_i \in C}$,*

$$\Pr[\,\Pi_C(a) = \langle s_i \rangle_{u_i \in C}\,] = \Pr[\,\Pi_C(b) = \langle s_i \rangle_{u_i \in C}\,].$$

In every secret-sharing scheme, the size of the domain of shares of each user is at least the size of the domain of the secrets [17], namely $|S_i| \geq |S|$ for all $i \in [n]$. This motivates the next definition.

Definition 3 (Ideal Access Structure). *A secret-sharing scheme with domain of secrets S is ideal if the domain of shares of each user is S. An access structure Γ is ideal if for some finite domain of secrets S there exists an ideal secret sharing scheme realizing it.*

Most previously known secret sharing schemes are *linear*. The concept of linear secret sharing schemes was introduced by Brickell [7] in the ideal setting and was latter generalized to non-ideal schemes. Linear schemes are equivalent to monotone span programs [16]. In an ideal linear secret sharing scheme, the secret is an element of a finite field, and each share is a linear combination of the secret and some additional random field elements.

In this paper we concentrate on special access structures, so-called weighted threshold access structures, that were already introduced in [26].

Definition 4 (Weighted Threshold Access Structure – WTAS). *Let $w : U \to \mathbb{N}$ be a weight function on U and $T \in \mathbb{N}$ be a threshold. Define $w(A) := \sum_{u \in A} w(u)$ and $\Gamma = \{A \subseteq U : w(A) \geq T\}$. Then Γ is called a weighted threshold access structure (WTAS) on U.*

Terminology and Notations. Throughout this paper we assume that the users are ordered in a nondecreasing order according to their weights, i.e., $w(u_1) \leq w(u_2) \leq \cdots \leq w(u_n)$. Let $A = \{u_{i_j}\}_{1 \leq j \leq k}$ be an ordered subset of U, where $1 \leq i_1 < \cdots < i_k \leq n$. In order to avoid two-levelled indices, we will denote the users in such a subset with the corresponding lower-case letter, namely, $A = \{a_j\}_{1 \leq j \leq k}$. We denote the first (lightest) and last (heaviest) users of A by $A_{\min} = a_1$ and $A_{\max} = a_k$ respectively. For an arbitrary ordered subset A we let $A_{s,t} = \{a_j\}_{s \leq j \leq t}$ denote a run-subset. If $s > t$ then $A_{s,t} = \emptyset$. Two types of runs that we shall meet frequently are prefixes and suffixes. A prefix of a subset A is a run-subset of the form $A_{1,\ell}$, while a suffix takes the form $A_{\ell,k}$, $1 \leq \ell \leq k$. A suffix $A_{\ell,k}$ is a proper suffix of $A_{1,k}$ if $\ell > 1$. We conclude this section by introducing the precedence relation $\prec$. When applied to users, $u_i \prec u_j$ indicates that $i < j$ (and, in particular, $w(u_i) \leq w(u_j)$). This relation induces a lexicographic order on subsets of U in the natural way.

3 Characterizing Ideal WTASs

The main result of this paper is a combinatorial characterization of ideal WTASs. We define in Definitions 5–7 the building blocks that play an essential role in this characterization. Using these definitions, we state Theorem 1, our main result, that characterizes ideal WTASs. We also outline the proof of that theorem, where the full proof is given in the subsequent sections.

3.1 Building Blocks

Definition 5 (Hierarchical Threshold Access Structure – HTAS). *Let m be an integer, $U = \bigcup_{i=1}^{m} L_i$ be a partition of the set of users into a hierarchy of m disjoint levels, and $\{k_i\}_{1 \leq i \leq m}$ be a sequence of decreasing thresholds, $k_1 > k_2 > \cdots > k_m$. These hierarchy and sequence of thresholds induce a hierarchical threshold access structure (HTAS) on U:*

$$\Gamma_H = \left\{ A \subseteq U : \text{There exists } i \in [m] \text{ such that } \left| A \cap \bigcup_{j=i}^{m} L_j \right| \geq k_i \right\}.$$

That is, a set $A \subseteq U$ is in Γ_H if and only if it contains at least k_i users from the ith level and above, for some $i \in [m]$. The family of HTASs was introduced by Simmons in [27] and further studied by Brickell who proved their ideality [7]. An explicit ideal scheme for these access structures was constructed in [33].

Remark 1. Without loss of generality, we assume that

$$|L_i| > k_i - k_{i+1} \text{ for every } i \in [m-1], \text{ and } |L_m| \geq k_m.$$

Indeed, if $|L_i| \leq k_i - k_{i+1}$ for some $i \in [m-1]$, then the ith threshold condition in the HTAS definition implies the $(i+1)$th threshold condition and, consequently, the ith condition is redundant.

Definition 6 (Tripartite Access Structure – TPAS). *Let U be a set of n users, such that $U = A \cup B \cup C$, where A, B, and C are disjoint, and A and C are nonempty. Let m, d, t be positive integers such that $m > t$. Then the following is a* tripartite access structure (TPAS) *on U:*

$$\Delta_1 = \{X \subseteq U : (|X| \geq m \text{ and } |X \cap (B \cup C)| \geq m - d) \text{ or } |X \cap C| \geq t\},$$

Namely, a set X is in Δ_1 if either it has at least m users, $(m - d)$ of which are from $B \cup C$, or it has at least t users from C. If $|B| \leq d + t - m$, then the following is also a tripartite access structure:

$$\Delta_2 = \{X \subseteq U : (|X| \geq m \text{ and } |X \cap C| \geq m - d) \text{ or } |X \cap (B \cup C)| \geq t\}.$$

That is, $X \in \Delta_2$ if either it has at least m users, $(m - d)$ of which are from C, or it has at least t users from $B \cup C$.

TPASs, introduced herein, generalize the concept of bipartite access structure that was presented in [23]. We show that TPASs are ideal by constructing a linear ideal secret sharing scheme that realizes them. Our scheme is a generalization of a scheme from [23] for bipartite access structures.

Definition 7 (Composition of Access Structures). *Let U_1 and U_2 be disjoint sets of users and let Γ_1 and Γ_2 be access structures over U_1 and U_2 respectively. Let $u_1 \in U_1$, and set $U = U_1 \cup U_2 \setminus \{u_1\}$. Then the* composition of Γ_1 and Γ_2 via u_1 *is*

$$\Gamma = \{X \subseteq U : X \cap U_1 \in \Gamma_1 \text{ or } (X \cap U_2 \in \Gamma_2 \text{ and } (X \cap U_1) \cup \{u_1\} \in \Gamma_1)\}.$$

3.2 The Characterization

Recall that the set U is viewed as a sequence which is ordered in a monotonic non-decreasing order according to the weights. Let M be the lexicographically minimal minterm of Γ (that is, $M \in \Gamma$ is a minterm and $M \prec M'$ for all other minterms $M' \in \Gamma$). It turns out that the form of M plays a significant role in the characterization of Γ.

If M is a prefix of U, namely $M = U_{1,k}$ for some $k \in [n]$, then, as we prove in Section 6.1, the access structure is a HTAS of at most three levels. If M is a lacunary prefix, in the sense that $M = U_{1,k} \setminus \{u_\ell\}$ for $1 \leq \ell < k \leq n$, then, as we discuss in Section 7, the access structure is a TPAS. Otherwise, if M is neither a prefix nor a lacunary prefix, the access structure is a composition of two weighted threshold access structures over smaller sets. More specifically, we identify a prefix $U_{1,k}$, where $1 < k < n$, that could be replaced by a single substitute user u, and then show that Γ is a composition of a WTAS on $U_{1,k}$ and another WTAS on $U_{k+1,n} \cup \{u\}$. Since Γ is ideal, so are the two smaller WTASs, as implied by Lemma 13. Hence, this result, which we prove in Section 8, completes the characterization of ideal WTASs in a recursive manner. Our main result in this paper is as follows.

Theorem 1 (Characterization Theorem). *Let U be a set of users, w be a weight function, T be a threshold, and Γ be the corresponding WTAS. Then Γ is ideal if and only if one of the following three conditions holds:*

- *The access structure Γ is a HTAS.*
- *The access structure Γ is a TPAS.*
- *The access structure Γ is a composition of Γ_1 and Γ_2, where Γ_1 and Γ_2 are ideal WTASs defined over sets of users smaller than U.*

In particular, if Γ is an ideal WTAS then there exists a linear ideal secret sharing scheme that realizes it.

4 Matroids and Ideal Secret Sharing Schemes

Ideal secret sharing schemes and matroids are strongly related [8]. If an access structure is ideal, there is a matroid that reflects its structure. On the other hand, every matroid that is representable over some finite field is the reflection of some ideal access structure. In this section we review some basic results from the theory of matroids and describe their relation to ideal secret sharing schemes. For more background on matroid theory the reader is referred to [22].

Matroids are a combinatorial structure that generalizes both linear spaces and the set of circuits in an undirected graph. They are a useful tool in several fields of theoretical computer science, e.g., optimization algorithms. A matroid $\mathcal{M} = \langle V, \mathcal{I} \rangle$ is a finite set V and a collection $\mathcal{I}$ of subsets of V that satisfy the following three axioms: **(I1)** $\emptyset \in \mathcal{I}$. **(I2)** If $X \in \mathcal{I}$ and $Y \subseteq X$ then $Y \in \mathcal{I}$. **(I3)** If X and Y are members of $\mathcal{I}$ with $|X| = |Y| + 1$ then there exists an element $x \in X \backslash Y$ such that $Y \cup \{x\} \in \mathcal{I}$. The elements of V are called the *points* of the matroid and the sets in $\mathcal{I}$ are called the *independent* sets of the matroid. A *dependent set* of the matroid is any subset of V that is not independent. The minimal dependent sets are called *circuits*. A matroid is said to be *connected* if for any two points there exists a circuit that contains both of them.

We now discuss the relations between ideal secret sharing schemes and matroids. Let Γ be an access structure over a set of users $U = \{u_1, \ldots, u_n\}$. If Γ is ideal, then, by the results of [8, 19], there exists a matroid $\mathcal{M}$ corresponding to Γ. The points of $\mathcal{M}$ are the users in U together with an additional point, denoted u_0, that could be thought of as representing the dealer. We denote hereinafter by $\mathcal{C}_0 = \{X \cup \{u_0\} : X \text{ is a minterm of } \Gamma\}$ the set of all Γ-minterms, supplemented by u_0.

Theorem 2 ([8, 19]). *Let Γ be a connected ideal access structure. Then there exists a connected matroid $\mathcal{M}$ such that $\mathcal{C}_0$ is exactly the set of circuits of $\mathcal{M}$ containing u_0.*

The next result implies the uniqueness of the matroid $\mathcal{M}$ that corresponds to a given connected ideal access structure, as discussed in Theorem 2, and it provides means to identify *all* the circuits of that matroid.

Lemma 1 ([22–Theorem 4.3.2]). *Let e be an element of a connected matroid $\mathcal{M}$ and let $\mathcal{C}_e$ be the set of circuits of $\mathcal{M}$ that contain e. Then all of the circuits of $\mathcal{M}$ that do not contain e are the minimal sets of the form $(C_1 \cup C_2)\setminus \bigcap \{C_3 : C_3 \in \mathcal{C}_e, C_3 \subseteq C_1 \cup C_2\}$ where C_1 and C_2 are distinct circuits in $\mathcal{C}_e$.*

The unique matroid whose existence and uniqueness are guaranteed by Theorem 2 and Lemma 1 is referred to as *the matroid corresponding to Γ*. The next definition will enable us to explicitly define the matroid corresponding to Γ using the authorized sets in Γ.

Definition 8 (Critical User). *Let M_1 and M_2 be distinct minterms of Γ. A user $x \in M_1 \cup M_2$ is* critical *for $M_1 \cup M_2$ if the set $M_1 \cup M_2 \setminus \{x\}$ is unauthorized. In addition, we define*

$$D(M_1, M_2) = (M_1 \cup M_2) \setminus \{x \in M_1 \cup M_2 : x \text{ is critical for } M_1 \cup M_2\}.$$

Corollary 1. *Let M_1 and M_2 be two distinct minterms of Γ. Then $D(M_1, M_2)$ is a dependent set of $\mathcal{M}$.*

Note that $D(M_1, M_2)$ is a dependent set of $\mathcal{M}$, but is not necessarily a circuit of $\mathcal{M}$.

Lemma 2 ([22–Lemma 1.1.3]). *Let C_1 and C_2 be two distinct circuits in a matroid and $e \in C_1 \cap C_2$. Then there exists a circuit $C_3 \subseteq (C_1 \cup C_2) \setminus \{e\}$.*

Finally, the next lemma is applicable when adding an element to an independent set results in a dependent set.

Lemma 3. *Let I be an independent set in a matroid $\mathcal{M}$ and let e be an element of $\mathcal{M}$ such that $I \cup \{e\}$ is dependent. Then $\mathcal{M}$ has a unique circuit contained in $I \cup \{e\}$ and that circuit contains e.*

Example 1. This example shows how to use the above statements in order to demonstrate that a given access structure is not ideal. Consider the WTAS Γ on the set $U = \{u_1, u_2, u_3, u_4\}$ with weights $w(u_1) = w(u_2) = w(u_3) = 1$ and $w(u_4) = 2$ and threshold $T = 3$. The minterms of Γ are $\{u_1, u_2, u_3\}$, $\{u_1, u_4\}$, $\{u_2, u_4\}$, and $\{u_3, u_4\}$. It follows from Benaloh and Leichter [5] that this access structure is not ideal.[1] Assume that it is ideal and consider the minterms $M_1 = \{u_1, u_4\}$ and $M_2 = \{u_2, u_4\}$. The set $\{u_1, u_2\}$ is unauthorized and thus u_4 is critical for $M_1 \cup M_2$. On the other hand, the users u_1 and u_2 are not critical for $M_1 \cup M_2$. Therefore, by Corollary 1, the set $D(M_1, M_2) = \{u_1, u_2\}$ is a dependent set of $\mathcal{M}$, the matroid corresponding to Γ. However, the set $\{u_1, u_2, u_3\}$ is a minterm of Γ and, consequently, it is independent in $\mathcal{M}$. Since $\{u_1, u_2\} \subset \{u_1, u_2, u_3\}$, we arrive at the absurd conclusion that a dependent set is contained in an independent set.

---

[1] In [10] it was shown that if the domain of secrets is S, the size of the domain of shares of at least one user in that access structure must be at least $|S|^{1.5}$. That result improved upon previous bounds that were derived in [9].

Definition 9 (Restriction). *Let $Y, X \subseteq U$ be two disjoint subsets of users. The restriction of Γ that is induced by Y on X is defined as the following access structure: $\Gamma_{Y,X} = \{Z \subseteq X : Z \cup Y \in \Gamma\}$.*

In other words, $\Gamma_{Y,X}$ consists of all subsets of X that complete Y to an authorized set in Γ. Since $\Gamma_{Y,X}$ is defined over a smaller set of users, restrictions can be helpful in recursively characterizing the structure of Γ. The following known result assures us that if Γ is ideal, $\Gamma_{Y,X}$ is ideal as well.

Lemma 4. *Let Γ be an access structure over a set of users U. Let $Y, X \subseteq U$ be sets such that $Y \notin \Gamma$ and $X \cap Y = \emptyset$. If Γ is ideal, then $\Gamma_{Y,X}$ is ideal. Furthermore, if Y is independent in the matroid corresponding to Γ and if a set $I \subseteq X$ is independent in the matroid corresponding to $\Gamma_{Y,X}$, then I is independent in the matroid corresponding to Γ.*

Lemma 5. *Let $X, Y \in U$ such that $X \cap Y = \emptyset$. If Γ is a WTAS, then $\Gamma_{Y,X}$ is a WTAS.*

5 WTASs and Matroids

In this section we prove several properties of matroids that are associated with ideal WTASs. These properties will serve us later in characterizing ideal WTASs. Let Γ be an ideal WTAS on $U = \{u_1, \ldots, u_n\}$ corresponding to a weight function $w : U \to \mathbb{N}$ and a threshold T. Let $\mathcal{M}$ be the matroid corresponding to Γ.

Lemma 6. *If $X = \{x_1, \ldots, x_k\} \in \Gamma$, it contains a suffix minterm, namely, there exists $i \in [k]$ such that $X_{i,k} = \{x_i, \ldots, x_k\}$ is a minterm.*

Lemma 7. *Let M be a minterm of Γ. Let $y \in U \setminus M$ be a user such that $w(M_{\min}) \le w(y)$. Then $M \cup \{y\}$ is a dependent set of $\mathcal{M}$.*

Proof. Let $X = (M \setminus \{M_{\min}\}) \cup \{y\}$ be the set that is obtained by replacing the minimal user in M with y. Since $w(X) \ge w(M)$, the set X is authorized and, thus, it contains a minterm M'. Moreover, $M \ne M'$ since $M_{\min} \in M \setminus M'$. Therefore, by Corollary 1, the set $M \cup M' = M \cup \{y\}$ is dependent in $\mathcal{M}$. □

We show in the next lemma that whenever two minterms have the same minimal member they must be of the same size.

Lemma 8. *Let X and Y be minterms of the access structure Γ such that $X_{\min} = Y_{\min}$. Then $|X| = |Y|$.*

Proof. As X and Y are minterms, they are independent sets of the matroid $\mathcal{M}$. Assume, w.l.o.g., that $|X| < |Y|$. Then, by Axiom **(I3)** of the matroid definition, there exists $y \in Y \setminus X$ such that the set $X \cup \{y\}$ is independent. However, as $X_{\min} = Y_{\min}$ is a user with the minimal weight in both X and Y, we have that $w(X_{\min}) \le w(y)$. Consequently, in view of Lemma 7, the set $X \cup \{y\}$ is dependent. This contradiction implies that $|X| = |Y|$. □

It turns out that the lexicographic order on the minterms of Γ, with respect to the relation $\prec$, is strongly related to dependence in $\mathcal{M}$. This is demonstrated through the following definition and lemmas.

Definition 10 (Canonical Complement). *Let P be a prefix of some minterm of Γ. Let $Y \subseteq U$ be the lexicographically minimal set such that: (1) $P_{\max} \prec Y_{\min}$, and (2) The set $P \cup Y$ is a minterm of Γ. Then the set Y is called the* canonical complement *of P.*

The following lemma shows that replacing the canonical complement by a user that precedes the first user of the canonical complement results in a dependent set.

Lemma 9. *Let P be a prefix of some minterm of Γ. Let $Y = \{y_1, \ldots, y_t\}$ be the canonical complement of P, and b be a user such $P_{\max} \prec b \prec y_1$. Then $P \cup \{b\}$ is dependent. Furthermore, the set $P \cup \{b\}$ includes a unique circuit that contains b.*

Proof. If $P = \emptyset$, then, since Γ is connected, there exists a minterm that starts with u_1, whence $y_1 = u_1$. Therefore, it cannot be that $b \prec y_1$ and thus the claim is trivially true. Otherwise, if $P \neq \emptyset$, denote by M_1 the minterm $M_1 = P \cup Y$. Let $X_2 = (M_1 \setminus \{P_{\max}\}) \cup \{b\}$ be the set resulting from replacing $P_{\max}$ with b in M_1. Since $w(P_{\max}) \leq w(b)$, the set X_2 is authorized (though not necessarily a minterm). Let M_2 be the suffix minterm contained in X_2 (such a minterm exists in view of Lemma 6). It must be that $b \in M_2$, since otherwise $M_2 \subseteq Y$, where Y is a proper subset of a minterm and thus is unauthorized.

Let $A = M_1 \cup M_2 = P \cup \{b\} \cup Y$. We proceed to show that every user in Y is critical for A. This will show that $D(M_1, M_2) \subseteq (M_1 \cup M_2) \setminus Y = P \cup \{b\}$. By Corollary 1, the set $D(M_1, M_2)$ is dependent, thus, this will imply that also $P \cup \{b\}$ is dependent. We also observe that it suffices to show that $Y_{\min} = y_1$ is critical for A; this will imply that also all other members of Y, having weight that is no smaller than $w(y_1)$, are also critical for A.

In view of the above, we show that y_1 is critical for A. Suppose this is not the case, namely, the set $A \setminus \{y_1\}$ is authorized. Since $A \setminus \{y_1\}$ results from M_1 by replacing y_1 by b where $w(b) \leq w(y_1)$, and since M_1 is a minterm, it must be that $A \setminus \{y_1\}$ is also a minterm. But this is a contradiction to the choice of y_1 as the first user in the canonical complement of P. Hence, all the elements of Y are critical for A, and, consequently, $P \cup \{b\}$ is dependent. Since P is part of a minterm, it must be that P is independent. Thus, by Lemma 3, the set $P \cup \{b\}$ must contain a unique circuit that contains b. □

The next lemma is a generalization of Lemma 9. Its proof, as well as the other missing proofs in this paper, can be found in the full version of this paper [2].

Lemma 10. *Let P be a prefix of some minterm of Γ. Let $Y = \{y_1, \ldots, y_t\}$ be the canonical complement of P, and $B = \{b_1, \ldots, b_j\}$ be a set such that $P_{\max} \prec B_{\min}$ and $b_j \prec y_j$. Then, the set $P \cup B$ is dependent.*

6 WTASs and HTASs

In this section we discuss the family of hierarchical threshold access structures (HTASs), from Definition 5, and their relation to WTASs. We show that if an ideal WTAS Γ has a minterm in the form of a prefix of U, then Γ is an HTAS.

When discussing a HTAS over some set $U = \{u_1, \ldots, u_n\}$, we shall assume that the users in U are ordered according to their position in the hierarchy, from the lowest level to the highest. Namely, that

$$L_i = U_{\ell_i, \ell_{i+1}-1} = \{u_{\ell_i}, \ldots, u_{\ell_{i+1}-1}\} \quad \forall i \in [m] \tag{1}$$

for some sequence $\ell_1 = 1 < \ell_2 < \cdots < \ell_m < \ell_{m+1} = n+1$. Given a nonempty subset $A \subseteq U$, if $A_{\min} \in L_i$, then A is said to be of level i and it is denoted by $L(A) = i$. Since any HTAS Γ_H is ideal [7,33], Theorem 2 implies that there exists a matroid $\mathcal{M}$ that is associated with it.

Lemma 11. *Let Γ_H and $\mathcal{M}$ be an HTAS and its associated matroid. Then U_{1,k_1+1} is a circuit of $\mathcal{M}$.*

Let U be a set of users and let Γ be a monotone access structure over U that is both a WTAS and a HTAS. Namely, on one hand, there exist a weight function $w : U \to \mathbb{N}$ and a threshold $T \in \mathbb{N}$ such that Γ is the corresponding WTAS, and, on the other hand, there exists a hierarchy in U, where $U = \bigcup_{i=1}^m L_i$, and thresholds $k_1 > k_2 > \cdots > k_m$ such that Γ is also the corresponding HTAS.

Lemma 12. *Let Γ be both a WTAS and an HTAS. Then the HTAS-parameters of Γ satisfy one of the following conditions: (1) $m = 1$. (2) $m = 2$ and $k_1 = k_2 + 1$. (3) $m = 2$ and $|L_1| = k_1 - k_2 + 1$. (4) $m \in \{2, 3\}$, the level L_m is trivial, and the restriction of Γ_H to the first $m-1$ levels is of the form that is described in cases (1)-(3).*

By constructing the appropriate weight function and threshold in each case, it can be shown that any HTAS with parameters as described in Lemma 12 is also a WTAS.

6.1 Ideal WTASs with a Prefix Minterm Are HTASs

In this section we make the first step towards proving Theorem 1. Let Γ be an ideal WTAS over a set U of n users, corresponding to a weight function $w : U \to \mathbb{N}$ and a threshold T. Assume that U possesses a prefix minterm $U_{1,k}$ for some $k \in [n]$ (namely, there exists $k \in [n]$ such that the k users of smallest weights form a minterm). We claim that Γ is an HTAS. We first describe the partition of U into levels and determine the corresponding thresholds. Denoting the resulting HTAS by Γ_H, we proceed to prove that $\Gamma = \Gamma_H$.

The decomposition of U to levels will respect the order of users according to their weights. Namely, each level will be a run of U and our goal is to determine the transition points between one level and the subsequent one. Since $U_{1,k}$ is a minterm, $U_{1,i}$ is authorized for every $i \in \{k, \ldots, n\}$. By Lemma 6, for every such

i there exists a run-minterm ending at u_i. Let us denote the length of that run-minterm by μ_i. By the non-decreasing monotonicity of the weights, we infer that the sequence of lengths $\boldsymbol{\mu} = (\mu_i)_{k \le i \le n}$ is monotonically non-increasing. Denote by m the number of distinct values assumed by the sequence $\boldsymbol{\mu}$, and let us denote those values by $k_1 > \cdots > k_m$. Then the HTAS Γ_H is defined as follows: m is the number of levels and k_i is the ith threshold. As for the levels, we denote by ℓ_i, where $i \in [m]$, the index of the first user in the first run-minterm of length k_i (e.g., $\ell_1 = 1$ since $U_{1,k}$ is the first run-minterm of length $k = k_1$ and its first user is u_1); then the ith level in the hierarchy is $L_i = U_{\ell_i, \ell_{i+1}-1}$, where $\ell_{m+1} = n+1$.

We denote by U_{s_i,t_i} the right-most run-minterm whose length is k_i, where $i \in [m]$, and consider the set $A_i = U_{s_i+1, \ell_{i+1}-1}$. As U_{s_i,t_i} is the last minterm that contains k_i users and U_{ℓ_{i+1}, t_i+1} is the first minterm that contains k_{i+1} users, the set A_i consists of the last $k_i - k_{i+1}$ users in L_i (where $k_{m+1} = 1$). An important observation is that given $i \in [m]$ and $u_h \in A_i$, there is no run-minterm of the WTAS Γ that starts with u_h; indeed, if $U_{h,j}$ was a run-minterm then it would be a proper subset of the minterm U_{s_i,t_i} if $j \le t_i$, or a proper superset of the minterm U_{ℓ_{i+1}, t_i+1} if $j \ge t_i + 1$. An illustration of the construction of the levels of the HTAS appears in Fig. 1. Next, we prove that the WTAS Γ coincides with the HTAS Γ_H described above.

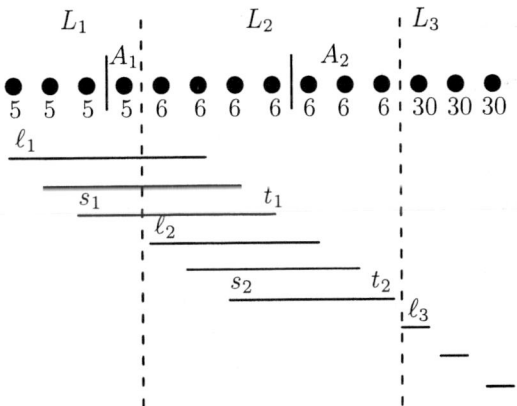

Fig. 1. A WTAS that is also an HTAS. The example is of a WTAS with 14 users of weights $5, 5, 5, 5, 6, 6, 6, 6, 6, 6, 6, 30, 30, 30$ and threshold $T = 30$. The vertical dashed lines indicate the three levels in the corresponding HTAS (the third one being a trivial one) and the horizontal lines indicate all of the run-minterms in that access structure

Theorem 3. *Let Γ be an ideal WTAS over U that has a prefix minterm. Then Γ is an HTAS.*

Proof. Let Γ_H be the HTAS as described above. We will prove that $\Gamma = \Gamma_H$, thus showing that Γ is an HTAS. We start with proving that $\Gamma_H \subseteq \Gamma$. Let $X \in \Gamma_H$. Then for some $i \in [m]$ the set X has at least k_i users from $\bigcup_{j=i}^{m} L_j$.

Letting $B_i = U_{\ell_i, \ell_i + k_i - 1}$ denote the set of the first k_i users from $\bigcup_{j=i}^{m} L_j$, the non-decreasing monotonicity of the weights implies that $w(X) \geq w(B_i)$. By the construction of levels in Γ_H, the set B_i is a minterm of Γ, whence $w(B_i) \geq T$. Therefore, $w(X) \geq T$ and, consequently, $X \in \Gamma$.

Conversely, assume that $X \notin \Gamma_H$. Then X has at most $k_i - 1$ users from $\bigcup_{j=i}^{m} L_j$, for every $i \in [m]$. Consider the set $A = \bigcup_{i=1}^{m} A_i$. By the definition of A, it has exactly $k_i - 1$ users from $\bigcup_{j=i}^{m} L_j$, for every $i \in [m]$. Moreover, A is the set with the maximal weight among the sets that are unauthorized in the HTAS and thus $w(X) \leq w(A)$. Therefore, it suffices to show that $A \notin \Gamma$ in order to conclude that $X \notin \Gamma$ and, thus, complete the proof.

To this end, assume that $A \in \Gamma$. Then A contains some minterm $M \in \Gamma$. Assume that M is of level i, $L(M) = i$, namely, i is the lowest level for which $M \cap L_i \neq \emptyset$. Then $M \cap A_i$ is a prefix of M. In order to arrive at a contradiction, we proceed to show that there can be no minterm that has a prefix which is a non-empty subset of A_i.

Assume, by contradiction, that there are such minterms, and let M' be the lexicographically minimal minterm of that sort. Let $u_h = M'_{\min}$ and let j be the maximal index such that $M' = U_{h,j} \cup Z$ for some $Z \subset U$. Since $u_h \in A_i$ and we observed earlier that no run-minterm starts in A_i, we conclude that $Z \neq \emptyset$ (and $u_{j+1} \prec Z_{\min}$ because of the maximality of j). We claim that $j < t_i$. Indeed, if $j \geq t_i$, then M' is a proper superset of $\hat{M} = U_{\ell_{i+1}, t_i} \cup \{Z_{\min}\}$. Since $w(\hat{M}) \geq w(U_{\ell_{i+1}, t_i + 1}) \geq T$, we get a contradiction since a minterm M' cannot be a proper superset of an authorized set $\hat{M}$.

Next, define $Q = M' \cup \{u_{j+1}\} \setminus \{u_j\}$. The set Q is authorized and, by Lemma 6, it contains a suffix minterm M'' that must contain u_{j+1}, for otherwise it would be a proper subset of M'. Therefore, $M' \cup M'' = M' \cup \{u_{j+1}\} = U_{h,j} \cup \{u_{j+1}\} \cup Z$. We claim that all members of Z are critical for this union. Assume, by contradiction, that $M^* = M' \cup \{u_{j+1}\} \setminus \{z\}$ is authorized, for some $z \in Z$. Since $w(u_{j+1}) \leq w(z)$ and M' was a minterm, also M^* is a minterm. But M^* is a minterm that starts within A_i and $M^* \prec M'$, thus contradicting our choice of M'. Hence, by Corollary 1, the set $U_{h,j+1}$ is dependent in $\mathcal{M}$. However, since $s_i < h$ and $j + 1 \leq t_i$, this dependent set is properly contained in the minterm U_{s_i, t_i}, leading to a contradiction. Hence, $A \notin \Gamma$. □

7 Ideal WTASs and TPASs

In the previous section we dealt with the case where the lexicographically minimal minterm of Γ is a prefix of U. Here, we handle the case where the lexicographically minimal minterm of Γ is a lacunary prefix, namely, it takes the form $M = U_{1,d} \cup U_{d+2,k}$ for some $1 \leq d \leq k-2$ and $k \leq n$. We assume that there is at least one minterm starting with the user u_2, and that there are no self-sufficient users. If this is not the case, then Γ is a simple composition of access structures as shown in Lemma 14. We show that under these conditions, Γ is a tripartite access structure, as defined in Definition 6. The idea of the proof is as follows:

first we show that $U_{2,k}$ must be a minterm of Γ. Thus, the restriction of Γ to $U_{2,n}$ has a prefix minterm, and consequently, by Theorem 3, it is an HTAS. This fact enables us to deduce that Γ is a TPAS.

Theorem 4. *Let Γ be an ideal WTAS with $M = U_{1,d} \cup U_{d+2,k}$ being its lexicographical minimal minterm for some $1 \leq d \leq k-2$ and $k \leq n$. If there exists a minterm in Γ that has u_2 as its minimal member and Γ has no self-sufficient users, then Γ is a TPAS.*

8 A Recursive Characterization of Ideal WTASs by Means of Composition

8.1 WTASs and Composition of Access Structures

We begin with the following lemma that asserts that a composition of two access structures is ideal if and only if those two access structures are ideal.

Lemma 13. *Let U_1 and U_2 be disjoint sets. Let $u_1 \in U_1$, and define $U = U_1 \cup U_2 \setminus \{u_1\}$. Suppose Γ_1 and Γ_2 are access structures over U_1 and U_2 respectively such that u_1 is not redundant in Γ_1 and $\Gamma_2 \neq \emptyset$. Furthermore, let Γ be the composition of Γ_1 and Γ_2 via u_1. Then Γ is ideal if and only if both Γ_1 and Γ_2 are ideal. Moreover, if both Γ_1 and Γ_2 have an ideal linear secret sharing schemes, then Γ has an ideal linear secret sharing scheme.*

The recursive characterization of ideal WTASs will be obtained by distinguishing between two types of users. Specifically, we shall identify a subset of so-called strong users that takes the form of a suffix, $S = U_{k,n}$, where $k \geq 3$, and then the complement subset will be thought of as the subset of weak users, $W = U_{1,k-1}$. A subset of strong users will be called S-cooperative if it is unauthorized, but it may become authorized if we add to it some weak users.

Definition 11 (Cooperative Set). *Given $Y \subseteq S$, if $Y \notin \Gamma$ but $W \cup Y \in \Gamma$, then Y is called an S-cooperative set.*

By Lemma 5, the access structure $\Gamma_{Y,W}$, the restriction of Γ induced by Y on W, is a WTAS for any partition $U = W \cup S$ and $Y \subseteq S$. We proceed to define a condition on the set S, such that if it is satisfied for some suffix $S = U_{k,n}$, where $k \geq 3$, the access structure Γ is a composition of two ideal WTASs that are defined on sets smaller than U.

Definition 12 (Strong Set). *If for any two S-cooperative sets $Y_1, Y_2 \subseteq S$, the corresponding restrictions of Γ to W coincide, i.e. $\Gamma_{Y_1,W} = \Gamma_{Y_2,W}$, the set S is called a strong set of users.*

If S is a strong set of users, there exists an access structure on W, denoted Γ_W, such that $\Gamma_W = \Gamma_{Y,W}$ for all cooperative subsets $Y \subset S$. In that case, every minterm $M \in \Gamma$ is either contained in S or $M \cap W \in \Gamma_W$. The following theorem shows that if S is a strong set of users, Γ is a composition of two ideal WTASs.

Theorem 5. *Let Γ be an ideal WTAS over U. Suppose $S = U_{k,n}$, for some $k \geq 3$, is a strong set of users. Then Γ is a composition of two ideal WTASs, where each access structure is defined on a set smaller than U.*

Simple Compositions. We identify two simple cases where an ideal WTAS is a composition of two ideal WTASs defined on sets smaller than U. If Γ has no minterm that starts with u_2 then every minterm that contains u_1 must contain also u_2 (otherwise, we could have replaced u_1 by u_2 in order to get a minterm that starts with u_2). Hence, for every $U_{3,n}$-cooperative set, $Y \subseteq U_{3,n}$, the access structure that Y induces on $U_{1,2}$ is the same, $\Gamma_{Y,U_{1,2}} = \{U_{1,2}\}$. Therefore, $U_{3,n}$ is a strong set of users in this case. Hence, by Theorem 5, the access structure Γ is a composition of two ideal WTASs that are defined on sets smaller than U. If u_n is a self-sufficient user, it can be shown easily that Γ is a composition of two ideal access structures, defined over smaller sets of users. To conclude, we get the following lemma:

Lemma 14. *Let Γ be an ideal WTAS over U. If Γ has self-sufficient users, or u_2 starts no minterm of Γ, then Γ is a composition of two ideal WTASs that are defined on sets smaller than U.*

8.2 Identifying Composition Structures

In this section we show that if Γ is an ideal WTAS, but it is not one of the access structures that were characterized in Sections 6.1 and 7, then it is a composition of two ideal WTASs as described in Section 8.1. In view of Lemma 14, we assume hereinafter that u_2 is the minimal user in some minterm of Γ.

Let M_1 be the lexicographically minimal minterm in Γ. Let u_r be the maximal user in M_1. Then since Γ is neither an HTAS nor a TPAS, there must be at least two users in $U_{1,r-1}$ that are not in M_1. Let u_ℓ be the *minimal* user in M_1 such that at least two users in $U_{1,\ell-1}$ are missing from M_1, and let u_d be the *maximal* user in M_1 such that $U_{1,d} \subset M_1$. We denote the users in $M_1 \cap U_{d+1,\ell-1}$, if there are any, by $Y = \{y_1, \ldots, y_t\}$. Note that if Y is not empty then Y is a run of U, and $y_1 = u_{d+2}$. Next, if $Y \neq \emptyset$ we denote the set of users of $U \setminus M_1$ between y_t and u_ℓ (excluding those two users) by $X = \{x_1, \ldots, x_s\}$. Otherwise, we denote the set $U_{d+2,\ell-1}$ as $X = \{x_1, \ldots, x_s\}$. Finally, we denote the users of $M_1 \cap U_{\ell,n}$ by $Z = \{z_1, \ldots, z_m\}$. Note that the sets X and Z are never empty, and that $z_1 = u_\ell$. The above notations are depicted in Fig. 2.

We claim that either $U_{\ell,n}$ or $U_{d+2,n}$ is a strong set of users. We start by partitioning U into $W = U_{1,\ell-1}$ and $S = U_{\ell,n}$. We show that if all the S-cooperative sets are of the same size, then $\Gamma_{Y_1,W} = \Gamma_{Y_2,W}$ for every two cooperative sets $Y_1, Y_2 \subseteq S$, namely, S is a set of strong users. If, however, that condition does not hold, we shall show that $U_{d+2,n}$ is a strong set.

Lemma 15. *Every minterm of the access structure Γ that intersects W contains at least m users from S.*

Proof. Assume towards contradiction that M is a minterm that intersects W such that $|M \cap S| < m$. The minterm M_1 is an independent set of size $d+t+m$,

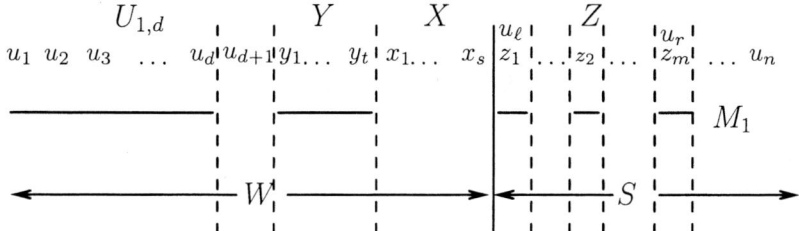

Fig. 2. Notations for the composition

and thus, by Axiom **(I3)**, every independent set of $\mathcal{M}$ that is smaller than $d+t+m$ can be expanded to an independent set of size $d+t+m$. Therefore, if $|M| < d+t+m$, the minterm M can be expanded to an independent set I of size $d+t+m$; otherwise, we set $I = M$. By Lemma 7, this expansion can only be done by adding to M users that precede $M_{\min}$. As M intersects W, the users in $I\setminus M$ are all from W. Hence, $|I \cap S| = |M \cap S| \leq m-1$. Therefore, $|I \cap W| = |I| - |I \cap S| \geq d+t+m-(m-1) = d+t+1$. Next, we view M_1 as the canonical complement of the empty set (see Definition 10). Its $(d+t+1)$th element is z_1. By Lemma 10 for $P = \emptyset$, $Y = M_1$, and $j = d+t+1$, any $d+t+1$ members of W form a dependent set. Hence I, which was assumed to be independent, contains a dependent set, a contradiction. □

When All S-Cooperative Sets are of the Same Size. Here we show that if all S-cooperative sets are of the same size, namely $|Z| = m$, the set S is a strong set. We accomplish this by showing that all the S-cooperative sets of size m induce the same access structure on W, which is the access structure induced by the S-cooperative set Z.

Lemma 16. *Let $V \subseteq S$ be an S-cooperative set. Then $w(V) \geq w(Z)$.*

Proof. Assume towards contradiction that $w(V) < w(Z)$ and consider the set $W \cup Z$. Since Z is S-cooperative, the set $W \cup Z$ is authorized. Thus, by Lemma 6, it must contain a suffix minterm of the form $B \cup Z$, where B is a suffix of W. There are two possible cases: either $B \cup V$ is authorized, or not.

If $B \cup V$ is authorized, then, since $w(V) < w(Z)$, the set $B \cup V$ is a minterm. Hence, as $B \cup V$ and $B \cup Z$ are two minterms that have the same minimal user, $B_{\min}$, Lemma 8 implies that $|V| = |Z| = m$. The set $B \cup V$ is independent in $\mathcal{M}$. If $|B \cup V| < d+t+m$, Axiom **(I3)** implies that $B \cup V$ can be expanded to an independent set I of size $d+t+m$; if $|B \cup V| \geq d+t+m$, we set $I = B \cup V$. By Lemma 7, all users in $I \setminus (B \cup V)$ must be from W. Hence, I includes at least $d+t$ users from W. On the other hand, since $|V| = |Z| = m$ and $w(V) < w(Z)$, there must be an index j such that $v_j \prec z_j$. Since M_1 is the canonical complement of the empty set $\emptyset$, we get from Lemma 10, applied to $P = \emptyset$, $Y = M_1$ and $B = I_{1,d+t+j}$, that the latter set is dependent. This is impossible since I is independent. Therefore, $B \cup V$ cannot be authorized.

If $B \cup V$ is unauthorized, we let Q be the canonical complement of B. Using Lemma 8, Lemma 15, and Lemma 10, we get that $B \cup V$ is dependent. On the other hand, since V is S-cooperative and B is a suffix of W, the set $B \cup V$ may be expanded to an authorized superset by adding to it users that precede $B_{\min}$, one by one, until the first time that we get an authorized set. This construction, where in each stage we add a new user that is smaller than all current users in the set, guarantees that we end up with a minterm. But a minterm of Γ cannot contain a dependent set. Therefore, this case is not possible either. We conclude that $w(V) \geq w(Z)$. □

Lemma 17. *Let V be an S-cooperative set of size m. Then $\Gamma_{V,W} = \Gamma_{Z,W}$.*

Proof. By Lemma 16, $w(V) \geq w(Z)$. If $w(V) = w(Z)$, the claim is trivial, since Γ is a WTAS. Therefore, we assume that $w(V) > w(Z)$. We first show that $U_{1,d} \cup Y \cup V$ is a minterm of Γ. Since $M_1 = U_{1,d} \cup Y \cup Z$ is authorized, the set $U_{1,d} \cup Y \cup V$ is authorized as well. Assume it is not a minterm. Then, by Lemma 6 it contains a suffix minterm of the form $B \cup V$, where B is a suffix of $U_{2,d} \cup Y$. Let Q be the canonical complement of B. Using Lemma 8, Lemma 15, and Lemma 10, we get that the set $B \cup Z$ is dependent. However, this set is contained in $U_{1,d} \cup Y \cup Z$, which is a minterm. This contradiction implies that $U_{1,d} \cup Y \cup V$ is a minterm of Γ. Consequently, since $U_{1,d} \cup Y \cup Z$ is a minterm and $U_{2,d} \cup Y \cup V$ is unauthorized (being a proper subset of a minterm), we infer that $w(Z) + w(u_1) > w(V)$.

We are now ready to prove that $\Gamma_{Z,W} = \Gamma_{V,W}$. Since we deal with the case where $w(Z) < w(V)$, the inclusion $\Gamma_{Z,W} \subseteq \Gamma_{V,W}$ is obvious. For the opposite inclusion, it is sufficient to concentrate on minterms of $\Gamma_{V,W}$. Let M be a minterm of $\Gamma_{V,W}$. Thus, $M \cup V \in \Gamma$, and since $M \cup V \setminus M_{\min} \notin \Gamma$, the set $M \cup V$ is a minterm in Γ. There are two possible cases: If $u_1 \in M$, the minterm $M \cup V$ must be of the same size as M_1 by Lemma 8. Since $|M_1| = d+t+m$ and $|V| = m$, we get that $|M| = d+t$. As $M_1 = U_{1,d} \cup Y \cup Z$ is the minimal minterm in Γ in terms of the precedence order $\prec$, the weight of $U_{1,d} \cup Y$ is minimal among all sets of size $d+t$ that are contained in a minterm. This implies that $w(M) \geq w(U_{1,d} \cup Y)$. This, in turn, implies that $M \cup Z \in \Gamma$ and thus $M \in \Gamma_{Z,W}$.

The second case is when $u_1 \notin M$. Assume, towards contradiction, that $M \cup Z \notin \Gamma$. Let Q be the canonical complement of M. Using Lemma 8, Lemma 15, and Lemma 10, we get that the set $M \cup Z$ is dependent. However, since $M \cup V \in \Gamma$ and $w(Z) + w(u_1) > w(V)$, we get that $\{u_1\} \cup M \cup Z \in \Gamma$. Moreover, it must be a minterm since any proper subset of $\{u_1\} \cup M \cup Z$ is of weight that does not exceed that of the unauthorized set $M \cup Z$. Hence, the dependent set $M \cup Z$ is contained in a minterm. This contradiction implies that $M \cup Z$ is authorized, and thus $M \in \Gamma_{Z,W}$. □

Corollary 2. *If there are no S-cooperative sets of size larger than m, then S is a strong set.*

Example 2. Consider the set $U = \{u_1, \ldots, u_8\}$, and let Γ be a WTAS where the weights are 1, 1, 1, 1, 1, 3, 3, 3 and the threshold is 6. The lexicographically minimal minterm is $\{u_1, u_2, u_3, u_6\}$, and so there is no prefix minterm and no lacunary minterm. In this example $W = U_{1,5}$ and $S = U_{6,8}$ and the access structure is a composition of a 3-out-of-5 threshold access structure on the week side W and a 2-out-of-4 threshold access structure on $S \cup \{u'\}$, where u' is an additional dummy user.

When Large S-Cooperative Sets Exist. The conclusion from Corollary 2 is that whenever all S-cooperative sets for $S = U_{\ell,n}$ are of the same size (i.e., $|Z| = m$), the set S is a strong set and, hence, by Theorem 5, the access structure Γ is a composition of ideal WTASs that are defined over two smaller sets. Here, we continue to deal with the case where there are S-cooperative sets of size larger than m. In that case we identify another strong set. Specifically, we show that $U_{d+2,n}$ is a strong set of users. The analysis of the structure of Γ when large S-cooperative sets exist is technically involved, and is omitted due to lack of space. It appears in the full version of the paper [2].

The following lemma summarizes the results in this case.

Lemma 18. *Suppose there is an S-cooperative set of size larger than m, and there is a minterm of Γ that starts with u_2. Then $U_{d+2,n}$ is a strong set of users.*

8.3 Proof of Theorem 1 – The Characterization Theorem

Let Γ be an ideal WTAS defined on a set of users U and let M_1 be its lexicographically minimal minterm. If either Γ has self-sufficient users or u_2 starts no minterm of Γ, then, by Lemma 14, the access structure Γ is a composition of two ideal WTASs on smaller sets of users.

If M_1 is a prefix then, by Theorem 3, the access structure Γ is an HTAS. If M_1 is a lacunary prefix, namely, $M_1 = U_{1,d} \cup U_{d+2,k}$ for some $1 \leq d \leq k-2$ and $k \leq n$, then, by Theorem 4, the access structure Γ is a TPAS. Otherwise, by Corollary 2 and Lemma 18, there exists within U a subset of strong users, and, by Theorem 5, the access structure Γ is a composition of two ideal WTASs that are defined on sets smaller than U.

As for the other direction, HTASs are ideal and may be realized by linear secret sharing schemes, as shown in [7, 33]. TPASs are also ideal and may be realized by linear secret sharing schemes, as shown in the full version of this paper [2]. Finally, given two ideal access structures, we showed in Lemma 13 how to construct an ideal secret sharing scheme for their composition. Hence, the composition is also ideal. Furthermore, by Lemma 13, if the secret sharing schemes for the two basic access structures are linear, so is the resulting scheme for the composition of the two access structures. This completes the proof of the characterization theorem. □

References

1. A. Beimel and B. Chor. Universally ideal secret sharing schemes. *IEEE Trans. on Information Theory*, 40(3):786–794, 1994.

2. A. Beimel, T. Tassa, and E. Weinreb. Characterizing ideal weighted threshold secret sharing. Technical Report 04-05, Dept. of Computer Science, Ben-Gurion University, 2004. Available at: www.cs.bgu.ac.il/~beimel/pub.html.
3. A. Beimel and E. Weinreb. Monotone circuits for weighted threshold functions, 2004. In preparation.
4. M. Ben-Or, S. Goldwasser, and A. Wigderson. Completeness theorems for non-cryptographic fault-tolerant distributed computations. In *20th STOC*, 1–10, 1988.
5. J. Benaloh and J. Leichter. Generalized secret sharing and monotone functions. In *CRYPTO '88*, volume 403 of *LNCS*, pages 27–35. 1990.
6. G. R. Blakley. Safeguarding cryptographic keys. *Proc. of the 1979 AFIPS National Computer Conference*, pages 313–317. 1979.
7. E. F. Brickell. Some ideal secret sharing schemes. *Journal of Combin. Math. and Combin. Comput.*, 6:105–113, 1989.
8. E. F. Brickell and D. M. Davenport. On the classification of ideal secret sharing schemes. *J. of Cryptology*, 4(73):123–134, 1991.
9. E. F. Brickell and D. R. Stinson. Some improved bounds on the information rate of perfect secret sharing schemes. *J. of Cryptology*, 5(3):153–166, 1992.
10. R. M. Capocelli, A. De Santis, L. Gargano, and U. Vaccaro. On the size of shares for secret sharing schemes. *J. of Cryptology*, 6(3):157–168, 1993.
11. D. Chaum, C. Crépeau, and I. Damgård. Multiparty unconditionally secure protocols. In *Proc. of the 20th ACM STOC*, pages 11–19, 1988.
12. R. Cramer, I. Damgård, and U. Maurer. General secure multi-party computation from any linear secret-sharing scheme. In *EUROCRYPT 2000*, volume 1807 of *LNCS*, pages 316–334. 2000.
13. Y. Desmedt and Y. Frankel. Shared generation of authenticators and signatures. In *CRYPTO '91*, volume 576 of *LNCS*, pages 457–469. 1992.
14. M. Ito, A. Saito, and T. Nishizeki. Secret sharing schemes realizing general access structure. In *Proc. of Globecom 87*, pages 99–102, 1987.
15. W. Jackson, K. M. Martin, and C. M. O'Keefe. Ideal secret sharing schemes with multiple secrets. *J. of Cryptology*, 9(4):233–250, 1996.
16. M. Karchmer and A. Wigderson. On span programs. In *Proc. of the 8th IEEE Structure in Complexity Theory*, pages 102–111, 1993.
17. E. D. Karnin, J. W. Greene, and M. E. Hellman. On secret sharing systems. *IEEE Trans. on Information Theory*, 29(1):35–41, 1983.
18. J. Martí-Farré and C. Padró. Secret sharing schemes on access structures with intersection number equal to one. *3rd SCN*, vol. 2576 of *LNCS*, pp. 354–363. 2002.
19. K. M. Martin. *Discrete Structures in the Theory of Secret Sharing*. PhD thesis, University of London, 1991.
20. K. M. Martin. New secret sharing schemes from old. *J. Combin. Math. Combin. Comput.*, 14:65–77, 1993.
21. P. Morillo, C. Padró, G. Sáez, and J. L. Villa. Weighted threshold secret sharing schemes. *Inform. Process. Lett.*, 70(5):211–216, 1999.
22. J. G. Oxley. *Matroid Theory*. Oxford University Press, 1992.
23. C. Padró and G. Sáez. Secret sharing schemes with bipartite access structure. *IEEE Trans. on Information Theory*, 46:2596–2605, 2000.
24. M. O. Rabin. Randomized Byzantine generals. In *Proc. of the 24th IEEE Symp. on Foundations of Computer Science*, pages 403–409, 1983.
25. P. D. Seymour. On secret-sharing matroids. *J. of Combinatorial Theory, Series B*, 56:69–73, 1992.
26. A. Shamir. How to share a secret. *Communications of the ACM*, 22:612–613, 1979.

27. G. J. Simmons. How to (really) share a secret. In *CRYPTO '88*, volume 403 of *LNCS*, pages 390–448. 1990.
28. G. J. Simmons. An introduction to shared secret and/or shared control and their application. In *Contemporary Cryptology, The Science of Information Integrity*, pages 441–497. IEEE Press, 1992.
29. G. J. Simmons, W. Jackson, and K. M. Martin. The geometry of shared secret schemes. *Bulletin of the ICA*, 1:71–88, 1991.
30. J. Simonis and A. Ashikhmin. Almost affine codes. *Designs, Codes and Cryptography*, 14(2):179–197, 1998.
31. D. R. Stinson. An explication of secret sharing schemes. *Designs, Codes and Cryptography*, 2:357–390, 1992.
32. D. R. Stinson. New general lower bounds on the information rate of secret sharing schemes. In *CRYPTO '92*, volume 740 of *LNCS*, pages 168–182. 1993.
33. T. Tassa. Hierarchical threshold secret sharing. In M. Naor, editor, *First Theory of Cryptography Conference, TCC 2004*, volume 2951 of *LNCS*, pages 473–490. 2004.

Author Index

Backes, Michael 210
Beimel, Amos 600
Ben-Or, Michael 386
Boneh, Dan 325

Cachin, Christian 210
Canetti, Ran 17, 150
Chawla, Shuchi 363
Cramer, Ronald 342

Damgård, Ivan 342
Datta, Anupam 476
Dedić, Nenad 227
Ding, Yan Zong 578
Dodis, Yevgeniy 188, 556
Dwork, Cynthia 363

Freedman, Michael J. 303

Goh, Eu-Jin 325
Goldwasser, Shafi 529
González Vasco, María Isabel 495
Groth, Jens 50

Halevi, Shai 17, 150
Hofheinz, Dennis 86
Hohenberger, Susan 264
Horodecki, Michał 386

Ishai, Yuval 303, 342, 445
Itkis, Gene 227

Jarecki, Stanisław 510

Katz, Jonathan 128, 150, 188
Kharchenko, Dmitriy 529
Kiltz, Eike 283
König, Robert 407
Küsters, Ralf 476
Kushilevitz, Eyal 445

Leander, Gregor 283
Lepinski, Matt 245
Leung, Debbie W. 386
Lin, Henry 34
Lindell, Yehuda 128
Lysyanskaya, Anna 264

Malone-Lee, John 283
Manabe, Yoshifumi 426

Martínez, Consuelo 495
Mayers, Dominic 386
McSherry, Frank 363
Micali, Silvio 1, 245
Micciancio, Daniele 169
Mitchell, John C. 476

Nagao, Waka 426
Naor, Moni 66
Nguyen, Minh-Huyen 457
Nissim, Kobbi 325
Nussboim, Asaf 66

Okamoto, Tatsuaki 426
Oppenheim, Jonathan 386
Ostrovsky, Rafail 445

Panjwani, Saurabh 169
Peikert, Chris 1
Pinkas, Benny 303
Prabhakaran, Manoj 104

Ramanathan, Ajith 476
Reingold, Omer 303
Renner, Renato 407
Reyzin, Leonid 227
Russell, Scott 227

Sahai, Amit 104
Saxena, Nitesh 510
Shelat, Abhi 245
Smith, Adam 363, 556
Steiner, Michael 17
Steinwandt, Rainer 495
Sudan, Madhu 1

Tassa, Tamir 600
Trevisan, Luca 34
Tromer, Eran 66

Unruh, Dominique 86

Villar, Jorge L. 495

Wee, Hoeteck 34, 363
Weinreb, Enav 600
Wilson A. David 1

Lecture Notes in Computer Science

For information about Vols. 1–3290

please contact your bookseller or Springer

Vol. 3412: X. Franch, D. Port (Eds.), COTS-Based Software Systems. XVI, 312 pages. 2005.

Vol. 3406: A. Gelbukh (Ed.), Computational Linguistics and Intelligent Text Processing. XVII, 829 pages. 2005.

Vol. 3403: B. Ganter, R. Godin (Eds.), Formal Concept Analysis. XI, 419 pages. 2005. (Subseries LNAI).

Vol. 3398: D.-K. Baik (Ed.), Systems Modeling and Simulation: Theory and Applications. XIV, 733 pages. 2005. (Subseries LNAI).

Vol. 3397: T.G. Kim (Ed.), Artificial Intelligence and Simulation. XV, 711 pages. 2005. (Subseries LNAI).

Vol. 3391: C. Kim (Ed.), Information Networking. XVII, 936 pages. 2005.

Vol. 3388: J. Lagergren (Ed.), Comparative Genomics. VIII, 133 pages. 2005. (Subseries LNBI).

Vol. 3387: J. Cardoso, A. Sheth (Eds.), Semantic Web Services and Web Process Composition. VIII, 148 pages. 2005.

Vol. 3386: S. Vaudenay (Ed.), Public Key Cryptography - PKC 2005. IX, 436 pages. 2005.

Vol. 3385: R. Cousot (Ed.), Verification, Model Checking, and Abstract Interpretation. XII, 483 pages. 2005.

Vol. 3382: J. Odell, P. Giorgini, J.P. Müller (Eds.), Agent-Oriented Software Engineering V. X, 239 pages. 2004.

Vol. 3381: P. Vojtáš, M. Bieliková, B. Charron-Bost, O. Sýkora (Eds.), SOFSEM 2005: Theory and Practice of Computer Science. XV, 448 pages. 2005.

Vol. 3379: M. Hemmje, C. Niederee, T. Risse (Eds.), From Integrated Publication and Information Systems to Information and Knowledge Environments. XXIII, 321 pages. 2005.

Vol. 3378: J. Kilian (Ed.), Theory of Cryptography. XII, 621 pages. 2005.

Vol. 3376: A. Menezes (Ed.), Topics in Cryptology – CT-RSA 2005. X, 385 pages. 2004.

Vol. 3375: M.A. Marsan, G. Bianchi, M. Listanti, M. Meo (Eds.), Quality of Service in Multiservice IP Networks. XIII, 656 pages. 2005.

Vol. 3368: L. Paletta, J.K. Tsotsos, E. Rome, G. Humphreys (Eds.), Attention and Performance in Computational Vision. VIII, 231 pages. 2005.

Vol. 3366: I. Rahwan, P. Moraitis, C. Reed (Eds.), Argumentation in Multi-Agent Systems. XII, 263 pages. 2005. (Subseries LNAI).

Vol. 3363: T. Eiter, L. Libkin (Eds.), Database Theory - ICDT 2005. XI, 413 pages. 2004.

Vol. 3362: G. Barthe, L. Burdy, M. Huisman, J.-L. Lanet, T. Muntean (Eds.), Construction and Analysis of Safe, Secure, and Interoperable Smart Devices. IX, 257 pages. 2005.

Vol. 3361: S. Bengio, H. Bourlard (Eds.), Machine Learning for Multimodal Interaction. XII, 362 pages. 2005.

Vol. 3360: S. Spaccapietra, E. Bertino, S. Jajodia, R. King, D. McLeod, M.E. Orlowska, L. Strous (Eds.), Journal on Data Semantics II. XI, 223 pages. 2004.

Vol. 3359: G. Grieser, Y. Tanaka (Eds.), Intuitive Human Interfaces for Organizing and Accessing Intellectual Assets. XIV, 257 pages. 2005. (Subseries LNAI).

Vol. 3358: J. Cao, L.T. Yang, M. Guo, F. Lau (Eds.), Parallel and Distributed Processing and Applications. XXIV, 1058 pages. 2004.

Vol. 3357: H. Handschuh, M.A. Hasan (Eds.), Selected Areas in Cryptography. XI, 354 pages. 2004.

Vol. 3356: G. Das, V.P. Gulati (Eds.), Intelligent Information Technology. XII, 428 pages. 2004.

Vol. 3355: R. Murray-Smith, R. Shorten (Eds.), Switching and Learning in Feedback Systems. X, 343 pages. 2005.

Vol. 3353: J. Hromkovič, M. Nagl, B. Westfechtel (Eds.), Graph-Theoretic Concepts in Computer Science. XI, 404 pages. 2004.

Vol. 3352: C. Blundo, S. Cimato (Eds.), Security in Communication Networks. XI, 381 pages. 2004.

Vol. 3350: M. Hermenegildo, D. Cabeza (Eds.), Practical Aspects of Declarative Languages. VIII, 269 pages. 2005.

Vol. 3349: B.M. Chapman (Ed.), Shared Memory Parallel Programming with Open MP. X, 149 pages. 2005.

Vol. 3348: A. Canteaut, K. Viswanathan (Eds.), Progress in Cryptology - INDOCRYPT 2004. XIV, 431 pages. 2004.

Vol. 3347: R.K. Ghosh, H. Mohanty (Eds.), Distributed Computing and Internet Technology. XX, 472 pages. 2004.

Vol. 3346: R.H. Bordini, M. Dastani, J. Dix, A.E.F. Seghrouchni (Eds.), Programming Multi-Agent Systems. XIV, 249 pages. 2005. (Subseries LNAI).

Vol. 3345: Y. Cai (Ed.), Ambient Intelligence for Scientific Discovery. XII, 311 pages. 2005. (Subseries LNAI).

Vol. 3344: J. Malenfant, B.M. Østvold (Eds.), Object-Oriented Technology. ECOOP 2004 Workshop Reader. VIII, 215 pages. 2004.

Vol. 3342: E. Şahin, W.M. Spears (Eds.), Swarm Robotics. IX, 175 pages. 2004.

Vol. 3341: R. Fleischer, G. Trippen (Eds.), Algorithms and Computation. XVII, 935 pages. 2004.

Vol. 3340: C.S. Calude, E. Calude, M.J. Dinneen (Eds.), Developments in Language Theory. XI, 431 pages. 2004.

Vol. 3339: G.I. Webb, X. Yu (Eds.), AI 2004: Advances in Artificial Intelligence. XXII, 1272 pages. 2004. (Subseries LNAI).

Vol. 3338: S.Z. Li, J. Lai, T. Tan, G. Feng, Y. Wang (Eds.), Advances in Biometric Person Authentication. XVIII, 699 pages. 2004.

Vol. 3337: J.M. Barreiro, F. Martin-Sanchez, V. Maojo, F. Sanz (Eds.), Biological and Medical Data Analysis. XI, 508 pages. 2004.

Vol. 3336: D. Karagiannis, U. Reimer (Eds.), Practical Aspects of Knowledge Management. X, 523 pages. 2004. (Subseries LNAI).

Vol. 3335: M. Malek, M. Reitenspieß, J. Kaiser (Eds.), Service Availability. X, 213 pages. 2005.

Vol. 3334: Z. Chen, H. Chen, Q. Miao, Y. Fu, E. Fox, E.-p. Lim (Eds.), Digital Libraries: International Collaboration and Cross-Fertilization. XX, 690 pages. 2004.

Vol. 3333: K. Aizawa, Y. Nakamura, S. Satoh (Eds.), Advances in Multimedia Information Processing - PCM 2004, Part III. XXXV, 785 pages. 2004.

Vol. 3332: K. Aizawa, Y. Nakamura, S. Satoh (Eds.), Advances in Multimedia Information Processing - PCM 2004, Part II. XXXVI, 1051 pages. 2004.

Vol. 3331: K. Aizawa, Y. Nakamura, S. Satoh (Eds.), Advances in Multimedia Information Processing - PCM 2004, Part I. XXXVI, 667 pages. 2004.

Vol. 3330: J. Akiyama, E.T. Baskoro, M. Kano (Eds.), Combinatorial Geometry and Graph Theory. VIII, 227 pages. 2005.

Vol. 3329: P.J. Lee (Ed.), Advances in Cryptology - ASIACRYPT 2004. XVI, 546 pages. 2004.

Vol. 3328: K. Lodaya, M. Mahajan (Eds.), FSTTCS 2004: Foundations of Software Technology and Theoretical Computer Science. XVI, 532 pages. 2004.

Vol. 3327: Y. Shi, W. Xu, Z. Chen (Eds.), Data Mining and Knowledge Management. XIII, 263 pages. 2004. (Subseries LNAI).

Vol. 3326: A. Sen, N. Das, S.K. Das, B.P. Sinha (Eds.), Distributed Computing - IWDC 2004. XIX, 546 pages. 2004.

Vol. 3323: G. Antoniou, H. Boley (Eds.), Rules and Rule Markup Languages for the Semantic Web. X, 215 pages. 2004.

Vol. 3322: R. Klette, J. Žunić (Eds.), Combinatorial Image Analysis. XII, 760 pages. 2004.

Vol. 3321: M.J. Maher (Ed.), Advances in Computer Science - ASIAN 2004. XII, 510 pages. 2004.

Vol. 3320: K.-M. Liew, H. Shen, S. See, W. Cai (Eds.), Parallel and Distributed Computing: Applications and Technologies. XXIV, 891 pages. 2004.

Vol. 3319: D. Amyot, A.W. Williams (Eds.), Telecommunications and beyond: Modeling and Analysis of Reactive, Distributed, and Real-Time Systems. XII, 301 pages. 2005.

Vol. 3318: E. Eskin, C. Workman (Eds.), Regulatory Genomics. VIII, 115 pages. 2005. (Subseries LNBI).

Vol. 3317: M. Domaratzki, A. Okhotin, K. Salomaa, S. Yu (Eds.), Implementation and Application of Automata. XII, 336 pages. 2005.

Vol. 3316: N.R. Pal, N.K. Kasabov, R.K. Mudi, S. Pal, S.K. Parui (Eds.), Neural Information Processing. XXX, 1368 pages. 2004.

Vol. 3315: C. Lemaître, C.A. Reyes, J.A. González (Eds.), Advances in Artificial Intelligence – IBERAMIA 2004. XX, 987 pages. 2004. (Subseries LNAI).

Vol. 3314: J. Zhang, J.-H. He, Y. Fu (Eds.), Computational and Information Science. XXIV, 1259 pages. 2004.

Vol. 3313: C. Castelluccia, H. Hartenstein, C. Paar, D. Westhoff (Eds.), Security in Ad-hoc and Sensor Networks. VIII, 231 pages. 2004.

Vol. 3312: A.J. Hu, A.K. Martin (Eds.), Formal Methods in Computer-Aided Design. XI, 445 pages. 2004.

Vol. 3311: V. Roca, F. Rousseau (Eds.), Interactive Multimedia and Next Generation Networks. XIII, 287 pages. 2004.

Vol. 3310: U.K. Wiil (Ed.), Computer Music Modeling and Retrieval. XI, 371 pages. 2005.

Vol. 3309: C.-H. Chi, K.-Y. Lam (Eds.), Content Computing. XII, 510 pages. 2004.

Vol. 3308: J. Davies, W. Schulte, M. Barnett (Eds.), Formal Methods and Software Engineering. XIII, 500 pages. 2004.

Vol. 3307: C. Bussler, S.-k. Hong, W. Jun, R. Kaschek, D.. Kinshuk, S. Krishnaswamy, S.W. Loke, D. Oberle, D. Richards, A. Sharma, Y. Sure, B. Thalheim (Eds.), Web Information Systems – WISE 2004 Workshops. XV, 277 pages. 2004.

Vol. 3306: X. Zhou, S. Su, M.P. Papazoglou, M.E. Orlowska, K.G. Jeffery (Eds.), Web Information Systems – WISE 2004. XVII, 745 pages. 2004.

Vol. 3305: P.M.A. Sloot, B. Chopard, A.G. Hoekstra (Eds.), Cellular Automata. XV, 883 pages. 2004.

Vol. 3303: J.A. López, E. Benfenati, W. Dubitzky (Eds.), Knowledge Exploration in Life Science Informatics. X, 249 pages. 2004. (Subseries LNAI).

Vol. 3302: W.-N. Chin (Ed.), Programming Languages and Systems. XIII, 453 pages. 2004.

Vol. 3300: L. Bertossi, A. Hunter, T. Schaub (Eds.), Inconsistency Tolerance. VII, 295 pages. 2005.

Vol. 3299: F. Wang (Ed.), Automated Technology for Verification and Analysis. XII, 506 pages. 2004.

Vol. 3298: S.A. McIlraith, D. Plexousakis, F. van Harmelen (Eds.), The Semantic Web – ISWC 2004. XXI, 841 pages. 2004.

Vol. 3296: L. Bougé, V.K. Prasanna (Eds.), High Performance Computing - HiPC 2004. XXV, 530 pages. 2004.

Vol. 3295: P. Markopoulos, B. Eggen, E. Aarts, J.L. Crowley (Eds.), Ambient Intelligence. XIII, 388 pages. 2004.

Vol. 3294: C.N. Dean, R.T. Boute (Eds.), Teaching Formal Methods. X, 249 pages. 2004.

Vol. 3293: C.-H. Chi, M. van Steen, C. Wills (Eds.), Web Content Caching and Distribution. IX, 283 pages. 2004.

Vol. 3292: R. Meersman, Z. Tari, A. Corsaro (Eds.), On the Move to Meaningful Internet Systems 2004: OTM 2004 Workshops. XXIII, 885 pages. 2004.

Vol. 3291: R. Meersman, Z. Tari (Eds.), On the Move to Meaningful Internet Systems 2004: CoopIS, DOA, and ODBASE, Part II. XXV, 824 pages. 2004.